Hurdles for Phage Therapy (PT) to Become a Reality

Hurdles for Phage Therapy (PT) to Become a Reality

Special Issue Editor

Harald Brüssow

MDPI • Basel • Beijing • Wuhan • Barcelona • Belgrade

Special Issue Editor
Harald Brüssow
KU Leuven
Belgium

Editorial Office
MDPI
St. Alban-Anlage 66
4052 Basel, Switzerland

This is a reprint of articles from the Special Issue published online in the open access journal *Viruses* (ISSN 1999-4915) from 2018 to 2019 (available at: https://www.mdpi.com/journal/viruses/special_issues/Phagetherapy).

For citation purposes, cite each article independently as indicated on the article page online and as indicated below:

LastName, A.A.; LastName, B.B.; LastName, C.C. Article Title. *Journal Name* **Year**, *Article Number*, Page Range.

ISBN 978-3-03921-391-7 (Pbk)
ISBN 978-3-03921-392-4 (PDF)

Contents

About the Special Issue Editor ix

Harald Brüssow
Hurdles for Phage Therapy to Become a Reality—An Editorial Comment
Reprinted from: *Viruses* **2019**, *11*, 557, doi:10.3390/v11060557 1

Damien Thiry, Virginie Passet, Katarzyna Danis-Wlodarczyk, Cédric Lood, Jeroen Wagemans, Luisa De Sordi, Vera van Noort, Nicolas Dufour, Laurent Debarbieux, Jacques G. Mainil, Sylvain Brisse and Rob Lavigne
New Bacteriophages against Emerging Lineages ST23 and ST258 of *Klebsiella pneumoniae* and Efficacy Assessment in *Galleria mellonella* Larvae
Reprinted from: *Viruses* **2019**, *11*, 411, doi:10.3390/v11050411 9

Dominique Holtappels, Rob Lavigne, Isabelle Huys and Jeroen Wagemans
Protection of Phage Applications in Crop Production: A Patent Landscape
Reprinted from: *Viruses* **2019**, *11*, 277, doi:10.3390/v11030277 18

Sarah Djebara, Christiane Maussen, Daniel De Vos, Maya Merabishvili, Benjamin Damanet, Kim Win Pang, Peggy De Leenheer, Isabella Strachinaru, Patrick Soentjens and Jean-Paul Pirnay
Processing Phage Therapy Requests in a Brussels Military Hospital: Lessons Identified
Reprinted from: *Viruses* **2019**, *11*, 265, doi:10.3390/v11030265 34

Susan M. Lehman, Gillian Mearns, Deborah Rankin, Robert A. Cole, Frenk Smrekar, Steven D. Branston and Sandra Morales
Design and Preclinical Development of a Phage Product for the Treatment of Antibiotic-Resistant *Staphylococcus aureus* Infections
Reprinted from: *Viruses* **2019**, *11*, 88, doi:10.3390/v11010088 45

Julien Lossouarn, Arnaud Briet, Elisabeth Moncaut, Sylviane Furlan, Astrid Bouteau, Olivier Son, Magali Leroy, Michael S. DuBow, François Lecointe, Pascale Serror and Marie-Agnès Petit
Enterococcus faecalis Countermeasures Defeat a Virulent *Picovirinae* Bacteriophage
Reprinted from: *Viruses* **2019**, *11*, 48, doi:10.3390/v11010048 61

Han Lin, Matthew L. Paff, Ian J. Molineux and James J. Bull
Antibiotic Therapy Using Phage Depolymerases: Robustness Across a Range of Conditions
Reprinted from: *Viruses* **2018**, *10*, 622, doi:10.3390/v10110622 83

Casandra W. Philipson, Logan J. Voegtly, Matthew R. Lueder, Kyle A. Long, Gregory K. Rice, Kenneth G. Frey, Biswajit Biswas, Regina Z. Cer, Theron Hamilton and Kimberly A. Bishop-Lilly
Characterizing Phage Genomes for Therapeutic Applications
Reprinted from: *Viruses* **2018**, *10*, 188, doi:10.3390/v10040188 100

Dana Štveráková, Ondrej Šedo, Martin Benešík, Zbyněk Zdráhal, Jiří Doškař and Roman Pantůček
Rapid Identification of Intact Staphylococcal Bacteriophages Using Matrix-Assisted Laser Desorption Ionization-Time-of-Flight Mass Spectrometry
Reprinted from: *Viruses* **2018**, *10*, 176, doi:10.3390/v10040176 120

Jean-Paul Pirnay, Gilbert Verbeken, Pieter-Jan Ceyssens, Isabelle Huys, Daniel De Vos, Charlotte Ameloot and Alan Fauconnier
The Magistral Phage
Reprinted from: *Viruses* **2018**, *10*, 64, doi:10.3390/v10020064 . . . . **139**

Katarzyna Leskinen, Henni Tuomala, Anu Wicklund, Jenni Horsma-Heikkinen, Pentti Kuusela, Mikael Skurnik and Saija Kiljunen
Characterization of vB_SauM-fRuSau02, a Twort-Like Bacteriophage Isolated from a Therapeutic Phage Cocktail
Reprinted from: *Viruses* **2017**, *9*, 258, doi:10.3390/v9090258 . . . . **146**

Andrei S. Bolocan, Aditya Upadrasta, Pedro H. de Almeida Bettio, Adam G. Clooney, Lorraine A. Draper, R. Paul Ross and Colin Hill
Evaluation of Phage Therapy in the Context of *Enterococcus faecalis* and Its Associated Diseases
Reprinted from: *Viruses* **2019**, *11*, 366, doi:10.3390/v11040366 . . . . **165**

Shawna McCallin, Jessica C. Sacher, Jan Zheng and Benjamin K. Chan
Current State of Compassionate Phage Therapy
Reprinted from: *Viruses* **2019**, *11*, 343, doi:10.3390/v11040343 . . . . **183**

Sandra-Maria Wienhold, Jasmin Lienau and Martin Witzenrath
Towards Inhaled Phage Therapy in Western Europe
Reprinted from: *Viruses* **2019**, *11*, 295, doi:10.3390/v11030295 . . . . **197**

Jonas D. Van Belleghem, Krystyna Dąbrowska, Mario Vaneechoutte, Jeremy J. Barr and Paul L. Bollyky
Interactions between Bacteriophage, Bacteria, and the Mammalian Immune System
Reprinted from: *Viruses* **2019**, *11*, 10, doi:10.3390/v11010010 . . . . **210**

Frank Oechslin
Resistance Development to Bacteriophages Occurring during Bacteriophage Therapy
Reprinted from: *Viruses* **2018**, *10*, 351, doi:10.3390/v10070351 . . . . **232**

Tobi E. Nagel
Delivering Phage Products to Combat Antibiotic Resistance in Developing Countries: Lessons Learned from the HIV/AIDS Epidemic in Africa
Reprinted from: *Viruses* **2018**, *10*, 345, doi:10.3390/v10070345 . . . . **255**

Marta Lourenço, Luisa De Sordi and Laurent Debarbieux
The Diversity of Bacterial Lifestyles Hampers Bacteriophage Tenacity
Reprinted from: *Viruses* **2018**, *10*, 327, doi:10.3390/v10060327 . . . . **266**

Andrzej Górski, Ryszard Międzybrodzki, Małgorzata Łobocka, Aleksandra Głowacka-Rutkowska, Agnieszka Bednarek, Jan Borysowski, Ewa Jończyk-Matysiak, Marzanna Łusiak-Szelachowska, Beata Weber-Dąbrowska, Natalia Bagińska, Sławomir Letkiewicz, Krystyna Dąbrowska and Jacques Scheres
Phage Therapy: What Have We Learned?
Reprinted from: *Viruses* **2018**, *10*, 288, doi:10.3390/v10060288 . . . . **277**

Antonet Svircev, Dwayne Roach and Alan Castle
Framing the Future with Bacteriophages in Agriculture
Reprinted from: *Viruses* **2018**, *10*, 218, doi:10.3390/v10050218 . . . . **305**

Zachary D. Moye, Joelle Woolston and Alexander Sulakvelidze
Bacteriophage Applications for Food Production and Processing
Reprinted from: *Viruses* **2018**, *10*, 205, doi:10.3390/v10040205 . . . . **318**

David R. Harper
Criteria for Selecting Suitable Infectious Diseases for Phage Therapy
Reprinted from: *Viruses* **2018**, *10*, 177, doi:10.3390/v10040177 . . . . **340**

Eoghan Casey, Douwe van Sinderen and Jennifer Mahony
In Vitro Characteristics of Phages to Guide 'Real Life' Phage Therapy Suitability
Reprinted from: *Viruses* **2018**, *10*, 163, doi:10.3390/v10040163 . . . . **351**

Alan Fauconnier
Phage Therapy Regulation: From Night to Dawn
Reprinted from: *Viruses* **2019**, *11*, 352, doi:10.3390/v11040352 . . . . **371**

Olivier Patey, Shawna McCallin, Hubert Mazure, Max Liddle, Anthony Smithyman and Alain Dublanchet
Clinical Indications and Compassionate Use of Phage Therapy: Personal Experience and Literature Review with a Focus on Osteoarticular Infections
Reprinted from: *Viruses* **2019**, *11*, 18, doi:10.3390/v11010018 . . . . **379**

Brigitte Roy, Cécile Philippe, Martin J. Loessner, Jacques Goulet and Sylvain Moineau
Production of Bacteriophages by Listeria Cells Entrapped in Organic Polymers
Reprinted from: *Viruses* **2018**, *10*, 324, doi:10.3390/v10060324 . . . . **400**

Clara Torres-Barceló
Phage Therapy Faces Evolutionary Challenges
Reprinted from: *Viruses* **2018**, *10*, 323, doi:10.3390/v10060323 . . . . **410**

Thomas Häusler
Phages Make for Jolly Good Stories
Reprinted from: *Viruses* **2018**, *10*, 209, doi:10.3390/v10040209 . . . . **418**

Christine Rohde, Grégory Resch, Jean-Paul Pirnay, Bob G. Blasdel, Laurent Debarbieux, Daniel Gelman, Andrzej Górski, Ronen Hazan, Isabelle Huys, Elene Kakabadze, et al.
Expert Opinion on Three Phage Therapy Related Topics: Bacterial Phage Resistance, Phage Training and Prophages in Bacterial Production Strains
Reprinted from: *Viruses* **2018**, *10*, 178, doi:10.3390/v10040178 . . . . **427**

Elene Kakabadze, Khatuna Makalatia, Nino Grdzelishvili, Nata Bakuradze, Marina Goderdzishvili, Ia Kusradze, Marie-France Phoba, Octavie Lunguya, Cédric Lood, Rob Lavigne, et al.
Selection of Potential Therapeutic Bacteriophages that Lyse a CTX-M-15 Extended Spectrum β-Lactamase Producing *Salmonella enterica* Serovar Typhi Strain from the Democratic Republic of the Congo
Reprinted from: *Viruses* **2018**, *10*, 172, doi:10.3390/v10040172 . . . . **442**

Irene Huber, Katerina Potapova, Andreas Kuhn, Herbert Schmidt, Jörg Hinrichs, Christine Rohde and Wolfgang Beyer
1st German Phage Symposium—Conference Report
Reprinted from: *Viruses* **2018**, *10*, 158, doi:10.3390/v10040158 . . . . **451**

About the Special Issue Editor

Harald Brüssow earned a PhD in virology at the Max-Planck Institute of Biochemistry in Martinsried/ Munich/Germany, and then worked for 36 years at the Nestlé Research Center in Lausanne/Switzerland on various projects exploring, in numerous clinical trials, the potential of passive antibodies, probiotics, prebiotics, and bacteriophages for the treatment of infectious diseases. He has also worked on malnutrition and immunity in field studies, and has experience studying phage infections in industrial food fermentation. He has authored 160 scientific publications and one book (The Quest for Food—a Natural History of Eating; Springer Publisher). After his retirement, he joined the group of Gene Technology at KU Leuven/Belgium as visiting scientist. He has served for over 20 years on the editorial boards of various journals from the American Society for Microbiology (*Clinical and Vaccine Immunology*; *Journal of Bacteriology*; and, currently, *Applied and Environmental Microbiology*) and is editor of *Microbial Biotechnology* (Wiley, Society for Applied Microbiology).

Editorial

Hurdles for Phage Therapy to Become a Reality—An Editorial Comment

Harald Brüssow

KU Leuven, Group of Gene Technology, 3001 Leuven, Belgium; haraldbruessow@yahoo.com

Received: 5 June 2019; Accepted: 9 June 2019; Published: 17 June 2019

This special issue of *Viruses* asks experts in the field about "Hurdles to phage therapy (PT) to become a reality". Their answers came as reviews, perspectives and opinions, along with a number of research papers. No singular hurdle was identified by the authors. According to the specialization of the contacted scientists, various different hurdles or gaps in knowledge impeding progress with PT were described. Collectively, the analyses give, however, a valuable description of the status quo and hopefully provide some direction for future fundamental and clinical research in PT. In view of the grim specter of a possible return to a pre-antibiotic era for a number of bacterial infections, exploring alternatives or adjuncts to antibiotics are of high public health importance and need no further justification. PT is without doubt an interesting approach to the antibiotic resistance problem and merits intensified research to get out of the fruitless confrontation between enthusiasm from the East and lingering Western skepticism.

1. Overview on the Contributions to This Issue

In this special issue, I invited a wide range of authors covering a science journalist who is author of a well-documented book on the history of PT [1] and a representative from a non-governmental organization [2], representatives of industry and opinion leaders in academic PT research and its clinical and agronomical application. Societal awareness of the problem is necessary to assure sufficient political support, which is needed to finance the development of phage products and costly clinical trials for the regulatory acceptance of PT. In my opinion, the currently available evidence of PT seen through clinical trials is not yet a sufficiently strong incentive for the private sector to invest heavily in this field. It is therefore likely that the public sector needs to take the lead to prove the value of the PT approach. This is not an unfair request, since exploring the potential of alternative antimicrobial agents is a task of the public health sector in view of the challenge of untreatable bacterial infections, which might in the near future dwarf past challenges, even that of the HIV epidemic. Once the scientific and clinical evidence is published for PT, it is likely that the private sector will follow with more investments.

Patent and regulatory issues still cause some hesitation in the private sector. Official organizations such as the World health Organization (WHO), European Medicines Agency (EMA) and the US Food and drug Administration (FDA) did not want to define their position towards regulatory aspects of PT in this special issue. This resulted in an over-representation of national, particularly Belgian, personalized medicine approaches with two contributions from the military hospital in Brussels, where the magistral phage approach was developed [3]. Large pharmaceutical industries have not shown much interest in the PT approach so far. There might be a number of reasons for this situation. On one side, there is no economic incentive to develop new antibiotics and even less developing non-antibiotic alternatives to their current antibiotic business. In addition, the classical pharmaceutical industry deals with small chemical drugs—or at most proteins that are molecularly well-defined—while phages represent large, replication-competent, biological material that is subject to biological variation and evolution. Thus, defining the composition of a phage product is not trivial and several contributions to this issue address this problem [4–8]. The pharmacokinetic properties of phages raise issues unknown to standard pharmacology approaches.

The fact that one of the few randomized controlled trials (RCT) with PT conducted by the private sector was organized by a food company (Nestlé, Vevey, Switzerland) should not be a surprise. Fermentation using bacterial starter cultures is used in various food production processes. These processes are always threatened by phage attacks, necessitating substantial phage research in the food industry, particularly in the field of phage-resistance. Food companies have therefore maintained active phage research groups. With the extension of several food companies into the nutrition and health area, the human microbiome research has come into focus, and with it again, phages as modulators of the bacterial microbiome. Using phages to correct microbiome dysbiosis is a potentially interesting application beyond just targeting single bacterial pathogens.

Fauconnier [9] proposes in his contribution an adapted approach to the regulation of PT. This point of view was not at all shared by one reviewer of the paper, who adamantly claimed that the current drug legislation both in Europe and North America is sufficient for PT introduction and that no alternatives exist to RCT demonstrating safety and efficacy if PT wants to see the market place. These two opposing views describe an unsettled controversy, although I personally think that both approaches are not mutually exclusive. A personalized medicine approach with phages, as currently under development in Belgium, will fulfill a pioneer function for PT in Western countries. Once sufficient efficacy data has been accumulated with that approach and with numbers of untreatable bacterial infections going into the several hundred of thousands, personalized medicine approaches will no longer be practical and phages would need to be developed as common drugs, provided that they show efficacy in RCT. The paper of Philipson et al. [7] describes how phages can be produced to FDA standards. Other frequently quoted issues hampering the introduction of PT, such as the difficulty of patenting approaches [10] or the problem of rapid phage-resistance development [11–13] are discussed and found to be less critical than commonly assumed. Oechslin even raises the possibility to explore Darwinian medicinal approaches, where phage treatment can induce virulence attenuation or reestablish antibiotic sensitivity [12]. Casey et al. argue that part of the clinical problems with PT can already be settled by careful selection with in vitro tests better suited for reflecting real life situations [14]. However, other contributions point to complications of PT that can only be assessed in a realistic in vivo context reflecting ecological [15] or evolutionary constraints [16] encountered at organismal or even population levels. I agree strongly with the notion that the lack of detailed in vivo knowledge of phages currently limits our capacity to design and eventually assure successful clinical trials. This in vitro orientation of phage research has historical reasons: phages were investigated by scientists under the perspective of the reductionist principle, which led to the molecular biology revolution [17]. This situation is likely to change with phages now returning on the scene when microbiome research has discovered the importance of phages in regulating microbial ecosystems as different as the oceans and the gut.

This special issue solicited insights from major stakeholders in the medical PT field, including a lead scientist from an industrial group that conducted the only successful RCT in PT [18], or the Polish [19] and Georgian PT centers; the latter with a contribution demonstrating how they select a therapeutic phage against a specific emerging pathogen in the field [20]. Compassionate phage use in the USA [21], in France [22] and with the Belgian Magistral Phage preparation [23] are described. Phage use in food production is reviewed by scientists from Intralytics [24] and phage in the service of agriculture is described by Svircev et al [25].

Considering that PT is a wide field, some subjects are only represented with a single paper: van Belleghem et al. [26] explore the impact of phage on the immune system, Thiry et al. [27] the use of a simple animal model for screening large numbers of phages for simplified in vivo phenotypes; Roy et al. [28] explore an interesting phage production system; Lin et al. [29] investigate the use of phage enzymes for infection treatment. However, some features important for the assessment of PT are missing, such as a failure analysis of the Phagoburn clinical trial [30] that had been supported by a grant from the European Community. A thorough microbiological work-up of failed RCTs are of substantial importance for future PT trial planning, such as that sponsored by the German government, where suitable phages against lower respiratory tract infections will be selected at the Leibniz Institute, produced to GMP standards at the Fraunhofer Institute and clinically tested at the Charité hospital as described by Wienhold et al. [31] in this issue.

2. Failure Analysis of the Bangladesh Diarrhea PT Trial

Since I was actively involved in a failed PT trial, the Nestlé diarrhea trial in Bangladesh [32], I will here summarize my personal evaluation of hurdles to phage therapy.

Perhaps it is best to start with what was not a hurdle in that RCT. Two aqueous phage products, a commercial Russian phage cocktail [33] and a phage cocktail specifically produced for this trial at the Nestlé Research Center [34] were tested. While maintained for the RCT over several years under refrigeration conditions, no decline in phage titer was seen [35], in contrast to initial experiences in the Phagoburn trial. The International Center for Diarrhoeal Disease Research in Bangladesh (icddr,b), the world's leading diarrhea research hospital, has a straightforward review process for clinical protocols consisting of four steps: in-house evaluation, external review, a research, followed by an ethical committee in Bangladesh. In fact, it was more difficult to get the export permit for phages from Russia than to get to their import permit into Bangladesh, once the protocol was approved by the ethical committee. Since oral phage use was planned, we only needed a food-grade phage preparation. Establishing a RCT for PT was not a difficulty, provided that all patients got the most efficient standard treatment consisting of oral rehydration solution supplemented with zinc. Since zinc already has a shortening effect on diarrhea duration, PT had to show an advantage over zinc treatment alone; this is a fair request in view of the low cost and risk of zinc supplementation. The start of the efficacy trial was delayed because the icddr,b clinicians asked for supplementary safety tests in healthy subjects of gradually decreasing age from Bangladesh [36] in addition to a safety test in adult Swiss healthy subjects [37]. Interestingly, external reviewers argued that healthy subjects would carry the risk of phage exposure without the possible therapeutic benefit of phage. Phage has been applied to many healthy subjects in Bangladesh and elsewhere without observing adverse events. As phage is not toxic as virion, but only when lysing the bacterial host during infection and releasing toxic bacterial products, the ethical committee in Bangladesh has subsequently also approved nasal application of commercial staphylococcal phage products from the Eliava Institute in Georgia [38]. The quality of clinical follow-up is very good at icddr,b, as documented by many influential publications coming from this research hospital. There is thus no objective hurdle to conduct RCTs with PT in Bangladesh to obtain scientific evidence for PT efficacy.

Now to the hurdles: there are indeed physico-chemical hurdles to phage use. In vitro experiments suggested heavy phage loss during simulated gastric passage conditions [34]. The ethical committee in Bangladesh did not allow buffering of gastric acidity in patients for concern of increased nosocomial infection risk in a diarrhea hospital with heavy pathogen load. We therefore probably lost a substantial amount of the orally applied phages in gastric passage. There are solutions to this problem (increasing the oral dose, microencapsulation), but we did not anticipate this difficulty since we had observed good oral phage transit in adult Swiss volunteers [37]. Since children and adults from developing countries produce less stomach acidity (hypochlorhydria) than Western adults [39], we anticipated an even better gut transit, which was not the case, therefore indicating limitations in our knowledge about the pharmacokinetics of oral phage products in subjects of the developing world. Apparently, more attention has to be paid to galenic preparations of phages to get phages at sufficient titers to the site of action of the targeted bacterial pathogen.

Laboratory analysis of the clinical samples also identified other factors that prevented clinical efficacy of the oral phages. As acute *Escherichia coli* diarrhea was the target for PT in this trial, phage treatment was started after rapid exclusion of non-*E. coli* diarrhea (rotavirus, cholera, shigellosis). However, further analysis revealed that only half of the enrolled cases showed a confirmed *E. coli* infection [32]. Many pathogens are involved in diarrhea, and a given pathogen might represent only a moderate share of all acute diarrhea cases. This observation is not restricted to diarrhea, but also applies to pneumonia, the major killer of children in developing countries. Under this condition, only a fraction of the treated patients would profit from a treatment with a phage preparation targeting a single pathogen. This problem can of course be addressed by using complex phage cocktails like Intestiphage preparations from Russia (Microgen) or Georgia (Eliava) containing phages against many enteropathogen species. Even then, two problems remained: first, even in confirmed cases of *E. coli* infection, *E. coli* did not represent the dominant bacterium in the stool [32]. Acute diarrhea cases showed a dominance of intestinal streptococci independent of their etiology in the stool [40], and this dysbiosis normalized with recovery from diarrhea. Diarrhea output correlated with streptococcal, but not *E. coli* stool abundance. In fact, the concentration of fecal pathogenic *E. coli* was near or below the replication threshold determined for T4-like coliphages to maintain an infection chain in the laboratory [32]. Second, acute diarrhea in children from developing countries is typically a polymicrobial infection [41], and this was also our observation. In addition, several *E. coli* pathotypes showed a low pathogenicity index in epidemiological surveys of children from developing countries [42], raising doubts about their role as pathogens. Due to this complexity, acute diarrhea is unlikely to represent a suitable target for PT. The problem is further compounded by the genetic variability of *E. coli*. Even with phage cocktails containing 10 phage strains, we achieved only about 50 per cent coverage (i.e., in vitro lysis) of the fecal *E. coli* isolates from the patients [32,43]. When including more phage strains, we encountered interference problems, where the cocktail showed less coverage than the sum of the individual phages.

3. Recommendations

The take home lessons from our PT experience are thus: successful PT trials are more likely with infections where:

(1) The disease-causing role of the bacterial pathogen is clearly established. Do not rely on textbook knowledge and confirm the role of the pathogen in your targeted patient population.
(2) Polymicrobial infections should be avoided or addressed with a multi-pronged approach.
(3) The pathogen is present with a sufficiently elevated concentration to allow productive phage infection chains to occur in the patient.
(4) Suitable phages are available to cover the genetic diversity of the pathogen.

Suitable phages are not always at hand. For example, when researchers screened a collection containing more than 10,000 mycobacteriophages (the largest collection of characterized phages directed against a single bacterial genus) for the treatment of two cystic fibrosis patients infected with *Mycobacterium abscessus*, they found only one lytic phage for one patient [44]. By genetic engineering they could transform a second temperate phage into a suitable lytic phage by deletion of the phage repressor. For two other phages, suitable host range mutants containing spontaneous point mutations were selected. The good news is that a cocktail of three phages, containing a genetically-engineered phage, was approved for clinical use and rescued one patient. This point proves that even a genetically modified phage was approved for patient use in Europe and this fact extends the possibilities offered to PT substantially. However, the bad news was that for the other patient, infected with another *M. abscessus* strain, no suitable phage could be found and the patient died. In contrast, some phage types have an extremely wide host range on *S. aureus*, including methicillin-resistant and to a lesser extent vancomycin-resistant strains; however, they also infect *S. epidermidis*, which represents a potential collateral damage on a skin commensal in skin application.

RCT of PT are more difficult to organize with acute rather than with chronic infections, since short disease durations need an early phage intervention frequently before the microbiological diagnosis becomes available, resulting in the enrolment of many uninformative patients. In contrast, prevention of acute diarrhea might be more attractive when the epidemiological situation is clear: for example, in case of prophylactic phage treatment of contact persons from cholera patients or outbreaks of cholera epidemics in refugee camps. In fact, the large successful prevention clinical trial of Shigella diarrhea conducted by the Eliava Institute in 1963 supports this point [45]. However, prevention trials depend on a careful follow-up causing logistic problems, thus making them frequently more costly than treatment trials of PT.

An additional hurdle is the fact that the targeted pathogen must be accessible to the applied phage. While oral phage application seems, at first view, an appropriate way to treat a gastro-intestinal infection, there are barriers beyond phage inactivation in the stomach. Gut peristalsis is accelerated in diarrhea and it becomes questionable if oral phage has long enough contact times to infect a pathogen like *Vibrio cholerae* [46]. Furthermore, it is not clear where the enteropathogen is actually located; is it in the lumen, in the mucus layer or epithelium-associated? Enteropathogens display a variety of virulence genes that allow them to penetrate the mucus layer and to adhere to gut epithelia. Some phages display depolymerase enzymes at their tail fibers, which allow penetration of bacterial capsular layers and sometimes bacterial biofilms. It is less clear whether phages are able to follow bacteria that adhere to the epithelia through the mucus layer. Mouse experiments showed that an in vitro fully-susceptible bacterial host could escape infection in the gut without developing genetically determined phage resistance. In this case, phage replicated in vivo only on a subpopulation of the host bacteria [47–49]. We still do not know enough about the physiological differentiation of bacteria in the mammalian gut. While clinical sampling is principally possible to study phage-pathogen interactions in at least some accessible gut segments of patients, the procedures are invasive and ethical committees will not allow invasive sampling that is not clinically indicated. It is thus preferable to target infections on more accessible body sites in future PT trials where sampling is easier than the gut. Purulent bacterial skin infections with *Staphylococcus aureus* or *Streptococcus pyogenes* come to mind.

Microbiome studies on the skin have demonstrated a substantial depth differentiation for bacterial colonization of the skin. Even in such "easy" sites for topical phage application like the skin, it remains to be shown in what epidermal cell layer the pathogen resides and whether phage can reach them. In fact, phages are commonly selected for vigorous in vitro planktonic growth on their target bacterium maintained under optimal nutrition. However, these are idealized laboratory conditions. In vivo, many bacteria grow very slowly in biofilms or in mucus layers. One might therefore ask whether we should not select phages for PT that are able to infect bacteria in biofilms or under simulated slow in vivo growth conditions. Complex biofilms consisting of different bacterial species are difficult to realize in the laboratory and not suitable for testing large numbers of source material containing phages (but see Thiry et al. [27] in this issue). Some in vivo properties can be predicted from in vitro observations (see Casey et al. [14] in this issue). For example, T4-like coliphages only replicate on exponentially growing *E. coli* cells, while T7-like phages replicate also on *E. coli* in stationary phase [47].

4. Outlook

Clearly, we need more ecophysiological data on in vivo phage-bacterium interaction in relevant animal models to select suitable phages for clinical application. As argued by Torres-Barceló [16] in this issue, evolutionary thinking should be included in this reasoning. A phage that kills off its host bacterium, present at low concentrations, wipes out its growth substrate and is unlikely to be maintained in evolution. Based on theoretical reasoning, phages should be active on expanding bacterial populations that shift the ecosystem to a state dominated by one or few bacteria. Phages might therefore play a positive role in ecology by maintaining bacterial genetic diversity in the environment [50]. This argument meets the threshold concept for phage replication and might suggest that PT could be more effective in fighting microbial dysbiosis due to an outgrowth of undesired bacteria as in antibiotic-associated diarrhea than against pathogens which mediate clinical effects while present in low numbers. If low-level food contaminants were to be eliminated, very high phage titers were needed to achieve enzymatic "lysis from without" rather than by phage replication.

From these arguments, one might conclude that we need more fundamental knowledge on phage-bacterium interaction in pertinent animal models before successful clinical application can be envisioned for PT. A possible short-cut to successful PT could be the careful evaluation of past personal experience [18,22], compassionate phage use [21], systematic evaluations of case reports [19] and patient follow-up with magistral phage preparation [3,23], all discussed in this issue. Case reports combining clinical observation with state-of-the-art laboratory investigation of in situ phage–bacterium–host interactions might pave the way to successful RCT with PT. We should avoid to target infections for PT according to the scientific background of the research group and their "favorite infection". The EMA and FDA have already called conferences for stakeholders of PT, without much concrete recommendations. Perhaps public health authorities should convene a consensus finding conference for the best target of PT for a RCT sponsored by the Horizon 2020 calls of the EU.

Acknowledgments: I thank the authors for their contributions to this special issue. I am also pleased to acknowledge the dedicated support by the staffs of *Viruses* Editorial Office and that of many reviewers assuring a fair, thorough and quick review process. For this editorial, I thank S. McCallin for critical reading.

Conflicts of Interest: The author declares no conflict of interest.

References

1. Häusler, T. Phages make for jolly good stories. *Viruses* **2018**, *10*, 209. [CrossRef] [PubMed]
2. Nagel, T.E. Delivering phage products to combat antibiotic resistance in developing countries: Lessons learned from the HIV/AIDS epidemic in Africa. *Viruses* **2018**, *10*, 345. [CrossRef] [PubMed]
3. Pirnay, J.P.; Verbeken, G.; Ceyssens, P.J.; Huys, I.; De Vos, D.; Ameloot, C.; Fauconnier, A. The magistral phage. *Viruses* **2018**, *10*, 64. [CrossRef] [PubMed]
4. Štveráková, D.; Šedo, O.; Benešík, M.; Zdráhal, Z.; Doškař, J.; Pantůček, R. Rapid identification of intact staphylococcal bacteriophages using matrix-assisted laser desorption ionization-time-of-flight mass spectrometry. *Viruses* **2018**, *10*, 176. [CrossRef] [PubMed]
5. Leskinen, K.; Tuomala, H.; Wicklund, A.; Horsma-Heikkinen, J.; Kuusela, P.; Skurnik, M.; Kiljunen, S. Characterization of vB_SauM-fRuSau02, a twort-like bacteriophage isolated from a therapeutic phage cocktail. *Viruses* **2017**, *9*, 258. [CrossRef] [PubMed]
6. Bolocan, A.S.; Upadrasta, A.; Bettio, P.H.A.; Clooney, A.G.; Draper, L.A.; Ross, R.P.; Hill, C. Evaluation of phage therapy in the context of *Enterococcus faecalis* and its associated diseases. *Viruses* **2019**, *11*, 366. [CrossRef] [PubMed]
7. Philipson, C.W.; Voegtly, L.J.; Lueder, M.R.; Long, K.A.; Rice, G.K.; Frey, K.G.; Biswas, B.; Cer, R.Z.; Hamilton, T.; Bishop-Lilly, K.A. Characterizing phage genomes for therapeutic applications. *Viruses* **2018**, *10*, 188. [CrossRef] [PubMed]
8. Lehman, S.M.; Mearns, G.; Rankin, D.; Cole, R.A.; Smrekar, F.; Branston, S.D.; Morales, S. Design and preclinical development of a phage product for the treatment of antibiotic-resistant *Staphylococcus aureus* infections. *Viruses* **2019**, *11*, 88. [CrossRef]
9. Fauconnier, A. Phage therapy regulation: From night to dawn. *Viruses* **2019**, *11*, 352. [CrossRef]
10. Holtappels, D.; Lavigne, R.; Huys, I.; Wagemans, J. Protection of phage applications in crop production: A patent landscape. *Viruses* **2019**, *11*, 277. [CrossRef]
11. Lossouarn, J.; Briet, A.; Moncaut, E.; Furlan, S.; Bouteau, A.; Son, O.; Leroy, M.; DuBow, M.S.; Lecointe, F.; Serror, P.; et al. *Enterococcus faecalis* countermeasures defeat a virulent *Picovirinae* bacteriophage. *Viruses* **2019**, *11*, 48. [CrossRef] [PubMed]
12. Oechslin, F. Resistance development to bacteriophages occurring during bacteriophage therapy. *Viruses* **2018**, *10*, 351. [CrossRef] [PubMed]
13. Rohde, C.; Resch, G.; Pirnay, J.P.; Blasdel, B.G.; Debarbieux, L.; Gelman, D.; Górski, A.; Hazan, R.; Huys, I.; Kakabadze, E.; et al. Expert opinion on three phage therapy related topics: Bacterial phage resistance, phage training and prophages in bacterial production strains. *Viruses* **2018**, *10*, 178. [CrossRef] [PubMed]
14. Casey, E.; van Sinderen, D.; Mahony, J. In vitro characteristics of phages to guide 'real life' phage therapy suitability. *Viruses* **2018**, *10*, 163. [CrossRef]
15. Lourenço, M.; De Sordi, L.; Debarbieux, L. The diversity of bacterial lifestyles hampers bacteriophage tenacity. *Viruses* **2018**, *10*, 327. [CrossRef]
16. Torres-Barceló, C. Phage therapy faces evolutionary challenges. *Viruses* **2018**, *10*, 323. [CrossRef]
17. Brüssow, H. Environmental microbiology: Too much food for thought?—An argument for reductionism. *Environ. Microbiol.* **2018**, *20*, 1929–1935. [CrossRef]

18. Harper, D.R. Criteria for selecting suitable infectious diseases for phage therapy. *Viruses* **2018**, *10*, 177. [CrossRef]
19. Górski, A.; Międzybrodzki, R.; Łobocka, M.; Głowacka-Rutkowska, A.; Bednarek, A.; Borysowski, J.; Jończyk-Matysiak, E.; Łusiak-Szelachowska, M.; Weber-Dąbrowska, B.; Bagińska, N. Phage therapy: What have we learned? *Viruses* **2018**, *10*, 288. [CrossRef]
20. Kakabadze, E.; Makalatia, K.; Grdzelishvili, N.; Bakuradze, N.; Goderdzishvili, M.; Kusradze, I.; Phoba, M.F.; Lunguya, O.; Lood, C.; Lavigne, R. Selection of potential therapeutic bacteriophages that lyse a CTX-M-15 extended spectrum β-lactamase producing *Salmonella enterica* serovar typhi strain from the democratic republic of the Congo. *Viruses* **2018**, *10*, 172. [CrossRef]
21. McCallin, S.; Sacher, J.C.; Zheng, J.; Chan, B.K. Current state of compassionate phage therapy. *Viruses* **2019**, *11*, 343. [CrossRef] [PubMed]
22. Patey, O.; McCallin, S.; Mazure, H.; Liddle, M.; Smithyman, A.; Dublanchet, A. Clinical indications and compassionate use of phage therapy: Personal experience and literature review with a focus on osteoarticular infections. *Viruses* **2018**, *11*, 18. [CrossRef] [PubMed]
23. Djebara, S.; Maussen, C.; De Vos, D.; Merabishvili, M.; Damanet, B.; Pang, K.W.; De Leenheer, P.; Strachinaru, I.; Soentjens, P.; Pirnay, J.P. Processing phage therapy requests in a Brussels military hospital: Lessons identified. *Viruses* **2019**, *11*, 265. [CrossRef] [PubMed]
24. Moye, Z.D.; Woolston, J.; Sulakvelidze, A. Bacteriophage applications for food production and processing. *Viruses* **2018**, *10*, 205. [CrossRef] [PubMed]
25. Svircev, A.; Roach, D.; Castle, A. Framing the future with bacteriophages in agriculture. *Viruses* **2018**, *10*, 218. [CrossRef] [PubMed]
26. Van Belleghem, J.D.; Dąbrowska, K.; Vaneechoutte, M.; Barr, J.J.; Bollyky, P.L. Interactions between bacteriophage, bacteria, and the mammalian immune system. *Viruses* **2018**, *11*, 10. [CrossRef] [PubMed]
27. Thiry, D.; Passet, V.; Danis-Wlodarczyk, K.; Lood, C.; Wagemans, J.; De Sordi, L.; van Noort, V.; Dufour, N.; Debarbieux, L.; Mainil, J.G.; et al. New bacteriophages against emerging lineages ST23 and ST258 of *Klebsiella pneumoniae* and efficacy assessment in *Galleria mellonella* larvae. *Viruses* **2019**, *11*, 411. [CrossRef]
28. Roy, B.; Philippe, C.; Loessner, M.J.; Goulet, J.; Moineau, S. Production of bacteriophages by listeria cells entrapped in organic polymers. *Viruses* **2018**, *10*, 324. [CrossRef]
29. Lin, H.; Paff, M.L.; Molineux, I.J.; Bull, J.J. Antibiotic therapy using phage depolymerases: Robustness across a range of conditions. *Viruses* **2018**, *10*, 622. [CrossRef]
30. Jault, P.; Leclerc, T.; Jennes, S.; Pirnay, J.P.; Que, Y.A.; Resch, G.; Rousseau, A.F.; Ravat, F.; Carsin, H.; Le Floch, R.; et al. Efficacy and tolerability of a cocktail of bacteriophages to treat burn wounds infected by *Pseudomonas aeruginosa* (PhagoBurn): A randomised, controlled, double-blind phase 1/2 trial. *Lancet Infect. Dis.* **2019**, *19*, 35–45. [CrossRef]
31. Wienhold, S.M.; Lienau, J.; Witzenrath, M. Towards inhaled phage therapy in Western Europe. *Viruses* **2019**, *11*, 295. [CrossRef] [PubMed]
32. Sarker, S.A.; Sultana, S.; Reuteler, G.; Moine, D.; Descombes, P.; Charton, F.; Bourdin, G.; McCallin, S.; Ngom-Bru, C.; Neville, T.; et al. Oral phage therapy of acute bacterial diarrhea with two coliphage preparations: A randomized trial in children from Bangladesh. *EBioMedicine* **2016**, *4*, 124–137. [CrossRef] [PubMed]
33. McCallin, S.; Sarker, S.A.; Barretto, C.; Sultana, S.; Berger, B.; Huq, S.; Krause, L.; Bibiloni, R.; Schmitt, B.; Reuteler, G.; et al. Safety analysis of a Russian phage cocktail: From metagenomic analysis to oral application in healthy human subjects. *Virology* **2013**, *443*, 187–196. [CrossRef] [PubMed]
34. Sarker, S.A.; McCallin, S.; Barretto, C.; Berger, B.; Pittet, A.C.; Sultana, S.; Krause, L.; Huq, S.; Bibiloni, R.; Bruttin, A.; et al. Oral T4-like phage cocktail application to healthy adult volunteers from Bangladesh. *Virology* **2012**, *434*, 222–232. [CrossRef] [PubMed]
35. Bourdin, G.; Schmitt, B.; Guy, L.M.; Germond, J.E.; Zuber, S.; Michot, L.; Reuteler, G.; Brüssow, H. Amplification and purification of T4-like escherichia coli phages for phage therapy: From laboratory to pilot scale. *Appl. Environ. Microbiol.* **2014**, *80*, 1469–1476. [CrossRef] [PubMed]
36. Sarker, S.A.; Berger, B.; Deng, Y.; Kieser, S.; Foata, F.; Moine, D.; Descombes, P.; Sultana, S.; Huq, S.; Bardhan, P.K.; et al. Oral application of Escherichia coli bacteriophage: Safety tests in healthy and diarrheal children from Bangladesh. *Environ. Microbiol.* **2017**, *19*, 237–250. [CrossRef] [PubMed]

37. Bruttin, A.; Brüssow, H. Human volunteers receiving Escherichia coli phage T4 orally: A safety test of phage therapy. *Antimicrob. Agents Chemother.* **2005**, *49*, 2874–2878. [CrossRef]
38. McCallin, S.; Sarker, S.A.; Sultana, S.; Oechslin, F.; Brüssow, H. Metagenome analysis of Russian and Georgian Pyophage cocktails and a placebo-controlled safety trial of single phage versus phage cocktail in healthy *Staphylococcus aureus* carriers. *Environ. Microbiol.* **2018**, *20*, 3278–3293. [CrossRef] [PubMed]
39. Sarker, S.A.; Ahmed, T.; Brüssow, H. Hunger and microbiology: Is a low gastric acid-induced bacterial overgrowth in the small intestine a contributor to malnutrition in developing countries? *Microb. Biotechnol.* **2017**, *10*, 1025–1030. [CrossRef] [PubMed]
40. Kieser, S.; Sarker, S.A.; Sakwinska, O.; Foata, F.; Sultana, S.; Khan, Z.; Islam, S.; Porta, N.; Combremont, S.; Betrisey, B.; et al. Bangladeshi children with acute diarrhoea show faecal microbiomes with increased *Streptococcus* abundance, irrespective of diarrhoea aetiology. *Environ. Microbiol.* **2018**, *20*, 2256–2269. [CrossRef]
41. Taniuchi, M.; Sobuz, S.U.; Begum, S.; Platts-Mills, J.A.; Liu, J.; Yang, Z.; Wang, X.Q.; Petri, W.A., Jr.; Haque, R.; Houpt, E.R. Etiology of diarrhea in Bangladeshi infants in the first year of life analyzed using molecular methods. *J. Infect. Dis.* **2013**, *208*, 1794–1802. [CrossRef] [PubMed]
42. Kotloff, K.L.; Nataro, J.P.; Blackwelder, W.C.; Nasrin, D.; Farag, T.H.; Panchalingam, S.; Wu, Y.; Sow, S.O.; Sur, D.; Breiman, R.F.; et al. Burden and aetiology of diarrhoeal disease in infants and young children in developing countries (the Global Enteric Multicenter Study, GEMS): A prospective, case-control study. *Lancet* **2013**, *382*, 209–222. [CrossRef]
43. Bourdin, G.; Navarro, A.; Sarker, S.A.; Pittet, A.C.; Qadri, F.; Sultana, S.; Cravioto, A.; Talukder, K.A.; Reuteler, G.; Brüssow, H. Coverage of diarrhoea-associated *Escherichia coli* isolates from different origins with two types of phage cocktails. *Microb. Biotechnol.* **2014**, *7*, 165–176. [CrossRef] [PubMed]
44. Dedrick, R.M.; Guerrero-Bustamante, C.A.; Garlena, R.A.; Russell, D.A.; Ford, K.; Harris, K.; Gilmour, K.C.; Soothill, J.; Jacobs-Sera, D.; Schooley, R.T.; et al. Engineered bacteriophages for treatment of a patient with a disseminated drug-resistant *Mycobacterium abscessus*. *Nat. Med.* **2019**, *25*, 730–733. [CrossRef] [PubMed]
45. Sulakvelidze, A.; Alavidze, Z.; Morris, J.G., Jr. Bacteriophage therapy. *Antimicrob. Agents Chemother.* **2001**, *45*, 649–659. [CrossRef] [PubMed]
46. Brüssow, H. Phage therapy for the treatment of human intestinal bacterial infections: Soon to be a reality? *Expert Rev. Gastroenterol. Hepatol.* **2017**, *11*, 785–788. [CrossRef]
47. Weiss, M.; Denou, E.; Bruttin, A.; Serra-Moreno, R.; Dillmann, M.L.; Brüssow, H. In vivo replication of T4 and T7 bacteriophages in germ-free mice colonized with *Escherichia coli*. *Virology* **2009**, *393*, 16–23. [CrossRef]
48. Maura, D.; Morello, E.; du Merle, L.; Bomme, P.; Le Bouguénec, C.; Debarbieux, L. Intestinal colonization by enteroaggregative Escherichia coli supports long-term bacteriophage replication in mice. *Environ. Microbiol.* **2012**, *14*, 1844–1854. [CrossRef]
49. De Sordi, L.; Lourenço, M.; Debarbieux, L. The battle within: Interactions of bacteriophages and bacteria in the gastrointestinal tract. *Cell Host Microbe* **2019**, *25*, 210–218. [CrossRef]
50. Wommack, K.E.; Colwell, R.R. Virioplankton: Viruses in aquatic ecosystems. *Microbiol. Mol. Biol. Rev.* **2000**, *64*, 69–114. [CrossRef]

Communication

New Bacteriophages against Emerging Lineages ST23 and ST258 of *Klebsiella pneumoniae* and Efficacy Assessment in *Galleria mellonella* Larvae

Damien Thiry [1,2,3,*], Virginie Passet [2], Katarzyna Danis-Wlodarczyk [3,‡], Cédric Lood [3,4], Jeroen Wagemans [3], Luisa De Sordi [5,6], Vera van Noort [4,7], Nicolas Dufour [5,8], Laurent Debarbieux [5], Jacques G. Mainil [1], Sylvain Brisse [2,†] and Rob Lavigne [3,†]

1 Bacteriology, Department of Infectious and Parasitic Diseases, FARAH and Faculty of Veterinary Medicine, ULiège, 4000 Liège, Belgium; jg.mainil@uliege.be
2 Biodiversity and Epidemiology of Bacterial Pathogens, Institut Pasteur, 75015 Paris, France; virginie.passet@pasteur.fr (V.P.); sylvain.brisse@pasteur.fr (S.B.)
3 Laboratory of Gene Technology, Department of Biosystems, KU Leuven, 3001 Heverlee, Belgium; danis@wp.pl (K.D.-W.); cedric.lood@kuleuven.be (C.L.); jeroen.wagemans@kuleuven.be (J.W.); rob.lavigne@kuleuven.be (R.L.)
4 Centre of Microbial and Plant Genetics, Department of Microbial and Molecular Systems, KU Leuven, 3001 Heverlee, Belgium; vera.vannoort@kuleuven.be
5 Department of Microbiology, Institut Pasteur, 75015 Paris, France; luisa.de_sordi@sorbonne-universite.fr (L.D.S.); nicolas.dufour@ght-novo.fr (N.D.); laurent.debarbieux@pasteur.fr (L.D.)
6 Laboratoire des Biomolécules, Hôpital Saint-Antoine, Sorbonne Université, 75012 Paris, France
7 Institute of Biology Leiden, Leiden University, 2311 Leiden, The Netherlands
8 Service de Réanimation Médico-Chirurgicale, Centre Hospitalier René Dubos, 95300 Pontoise, France
* Correspondence: damien.thiry@uliege.be; Tel.: +32-4-3669388
† These authors contributed equally to this work.
‡ Current Address: Department of Microbiology and Microbial Infection and Immunity, The Ohio State University, 43210 Columbus, OH, USA.

Received: 14 February 2019; Accepted: 29 April 2019; Published: 3 May 2019

Abstract: *Klebsiella pneumoniae* is a bacterial pathogen of high public health importance. Its polysaccharide capsule is highly variable but only a few capsular types are associated with emerging pathogenic sublineages. The aim of this work is to isolate and characterize new lytic bacteriophages and assess their potential to control infections by the ST23 and ST258 *K. pneumoniae* sublineages using a *Galleria mellonella* larvae model. Three selected bacteriophages, targeting lineages ST258 (bacteriophages vB_KpnP_KL106-ULIP47 and vB_KpnP_KL106-ULIP54) and ST23 (bacteriophage vB_KpnP_K1-ULIP33), display specificity for capsular types KL106 and K1, respectively. These podoviruses belong to the *Autographivirinae* subfamily and their genomes are devoid of lysogeny or toxin-associated genes. In a *G. mellonella* larvae model, a mortality rate of 70% was observed upon infection by *K. pneumoniae* ST258 and ST23. This number was reduced to 20% upon treatment with bacteriophages at a multiplicity of infection of 10. This work increases the number of characterized bacteriophages infecting *K. pneumoniae* and provides information regarding genome sequence and efficacy during preclinical phage therapy against two prominent sublineages of this bacterial species.

Keywords: antimicrobial resistance; capsule; *Galleria mellonella*; *Klebsiella pneumoniae*; phage therapy

---

Klebsiella pneumoniae, a member of the Enterobacteriaceae family, causes a variety of human and animal infections including pneumonia, infections of the urinary tract, bacteremia, and liver abscess. *K. pneumoniae* infections are becoming increasingly difficult, and sometimes impossible [1], to treat due to the continuous emergence of multidrug-resistant strains [2–4]. Cells of *K. pneumoniae* are characteristically surrounded by a thick capsule of variable chemical composition, which translates into a large number of classically defined capsular serotypes [5] and an even larger number of in silico-defined *wzi*, *wzc*, or KL-types [6,7]. These three molecular classifications denote the diversity of the capsular polysaccharide synthesis gene cluster and serve as proxies of capsular antigen variation. *K. pneumoniae* isolates can be roughly classified into two pathotypes: opportunistic *K. pneumoniae*, which are often multidrug-resistant (mdrKp), and hypervirulent *K. pneumoniae* (hvKp) [8,9], which are able to infect healthy individuals and cause invasive infections including pyogenic liver abscess. The majority of clinical mdrKp and hvKp isolates are part of a small number of genetic lineages (also called clonal groups). Prominent lineages include mdrKpST258, which is frequently associated with specific carbapenemases (i.e., those of the KPC family) and resistant to multiple other antimicrobials, and the ST23 lineage, which is the most frequent cause of liver abscess [1] and can also acquire clinically significant antibiotic resistance genes [10]. Recently, there has been has a sharp increase in the clinical significance of mdrKp and hvKp infections [4,9,11].

New therapeutic strategies are critically needed against *K. pneumoniae* infections. Phage therapy is increasingly recognized as an attractive approach [12]. Previous work has shown that bacteriophages (phages) against *Klebsiella* can be readily isolated from diverse sources and are a promising tool against *K. pneumoniae* infections in *Galleria mellonella* models [13,14].

The aim of this study was to contribute to developments of the phage therapy approach against *K. pneumoniae* and, more specifically, against its two prominent lineages ST23 and ST258. Specifically, our objectives were (i) to isolate and characterize phages against bacteria in these lineages and to sequence the genome of these phages; (ii) to implement an infection model of *G. mellonella* larvae with *K. pneumoniae* strains of interest; and (iii) to test phages against *K. pneumoniae* in this model.

Two clinical *K. pneumoniae* strains were selected for phage isolation [6,15,16]. The first was the 2198 (SB4551) strain, a *K. pneumoniae* carbapenemase-producing isolate from an outbreak in Ireland [15]. This strain, characterized by *wzc*-921 and *wzi*-29 alleles, belongs to ST258 clade 1 [17,18] or ST258a [19] associated with the production of a newly described capsular polysaccharide [20]. It carries $bla_{\text{KPC-2}}$ and $bla_{\text{TEM-1}}$ genes, as well as a chromosomal $bla_{\text{SHV-11}}$ gene; aminoglycoside resistance genes *aac6-Ib* and *aadA2*; mutations in the QRDR region of quinolone targets (ParC-80I, GyrA-83I); genes conferring resistance to phenicols, sulfonamide, tetracycline, and trimethoprim (*catA1, sulI, tetB, dfrA12*); and has no virulence genes. The second strain was SA12 (SB4385), an ST23, K1 capsular-type isolate from a human liver abscess infection in France [9]. It carried virulence genes for yersiniabactin (*ybt 1; ICEKp10*), colibactin (*clb 2*), aerobactin (*iuc 1*), salmochelin (*iro 1*), and the regulator of mucoid phenotype genes *rmpA* and *rmpA2*; it has no resistance genes except for the chromosomal gene $bla_{\text{SHV-11}}$. Phages vB_KpnP_KL106-ULIP47 and vB_KpnP_KL106-ULIP54 were isolated against 2198 and phage vB_KpnP_K1-ULIP33 was isolated against SA12; all three from wastewater collected in France (Clichy, Saint-Denis, and Rueil-Malmaison, respectively) in 2015 using standard procedures [21]. Briefly, the wastewater samples were centrifuged at 4000 rpm for 10 min to remove large particles, then filtered and sterilized (0.45 μm). A first enrichment step was performed at 37 °C for 24 h with gentle agitation (50 rpm). When a clarification of the medium was observed, it was then centrifuged at 5000 *g* for 10 min and 20 μL of supernatant was spread on the surface of LB agar and then covered by a liquid culture of the target bacteria. After incubation for 18 h at 37 °C, individual plaques were selected and purified three times following the same procedure. These three phages produced large, clear plaques surrounded by a halo zone (Figure S1) reflecting the potential presence of an exopolysaccharide depolymerase [22]. The pH, temperature, storage stability, and the lysis kinetic curves were assessed (Figures S2–S5). The host range of the isolated phages was determined using a set of 23 *Klebsiella* spp. strains representative of diverse species and capsular serotypes (Table S1). Based on standard spot assays [23], the three phages showed specificity for the capsular type of their original bacterial host. vB_KpnP_K1-ULIP33 showed clear lysis specifically against the K1 strains, whereas vB_KpnP_KL106-ULIP47 and vB_KpnP_KL106-ULIP54 were specific for the "undefined" capsular type of their parental strain (KL106, *wzi* 29) (Table S1). This capsular specificity probably reflects the need for phages to first adsorb to and depolymerize the thick capsule. The depolymerases allowing the disruption of the polysaccharide capsule are generally K-type specific in *Klebsiella* [24–26].

To analyze the genome of these phages, polyethylene glycol (PEG) precipitation was performed, followed by CsCl density gradient (layers of 1.33, 1.45, 1.50, and 1.70 g/cm^3) ultracentrifugation (28,000 *g*; 3 h; 4 °C), dialysis using Slide-A-Lyzer dialysis cassettes G2 (Thermo Fisher Scientific Inc., Merelbeke, Belgium) and, finally, DNA extraction [27,28]. A sequencing library was obtained using the NEBNext Ultra DNA kit (New England Biolabs, Ipswich, MA, USA) and sequenced using an Illumina MiSeq instrument equipped with a nanoFlowcell (Illumina MiSeq Reagent Nano Kit v2, Brussels, Belgium, paired-end 2*250 bp reads). After correction of reads (Trimmomatic v0.38) [29], assembly (SPAdes v3.9) [30], and analysis of the genome ends (PhageTerm v1.0.11) [31], the average read coverage depths of the assemblies were 550×, 423×, and 815× for phages vB_KpnP_K1-ULIP33, vB_KpnP_KL106-ULIP47, and vB_KpnP_KL106-ULIP54, respectively. Annotation was performed with the RAST server using the virus domain option [32] followed by manual curation. All genomic data related to this project, including raw Illumina read and GenBank annotation, are available via the NCBI BioProject PRJNA488998. GenBank accession numbers are MK380014 (vB_KpnP_K1-ULIP33), MK380015 (vB_KpnP_KL106-ULIP47), and MK380016 (vB_KpnP_KL106-ULIP54). All three phages carry a linear dsDNA genome with predicted direct repeats, totaling 44,122 bp (vB_KpnP_K1-ULIP33), 41,397 bp (vB_KpnP_KL106-ULIP47), and 41,109 bp (vB_KpnP_KL106-ULIP54). Phage vB_KpnP_K1-ULIP33 has direct repeats of length 163 nt, whereas phages vB_KpnP_KL106-ULIP47 and vB_KpnP_KL106-ULIP54 have direct repeats of 180 nt. Comparative genomics of vB_KpnP_K1-ULIP33 with Enterobacteria phage Sp6, and of vB_KpnP_KL106-ULIP47 and vB_KpnP_KL106-ULIP54 with *Klebsiella* phage KP32, illustrate their genetic relatedness to reference phages [33,34] and the conserved genome organization of the *Autographivirinae* subfamily (Figure 1a–c). Distinguishing features of this phage subfamily include a unidirectional and progressive transcriptional scheme, regulated by the presence of a single subunit RNAP driving the middle/late expression. Analysis of the tailspike proteins with HMMER and HHPRED suggested the presence of tailspike-associated depolymerases present in phages vB_KpnP_KL106-ULIP47 (locus D3A56_0040) and vB_KpnP_KL106-ULIP54 (locus D3A57_0040), consistent with the presence of expanding halos in the plaques [35]. These proteins typically show a conserved (T7-related, gp17) N-terminal connector (aa1–154 pfam03906) and diverse C-terminal domains, associated with predicted pectate lyase domains. Pectate lyase domains were previously shown to have depolymerase activity against *Acinetobacter baumannii* polysaccharide capsules and against extracted exopolysaccharides [36,37]. These domains are likely associated with the capsular specificity of these phages [38,39]. Although vB_KpnP_K1-ULIP33 also induced a halo zone around the clear region of plaque lysis, suggestive of a putative depolymerase activity, no depolymerase domain was predicted. However, a tailspike protein (locus D3A55_0041) was found to have a conserved N-terminal phage_T7 connector domain (aa3–171 pfam03906). No gene related to phage lysogeny was predicted, suggesting that these phages are strictly lytic, which is an important prerequisite for phage therapy [40]. The location of the lysis cassette genes in vB_KpnP_KL106-ULIP47 and vB_KpnP_KL106-ULIP54 suggests a typical T7-related genome organization in which the endolysin is located among the middle genes, presumably having a secondary function as a regulator of the phage-encoded RNAP.

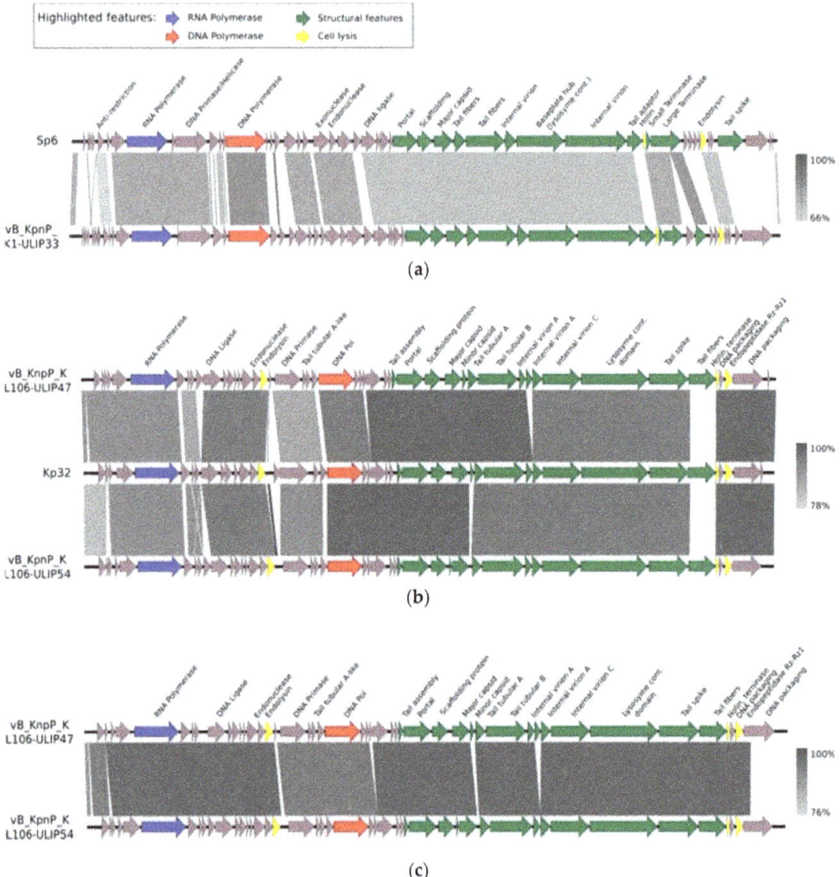

Figure 1. Comparative genomics (nucleic acid sequence) of (**a**) vB_KpnP_K1-ULIP33 with Enterobacteria phage Sp6 (Genus Sp6virus, AY288927), (**b**) vB_KpnP_KL106-ULIP47, vB_KpnP_KL106-ULIP54 with *Klebsiella* phage *Klebsiella pneumoniae* 32 (Genus *K. pneumoniae* 32virus, MH172262); and (**c**) vB_KpnP_KL106-ULIP47 with vB_KpnP_KL106-ULIP54.

To assess the potential in vivo efficacy of phages against *K. pneumoniae* in a preclinical setting with an emphasis on the prevention of infection, a *G. mellonella* larvae model was used. This model allows testing phages within a more complex system than Petri dishes and has interesting features, including similarities between the systemic cellular and humoral immune responses of these larvae and the inflammatory responses of the mammalian innate immune system [41]. Previous reports have found this model to be useful for studies of the virulence of *K. pneumoniae* and for therapeutic approaches [14,42–44]. We first determined that the optimal inoculum concentration for *K. pneumoniae* infection was 10^4 CFU/10 μL, as this dose induced a mortality rate of 70–90% in 4 days, both for strain 2198 and for strain SA12. We confirmed (data not shown) that the mortality of larvae infected with *K. pneumoniae* was dose-dependent [43]. We next assessed phage efficacy against *K. pneumoniae* infection in two independent experimental setups.

In the first experiment, we assessed the efficacy of phage vB_KpnP_K1-ULIP33 against infection by strain SA12. A total of 150 larvae were divided into five groups of 10 larvae with technical triplicates (Table S2a). In a second experiment, we analyzed the individual or combined effect of phages vB_KpnP_KL106-ULIP47 and vB_KpnP_KL106-ULIP54 on strain 2198. Here, a total of 330 larvae were divided into five groups and 11 subgroups of 10 larvae in technical triplicates (Table S2b). In both experiments, phages were administered either 1 h prior to bacterial infection (group A) or 1 h post-bacterial inoculation (group B). The timing of phage inoculation was selected in order to allow the spread of bacteria within the larvae but without allowing enough time for the infection to develop. Groups C, D, and E corresponded to assays of phage toxicity, infectivity control, and injection safety, respectively. Phages were inoculated with a multiplicity of infection (MOI) close to 10 on the left last proleg and the bacterial inoculation was performed on the right last proleg. The concentrations of the inoculated *K. pneumoniae* SA12 and 2198 were verified and were, respectively, 2×10^4 CFU/10 μL and 7×10^3 CFU/10 μL. The titers of the phage inoculums were also verified after inoculation and were 2×10^5 PFU/10 μL for vB_KpnP_K1-ULIP33, 2×10^5 PFU/10 μL for vB_KpnP_KL106-ULIP47, and 7×10^4 PFU/10 μl for vB_KpnP_KL106-ULIP54. Data from each independent experiment were pooled and the protection of the *G. mellonella* larvae by the phages was assessed with the log-rank test (p-values < 0.005 were considered as statistically significant). The Kaplan–Meier analyses were performed with the LIFETEST procedure of SAS version 9.4 for Windows and graphs were designed with SAS® ODS Graphics Editor.

Considering the different technical replicates, the first experiment, which tested the in vivo efficacy of vB_KpnP_K1-ULIP33 against SA12, showed that only 0–30% of the larvae survived in the infected groups at 4 days post-inoculation (DPI), whereas the survival rates of prophylactic and treatment groups ranged from 90% to 100%. In the second experiment, which tested the in vivo efficacy of vB_KpnP_KL106-ULIP47 and vB_KpnP_KL106-ULIP54 against strain 2198, 0–10% of the larvae survived in the infected groups at 4 DPI, whereas the survival rates of prophylactic and treatment groups ranged from 80% to 100%. In both experiments, groups of larvae inoculated with phage (but not bacteria) showed comparable survival rates as the PBS control groups, ranging from between 70% and 100% (Figure S1a,b). The survival curves are presented in Figure 2; data from the triplicate experiments were pooled. Protection of the *G. mellonella* larvae by the phages was found to be statistically significant (p-values < 0.0001 for each experiment). No significant difference was observed between the cocktail and the monophage groups. Note that despite their different stabilities (Figures S2–S5), these phages have high genetic relatedness and similar host ranges, and may therefore not be the best candidates for a phage cocktail.

These data show that the three studied phages could efficiently prevent a *K. pneumoniae* infection induced by their host strains. Both phage only and PBS control groups showed similar survival rates, demonstrating the safety of the phages in this model. A recent report indicated protection against *K. pneumoniae* ST258 infection in *G. mellonella* with another phage [14]. The present study confirms that strains belonging to this ST can be targeted by phages and reports, for the first time, on phage efficacy against ST23 *K. pneumoniae* in *G. mellonella*. The very low MOI used in this study allowed for assessment of the efficacy of the phages while avoiding the phenomenon of "lysis from without". Overall, this study confirms that the *G. mellonella* is a flexible and rapid tool to assess phage efficacy. Indeed, it accommodates many human pathogenic strains in contrast to rodent models and it allows a quick (less than 48 h in this study) evaluation of the killing activity of phages in vivo. However, the relevance of the *G. mellonella* model to predict the phage efficacy in higher animals including humans, and in particular, with higher MOIs and timings of phage administration, remains to be determined [44].

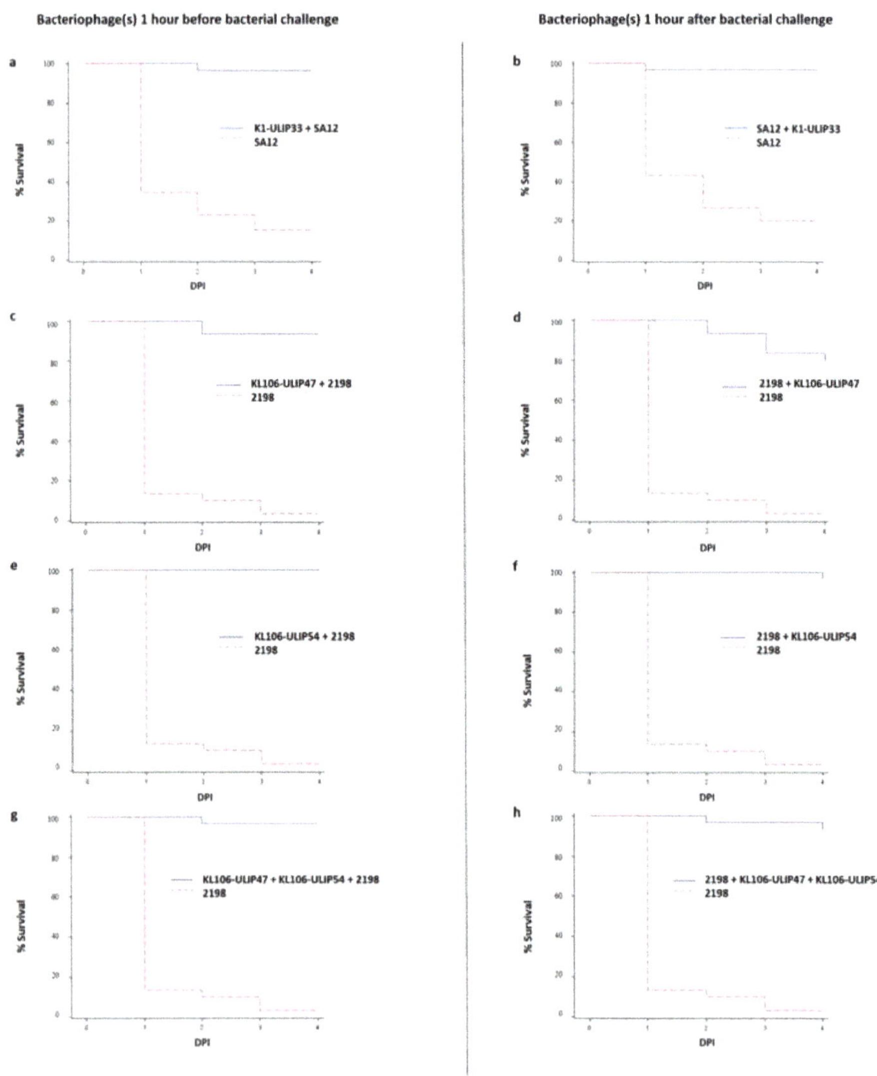

Figure 2. Kaplan–Meier survival curves of the *Galleria mellonella* larvae inoculated with *K. pneumoniae* SA12 (ST23) (**a**,**b**) and *K. pneumoniae* 2198 (ST258) (**c**–**h**) with, respectively, phage vB_KpnP_K1-ULIP33 (K1-ULIP33), and phages vB_KpnP_KL106-ULIP47 (KL106-ULIP47) and vB_KpnP_KL106-ULIP54 (KL106-ULIP54), one hour before or one hour after bacterial inoculation.

Supplementary Materials: The following are available online at http://www.mdpi.com/1999-4915/11/5/411/s1, Figure S1: Picture of halo zones of Phages (a) vB_KpnP_K1-ULIP33, (b) vB_KpnP_KL106-ULIP47 and (c) vB_KpnP_KL106-ULIP54. Figure S2: The temperature stability of phages vB_KpnP_K1-ULIP33 (A), vB_KpnP_KL106-ULIP47 (B), and vB_KpnP_KL106-ULIP54 (C). Figure S3: The pH stability of phages vB_KpnP_K1-ULIP33 (A), vB_KpnP_KL106-ULIP47 (B), and vB_KpnP_KL106-ULIP54 (C). Figure S4: The storage stability of phages vB_KpnP_K1-ULIP33 (A), vB_KpnP_KL106-ULIP47 (B), and vB_KpnP_KL106-ULIP54 (C) at 4 °C. Figure S5: Lysis kinetic curves of vB_KpnP_K1-ULIP33 lysis on the SB4385 strain (A), vB_KpnP_KL106-ULIP47 (B), and vB_KpnP_KL106-ULIP54 (C) on the SB4551 strain. Table S1: Bacterial strains characteristics and phages spot assays results. Table S2: Experimental designs of the main *Galleria mellonella* experiments with (a) *K. pneumoniae* SA12 (ST23) and phage vB_KpnP_K1-ULIP33 and (b) *K. pneumoniae* 2198 (ST258), phage vB_KpnP_KL106-ULIP47, and vB_KpnP_KL106-ULIP54. Each group contains 10 larvae and each experiment condition was reproduced in technical triplicates.

Author Contributions: Conceptualization, S.B., R.L., L.D. and D.T. Methodology, S.B., R.L. and D.T. Manipulations, D.T., V.P., K.D.-W., N.D., J.W. and L.D.S. Genomic analysis, C.L., R.L. and D.T. Writing—Original Draft Preparation, D.T., S.B. and R.L.; Writing—Review & Editing, All authors. Supervision, S.B., R.L., V.v.N. and J.G.M.

Funding: We thank the Biermans-Lapôtre Foundation for a post-doctoral fellowship that supported D.T. work in Institut Pasteur. L.D.S. is founded by a Roux-Cantarini fellowship from the Institut Pasteur (Paris, France). R.L., K.D.-W. and J.W. are supported by a GOA grant from KU Leuven ('Phage Biosystems'), C.L. holds a SB PhD fellowship from FWO Vlaanderen (1S64718N). K.D.-W. was supported by a PDM mandate from KU Leuven.

Acknowledgments: We thank Dearbhaile Morris (NUIG, Galway, Ireland) for providing strain 2198 (SB4551), Dominique Decré (Saint-Antoine hospital, Paris, France) for providing strain SA12 (SB4385) and Carla Rodrigues for assistance with *wzi* and KL typing of the isolates.

Conflicts of Interest: The authors declare no conflict of interest.

References

1. Lee, C.R.; Lee, J.H.; Park, K.S.; Jeon, J.H.; Kim, Y.B.; Cha, C.J.; Jeong, B.C.; Lee, S.H. Antimicrobial Resistance of Hypervirulent *Klebsiella pneumoniae*: Epidemiology, Hypervirulence-Associated Determinants, and Resistance Mechanisms. *Front. Cell. Infect. Microbiol.* **2017**, *7*, 483. [CrossRef]
2. Pendleton, J.N.; Gorman, S.P.; Gilmore, B.F. Clinical relevance of the ESKAPE pathogens. *Expert. Rev. Anti. Infect. Ther.* **2013**, *11*, 297–308. [CrossRef]
3. Seiffert, S.N.; Marschall, J.; Perreten, V.; Carattoli, A.; Furrer, H.; Endimiani, A. Emergence of *Klebsiella pneumoniae* co-producing NDM-1, OXA-48, CTX-M-15, CMY-16, QnrA and ArmA in Switzerland. *Int. J. Antimicrob. Agents.* **2014**, *44*, 260–262. [CrossRef]
4. Davis, G.S.; Price, L.B. Recent Research Examining Links Among *Klebsiella pneumoniae* from Food, Food Animals, and Human Extraintestinal Infections. *Curr. Environ. Heal. Reports* **2016**, *3*, 128–135. [CrossRef] [PubMed]
5. Wyres, K.L.; Wick, R.R.; Gorrie, C.; Jenney, A.; Follador, R.; Thomson, N.R.; Holt, K.E. Identification of *Klebsiella capsule* synthesis loci from whole genome data. *Microb. Genom.* **2016**, *2*, e000102. [CrossRef] [PubMed]
6. Brisse, S.; Passet, V.; Haugaard, A.B.; Babosan, A.; Kassis-Chikhani, N.; Struve, C.; Decré, D. wzi Gene Sequencing, a Rapid Method for Determination of Capsular Type for Klebsiella Strains. *J. Clin. Microbiol.* **2013**, *51*, 4073–4078. [CrossRef]
7. Pan, Y.J.; Lin, T.L.; Chen, Y.H.; Hsu, C.R.; Hsieh, P.F.; Wu, M.C.; Wang, J.T. Capsular types of *Klebsiella pneumoniae* revisited by wzc sequencing. *PLoS ONE* **2013**, *8*, e80670. [CrossRef]
8. Shon, A.S.; Bajwa, R.P.S.; Russo, T.A. Hypervirulent (hypermucoviscous) *Klebsiella pneumoniae*: A new and dangerous breed. *Virulence* **2013**, *4*, 107–118. [CrossRef] [PubMed]
9. Bialek-Davenet, S.; Criscuolo, A.; Ailloud, F.; Passet, V.; Jones, L.; Delannoy-Vieillard, A.S.; Garin, B.; Le Hello, S.; Arlet, G.; Nicolas-Chanoine, M.H.; Decré, D.; Brisse, S. Genomic definition of hypervirulent and multidrug-resistant *Klebsiella pneumoniae* clonal groups. *Emerg. Infect. Dis.* **2014**, *20*, 1812–1820. [CrossRef]
10. Cheong, H.S.; Chung, D.R.; Park, M.; Kim, S.H.; Ko, K.S.; Ha, Y.E.; Kang, C.I.; Peck, K.R.; Song, J.H. Emergence of an extended-spectrum β-lactamase-producing serotype K1 *Klebsiella pneumoniae* ST23 strain from Asian countries. *Epidemiol. Infect.* **2017**, *145*, 990–994. [CrossRef] [PubMed]
11. Holt, K.E.; Wertheim, H.; Zadoks, R.N.; Baker, S.; Whitehouse, C.A.; Dance, D.; Jenney, A.; Connor, T.R.; Hsu, L.Y.; Severin, J.; et al. Genomic analysis of diversity, population structure, virulence, and antimicrobial resistance in *Klebsiella pneumoniae*, an urgent threat to public health. *Proc. Natl. Acad. Sci. USA* **2015**, *112*, E3574–E3581. [CrossRef]
12. Vandenheuvel, D.; Lavigne, R.; Brüssow, H. Bacteriophage Therapy: Advances in Formulation Strategies and Human Clinical Trials. *Annu. Rev. Virol.* **2015**, *2*, 599–618. [CrossRef]
13. Manohar, P.; Nachimuthu, R.; Lopes, B.S. The therapeutic potential of bacteriophages targeting gram-negative bacteria using *Galleria mellonella* infection model. *BMC Microbiol.* **2018**, *18*, 97. [CrossRef]

14. D'Andrea, M.M.; Marmo, P.; Henrici De Angelis, L.; Palmieri, M.; Ciacci, N.; Di Lallo, G.; Demattè, E.; Vannuccini, E.; Lupetti, P.; Rossolini, G.M.; et al. φBO1E, a newly discovered lytic bacteriophage targeting carbapenemase-producing *Klebsiella pneumoniae* of the pandemic Clonal Group 258 clade II lineage. *Sci. Rep.* **2017**, *7*, 2614. [CrossRef]
15. Morris, D.; Boyle, F.; Morris, C.; Condon, I.; Delannoy-Vieillard, A.S.; Power, L.; Khan, A.; Morris-Downes, M.; Finnegan, C.; Powell, J.; et al. Inter-hospital outbreak of *Klebsiella pneumoniae* producing KPC-2 carbapenemase in Ireland. *J. Antimicrob. Chemother.* **2012**, *67*, 2367–2372. [CrossRef] [PubMed]
16. Wick, R.R.; Heinz, E.; Holt, K.E.; Wyres, K.L. Kaptive Web: User-Friendly Capsule and Lipopolysaccharide Serotype Prediction for *Klebsiella* Genomes. *J. Clin. Microbiol.* **2018**, *56*, e00197-18. [CrossRef]
17. Deleo, F.R.; Chen, L.; Porcella, S.F.; Martens, C.A.; Kobayashi, S.D.; Porter, A.R.; Chavda, K.D.; Jacobs, M.R.; Mathema, B.; Olsen, R.J.; et al. Molecular dissection of the evolution of carbapenem-resistant multilocus sequence type 258 *Klebsiella pneumoniae*. *Proc. Natl. Acad. Sci. USA* **2014**, *111*, 4988–4993. [CrossRef] [PubMed]
18. Bowers, J.R.; Kitchel, B.; Driebe, E.M.; MacCannell, D.R.; Roe, C.; Lemmer, D.; de Man, T.; Rasheed, J.K.; Engelthaler, D.M.; Keim, P.; et al. Genomic Analysis of the Emergence and Rapid Global Dissemination of the Clonal Group 258 *Klebsiella pneumoniae* Pandemic. *PLoS ONE* **2015**, *10*, e0133727. [CrossRef] [PubMed]
19. Wright, M.S.; Perez, F.; Brinkac, L.; Jacobs, M.R.; Kaye, K.; Cober, E.; van Duin, D.; Marshall, S.H.; Hujer, A.M.; Rudin, S.D.; et al. Population structure of KPC-producing *Klebsiella pneumoniae* isolates from midwestern U.S. hospitals. *Antimicrob. Agents Chemother.* **2014**, *58*, 4961–4965. [CrossRef]
20. Bellich, B.; Ravenscroft, N.; Rizzo, R.; Lagatolla, C.; D'Andrea, M.M.; Rossolini, G.M.; Cescutti, P. Structure of the capsular polysaccharide of the KPC-2-producing *Klebsiella pneumoniae* strain KK207-2 and assignment of the glycosyltransferases functions. *Int. J. Biol. Macromol.* **2019**, *130*, 536–544. [CrossRef] [PubMed]
21. Van Twest, R.; Kropinski, A.M. Bacteriophage enrichment from water and soil. *Methods Mol. Biol.* **2009**, *501*, 15–21. [PubMed]
22. Hughes, K.A.; Sutherland, I.W.; Clark, J.; Jones, M.V. Bacteriophage and associated polysaccharide depolymerases–novel tools for study of bacterial biofilms. *J. Appl. Microbiol.* **1998**, *85*, 583–590. [CrossRef] [PubMed]
23. Kutter, E. Phage host range and efficiency of plating. *Methods Mol. Biol.* **2009**, *501*, 141–149.
24. Lin, T.L.; Hsieh, P.F.; Huang, Y.T.; Lee, W.C.; Tsai, Y.T.; Su, P.A.; Pan, Y.J.; Hsu, C.R.; Wu, M.C.; Wang, J.T. Isolation of a bacteriophage and its depolymerase specific for K1 capsule of *Klebsiella pneumoniae*: Implication in typing and treatment. *J. Infect. Dis.* **2014**, *210*, 1734–1744. [CrossRef]
25. Maciejewska, B.; Olszak, T.; Drulis-Kawa, Z. Applications of bacteriophages versus phage enzymes to combat and cure bacterial infections: an ambitious and also a realistic application? *Appl. Microbiol. Biotechnol.* **2018**, *102*, 2563–2581. [CrossRef] [PubMed]
26. Hsu, C.R.; Lin, T.L.; Pan, Y.J.; Hsieh, P.F.; Wang, J.T. Isolation of a bacteriophage specific for a new capsular type of *Klebsiella pneumoniae* and characterization of its polysaccharide depolymerase. *PLoS ONE* **2013**, *8*, e70092. [CrossRef]
27. Ceyssens, P.J.; Miroshnikov, K.; Mattheus, W.; Krylov, V.; Robben, J.; Noben, J.P.; Vanderschraeghe, S.; Sykilinda, N.; Kropinski, A.M.; Volckaert, G.; et al. Comparative analysis of the widespread and conserved PB1-like viruses infecting Pseudomonas aeruginosa. *Environ. Microbiol.* **2009**, *11*, 2874–2883. [CrossRef]
28. Green, M.R.; Sambrook, J. Preparation of Single-Stranded Bacteriophage M13 DNA by Precipitation with Polyethylene Glycol. *Cold Spring Harb. Protoc.* **2017**, *2017*, pdb.prot093419. [CrossRef]
29. Bolger, A.M.; Lohse, M.; Usadel, B. Trimmomatic: A flexible trimmer for Illumina sequence data. *Bioinformatics* **2014**, *30*, 2114–2120. [CrossRef]
30. Bankevich, A.; Nurk, S.; Antipov, D.; Gurevich, A.A.; Dvorkin, M.; Kulikov, A.S.; Lesin, V.M.; Nikolenko, S.I.; Pham, S.; Prjibelski, A.D.; et al. SPAdes: A New Genome Assembly Algorithm and Its Applications to Single-Cell Sequencing. *J. Comput. Biol.* **2012**, *19*, 455–477. [CrossRef]
31. Garneau, J.R.; Depardieu, F.; Fortier, L.-C.; Bikard, D.; Monot, M. PhageTerm: A tool for fast and accurate determination of phage termini and packaging mechanism using next-generation sequencing data. *Sci. Rep.* **2017**, *7*, 8292. [CrossRef] [PubMed]
32. Aziz, R.K.; Bartels, D.; Best, A.A.; DeJongh, M.; Disz, T.; Edwards, R.A.; Formsma, K.; Gerdes, S.; Glass, E.M.; Kubal, M.; et al. The RAST Server: Rapid Annotations using Subsystems Technology. *BMC Genomics.* **2008**, *9*, 75. [CrossRef] [PubMed]

33. Dobbins, A.T.; George, M., Jr.; Basham, D.A.; Ford, M.E.; Houtz, J.M.; Pedulla, M.L.; Lawrence, J.G.; Hatfull, G.F.; Hendrix, R.W. Complete genomic sequence of the virulent Salmonella bacteriophage SP6. *J. Bacteriol.* **2004**, *186*, 1933–1944. [CrossRef]
34. Kęsik-Szeloch, A.; Drulis-Kawa, Z.; Weber-Dąbrowska, B.; Kassner, J.; Majkowska-Skrobek, G.; Augustyniak, D.; Lusiak-Szelachowska, M.; Zaczek, M.; Górski, A.; Kropinski, A.M. Characterising the biology of novel lytic bacteriophages infecting multidrug resistant *Klebsiella pneumoniae*. *Virol. J.* **2013**, *10*, 100. [CrossRef] [PubMed]
35. Cornelissen, A.; Ceyssens, P.J.; Krylov, V.N.; Noben, J.P.; Volckaert, G.; Lavigne, R. Identification of EPS-degrading activity within the tail spikes of the novel Pseudomonas putida phage AF. *Virology* **2012**, *434*, 251–256. [CrossRef]
36. Oliveira, H.; Costa, A.R.; Konstantinides, N.; Ferreira, A.; Akturk, E.; Sillankorva, S.; Nemec, A.; Shneider, M.; Dötsch, A.; Azeredo, J. Ability of phages to infect Acinetobacter calcoaceticus-Acinetobacter baumannii complex species through acquisition of different pectate lyase depolymerase domains. *Environ. Microbiol.* **2017**, *19*, 5060–5077. [CrossRef] [PubMed]
37. Liu, Y.; Mi, Z.; Mi, L.; Huang, Y.; Li, P.; Liu, H.; Yuan, X.; Niu, W.; Jiang, N.; Bai, C.; et al. Identification and characterization of capsule depolymerase Dpo48 from *Acinetobacter baumannii* phage IME200. *PeerJ* **2019**, *7*, e6173. [CrossRef] [PubMed]
38. Olszak, T.; Shneider, M.M.; Latka, A.; Maciejewska, B.; Browning, C.; Sycheva, L.V.; Cornelissen, A.; Danis-Wlodarczyk, K.; Senchenkova, S.N.; Shashkov, A.S.; et al. The O-specific polysaccharide lyase from the phage LKA1 tailspike reduces Pseudomonas virulence. *Sci. Rep.* **2017**, *7*, 16302. [CrossRef]
39. Oliveira, H.; Costa, A.R.; Ferreira, A.; Konstantinides, N.; Santos, S.B.; Boon, M.; Noben, J.P.; Lavigne, R.; Azeredo, J. Functional Analysis and Antivirulence properties of a new depolymerase from a Myovirus that infects Acinetobacter baumannii capsule K45. *J. Virol.* **2019**, *93*, e01163-18. [CrossRef]
40. Pirnay, J.-P.; Verbeken, G.; Ceyssens, P.-J.; Huys, I.; De Vos, D.; Ameloot, C.; Fauconnier, A. The Magistral Phage. *Viruses* **2018**, *10*, 64. [CrossRef]
41. Desbois, A.P.; Coote, P.J. Utility of Greater Wax Moth Larva (*Galleria mellonella*) for Evaluating the Toxicity and Efficacy of New Antimicrobial Agents. *Adv. Appl. Microbiol.* **2012**, *78*, 25–53. [PubMed]
42. Wei, W.J.; Yang, H.F.; Ye, Y.; Li, J.B. *Galleria mellonella* as a model system to assess the efficacy of antimicrobial agents against *Klebsiella pneumoniae* infection. *J. Chemother.* **2017**, *29*, 252–256. [CrossRef] [PubMed]
43. Insua, J.L.; Llobet, E.; Moranta, D.; Pérez-Gutiérrez, C.; Tomás, A.; Garmendia, J.; Bengoechea, J.A. Modeling *Klebsiella pneumoniae* Pathogenesis by Infection of the Wax Moth *Galleria mellonella*. *Infect. Immun.* **2013**, *81*, 3552–3565. [CrossRef] [PubMed]
44. Forti, F.; Roach, D.R.; Cafora, M.; Pasini, M.E.; Horner, D.S.; Fiscarelli, E.V.; Rossitto, M.; Cariani, L.; Briani, F.; Debarbieux, L.; et al. Design of a Broad-Range Bacteriophage Cocktail That Reduces *Pseudomonas aeruginosa* Biofilms and Treats Acute Infections in Two Animal Models. *Antimicrob. Agents Chemother.* **2018**, *62*, e02573-17. [CrossRef]

Article

Protection of Phage Applications in Crop Production: A Patent Landscape

Dominique Holtappels [1], Rob Lavigne [1], Isabelle Huys [2,*] and Jeroen Wagemans [1,*]

1 Laboratory of Gene Technology, KU Leuven, 3001 Heverlee, Belgium; dominique.holtappels@kuleuven.be (D.H.); rob.lavigne@kuleuven.be (R.L.)

2 Clinical Pharmacology and Pharmacotherapy, KU Leuven, 3000 Leuven, Belgium

* Correspondence: Isabelle.huys@kuleuven.be (I.H.); Jeroen.wagemans@kuleuven.be (J.W.)

Received: 19 February 2019; Accepted: 16 March 2019; Published: 19 March 2019

Abstract: In agriculture, the prevention and treatment of bacterial infections represents an increasing challenge. Traditional (chemical) methods have been restricted to ensure public health and to limit the occurrence of resistant strains. Bacteriophages could be a sustainable alternative. A major hurdle towards the commercial implementation of phage-based biocontrol strategies concerns aspects of regulation and intellectual property protection. Within this study, two datasets have been composed to analyze both scientific publications and patent documents and to get an idea on the focus of research and development (R&D) by means of an abstract and claim analysis. A total of 137 papers and 49 patent families were found from searching public databases, with their numbers increasing over time. Within this dataset, the majority of the patent documents were filed by non-profit organizations in Asia. There seems to be a good correlation between the papers and patent documents in terms of targeted bacterial genera. Furthermore, granted patents seem to claim rather broad and cover methods of treatment. This review shows that there is indeed growing publishing and patenting activity concerning phage biocontrol. Targeted research is needed to further stimulate the exploration of phages within integrated pest management strategies and to deal with bacterial infections in crop production.

Keywords: phage biocontrol; patent landscape; crop production

---

1. Introduction

Soon after the discovery of bacteriophages by d'Herelle and Twort at the beginning of the 20th century [1,2], the potential to use these bacterial viruses as a therapeutic was recognized. Although the first applications of phages focused on human medicine [3], other fields including agriculture soon began to explore the potential of bacteriophages as biocontrol agents [4]. The first isolation of phages infecting plant pathogenic bacteria (PPB) dates back to 1924, when it was shown that *Xanthomonas campestris* pv. *campestris*, causing black rot in *Brassicaceae,* could be lysed by the filtrate of diseased cabbages [4,5]. In the following years, interest in phages as biocontrol agents remained relatively high [4]. However, the discovery of broad-spectrum antibiotics and other bactericidal chemicals resulted in a dwindling interest in phage therapy research in general [6].

Within the agricultural sector, the prevention and treatment of bacterial infections represents an increasing challenge. For farmers, devastating losses by bacterial pathogens are generally estimated to reach 10% of the total production [7] and for some bacterial species like *Xanthomonas campestris* pv. *campestris*, crop yield can be reduced by 25% [8]. Other major threats include *Ralstonia solanacearum, Xylella fastidiosa*, and *Pseudomonas syringae* pathovars [9]. In recent years, general antibiotics like streptomycin, as well as copper-based chemicals, have been restricted in crop production to ensure public health and to limit the occurrence of resistant strains [10–13]. Because of these restrictions, the search for sustainable, natural biocontrol of PPBs has reached a critical stage, especially considering the increased food production needs [14]. Governments have decided to implement integrated pest management strategies (IPMs) as the standard for crop protection (e.g., the European Parliament and the Council of the European Union (2009) Directive 2009/128/EC of the European Parliament and of the Council of 21 October 2009) [15]. These strategies are based on the implementation of sustainable pest control strategies with the emphasis on biological control not to eradicate pests, but to maintain their populations to avoid economical losses [15]. As phages are the natural predators of bacteria, and thus fit into this framework of IPM, research on phage-based biocontrol is coming back into the picture. Some critical proofs of concept on the efficacy of phages as biocontrol agents have been demonstrated in recent years, as reviewed elsewhere [4].

A major hurdle towards the commercial implementation of these phage-based biocontrol strategies in crop protection concerns aspects of regulation and intellectual property protection. The cost to develop new crop protectants can reach $286 million and may take eleven years [16,17]. Therefore, patents can act as a tool to stimulate innovations in the field. They provide the applicant(s) the right to prohibit others to use their invention for a time-span of twenty years in a specific geographical area in which the patent has been filed. As a consequence, when an applicant wants to have an exclusive right for the invention in different countries, the patent application should be filed for each country individually (full application procedure available [18]). This exclusive right could assure a return-on-investment to the patent holder and hence serve as an incentivizing tool for research and innovation [19,20]. However, patenting biological substances like bacteriophages has been shown to be difficult [21,22]. Nevertheless, the question remains as to whether phages and phage-containing products are protectable by patents and what the scope is of these patents. Here, we present a patent landscape on the existing patents within the field of phage biocontrol using natural phages in agriculture and correlate this with a systematic survey of the scientific literature. A database containing the relevant patent documents within this research area has been set up and analyzed. The number, geographical distribution, and legal status were examined and the scope of the patents was investigated by means of a claim analysis. In parallel, a database of scientific publications within the same research topic was analyzed to enable comparisons with the patent information. This analysis allows us to draw conclusions on the amount and scope of protection of phage-based biocontrol applications in crop production.

2. Materials and Methods

2.1. Dataset Scientific Publications

A dataset consisting of scientific publications on the topic of phage biocontrol in crop production was created by searching Web of Science and PubMed using Boolean search operators combined with a set of keywords relevant for this topic ((("phage"[Title/Abstract] OR "phage therapy" [Title/Abstract] OR "bacteriophage" [Title/Abstract] OR "phage biocontrol" [Title/Abstract] OR "bacteriophage therapy" [Title/Abstract])) AND (("plant"[Title/Abstract] OR "crop"[Title/Abstract])). The database was manually curated by eliminating non-relevant papers (papers discussing human phage therapy or phage biocontrol in food). Therefore, all abstracts of the scientific publications were manually curated to verify the relevance to the topic. The last update of the dataset was on 17 February 2019 (Supplementary Table S1) and can be considered as an up-to-date snapshot of the situation.

2.2. Patent Search, Legal Status, and Geographical Distribution

An algorithm based on Verbeure et al. (2006) [23] was used to set up a dataset of relevant patent applications and granted patents. In short, a classification search was performed using the International Patent Classification system (IPC) (A01N63—C12N7) combined with a set of keywords relevant for phage applications in the agricultural sector (bacteriophage, phage, phage biocontrol, phage therapy, bacteriophage therapy, plant, crop) in public databases (EspaceNet, Patent Scope, Google Patents). The dataset was manually evaluated and was last updated 17 February 2019 (Supplementary Table S2).

The patent landscape was performed according to Huys et al. (2009) [24]. Three different categories of the legal status were used as retrieved from EspaceNet by evaluating the "Global dossier" in the case of non-European patents or the "INPADOC legal status" and "EP register" for European patents: pending (patent application in consideration), granted (patent that is active in a specific territory), dead (meaning that the application or patent is abandoned) (e.g., WO2014177996 (A1)), refused (e.g., JP2005073562 (A)), withdrawn or deemed to be withdrawn (e.g., CN103430973 (A)), or lapsed (e.g., US19840662065). A patent and/or application is considered "active" when the application is still pending or when the patent is granted. On the contrary, if the patent and/or application is dead, it is considered "non-active".

The applicants and the geographical span of the patents were evaluated by looking at the patent document and its attributed number. The documents were categorized according to continent to have a better overview (Africa (ZA), Asia (CN, IN, JP, KR), Oceania (AU, NZ), Europe (DE, EA, EP, ES, GB, HU, IT), North America (CA, US), South America (AR, BR, CL, CR, GT, MX, PE)).

2.3. Categorization of Patent Documents and Scientific Publications

To have an understanding of the core focus of patent documents and scientific publications in terms of genera of bacteria being addressed, both were categorized among the most prominent bacterial genera causing plant bacterial diseases, as determined by Mansfield et al. 2012: *Agrobacterium, Dickeya, Erwinia, Pectobacterium, Pseudomonas, Ralstonia, Xanthomonas, Xylella*, and Other [9]. Here, "Other" was defined as any other bacterial genus or if the document did not specify the bacterium addressed. The categorization was based on a claim analysis (both dependent and independent claims) of the patent documents and an abstract analysis of the scientific publications.

A detailed independent claim analysis was performed for the granted, active patents to determine the scope of the patent. Independent claims were analyzed as these generally define the broadest scope. Claims were categorized into four categories: (i) the phage itself ("phage"), (ii) the composition of the phage cocktail and the final product ("composition"), (iii) methods for producing and/or obtaining phages ("production"), and (iv) the use of phages to treat (non-human) bacterial diseases ("treatment"). One limitation of this research is that the authors had to rely on automatic translations of the claims in Asian patents in order to understand their scope. The same categories were applied on the scientific publications by interpreting the abstracts of the papers, enabling a comparison between the scope of patents and scientific papers.

Moreover, the impact of the independent claims was evaluated based on Huys et al. 2009 [24] according to Art. 69, Art. 83, and the Protocol of the European Patent Convention (EPC) ("fair protection for the patentee with a reasonable degree of certainty for third parties"). In the case of U.S. patents, the claim interpretation was based on U.S. Utility Patent Act §112, demanding a "clear written description" and "best mode for carrying out the invention". Non-European and non-U.S. patents were interpreted by the authors in a similar way. Using this methodology, three impact levels could be defined: narrow, intermediate, and broadly-defined claims. The circumvention of the claims was estimated according to the author's appraisal. Narrow defined claims (green) cover specific details of the invention and can easily be circumvented, for example, changes in the genomic sequence of the phage, adaptations to the product composition, different protocols, and methods of treatment not covered by the claims. Intermediate claims (orange) cover the invention as such without describing details (e.g., a specific phage for a specific bacterium, a composition of a product or production process). These claims can be circumvented, though this would require substantial inventiveness. The authors categorized production and treatment claims as intermediate when these claims cover broad methods of production of treatment, but only for a particular phage. On the other hand, broadly defined claims (red) cover every aspect of the invention, but are vaguely described and hence claim outside the invention. However, no full freedom-to-operate analysis for each individual invention was conducted, as this is beyond the scope of this study.

3. Results

3.1. Patent Search, Scientific Papers, and Legal Status

Searching the different databases resulted in the identification of 49 different patent families within the field of phage biocontrol in crop production. A patent family is defined as a group of patents and/or patent applications that have been filed in different countries, but protect one and the same invention and have the same inventor(s). To highlight specific patent families, the first attributed application number was chosen to represent the family. Within the database, the families comprise in total 97 patents and applications (both active and non-active). Figure 1 shows the different patent families organized per year (priority year—dark blue area chart). It shows that the number of patent families slightly increases in time (peak at 2013) and decreases again as less patents were filed in 2016. The information on patent applications from 2017, 2018, and 2019 is not complete as these applications may not have been published yet.

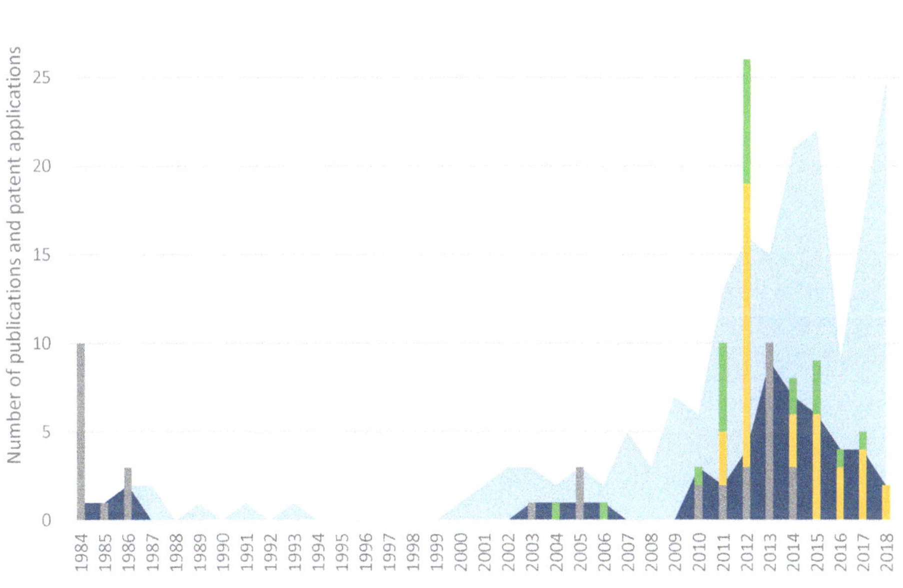

Figure 1. An overview of the number of scientific publications, patent families, patents, and patent applications in the field of phage biocontrol in crop production organized by year. The light blue area chart represents the number of scientific publications and the dark blue area chart the number of patent families per priority year. The bars represent the number of patent applications per priority year: green ("Granted patents") corresponds to the number of granted patents, yellow ("Pending applications") to pending applications, and grey ("Dead applications and patents") to dead patents and applications. This last group consists of patents and applications that are abandoned, refused, withdrawn, deemed to be withdrawn, or lapsed.

In sharp contrast, 137 scientific publications (1984–2019) were found within Web of Science and PubMed (Figure 1—light blue area chart). Publications on phage biocontrol were scarce in the eighties and nineties of the 20th century. However, from the year 2000 onwards, this number has increased steadily (25 peer reviewed publications in 2018). In other words, there is an increasing trend in the number of publications over the past decade, demonstrating a discrepancy between scientific publishing and patent filing. However, not all patent applications from 2017, 2018, and 2019 are publically available, as the 18-month period before publishing has not passed yet.

When looking at the patent documents in more detail (Figure 1—bar charts), a distinction should be made between "Granted", "Pending", and "Dead", based on their legal status as derived from Espacenet. In total, 59 patents and patent applications (61%) are active, 22 patents (23%) have been granted, 37 applications (38%) are pending, and 38 patent documents (39%) can be considered dead. From the latter, 39% are applications that were deemed to be withdrawn, 16% are rejected applications, and 16% are lapsed patents. Figure 1 shows an overview of the percentages of granted, pending, and dead documents with the same priority year. The first active patent from the dataset dates from 2004 (JP4532959 (B2)). In 2011, there were two families filed, represented by GB20110010647 and JP20110102153, containing in total six patents and four applications, respectively. All the applications and patents within the first family remain active—50% is granted and 50% is pending—whereas two out of four patent applications from the latter family have been rejected. The year 2012 has the most patents—26 in total, divided among four different families represented by GB20120017097, KR101584214, MX2012011440, and US201261716245. This last family contains four granted patents and sixteen applications. The family represented by GB20120017097 consists of one pending and two dead applications and one granted patent (US9278141 (B2)). In 2013, there was a peak in the amount of patent families filed as the number reached nine families. These families consist in total of ten applications, all of which are dead. The majority of these applications were deemed to be withdrawn (80%), meaning that the designation fee was not paid.

3.2. Applicants and Geographical Distribution

Analyzing the applicants of the different patent documents, it shows that 56% (54) of the 97 patent documents have been filed by academia, whereas 37% (36) are linked to industry (without joint applicants), and 7% (7) are joint applicants. This means that the family, or patent documents belonging to a family, have more than one applicant and thus the rights of the patent are distributed among the different partners. The most prominent academic applicants based on the amount of patent families filed by these applicants are the University of Hiroshima (JP) (21% of patent families –6/28) and the Rural Development Administration (KR) (14%, 4/28). The most prolific company in terms of patent filing is Qingdao Biological Technology Corporation LTD (CN), accounting for 50% (8/16). However, it is worthwhile to mention that all patents within the database from the Qingdao Biological Technology Corporation LTD have been withdrawn. Remarkably, only 4 out of 21 granted patents are filed by applicants from the private sector and all of these patents belong to the same family (represented by GB20110010647—Fixed Phage).

Figure 2 provides an overview of the relative contributions by countries in terms of filed and granted patent applications. One percent of all patent documents were filed in Africa (ZA), 3% in Oceania (AU, NZ), 13% in Europe (DE, EA, EP, ES, GB, HU, IT), and 13% worldwide. The majority of the patent documents have been filed in North America and Asia—16% in the USA and Canada and 43% in Asia. These Asian documents consist of 45% (19/42) Chinese, 31% (13/42) Japanese, 21% (9/42) Korean, and 2% (1/42) Indian applications and patents. When analyzing the legal status of the different patents organized per continent, 33% (14/42) of the Asian patents have been granted, 26% (11/42) of the applications are pending, and 40% (17/42) are dead. The large portion of dead patents and applications is also visible among the European (54%, 7/13) and North American (44%, 7/16) documents. Only 8% (1/13) of European patents have been granted, while this figure is 25% (4/16) for North American patents. The amount of pending applications is similar at 38% (5/13) and 31% (5/16) of the European and North American applications, respectively. On the other hand, in the case of the South American patent applications, 89% (8/9) are pending and 11% are dead. Notably, 7/9 of these South American patents are part of the same patent family (US2016309723).

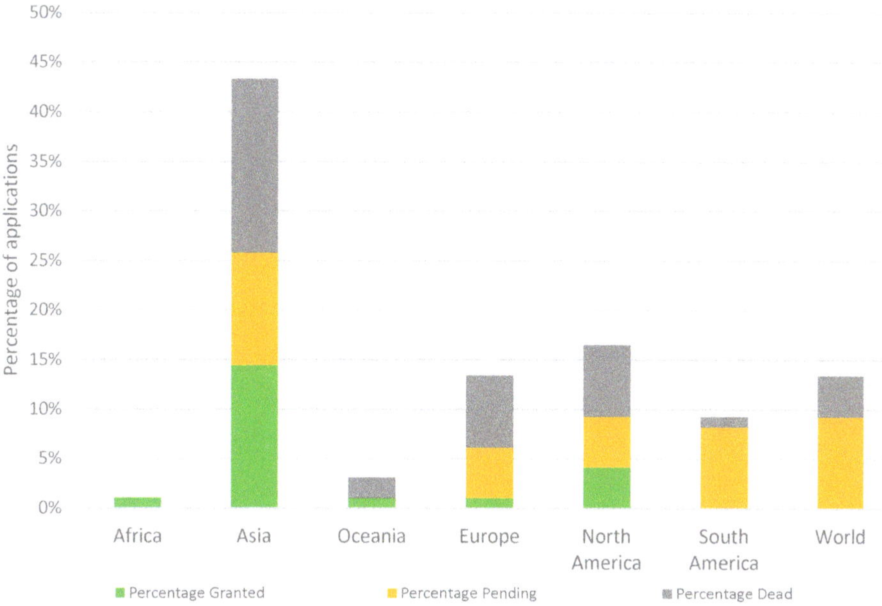

Figure 2. Percentage of patents and applications organized per continent. The total height of the bars indicate the percentage of patents and applications per continent. Africa (ZA), Asia (CN, IN, JP, KR), Oceania (AU, NZ), Europe (DE, EA, EP, ES, GB, IT), North America (CA, US), South America (AR, BR, CL, CR, GT, MX, PE), and world applications. In green, the percentage of granted patents; in yellow, the percentage of pending applications; and in grey, the percentage of dead applications and dead patents.

Furthermore, China has the highest percentage of dead patents, with 63% (12/19). The majority of these patent were deemed to be withdrawn after the admission fee was not paid (92%). On the contrary, Japan and Korea have the highest amount of granted patents, 38% (5/13) and 67% (6/9) of the patents filed, respectively, are granted and still active.

3.3. Categorization of Patents and Scientific Publications

As the most prominent bacterial species belong to the genera of *Agrobacterium, Dickeya, Erwinia, Pectobacterium, Pseudomonas, Ralstonia, Xanthomonas,* and *Xylella* [9], patent families and scientific publications were categorized and quantified among these groups according to (in)dependent claims (patent applications) and abstracts (scientific publications) (Figure 3). Combinations of genera were created when phages against certain pathogens were combined in a single cocktail (*Dickeya/Pectobacterium* and *Xanthomonas/Xylella*). A patent family or publication was classified as "Other" if it concerns a phage infecting bacteria of other genera (e.g., *Streptomyces*—"McKenna, F., et al. 2001. Novel In vivo use of a polyvalent Streptomyces phage to disinfest Streptomyces scabies-infected seed potatoes. Plant Pathol. 50:666–675." [25]) or if the paper/publication has no specification of the type of phage neither in the independent nor in the dependent claims (e.g., WO2016154602—"A method of preparing a phage composition for killing or degrading fitness of a pest").

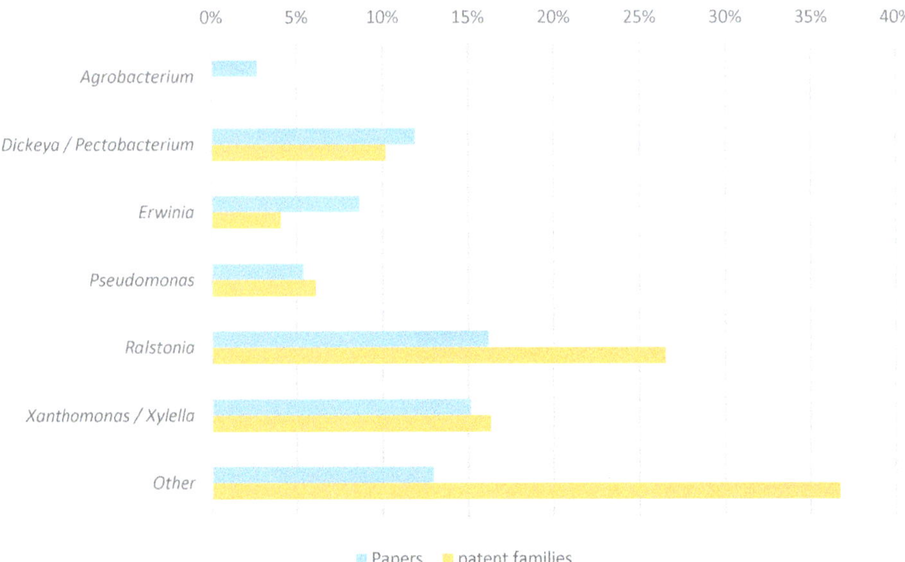

Figure 3. Distribution of patent families and scientific publications classified according to the bacterial genera that is tackled by the phage product. On the basis of Mansfield et al., 2012, seven categories (*Agrobacterium, Dickeya/Pectobacterium, Erwinia, Pseudomonas, Ralstonia, Xanthomonas/Xylella,* and Other) were made to classify patent families and publications. Groupings of bacterial genera (*Dickeya/Pectobacterium* and *Xanthomonas/Xylella*) were created as phage cocktails were created to tackle both bacterial genera. The last category "Other" consists of patent families and publications that do not specify the bacterial pathogen that is being targeted or that tackle an alternative bacterial genera.

The least researched bacterial genera within both databases are *Agrobacterium* (0% patent families, 3% publications), *Pseudomonas* (6% patent families, 5% publications), and *Erwinia* (4% patent families, 9% publications). The most represented bacterial genera within the datasets as suggested by Figure 3 are *Ralstonia, Xanthomonas/Xylella,* and *Dickeya/Pectobacterium*. Sixteen percent of the scientific publications discuss phages infecting *Ralstonia*; 15% *Xanthomonas, Xylella,* or a combination of *Xanthomonas and Xylella*; and 12% *Dickeya, Pectobacterium,* or a combination of *Dickeya* and *Pectobacterium*. In the case of the patent applications, a similar trend can be observed—27% *Ralstonia*, 16% *Xanthomonas / Xylella*, and 10% *Dickeya / Pectobacterium*.

A closer inspection of the different categories reveals that three patent families within the category of *Ralstonia* have included a private company as co-applicant (23%, 3/13). On the other hand, inventions from all patent families filed by public institutes have also been published in scientific papers (Supplementary Table S3). In the case of the category of *Xanthomonas / Xylella*, a similar trend can be observed, as there are three families filed by industrial applicants represented by WO2015200519 by Auxergen (US), US2017142976 by Fairhaven Vineyards (US), and HU1700178 Envirolnvest Koernyezetvedelmi es biotechnologiai (HU).

The other families are filed by the University of Hiroshima (JP), Texas A&M University (US), University Huazhong (CN), and the National institute for Agro-Environmental Sciences (JP), all situated in the public sector. This is also the case for *Dickeya* and *Pectobacterium*, as no patents were filed by private institutes. The applicants here include the Rural Development Administration (KR) and the national university of Seoul (KR). The majority of the families filed by private institutes are classified as "Other" (37%), indicating that the patents are not discussing the major genera of bacteria, but rather do not specify the phages nor their host. Hence, the claims are defined broadly.

Within this group, the majority of the families are filed by industrial applicants (88%): Fixed phage (GB), Epibiome (US), Qingdao Biological Technology Corporation LTD (CN), and Internalle (US).

3.4. Claim and Abstract Analysis

A claim analysis was performed to determine the scope of the individual patents along with an analysis of the scientific publications. Among the dataset of 91 patent documents analyzed, 21 were active, granted patents. In total, 79 independent claims were systematically analyzed and classified into four different categories (Figure 4): (1) Phage (e.g., "*A bacteriophage able to lyse cells of Ralstonia solanacearum selected from the group of the following: (a) vRsoP-WF2 (DSM 32039), vRsoP-WM2 (DSM32040), vRsoP-WR2 (DSM32041); or (b) a podovirus whose genome has the sequence of SEQ ID NO: 1 (corresponding to vRsoP-WF2), SEQ ID NO: 2 (corresponding to vRsoP- WM2), or SEQ ID NO: 3 (corresponding to vRsoP-WR2)*"—ES2592352), (2) Composition (e.g., "*A composition for inhibiting or preventing the growth of Pectobacterium carotovorum, which comprises as an active ingredient bacteriophage PM-2 (KACC97022P) having an entire genome sequence consisting of the nucleotide sequence of SEQ ID NO: 1.*"—KR101797463), (3) Production (e.g., "*A method of propagating a virulent bacteriophage that includes X fastidiosa in its host range, comprising the steps of the following: (a) infecting a culture of Xanthomonas bacteria with said virulent bacteriophage; (b) allowing said bacteriophage to propagate; and (c) isolating virulent bacteriophage particles from the culture.*"—"Production claim": AU2013331060), and (4) Treatment (e.g., "*A method of preventing or reducing symptoms or disease associated with Xylella fastidiosa or Xanthomonas in a plant, comprising contacting said plant with particles of at least one virulent bacteriophage, wherein Xylella fastidiosa and/or Xanthomonas axonopodis are hosts of the bacteriophage, wherein the bacteriophage is a Xfas 300-type bacteriophage and displays the following characteristics: (a) the bacteriophage is capable of lysing said Xylella fastidiosa and/or Xanthomonas bacteria; (b) the bacteriophage infects a cell by binding to a Type IV pilus; (c) the bacteriophage comprises a non-contractile tail with a capsid size ranging from 58–68 nm in diameter and belongs to the Podoviridae family; (d) the genomic size of the bacteriophage is about 43,300 bp to 44,600 bp; and (e) the bacteriophage prevents or reduces symptoms associated with Pierce's disease in a plant or plants.*"—US9357785). The same categories were used to classify the scientific publications based on the abstract of the publication. Both scientific papers and patents can fall in different categories as they may discuss one or more phages and their basic characterization, the composition of a cocktail, and the testing of this cocktail in bioassays and/or field trials. As Figure 4 shows, there are differences between the relative contributions of scientific papers and patents that address a specific topic. The majority of the scientific papers focus on the isolation and basic characterization of one or more phages (84%). Thirty-six percent of the manuscripts discuss the composition of a phage cocktail and 39% use this cocktail in bioassays and/or field trials. The least represented topic discusses production methods of phage (4%). In contrast, the majority of the patents (90%) protect the use of phages to treat a plant in one way or another. Seventy-six percent of the patents claim the composition of a product containing phage, 62% protect the phage itself, and 24% contain claims protecting the production of the phage product. On the other hand, no patents within the database contain claims to protect methods of detection nor of the phage nor of the host.

When evaluating the 77 independent claims of all active, granted patents, the majority of the claims are process claims protecting different methods to use phages as a treatment for bacterial infection (40%—Supplementary Figure S1). Furthermore, this data shows that 30% of the claims protect the composition of the phage product. This means that a combination of phages is protected and/or the formulation of the final product. Nineteen percent can be considered as compound claims as the claim is protecting the phage(s) as an active ingredient and only 10% of the claims are production claims.

The 77 independent claims were also categorized within three classes: narrow, intermediate, and broadly-defined. This gives an indication of whether a specific claim can be easily circumvented (narrowly-defined) or not (broadly-defined). Table 1 gives an overview of all the granted patents and the independent claims that belong to these patents. Twenty-six percent of the claims are narrow claims (e.g., "*The invention relates to a method for applying a bacteriophage of R. solanacearum, characterized in that the bacteriophage liquid of R. solanacearum is placed in a sterile syringe needle obliquely inserted into the stem of the tobacco plant, and then the bacterial phage liquid is covered with sterile mineral oil to prevent evaporation. And pollution, so that the R. solanaceans phage directly enters the stem of the tobacco through a sterile syringe needle; the sterile mineral oil is prepared by pouring 300 mL of mineral oil into a 500 mL screw bottle and sterilizing at 121 °C 30 min, then stored and reserved after cooling; the sterile syringe needle is prepared by using a sterile blister needle overnight, rinsing twice with sterile water, drying at 50 °C after autoclaving, and after cooling to room temperature, Packed and stored for use; each tobacco stem is injected with 50–100 µL of R. solanacearum phage solution; in the sterile syringe needle inserted into the stem of the tobacco plant, the amount of sterile mineral oil is 50–100 µL*"—"Treatment claim": CN104542717). Thirty percent of the claims are intermediate as they do not cover specific details on the invention (e.g., "*A method for controlling bacterial wilt disease bacteria, which comprises spraying the bacteriophage strain according to any one of claims 1 to 3 to a plant or soil.*"—"Treatment claim": JP4532959). Finally, 44% can be defined as broad claims (e.g., "*An isolated bacteriophage which is toxic to X. fastidiosa and Xanthomonas species, wherein said bacteriophage is at least one member selected from the group consisting of Xfas 100 phage type bacteriophage and Xfas 300 phage type bacteriophage Of the bacteriophage, wherein the Xfas 100 phage type is selected from the group consisting of SEQ ID NO: 11, SEQ ID NO: 12, SEQ ID NO: 13, SEQ ID NO: 14, SEQ ID NO: 15, SEQ ID NO: 16, SEQ ID NO: 17, and SEQ ID NO: 18 Wherein the Xfas 300 phage type comprises a genome having a DNA sequence that is 99% or more identical to the sequence of SEQ ID NO: 19, SEQ ID NO: 20, SEQ ID NO: 21, SEQ ID NO: 22, SEQ ID NO: 23 and SEQ ID NO: 24 Select from It comprises a sequence a genome having the same DNA sequence 99% or more that is, the bacteriophage.*"—"Phage claim": JP6391579).

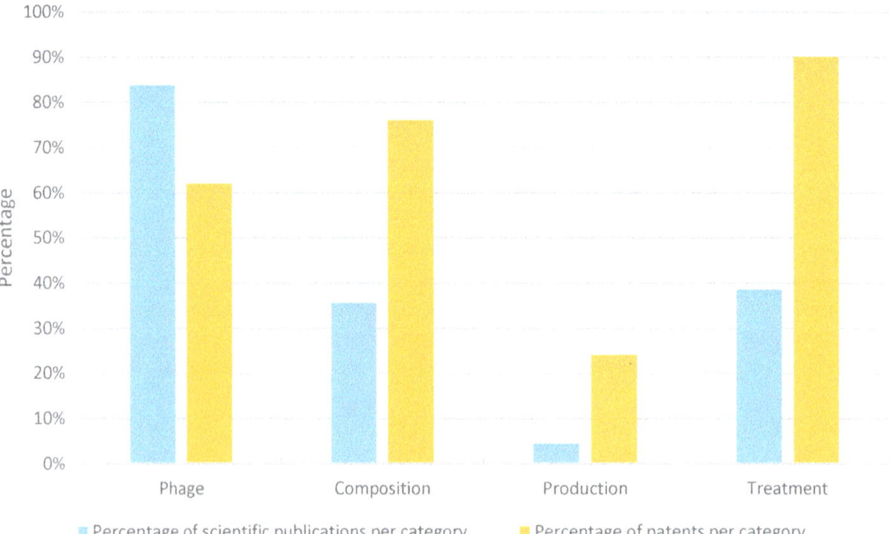

Figure 4. Claim and abstract analysis of the active, granted patents and scientific publications. In total, 79 independent claims from 21 patents and 137 abstracts from scientific papers were categorized among four different categories: (1) Phage—here the phage was described as the active ingredient or the isolation of a phage was described, (2) Cocktail—this category contains claims that protect the combination of phages and publications that describe a phage cocktail, (3) Production—ways of how the phage is produced, and (4) Treatment—claims that protect the use of phages to fight a specific bacterial infection or methods and application strategies for using the phage (e.g., bioassays, field trials). In blue, the percentages of publications are shown; in yellow, the percentage of claims are shown. Note: one publication and patent can be categorized in multiple categories.

Table 1. Claim analysis of the granted patents. Within this table, all granted patents have been organized according to the bacterial genera they describe, their patent number, the applicant (private sector in blue and public sector in yellow) and the nationality of the applicant (CN China, ES Spain, GB United Kingdom, JP Japan, KR Korea, and US United States), and the filing year. Moreover, the claims were categorized according the four different categories: (1) Phage: claims protecting one or multiple phages and their genome sequence, (2) Composition: claims discussing the composition of a phage product, (3) Production: methods to produce phages, and (4) Treatment: methods to use phages as a treatment for plant diseases. If a patent contains claims that belong to a specific category, it is quantified. The colors indicate the scope of a claim or set of claims: very narrow, intermediate, or broadly-defined (green, orange, and red, respectively).

Bacterial Genera	Patent Number	Applicant	Filing Year	Type of Claim			
				Phage	Composition	Production	Treatment
Other	US9539343	Fixed Phage (GB)	2012		1		1
	CN103747792	Fixed Phage (GB)	2012		1		1
	JP6230994	Fixed Phage (GB)	2012		1		1
	US9278141	Fixed Phage (GB)	2013		5		
	CN104630154	University of Zhejiang (CN)	2015	1	1	1	2
	KR101887987	Pukyong National University (KR)	2016	1	1		1
Dickeya/ Pectobacterium	KR101368328	Rural development Administration (KR)	2010	1	1		
	KR101790019	Rural development administration (KR)	2014	1	1		1
	KR101797463	Rural development Administration (KR)	2014	1	1		1
	KR101891298	Seoul National University (KR)	2016	1	2		1
Pseudomonas	KR101584214	Chungbuk National University (KR)	2012		1		1
Ralstonia	JP4532959	Sanin Kensetsu Kogyo (JP)	2004	3	2		2
	JP4862154	University of Hiroshima (JP)	2006	1	1		3
	US9380786	University of Hiroshima (JP)	2012				1
	JP5812466	University of Hiroshima (JP)	2011	1			3
	CN104542717	Guizhou Tobacco science research institute (CN)	2015			1	1
	ES2592352	University of Valencia (ES)	2015	1	1		2
Xanthomonas/ Xylella	AU2013331060	Texas A&M University (US)	2013	1	1	2	4
	JP6391579	Texas A&M University (US)	2013	1	1	2	2
	US9357785	Texas A&M University (US)	2013				1
	ZA201502230	Texas A&M University (US)	2013	1	1	2	2

The claims categorized as "Phage claims" are generally narrowly-defined (41%), as the claims protect the phage and its genomic sequence. However, highly similar phages with differences in their genomic sequence fall outside the scope of these claims. Thus, the claims can be circumvented. A judge, however, may interpret by the claim as equivalent by the doctrine of equivalence. The isolation of a different phage that targets the same bacterium also circumvents the claim. On the other hand, the majority of the "Composition claims" and "Production claims" are difficult to circumvent as they are broadly defined (56% and 75%, respectively).

As mentioned above, 4 out of 21 granted patents have been filed by a private organization (Fixed Phage) and these patents belong to the same patent family. When looking at the claims for these patents, one notices that all the claims are broadly-defined. As the invention discusses the use of phages in a composition to fight bacterial infections without defining the phage itself, the majority of the claims protect the composition of the product rather than the phage or the production of the product.

4. Discussion

Within this study, we analyzed publishing and patenting activities in phage biocontrol as a crop protectant using the number of scientific papers and patent documents as a measuring tool for assessing interest in this area. Furthermore, we looked into the applicants, geographical distribution, legal status, and scope of the patent documents by categorizing both the patents and scientific manuscripts by claims and abstracts. This allowed us to make a general comparison between the focus of patents and scientific literature and the willingness of R&D to look into the potential of phages as a part of IPM.

4.1. Despite a Growing Interest in Phage Biocontrol, Patenting Activities Remain Limited

The increasing number of scientific publications (Figure 1) shows that there is indeed a growing interest in using phages as an alternative to existing plant protecting products. The number of patent filings seems to fall behind in 2016, but increases slightly in 2017. However, it is premature to determine a trend because not all the applications filed in 2017 and 2018 have passed the 18-month period before publication and thus are not made publicly available [26]. Fifty-six percent of the patent documents were filed by non-profit organizations like the University of Hiroshima (JP), the Rural Development Administration of South-Korea (KR), and Texas A&M University (US). This confirms the slightly stronger interest from non-profit organizations in protecting phage biocontrol inventions compared with industry. It might also suggest that private companies are still dangling in the start-up phase or have not picked up the topic yet (apart from a few early adopters). This could also indicate that the expertise still remains in non-profit and needs to be transferred to the private sector.

The assessment of the legal status of the documents shows a high fraction of dead documents (39%). Within this group of dead applications, 16% of the applications were rejected as the invention was not considered novel (US2009053179) and/or did not include an inventive step (JP2005073562). The majority of the dead documents are applications that were deemed to be withdrawn (39%) and patents that are lapsed (16%—no oppositions were filed, e.g., EP0182106). In both cases, the applicants did not pay the fees needed to maintain the rights of the patent. For some patent applications, this is because of a negative search report (GB2519913 and WO2007044428), in which the invention was not found novel nor inventive. On the other hand, not paying the fees may also imply internal shifts of interest for commercial or other reasons. Hurdles in the regulation of biopesticides may influence such decisions as well. In Europe, for instance, the registration exists for two phases: registration of the active ingredient at European level by the European Food Safety Authority (EFSA), and the authorization of the formulated product at member state level leading to bureaucratic difficulties and high cost [17]. In the United States, a similar procedure should be taken as the product needs to be registered by the Environmental Protection Agency (EPA) [27]. Other countries including China and South Korea have similar agencies [28,29]. As phages are highly specific and can only infect a few strains of one bacterial species, they are often combined in a cocktail [30]. In terms of registrations, this means that every phage in the cocktail should be registered as an active ingredient (Regulation (EC) No

1107/2009) and reformulations of the product require re-registrations [4]. The cost of such registrations may be high, for example, the registration costs of products corresponding to a "New Active Ingredient, Non-food use; outdoor; reduced risk" can reach $436,004 in the United States [31] and in Europe, the total R&D costs (including registration fees) are estimated to reach $286 million [16]. Moreover, phage genomes are variable as a result of evolutionary fluxes [32], thus phages cannot be considered as stable, fixed products that could impede registration [33]. Luckily, this can partially be addressed by recent insights on genome-based taxonomy that phages sharing 95% nucleotide similarity are considered as isolates the same species [34]. Changes in the regulation including fast track registration, priority registration, and zonal authorization (i.e., authorization for all of the EU instead of registration per country) could function as an incentive to further develop innovations and promote patenting in biocontrol [35]. These reasons could also explain the low number of filed applications in Europe and North America, as observed in Figure 2. Nevertheless, a first phage product line, AgriPhage™, was registered in the United States (2005) by Omnilytics (part of Phagelux). AgriPhage™ is approved by the Environmental Protection Agency (EPA) and contains products with phages against *Xanthomonas campestris* pv. *vesicatoria* and *Pseudomonas syringae* pv. *tomato*. Currently, Omnilytics has added cocktails against *Erwinia amylovora* and *Clavibacter michiganensis* subsp. *michiganensis* to their product line [36]. Notably, none of these products or phages were found by the authors in the public assessed databases. This may imply that the patents could be licensed (e.g., from an academic partner); the applications are not publicly available yet; or the product is not patented, opening the discussion of whether patenting is indeed crucial for commercial activities of phage applications in agriculture. Also, in Europe, a few phage-containing products are on the market. In Hungary, Enviroinvest has a product (available since March 2018) on the Hungarian market called Erwiphage PLUS, containing bacteriophages targeting *Erwinia amylovora* [37]. Moreover, the Scottish company APS Biocontrol distributes a product line Biolyse®, a food-packaging aid that contains phages that prevent soft rot disease in stored potatoes [38].

4.2. *Efforts in Asia to Protect Phage Biocontrol Preparations*

The geographical distribution of the patent documents (Figure 2) shows that the majority were filed in Asia (43%). This observation might be explained by local governmental stimuli to promote patenting to cause rapid economic growth [39,40]. China, for instance, has implemented so-called patent promotion policies. These policies stimulate patenting as tax incentives and subsidies are linked to patent ownership, and have caused a booming growth in the number of Chinese patent applications and patents [40]. Furthermore, Asia, in general, is taking measures to reduce the amount of chemical pesticides and fertilizers. China, for instance, has launched a "national research program on reduction in chemical pesticides and fertilizers in China" ($340 million) [35]. Within the study presented here, 19% of all documents were filed in China. In contrast, 63% of these documents are dead, from which 92% are applications that were deemed to be withdrawn as fees were not paid. This might implicate that the applicants lost their confidence in the invention. The number of granted, active patents is high in Japan and South Korea, further illustrating Asian efforts to develop new biocontrol strategies. A study from the Food and Agriculture Organization from the United Nations (FAO) and the Asia and Pacific Plant Protection Commission (APPPC) has shown that both Japan and South Korea are promoting integrated pest management strategies by setting pesticide reduction targets (KR) and appointing IPM expert groups (JP) [41].

4.3. *Scientific Publications and Patent Documents Show a Correlation in Terms of Targeted Pathogens*

To get a more profound idea about the focus of the patent and scientific literature, both datasets were classified according to the most prominent bacterial genera causing plant diseases [9] (Figure 3). In general, the percentages of scientific publications and patent families discussing a specific bacterial genus correspond quite well. Within the datasets presented here, the most studied bacterial genus is *Ralstonia*; not surprisingly, as these bacteria cause disease in a wide range of cash crops like tomato, potato, tobacco, eggplant, and banana [42]. In the dataset, tomato is the most researched crop.

The majority of the documents categorized within this group were filed by non-profit organizations rather than industrial applicants. This suggests that the actual interest from industry is still rather limited compared with that of academia. The second most prominent group is the combination of *Xanthomonas* and *Xylella*. Both pathogens can have a significant impact in crop production and hence, it is not remarkable that these bacteria are largely represented in phage biocontrol research [43,44]. Furthermore, the majority of the applications were filed by the public sector. The *Pseudomonas* genus, including that of *Pseudomonas syringae* and its pathovars, is less represented within both databases. This is striking as *P. syringae* pathovars are considered as one of the most important PPBs causing disease in different crops (tomato, bean, kiwi, leek, and others) [45]. As Figure 3 shows, there is a discrepancy within the "*Other*" group. This category combines documents that talk about other bacterial genera or do not specify the bacterium that is being addressed. The majority of the patents and patent applications were classified in this category because these do not specify a particular bacterial genus. From these patent documents, we observe that all the granted patents filed by industrial applicants have chosen this option (Table 1). This might suggest that industry defines claims containing bacteriophages as broad as possible to get patent exclusive rights to any phage product that falls under the protection of the patent. By applying this strategy, companies are taking risks in their patenting strategy because these broad claims are more susceptible in terms of novelty anticipation, which may invalidate them [46].

4.4. Granted Patents Include Broad Claims

The main focus of the scientific publications and the granted patents indicate a difference in emphasis (Figure 4). While the majority of the manuscripts discuss the isolation of one or more bacteriophages and their basic characterization, the majority of the patents claim a method of treating a bacterial infection by means of a phage product. Only 39% of scientific papers address the latter, although this could be considered as the ultimate goal of phage biocontrol. This illustrates that first efforts are made in the field, but that there are still opportunities for further innovations.

Limited efforts have been made towards patenting detection strategies for and by phages in the field of phage biocontrol in crop production. As phages are highly specific to a specific bacterium and can locate their host in a complex matrix of bacteria, they can easily be used as a detection tool [47]. However, this study suggests that little published evidence is available in this area by both the scientific community and the industrial early adopters, as there are no claims protecting possible methods of detection and only a small minority (8%) of scientific papers discuss this matter.

The claims that belong to the "Phage" category consist of claims that protect a natural occurring phage. According to Art. 3 of the Biotechnology Directive 98/44/EC (European Commission), natural phages can be patented because the phage is isolated from the environment [22]. The techniques to isolate phages, however, are similar to those used in 1920, which makes the patentability of natural phages fragile [21]. Nevertheless, claims for isolated phages can be of great value as patents covering such claims may give the right to prevent others from using that particular phage. On the contrary, as phages are abundantly present in the environment, the chances of finding a similar, yet different phage, which may circumvent the claim, exists. However, as the literature on phages applicable for phage biocontrol increases, anticipation of novelty as a patentability requirement of a particular phage may become an issue. In the United States, the requirement of novelty is more complicated, as the phage will be excluded for patentability if on the one side it is known or used within the United States, or on the other side, published or patented wherever in the world [22].

Figure 4 also shows that there is limited evidence on the optimization of the production of phage cocktails based on both scientific and patent literature. One could argue that the production of phages is phage-dependent (specific bacterial strains, media, temperature) and thus keeping the production of phages secret could be a valid strategy to maintain a competitive edge [19].

On the other hand, many patents claim the combination of phages or phages as part of a formulation. This is illustrated by Supplementary Figure S1. Claiming a combination of phages can act as a buffer to reduce the risk of resistance development of the target bacterium. Phages are known to have different infection strategies and highly diverse genomes, leading to the chance that a certain bacterium develops resistance (by altering receptors, CRISPRs, restriction enzymes) against two or even three phages in one cocktail are theoretically small [4,48,49]. The strength of these claims can be questioned as minor changes to the cocktail could already be sufficient to circumvent claims.

Table 1 depicts in which categories the individual patents can be classified. Here, it is clear that the different patents combine different types of claims. Combined with the scope of the claims, this might support the previous statement that companies employ a "throw everything at the wall to see what sticks" patenting strategy because of the uncertainty and the ignorance in how to achieve a strong protection for the invention [19].

4.5. Phages and Other Viruses as Part of an Integrated Pest Management Strategy

To safeguard public health and minimize impact on the environment, traditional pesticides used in crop production are currently being stigmatized. This triggers the research community to evaluate new, alternative methods to deal with different kinds of pests like fungi, bacteria, and insects. The use of relatively high concentrations of natural predators to eradicate a certain pest, also known as augmentative biocontrol, is gaining in popularity [35]. In this regard, bacteriophages and other viruses, like insect viruses and mycoviruses, are ideal candidates as biocontrol agents because they are sustainable, specific, and do not leave residues on the crop. Moreover, they fit into the framework of integrated pest management, where the use of biocontrol agents is heavily promoted [35]. Viruses as biocontrol agents are, however, sometimes overlooked in IPM [15], which indicates that more efforts need to be taken to integrate these viral strategies. This review shows that there is indeed a basis within the scientific community to investigate the potential of phages to be used as biocontrol agents. In 2018, a Horizon 2020 consortium was established to investigate viruses and their potential to serve as a possible solution against pests as both a probiotic and viral treatment strategy (https://viroplant.eu/). Initiatives like these, together with more in-depth research, are needed to provide fundamental insights to close the gap between academia and industry in this matter and to stimulate the industry to invest in phage biocontrol.

Supplementary Materials: Supplementary materials can be found at http://www.mdpi.com/1999-4915/11/3/277/s1.

Author Contributions: Conceptualization, D.H., R.L., I.H., and J.W.; methodology, D.H. and I.H.; writing—original draft preparation, D.H.; writing—review and editing, R.L., I.H., and J.W.

Funding: D.H., I.H., R.L., and J.W. are supported by the European Union's Horizon 2020 Research and Innovation Program (773567; www.viroplant.eu) and by a VLAIO LA-grant (150914). DH holds a predoctoral scholarship from FWO-strategic basic research (1S02518N).

Conflicts of Interest: The authors declare no conflict of interest.

References

1. Twort, F.W. An Investigation On The Nature Of Ultra-Microscopic Viruses. *Lancet* **1915**, *186*, 1241–1243. [CrossRef]
2. D'Herelle, F. Roux On an invisible microbe antagonistic toward dysenteric bacilli: Brief note by Mr. F. D'Herelle, presented by Mr. Roux. *Res. Microbiol.* **2007**, *158*, 553–554. [PubMed]
3. Pirnay, J.P.; De Vos, D.; Verbeken, G.; Merabishvili, M.; Chanishvili, N.; Vaneechoutte, M.; Zizi, M.; Laire, G.; Lavigne, R.; Huys, I.; et al. The phage therapy paradigm: Prêt-à-porter or sur-mesure? *Pharm. Res.* **2011**, *28*, 934–937. [CrossRef]
4. Buttimer, C.; McAuliffe, O.; Ross, R.P.; Hill, C.; O'Mahony, J.; Coffey, A. Bacteriophages and bacterial plant diseases. *Front. Microbiol.* **2017**, *8*, 34. [CrossRef]
5. Mallmann, W.; Hemstreet, C. Isolation ofan inhibitory substance from plants. *Agric. Res.* **1924**, *28*, 599–602.
6. Okabe, N.; Goto, M. Bacteriophages of Plant Pathogens. *Annu. Rev. Phytopathol.* **1963**, *1*, 397–418. [CrossRef]

7. Strange, R.; Scott, P.R. Plant disease: A threat to global food security. *Annu. Rev. Phytopathol.* **2005**, *43*, 83–116. [CrossRef] [PubMed]
8. Inagro IWT—Beheersing bacteriële pathogeen opkweek kolen-prei. Available online: http://leden.inagro.be/Wie-is-Inagro/Projecten/project/13960 (accessed on 27 December 2018).
9. Mansfield, J.; Genin, S.; Magori, S.; Citovsky, V.; Sriariyanum, M.; Ronald, P.; Dow, M.; Verdier, V.; Beer, S.V.; Machado, M.A.; et al. Top 10 plant pathogenic bacteria in molecular plant pathology. *Mol. Plant Pathol.* **2012**, *13*, 614–629. [CrossRef] [PubMed]
10. Cooksey, D.A. Genetics of Bactericide. *Annu. Rev. Phytopathol.* **1990**, *28*, 201–219. [CrossRef]
11. Frampton, R.A.; Pitman, A.R.; Fineran, P.C. Advances in bacteriophage-mediated control of plant pathogens. *Int. J. Microbiol.* **2012**, *2012*, 326452. [CrossRef]
12. Pietrzak, U.; McPhail, D.C. Copper accumulation, distribution and fractionation in vineyard soils of Victoria, Australia. *Geoderma* **2004**, *122*, 151–166. [CrossRef]
13. Copping, L.G.; Duke, S.O. Natural products that have been used commercially as crop protection agents. *Pest Manag. Sci.* **2007**, *63*, 524–554. [CrossRef] [PubMed]
14. Ray, D.K.; Mueller, N.D.; West, P.C.; Foley, J.A. Yield Trends Are Insufficient to Double Global Crop Production by 2050. *PLoS ONE* **2013**, *8*, e66428. [CrossRef] [PubMed]
15. Stenberg, J.A. A Conceptual Framework for Integrated Pest Management. *Trends Plant Sci.* **2017**, *22*, 759–769. [CrossRef] [PubMed]
16. Phillips, M.D. *The Cost of New Agrochemical Product Discovery, Development and Resgistration in 1995, 2000, 2005–8 and 2010 to 2014. R&D Expenditure in 2014 and Expectations for 2019. A Consultancy Study for CropLife International, CropLife America and the European Crop*; Pathhead: Midlothian, UK, 2016.
17. European Crop Protection Registering Plant Protection Products in the EU. Available online: https://www.ecpa.eu/sites/default/files/7450_Registrationbrochure_3.pdf%0Ahttp://www.ecpa.eu/files/attachments/20110125_PPPBrochure_ECPA.pdf (accessed on 5 February 2019).
18. WIPO—Protecting Your Inventions Abroad: Frequently Asked Questions About the Patent Cooperation Treaty (PCT). Available online: https://www.wipo.int/pct/en/faqs/faqs.html (accessed on 12 March 2019).
19. Todd, K. The Promising Viral Threat To Bacterial Resistance: The Uncertain Patentability of Phage Therapeutics and the Necessity of Alternative Incentives. *J. D. Expect.* **2019**, *68*, 767–805.
20. Williams, H.L. How do patents affect research investments? *Annu. Rev. Econ.* **2017**, *9*, 441–469. [CrossRef]
21. Pirnay, J.-P.; Verbeken, G.; Rose, T.; Jennes, S.; Zizi, M.; Huys, I.; Lavigne, R.; Merabishvili, M.; Vaneechoutte, M.; Buckling, A.; et al. Introducing yesterday's phage therapy in today's medicine. *Future Virol.* **2012**, *7*, 379–390. [CrossRef]
22. Verbeken, G. *Towards An Adequate Regulatory Framework For Bacteriophage Therapy*; KU Leuven and Royal Military Academy: Leuven, Belgium, 2015.
23. Verbeure, B.; Matthijs, G.; Van Overwalle, G. Analysing DNA patents in relation with diagnostic genetic testing. *Eur. J. Hum. Genet.* **2006**, *14*, 26–33. [CrossRef]
24. Huys, I.; Berthels, N.; Matthijs, G.; Van Overwalle, G. Legal uncertainty in the area of genetic diagnostic testing. *Nat. Biotechnol.* **2009**, *27*, 903–909. [CrossRef]
25. McKenna, F.; El-Tarabily, K.A.; Hardy, G.S.J.; Dell, B. Novel in vivo use of a polyvalent Streptomyces phage to disinfest Streptomyces scabies-infected seed potatoes. *Plant Pathol.* **2001**, *50*, 666–675. [CrossRef]
26. EPO—Basic Definitions. Available online: https://www.epo.org/searching-for-patents/helpful-resources/first-time-here/definitions.html (accessed on 14 February 2019).
27. Braverman, M. United States. In *Use and Regulation of Microbial Pesticides in Representative Jurisdictions Worldwide*; 2015; Available online: https://www.IOBC-Global.org (accessed on 14 February 2019).
28. Wang, B.; Li, Z. Use and Regulation of Biopesticides in China. In *Use And Regulation Of Microbial Pesticides In Representative Jurisdictions Worldwide*; 2015; Available online: https://www.IOBC-Global.org (accessed on 14 February 2019).
29. Kim, J.J.; Lee, S.G.; Lee, S.; Jee, H.-J. South Korea. In *Use and Regulation of Microbial Pesticides in Representative Jurisdictions Worldwide*; 2015; Available online: https://www.IOBC-Global.org (accessed on 14 February 2019).

30. Rombouts, S.; Volckaert, A.; Venneman, S.; Declercq, B.; Vandenheuvel, D.; Allonsius, C.N.; Van Malderghem, C.; Jang, H.B.; Briers, Y.; Noben, J.P.; et al. Characterization of novel bacteriophages for biocontrol of bacterial blight in leek caused by Pseudomonas syringae pv. porri. *Front. Microbiol.* **2016**, *7*, 1–15. [CrossRef]
31. EPA PRIA Fee Category Table—Registration Division—New Active Ingredients. Available online: https://www.epa.gov/pria-fees/pria-fee-category-table-registration-division-new-active-ingredients (accessed on 15 February 2019).
32. Barbosa, C.; Venail, P.; Holguin, A.V.; Vives, M.J. Co-Evolutionary Dynamics of the Bacteria Vibrio sp. CV1 and Phages V1G, V1P1, and V1P2: Implications for Phage Therapy. *Microb. Ecol.* **2013**, *66*, 897–905. [CrossRef]
33. European Commission. *Guidance Document For The Assessment Of The Equivalence Of Technical Grade Active Ingredients For Identical Microbial Strains Or Isolates Approved Under Regulation (EC) No 1107/2009*; European Commission Health and Food safety: Brussels, Belgium, 2014.
34. Adriaenssens, E.M.; Rodney Brister, J. How to name and classify your phage: An informal guide. *Viruses* **2017**, *9*, 70. [CrossRef]
35. Van Lenteren, J.C.; Bolckmans, K.; Köhl, J.; Ravensberg, W.J.; Urbaneja, A. Biological control using invertebrates and microorganisms: Plenty of new opportunities. *BioControl* **2018**, *63*, 39–59. [CrossRef]
36. AgriPhageTM Product Info AgriPhage. Available online: https://www.agriphage.com/product-info/ (accessed on 9 February 2019).
37. Enviroinvest Zrt. Erwiphage PLUS. Available online: http://www.erwiphage.com/ (accessed on 12 March 2019).
38. APS Biocontrol Ltd. Biolyse Products. Available online: https://www.apsbiocontrol.com/products (accessed on 12 March 2019).
39. Erstling, J.; Strom, R. Korea's Patent Policy and Its Impact on Economic Development: A Model for Emerging Countries. *San Diego Int. Law J.* **2009**, *11*, 441–481.
40. Long, C.X.; Wang, J. China's patent promotion policies and its quality implications. *Sci. Pub. Policy* **2018**, *46*, 91–104. [CrossRef]
41. APPPC; FAO. Plant Protion Profiles from Asia-Pacific countries—Chapter 3 Intergrated Pest Management. Available online: http://www.fao.org/docrep/010/ag123e/AG123E22.htm (accessed on 13 February 2019).
42. Álvarez, B.; Biosca, E.G. Bacteriophage-Based Bacterial Wilt Biocontrol for an Environmentally Sustainable Agriculture. *Front. Plant Sci.* **2017**, *8*, 1–7. [CrossRef]
43. Ryan, R.P.; Vorhölter, F.J.; Potnis, N.; Jones, J.B.; Van Sluys, M.A.; Bogdanove, A.J.; Dow, J.M. Pathogenomics of Xanthomonas: Understanding bacterium-plant interactions. *Nat. Rev. Microbiol.* **2011**, *9*, 344–355. [CrossRef]
44. Janse, J.D.; Obradovic, A. Xylella Fastidiosa: Its Biology, Diagnosis, Control And Risks. *J. Plant Pathol.* **2010**, *92*, S1.35–S1.48.
45. Xin, X.F.; Kvitko, B.; He, S.Y. Pseudomonas syringae: What it takes to be a pathogen. *Nat. Rev. Microbiol.* **2018**, *16*, 316–328. [CrossRef]
46. WIPO. IP and Business: Quality Patents: Claiming what Counts. Available online: https://www.wipo.int/wipo_magazine/en/2006/01/article_0007.html (accessed on 14 February 2019).
47. Schofield, D.A.; Bull, C.T.; Rubio, I.; Wechter, W.P.; Westwater, C.; Molineux, I.J. Development of an engineered bioluminescent reporter phage for detection of bacterial blight of crucifers. *Appl. Environ. Microbiol.* **2012**, *78*, 3592–3598. [CrossRef] [PubMed]
48. Labrie, S.J.; Samson, J.E.; Moineau, S. Bacteriophage resistance mechanisms. *Nat. Rev. Microbiol.* **2010**, *8*, 317–327. [CrossRef] [PubMed]
49. Chan, B.K.; Abedon, S.T.; Loc-Carrillo, C. Phage cocktails and the future of phage therapy. *Future Microbiol.* **2013**, *8*, 769–783. [CrossRef] [PubMed]

Communication

Processing Phage Therapy Requests in a Brussels Military Hospital: Lessons Identified

Sarah Djebara [1,*], Christiane Maussen [1], Daniel De Vos [2], Maya Merabishvili [2], Benjamin Damanet [1], Kim Win Pang [1], Peggy De Leenheer [1], Isabella Strachinaru [1], Patrick Soentjens [1] and Jean-Paul Pirnay [2]

1 Center for Infectious diseases ID4C, Queen Astrid military hospital, Bruynstraat 1, B-1120 Brussels, Belgium; christiane.maussen@mil.be (C.M.); benjamin.damanet@mil.be (B.D.); winggopang@gmail.com (K.W.P.); peggy.deleenheer@mil.be (P.D.L.); isabella.strachinaru@mil.be (I.S.); Patrick.Soentjens@mil.be (P.S.)

2 Laboratory for molecular and cellular technology, Queen Astrid military hospital, Bruynstraat 1, B-1120 Brussels, Belgium; danielmarie.devos@mil.be (D.D.V.); maia.merabishvili@mil.be (M.M.); jean-paul.pirnay@mil.be (J.-P.P.)

* Correspondence: sarah.djebara@mil.be; Tel.: +32-2-264-4598

Received: 19 February 2019; Accepted: 14 March 2019; Published: 17 March 2019

Abstract: There is a growing interest in phage therapy as a complementary tool against antimicrobial resistant infections. Since 2007, phages have been used sporadically to treat bacterial infections in well-defined cases in the Queen Astrid military hospital (QAMH) in Brussels, Belgium. In the last two years, external requests for phage therapy have increased significantly. From April 2013 to April 2018, 260 phage therapy requests were addressed to the QAMH. Of these 260 requests, only 15 patients received phage therapy. In this paper, we analyze the phage therapy requests and outcomes in order to improve upon the overall capacity for phage therapy at the QAMH.

Keywords: bacteriophages; phage therapy; antibiotic resistance; *Pseudomonas aeruginosa*; *Escherichia coli*; *Staphylococcus aureus*; Brussels; Belgium

---

1. Introduction

Antibiotic resistance is an increasing threat not only to human health but also to the production of food and to sustainable development. By 2050, it is estimated that antimicrobial resistant infections will kill more than 10 million people per year (more than cancer), and the cost in terms of lost global production will amount to 100 trillion USD, if no action is undertaken [1]. In 2016, the United Nations acknowledged that the current antimicrobial resistance crisis is mainly due to the inappropriate use of antimicrobial medicines in the public health, animal, food, agriculture, and aquaculture sectors; a lack of access to health services (including to diagnostics and laboratory capacity); and the presence of antimicrobial residues in soil, crops, and water. They subsequently committed to work at national, regional, and global levels to support the development of new antimicrobial agents and therapies [2]. In 2017, the World Health Organization published a list of 12 drug-resistant bacteria for which new antibiotics are urgently needed. The critical priority category consisted of *Acinetobacter baumannii* (carbapenem-resistant), *Pseudomonas aeruginosa* (carbapenem-resistant), and *Enterobacteriaceae* (carbapenem-resistant, ESBL-producing) [3]. In the US, Rice coined the term "ESKAPE" pathogens (*Enterococcus faecium*, *Staphylococcus aureus*, *Klebsiella pneumoniae*, *Acinetobacter baumannii*, *Pseudomonas aeruginosa*, and *Enterobacter* spp.) to emphasize that these bacteria currently cause the majority of US hospital infections and effectively "escape" the effects of antibacterial drugs [4]. However, only a few new antibiotics are being developed, and none are expected to be effective against the most dangerous antibiotic-resistant bacteria called "superbugs" [5]. There is renewed interest in phage therapy as an alternative or addition to antibiotic therapy for the treatment of bacterial infections.

Bacteriophages (phages for short) are viruses that target and infect a subset of bacteria with almost no collateral damage to the commensal flora (e.g., the gut and skin microbiomes). Phage therapy was first introduced in Western medicine in the 1920s. Upon the widespread marketing of antibiotics, which had the advantage of exhibiting a broad spectrum antimicrobial activity, phage therapy was abandoned in the West by about the 1940s. It continued to be developed and used in Eastern Europe and in former Soviet Republics, with the Eliava Institute in Tbilisi (Georgia) as the epicenter [6]. In 2007, the Queen Astrid military hospital (QAMH) in Brussels was the first Belgian hospital to reinitiate a focus on phage therapy, and this under the umbrella of article §37 (unproven interventions in clinical practice) of the Declaration of Helsinki, which was developed by the World Medical Association [7].

Article §37. In the treatment of an individual patient, where proven interventions do not exist or other known interventions have been ineffective, the physician, after seeking expert advice, with informed consent from the patient or a legally authorised representative, may use an unproven intervention if, in the physician's judgement, it offers hope of saving life, re-establishing health, or alleviating suffering. This intervention should subsequently be made the object of research, designed to evaluate its safety and efficacy. In all cases, new information must be recorded and, where appropriate, made publicly available.

Since then, patients have been occasionally treated by phage therapy at the QAMH. Last year, we reported the case of a patient treated with intravenous bacteriophage monotherapy (no antibiotics were used) against colistin-only-sensitive *P. aeruginosa* [8]. Belgium is now implementing a pragmatic phage therapy framework that centers on the magistral preparation (compounding pharmacies in the US) of tailor-made phage medicines [9], which can pave the way for a broader and more structured application of phages in Belgium. Most requests for phage therapy used to originate from within the QAMH, and more specifically from the burn wound center. Since 2017, however, a spectacular increase in external phage therapy requests to the QAMH has been observed, most of them related to the broadcast of two phage therapy prime time documentaries on Dutch television: *Bacteriofagen: een alternatief voor antibiotica?* (Bacteriophages: an alternative to antibiotics?) on the 21st of March 2017 [10] and *Dokters van Morgen over bacteriën* (Doctors of Tomorrow on bacteria) on the 24th of October 2017 [11]. One hundred and fifty-one phage therapy requests were registered in 2017, with increases in requests following the documentaries' broadcast dates. A third Dutch documentary aired on 5 February 2019 [12] was again followed by a considerable increase in phage therapy request. Between April 2013 and April 2018, 260 phage therapy requests were addressed to QAMH medical staff by e-mail, post, or telephone. All these requests were re-directed to a centralized e-mail address (pt@mil.be), upon the receipt of which a reply was sent to request standardized medical information (Supplementary File 1) with regard to the particular case. One hundred and ninety applicants (73.1%) responded (Figure 1). The received medical files were stored in a dedicated access database to ensure a uniform follow-up.

The aim of this paper was to describe, analyze, and discuss the 260 phage therapy requests addressed to the QAMH to raise awareness for the increased interest in phage therapy in Northwest Europe and to guide future phage therapy R&D in and outside the QAMH.

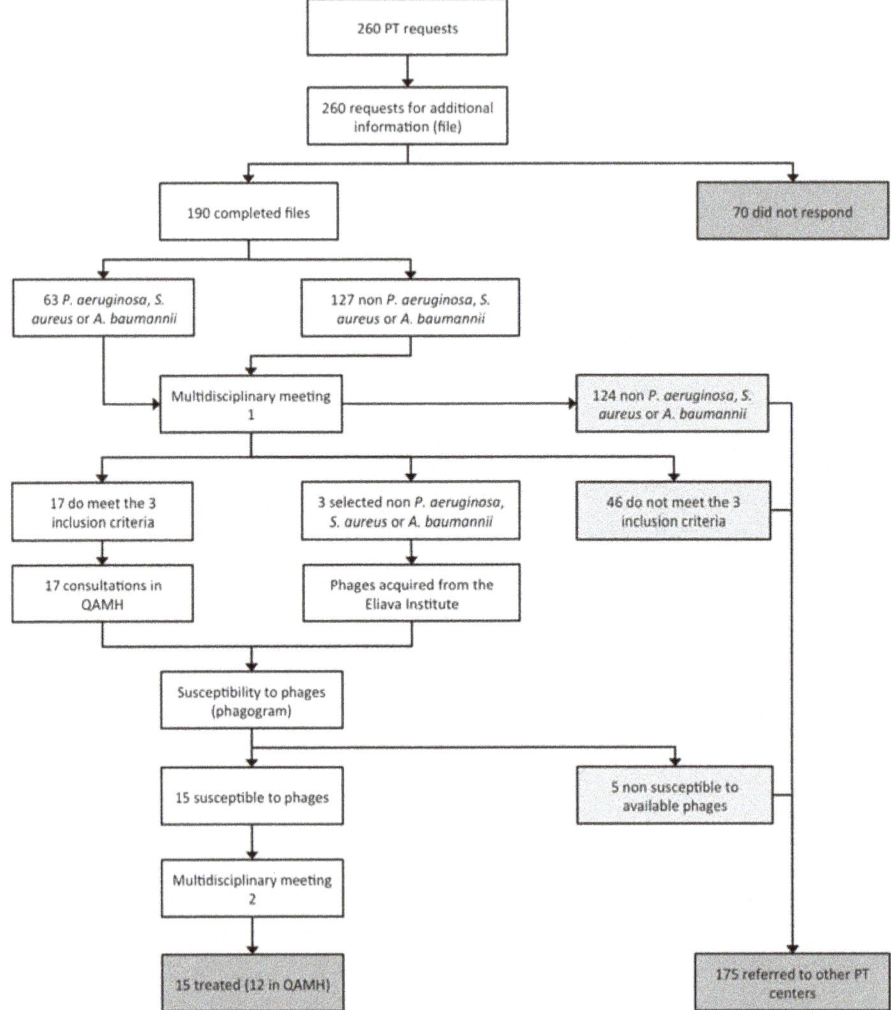

Figure 1. Patient care workflow in phage therapy at the Queen Astrid military hospital in Brussels (Belgium). PT, phage therapy; QAMH, Queen Astrid military hospital.

2. Demographics

Most phage therapy requests were initiated by the patients themselves (70.8%), followed by physicians (26.1%) and the patient's family (3.1%) (Figure 2). The increased attention for phage therapy in the popular media seems to have raised the awareness of patients to this new therapeutic alternative. Not surprisingly, the majority of phage therapy requests to the QAMH in Brussels (Belgium) originated from The Netherlands (66.9%), one of Belgium's neighboring countries. The other countries of patients' origins were, in decreasing order, Belgium (19.2%), France (7.3%), Germany (2.3%), and Luxembourg (1.1%) (Table 1). Fifty-three percent of the patients requesting phage therapy were male, and the mean age (SD) of the 132 patients, who communicated their age, was 57.9 (20.7) years (Table 1). Patients older than 60 years were more prevalent (61.4%) (Table 1).

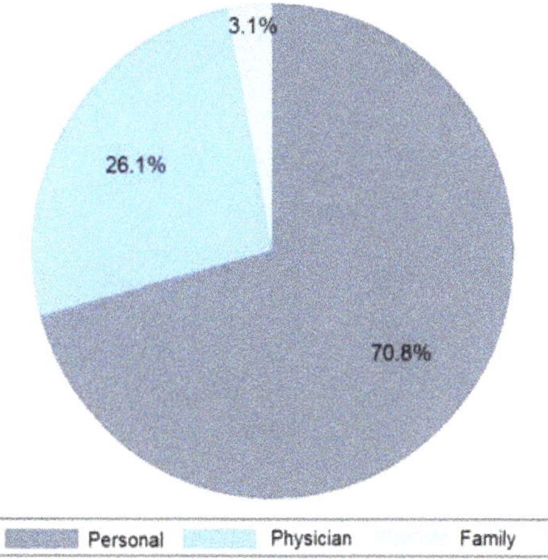

Figure 2. Initiators of the 260 phage therapy requests.

Table 1. Demographics and microbiology of patients requesting phage therapy at the Queen Astrid military hospital (n = 260).

Infection Types		**LRTI**	**UTI**	**SSTI**	**ENTI**	**BoneI**	**OPI**	**AbdI**	**ND**	**Other**	**Total**
Demographics											
Number of requesters		59	79	21	22	16	14	9	12	28	260
Age	≤14	1	3	3	1				1		9
	15–29	2	1			1			1		5
	30–59	5	11	5	3	4	2	1	1	5	37
	≥60	21	26	9	6	4	10	2	1	2	81
	ND	30	38	4	12	7	2	6	8	21	128
Gender	Male	23	33	12	13	10	11	5	7	20	134
	Female	36	46	9	9	6	3	4	5	8	126
Countries	The Netherlands	38	69	5	14	5	7	5	10	21	174
	Belgium	9	4	12	2	5	6	3	2	7	50
	France	5	3	3	4	3	1				19
	Germany	3			2	1					6
	Luxembourg	1	2								3
	Italy			1							1
	Spain	1									1
	United States					2		1			3
	Israel	2									2
	Unknown		1								1
Bacterial pathogens											
Pseudomonas aeruginosa		36	7	8	9	3		2	1	4	70
Escherichia coli		3	29	2	3	2	2	2		1	44
Staphylococcus aureus		11		6	5	2	5	1	1		31
Klebsiella pneumoniae			18	1		2	1		2		24
Enterococcus faecalis			12		1					4	17
Proteus mirabilis		4	5		1	1					11
Enterobacter cloacae			6								6
Mycobacterium avium		3									3
Streptococcus pyogenes		1	2		1	1					5
Staphylococcus epidermidis						3	2		1		6

Table 1. *Cont.*

Infection Types	LRTI	UTI	SSTI	ENTI	BoneI	OPI	AbdI	ND	Other	Total
Staphylococcus dysgalactiae				1	2					3
Acinetobacter baumannii	2		1					1		4
Serratia marcescens	1		1							2
Staphylococcus capitis	1									1
Staphylococcus warneri						2				2
Borrelia burgdorferi									2	2
Burkholderia cenocepacia	1									1
Burkholderia multivorans	1									1
Enterobacter aerogenes			2							2
Granulicatella adiacens									2	2
Haemophilus influenzae	1									1
Morganella morganii					1				1	2
Moraxella catarrhalis	1									1
Cutibacterium acnes									2	2
Stenotrophomonas maltophilia	1									1
Yersinia enterocolitica								1		1
Coxiella burnetii									1	1
Clostridium hathewayi			1							1
Helicobacter pylori							1			1
Corynebacterium amycolatum				1						1
ND	8	22	3	5	2	2	3	5	12	62
Total	75	101	25	27	19	14	9	12	29	311
Polymicrobial (caused by a combination of bacteria)	10	14	4	4	3				2	37

AbdI, abdominal infection; BoneI, bone infection; ENTI, ear-nose-throat infection; LRTI, lower respiratory tract infection; ND, no data; OPI, orthopedic prosthesis infection; SSTI, skin and soft tissue infection; UTI, urinary tract infection.

3. Infection Types and Bacterial Pathogens

The infection types and their causative bacterial pathogens are shown in Table 1. Urinary tract infection (UTI) was the most common type of infection (31.8% of all requests), with chronic bladder infection as the most frequent UTI type. Less frequent were responders with neurogenic bladder and recurrent UTIs. The leading causative infectious agents in UTI were (in descending order) *Escherichia coli*, *Enterococcus faecalis*, *K. pneumoniae*, and *Enterobacter cloacae*.

Lower respiratory tract infection (LRTI) was the second most frequent type of infection reported to request phage therapy (23.8%), with *P. aeruginosa* as the predominant respiratory pathogen. In this category, cystic fibrosis, bronchiectasis, chronic obstructive pulmonary disease (COPD), and asthma were the most common underlying pathologies. Third in line were bone infections (Bone Is) and orthopedic prosthesis infections (OPIs) with 12.1% of requests, including osteomyelitis, osteitis in diabetic foot, infected traumatic fractures, and hip and knee prosthesis infections. Ear, nose, and throat infections (ENTIs) came in fourth position (8.9%), with chronic sinusitis and chronic otitis as the main pathologies. Skin and soft tissue infections (SSTIs) were fifth, with 8.5% of requests. Burn and chronic wound infections (including postoperative surgical wounds and diabetic foot ulcers) were the most common SSTIs. Surprisingly, only few phage therapy requests concerned patients with abdominal infections (AbdIs) (3.6%). There were also 11 requests from patients seeking phage therapy for non-bacterial or non-infectious medical conditions such as arthritis, interstitial cystitis, cirrhosis, collage colitis, and irritable bowel disease. No fewer than 30 bacterial species were at the basis of the reported infections, with *P. aeruginosa* (22.5%) as the leading causative agent (Figure 3). This pathogen was found mostly in LRTIs (51.4%), and to a lesser extent in ENTIs, SSTIs, and UTIs (Figure 4). The second most prevalent bacterium was *E. coli* (14.1%), found mostly in UTI patients (66.1%)

(Figure 4). The third one was *S. aureus* (10%), mostly found in LRTIs, ENTIs, OPIs, and SSTIs. Other frequently encountered bacteria were *Enterobacteriaceae*, including *K. pneumoniae* (7.7%, mainly in UTI), *E. faecalis* (5.5%), and *Proteus mirabilis* (3.5%) (Figure 3). Interestingly, *E. faecium*, which is often considered as the leading cause of multi-drug resistant enterococcal infections (over *E. faecalis*), was not represented. From the 190 requests with completed files, only 102 antibiograms could be retrieved and analyzed (Supplementary Table S1). Bacterial strains were classified in five different categories of acquired antibiotic resistance according to Magiorakos' classification proposal [13]. Multidrug-resistant (MDR) was defined as non-susceptible to at least one agent in three or more antimicrobial categories, which Magiorakos and colleagues had previously constructed, for each of the organisms with the intent of placing antimicrobial agents into more therapeutically relevant groups. Extensively drug-resistant (XDR) was defined as non-susceptible to at least one agent in all but two or fewer antimicrobial categories (i.e., bacterial isolates remain susceptible to only one or two categories), and pandrug-resistant (PDR) was defined as non-susceptible to all agents in all antimicrobial categories (i.e., no agents tested as susceptible for that organism). Non-defined (ND) was used when the information given by the antibiogram (the result of an antibiotic susceptibility test) was incomplete to classify the germ into one of the five antibiotic resistance categories. We chose to focus on the three most encountered pathogens in our cohort (*P. aeruginosa*, *E. coli*, and *S. aureus*), for which, respectively, 28, 19, and 14 antibiograms were collected and analyzed (Figure 5). For *P. aeruginosa*, 7.1% of strains were classified as MDR, 10.7% as XDR, and 7.1% as PDR. The proportion of MDR *E. coli* strains was no less than 47.3% with 5.2% of XDR, but no PDR strains were observed. Approximately a fifth (21.4%) of *S. aureus* strains were MDR, and none were XDR or PDR. Notwithstanding the fact that these statistics are more or less in line with the literature with regard to the current antibacterial resistance crisis, we observed that—with the exception of *E. coli*—the majority of phage therapy requests concerned non-MDR organisms. Technically speaking, under the umbrella of article 37 of the Declaration of Helsinki, phage therapy can only be applied when proven (e.g., antibiotic) therapies are ineffective. So, when there are indications (e.g., based on an antibiogram) that the infection can be treated with an antibiotic, phage therapy should not be considered.

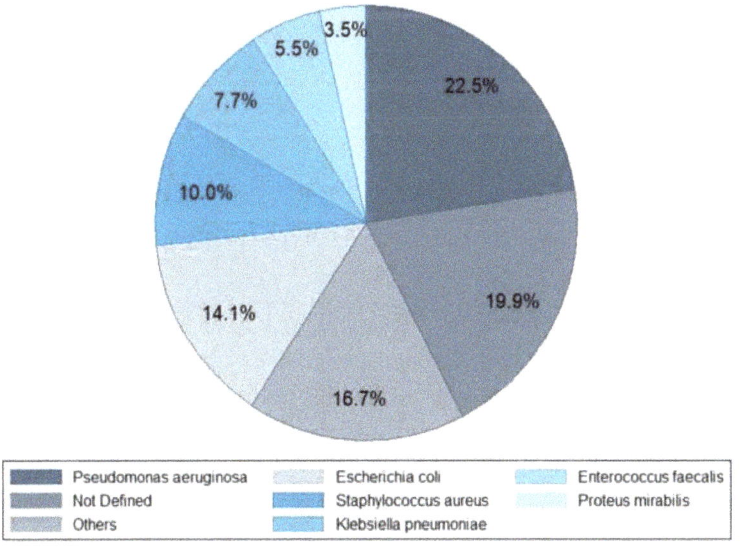

Figure 3. Relative prevalence of 311 reported bacterial pathogens (Table 1).

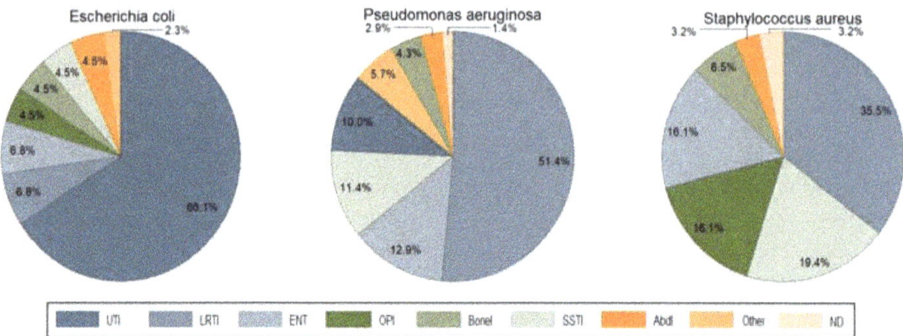

Figure 4. The proportion of the three most prevalent bacteria in the different infection types. AbdI, abdominal infection; BoneI, bone infection; ENT, ear-nose-throat; LRTI, lower respiratory tract infection; OPI, Orthopedic prosthesis infection; SSTI, skin and soft tissue infection; UTI, urinary tract infection.

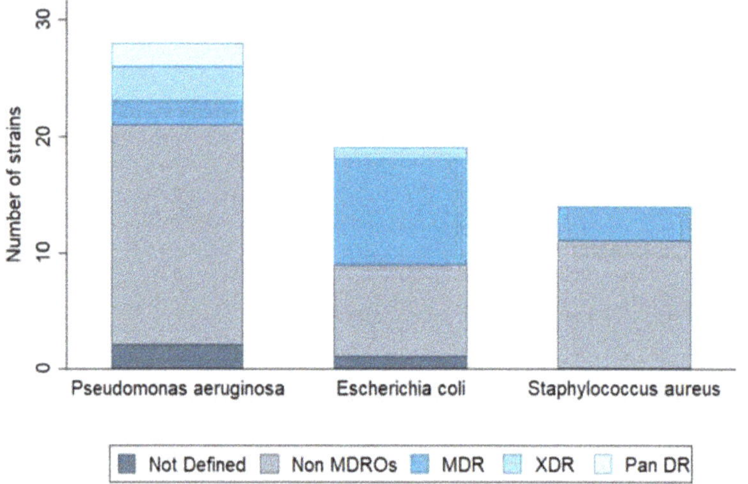

Figure 5. Proportion of drug-resistant strains in the three most prevalent bacterial pathogens (see also Supplementary Table S1). MDR, multidrug-resistant; non-MDROs, non-multidrug-resistant organisms; Pan DR, pandrug-resistant; XDR, extensively drug-resistant.

4. Patient Care Workflow

A dedicated patient care workflow in phage therapy was created to ensure an accurate and systematic monitoring of phage therapy requests, treatments, and follow-up (Figure 1).

When the medical dossier was complete, which was the case for 190 responders; the case was discussed by dedicated infectious disease specialists and microbiologists during a first multidisciplinary meeting. Three inclusion criteria were taken into account:

- Infection with *S. aureus*, *P. aeruginosa*, and/or *A. baumannii*, the three bacterial pathogens against which the QAMH possessed potent phages [14,15];
- Bacterial infection associated with antibiotic treatment failure;
- The absence of other therapeutic options.

When eligibility criteria were met, which was the case for only 20 patients (Figure 1), a consultation with an infectious disease specialist was scheduled, during which a physical examination and an anamnesis of the patient were performed and bacterial samples of the infection site(s) were taken. Consecutively, these bacterial samples were sent to the clinical laboratory for standard bacterial culture, isolation, and identification (using VITEK II, bioMérieux, Marcy l'Etoile, France). For *P. aeruginosa*, *S. aureus*, and *A. baumannii* isolates, a "phagogram" (by analogy with an antibiogram) was performed, based on the spot-test and the double-agar overlay method (both methods are described in Kakabadze et al. [16]), to determine their susceptibility to the phages available (for non-commercial R&D purposes) in the QAMH and described in Merabishvili et al. [14,15]. For the spot test, 100 µL of bacterial suspension at a concentration of 10^8–10^9 cfu/mL was mixed with 3.5 mL of LB (Lysogeny Broth, Becton Dickinson, Franklin Lakes, NJ, USA) medium with 0.6% agar (Becton Dickinson) at 45 °C and poured on petri plates containing a solidified bottom layer of LB medium with 1.5% agar. After air-drying, 10 µL of serial 100-fold dilutions of phage cocktails were spotted on the bacterial lawn. Plates were incubated overnight at 37 °C. Clearance zones (when present) were examined the next day. In case phage activity was observed against the tested bacterial strain, the double-agar overlay method was applied to determine the Efficiency of Plating (EOP) of the phage cocktails, calculated as the ratio of activity in the test strain (i.e., the patient's strain) to the activity on the host strain (i.e., the production strain). In 15 cases, the phagogram indicated susceptibility of the infecting bacterial pathogen(s) to the tested phage(s) and treatment was proposed to the patient and to his treating physician (Figure 1). Treatment protocols and patient follow up were discussed during a second multidisciplinary meeting. Details with regard to the phage therapy protocol and outcome will be the subject of separate publications (grouped according to medical indications and authored by the different treating physicians) and will not be discussed in this article. However, we can disclose that no serious adverse events were observed and that, in general, phage therapy seemed helpful in a considerable number of the cases. We also would like to stress that, with the exception of one case that was recently published [8], other antimicrobial agents (e.g., antibiotics) were applied simultaneously with the phages.

5. Implications for Future Activities

Most phage therapy requests were initiated by the patients themselves, which in part could explain the low proportion of MDR infections and the occurrence of requests for phage therapy against non-bacterial infections. The role of the media was non-negligible in the patients' self-management of their disease, as demands increased spectacularly immediately after two prime-time TV-shows promoting phage therapy, but it probably also reflected the increasing will of patients to find alternatives to (effective or ineffective) antibiotics. It is therefore important to understand that desperate patients take the matter in their own hands and try to find alternative therapeutic options.

Only 15 (5.8%) of the 260 phage therapy requests resulted in actual phage therapy. Two hundred and forty five requests were rejected for diverse reasons (Figure 1):

- 70 applicants (26.9%) did not respond to the email request for more information;
- 124 requests (47.7%) concerned bacterial pathogens against which the QAMH had no potent phages available;
- 46 applications (17.7%) did not meet the other two eligibility criteria (antibiotic treatment failure and/or absence of other therapeutic options);
- 5 (25%) out of the 20 infecting bacterial strains for which a phagogram was performed were found to be non-susceptible to the available phages.

In most cases, the rejected applications were referred to reputable phage therapy centers abroad.

The high frequency of non-responsiveness of applicants to the initial information request could have been partly due to an inability to provide the necessary information. For instance, access to medical data could have been hampered by a lack of confidence from the treating physicians. This could have been partially remedied by conducting randomized controlled trials to demonstrate phage therapy efficacy and by awareness campaigns. The QAMH is developing tools, such as a comprehensive website with information and instructions, to dispense phage therapy information to health care professionals.

Almost half of the requests (47.7%) had to be dismissed because no suitable phages were available to treat the causative bacterial agents, which increasingly belong to the family of the *Enterobacteriaceae*. This observation prompted us to initiate research programs to isolate and characterize potent phages against, amongst others, problematic *E. coli* and *K. pneumoniae* strains. For this, we will need to address the remarkably high phage specificity within the *Enterobacteriaceae* family. Of importance, in this cohort 30 bacterial species were at the basis of the infections for which phage therapy was pursued. This observation highlights the main difficulty with which a phage therapy center is confronted. Indeed, to be able to cope with these 30 bacterial species, hundreds of potent and characterized phages need to be readily available or isolated de novo from the environment and produced to a quality acceptable for human application. As a consequence, big phage repositories will be required. This observation shows the importance of directing a considerable part of R&D efforts towards new technologies (e.g., synthetic biology) [17], which would allow the accelerated selection and production of potent therapeutic phages, for every possible pathogen.

Forty-six requests (17.7%) did not involve antibiotic treatment failure and/or the absence of other therapeutic options than phage therapy. Indeed, to be able to apply article §37 of the Declaration of Helsinki, a physician must be certain that "proven prophylactic, diagnostic and therapeutic methods do not exist or have been ineffective". This is also a condition for phage therapy in the Wroclaw Ludwik Hirszfeld Institute. However, we found that the majority of cases did not involve MDR infection. If phage therapy requests had predominantly been made by (university) hospitals instead of individuals, the proportion of MDR cases would likely have been greater.

It must be said that some medical conditions such as COPD, bronchiectasis, and diabetic foot infection can be very difficult to treat due to underlying complications such as poor blood flow, a weakened immune system, or the presence of highly protected bacterial communities (in biofilms), even when the infecting bacterium is susceptible to common antibiotics. The difficulty is to demonstrate this unambiguously. The implementation of the magistral phage framework, which does not require the proven ineffectiveness of conventional therapies, should solve these issues in the near future in Belgium [9].

Currently, the limited (as compared to renown phage therapy centers, such as the Eliava Institute in Tbilisi) phage therapy expertise in the QAMH mainly concerns military- or mass casualty-associated indications such as burn wound and orthopedic infections, and to some extent respiratory diseases. Based on this analysis of phage therapy requests, we will expand our ability, capacity, and experience (including adequate treatment protocols) to treat other pathologies such as urological infection with phages.

Finally, we must keep in mind that phages are not "miracle drugs", as antibiotics were once presented, but that they are probably only useful as additional tools in certain indications and conditions, which still need to be determined. As such, more phage therapy randomized controlled trials are needed, and phage antibiotic synergy (PAS) should be further explored [18,19].

With this report, we hope to help and guide the "phage therapy centers in the making", which are slowly emerging from the "phage-averse" setting called Western medicine.

Supplementary Materials: The following are available online at http://www.mdpi.com/1999-4915/11/3/265/s1, Supplementary Table S1: Antibiograms reported for 102 infecting bacterial pathogens and Supplementary File 1: Phage therapy questionnaire.

Author Contributions: Conceptualization, S.D., D.D.V., P.S. and J.-P.P.; methodology, S.D., D.D.V., M.M., K.W.P., P.D.L., I.S., P.S. and J.-P.P.; investigation, S.D., D.D.V., M.M., K.W.P., P.D.L., I.S., P.S. and J.-P.P.; software, C.M. and B.D.; data curation, S.D., C.M., M.M., P.D.L.; writing – original draft preparation, S.D. and J.-P.P.; writing – review and editing, S.D., D.D.V., M.M., P.S. and J.-P.P.; visualization, B.D. and J.-P.P.; supervision, S.D., P.S. and J.-P.P.; project administration, P.D.L. and C.M.; funding acquisition, P.S. and J.-P.P.

Funding: M.M. was supported by grant HFM 15-5 of the Royal Higher Institute for Defence.

Acknowledgments: The authors express their thanks to the personnel of the clinical laboratory and to the hospital staff and patients.

Conflicts of Interest: The authors declare no conflict of interest.

References

1. Tackling Drug-Resistant Infections Globally: Final Report and Recommendations. The Review on Antimicrobial Resistance. 2016 Release. Available online: https://amr-review.org/sites/default/files/160525_Final%20paper_with%20cover.pdf (accessed on 10 February 2019).
2. United Nations. Draft political declaration of the high-level meeting of the General Assembly on antimicrobial resistance (16-16108 (E)). 2016 Release. Available online: https://www.un.org/pga/71/wp-content/uploads/sites/40/2016/09/DGACM_GAEAD_ESCAB-AMR-Draft-Political-Declaration-1616108E.pdf (accessed on 10 February 2019).
3. WHO Publishes List of Bacteria for Which New Antibiotics Are Urgently Needed. Available online: https://www.who.int/news-room/detail/27-02-2017-who-publishes-list-of-bacteria-for-which-new-antibiotics-are-urgently-needed (accessed on 10 February 2019).
4. Rice, L.B. Federal funding for the study of antimicrobial resistance in nosocomial pathogens: no ESKAPE. *J. Infect. Dis.* **2009**, *197*, 1079–1081. [CrossRef] [PubMed]
5. Boucher, H.W.; Talbot, G.H.; Bradley, J.S.; Edwards, J.E.; Gilbert, D.; Rice, L.B.; Scheld, M.; Spellberg, B.; Bartlett, J. Bad bugs, no drugs: no ESKAPE! An update from the Infectious Diseases Society of America. *Clin. Infect. Dis.* **2009**, *48*, 1–12. [CrossRef] [PubMed]
6. Thiel, K. Old dogma, new tricks—21st Century phage therapy. *Nat. Biotechnol.* **2004**, *22*, 31–36. [CrossRef] [PubMed]
7. World Medical Association Declaration of Helsinki: Ethical principles for medical research involving human subjects. *JAMA* **2013**, *10*, 2191–2194.
8. Jennes, S.; Merabishvili, M.; Soentjens, P.; Pang, K.W.; Rose, T.; Keersebilck, E.; Soete, O.; François, P.M.; Teodorescu, S.; Verween, G.; Verbeken, G.; De Vos, D.; Pirnay, J.P. Use of bacteriophages in the treatment of colistin-only-sensitive *Pseudomonas aeruginosa* septicaemia in a patient with acute kidney injury-a case report. *Crit. Care.* **2017**, *21*, 129. [CrossRef] [PubMed]
9. Pirnay, J.P.; Verbeken, G.; Ceyssens, P.J.; Huys, I.; De Vos, D.; Ameloot, C.; Fauconnier, A. The Magistral Phage. *Viruses* **2018**, *10*, 64. [CrossRef] [PubMed]
10. Zorg.nu. Bacteriofagen: Een Alternatief Voor Antibiotica? Available online: https://zorgnu.avrotros.nl/uitzending/21-03-2017/bacteriofagen-een-alternatief-voor-antibiotica/ (accessed on 10 February 2019).
11. Dokters van Morgen over bacteriën. Available online: https://zorgnu.avrotros.nl/uitzending/24-10-2017/ (accessed on 10 February 2019).
12. Dokters van Morgen: Bacteriofagen. Available online: https://zorgnu.avrotros.nl/uitzending/05-02-2019/ (accessed on 10 February 2019).
13. Magiorakos, A.P.; Srinivasan, A.; Carey, R.B.; Carmeli, Y.; Falagas, M.E.; Giske, C.G.; Harbarth, S.; Hindler, J.F.; Kahlmeter, G.; Olsson-Liljequist, B.; et al. Multidrug-resistant, extensively drug-resistant and pandrug-resistant bacteria: an international expert proposal for interim standard definitions for acquired resistance. *Clin. Microbiol. Infect.* **2012**, *18*, 268–281. [CrossRef] [PubMed]
14. Merabishvili, M.; Pirnay, J.P.; Verbeken, G.; Chanishvili, N.; Tediashvili, M.; Lashkhi, N.; Glonti, T.; Krylov, V.; Mast, J.; Van Parys, L.; et al. Quality-controlled small-scale production of a well-defined bacteriophage cocktail for use in human clinical trials. *PLoS ONE* **2009**, *4*, e4944. [CrossRef] [PubMed]
15. Merabishvili, M.; Vandenheuvel, D.; Kropinski, A.M.; Mast, J.; De Vos, D.; Verbeken, G.; Noben, J.P.; Lavigne, R.; Vaneechoutte, M.; Pirnay, J.P. Characterization of newly isolated lytic bacteriophages active against *Acinetobacter baumannii*. *PLoS ONE* **2014**, *9*, e104853. [CrossRef] [PubMed]
16. Kakabadze, E.; Makalatia, K.; Grdzelishvili, N.; Bakuradze, N.; Goderdzishvili, M.; Kusradze, I.; Phoba, M.F.; Lunguya, O.; Lood, C.; Lavigne, R.; et al. Selection of Potential Therapeutic Bacteriophages that Lyse a CTX-M-15 Extended Spectrum β-Lactamase Producing *Salmonella enterica* Serovar Typhi Strain from the Democratic Republic of the Congo. *Viruses* **2018**, *10*, 172. [CrossRef] [PubMed]
17. Barbu, E.M.; Cady, K.C.; Hubby, B. Phage Therapy in the Era of Synthetic Biology. *Cold Spring Harb. Perspect. Biol.* **2016**, *8*, a023879. [CrossRef] [PubMed]

18. Torres-Barceló, C.; Arias-Sánchez, F.I.; Vasse, M.; Ramsayer, J.; Kaltz, O.; Hochberg, M.E. A window of opportunity to control the bacterial pathogen *Pseudomonas aeruginosa* combining antibiotics and phages. *PLoS ONE* **2014**, *9*, e106628. [CrossRef] [PubMed]
19. Kirby, A.E. Synergistic Action of Gentamicin and Bacteriophage in a Continuous Culture Population of *Staphylococcus aureus*. *PLoS ONE* **2012**, *7*, e51017. [CrossRef] [PubMed]

Article

Design and Preclinical Development of a Phage Product for the Treatment of Antibiotic-Resistant *Staphylococcus aureus* Infections

Susan M. Lehman [1], Gillian Mearns [2], Deborah Rankin [2], Robert A. Cole [2], Frenk Smrekar [3], Steven D. Branston [2] and Sandra Morales [2,*]

1 AmpliPhi Biosciences, San Diego, CA 92130, USA; sml@ampliphibio.com
2 AmpliPhi Australia, Sydney, NSW 2100, Australia; gm@ampliphibio.com (G.M.); dar@ampliphibio.com (D.R.); rac@ampliphibio.com (R.A.C.); sb@ampliphibio.com (S.D.B.)
3 AmpliPhi d.o.o., 1261 Ljubljana-Dobrunje, Slovenia; frenk.smrekar@jafral.com
* Correspondence: spm@ampliphibio.com

Received: 20 December 2018; Accepted: 16 January 2019; Published: 21 January 2019

Abstract: Bacteriophages, viruses that only kill specific bacteria, are receiving substantial attention as nontraditional antibacterial agents that may help alleviate the growing antibiotic resistance problem in medicine. We describe the design and preclinical development of AB-SA01, a fixed-composition bacteriophage product intended to treat *Staphylococcus aureus* infections. AB-SA01 contains three naturally occurring, obligately lytic myoviruses related to *Staphylococcus* phage K. AB-SA01 component phages have been sequenced and contain no identifiable bacterial virulence or antibiotic resistance genes. In vitro, AB-SA01 killed 94.5% of 401 clinical *Staphylococcus aureus* isolates, including methicillin-resistant and vancomycin-intermediate ones for a total of 95% of the 205 known multidrug-resistant isolates. The spontaneous frequency of resistance to AB-SA01 was $\leq 3 \times 10^{-9}$, and resistance emerging to one component phage could be complemented by the activity of another component phage. In both neutropenic and immunocompetent mouse models of acute pneumonia, AB-SA01 reduced lung *S. aureus* populations equivalently to vancomycin. Overall, the inherent characteristics of AB-SA01 component phages meet regulatory and generally accepted criteria for human use, and the preclinical data presented here have supported production under good manufacturing practices and phase 1 clinical studies with AB-SA01.

Keywords: bacteriophage; phage therapy; *Staphylococcus aureus*; biofilm; antimicrobial; frequency of resistance; phage sensitivity; resistance management; nontraditional antibacterial

1. Introduction

The use of bacteriophages (phages) as antibacterial drugs, frequently referred to as "phage therapy" has been discussed and deployed since these bacterial viruses were discovered in the early 1900s. Interest in phage therapy has waxed and waned in various parts of the world, heavily influenced by the availability, affordability, and efficacy of potent small-molecule antibiotics [1]. The current resurgence of interest is persisting in light of the growing urgency of the antimicrobial resistance (AMR) crisis, which predicts that AMR will be the leading cause of human death by 2050, causing 10 million global deaths per year [2].

Staphylococcus aureus, one of the ESKAPE pathogens (*Enterococcus faecium*, *Staphylococcus aureus*, *Klebsiella pneumoniae*, *Acinetobacter baumannii*, *Pseudomonas aeruginosa*, and *Enterobacter spp.*) [3], is a problem in both hospital-associated and community-associated infections [4,5]. It is a significant problem in many clinical settings and the antibiotic-resistant forms are classified as a "High Priority" pathogen by the World Health Organization (WHO) [6] and a "Serious Threat" by the U.S. Centers for

Disease Control and Prevention (CDC) [7]. Since 1999, nine antibiotics targeting methicillin-resistant *S. aureus* (MRSA) have been approved (linezolid, daptomycin, tigecycline, ceftobiprole, telavancin, ceftaroline, dalbavancin, oritavancin, tedizolid) [8]. Of these, only the oxazolidinones, now nearly 20 years old, were a completely new class [9]. Clinical resistance has already been observed for all nine of these drugs, though it can be difficult to predict how quickly or widely some of these resistances will spread [10–19]. Side effects such as renal toxicity and cross-resistance (e.g. among glyo- and lipoglycopeptides) can limit clinical use [8]. Moreover, many antibiotics have reduced efficacy against *Staphylococcus* spp. when it grows in biofilms [20], as is often the case with device-associated infections or endocarditis. Thus, there remains an urgent need for anti-staphylococcal drugs, especially ones with fundamentally different mechanisms of action. Here, we describe the design and composition of AB-SA01, a highly characterized anti-*Staphylococcus* phage product that is being developed to treat acute and chronic *S. aureus* infections in humans, including those caused by MRSA.

Phage therapy has frequently been cited as a form of technology that could help address the AMR problem, provided that high-quality evidence can be gathered in controlled studies focused on testing product efficacy in well-defined clinical indications and administration parameters [7,21–23]. Phages are unable to infect mammalian cells and are usually specific for one or a few bacterial species or strains. Obligately lytic phages would comprise a self-replicating, self-limiting antimicrobial that can be administered by a variety of routes, and that functions via an entirely different mechanism of action compared to small-molecule antibiotics. Humans are continuously exposed to phages present in the environment and as part of the human microbiome, and there is no evidence of any direct toxicity resulting from intentionally administered phages as long as non-phage contaminants such as endotoxins are removed [24–26].

While there remains some debate about the optimal features of therapeutic phages, there is widespread agreement that the traits listed below are either required or particularly desirable for phage products and their individual components [27–31]. Individual phages should be:

- Obligately lytic, to avoid specialized transduction of bacterial genes, and maximize chances for bacterial killing;
- Not known, by empirical testing and/or inference from genomics, to be prone to generalized transduction; and,
- Fully sequenced, to avoid phages with known antibiotic resistance or bacterial virulence genes, and to help assess other lifestyle traits.

Collectively, phages used together to treat a patient should:

- Have broad activity against the target pathogen but not other species, to maximize potential utility and minimize off-target effects; and,
- Be capable of complementation, in which resistant mutants arising to one phage are sensitive to another phage.

In addition to characteristics of the phages themselves, material for clinical use should be produced in such a way as to give confidence that the final product retains these characteristics (i.e., are still the same phages) and does not contain potentially harmful (or harmful amounts) of impurities such as endotoxin or host cell proteins. AB-SA01 satisfies these criteria and has entered clinical development.

2. Methods

2.1. Bacteriophages, Source and Propagation

Each of the selected phages was isolated from an environmental source and subsequently paired to a well-characterized *S. aureus* strain that serves as its manufacturing host. Host-paired phages were purified to ensure that the resulting master stocks produced genetically and phenotypically consistent batches of each phage. Unless otherwise stated, all data is derived from the host-paired, plaque-purified phages. Phages were propagated in liquid culture using vegetable peptone media (VP0101, Oxoid, Hampshire, UK). Lysates were passed through a 0.2-μm filter to remove large cellular debris and, depending on the needs of subsequent testing, optionally subjected to a proprietary process of column-based purification steps to further remove host cell proteins and other bacterial debris and to replace growth medium with phosphate-buffered saline (PBS; Oxoid, Hampshire, UK) containing 10 mM magnesium sulfate (PBS+Mg).

2.2. Bacteria

AB-SA01 manufacturing hosts are *S. aureus* strains originally isolated from humans. The *S. aureus* diversity panel and the species-specificity panel were sourced from the American Type Culture Collection (Manassas, VA, USA), the Walter Reed Army Institute of Research Multidrug-resistant Organism Repository and Surveillance Network ("MRSN", Silver Spring, MD, USA), and clinical sites in Australia and the United Kingdom. Global surveillance panels of *S. aureus* strains were obtained from JMI Laboratories (North Liberty, IA, USA). Targeted interest panels included chronic rhinosinusitis (CRS) strains from Belgium, and vancomycin intermediate (VISA) strains from the CDC and Food and Drug Administration (FDA) Antimicrobial Resistance Isolate Bank (Atlanta, GA, USA). The definition of multidrug resistant (MDR) strains is according to Magiorakos et al [32].

2.3. Phage Sensitivity Assays

Testing on the *S. aureus* panels used Heart Infusion Broth (BD, Franklin Lakes, NJ, USA), amended with 1.5% agar (Oxoid, Hampshire, UK) for plates or 0.7% agar for overlays. Phage activity was assessed using a modification of the small drop agar overlay method [33]. Briefly, 100 μL of 16–18 h planktonic bacterial culture was mixed with molten 0.7% top agar and poured evenly over an agar plate. When the top agar layer was set, serial dilutions of standardized phage solutions were spotted onto the overlay and plates incubated overnight at 37 °C. Phage activity was indicated by clearing of the bacterial lawn at the site of phage application, and by the development of individual plaques as the phage sample is diluted. Strains were only considered sensitive if discrete plaques could be observed as the sample was diluted, indicating phage replication. The titer for each phage+bacteria combination tested was calculated from the drop dilutions. Testing on the species-specificity panel was conducted similarly, using media recommended for the specific bacterial species and bacterial culture volumes suitable to produce a uniform lawn.

2.4. Frequency of Resistance and Complementation

Complementation studies conducted during product selection used apparent bacteriophage-insensitive mutant (BIM) colonies that were isolated after infecting a sensitive *S. aureus* strain with the individual candidate phages. Surviving colonies were streak-purified once on agar plates. The double-drop method was then used to screen for phage sensitivity: 10 μL spots of PBS or phage (~1×10^9 plaque-forming units (PFU)/mL) were spotted onto nutrient agar plates and after 10 min, 5 μL of overnight nutrient broth culture from each BIM or the parental strain was applied to each phage spot. After 24 h incubation at 37 °C, phage+bacteria spots were compared to PBS+bacteria controls and scored as R (resistant, no difference from control spot), I (intermediate, phage activity seen within bacterial spot), or S (sensitive, <10 bacterial colonies in spot).

For the final AB-SA01 composition, the frequency of spontaneous phage resistance in triplicate populations of the same *S. aureus* strain was assessed using a modification of the method of O'Flynn et al. [34]. In a final volume of 200 μL, 6–8 $\times 10^8$ colony forming units (CFU) in nutrient broth was mixed with 2–3 $\times 10^9$ PFU of purified phage (AB-SA01 or individual components), incubated for 10 min at 37 °C, then mixed with 3 mL molten 0.4% nutrient agar and poured over a 90-mm round 1.5% nutrient agar plate. Bacterial colonies were counted after 24 h and 48 h of incubation at 37 °C. The apparent frequency of BIMs was calculated as the number of colonies on each test plate divided by the input number of bacteria in that replicate. Results were compared using a repeated measures

ANOVA and a priori planned comparisons between AB-SA01 and the three component phages were conducted using paired *t*-tests. Up to 10 BIMs from each phage+host combination (all BIMs if <10) were picked and streak-purification was attempted on agar plates.

2.5. Genome Sequencing and Analysis

Phage genomic DNA was purified from filtered lysates or purified preparations and sequenced by Illumina paired-end (ACGT, Wheeling, IL, USA) or PacBio technologies (Expression Analysis, Durham, NC, USA), using PCR-free libraries (Illumina TruSeq PCR-free Library Prep kit, PacBio SMRTbell library). Nucleotide sequences have been deposited in GenBank (Sa83: MK417514, Sa87: MK417515, J-Sa36: MK417516). Annotation was conducted using myRAST v36 (http://blog.theseed.org/servers/). Similarities of (1) annotated proteins to all *Staphylococcus* integrases in GenBank and (2) annotated genes to a proprietary database of bacterial virulence and antibiotic resistance genes were assessed using BLAST searches requiring at least 30% identity across 50% of the sequence, and $E \leq 0.05$; any hits were manually inspected for validity based on factors such as the likely accuracy of the hit's original annotation and evidence from secondary structure predicted by HHPred (https://toolkit.tuebingen.mpg.de) [35,36]. Genome alignments were constructed using Progressive Mauve (http://darlinglab.org/mauve/mauve.html) with the default parameters [37].

2.6. Animal Studies

Purified phage material was used for all animal studies. **(1) Prototype 4-phage product:** Six groups of five female CD-1 mice (Harlan Laboratories, Houston, TX, USA) were rendered neutropenic by administering 150 and 100 mg/kg cyclophosphamide on day -4 and day -1 prior to infection, respectively. Mice were anaesthetized by intraperitoneal (IP) injection of 0.15 mL of a mixture of ketamine HCl (100 mg/kg body weight) plus xylazine (10 mg/kg body weight). Once anaesthetized, an inoculum of 9.5×10^6 CFU of MRSA strain UNT144-3 was delivered intranasally (IN) in a 50 μL volume. At 2 and 6 hours post-infection (hpi), untreated controls received 50 μL PBS+Mg IN, antibiotic controls received 100 mg/kg vancomycin as a subcutaneous (SC) injection, and the three phage treatment groups received 1×10^9 PFU per phage, 1×10^8 PFU per phage, or 1×10^7 PFU per phage in a 50-μL IN dose. At 24 hpi, mice were euthanized by CO_2 inhalation and lungs processed for bacterial load. Bacterial counts were enumerated on Brain Heart Infusion agar with 0.5% activated charcoal. **(2) AB-SA01:** Three groups of five female BALB/c mice were anesthetized and an inoculum of 3.0×10^8 CFU of methicillin-sensitive *S. aureus* strain Xen29 was delivered IN in a volume of 35 μL. At 2 and 6 hpi, untreated controls received 50 μL PBS-Mg IN and the phage treatment group received 5×10^8 PFU per phage in a 50 μL IN dose. At 2, 6, and 12 hpi, the antibiotic controls received 110 mg/kg vancomycin as a SC injection. At 24 hpi, mice were euthanized by CO_2 inhalation and lungs processed for bacterial load. Care was taken to ensure tissue samples were kept cold and processed promptly for bacterial presence. Bacterial counts were enumerated on Mueller Hinton agar. After both mouse studies, bacteria recovered from mouse lung tissue were tested for phage sensitivity according to the method of 2.3. **Statistical Analysis:** Treatments were compared by one-way ANOVA on $\log_{10}$-transformed values with Tukey's test for all pairwise comparisons. Adjusted *p*-values are reported. **Bacterial strains:** *S. aureus* strains were provided by the vendors conducting the studies. UNT144-3 is MRSA and carries the *tetM* gene. Xen29 [38] is available from Perkin Elmer, Inc. Media choice for bacterial enumeration was per each vendor's standard practice. Each vendor had previously established both the dosing for their vancomycin control groups and the bacterial inoculation methods yielding consistent infection outcomes for each specific mouse and *S. aureus* strain combination.

2.7. Animal Welfare

Mouse studies were conducted by external vendors. Study "AmpliPhi 2014-01": The University of North Texas Health Science Center Animal Facility is a member in good standing with the Association for Assessment and Accreditation of Laboratory Animal Care International. Study "APP004-2" (study approved21 March, 2016): KWS BioTest conducts all in-life experimental procedures in accordance with United Kingdom Animals (Scientific Procedures) Act 1986. Their local Ethical Review Process occurred under the auspices of the University of Bristol's Animal Welfare and Ethical Review Body (AWERB).

3. Results

3.1. Physicochemical Characteristics of AB-SA01 Component Phages

All three AB-SA01 component phages produce small, clear plaques when plated on their paired *S. aureus* hosts. Transmission electron microscopy (TEM) images of the three AB-SA01 component phages show the straight, contractile tail and narrow neck that are characteristic of phages belonging to the order *Caudovirales*, family *Myoviridae* (Figure 1).

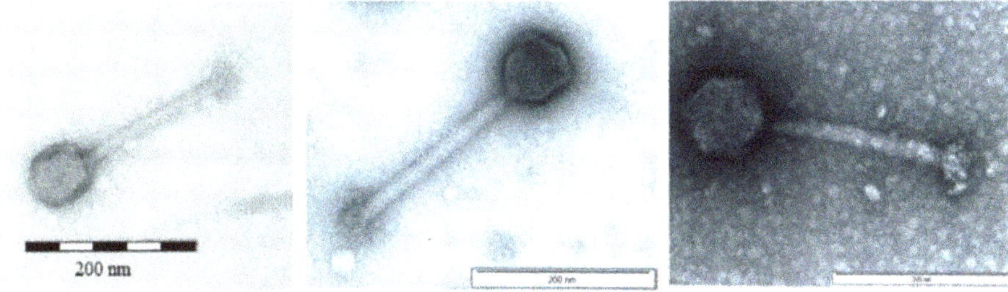

Figure 1. Transmission electron microscopy images of (left to right) Sa83, Sa87, and J-Sa36. Scale bars: 200 nm. Filtered lysates were PEG8000 precipitated, suspended in salt-magnesium buffer, stained with 2% uranyl acetate, and imaged at 80–100kV [39].

All AB-SA01 component phages were sequenced from amplification-free libraries capable of revealing the relative frequencies of genome regions. Read-mapping data showed regions of approximately doubled coverage identifying the genome termini and associated fixed direct terminal repeats between approximately 8 and 10 kb. These genome structures indicate a sequence-specific packaging mechanism not associated with generalized transduction. The pairwise relatedness of the collinear single-copy component phage genomes ranges from 93 to 97% nucleotide identity (Figure 2) and all are related to well-studied *S. aureus* myovirus phage K. No identifiable integrases were found in the AB-SA01 component phage genomes and none of the ca. 200 predicted phage genes in each of the three phages were similar to known bacterial virulence or antibiotic resistance genes.

3.2. In Vitro Activity of AB-SA01

The target species for AB-SA01 is *S. aureus*. Overall, 94.5% of 401 clinical *S. aureus* isolates were sensitive to AB-SA01 (Table 1), including 95% of the 205 total isolates known to be MDR, and with little apparent variation by genetic lineage (Supplementary Table S1), year of isolation, or infection type. When tested on representatives of normal human microflora and related staphylococci, AB-SA01 and its component phages showed some activity against two of five tested *S. epidermidis* strains, but no cross-genus activity (Table 2). When tested on *S. aureus* strains, no evidence of interference among the component phages was observed. The titers observed for AB-SA01 were mostly consistent with the component phage activities, except for a few cases of apparent synergy in which AB-SA01 generated plaques on the *S. aureus* strain even though none of the individual component phages did so. Since testing was conducted in triplicate, this observation of synergy is likely to be real, as opposed to a case of borderline results in which plaques were a bit more obvious with AB-SA01 by simple chance.

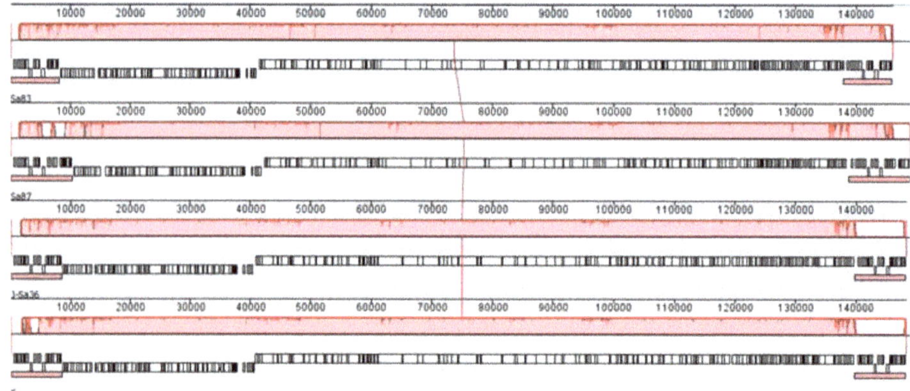

Figure 2. A Progressive Mauve alignment of (top to bottom) Sa83, Sa87, J-Sa36, and phage K (GenBank K766114), each showing annotated genes (white boxes) and long terminal repeats (small red boxes immediately below white gene blocks). The large red blocks above each annotated genome (connected by the red vertical line at approximately 75 kb) represent local collinear blocks of genomes identity; interruptions in these red blocks indicate differences among the four aligned nucleotide sequences.

Table 1. In vitro antibacterial activity of AB-SA01 and its component phages on *Staphylococcus aureus*.

Panel Type	Panel \ Phage	Percentage of Total Isolates Sensitive to Indicated Phage				% of MDR Isolates Sensitive to AB-SA01
		Sa83	Sa87	J-Sa36	AB-SA01	
Selection	AmpliPhi Reference Panel (n = 68) [1]	85.2%	86.8%	76.4%	94.1%	94% (61/65)
Prevalence [2]	2013 Global Panel (n = 53)	96.2%	96.2%	86.8%	100%	100% (38/38)
	2015 Global Panel (n = 60)	85.0%	93.3%	75.0%	96.7%	100% (28/28)
	2016 Global Panel (n = 60)	80.0%	83.3%	63.3%	88.3%	94% (30/32)
Targeted	CDC VISA Panel (n = 14)	64.3%	64.3%	64.3%	64.3%	69% (9/13)
	Regional USA300 Panel (n = 29) [3]	100%	100%	100%	100%	100% (29/29)
	Ghent CRS Panel (n = 90)	NT	NT	NT	96.7%	Insufficient AST data
NA	Expanded Access Requests (n = 27) [4]	85.2%	92.6%	88.9%	96.3%	Insufficient AST data
Summary Values		Diversity Panels: Selection and Prevalence (n = 241)			94.6%	-
		All Panels (n = 401)			94.5%	95% (n = 205)

Abbreviations: AST: Antibiotic Sensitivity Testing; CDC: Centers for Disease Control and Prevention; CRS: chronic rhinosinusitis; NA: not applicable; NT: not tested; MDR: multidrug resistant; VISA: vancomycin intermediate. [1] Includes all major hospital-acquired methicillin-resistant *S. aureus* (HA-MRSA) and community-acquired methicillin-resistant *S. aureus* (CA-MRSA) lineages. [2] Nearly random samples fitting geographic distribution 45% North America, 45% Europe., 10% Asia-Pacific, obtained from JMI Laboratories SENTRY program for antimicrobial surveillance. [3] Isolates selected from [40–43], each having a different Pulsed Field Gel Electrophoresis pattern. [4] Initial patient isolates submitted for sensitivity testing as part of requests for product use under U.S. Individual Patient Expanded Access or Australian Special Access Scheme policies between August 2017 and September 2018, inclusive. These policies allow patients with serious or life-threatening infections that are not responding to existing approved therapies to access investigational products on an emergency basis. AB-SA01 is listed as NCT03395769 for Expanded Access use in the United States.

The AB-SA01 component phages were selected partly based on the 68-member diversity panel, which included representatives of all major community-acquired (CA-) and hospital-acquired (HA-) MRSA lineages [4,5]. Each component phage had a different host range, with most bacterial strains being sensitive to more than one of the AB-SA01 phages. AB-SA01 activity was similarly high across panels of isolates that represent globally prevalent *S. aureus* from blood, wound, lung, urinary, and other infections in later years. Using targeted interest panels, AB-SA01 was also shown to have activity on strains being relatively rare but concerning the VISA phenotype, a panel of exclusively CRS isolates, and a variety of the clinically significant USA300 lineage.

Table 2. In vitro activity of AB-SA01 and its component phages on bacterial species other than *S. aureus*.

Bacteria		Number of Strains Tested	Number of Strains Productively Infected			
Order	**Genus, Species**		**Sa83**	**Sa87**	**J-Sa36**	**AB-SA01**
Bacillales	*Staphylococcus epidermidis*	5	2	2	2	2
Lactobacillales	*Streptococcus spp.*	3	0	0	0	0
Corynebacteriales	*Corynebacterium spp.*	4	0	0	0	0
Micrococcales	*Micrococcus luteus*	1	0	0	0	0
Burkholderiales	*Achromobacter xylosoxidans*	1	0	0	0	0
	Burkholderia cepacia	1	0	0	0	0
Pseudomonales	*Acinetobacter baumannii*	1	0	0	0	0
	Pseudomonas aeruginosa	3	0	0	0	0
	Pseudomonas oryzihabitans	1	0	0	0	0
Enterobacteriales	*Enterobacter cloacae*	1	0	0	0	0
	Escherichia coli	1	0	0	0	0
	Klebsiella pneumoniae	1	0	0	0	0
	Pantoea agglomerans	1	0	0	0	0
Xanthamonadales	*Stenotrophomonas maltophilia*	1	0	0	0	0

3.3. *Frequency of Resistance and Complementation*

The potential for phages to complement each other in the event that bacterial resistance arises was considered as part of AB-SA01 development. During product selection, six candidate phages with broad or differing host ranges were assessed on a sensitive *S. aureus* strain. Using the double-drop method of 2.4, BIMs that were generated using one phage were first tested to confirm whether they truly exhibited reduced phage sensitivity after streak-purification, then cross-resistance to other phages was tested (Table 3). Sa83, Sa81, and Sa76 were similarly able to complement Sa87-induced resistance and had previously shown very similar host ranges. Of these, only Sa83 was retained because it made a slightly better contribution to the total host range and complementation profile of AB-SA01. J-Sa37 more often exhibited cross-resistance than complementation and was not included in AB-SA01. J-Sa36 exhibited different complementation behavior as compared to Sa87 or Sa83.

Table 3. Complementation among candidate phages.

Phage Used to Generate BIM	Bacterial Lawn	BIM Confirmation [1]	Test for Complementation					
			Sa83	**Sa87**	**J-Sa36**	**Sa76**	**Sa81**	**J-Sa37**
Sa87	parental	S	S	S	S	S	S	S
	BIM 1	I	S	-	S	S	S	R
	BIM 2	I	S	-	S	S	S	R
	BIM 3	NG [2]	-	-	-	-	-	-
	BIM 4	I	S	-	S	S	S	R
	BIM 5	I	S	-	I	S	S	R
	BIM 6	I	S	-	S	S	S	R
	BIM 7	I	S	-	S	S	S	R
	BIM 8	I	S	-	I	S	S	R
	BIM 9	I	S	-	I	S	S	R
	BIM 10	I	S	-	I	S	S	R
J-Sa36	Parental	S	S	S	S	-	-	S
	BIM 1	I	S	I	-	-	-	R
	BIM 2	I	I	I	-	-	-	R
	BIM 3	I	I	S	-	-	-	S
	BIM 4	I	I	I	-	-	-	S

[1] Bacteriophage-insensitive mutant (BIM) confirmation conducted using same phage as in column 1. R (red): resistant (no phage activity seen within bacterial spot); I (yellow): intermediate (phage activity seen within bacterial spot); S (green): sensitive (<10 colonies within bacterial spot), -: not tested. [2] NG: no growth; BIM was not recovered during single colony purification and is therefore presumed to be sensitive; no other testing possible.

After AB-SA01 composition was finalized, the mean apparent frequency of resistance to AB-SA01 was lower than the values observed for the individual phages, both at 24 h and 48 h (Table 4). However, this trend was not statistically significant ($p > 0.05$), possibly because the values observed in this study were close to the limit of detection. This suggests that the spontaneous frequency of AB-SA01 resistance among sensitive *S. aureus* populations is no greater than ~3×10^{-9}. None of the BIM colonies observed in this study could be recovered by picking and re-streaking on agar to isolate them away from phages on the original test plate, implying that their growth on the test plates was not due to stable, heritable phage resistance, but was instead a temporary phenotype or a spatial phenomenon in which cells escape contact with the phages during incubation of the phage-bacteria mixture before plating.

Table 4. Apparent frequency of intrinsic phage resistance in populations of *S. aureus* sensitive to AB-SA01 and its component phages.

Phage	After 24 h Plate Incubation			After 48 h Plate Incubation		
	Replicate 1 [1]	Replicate 2	Replicate 3	Replicate 1	Replicate 2	Replicate 3
Sa83	1.1E-8	3.8E-9	5.0E-9	7.1E-9	3.8E-9	6.7E-9
Sa87	2.0E-8	5.0E-9	5.0E-9	1.7E-8	5.0E-9	5.0E-9
J-Sa36	2.9E-9	2.5E-9	1.2E-8	2.9E-9	1.3E-9	5.0E-9
AB-SA01 [2]	1.4E-9	3.8E-9	3.3E-9	2.9E-9	0 [3]	3.3E-9

[1] Within each replicate, all aliquots of the same culture were exposed to each phage test sample. Since replicates contained a slightly different initial bacterial concentration, each replicate is displayed separately to allow for more accurate comparisons among the different phages. [2] Prepared as equal volume mixture of the three component stocks. [3] Limit of detection is 1.0E-9.

3.4. *In Vivo Activity of AB-SA01*

AB-SA01 showed efficacy equivalent to vancomycin in two murine acute lung infection models, each of which used a different *S. aureus* challenge strain, murine genetic background, and immune status. In the first murine pneumonia model (Figure 3A), three doses of the AB-SA01 prototype were tested in neutropenic CD-1 mice. This prototype contained the three phage components of AB-SA01 plus the J-Sa37 phage that was later removed from the product because its fractional contribution to in vitro host range and complementation were deemed insufficient to justify manufacturing a fourth component phage. At 24 hpi, lung homogenates from mice treated with 4×10^9 or 4×10^8 total PFU contained significantly fewer bacteria than mice treated with buffer and were statistically equivalent (all $p > 0.79$) to mice that had been treated with vancomycin at the same time points. The mean reductions in lung bacterial load relative to untreated mice were 3.63 $\log_{10}$CFU ($p < 0.0001$) for the vancomycin group, 3.09 $\log_{10}$CFU ($p < 0.0001$) for the highest AB-SA01 dose group, and 3.02 $\log_{10}$CFU ($p < 0.0001$) for the medium AB-SA01 dose groups. These results suggested that doses higher than 4×10^7 PFU were required for efficacy in this model.

In a follow-up experiment using the final AB-SA01 composition (Figure 3B), a 1.5×10^9 total PFU dose group was tested in immunocompetent BALB/c mice. At 24 hpi, lung homogenates from mice that had received two doses of AB-SA01 contained statistically fewer bacteria than those from untreated mice ($p = 0.0058$), and were statistically equivalent to mice that had received three doses of vancomycin ($p = 0.9172$). The mean reductions in lung bacterial load relative to the untreated control were 1.64 $\log_{10}$CFU for AB-SA01-treated mice and 1.80 $\log_{10}$CFU for vancomycin-treated mice.

In both mouse studies, *S. aureus* colonies recovered from infected animals showed patterns of sensitivity to AB-SA01 and its component phages that were similar to their respective parental strains, and no phage-resistant colonies were observed.

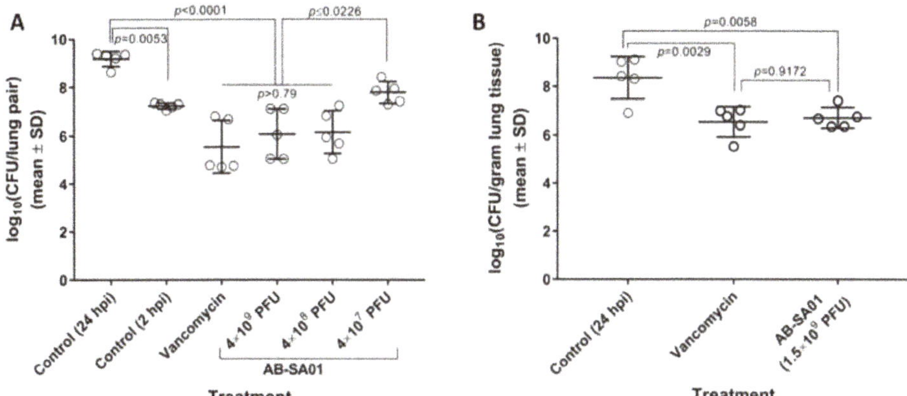

Figure 3. AB-SA01 reduces lung bacterial burden in (**A**) neutropenic CD-1 mice and (**B**) immunocompetent BALB/c mice. Phage doses are given as total plaque-forming units (PFU) per dose.

4. Discussion

The suitability of a medicinal product for human administration depends in part on the intrinsic characteristics of its active components. While no phage product to treat human infections has yet received market approval from the FDA or most of its global equivalents, the characteristics that make individual phages suitable for human use are commonly accepted within the phage research community [27–29,31] and generally supported by the FDA in public commentary on the subject [30]. AB-SA01, which is being developed to treat *S. aureus* infections, consists of three component phages that each meet these criteria in that they are: obligately lytic (not temperate), kill a wide range of clinical *S. aureus* strains, are incapable of specialized transduction and likely incapable of generalized transduction, and no bacterial virulence factors or drug resistance genes were identified by whole genome sequence analysis. Since a potential advantage of phage therapy is that it can be targeted to a pathogen of interest and therefore cause less disruption of the patient's commensal flora than a broad-spectrum antibiotic, it is relevant that the AB-SA01 component phages appear to be specific to *Staphylococcus spp.*, exhibiting no in vitro cross-genus activity.

In addition to the characteristics of individual phages, there is a rationale for the specific combination of phages that makes up AB-SA01. Within AB-SA01, phages Sa83, Sa87, and J-Sa36 each contribute different anti-*S. aureus* activity to AB-SA01; there is evidence of occasional synergy to kill otherwise non-susceptible *S. aureus* strains, the intrinsic frequency of resistance within populations of sensitive bacteria is low, and complementation is possible when resistance does develop.

The clinical utility of an antibacterial agent depends in large part on its spectrum of activity against target pathogens and non-target bacteria. The in vitro activity of AB-SA01 is high and the percentage of susceptible isolates is nearly identical on MDR and non-MDR *S. aureus* strains. This is similar to results from an external study that looked at two of the three AB-SA01 component phages and found no significant association between phage susceptibility and antibiotic resistance among 65 clinical *S. aureus* isolates [44]. The apparently lower activity of AB-SA01 on the VISA strains is difficult to interpret because this panel represents a diversity of vancomycin resistance determinants and not a diversity of *S. aureus* strain backgrounds. For *S. aureus* strains with a known multilocus sequence type there was no apparent association between genetic lineage and phage sensitivity, which is not unexpected given that the housekeeping genes on which bacterial strain typing systems are based are not expected to affect phage adsorption, replication, or lysis. It is possible for bacteria to become resistant to phages by a variety of mechanisms such as mutations in cell surface receptors and CRISPR, restriction-modification, or abortive infection systems [45]. The AB-SA01 component phages were partially chosen based on empirical evidence that the individual phages can complement resistance that may arise to another component. This is somewhat analogous to antibiotics that target multiple critical points in bacterial metabolism. The frequency of spontaneous resistance to AB-SA01 was measured as no greater than 3×10^{-9}. This value is less frequent than for rifampicin [22] and approximately 10-fold higher than for daptomycin and linezolid [46,47], though it may be an overestimate since none of the counted colonies proved to be heritably resistant to AB-SA01. Unlike static small molecules, phages also have the potential to evolve in situ, adapting to local bacterial populations and undergoing antagonistic co-evolution to bypass newly developed resistance [48,49]. How this will play out clinically remains to be seen. In vitro, mutual adaptation often leads to long-term maintenance of both phage and bacterial populations [49], but patterns of in vitro and in vivo mutation have been shown to differ [48]. The collective global experience treating single patients, including with AB-SA01 [50], strongly suggests that phage administration can lead to clinical resolution of infection, sometimes with confirmed pathogen eradication [51–60].

The rare instances in which AB-SA01 formed plaques on a *S. aureus* strain, even though the individually tested component phages did not, are intriguing. Between-phage synergy has not been extensively studied. Commonly proposed mechanisms tend to focus on combinations of unrelated or distantly related phages in which, for example, two phages use different receptors [61] or one phage has a tailspike protein with depolymerase activity that degrades bacterial capsule and increases the accessibility of a cell surface receptor to a second phage that does possess such enzymatic activity [62]. This type of mechanism seems unlikely for AB-SA01 given the high degree of relatedness among its component phages. Our observations could conceivably be the result of interactions downstream of phage adsorption, e.g. an in trans effect in which each phage in a co-infected cell expresses gene(s) necessary for both phages to bypass an intracellular resistance mechanism that would otherwise have prevented the second phage from completing replication and lysis. However, this is hypothetical and would need to be investigated further.

A frequent point of discussion for phage therapy is whether a fixed composition phage product will remain active against globally circulating strains of bacteria for long enough to be useful. It has sometimes been postulated that the rapid pace of bacterial evolution might cause the clinical populations of a target pathogen to change rapidly enough that a phage product might no longer be relevant by the time it obtains market approval, or that once in use, resistance may develop too quickly for the phage product to remain useful. While resistance development is a relevant issue for any novel antibacterial, we are not aware of evidence that this risk or rate would be higher for phage products than for other antibacterial agents being developed with a similar focus on novel mechanisms of action and resistance management. On the contrary, traits such as complementation among component phages and phage evolution offer a means of combating this and the evolution of phage resistance often carries other fitness costs [63]. The data presented here show that, at least for *S. aureus*, it appears possible to create a fixed-composition phage product that has activity against the vast majority of circulating clinical strains over several years, including MDR strains. When looking only at in vitro AB-SA01 activity on the 2013, 2015, and 2016 Global Panels, it is possible to suppose that activity has been gradually decreasing over time. However, it is equally possible, especially considering the aggregate results shown in Table 1, that the three Global Panels represent a mean of approximately 94% with one result each above and below this percentage. Notably, 96.3% of the 27 contemporary *S. aureus* isolates received between 2017 and 2018 by AmpliPhi from physicians requesting AB-SA01 to treat individual patients with refractory *S. aureus* infections were sensitive to AB-SA01, offering "real-world" support for the expectation that AB-SA01 will be active against the isolates of patients not responding to antibiotics.

Murine models of acute pneumonia showed that AB-SA01 exhibits antibacterial activity in a vertebrate infection. Both the prototype product and AB-SA01 were as effective as vancomycin in reducing lung bacterial burdens. The efficacy of *S. aureus* phages was observed in both neutropenic and immunocompetent mice. In *Pseudomonas aeruginosa* pneumonia models, phages were observed to be ineffective in neutropenic mice even if the same phages had successfully controlled a similar infection in mice with different or no immune deficiencies [64]. This likely reflects a genuine difference between *S. aureus* and *P. aeruginosa* pathogenesis. Skerrett et al [65] reported that myeloid differentiation factor 88, which is required for neutrophil production, is essential for host defense against *P. aeruginosa* but not *S. aureus* pneumonia. Neutrophil elastase is important for eradication of *P. aeruginosa* by the host's innate immune system [66], whereas *S. aureus* produces neutrophil elastase inhibitors and appears particularly resistant to neutrophil killing [67,68].

Most staphylococcal phages fall into three broad categories, temperate siphoviruses, obligately lytic myoviruses, and obligately lytic podoviruses [69]. The myoviruses have historically been grouped together and described as K-like or Twort-like [69], though recent taxonomic proposals divide them into four genera within a proposed *Twortvirinae* subfamily [70,71]. Collectively the staphylococcus myoviruses tend to have broad host ranges and are frequently discussed as actual and proposed components of therapeutic phage preparations [72,73]. Previous studies have also shown the potential of K-like *S. aureus* phages to treat biofilm-associated infections. Guimin et al. [44] studied two of the three AB-SA01 component phages and showed that they can reduce in vitro *S. aureus* biofilm. A four-phage mix containing the precursor of an AB-SA01 component phage also significantly degraded in vitro biofilm [74] and was used to treat mature *S. aureus* biofilm in a sheep sinus infection model [75]. After 3 days of treatment, the phage-treated sheep had significantly lower mucosal biofilm mass. Compared to the controls, the sheep were healthy, showed comparable levels of sinus mucosal inflammation and had healthy looking cilia.

Randomized, controlled clinical trials are needed to show that single-patient clinical observations and systematic preclinical data collected in a research environment will translate into broad clinical efficacy. The chemistry, manufacturing, and control aspects of AB-SA01 production (including but not limited to production, purification, quality control, storage, and stability) are beyond the scope of this manuscript. However, when added to the preclinical characterization data presented here, AB-SA01 and its associated data package enabled clinical studies under the oversight of the U.S. Food and Drug Administration and Australia's Therapeutic Goods Administration (TGA). In 2016, the safety and tolerability of AB-SA01 was tested in two clinical trials: one healthy volunteer study in the United States under an Investigational New Drug (IND) application (NCT02757755) and one open-label investigator-initiated study in Australia among post-rhinoplasty CRS patients (ACTRN12616000002482). AB-SA01 was safe and well tolerated in both study populations. Among CRS patients, there were preliminary indications of efficacy that will need to be confirmed in placebo-controlled studies, such as reductions in sinus bacterial load, improved endoscopic findings, and general symptom improvement [76]. Finally, 15 patients with serious or life-threatening *S. aureus* infections not responding to antibiotics have received a cumulative total of more than 400 doses AB-SA01, including more than 300 administered intravenously under Individual Patient Expanded Access INDs in the United States or Australia's Special Access Scheme. No serious adverse events attributed to AB-SA01 were reported and observations from these patients suggest that it may be fruitful to investigate the efficacy of AB-SA01 in randomized controlled trials involving indications such as bacteremia, native and prosthetic valve endocarditis, prosthetic joint infections, and ventricular assist device infections.

AB-SA01 is a well-characterized phage investigational product that has entered clinical development for the treatment of *S. aureus* infections. While it is frequently suggested that existing regulatory structures are not compatible with the clinical development of phage products or with the timely emergency treatment of patients not responding to antibiotics, AB-SA01 has thus far satisfied FDA and TGA requirements to conduct clinical trials and single-patient emergency treatment. As with any antibacterial, epidemiological shifts might eventually necessitate a compositional update. At that point, the accumulated clinical and regulatory experience that will hopefully have been established with fixed-composition products should pave the way for data-driven strategies to streamline updates to phage products.

Supplementary Materials: The following are available online at http://www.mdpi.com/1999-4915/11/1/88/s1, Table S1: Phage sensitivity of *S. aureus* strains with known MLST.

Author Contributions: Conceptualization, S.M., S.M.L.; Methodology, S.M., S.M.L., G.M., D.R., R.A.C., F.S., S.D.B.; Writing—Original Draft Preparation, S.M.L., S.M.; Writing—Review and Editing, S.M.L., G.M., D.R., R.A.C., F.S., S.D.B., S.M.

Acknowledgments: We thank Mikeljon Nikolich, Andrey Filippov, and Kirill Sergueev at Walter Reed Army Institute of Research, as well as Joseph Bertsche and Sam Boundy at AmpliPhi Biosciences for their contributions to host range testing. We thank Jon Iredell, Bernard Hudson, Peter Hawkey, and H.C. Claus Bachert for generously sharing clinical bacterial isolates. We thank Anthony Smithyman for his insightful scientific discussions and input over the years. We thank the University of North Texas Health Sciences Center pre-clinical services lab (Fort Worth, TX) and Karen J Shaw (Hearts Consulting) for contributions to the neutropenic mouse study and KWS Biotest (Bristol, UK) for the immunocompetent mouse study.

Conflicts of Interest: The authors are employees of AmpliPhi Biosciences, which is developing AB-SA01. Following collaborative study design, all animal studies were independently conducted and analyzed by external vendors.

References

1. Summers, W.C. The strange history of phage therapy. *Bacteriophage* **2012**, *2*, 130–133. [CrossRef] [PubMed]
2. O'Neill, J. *Tackling Drug-Resistance Infections Globally: Final Report and Reccomendations*; Review on Antimicrobial Resistance: London, UK, 2016.
3. Rice, L.B. Federal funding for the study of antimicrobial resistance in nosocomial pathogens: No ESKAPE. *J. Infect. Dis.* **2008**, *197*, 1079–1081. [CrossRef] [PubMed]
4. Stefani, S.; Chung, D.R.; Lindsay, J.A.; Friedrich, A.W.; Kearns, A.M.; Westh, H.; Mackenzie, F.M. Methicillin-resistant *Staphylococcus aureus* (MRSA): Global epidemiology and harmonisation of typing methods. *Int. J. Antimicrob. Agents* **2012**, *39*, 273–282. [CrossRef] [PubMed]
5. Otter, J.A.; French, G.L. Molecular epidemiology of community-associated meticillin-resistant *Staphylococcus aureus* in Europe. *Lancet Infect. Dis.* **2010**, *10*, 227–239. [CrossRef]
6. WHO. *Prioritization of Pathogens to Guide Discovery, Research and Development Of New Antibiotics for Drug-Resistant Bacterial Infections, Including Tuberculosis*; WHO: Geneva, Switzerland, 2017.
7. CDC. *Antibiotic Resistance Threats in the United States*; CDC: Atlanta, GA, USA, 2013.
8. Kallberg, C.; Ardal, C.; Salvesen Blix, H.; Klein, E.; Martinez, E.M.; Lindbaek, M.; Outterson, K.; Rottingen, J.A.; Laxminarayan, R. Introduction and geographic availability of new antibiotics approved between 1999 and 2014. *PLoS ONE* **2018**, *13*, e0205166. [CrossRef] [PubMed]
9. Senior, K. FDA approves first drug in new class of antibiotics. *Lancet* **2000**, *355*, 1523. [CrossRef]
10. Ikeda-Dantsuji, Y.; Hanaki, H.; Nakae, T.; Takesue, Y.; Tomono, K.; Honda, J.; Yanagihara, K.; Mikamo, H.; Fukuchi, K.; Kaku, M.; et al. Emergence of Linezolid-Resistant Mutants in a Susceptible-Cell Population of Methicillin-Resistant *Staphylococcus aureus*. *Antimicrob. Agents Chemother.* **2011**, *55*, 2466–2468. [CrossRef] [PubMed]
11. Morales, G.; Picazo, J.J.; Baos, E.; Candel, F.J.; Arribi, A.; Peláez, B.; Andrade, R.; de la Torre, M.-Á.; Fereres, J.; Sánchez-García, M. Resistance to Linezolid Is Mediated by the *cfr* Gene in the First Report of an Outbreak of Linezolid-Resistant *Staphylococcus aureus*. *Clin. Infect. Dis.* **2010**, *50*, 821–825. [CrossRef] [PubMed]
12. Bayer, A.S.; Schneider, T.; Sahl, H.-G. Mechanisms of daptomycin resistance in *Staphylococcus aureus*: Role of the cell membrane and cell wall. *Ann. N. Y. Acad. Sci.* **2013**, *1277*, 139–158. [CrossRef]
13. Dortet, L.; Anguel, N.; Fortineau, N.; Richard, C.; Nordmann, P. In vivo acquired daptomycin resistance during treatment of methicillin-resistant *Staphylococcus aureus* endocarditis. *Int. J. Infect. Dis.* **2013**, *17*, e1076–e1077. [CrossRef]
14. Stein, G.E.; Babinchak, T. Tigecycline: An update. *Diagn. Microbiol. Infect. Dis.* **2013**, *75*, 331–336. [CrossRef] [PubMed]
15. Long, S.W.; Olsen, R.J.; Mehta, S.C.; Palzkill, T.; Cernoch, P.L.; Perez, K.K.; Musick, W.L.; Rosato, A.E.; Musser, J.M. PBP2a mutations causing high-level ceftaroline resistance in clinical methicillin-resistant *Staphylococcus aureus* isolates. *Antimicrob. Agents Chemother.* **2014**, *58*, 6668–6674. [CrossRef] [PubMed]
16. Karlowsky, J.A.; Nichol, K.; Zhanel, G.G. Telavancin: Mechanisms of action, in vitro activity, and mechanisms of resistance. *Clin. Infect. Dis* **2015**, *61* (Suppl. 2), S58–S68. [CrossRef]

17. Jones, R.N.; Moeck, G.; Arhin, F.F.; Dudley, M.N.; Rhomberg, P.R.; Mendes, R.E. Results from Oritavancin Resistance Surveillance Programs (2011 to 2014): Clarification for Using Vancomycin as a Surrogate To Infer Oritavancin Susceptibility. *Antimicrob. Agents Chemother.* **2016**, *60*, 3174–3177. [CrossRef] [PubMed]
18. Morroni, G.; Brenciani, A.; Brescini, L.; Fioriti, S.; Simoni, S.; Pocognoli, A.; Mingoia, M.; Giovanetti, E.; Barchiesi, F.; Giacometti, A.; et al. A high rate of ceftobiprole resistance among clinical MRSA from a hospital in central Italy. *Antimicrob. Agents Chemother.* **2018**, *62*, e01663-1810. [CrossRef] [PubMed]
19. Werth, B.J.; Jain, R.; Hahn, A.; Cummings, L.; Weaver, T.; Waalkes, A.; Sengupta, D.; Salipante, S.J.; Rakita, R.M.; Butler-Wu, S.M. Emergence of dalbavancin non-susceptible, vancomycin-intermediate *Staphylococcus aureus* (VISA) after treatment of MRSA central line-associated bloodstream infection with a dalbavancin- and vancomycin-containing regimen. *Clin. Microbiol. Infect.* **2018**, *24*, e421–e429. [CrossRef] [PubMed]
20. Molina-Manso, D.; del Prado, G.; Ortiz-Perez, A.; Manrubia-Cobo, M.; Gomez-Barrena, E.; Cordero-Ampuero, J.; Esteban, J. In vitro susceptibility to antibiotics of staphylococci in biofilms isolated from orthopaedic infections. *Int. J. Antimicrob. Agents* **2013**, *41*, 521–523. [CrossRef] [PubMed]
21. Czaplewski, L.; Bax, R.; Clokie, M.; Dawson, M.; Fairhead, H.; Fischetti, V.A.; Foster, S.; Gilmore, B.F.; Hancock, R.E.; Harper, D.; et al. Alternatives to antibiotics-a pipeline portfolio review. *Lancet Infect. Dis.* **2016**, *16*, 239–251. [CrossRef]
22. O'Neill, A.J.; Cove, J.H.; Chopra, I. Mutation frequencies for resistance to fusidic acid and rifampicin in *Staphylococcus aureus*. *J. Antimicrob. Chemother.* **2001**, *47*, 647–650. [CrossRef] [PubMed]
23. Gottlieb, S.; Washington, DC, USA. FDA's Strategic Approach for Combating Antimicrobial Rsistance. Personal Communication, 2018.
24. Loc-Carrillo, C.; Abedon, S.T. Pros and cons of phage therapy. *Bacteriophage* **2011**, *1*, 111–114. [CrossRef]
25. Gorski, A.; Dabrowska, K.; Switala-Jelen, K.; Nowaczyk, M.; Weber-Dabrowska, B.; Boratynski, J.; Wietrzyk, J.; Opolski, A. New insights into the possible role of bacteriophages in host defense and disease. *Med. Immunol.* **2003**, *2*, 2. [CrossRef] [PubMed]
26. Nguyen, S.; Baker, K.; Padman, B.S.; Patwa, R.; Dunstan, R.A.; Weston, T.A.; Schlosser, K.; Bailey, B.; Lithgow, T.; Lazarou, M.; et al. Bacteriophage Transcytosis Provides a Mechanism To Cross Epithelial Cell Layers. *mBio* **2017**, *8*, e01874–e01817. [CrossRef] [PubMed]
27. Gill, J.J.; Hyman, P. Phage choice, isolation, and preparation for phage therapy. *Curr. Pharm. Biotechnol.* **2010**, *11*, 2–14. [CrossRef] [PubMed]
28. Merabishvili, M.; Pirnay, J.P.; Verbeken, G.; Chanishvili, N.; Tediashvili, M.; Lashkhi, N.; Glonti, T.; Krylov, V.; Mast, J.; Van Parys, L.; et al. Quality-controlled small-scale production of a well-defined bacteriophage cocktail for use in human clinical trials. *PLoS ONE* **2009**, *4*, e4944. [CrossRef] [PubMed]
29. Casey, E.; van Sinderen, D.; Mahony, J. In Vitro Characteristics of Phages to Guide 'Real Life' Phage Therapy Suitability. *Viruses* **2018**, *10*, 163. [CrossRef] [PubMed]
30. Reindel, R.; Fiore, C.R. Phage Therapy: Considerations and Challenges for Development. *Clin. Infect. Dis.* **2017**, *64*, 1589–1590. [CrossRef] [PubMed]
31. Carlton, R.M. Phage therapy: Past history and future prospects. *Arch. Immunol. Ther. Exp.* **1999**, *47*, 267–274.
32. Magiorakos, A.P.; Srinivasan, A.; Carey, R.B.; Carmeli, Y.; Falagas, M.E.; Giske, C.G.; Harbarth, S.; Hindler, J.F.; Kahlmeter, G.; Olsson-Liljequist, B.; et al. Multidrug-resistant, extensively drug-resistant and pandrug-resistant bacteria: an international expert proposal for interim standard definitions for acquired resistance. *Clinic. Microbiol. Infect.* **2012**, *18*, 268–281. [CrossRef] [PubMed]
33. Mazzocco, A.; Waddell, T.E.; Lingohr, E.; Johnson, R.P. Enumeration of bacteriophages using the small drop plaque assay system. *Methods mol. Biol.* **2009**, *501*, 81–85. [PubMed]
34. O'Flynn, G.; Ross, R.P.; Fitzgerald, G.F.; Coffey, A. Evaluation of a cocktail of three bacteriophages for biocontrol of *Escherichia coli* O157:H7. *Appl. Environ. Microbiol.* **2004**, *70*, 3417–3424. [CrossRef]
35. Soding, J. Protein homology detection by HMM-HMM comparison. *Bioinformatics* **2005**, *21*, 951–960. [CrossRef] [PubMed]
36. Soding, J.; Biegert, A.; Lupas, A.N. The HHpred interactive server for protein homology detection and structure prediction. *Nucleic Acids Res.* **2005**, *33*, W244–W248. [CrossRef] [PubMed]
37. Darling, A.C.E.; Mau, B.; Blattner, F.R.; Perna, N.T. Mauve: Multiple alignment of conserved genomic sequence with rearrangements. *Genome Res.* **2004**, *14*, 1394–1403. [CrossRef] [PubMed]

38. Kadurugamuwa, J.L.; Sin, L.V.; Yu, J.; Francis, K.P.; Kimura, R.; Purchio, T.; Contag, P.R. Rapid Direct Method for Monitoring Antibiotics in a Mouse Model of Bacterial Biofilm Infection. *Antimicrob. Agents Chemother.* **2003**, *47*, 3130–3137. [CrossRef] [PubMed]
39. Carlson, K. Working with bacteriophages: Common techniques and methodological approaches. In *Bacteriophages: Biology and Applications*; CRC Press: Boca Raton, FL, USA, 2005.
40. Tattevin, P.; Diep, B.A.; Jula, M.; Perdreau-Remington, F. Long-term follow-up of methicillin-resistant *Staphylococcus aureus* molecular epidemiology after emergence of clone USA300 in San Francisco jail populations. *J. Clin. Microbiol.* **2008**, *46*, 4056–4057. [CrossRef]
41. Diep, B.A.; Carleton, H.A.; Chang, R.F.; Sensabaugh, G.F.; Perdreau-Remington, F. Roles of 34 virulence genes in the evolution of hospital- and community-associated strains of methicillin-resistant Staphylococcus aureus. *J. Infect. Dis.* **2006**, *193*, 1495–1503. [CrossRef] [PubMed]
42. Diep, B.A.; Chambers, H.F.; Graber, C.J.; Szumowski, J.D.; Miller, L.G.; Han, L.L.; Chen, J.H.; Lin, F.; Lin, J.; Phan, T.H.; et al. Emergence of multidrug-resistant, community-associated, methicillin-resistant Staphylococcus aureus clone USA300 in men who have sex with men. *Ann. Intern. Med.* **2008**, *148*, 249–257. [CrossRef] [PubMed]
43. Diep, B.A.; Gill, S.R.; Chang, R.F.; Phan, T.H.; Chen, J.H.; Davidson, M.G.; Lin, F.; Lin, J.; Carleton, H.A.; Mongodin, E.F.; et al. Complete genome sequence of USA300, an epidemic clone of community-acquired meticillin-resistant Staphylococcus aureus. *Lancet* **2006**, *367*, 731–739. [CrossRef]
44. Guimin, Z.; Yin, Z.; Paramasivan, S.; Richter, K.; Morales, S.; Wormald, P.J.; Vreugde, S. Bacteriophage effectively kills multidrug resistant *Staphylococcus aureus* clinical isolates from chronic rhinosinusitis patients. *Int. Forum Allergy Rhinol.* **2018**, *8*, 406–414.
45. Labrie, S.J.; Samson, J.E.; Moineau, S. Bacteriophage resistance mechanisms. *Nat. Rev. Microbiol.* **2010**, *8*, 317–327. [CrossRef]
46. Silverman, J.A.; Oliver, N.; Andrew, T.; Li, T. Resistance Studies with Daptomycin. *Antimicrob. Agents Chemother.* **2001**, *45*, 1799–1802. [CrossRef] [PubMed]
47. Zurenko, G.E.; Yagi, B.H.; Schaadt, R.D.; Allison, J.W.; Kilburn, J.O.; Glickman, S.E.; Hutchinson, D.K.; Barbachyn, M.R.; Brickner, S.J. In vitro activities of U-100592 and U-100766, novel oxazolidinone antibacterial agents. *Antimicrob. Agents Chemother.* **1996**, *40*, 839–845. [CrossRef] [PubMed]
48. De Sordi, L.; Khanna, V.; Debarbieux, L. The Gut Microbiota Facilitates Drifts in the Genetic Diversity and Infectivity of Bacterial Viruses. *Cell Host Microbe* **2017**, *22*, 801–808. [CrossRef] [PubMed]
49. Buckling, A.; Rainey, P.B. Antagonistic coevolution between a bacterium and a bacteriophage. *Proc. Biol. Sci.* **2002**, *269*, 931–936. [CrossRef] [PubMed]
50. Fabijan, A.; Ho, J.; Lin, R.C.Y.; Maddocks, S.; Gilbey, T.; Sandaradura, I.; Chan, S.; Morales, S.; Venturini, C.; Branston, S.; et al. Safety and tolerability of bacteriophage therapy in the treatment of severe *Staphylococcus aureus* bacteremia. Manuscript in preparation.
51. Slopek, S.; Weber-Dabrowska, B.; Dabrowski, M.; Kucharewicz-Krukowska, A. Results of bacteriophage treatment of suppurative bacterial infections in the years 1981-1986. *Arch. Immunol. Ther. Exp.* **1987**, *35*, 569–583.
52. Weber-Dabrowska, B.; Mulczyk, M.; Gorski, A. Bacteriophage therapy of bacterial infections: An update of our institute's experience. *Arch. Immunol. Ther. Exp.* **2000**, *48*, 547–551.
53. Kutateladze, M.; Adamia, R. Phage therapy experience at the Eliava Institute. *Med. Mal. Infect.* **2008**, *38*, 426–430. [CrossRef]
54. Wright, A.; Hawkins, C.H.; Anggard, E.E.; Harper, D.R. A controlled clinical trial of a therapeutic bacteriophage preparation in chronic otitis due to antibiotic-resistant *Pseudomonas aeruginosa*; a preliminary report of efficacy. *Clin. Otolaryngol.* **2009**, *34*, 349–357. [CrossRef]
55. Chanishvili, N. Phage therapy–history from Twort and d'Herelle through Soviet experience to current approaches. *Adv. Virus Res.* **2012**, *83*, 3–40.
56. Miedzybrodzki, R.; Borysowski, J.; Weber-Dabrowska, B.; Fortuna, W.; Letkiewicz, S.; Szufnarowski, K.; Pawelczyk, Z.; Rogoz, P.; Klak, M.; Wojtasik, E.; et al. Clinical aspects of phage therapy. *Adv. Virus Res.* **2012**, *83*, 73–121.

57. Schooley, R.T.; Biswas, B.; Gill, J.J.; Hernandez-Morales, A.; Lancaster, J.; Lessor, L.; Barr, J.J.; Reed, S.L.; Rohwer, F.; Benler, S.; et al. Development and Use of Personalized Bacteriophage-Based Therapeutic Cocktails To Treat a Patient with a Disseminated Resistant *Acinetobacter baumannii* Infection. *Antimicrob. Agents Chemother.* **2017**, *61*, e00954-17. [CrossRef] [PubMed]
58. Jennes, S.; Merabishvili, M.; Soentjens, P.; Pang, K.W.; Rose, T.; Keersebilck, E.; Soete, O.; François, P.-M.; Teodorescu, S.; Verween, G.; et al. Use of bacteriophages in the treatment of colistin-only-sensitive *Pseudomonas aeruginosa* septicaemia in a patient with acute kidney injury-a case report. *Crit. Care* **2017**, *21*, 129. [CrossRef] [PubMed]
59. Chan, B.K.; Turner, P.E.; Kim, S.; Mojibian, H.R.; Elefteriades, J.A.; Narayan, D. Phage treatment of an aortic graft infected with *Pseudomonas aeruginosa*. *Evolut. Med. Public Health* **2018**, *2018*, 60–66. [CrossRef] [PubMed]
60. Gorski, A.; Miedzybrodzki, R.; Borysowski, J.; Weber-Dabrowska, B.; Lobocka, M.; Fortuna, W.; Letkiewicz, S.; Zimecki, M.; Filby, G. Bacteriophage therapy for the treatment of infections. *Curr. Opin. Investig. Drugs* **2009**, *10*, 766–774. [PubMed]
61. Chaudhry, W.N.; Concepcion-Acevedo, J.; Park, T.; Andleeb, S.; Bull, J.J.; Levin, B.R. Synergy and Order Effects of Antibiotics and Phages in Killing *Pseudomonas aeruginosa* Biofilms. *PLoS ONE* **2017**, *12*, e0168615. [CrossRef] [PubMed]
62. Schmerer, M.; Molineux, I.J.; Bull, J.J. Synergy as a rationale for phage therapy using phage cocktails. *PeerJ* **2014**, *2*, e590. [CrossRef] [PubMed]
63. Oechslin, F. Resistance Development to Bacteriophages Occurring during Bacteriophage Therapy. *Viruses* **2018**, *10*, 351. [CrossRef]
64. Roach, D.R.; Leung, C.Y.; Henry, M.; Morello, E.; Singh, D.; Di Santo, J.P.; Weitz, J.S.; Debarbieux, L. Synergy between the Host Immune System and Bacteriophage Is Essential for Successful Phage Therapy against an Acute Respiratory Pathogen. *Cell Host Microbe* **2017**, *22*, 38–47.e4. [CrossRef]
65. Skerrett, S.J.; Liggitt, H.D.; Hajjar, A.M.; Wilson, C.B. Cutting edge: Myeloid differentiation factor 88 is essential for pulmonary host defense against *Pseudomonas aeruginosa* but not *Staphylococcus aureus*. *J. Immunol.* **2004**, *172*, 3377–3381. [CrossRef]
66. Hirche, T.O.; Benabid, R.; Deslee, G.; Gangloff, S.; Achilefu, S.; Guenounou, M.; Lebargy, F.; Hancock, R.E.; Belaaouaj, A. Neutrophil elastase mediates innate host protection against *Pseudomonas aeruginosa*. *J. Immunol.* **2008**, *181*, 4945–4954. [CrossRef]
67. Stapels, D.A.; Geisbrecht, B.V.; Rooijakkers, S.H. Neutrophil serine proteases in antibacterial defense. *Curr. Opin. Microbiol.* **2015**, *23*, 42–48. [CrossRef] [PubMed]
68. Stapels, D.A.; Ramyar, K.X.; Bischoff, M.; von Kockritz-Blickwede, M.; Milder, F.J.; Ruyken, M.; Eisenbeis, J.; McWhorter, W.J.; Herrmann, M.; van Kessel, K.P.; et al. *Staphylococcus aureus* secretes a unique class of neutrophil serine protease inhibitors. *Proc. Natl. Acad. Sci. USA* **2014**, *111*, 13187–13192. [CrossRef] [PubMed]
69. Ajuebor, J.; Buttimer, C.; Arroyo-Moreno, S.; Chanishvili, N.; Gabriel, E.M.; O'Mahony, J.; McAuliffe, O.; Neve, H.; Franz, C.; Coffey, A. Comparison of *Staphylococcus* Phage K with Close Phage Relatives Commonly Employed in Phage Therapeutics. *Antibiotics* **2018**, *7*, 37. [CrossRef] [PubMed]
70. McCallin, S.; Sarker, S.A.; Sultana, S.; Oechslin, F.; Brussow, H. Metagenome analysis of Russian and Georgian Pyophage cocktails and a placebo-controlled safety trial of single phage versus phage cocktail in healthy *Staphylococcus aureus* carriers. *Environ. Microbiol.* **2018**, *20*, 3278–3293. [CrossRef] [PubMed]
71. Łobocka, M.; Hejnowicz, M.S.; Dabrowski, K.; Gozdek, A.; Kosakowski, J.; Witkowska, M.; Ulatowska, M.I.; Weber-Dabrowska, B.; Kwiatek, M.; Parasion, S.; et al. Genomics of staphylococcal Twort-like phages–potential therapeutics of the post-antibiotic era. *Adv. Virus Res.* **2012**, *83*, 143–216. [PubMed]
72. Adriaenssens, E.M.; Wittmann, J.; Kuhn, J.H.; Turner, D.; Sullivan, M.B.; Dutilh, B.E.; Jang, H.B.; van Zyl, L.J.; Klumpp, J.; Lobocka, M.; et al. Taxonomy of prokaryotic viruses: 2017 update from the ICTV Bacterial and Archaeal Viruses Subcommittee. *Arch. Virol.* **2018**, *163*, 1125–1129. [CrossRef] [PubMed]
73. Barylski, J.; Enault, F.; Dutilh, B.E.; Schuller, M.B.P.; Edwards, R.A.; Gillis, A.; Klumpp, J.; Knezevic, P.; Krupovic, M.; Kuhn, J.H.; et al. Analysis of Spounaviruses as a Case 3 Study for the Overdue Reclassification of 4 Tailed Bacteriophages. *bioRxiv* **2017**. [CrossRef]
74. Drilling, A.; Morales, S.; Jardeleza, C.; Vreugde, S.; Speck, P.; Wormald, P.J. Bacteriophage reduces biofilm of *Staphylococcus aureus* ex vivo isolates from chronic rhinosinusitis patients. *Am. J. Rhinol. Allergy* **2014**, *28*, 3–11. [CrossRef] [PubMed]

75. Drilling, A.; Morales, S.; Boase, S.; Jervis-Bardy, J.; James, C.; Jardeleza, C.; Tan, N.C.; Cleland, E.; Speck, P.; Vreugde, S.; et al. Safety and efficacy of topical bacteriophage and ethylenediaminetetraacetic acid treatment of *Staphylococcus aureus* infection in a sheep model of sinusitis. *Int. Forum Allergy Rhinol.* **2014**, *4*, 176–186. [CrossRef]
76. Ooi MLD, A.J.; Morales, S.; Fong, S.; Moraitis, S.; Macias-Valle, L.; Vreugde, S.; Psaltis, A.; Wormald, P.-J. Phage Therapy for *S. aureus* Chronic Rhinosinusitis: A Phase 1 First-in-Human Study. Manuscript in preparation.

Article

Enterococcus faecalis Countermeasures Defeat a Virulent *Picovirinae* Bacteriophage

Julien Lossouarn [1,*], Arnaud Briet [1], Elisabeth Moncaut [1], Sylviane Furlan [1], Astrid Bouteau [1], Olivier Son [1], Magali Leroy [2], Michael S. DuBow [2], François Lecointe [1], Pascale Serror [1] and Marie-Agnès Petit [1]

1 Micalis Institute, INRA, AgroParisTech, Université Paris-Saclay, 78350 Jouy-en-Josas, France; Arnaud.Briet@ANSES.FR (Ar.B.); elisabeth.moncaut@inra.fr (E.M.); sylviane.furlan@inra.fr (S.F.); astridbouteau@gmail.com (As.B.); olivier.son@inra.fr (O.S.); françois.lecointe@inra.fr (F.L.); pascale.serror@inra.fr (P.S.); marie-agnes.petit@inra.fr (M.-A.P.)

2 Institute for Integrative Biology of the Cell (I2BC), CEA, CNRS, Université Paris-Sud, Université Paris-Saclay, 91198 Gif-sur-Yvette, France; magali.o.leroy@gmail.com (M.L.); micheal.dubow@igmors.u-psud.fr (M.S.D.)

* Correspondence: julien.lossouarn@inra.fr; Tel.: +33-(0)1-34-65-28-67

Received: 14 December 2018; Accepted: 31 December 2018; Published: 10 January 2019

Abstract: *Enterococcus faecalis* is an opportunistic pathogen that has emerged as a major cause of nosocomial infections worldwide. Many clinical strains are indeed resistant to last resort antibiotics and there is consequently a reawakening of interest in exploiting virulent phages to combat them. However, little is still known about phage receptors and phage resistance mechanisms in enterococci. We made use of a prophageless derivative of the well-known clinical strain *E. faecalis* V583 to isolate a virulent phage belonging to the *Picovirinae* subfamily and to the P68 genus that we named Idefix. Interestingly, most isolates of *E. faecalis* tested—including V583—were resistant to this phage and we investigated more deeply into phage resistance mechanisms. We found that *E. faecalis* V583 prophage 6 was particularly efficient in resisting Idefix infection thanks to a new abortive infection (Abi) mechanism, which we designated Abiα. It corresponded to the Pfam domain family with unknown function DUF4393 and conferred a typical Abi phenotype by causing a premature lysis of infected *E. faecalis*. The *abiα* gene is widespread among prophages of enterococci and other Gram-positive bacteria. Furthermore, we identified two genes involved in the synthesis of the side chains of the surface rhamnopolysaccharide that are important for Idefix adsorption. Interestingly, mutants in these genes arose at a frequency of ~10^{-4} resistant mutants per generation, conferring a supplemental bacterial line of defense against Idefix.

Keywords: abortive infection; prophage; adsorption; *Enterococcus*; rhamnopolysaccharide

1. Introduction

Enterococci are ubiquitous Gram-positive facultative anaerobic bacteria that colonize the mammalian gastrointestinal tract [1]. In particular, the two species *Enterococcus faecalis* and *Enterococcus faecium* are part of the normal human gut microbiota and generally have no adverse effects on healthy individuals. However, they also represent opportunistic pathogens that have emerged as a leading source of nosocomial infections, particularly in immunocompromised patients [2]. *E. faecalis* and *E. faecium* mostly cause urinary tract infections, peritonitis, bacteraemia, and endocarditis [2]. The clinical importance of these bacterial species is directly related to their antibiotic resistance. The rapid spread of clinical isolates resistant to last resort antibiotics such as vancomycin and daptomycin has been of particular concern and associated hospital acquired infections have become a growing problem [3].

Virulent bacteriophages, i.e., viruses that infect and obligatorily lyse bacteria, have long held promise to treat bacterial infections and combat multi-drug resistant (MDR) bacteria [4]. In the recent years, the complete genome and phage cycle characteristics of a dozen virulent *E. faecalis* and *E. faecium* phages have been reported, emphasizing the growing interest for phage therapy in the future [5–15]. However, similar to the phenomenon of antibiotic resistance, bacterial resistance to phage is also taking place. For instance, mutations in the cell wall protein PIP_{EF} were recently found to provide resistance to two siphophages by limiting phage DNA entry in *E. faecalis* [16]. Apart from this, the modes of defence of enterococci against phages largely remain a *terra incognita*.

Keeping in mind that, in most ecosystems, bacteria co evolve with a plethora of bacteriophages, which imperatively depend on them for their reproduction, it is no surprise that bacteria invest efforts in fighting against them. They deploy for this goal different functions that have been studied for decades (reviewed in [17,18]), and still led to new discoveries [19–23]. The *E. faecalis* species, however, seems to have few acquired and innate immune systems. The only common clustered regularly interspaced short palindromic repeat (CRISPR)-CRISPR associated (Cas) locus found in *E. faecalis* isolates has lost its *cas* genes, and two complete type II CRISPR-Cas systems occur variably across the species [24–26]. Furthermore, no restriction-modification (R-M) system is commonly found within the species, and to date only a single type II R-M system has been described in three *E. faecalis* chromosomes [27,28]. Finally, no complete defense island system associated with restriction-modification was detected in *E. faecalis* to date [22]. The relative scarcity of R-M and CRISPR-Cas systems in MDR strains may facilitate plasmid-encoded antibiotic resistance genes acquisition [29,30].

Although less studied, phages themselves compete for their hosts on the bacterial battlefield [31]. Phage genes involved in anti-phage mechanisms are mainly found in temperate or defective prophages, rather than in virulent phages. During the prophage state (where the temperate phage is stably associated with its host), in addition to repressor-dependent immunity against similar phages, some phage express genes conferring resistance to infection by more or less unrelated phages. These genes can encode generalist functions such as R-M systems [32] or CRISPR-Cas loci [33], or specific ones. For instance, the *Escherichia coli* temperate phage P2 has three genes, *fun*, *tin*, and *old*, preventing growth of T5, T-even, and lambda respectively [34]. Furthermore, the O-antigen acetylase of *Salmonella typhimurium* phage BTP1 is reported to prevent adsorption of other phages [35]. Other typical prophage-encoded superinfection exclusion (Sie) systems act to block phage DNA injection into *E. coli* [36], *Lactococcus lactis* [37], and *Streptococcus thermophilus* [38]. Similar systems have also been recently discovered in *Pseudomonas* [39] and *Mycobacterium* prophages [40].

Prophages can also encode abortive infection (Abi) mechanisms that cause an interruption of invasive phage development and a premature death of the infected bacteria. This leads to the release of few or no progeny particles and thus prevents the expansion of the infection to the neighboring bacteria [17,18]. Abi are very diverse; among those carried by prophages, some have been well characterized such as the two component Rex system preventing lambda-lysogenic *E. coli* strain infection by T4 phage [41] or the tyrosine kinase Stk of coliphage 933W that blocks the replication cycle of HK97 [41,42].

Our earlier work showed that the vancomycin-resistant *E. faecalis* V583 clinical isolate hosts seven prophage elements. One prophage is remnant and completely domesticated (prophage 2) while the six others have various degrees of autonomy and different levels of interference with each other [43]. Only three of them—prophages 1, 3, and 5—are fully active and grow as plaques on a V583 derivative cured of all six prophages. However, prophage 1 is parasitized during the induction of it lytic cycle by the satellite prophage 7 also named a phage inducible chromosomal island. Prophages 4 and 6 finally retain some phage-like behavior but no longer produce infective virions. They excise from the bacterial chromosome in a process controlled by the other prophages, and once excised, prophage 4 replicates, while prophage 6 cannot [43].

To investigate putative anti-phage roles played by V583 prophages, we performed a phage screening using the V583 derivative strain cured of all plasmids and the six active prophages. We isolated a virulent phage belonging to the *Picovirinae* subfamily and to the P68 genus that contains phages with small size genome (19–20 kb). This prompted us to name this phage Idefix, after the French name of the small dog from Asterix comics. The genome sequence of two phages similar to Idefix have been reported recently [13,44]. We showed that *E. faecalis* prophage 6, which apparently belongs to the *Siphoviridae* family, is especially efficient in resisting to Idefix infection due to a new abortive infection system that we call Abiα. This latter confers a typical Abi phenotype, by causing a premature lysis of infected *E. faecalis*. The *abiα* gene is notably widespread among prophages integrated in enterococci and other Gram-positive bacteria. Furthermore, the bacterium itself provides another line of defence against Idefix through efficient mutagenesis of the phage bacterial receptor encoded within the variable part of the *epa* locus responsible for the building up of the surface rhamnopolysaccharide of *E. faecalis*.

2. Materials and Methods

2.1. Sample for Bacteriophage Isolation, Bacterial Strains, Plasmids, and Growth Conditions

One raw sewage water sample from the Sèvres wastewater treatment plant (Paris area, France) was used as a source of phage. *E. faecalis* indicator strain (cured of active prophages (strain VE18590)) used for phage Idefix isolation, as well as strains and plasmids used for *E. faecalis* phage-resistance characterization are listed in Table S1 (with supplementary references [45–50]). *Enterococcus* strains tested to determine phage Idefix host range are detailed in Table S2 (with supplementary references [51,52]). *E. coli* was cultivated at 37 °C in LB medium with shaking. *L. lactis* was grown statically at 30 °C in M17 medium supplemented with glucose 0.5% (M17G). *Enterococcus* strains were cultivated statically at 37 °C either in BHI medium or in M17G medium. Following antibiotics were added when necessary: erythromycin, 10 $\mu g \cdot mL^{-1}$ for *E. faecalis* and 100 $\mu g \cdot mL^{-1}$ for *L. lactis* and *E. coli* strains; chloramphenicol, 20 $\mu g \cdot mL^{-1}$ and 10 $\mu g \cdot mL^{-1}$ for *E. coli* and *E. faecalis* strains, respectively; kanamycin, 50 $\mu g \cdot mL^{-1}$ for *E. coli* strains.

2.2. Phage Isolation

Phage Idefix was isolated using the standard double overlay plaque assay technique as previously described [44] with minor modifications. A standard Petri dish was filled with 20–30 mL of BHI medium containing agar 1.5% and $MgSO_4$ 10 mM. One hundred microliters of 0.45 µm filtered sewage was mixed with 500 µL of an overnight culture of the indicator strain, then added to 5 mL of BHI medium containing agarose 0.4% and $MgSO_4$ 10 mM, and poured onto the bottom agar. The double agar/agarose plate was incubated 24 h at 37 °C and screened for plaque appearance. Many different plaques were obtained, and among them, a large clear plaque was picked and streaked on an agar base before applying the second layer of BHI top agarose mixed with the overnight culture of the indicator strain. The plate was incubated 24 h at 37 °C and a large clear plaque was streaked again twice, to ensure phage purity.

2.3. Phage Concentration and Purification

Starting from a large clear plaque, virions were resuspended in 1 mL of SM buffer (Tris-HCl 10 mM pH 7.5, $MgSO_4$ 10 mM, NaCl 300 mM) and the suspension was centrifuged 10 min at 7500× *g* and room temperature. Phage titer in the supernatant was determined by preparing serial dilutions in SM buffer and using the double overlay method. High-titer phage stock was obtained as previously described [53] with slight modifications. Five hundred to one thousand phages were plated using the double overlay method, which led to a confluent lysis after 12 h at 37 °C. The plate was then flooded with 5 mL of SM buffer and incubated 2 h at 4 °C. The overlay was carefully collected and centrifuged 20 min at 7500× *g* at room temperature and the phages-containing supernatant was passed through a 0.45 µm filter. Phage stocks were titrated (~10^{11} $PFU \cdot mL^{-1}$) and stored at 4 °C for further experiments. Ten milliliters of filtered phage stock was incubated 12 h at 4 °C under stirring in the presence of NaCl 1M and PEG 8000 10%. Precipitated phages were then centrifuged 20 min at 10,000× *g* and 4 °C and slowly resuspended in 1 mL of SM buffer for 1 h at 4 °C prior to be purified by centrifugation in a CsCl buoyant gradient. The purified phage-containing fraction was then recovered (density 1.4), titrated and stored at 4 °C for further experiments.

2.4. Phage Examination in Transmission Electron Microscopy

Ten microliters of purified phage Idefix fraction were directly spotted onto a Formwar carbon coated copper grid. Phages were allowed to adsorb to the carbon layer for 5 min and excess of liquid was removed. Ten microliters of a staining uranyl acetate solution (1%) was then spotted to the grid for 10 s and excess of liquid was removed again. The grid was imaged at 80 kV in a Hitachi HT7700 transmission electron microscope.

2.5. Phage Genomic Nucleic Acid Extraction, Whole Genome Sequencing, and Bioinformatic Analysis

Total DNA was extracted as described in [54,55]. Prior to DNA extraction, 10 mL of the phage stock (10^{11} PFU·mL^{-1}) was treated with 40 µL of nuclease mix (50% glycerol, 0.25 mg·mL^{-1} RNAse A, 0.25 mg·mL^{-1} Dnase I, 150 mM NaCl), for 15 min at 37 °C. Particles were then precipitated by adding PEG 8000 (10%, *w*/*v*) and NaCl to 1 M, and let sit overnight at 4 °C once PEG was solubilized. Phages were centrifuged 10 min at 10,000× *g* and room temperature, and resuspended into 500 µL of SM buffer. Insoluble particles were removed by centrifugation (20 s at 12,000× *g*), and the clarified supernatant was used for DNA extraction. The PROMEGA Wizard DNA Clean up kit (ref A7280) was then used, following essentially the manufacturer instructions. Prior to DNA elution from the column, a washing step with 5.4 M guanidium thiocyanate (resin solution) was applied. DNA was sent to a 454-sequencing platform, and reads were assembled with Newbler [56]. The phage genome was annotated using RAST [57], followed by manual inspection. The genome sequence is available under the accession number LT630001.1. All genomic figures, including Idefix genome comparison, were generated using Easyfig [58].

2.6. Determination of Phage Burst Size

The one-step growth kinetic curve of phage Idefix was measured using a standard method [59] with minor modifications. One milliliter of a log-phase culture of indicator strain was centrifuged 5 min at 10,000× *g* and room temperature, and resuspended in 100 µL of prewarmed BHI medium with $MgSO_4$ (10 mM). Phage from the high-titer phage stock was added at a MOI of 0.001 and allowed to adsorb for 5 min at 37 °C. The phage/bacteria mix was centrifuged 2 min at 10,000× *g* and room temperature. The supernatant was titrated to count unadsorbed phage particles, whereas the bacterial pellet was washed in 100 µL of prewarmed BHI medium with $MgSO_4$ (10 mM) and recentrifuged. The pellet was suspended and diluted in 10 mL of prewarmed BHI medium with $MgSO_4$ (10 mM) and cultured at 37 °C. Samples were taken at regular intervals and plated at the correct dilution for phage titration. A second set of samples from a synchronized 100-fold diluted culture was taken at same intervals and titrated. The values indicate the means and standard deviations of three independent experiments.

2.7. Determination of Phage Host Range

Ten microliters of serially diluted high-titer phage stock were spotted (10 µL) on top of agarose overlays containing overnight culture of enterococci, as described above (Table S2). Plates were incubated at 37 °C and examined for plaque appearance 6, 12, and 24 h after spotting.

2.8. Determination of Phage Efficiency of Plaquing

Efficiency of plaquing (EOP) was determined for *E. faecalis* derivative strains (Table S1). One hundred phages were plated using the double overlay plaque assay technique. EOP were calculated as ((phage titer on tested strain) × (phage titer on the indicator strain)$^{-1}$) × 100. These experiments were independently performed three times and average values are reported with standard deviations.

2.9. Phage Adsorption Assay

Adsorption of phage to *E. faecalis* derivative strains were determined as reported previously [60] with minor modifications. One milliliter of log-phase cultures was harvested and resuspended in 100 µL of BHI medium with $MgSO_4$ (10 mM), and then phage from the high-titer phage stock was added at a MOI of 0.001. Following incubation for 10 min at 37 °C, the phage/bacteria mixtures were centrifuged 2 min at 10,000× *g* and room temperature. Supernatants were plated at the correct dilution and titrated for phage. Percentages of adsorption were calculated as ((control titer - residual titer) × (control titer)$^{-1}$) × 100. These experiments were independently conducted three times and average values are given with standard deviations.

2.10. Determination of Phage Efficiency of Center of Infection

Efficiency of center of infection (ECOI) was determined for *E. faecalis* derivative strains as detailed by [61] with minor modifications. One milliliter of log-phase cultures was harvested and resuspended as previously, and then phage from the high-titer phage stock was added at a MOI of 0.001. Following incubation for 5 min at 37 °C, the phage/bacteria mixtures were centrifuged 1 min at 10,000× *g* and 4 °C, washed twice, diluted, and assayed for infective centers. An *E. faecalis* strain which does not adsorb Idefix was used as control to monitor the effectiveness of phage removal during washing. Percentages of ECOI were calculated as ((number of centers on indicator strain) × (number of centers on tested strain)$^{-1}$) × 100. These experiments were independently performed three times and average values are reported with standard deviations.

2.11. Bacterial Survival Assay

E. faecalis derivative strains survival was assayed as essentially described in [62] with slight modifications. Briefly, one milliliter of log-phase bacterial cultures was harvested and resuspended as previously described, and then phage from the high-titer phage stock was added at both MOI of 1 and 10. Following incubation for 20 min at 37 °C, bacterial suspensions were plated at the correct dilution on agar plates and surviving bacteria were enumerated as CFU. Percentages of bacterial death were calculated as ((CFU·mL^{-1} in cultures without phage − CFU·mL^{-1} in cultures with phage) × (CFU·mL^{-1} in cultures without phage)$^{-1}$) × 100. These experiments were independently conducted three times and average values are shown with standard deviations.

2.12. Lysis Curve Experiments

One milliliter of log-phase bacterial cultures was harvested and resuspended as previously described, and then phage from the high-titer phage stock was added at an MOI of 10. Following incubation for 10 min at 37 °C, the phage/bacteria mixtures were centrifuged 2 min at 10,000× *g* and room temperature. Pellets were resuspended in 200 µL of BHI medium with $MgSO_4$ (10 mM) and the suspensions were transferred in wells of a 96-wells plate. Empty wells are filled with BHI medium or uninfected bacterial suspensions to have negative and positive growth control respectively. Bacterial growths were then monitored during 30–50 min at 37 °C using a Tecan plate reader (OD 600 nm, measurements at 2 min intervals after shakings). During the experiment, each sample was prepared and monitored in triplicates. Kinetics shown are representative of three independent experiments.

2.13. Luria–Delbrück Fluctuation Tests

An exponentially growing *E. faecalis* strain cured of prophage 6 (strain VE18306) culture supplemented with 10 mM $MgSO_4$ was distributed in a 96-wells plate (180 µL, corresponding to ~2 × 10^4 CFU, in 93 wells), and infected with phage Idefix (20 µL, corresponding to ~2 × 10^9 PFU, which are dispensed before bacteria, in 90 of the 96 wells). Remaining wells contained either BHI medium or uninfected VE18306 cultures. Following static incubation of the plates for 12 h at 37 °C, bacterial growth was evaluated using a Tecan plate reader, after a 1 min shaking step (OD 600 nm). If mutations occur at random in the bacterial population, the number of mutational events per well follows a Poisson distribution [63], and the proportion of wells in which no mutant resisting to Idefix was present at the time of infection P_0, is related to h, the expectation of this law, by the formula $P_0 = e^{-h}$. Knowing the average number of bacteria per well at infection, N, mutation frequency $f = h/N$, so $f = -\ln P_0/N$. These experiments were independently performed three times giving a similar value of f.

2.14. General Molecular Biology Methods

All PCR reactions to clone or sequence *IDF_13*, *IDF_15*, *ef2833*, *ef2847*, *ef2850*, *ef2169*, and *ef2170* genes were performed in a Mastercycler Eppendorf with Phusion high-fidelity DNA polymerase (NEB) and according to manufacturer's instructions. PCR products and DNA restriction fragments were purified with QIAquick kits (QIAGEN, Hilden, Germany) when necessary. Electro-transformation of *E. coli*, *L. lactis*, and *E. faecalis* were carried out as previously described [64,65] using a Gene Pulser apparatus (Bio-Rad Laboratories, Inc., Hercules, CA, USA). Transformations of chemically competent *E. coli* ER2566 were carried out by a heat shock procedure.

2.15. Cloning of IDF_13 and IDF_15, Expression and Preparation of Extracts of E. coli, and Spot Assay of the Extracts on the Indicator Strain

PCR amplification of *IDF_13* and *IDF_15* were performed using primer pairs OFL264/OFL265 and OFL266/267 respectively, with Idefix genomic DNA as template. PCR products were purified and digested by *NdeI* and *BamHI* restriction enzymes. Digested products were purified and cloned between the *NdeI* and *BamHI* sites of a linearized pJ411 derivative (Table S1). JM105 transformants were selected on LB plates supplemented with kanamycin. The resulting plasmids, pAB1and pAB2, contained the corresponding ORF fused in 5′ to a sequence coding for a His6-tag, placed under the control of a T7 promoter. His6-tagged IDF_13 and IDF_15 were produced in *E. coli* ER2566 transformed with pAB1 and pAB2 respectively. Cells were grown in 50 mL of LB supplemented with kanamycin at 37 °C. At $OD_{600} = 0.6$, production of the phage proteins was induced by addition of IPTG (0.5 mM final concentration) to the culture for 2.5 h. Cells were then harvested by centrifugation at 5200× *g* for 7 min at 4 °C and resuspended in 1 mL of lysis buffer (Tris-HCl 50 mM pH 8, NaCl 150 mM). Cells suspensions were stored at −20 °C until preparation of crude extracts. After thawing, bacteria were lyzed by sonication on ice. The lysates were centrifuged at 4 °C for 20 min at 20,000× *g* and the supernatants were stored at −20 °C. Production and solubility of IDF_13 and IDF_15 were verified by SDS-PAGE. Lysis activity of IDF_13 and IDF_15 was assessed by spotting 5 μL of the supernatants alone or a mix of both fractions on an indicator strain bacterial lawn as previously described.

2.16. Construction of E. faecalis VE18306 ef2833-, ef2847- and ef2850-Complemented Strains and E. faecalis VE18590 ef2833- Complemented Strain

Complementations of the prophage 6 genes *ef2833*, *ef2847*, and *ef2850* were done using the pJIM2246 vector (Table S1). The three genes were amplified including their constitutive promoters and ribosome binding sites from DNA of *E. faecalis* strain containing all prophages (strain VE14089) with primer pairs JL12/JL13, JL10/JL11, and JL8/JL9 respectively (Table S1). The three purified products were separately cloned into pJIM2246 yielding plasmids pJL1, pJL2, and pJL3 after respective transformation into *E. coli* JM105 and chloramphenicol selection. pJL1, pJL2, and pJL3 were then separately electroporated into *E. faecalis* cured of prophage 6 (strain VE18306). The *ef2833-*, *ef2847-*, and *ef2850*-complemented strains were respectively selected on chloramphenicol. The nucleotide sequences of cloned PCR products were systematically confirmed by sequencing using primer pairs OEF879/1233 (Table S1). pJL3 and pJIM2246 were also separately electroporated into the indicator strain and selected as previously described.

2.17. PCR Amplification in epaX Region

epaX region was amplified from total DNA of nine Idefix spontaneous resisting mutants using primer pairs OEF394/OEF397 and OEF527/OEF397 (Table S1). PCR analysis was extended for the three mutants for which no IS insertion was discovered with larger amplifications downstream and upstream *epaX* region using primer pairs OEF857/OEF527 and OEF885/OEF858; and OEF528/OEF856 and OEF859/OEF823 (Table S1) respectively.

2.18. Construction of E. faecalis VE18306ΔepaX Strain

A deletion of *epaX* in the VE18306 background was constructed by double homologous recombination using pVE14283 plasmid (Table S1). pVE14283 was electroporated into strain VE18306, and the *epaX* deletion (strain VE18393, Table S1) was selected as described in [49,66].

2.19. Construction of E. faecalis VE18393 epaX—Complemented Strain

A complementation of *epaX* was constructed using pVE14297 plasmid (Table S1). This latter, containing *epaX* under the control of the constitutive promoter P_{aphA3}, was electroporated into the VE18306 Δ*epaX* strain (strain VE18393) and the complementation (strain VE18945, Table S1) was selected as described in [49]. A control strain, harboring pVE14176 vector devoid of *epaX*, was also obtained (strain VE18944, Table S1).

3. Results

3.1. Characterization of the Enterococcus Phage Idefix

3.1.1. Isolation, Morphological Characterization, and Phage–Host Relationship

One municipal sewage water sample from Sèvres (Paris area, Ile de France region) was screened by direct plating without enrichment for phages forming plaques on the indicator strain. This latter, VE18590 (here below pp^-, complete names of all strains used are listed in Table S1), corresponds to an *E. faecalis* V583 derivative deleted from its endogenous plasmids and six active prophages [43]. A phage making particularly large and clear plaques was isolated with the double-layer technique. After purification and amplification, transmission electron microscopy revealed a virion with an icosahedral head ~40 nm in diameter and a very short non-contractile tail (Figure 1A). This phage, Idefix, therefore belongs to the *Caudovirales* order and the *Podoviridae* family. A one-step growth kinetic indicated that Idefix has a latent period of 15 min and a burst size ~50 PFUs per infected pp^- bacterium (Figure 1B).

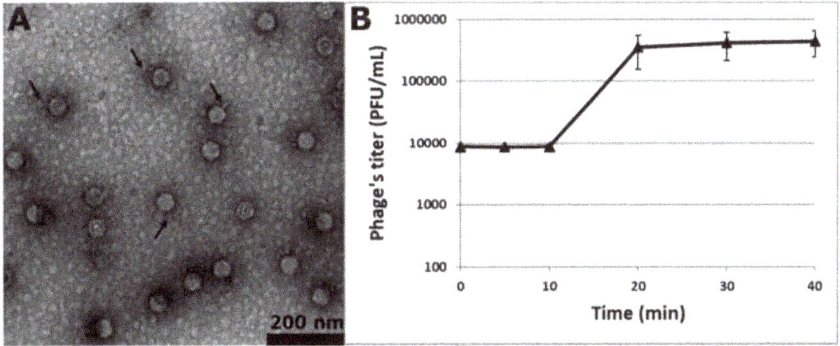

Figure 1. Characterization of the virulent phage Idefix. (**A**) Electron micrograph of Idefix particles negatively stained with 2% uranyl acetate (some Idefix tails are shown by arrows). (**B**) One step growth kinetic of phage Idefix determined in its host strain *E. faecalis* pp^-. The values indicate the means and standard deviations of three independent experiments.

3.1.2. Genomic Characterization

The genome of phage Idefix is a double stranded linear DNA, consisting of 18,168 bp with inverted terminal repeats 61 bp long and an average GC content of 33.2%. It is highly homologous to *Enterococcus* phages vB_EfaP_IME195 (95% coverage and 92% nt identity) and vB_Efae230P-4 (75% coverage and 85% identity) (Figure 2). These genomes belong to virulent podophages, which were isolated from sewage samples in China and Poland, respectively [13,44]. A close relative phage, vB_EfaP_IME199, infecting an *E. faecium* strain has also been described [14]. Genbank accession numbers of these genomes are listed in Table S3.

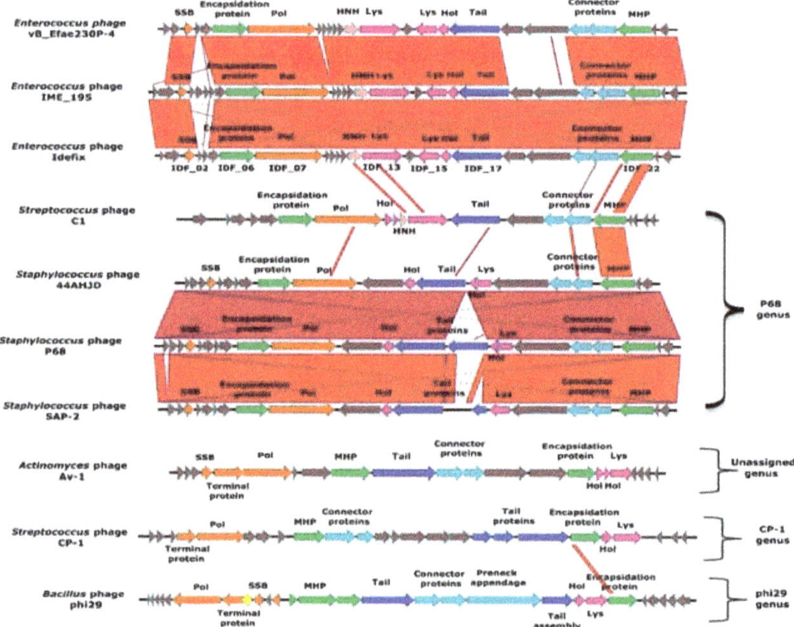

Figure 2. Annotation and comparison of the phage Idefix genome. *Enterococcus* phage genomes from vB_Efae230P-4, vB_EfaP_IME195, and Idefix compared to *Streptococcus* phage C1, *Staphylococcus* phages 44AHJD, P68, SAP-2, 66, *Actinomyces* phage Av-1, *Streptococcus* phage CP-1 and *Bacillus* phage phi29 all belonging to the *Picovirinae* subfamily. Gene functions are color-coded and detailed (yellow: transcriptional regulation, orange: DNA metabolism, green: DNA packaging and head, light blue: head to tail, dark blue: tail, pink: HNH endonuclease, fuchsia: lysis, grey: hypothetical proteins).

The Idefix genome encodes 25 ORFs, 12 of which were assigned a function (Figure 2 and Table S4). The replication module contains a single strand DNA binding protein (SSB, IDF_02) and a DNA polymerase belonging to the B type superfamily (Pol, IDF_07). This polymerase presents all the conserved motifs typical of phage phi29 polymerase, including the specific regions of the "protein priming" subfamily [67–71] (Figure S1).We next searched for the gene coding the terminal protein used to initiate the replication at both ends of *Picovirinae* linear genomes. Such genes are generally located near the polymerase gene but very poorly conserved. Terminal proteins share features like a small size and a high isoelectric point [72,73]. Based on these criteria, *IDF_05* appears the best candidate (Table S4) but additional investigations are needed to confirm this hypothesis. A gene encoding an encapsidation protein (IDF_06) separates this putative terminal gene and the polymerase gene. The lysis module is composed of a holin (Hol, IDF_16) and two different putative endolysins (Lys): IDF_15 and IDF_13. IDF_15 displays homology with *N*-acetylmuramoyl-L-alanine amidases (Figure S2A,B). To test whether this gene expresses an active endolysin, *IDF_15* was cloned in an expression vector based on a T7 promoter (plasmid pAB2) and expressed in *E. coli* ER2566 (Figure S4A). Spotting of a soluble cell extract (containing IDF_15, Figure S4A) on a pp- bacterial lawn revealed a lysis zone (Figure S4B). This result tends to show that IDF_15 has an endolysin activity. IDF_13 displays homology with PlyCA [74–76], a subunit of the PlyC endolysin synthetized by the *Streptococcus* podophage C1 (Figure S3A). Indeed *IDF_13*, like *plyCA*, encodes a protein with two catalytic domains. A putative glycosidase domain (homologous to the one of PlyCA) and a putative "CHAP" domain (more divergent compared with the one found in PlyCA) are respectively located in the N and the C terminus of the protein (Figure S3B,C). In contrast to IDF_15, a soluble cell extract of an *E. coli* strain expressing the *IDF_13* (via plasmid pAB1) did not lead to a bacterial lysis (Figure S4A,B). Moreover, a mix of extracts containing IDF_13 and IDF_15 did not produce increased lysis (Figure S4B). C1 is the only reported phage whose endolysin PlyC is synthesized from two genes: *plyCA* and *plyCB*. To form the complete enzyme, the catalytic subunit PlyCA and eight subunits PlyCB harboring the cell wall binding domain (CBD) are associated [75,77]. However, we were not able to find an Idefix ORF displaying any significant similarity with PlyCB. We concluded that *IDF_15* seems to encode a canonical monomeric endolysin, and that more experiments are required to determine whether *IDF_13* is just remnant of a former endolysin module, or retains activity combined with another Idefix protein to form a multimeric endolysin. Finally, the structural module includes genes encoding a tail protein (IDF_17), connector proteins (IDF_20 and IDF_21) and a major head protein (MHP, IDF_22) (Table S4).

Comparative genomic analyses clearly show that Idefix and the other homologous *Enterococcus* phages belong to the *Picovirinae* subfamily [78] composed of small virulent podophages infecting Gram-positive bacteria and encoding a phi29-like DNA polymerase. The *Picovirinae* subfamily is subdivided into P68, phi29, Cp-1 genera and one unassigned genus (Figure 2). The overall synteny and the MHP similarities [14] both lead to assign Idefix and all other *Enterococcus Picovirinae* to the P68 genus, including one *Streptococcus* and some *Staphylococcus* phages (Figure 2). Within this genus, a hallmark of all *Enterococcus* phages is the presence of two genes coding putative endolysins (IDF_13 and IDF_15 in Idefix).

3.1.3. Phage Host Range

Plaque assays of Idefix were performed on fifty-nine *Enterococcus* strains, belonging to the *E. faecalis* (47) and to the *E. faecium* (12) species. *E. faecalis* strains had sewage, clinical, commensal, or food origins and represented a range of different clonal complexes and capsule types, whereas *faecium* strains essentially belonged to different sequence types from the most prevalent clinical clonal complex (Table S2). Idefix did not propagate on any of them, including the wild type strain V583 (wt), from which our indicator sensitive strain pp^- is derived (Figure 3A,B).

3.2. Characterization of E. faecalis V583 Resistance to Idefix

3.2.1. Novel Abi System Encoded by V583 Prophage 6

To investigate the mechanism by which *E. faecalis* V583 resists Idefix infection, we made use of our collection of V583 derivatives deleted for either plasmids or prophages [43]. The plasmids hosted by V583 were not responsible for the resistance to Idefix, as the use of the plasmidless V583 derivative strain VE14089 (pp^+, Table S1), did not permit Idefix growth (Figure 3C). Plaque assays were next performed on V583 plasmidless derivatives where only one of the prophages remains. Idefix grew on all derivatives (Figure 3D–H) but one, VE18581 ($pp6^+$) in which prophage 6 is present (Figure 3I). We concluded that prophage 6 is necessary to confer resistance to Idefix. Plaque assays on the V583 derivative VE18306 ($pp6^-$) in which only prophage 6 is deleted, led to clear plaque appearance (Figure 3J) with efficiency of plaquing (EOP) similar to pp^- (Table 1). We concluded that prophage 6, which structural genes are notably related to the siphophage HK97, is sufficient for Idefix resistance. Interestingly, we noticed that plaque size was reduced (~three times smaller) whenever prophage 3 was present (Figure 3G,H,J) (Table 1), suggesting that prophage 3 slightly interferes during Idefix infection.

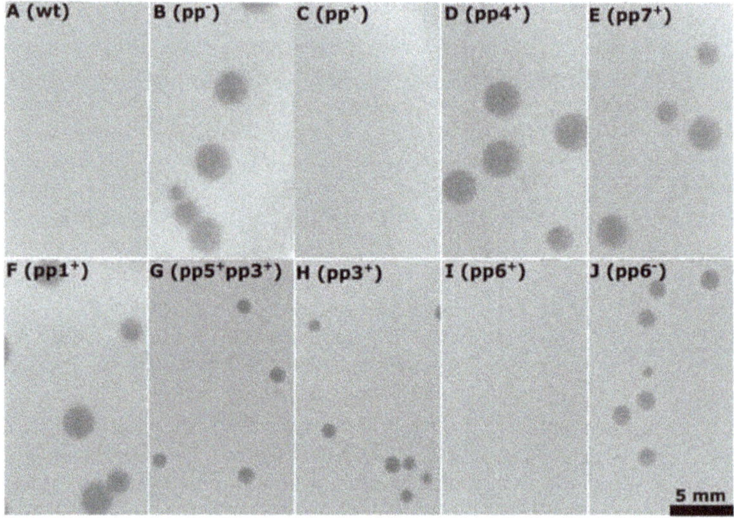

Figure 3. V583 endogenous prophages (pps) interference with Idefix infection, plaque assays performed on strains: A. wt (VE14002), V583 clinical isolate. B. pp- (VE18590), a V583 derivative deleted for all its plasmids and prophages. C. pp^+ (VE14089), a V583 derivative only deleted from all its plasmids. D. $pp4^+$ (VE18582), a VE14089 derivative deleted for all its pps but pp4. E. $pp7^+$ (VE18589), a VE14089 derivative deleted for all its pps but pp7. F. $pp1^+$ (VE18562), a VE14089 derivative deleted for all its pps but pp1. G. $pp3^+pp5^+$ (VE18583), a VE14089 derivative deleted for all its pps but pp3 and pp5. H. $pp3^+$ (VE18584), a VE14089 derivative deleted for all its pps but pp3. I. $pp6^+$ (VE18581), a VE14089 derivative deleted for all its pps but pp6. J. $pp6^-$ (VE18306), a VE14089 derivative only deleted from pp6.

Table 1. Efficiency of plaquing (EOP) of phage Idefix on *E. faecalis* V583 derivatives deleted from different prophages.

Strain (Genotype)	Average EOP of Idefix (sd)	Plaque (Diameter)
VE18590 (pp^-)	1	Big clear plaque (3–4 mm)
VE18562 ($pp1^+$)	1.1 (0.17)	Big clear plaque (3–4 mm)
VE18582 ($pp4^+$)	0.92 (0.14)	Big clear plaque (3–4 mm)
VE18589 ($pp7^+$)	1.03 (0.16)	Big clear plaque (3–4 mm)
VE18583 (pp3+, $pp5^+$)	0.97 (0.16)	Reduction in size (1 mm)
VE18584 ($pp3^+$)	0.92 (0.17)	Reduction in size (1 mm)
VE18581 ($pp6^+$)	0 *	No plaque visible
VE18306 ($pp6^-$)	1 (0.13)	Reduction in size (1 mm)

* Below the detectable limit of the assay, at least $< 1.2 \times 10^{-8}$.

To characterize the resistance conferred by prophage 6, adsorption assays were first performed on a panel of strains differing for the presence or absence of this prophage. Idefix adsorption was always efficient, regardless of the presence of prophage 6 (Table 2), showing therefore that prophage 6 does not prevent Idefix adsorption. We next assayed efficiency of center of infection (ECOI) by Idefix on the same panel of strains, which can detect infective events, regardless of phage burst size. It was reduced to 39% upon infection of strain $pp6^+$. This reduction was even more drastic when the strain pp^+, carrying all prophages, was tested, with only 15% of the infections leading to phage production. This suggests again that some other prophage gene(s) impede Idefix growth. However, infection of strain $pp6^-$ permitted to recover plaques for nearly 70% of all infecting particles, suggesting that prophage

6 contribution to interference is major (Table 2). We finally tested whether prophage 6 affected cell survival upon Idefix infection. Cell survival following exposure to Idefix was similar irrespective of the presence of prophage 6: even in the Idefix resistant strains, percentage of cell death was around 50–70% at MOI of 1, and around 100% at MOI of 10, like in Idefix sensitive strains (Table 2). We concluded that prophage 6 encodes a typical Abi mechanism affecting the production of virions while leading simultaneously to host cell death upon Idefix infection.

Table 2. Parameters of phage Idefix proliferation on *E. faecalis* strains differing by the presence or absence of prophage 6 or prophage 6-encoded *ef2833*.

Strain (Genotype)	% EOP (Plaques Diameter)	% Adsorption (sd)	% ECOI (sd)	% Bacterial Death at MOI 1 (sd) and MOI 10 (sd)
Indicator strain VE18590 (pp^-)	100 (3–4 mm)	96.5 (1.63)	100	69.0 (0.39) and 99.2 (0.78)
VE18306 ($pp6^-$)	100 (1 mm)	97.1 (1.35)	69.3 (3.10)	52.8 (3.00) and 98.2 (0.98)
VE18581 ($pp6^+$)	0 None plaques visible	97.5 (0.92)	39.5 (3.94)	66.8 (2.78) and 98.2 (1.32)
VE14089 (pp^+)	0 None plaques visible	98.1 (2.36)	15.9 (4.01)	60.2 (3.46) and 98.9 (0.19)
VEJL3 ($pp6^-$, *ef2833*$^+$)	0 None plaques visible	98.4 (0.68)	9.46 (2.68)	67.7 (1.65) and 98.6 (0.46)
VEJL5 (pp^-, *ef2833*$^+$)	0 None plaques visible	98.5 (1.33)	13.8 (0.58)	63.2 (1.98) and 96.6 (1.36)

To search for prophage 6 candidate genes involved in this Abi mechanism, we analyzed transcriptomic data of strain pp^+ [48]. Eight of the prophage 6 encoded genes are expressed constitutively during normal growth conditions, in contrast to the generally low expression level of the remaining 50 genes carried by the prophage (Figure 4A). Two of these genes, *ef2855* and *ef2852*, encode the integrase and repressor proteins, which are implicated in the lysogenic control of the prophage. Between these two genes, and probably forming an operon with them, genes *ef2854* and *ef2853* encode a putative membrane protein and a putative metallo-peptidase, respectively. A similar gene pair, placed between an integrase and a repressor, is found in several Sie prophage-encoded phage resistance systems within the *Lactococcus* genus [37,79]. These systems block the phage DNA injection step, due to the membrane protein. The nearby metallo-peptidase is dispensable, but sometimes enhances the resistance efficiency [37,79]. Given that Sie systems do not affect bacterial survival, we did not consider the gene pair *ef2853* and *ef2854* as responsible for the observed resistance to Idefix infection. Among the four last candidate genes, *ef2801* is disrupted by a transposable element invalidating a putative glycosyltransferase, so that our investigation was restricted to the three remaining genes *ef2833*, *ef2847*, and *ef2850*.

All of them are preceded by a promoter region including an experimentally mapped transcription-starting site [80] and encode proteins with unknown function. We thus cloned these three genes individually with their own promoters on pJIM2246 vector (Table S1), and performed Idefix infections in $pp6^-$ transformed by each of the plasmids. Plaque assays revealed that Idefix was able to infect all strains (Figure 4B(a,b) but the one hosting plasmid pJL3, in which *ef2833* is expressed (Figure 4B(c)). We additionally tested plasmid pJL3 in the pp^- background and found that this strain was also resistant to Idefix (Figure 4B(d)) whereas the phage grew on the isogenic strain hosting the empty vector (Figure 4B(e)) as well as on pp^- (Figure 4B(f)).

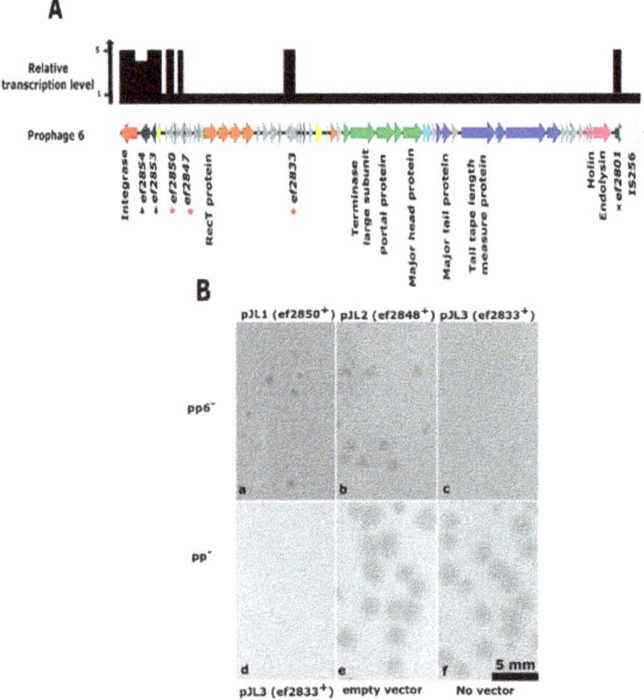

Figure 4. Identification of the V583 endogenous prophage 6 (pp6) gene responsible for Idefix resistance. (**A**) Genomic organization of pp6 and relative transcription level of each gene. The genomic organization of pp6 is modified from [43]. Gene functions are color-coded and detailed wherever possible (red: integrase, yellow: transcriptional regulation, orange: DNA metabolism, green: DNA packaging and head, light blue: head to tail, dark blue: tail, fuchsia: lysis, grey: hypothetical proteins, white: transposable element). Relative transcriptional level of each gene is indicated in black and designed from a previously published transcriptomic study [48]. Eight pp6 genes are constitutively expressed during V583 normal growth conditions. Among those not involved in pp6 lysogenic decision: a membrane protein and putative metallo-peptidase genes are indicated with triangles, a putative glycosyltransferase gene disrupted by a transposable element is indicated with a cross and the three other genes tested as *abi* candidates are indicated with red stars. (**B**) Interference with Idefix infection following the expression of the three pp6 *abi* candidates, plaque assays performed on strain pp6$^-$ (A to C) or pp$^-$ (D to F) transformed by: **a.** pJL1. **b.** pJL2. **c.** pJL3. **d.** pJL3. **e.** the pJIM2246 empty vector. **f.** no vector.

We then checked the other proliferation parameters of Idefix on the two strains hosting the pJL3 plasmid. The phage adsorbed as efficiently on these strains, its ECOI was reduced to ~10% and the percentage of cell death were around 60–70% at MOI 1 and around 100% at MOI 10 in both cases (Table 2). These results confirmed that *ef2833* is responsible for the abortive mechanism. As letters from *abiA* to *abiZ* have been used once to designate over 20 *L. lactis* Abi systems [18,81], we renamed *ef2833* as "*abiα*". The corresponding encoded protein contains 272 residues, and does not exhibit similarity with any domain or protein of known function. Abiα nevertheless constitutes a PFAM domain referred to as DUF4393 (158 sequences, 140 species, both Gram$^+$ and Gram$^-$), suggesting already its broad distribution (see below). Abiα also shares similarity with Pfam entry PF10987 and PDB entry 3H35 without providing any substantial information about Abiα function.

To characterize the Abiα mode of action, lysis curves were conducted and indicated that strain pp$^-$ containing plasmid pJL3 lyses ~10 min earlier than its isogenic strain devoid of *abiα* after infection by Idefix (MOI 10) (Figure 5A). The same results were obtained when comparing lysis curves of the prophage positive and prophageless strains pp6$^+$ and pp$^-$, respectively (Figure 5B). We concluded that Abiα provokes a lysis asynchrony.

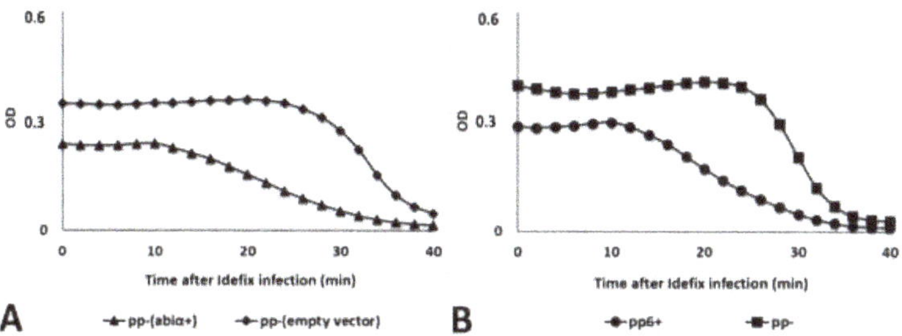

Figure 5. Lysis curves after Idefix infection at MOI 10. (**A**) Comparison between pp$^-$pJIM2246*abiα*$^+$ (VEJL5) and pp$^-$ pJIM2246 (VEJL4) in triangles and diamonds, respectively. (**B**) Comparison between pp6$^+$ and pp$^-$ in circles and squares, respectively. The results shown are representative of three biological replicates (see Figure S5).

To evaluate the prevalence of *abiα* and its genetic context, the protein sequence was searched in the JGI database with the IMG interface ([82]; BLASTP, E-value < 3.10^{-8}, Identity $\geq$ 25%). Homologs of Abiα were found mostly on prophages, with some of them shown in Figure S6: the closer relatives (61–59% sequence identity) came from various enterococci, such as *E. faecium* and *Enterococcus villorum* isolates, and subsequent analyses with the NCBI nt database also revealed a homolog in an *Enterococcus hirae* prophage (not shown in Figure S6). Other Abiα were found in about 50 *Lactobacillus* strains (27–42% sequence identity). Roughly, half of these genes are encoded on prophages (as in *L. johnsonii*, *L. salivarius* or *L. plantarum* isolates) whereas the other half are located on the chromosomes (as in *L. lindneri* and *L. helveticus* isolates). Interestingly, in *L. helveticus*, and in two *Oenococcus* species (*O. kitaharae* and *O. oeni*), the *abiα* gene is close to other putative Abi systems, and may belong to a phage resistance island. Sporadic occurrences (31–37% sequence identity) were also found in three *Streptococcus* species (*S. suis*, *S. bovis*, and *S. parasanguinis*), and one *Carnobacterium* isolate. A prophage context was detectable in this latter and in *S. parasanguinis*. Additional prophage-encoded genes with 25–27% sequence identity with *abiα* were located in one *Lactococcus lactis* ssp. *lactis* (not shown in Figure S6), *Bacillus velezensis*, *Staphylococcus epidermidis*, *Staphylococcus scuiri* and *Staphylococcus aureus*. We conclude that Abiα is widely distributed, and hypothesize it has a phage origin.

3.2.2. Mutagenesis in V583 *epa* Variable Region as Potential Additional Line of Defense

During this work, we regularly observed phage-resistant colonies growing within lysis zones of the pp6$^-$ strain. Fluctuation tests of Luria-Delbrück allowed estimating that the frequency of mutations resulting in resistance of the strain was 1.56 ($\pm$ 0.54) $\times$ 10^{-4} per cellular division. We conclude that spontaneous resistance to Idefix arises at high frequency in *E. faecalis* strain devoid of *abiα*.

To gain further insights into the bacterial mechanisms of resistance to Idefix, we tested nine spontaneous Idefix resistant mutants randomly isolated from fluctuation tests. Idefix adsorption was reduced for all mutants: three of them had a 2-to-4-fold reduced adsorption efficiency and the remaining six, rather a 10-fold defect (Table S5). We conclude that bacterial resistance to Idefix is due to an adsorption defect.

More than 10 years ago, a study of the podophage C1 distantly related to Idefix (see Figure 2) had proposed that the *N*-acetylgalactosamine (GalNac) composing the side chains of the group C streptococci surface rhamnopolysaccharide should be a crucial element of the bacterial receptor [83]. Enterococci, like group C streptococci, harbor a surface rhamnopolysaccharide or Epa (Enterococcal polysaccharide antigen), which is synthesized by more than thirty proteins encoded within a single cluster of genes. The upstream part of this gene cluster is shared among *E. faecalis* strains, while the immediately downstream region is variable between strains [84]. Within this *epa* variable region, *epaX* encodes a glycosyltransferase involved in the incorporation of galactose and/or GalNac proposed to form the side chains of the rhamnopolysaccharide in *E. faecalis* [49]. We therefore started by examining the *epaX* region in the nine mutants. PCR amplification of *epaX* led to a DNA product of increased size in six out of nine mutants (~4000 instead of ~2500 bp). Sequencing revealed that *epaX* gene was disrupted by an IS256 insertion sequence, positioned in either direction, and at six different locations, all of them in the very distal 3′ region of *epaX* (Figure 6 and Table S5).

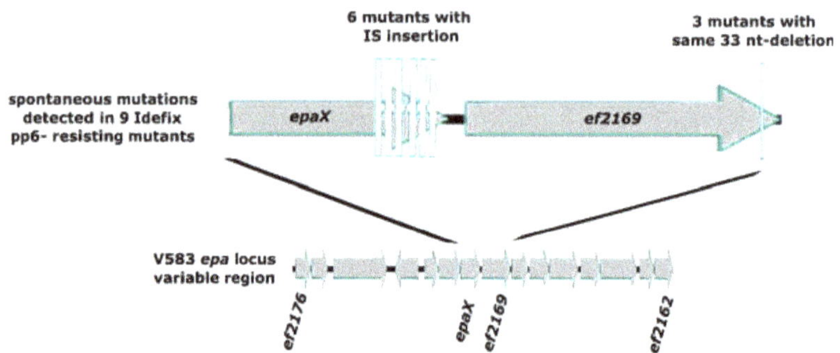

Figure 6. Genomic organization of V583 variable *epa* locus and focus on spontaneous mutations found in *epaX* region for nine Idefix pp6$^-$ resisting mutants.

We next reconstructed a deletion of *epaX* in the pp6$^-$ background. Again, Idefix was almost unable to adsorb to the resulting strain (VE18393, Table S1) with only 5% (± 8) of adsorption, nor to lyse it (Figure 7A). Complementation of this *epaX* mutant with plasmid pVE14176 expressing constitutively *epaX* under a strong promoter (strain VE18945, Table S1) restored Idefix adsorption, which reached 96% (± 2), as well as the capacity of plaquing (Figure 7B). Results on the isogenic strain (VE18944, Table S1) hosting the empty vector were 5% (± 6) of phage adsorption and no plaque formation (Figure 7C). We concluded that *epaX* is required for Idefix adsorption.

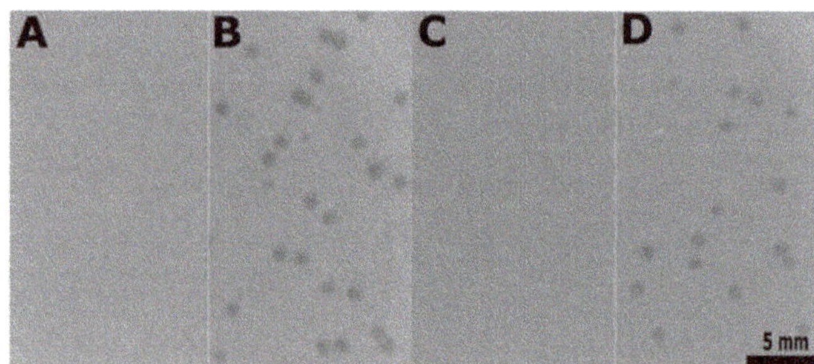

Figure 7. *epaX* gene expression is required for Idefix infection of pp6$^-$. Plaque assays performed on strains: (**A**) pp6$^-$*epaX*$^-$ (VE18393), a VE18306 derivative deleted from *epaX*. (**B**) pp6$^-$*epaX*$^-$pVE14176*epaX*$^+$ (VE18945), a VE18393 derivative complemented with *epaX*. (**C**) pp6$^-$*epaX*$^-$pVE14176 (VE18944), a VE18393 derivative carrying the empty vector. (**D**) pp6$^-$ (VE18306), a V583 derivative deleted from its plasmids and pp6.

To characterize the three remaining Idefix-resistant mutants, in which no IS was detected in the *epaX* region by PCR, we sequenced the genome of one of them, and mapped the reads against the pp6$^-$ reference genome. We identified a 33 nt-long deletion, flanked by 11 nt-long direct repeats, within ORF *ef2169* encoding a predicted O antigen polymerase, and located immediately downstream of *epaX*.

PCR amplification and sequencing of *ef2169* gene in the two last mutants allowed to detect the same 33 nt deletion in both of them (Figure 6 and Table S5). The observed deletion results in an in-frame 11 amino acids deletion in the C-terminal region of the putative O antigen polymerase. Additional single nucleotide polymorphisms (introducing amino acid substitution and a premature stop codon, respectively) were also observed in these two last *ef2169* mutants (Table S5). Complementation of the *ef2169* mutation could not be checked, due to our failure to clone the *ef2169* gene on a plasmid, despite repeated attempts in *E coli*, *L. lactis*, and *E. faecalis*.

4. Discussion

Enterococci are particularly successful at rapidly acquiring resistance to virtually any antibiotic used in therapy, with vancomycin-resistant enterococci (VRE) being a major clinical problem. If phage therapy is consequently nowadays (re)considered as an alternative to combat VRE infections, we are still nowhere near using enterococci phages as therapeutic agents in routine. Before that, we need to better characterize (i) enterococci phage receptors that define phage strain specificity, and (ii) enterococci resistance mechanisms that constitute the major constraint of phage therapy [7,16].

The present study with the new virulent podophage, Idefix, fulfills these goals. The phage was isolated by using the reference vancomycin-resistant *E. faecalis* V583 clinical isolate, deleted from its endogenous plasmids and prophages, as a recipient for phage infections. Idefix belongs to the *Picovirinae* subfamily and P68 genus and is closely related to three phages recently isolated on *E. faecalis* [13,44] or *E. faecium* [14]. They encode a DNA polymerase from the B type superfamily, which is a hallmark of the *Picovirinae*. As an additional hallmark of these enterococci podophages, they all encode two types of predicted endolysins. Endolysin engineering has emerged as a suitable strategy for food safety, environment decontamination, and infections control [85]. Some of the characterized enterococci phage endolysins show potential to combat VRE in vivo [15,86] and could represent an alternative to resolve the VRE problem. Idefix both encodes endolysin IDF_15, a predicted N-acetylmuramoyl-L-alanine amidase for which we could presume activity by spot assays on bacterial lawns, and IDF_13, similar to the catalytic subunit of the endolysin PlyC encoded by *Streptococcus* phage C1, which was inactive in the same spot assay. PlyC is the unique example of multimeric endolysin and represents the more active peptidoglycan hydrolase reported to date [85]. We were unable to identify by similarity an Idefix gene that could encode the CBD of this complex, and allow the formation of a putative multimeric endolysin. CBD of endolysins bind to specific substrates on the bacterial surfaces, often giving rise to near-species-specific binding, so that endolysins of the same class often share very little sequence similarity in the binding region [85]. Identification of the CBD subunit homolog in Idefix and complete characterization of this intriguing endolysin requires further investigation.

Idefix was unable to lyse any of our fifty-nine strains of *E. faecalis* and *E. faecium*, suggesting it could originate from another enterococcal species. Plaques were obtained against the indicator strain used for its isolation, and not on the wild type isolate V583 from which the indicator strain is derived. This apparently narrow host range could be explained by the reported Abiα-sensitivity and perhaps more importantly by the receptor-specificity of Idefix.

In V583, resistance to the phage was mainly due to the gene *abiα* (formerly ef2833) encoded by prophage 6. Upon Idefix infection, cells are dying while the Idefix lytic cycle is perturbed (ECOI of ~40% and ~10% whenever *abiα* is expressed on prophage 6 or overexpressed on cloning vectors, respectively) so that its efficiency of plaquing is below 10^{-8} and the detectable limits of the assay for Idefix. To our knowledge, it is the first time that an Abi system is described within enterococci, and the first one targeting a *Picovirinae*. This new Abi protein is widely distributed, it corresponds to the PFAM DUF 4393, and we report here that it is mostly found in prophages, arguing that it is essentially a temperate phage weapon to fight against other phages. Abi are specific anti-phage defence mechanisms present in many bacterial species including *Shigella dysenteriae*, *Streptococcus pyogenes*, *Vibrio cholerae*, *Bacillus subtilis*, and *Bacillus licheniformis* [62,87–90]. They have been more extensively studied in *L. lactis* and *E. coli* [17,81]. Most *abi* genes are encoded by a single gene on mobile genetic elements (MGE), on prophages in *E. coli* [91], or plasmids in *L. lactis* [81]. The fact that most *abi* genes are plasmidic in *L. lactis* may result however from a bias of human selection for plasmid-encoded systems in the dairy industry. Abi actually include a large collection of diverse mechanisms acting at any stage of the phage development to decrease or completely block virion production and cause host cell death. All these mechanisms display little or no known evolutionary relationship, apart from a very similar phenotype [18,81,92].

Lysis curve experiments show that the expression of *abiα* triggers a premature lysis of Idefix-infected *E. faecalis*. A similar phenotype was described in *L. lactis*, in which lysis, upon infection by siphophages from the P335 group, occurs earlier in the presence of AbiZ [93]. AbiZ speeds up the lysis of *L. lactis* bacteria in which the phage holin and endolysin are expressed, and enhances membrane permeability of bacteria expressing the phage holin. Durmaz and Klaenhammer propose that AbiZ interacts with phage holin to accelerate the lysis clock and prevent normal phage multiplication [93]. Abiα and AbiZ do not share any sequence similarity. We were unable to obtain Idefix mutants that were resistant to Abiα, suggesting a very high efficiency of this Abi system. Based on our results, we can only speculate that Abiα interferes with lysis, possibly by preventing timely holin and consequently endolysin(s) actions, as described for AbiZ. Another possibility is that Abiα targets a putative Idefix holin inhibitor. The holin triggering, which is determinant for optimal burst size, is critically regulated by the expressed ratio between the holin and its inhibitor. This latter is encoded within the lambda holin gene, as a separate transcript initiated at a dual-start motif [94,95]. Unlike lambda (and phi29), the Idefix holin gene does not harbor a dual-start motif and its regulation may be mediated by the expression of another non-identified Idefix gene, as described in other phages [94,95]. Finally, premature triggering of lambda holin can also be induced by any poison that efficiently reduces the proton-motrice force (pmf) of the cytoplasmic membrane [94,95]. Thus, we cannot exclude Abiα causes a holin-independent reduction of the pmf underlying an early trigger of the holin and/or a permeabilization of the membrane leading to endolysin release. If this is the case, you might imagine that Abiαcould be active against a wider range of phages and not only Idefix or related phages. Further experiments will be needed to test these different hypotheses and complete the characterization of this defence mechanism.

Genomes of V583 and other MDR *E. faecalis* isolates have few generalist anti-MGE systems, which favor horizontal gene transfer and polylysogeny [29]. MDR enterococci may thus provide a suitable ground for bacteriophages confrontation. Whereas the spread of Idefix infection is thus limited by V583 prophage 6, it is interesting to note this latter is itself strictly controlled by V583 prophages 3 and 5 that block its excision [43]. This 'domestication' could be seen as a way for both prophages to sustain the specific line of defense conferred by prophage 6 and thus protect themselves and their host from an external phage 'attack'.

Beyond the intra-bacteriophage warfare, bacteria are also able to evolve resistance to phages. We observed that *E. faecalis* mutants resisting to Idefix infection arose at a high frequency of 1.5×10^{-4} per generation. The nine Idefix-resistant mutants analyzed had either acquired an IS256 or recombined between eleven base paires long direct repeats in the *epa* variable region, indicating that *E. faecalis* has a high potential for evolution by recombination. Mutants in glycosyltransferase gene *epaX* had a 10-fold defect in Idefix adsorption, whereas mutants in the next o-polymerase gene had a milder adsorption defect. The fact that all mutations cluster into two genes of the *epa* variable region underlines the role of the rhamnopolysaccharide for Idefix adsorption. Epa was recently confirmed as a receptor of *E. faecalis* virulent phage NPV1 [96,97], and we provide here the first evidence for a role of Epa decoration chains in phage/host recognition in enterococci. Rhamnose-rich cell wall polysaccharides (CWPS) and their structural diversity are important in phage adsorption within lactococci and streptococci [98,99]. In *L. lactis* as well, CWPS biosynthesis is encoded on a large chromosomal gene cluster also including a conserved and a variable region. Based on sequence similarity and difference in the variable region, *L. lactis* strains were divided into three groups, and one of them in several subtypes [100]. These latter are distinguished by their glycosyltransferase-encoding genes composition and consequent differences between CWPS structures are critical in determining phage sensitivity [100]. The use of unconserved sugar decorations of the saccharidic chains as a receptor combined with the presence of a prophage-encoded Abi system targeting Idefix may explain Idefix very low success in *E. faecalis*.

5. Conclusions

This study of the *Enterococcus* infecting *Picovirinae* Idefix permitted to unveil two resistance mechanisms against it, one bacterial and the other viral. Bacterial mutations suppressed the Epa decoration needed for phage adsorption, and the prophage-encoded product Abiα interfered with Idefix timing of lysis. However, the bacterial line of defence is likely to be counterselected in vivo as V583 *epaX* mutants have a defect in mouse gut colonization [49]. In fact, phage selective pressure might be one of the actors of the observed diversification of the *epa* locus in enterococci. The viral line of defence based on Abiα is likely to be more robust, given that no escape mutant could be isolated. Indeed, Abiα is widespread both in enterococci and in *Firmicutes* and it may allow to fight against *Picovirinae* or other ranges of phages. Somehow, the different families of phages infecting the same bacterial species are in competition, so that their host resembles a battlefield. It is especially true if bacteria are devoid of generalist defense systems, as MDR *E. faecalis* strains tend to be. In this context, it might be expected that Abi and other specific defensive systems pave prophage genomes, allowing temperate phages to align their strategy with their host, and fight against virulent phage invaders.

Supplementary Materials: The following are available online at http://www.mdpi.com/1999-4915/11/1/48/s1, Figure S1: Structural and functional map of family of B phage DNA polymerases belonging to the "protein priming" subfamily. Figure S2: IDF_15 alignments. Figure S3: Alignments of IDF_13. Figure S4: Activities of *E. coli* extracts containing putative Idefix endolysins (IDF_15 and IDF_13). Figure S5: Biological replicates relative to lysis curves experiments after Idefix infection at MOI 10. Figure S6: Putative *abiα* homologs identified on several prophages of Gram-positive bacteria. Table S1: Bacterial strains, plasmids, and oligonucleotide primers used. Table S2: Description of *Enterococcus* isolates tested as potential Idefix hosts. Table S3: Respective genome accession numbers of the *Enterococcus Picovirinae* phages. Table S4: Features of Idefix ORFs and predicted functions of their products. Table S5: Spontaneous mutation in *epaX* (ef2170) or *ef2169* in pp6$^-$ and resistance to Idefix infection.

Author Contributions: Conceptualization, J.L., F.L., P.S., and M.-A.P.; Investigation, J.L., Ar.B., E.M., S.F., As.B., O.S., M.L., M.D., and F.L.; Supervision, M.-A.P.; Writing—original draft, J.L., F.L., P.S., and M.-A.P.; Writing—review & editing, J.L.

Acknowledgments: We thank Marianne De Paepe and Lionel Rigottier-Gois for their help during this study; Christine Longin from the MIMA2 facilities (UMR 1313 GABI, INRA) for the TEM observations; Laurent Debarbieux and Sylwia Bloch, Alicja Węgrzyn as well as Vincent Cattoir who respectively provided sewage water samples and enterococci isolates.

Conflicts of Interest: The authors declare no conflict of interest.

References

1. Lebreton, F.; Willems, R.J.; Gilmore, M.S. *Enterococcus* diversity, origins in nature, and gut colonization. In *Enterococci from Commensals to Leading Causes of Drug Resistant Infection*; Gilmore, M.S., Clewell, D.B., Ike, Y., Shankar, N., Eds.; Massachusetts Eye and Ear Infirmary: Boston, MA, USA, 2014; pp. 3–46.
2. Higuita, N.I.A.; Huycke, M.M. Enterococcal disease, epidemiology, and implications for treatment. In *Enterococci from Commensals to Leading Causes of Drug Resistant Infection*; Gilmore, M.S., Clewell, D.B., Ike, Y., Shankar, N., Eds.; Massachusetts Eye and Ear Infirmary: Boston, MA, USA, 2014; pp. 47–72.
3. Kristich, C.J.; Rice, L.B.; Arias, C.A. Enterococcal infection—Treatment and antibiotic resistance. In *Enterococci from Commensals to Leading Causes of Drug Resistant Infection*; Gilmore, M.S., Clewell, D.B., Ike, Y., Shankar, N., Eds.; Massachusetts Eye and Ear Infirmary: Boston, MA, USA, 2014; pp. 89–134.
4. Sulakvelidze, A.; Alavidze, Z.; Morris, J.G. Bacteriophage therapy. *Antimicrob. Agents Chemother.* **2001**, *45*, 649–659. [CrossRef] [PubMed]
5. Cheng, M.; Liang, J.; Zhang, Y.; Hu, L.; Gong, P.; Cai, R.; Zhang, L.; Zhang, H.; Ge, J.; Ji, Y.; et al. The bacteriophage EF-P29 efficiently protects against lethal vancomycin-resistant *Enterococcus faecalis* and alleviates gut microbiota imbalance in a murine bacteremia model. *Front. Microbiol.* **2017**, *8*, 837. [CrossRef]

6. Cheng, S.; Xing, S.; Zhang, X.; Pei, G.; An, X.; Mi, Z.; Huang, Y.; Tong, Y. Complete genome sequence of a new *Enterococcus faecalis* bacteriophage, vB_EfaS_IME197. *Genome Announc.* **2016**, *4*, e00827-16. [CrossRef] [PubMed]
7. Duerkop, B.A.; Palmer, K.L.; Horsburgh, M.J. Enterococcal bacteriophages and genome defense. In *Enterococci from Commensals to Leading Causes of Drug Resistant Infection*; Gilmore, M.S., Clewell, D.B., Ike, Y., Shankar, N., Eds.; Massachusetts Eye and Ear Infirmary: Boston, MA, USA, 2014; pp. 309–336.
8. Gelman, D.; Beyth, S.; Lerer, V.; Adler, K.; Poradosu-Cohen, R.; Coppenhagen-Glazer, S.; Hazan, R. Combined bacteriophages and antibiotics as an efficient therapy against VRE *Enterococcus faecalis* in a mouse model. *Res. Microbiol.* **2018**, *169*, 531–539. [CrossRef]
9. Gong, P.; Cheng, M.; Li, X.; Jiang, H.; Yu, C.; Kahaer, N.; Li, J.; Zhang, L.; Xia, F.; Hu, L.; et al. Characterization of *Enterococcus faecium* bacteriophage IME-EFm5 and its endolysin LysEFm5. *Virology* **2016**, *492*, 11–20. [CrossRef]
10. Khalifa, L.; Brosh, Y.; Gelman, D.; Coppenhagen-Glazer, S.; Beyth, S.; Poraduso-Cohen, R.; Que, Y.-A.; Beyth, N.; Hazan, R. Targeting *Enterococcus faecalis* biofilm using phage therapy. *Appl. Environ. Microbiol.* **2015**, *81*, 2696–2705. [CrossRef] [PubMed]
11. Khalifa, L.; Gelman, D.; Shlezinger, M.; Dessal, A.L.; Coppenhagen-Glazer, S.; Beyth, N.; Hazan, R. Defeating antibiotic-and phage-resistant *Enterococcus faecalis* using a phage cocktail in vitro and in a clot model. *Front. Microbiol.* **2018**, *9*, 326. [CrossRef]
12. Rahmat Ullah, S.; Andleeb, S.; Raza, T.; Jamal, M.; Mehmood, K. Effectiveness of a lytic Phage SRG1 against vancomycin-resistant *Enterococcus faecalis* in compost and soil. *Biomed. Res. Int.* **2017**, *2017*. [CrossRef]
13. Wang, R.; Xing, S.; Zhao, F.; Li, P.; Mi, Z.; Shi, T.; Liu, H.; Tong, Y. Characterization and genome analysis of novel phage vB_EfaP_IME195 infecting *Enterococcus faecalis*. *Virus Genes* **2018**, *54*, 804–811. [CrossRef]
14. Xing, S.; Zhang, X.; Sun, Q.; Wang, J.; Mi, Z.; Pei, G.; Huang, Y.; An, X.; Fu, K.; Zhou, L.; et al. Complete genome sequence of a novel, virulent Ahjdlikevirus bacteriophage that infects *Enterococcus faecium*. *Arch.Virol.* **2017**, *162*, 3843–3847. [CrossRef]
15. Zhang, W.; Mi, Z.; Yin, X.; Fan, H.; An, X.; Zhang, Z.; Chen, J.; Tong, Y. Characterization of *Enterococcus faecalis* phage IME-EF1 and its endolysin. *PLoS ONE* **2013**, *8*, e80435. [CrossRef] [PubMed]
16. Duerkop, B.A.; Huo, W.; Bhardwaj, P.; Palmer, K.L.; Hooper, L.V. molecular basis for lytic bacteriophage resistance in enterococci. *MBio* **2016**, *7*. [CrossRef] [PubMed]
17. Labrie, S.J.; Samson, J.E.; Moineau, S. Bacteriophage resistance mechanisms. *Nat. Rev. Microbiol.* **2010**, *8*, 317–327. [CrossRef] [PubMed]
18. Seed, K.D. Battling Phages: How bacteria defend against viral attack. *PLoS Pathog.* **2015**, *11*, e1004847. [CrossRef] [PubMed]
19. Goldfarb, T.; Sberro, H.; Weinstock, E.; Cohen, O.; Doron, S.; Charpak-Amikam, Y.; Afik, S.; Ofir, G.; Sorek, R. BREX is a novel phage resistance system widespread in microbial genomes. *EMBO J.* **2014**, *34*, 169–183. [CrossRef]
20. Doron, S.; Melamed, S.; Ofir, G.; Leavitt, A.; Lopatina, A.; Keren, M.; Amitai, G.; Sorek, R. Systematic discovery of antiphage defense systems in the microbial pangenome. *Science* **2018**, *359*, eaar4120. [CrossRef]
21. Depardieu, F.; Didier, J.-P.; Bernheim, A.; Sherlock, A.; Molina, H.; Duclos, B.; Bikard, D. A eukaryotic-like serine/threonine kinase protects staphylococci against phages. *Cell Host Microbe* **2016**, *20*, 471–481. [CrossRef]
22. Ofir, G.; Melamed, S.; Sberro, H.; Mukamel, Z.; Silverman, S.; Yaakov, G.; Doron, S.; Sorek, R. DISARM is a widespread bacterial defence system with broad anti-phage activities. *Nat. Microbiol.* **2018**, *3*, 90–98. [CrossRef]
23. Kronheim, S.; Daniel-Ivad, M.; Duan, Z.; Hwang, S.; Wong, A.I.; Mantel, I.; Nodwell, J.R.; Maxwell, K.L. A chemical defence against phage infection. *Nature* **2018**, *564*, 283–286. [CrossRef]
24. Burley, K.M.; Sedgley, C.M. CRISPR-Cas, a prokaryotic adaptive immune system, in endodontic, oral, and multidrug-resistant hospital-acquired *Enterococcus faecalis*. *J. Endod.* **2012**, *38*, 1511–1515. [CrossRef]
25. Hullahalli, K.; Rodrigues, M.; Schmidt, B.D.; Li, X.; Bhardwaj, P.; Palmer, K.L. Comparative analysis of the orphan CRISPR2 locus in 242 *Enterococcus faecalis* strains. *PLoS ONE* **2015**, *10*, e0138890. [CrossRef] [PubMed]
26. Lindenstrauß, A.G.; Pavlovic, M.; Bringmann, A.; Behr, J.; Ehrmann, M.A.; Vogel, R.F. Comparison of genotypic and phenotypic cluster analyses of virulence determinants and possible role of CRISPR elements towards their incidence in *Enterococcus faecalis* and *Enterococcus faecium*. *Syst. Appl. Microbiol.* **2011**, *34*, 553–560. [CrossRef] [PubMed]

27. Price, V.J.; Huo, W.; Sharifi, A.; Palmer, K.L. CRISPR-cas and restriction-modification act additively against conjugative antibiotic resistance plasmid transfer in *Enterococcus faecalis*. *mSphere* **2016**, *1*. [CrossRef] [PubMed]
28. Huo, W.; Adams, H.M.; Zhang, M.Q.; Palmer, K.L. Genome modification in *Enterococcus faecalis* OG1RF assessed by bisulfite sequencing and single-molecule real-time sequencing. *J. Bacteriol.* **2015**, *197*, 1939–1951. [CrossRef] [PubMed]
29. Palmer, K.L.; Gilmore, M.S. Multidrug-resistant enterococci lack CRISPR-cas. *MBio* **2010**, *1*, e00227-10. [CrossRef] [PubMed]
30. Hullahalli, K.; Rodrigues, M.; Nguyen, U.T.; Palmer, K. An attenuated CRISPR-cas system in *Enterococcus faecalis* permits DNA acquisition. *MBio* **2018**, *9*, e00414-18. [CrossRef]
31. Villarreal, L.P. Viral ancestors of antiviral systems. *Viruses* **2011**, *3*, 1933–1958. [CrossRef]
32. Arber, W.; Dussoix, D. Host specificity of DNA produced by *Escherichia coli*. I. Host controlled modification of bacteriophage lambda. *J. Mol. Biol.* **1962**, *5*, 18–36. [CrossRef]
33. Seed, K.D.; Lazinski, D.W.; Calderwood, S.B.; Camilli, A. A bacteriophage encodes its own CRISPR/cas adaptive response to evade host innate immunity. *Nature* **2013**, *494*, 489–491. [CrossRef]
34. Nilsson, A.S.; Karlsson, J.L.; Haggård-Ljungquist, E. Site-specific recombination links the evolution of P2-like coliphages and pathogenic enterobacteria. *Mol. Biol. Evol.* **2004**, *21*, 1–13. [CrossRef]
35. Kintz, E.; Davies, M.R.; Hammarlöf, D.L.; Canals, R.; Hinton, J.C.D.; van der Woude, M.W. A BTP1 prophage gene present in invasive non-typhoidal *Salmonella* determines composition and length of the O-antigen of the lipopolysaccharide. *Mol. Microbiol.* **2015**, *96*, 263–275. [CrossRef] [PubMed]
36. Cumby, N.; Edwards, A.M.; Davidson, A.R.; Maxwell, K.L. The bacteriophage HK97 gp15 moron element encodes a novel superinfection exclusion protein. *J. Bacteriol.* **2012**, *194*, 5012–5019. [CrossRef] [PubMed]
37. Mahony, J.; McGrath, S.; Fitzgerald, G.F.; van Sinderen, D. Identification and characterization of lactococcal-prophage-carried superinfection exclusion genes. *Appl. Environ. Microbiol.* **2008**, *74*, 6206–6215. [CrossRef] [PubMed]
38. Bebeacua, C.; Lorenzo Fajardo, J.C.; Blangy, S.; Spinelli, S.; Bollmann, S.; Neve, H.; Cambillau, C.; Heller, K.J. X-ray structure of a superinfection exclusion lipoprotein from phage TP-J34 and identification of the tape measure protein as its target. *Mol. Microbiol.* **2013**, *89*, 152–165. [CrossRef] [PubMed]
39. Bondy-Denomy, J.; Qian, J.; Westra, E.R.; Buckling, A.; Guttman, D.S.; Davidson, A.R.; Maxwell, K.L. Prophages mediate defense against phage infection through diverse mechanisms. *ISME J.* **2016**, *10*, 2854–2866. [CrossRef] [PubMed]
40. Dedrick, R.M.; Jacobs-Sera, D.; Bustamante, C.A.G.; Garlena, R.A.; Mavrich, T.N.; Pope, W.H.; Reyes, J.C.C.; Russell, D.A.; Adair, T.; Alvey, R.; et al. Prophage-mediated defence against viral attack and viral counter-defence. *Nat. Microbiol.* **2017**, *2*, 16251. [CrossRef] [PubMed]
41. Parma, D.H.; Snyder, M.; Sobolevski, S.; Nawroz, M.; Brody, E.; Gold, L. The Rex system of bacteriophage lambda: Tolerance and altruistic cell death. *Genes Dev.* **1992**, *6*, 497–510. [CrossRef] [PubMed]
42. Friedman, D.I.; Mozola, C.C.; Beeri, K.; Ko, C.-C.; Reynolds, J.L. Activation of a prophage-encoded tyrosine kinase by a heterologous infecting phage results in a self-inflicted abortive infection. *Mol. Microbiol.* **2011**, *82*, 567–577. [CrossRef]
43. Matos, R.C.; Lapaque, N.; Rigottier-Gois, L.; Debarbieux, L.; Meylheuc, T.; Gonzalez-Zorn, B.; Repoila, F.; de Lopes, M.F.; Serror, P. *Enterococcus faecalis* prophage dynamics and contributions to pathogenic traits. *PLoS Genet.* **2013**, *9*, e1003539. [CrossRef]
44. Jurczak-Kurek, A.; Gąsior, T.; Nejman-Faleńczyk, B.; Bloch, S.; Dydecka, A.; Topka, G.; Necel, A.; Jakubowska-Deredas, M.; Narajczyk, M.; Richert, M.; et al. Biodiversity of bacteriophages: Morphological and biological properties of a large group of phages isolated from urban sewage. *Sci. Rep* **2016**, *6*, 34338. [CrossRef]
45. Dumoulin, R.; Cortes-Perez, N.; Gaubert, S.; Duhutrel, P.; Brinster, S.; Torelli, R.; Sanguinetti, M.; Posteraro, B.; Repoila, F.; Serror, P. Enterococcal Rgg-like regulator ElrR activates expression of the elrA operon. *J. Bacteriol.* **2013**, *195*, 3073–3083. [CrossRef]
46. Gasson, M.J. Plasmid complements of *Streptococcus lactis* NCDO 712 and other lactic streptococci after protoplast-induced curing. *J. Bacteriol.* **1983**, *154*, 1–9. [PubMed]
47. Renault, P.; Corthier, G.; Goupil, N.; Delorme, C.; Ehrlich, S.D. Plasmid vectors for gram-positive bacteria switching from high to low copy number. *Gene* **1996**, *183*, 175–182. [CrossRef]

48. Rigottier-Gois, L.; Alberti, A.; Houel, A.; Taly, J.-F.; Palcy, P.; Manson, J.; Pinto, D.; Matos, R.C.; Carrilero, L.; Montero, N.; et al. Large-scale screening of a targeted *Enterococcus faecalis* mutant library identifies envelope fitness factors. *PLoS ONE* **2011**, *6*, e29023. [CrossRef]
49. Rigottier-Gois, L.; Madec, C.; Navickas, A.; Matos, R.C.; Akary-Lepage, E.; Mistou, M.-Y.; Serror, P. The surface rhamnopolysaccharide Epa of *Enterococcus faecalis* is a key determinant of intestinal colonization. *J. Infect. Dis.* **2015**, *211*, 62–71. [CrossRef]
50. Sahm, D.F.; Marsilio, M.K.; Piazza, G. Antimicrobial resistance in key bloodstream bacterial isolates: Electronic surveillance with the Surveillance Network Database-USA. *Clin. Infect. Dis.* **1999**, *29*, 259–263. [CrossRef] [PubMed]
51. Fritzenwanker, M.; Kuenne, C.; Billion, A.; Hain, T.; Zimmermann, K.; Goesmann, A.; Chakraborty, T.; Domann, E. Complete Genome Sequence of the Probiotic *Enterococcus faecalis* Symbioflor 1 Clone DSM 16431. *Genome Announc.* **2013**, *1*, e00165-12. [CrossRef] [PubMed]
52. Jamet, E.; Akary, E.; Poisson, M.-A.; Chamba, J.-F.; Bertrand, X.; Serror, P. Prevalence and characterization of antibiotic resistant *Enterococcus faecalis* in French cheeses. *Food Microbiol.* **2012**, *31*, 191–198. [CrossRef]
53. Fortier, L.-C.; Moineau, S. Phage production and maintenance of stocks, including expected stock lifetimes. *Methods Mol. Biol.* **2009**, *501*, 203–219. [PubMed]
54. Pickard, D.J.J. Preparation of bacteriophage lysates and pure DNA. *Methods Mol. Biol.* **2009**, *502*, 3–9.
55. Summer, E.J. Preparation of a phage DNA fragment library for whole genome shotgun sequencing. *Methods Mol. Biol.* **2009**, *502*, 27–46. [PubMed]
56. Margulies, M.; Egholm, M.; Altman, W.E.; Attiya, S.; Bader, J.S.; Bemben, L.A.; Berka, J.; Braverman, M.S.; Chen, Y.-J.; Chen, Z.; et al. Genome sequencing in microfabricated high-density picolitre reactors. *Nature* **2005**, *437*, 376–380. [CrossRef] [PubMed]
57. Aziz, R.K.; Bartels, D.; Best, A.A.; DeJongh, M.; Disz, T.; Edwards, R.A.; Formsma, K.; Gerdes, S.; Glass, E.M.; Kubal, M.; et al. The RAST Server: Rapid annotations using subsystems technology. *BMC Genom.* **2008**, *9*, 75. [CrossRef] [PubMed]
58. Sullivan, M.J.; Petty, N.K.; Beatson, S.A. Easyfig: A genome comparison visualizer. *Bioinformatics* **2011**, *27*, 1009–1010. [CrossRef]
59. Hyman, P.; Abedon, S.T. Practical methods for determining phage growth parameters. *Methods Mol. Biol.* **2009**, *501*, 175–202.
60. Garvey, P.; Hill, C.; Fitzgerald, G.F. The lactococcal plasmid pNP40 encodes a third bacteriophage resistance mechanism, one which affects phage DNA penetration. *Appl. Environ. Microbiol.* **1996**, *62*, 676–679.
61. Sing, W.D.; Klahenhammer, T.R. Characteristics of phage abortion conferred in lactococci by the congugal plasmid pTR2030. *J. Mol. Biol.* **1990**, *136*, 1807–1815.
62. Behnke, D.; Malke, H. Bacteriophage interference in *Streptococcus pyogenes*. II. A25 mutants resistant to prophage-medicated interference. *Virology* **1978**, *85*, 129–136. [CrossRef]
63. Luria, S.E.; Delbrück, M. Mutations of bacteria from virus sensitivity to virus resistance. *Genetics* **1943**, *28*, 491–511.
64. Dower, W.J.; Miller, J.F.; Ragsdale, C.W. High efficiency transformation of *E. coli* by high voltage electroporation. *Nucleic Acids Res.* **1988**, *16*, 6127–6145. [CrossRef]
65. Dunny, G.M.; Lee, L.N.; LeBlanc, D.J. Improved electroporation and cloning vector system for gram-positive bacteria. *Appl. Environ. Microbiol.* **1991**, *57*, 1194–1201. [PubMed]
66. Brinster, S.; Furlan, S.; Serror, P. C-terminal WxL domain mediates cell wall binding in *Enterococcus faecalis* and other gram-positive bacteria. *J. Bacteriol.* **2007**, *189*, 1244–1253. [CrossRef] [PubMed]
67. Blanco, L.; Salas, M. Relating structure to function in phi29 DNA polymerase. *J. Biol. Chem.* **1996**, *271*, 8509–8512. [CrossRef] [PubMed]
68. Dufour, E.; Méndez, J.; Lázaro, J.M.; de Vega, M.; Blanco, L.; Salas, M. An aspartic acid residue in TPR-1, a specific region of protein-priming DNA polymerases, is required for the functional interaction with primer terminal protein1. *J. Mol. Biol.* **2000**, *304*, 289–300. [CrossRef] [PubMed]
69. Eisenbrandt, R.; Lázaro, J.M.; Salas, M.; Vega, M. de. Phi29 DNA polymerase residues Tyr59, His61 and Phe69 of the highly conserved ExoII motif are essential for interaction with the terminal protein. *Nucleic Acids Res.* **2002**, *30*, 1379–1386. [CrossRef]
70. Meijer, W.J.J.; Horcajadas, J.A.; Salas, M. φ29 Family of Phages. *Microbiol. Mol. Biol. Rev.* **2001**, *65*, 261–287. [CrossRef]

71. Rodríguez, I.; Lázaro, J.M.; Blanco, L.; Kamtekar, S.; Berman, A.J.; Wang, J.; Steitz, T.A.; Salas, M.; de Vega, M. A specific subdomain in phi29 DNA polymerase confers both processivity and strand-displacement capacity. *Proc. Natl. Acad. Sci. USA* **2005**, *102*, 6407–6412. [CrossRef]
72. Redrejo-Rodríguez, M.; Salas, M. Multiple roles of genome-attached bacteriophage terminal proteins. *Virology* **2014**, *468–470*, 322–329. [CrossRef]
73. Redrejo-Rodríguez, M.; Muñoz-Espín, D.; Holguera, I.; Mencía, M.; Salas, M. Nuclear and nucleoid localization are independently conserved functions in bacteriophage terminal proteins. *Mol. Microbiol.* **2013**, *90*, 858–868. [CrossRef] [PubMed]
74. Bateman, A.; Rawlings, N.D. The CHAP domain: A large family of amidases including GSP amidase and peptidoglycan hydrolases. *Trends Biochem. Sci.* **2003**, *28*, 234–237. [CrossRef]
75. McGowan, S.; Buckle, A.M.; Mitchell, M.S.; Hoopes, J.T.; Gallagher, D.T.; Heselpoth, R.D.; Shen, Y.; Reboul, C.F.; Law, R.H.P.; Fischetti, V.A.; et al. X-ray crystal structure of the streptococcal specific phage lysin PlyC. *Proc. Natl. Acad. Sci. USA* **2012**, *109*, 12752–12757. [CrossRef] [PubMed]
76. Rigden, D.J.; Jedrzejas, M.J.; Galperin, M.Y. Amidase domains from bacterial and phage autolysins define a family of gamma-D,L-glutamate-specific amidohydrolases. *Trends Biochem. Sci.* **2003**, *28*, 230–234. [CrossRef]
77. Nelson, D.; Schuch, R.; Chahales, P.; Zhu, S.; Fischetti, V.A. PlyC: A multimeric bacteriophage lysin. *Proc. Natl. Acad. Sci. USA* **2006**, *103*, 10765–10770. [CrossRef] [PubMed]
78. Lavigne, R.; Seto, D.; Mahadevan, P.; Ackermann, H.-W.; Kropinski, A.M. Unifying classical and molecular taxonomic classification: Analysis of the Podoviridae using BLASTP-based tools. *Res. Microbiol.* **2008**, *159*, 406–414. [CrossRef] [PubMed]
79. McGrath, S.; Fitzgerald, G.F.; van Sinderen, D. Identification and characterization of phage-resistance genes in temperate lactococcal bacteriophages. *Mol. Microbiol.* **2002**, *43*, 509–520. [CrossRef] [PubMed]
80. Innocenti, N.; Golumbeanu, M.; d' Hérouël, A.F.; Lacoux, C.; Bonnin, R.A.; Kennedy, S.P.; Wessner, F.; Serror, P.; Bouloc, P.; Repoila, F.; et al. Whole-genome mapping of 5′ RNA ends in bacteria by tagged sequencing: A comprehensive view in *Enterococcus faecalis*. *Rna* **2015**, *21*, 1018–1031. [CrossRef] [PubMed]
81. Chopin, M.-C.; Chopin, A.; Bidnenko, E. Phage abortive infection in lactococci: Variations on a theme. *Curr. Opin. Microbiol.* **2005**, *8*, 473–479. [CrossRef]
82. Paez-Espino, D.; Roux, S.; Chen, I.-M.A.; Palaniappan, K.; Ratner, A.; Chu, K.; Huntemann, M.; Reddy, T.B.K.; Pons, J.C.; Llabrés, M.; et al. IMG/VR v.2.0: An integrated data management and analysis system for cultivated and environmental viral genomes. *Nucleic Acids Res.* **2018**. [CrossRef]
83. Nelson, D.; Schuch, R.; Zhu, S.; Tscherne, D.M.; Fischetti, V.A. Genomic sequence of C1, the first streptococcal phage. *J. Bacteriol.* **2003**, *185*, 3325–3332. [CrossRef]
84. Palmer, K.L.; Godfrey, P.; Griggs, A.; Kos, V.N.; Zucker, J.; Desjardins, C.; Cerqueira, G.; Gevers, D.; Walker, S.; Wortman, J.; et al. Comparative genomics of enterococci: Variation in Enterococcus faecalis, clade structure in E. faecium, and defining characteristics of E. gallinarum and E. casseliflavus. *MBio* **2012**, *3*, e00318-11. [CrossRef]
85. Nelson, D.C.; Schmelcher, M.; Rodriguez-Rubio, L.; Klumpp, J.; Pritchard, D.G.; Dong, S.; Donovan, D.M. Endolysins as antimicrobials. *Adv. Virus Res.* **2012**, *83*, 299–365. [PubMed]
86. Cheng, M.; Zhang, Y.; Li, X.; Liang, J.; Hu, L.; Gong, P.; Zhang, L.; Cai, R.; Zhang, H.; Ge, J.; et al. Endolysin LysEF-P10 shows potential as an alternative treatment strategy for multidrug-resistant *Enterococcus faecalis* infections. *Sci. Rep.* **2017**, *7*, 10164. [CrossRef] [PubMed]
87. Chowdhury, R.; Biswas, S.K.; Das, J. Abortive replication of choleraphage phi 149 in *Vibrio cholerae* biotype el tor. *J. Virol.* **1989**, *63*, 392–397. [PubMed]
88. Rettenmier, C.W.; Hemphill, H.E. Abortive infection of lysogenic *Bacillus subtilis* 168(SPO2) by bacteriophage phi 1. *J. Virol.* **1974**, *13*, 870–880. [PubMed]
89. Smith, H.S.; Pizer, L.I.; Pylkas, L.; Lederberg, S. Abortive infection of *Shigella dysenteriae* P2 by T2 bacteriophage. *J. Virol.* **1969**, *4*, 162–168. [PubMed]
90. Tran, L.S.; Szabó, L.; Ponyi, T.; Orosz, L.; Sík, T.; Holczinger, A. Phage abortive infection of *Bacillus licheniformis* ATCC 9800; identification of the abiBL11 gene and localisation and sequencing of its promoter region. *Appl. Microbiol. Biotechnol.* **1999**, *52*, 845–852. [CrossRef] [PubMed]
91. Snyder, L. Phage-exclusion enzymes: A bonanza of biochemical and cell biology reagents? *Mol. Microbiol.* **1995**, *15*, 415–420. [CrossRef]

92. Stern, A.; Sorek, R. The phage-host arms race: Shaping the evolution of microbes. *Bioessays* **2011**, *33*, 43–51. [CrossRef]
93. Durmaz, E.; Klaenhammer, T.R. Abortive phage resistance mechanism AbiZ speeds the lysis clock to cause premature lysis of phage-infected Lactococcus lactis. *J. Bacteriol.* **2007**, *189*, 1417–1425. [CrossRef]
94. Young, R. Phage lysis: Three steps, three choices, one outcome. *J. Microbiol.* **2014**, *52*, 243–258. [CrossRef]
95. Wang, I.N.; Smith, D.L.; Young, R. Holins: The protein clocks of bacteriophage infections. *Annu. Rev. Microbiol.* **2000**, *54*, 799–825. [CrossRef] [PubMed]
96. Ho, K.; Huo, W.; Pas, S.; Dao, R.; Palmer, K.L. Loss-of-Function Mutations in epaR Confer Resistance to ϕNPV1 Infection in *Enterococcus faecalis* OG1RF. *Antimicrob. Agents Chemother.* **2018**, *62*. [CrossRef] [PubMed]
97. Teng, F.; Singh, K.V.; Bourgogne, A.; Zeng, J.; Murray, B.E. Further characterization of the *epa* gene cluster and Epa polysaccharides of *Enterococcus faecalis*. *Infect. Immun.* **2009**, *77*, 3759–3767. [CrossRef] [PubMed]
98. Chapot-Chartier, M.-P. Interactions of the cell-wall glycopolymers of lactic acid bacteria with their bacteriophages. *Front. Microbiol.* **2014**, *5*, 236. [CrossRef] [PubMed]
99. Mistou, M.-Y.; Sutcliffe, I.C.; van Sorge, N.M. Bacterial glycobiology: Rhamnose-containing cell wall polysaccharides in Gram-positive bacteria. *FEMS Microbiol. Rev.* **2016**, *40*, 464–479. [CrossRef]
100. Ainsworth, S.; Sadovskaya, I.; Vinogradov, E.; Courtin, P.; Guerardel, Y.; Mahony, J.; Grard, T.; Cambillau, C.; Chapot-Chartier, M.-P.; van Sinderen, D. Differences in lactococcal cell wall polysaccharide structure are major determining factors in bacteriophage sensitivity. *MBio* **2014**, *5*, e00880-14. [CrossRef]

Article

Antibiotic Therapy Using Phage Depolymerases: Robustness Across a Range of Conditions

Han Lin [1], Matthew L. Paff [1,†], Ian J. Molineux [2,3,4,*] and James J. Bull [1,2,5,*]

1 Department of Integrative Biology, University of Texas, Austin, TX 78712, USA; hanl@austin.utexas.edu (H.L.); matthew.paff@utexas.edu (M.L.P.)
2 Institute for Cellular and Molecular Biology, University of Texas, Austin, TX 78712, USA
3 Department of Molecular Biosciences, University of Texas, Austin, TX 78712, USA
4 LaMontagne Center for Infectious Disease, University of Texas, Austin, TX 78712, USA
5 Center for Computational Biology and Bioinformatics, University of Texas, Austin, TX 78712, USA
* Correspondence: molineux@austin.utexas.edu (I.J.M.); bull@utexas.edu (J.J.B.)
† Current address: Nano Vision, 1705 Guadalupe St, Austin, TX 78712, USA.

Received: 26 October 2018; Accepted: 10 November 2018; Published: 12 November 2018

Abstract: Phage-derived depolymerases directed against bacterial capsules are showing therapeutic promise in various animal models of infection. However, individual animal model studies are often constrained by use of highly specific protocols, such that results may not generalize to even slight modifications. Here we explore the robustness of depolymerase therapies shown to succeed in a previous study of mice. Treatment success rates were reduced by treatment delay, more so for some enzymes than others: K1- and K5 capsule-degrading enzymes retained partial efficacy on delay, while K30 depolymerase did not. Phage were superior to enzymes under delayed treatment only for K1. Route of administration (intramuscular versus intraperitoneal) mattered for success of K1E, possibly for K1F, not for K1H depolymerase. Significantly, K1 capsule-degrading enzymes proved highly successful when using immune-suppressed, leukopenic mice, even with delayed treatment. Evolution of bacteria resistant to K1-degrading enzymes did not thwart therapeutic success in leukopenic mice, likely because resistant bacteria were avirulent. In combination with previous studies these results continue to support the efficacy of depolymerases as antibacterial agents in vivo, but system-specific details are becoming evident.

Keywords: phage therapy; bacterial infection; capsule depolymerase; antibiotic; animal model; bacterial resistance

---

1. Introduction

In the increasingly urgent search for new treatments against bacterial infections, phages and phage products hold promise [1–3]. Offsetting the now many laboratory studies of phages and phage products showing positive results [4–6], the few clinical phage therapy trials conducted under standards of Western medicine have actually failed [7–9]. The most recent randomized clinical phage therapy trial was also stopped prematurely for lack of efficacy, although small effects were deemed encouraging [10]. The contrasting outcomes between actual trials and laboratory infections raise the possibility that experimental studies poorly represent applications. One obvious concern in generalizing from experimental infections to clinical settings is the potential sensitivity of results to specifics of the experimental protocol. A step toward generality of a therapeutic agent is thus to broaden the experimental protocol and measure the sensitivity of treatment success to experimental variables. Here we evaluate the robustness of phage depolymerase therapies previously demonstrated to rescue mice from experimental infections of capsulated *E. coli* [11].

Several classes of phage proteins have exhibited antibiotic potency, including endolysins [12], viron-associated lysins [13], holins [5], spanins [14] and bacterial polysaccharide depolymerases [15]. Some endolysins are already commercially available and are currently being tested in clinical trials [3,16]. Capsular depolymerases represent an interesting type of antibiotic: they do not kill per se, but merely strip the bacteria of protective polysaccharides and thus expose the bacteria to immune components [17]. They have a potential advantage over endolysins in that they do not lyse the bacteria, thereby minimizing inflammatory responses from endotoxins [18]. In vivo studies of capsule depolymerases are yet limited but appear to generalize across different animal models (Table 1). Furthermore, enzymes that disrupt biofilms, in part by degradation of polysaccharide, but do not kill bacteria are also showing promise [19,20].

Table 1. In vivo studies of capsule depolymerases in animal infection models.

Capsule Depolymerases	Animal Infection Model	References
EndoE (K1E)	*E. coli*; neonatal rats	Mushtaq et al., 2004 [21]; 2005 [22]
CapD; EnvD	*B. anthracis*; mice	Scorpio et al., 2008 [23]; Negus et al., 2015 [24]
K1-ORF34; K64dep	*K. pneumoniae*; mice	Lin et al., 2014 [25]; Pan et al., 2015 [26]
depoKP36	*K. pneumoniae*; moth larvae	Majkowska-Skrobek et al., 2016 [27]
K1F, K1H, K5, K30	*E. coli*; mice	Lin et al., 2017 [11]

Our purpose here is to explore the robustness of capsular depolymerase enzymes in treating experimentally infected mice. The bacteria are *E. coli* with K1, K5 or K30 capsules, and the enzymes were obtained as purified proteins expressed from clones of phage genes. Previous work demonstrated therapeutic success of several enzymes when using (i) simultaneous infection and treatment, (ii) intramuscular administration, and (iii) immunocompetent mice [11]. Here we explore treatment efficacy when relaxing these conditions. By measuring performance of depolymerase treatments under different conditions and with different enzymes, our work also exposes possible realms for improving therapeutic efficacy of depolymerases, as in structure/function properties of the enzymes.

2. Materials and Methods

2.1. Bacterial Strains and Cell Culture

The pathogenic bacterial strains used in this study were K1-capsulated *E. coli* RS218 [28], K5-capsulated ATCC 23506, and K30-capsulated E69 [29]. *E. coli* lab strains used only for phage propagation or cloning were the K1-capsulated K12 strain EV36 [30] and BL21(DE3). Cells were generally grown in LB broth (10 g tryptone, 5 g yeast extract, 10 g NaCl per liter) in 37 °C shakers. Cell density was determined by plating serial cell dilutions on LB agar (1.3% w/v) plates for colony counts.

Capsule-free isolates of RS218 were selected by culturing RS218 on LB plates containing K1H phage [31]. Colonies that grew on the plates were picked, diluted in LB medium and replated on LB plates containing K1 phage. The colonies grown on these second plates were picked and cultured in LB medium containing K1 phage before mouse injection. The addition of phage was a precaution against any possible revertants to a capsulated state.

2.2. Coliphage Strains and Culture

Phages K1E, K1F and K1H [32] were propagated on host *E. coli* EV36. K1–5 [33] and K30 [34] were grown with host *E. coli* ATCC 23506 and E69, respectively. Phages were purified by equilibrium CsCl-gradient centrifugation and dialyzed into SM buffer (50 mM Tris–HCl, 100 mM NaCl, 8 mM $MgSO_4$, pH 7.5) as previously described [11,35]. Phage titrations were performed by plaque counts on an appropriate host in LB soft agar (0.65%) overlay.

2.3. Plasmids, Protein Expression and Purification

The phage capsule depolymerases were expressed in *E. coli* BL21(DE3) with pET28b constructs containing cloned depolymerase genes as previously described [11]. Proteins were purified with HisPur Ni-NTA resin (Thermo Fisher Scientific, Rockford, IL, USA), and dialyzed into PBS buffer (137 mM NaCl, 2.7 mM KCl, 10 mM Na_2HPO_4, 1.8 mM KH_2PO_4, pH 7.5) with 3.5 kDa MWCO dialysis membranes (Spectrum-Repligen, Houston, TX, USA). Protein concentrations were estimated by absorption at 280 nm with a Nanodrop ND-1000 (Thermo Fisher Scientific, Wilmington, DE, USA).

2.4. Mouse Infection Model

Animal work was performed under NIH guidelines and the University of Texas IACUC protocols (AUP-2015-00035, AUP-2018-00010). Female NIH Swiss outbred mice (Envigo, Somerset, NJ, USA) aged 4–6 weeks with 20–25 g weights were used here in all studies. All intramuscular (IM) inoculations of bacteria used the left thigh; all IM inoculations of enzyme or phage used the right thigh. Mouse survival was monitored twice daily for 5 days.

The following experiments were undertaken.

Delayed treatment of normal mice with enzyme or phage. Mice received an IM injection of 100 µL bacteria, dosed at either 1.2–3.4 $\times$ 10^8 CFU of RS218, 3.1–6.1 $\times$ 10^8 CFU of ATCC 23506, or 1.1–2.8 $\times$ 10^8 CFU of E69 per mouse. Enzyme or phage inoculation in the contralateral thigh followed at 8 h, dosed at 20 µg enzyme in 100 µL PBS or 10^7 pfu phage in 100 µL SM buffer. A dose of 10^7 phage is 10-fold lower than used in the delayed treatment studies of [36], but much higher than doses used in studies with simultaneous treatment [37]. Given the rate of phage amplification in the host [37], 10^7 was expected to be highly effective.

Different administration routes for immediate enzyme treatment. Mice received IM 1.2–2.9 $\times$ 10^8 CFU of RS218 in 100 µL volume, and then 20 µg K1E, 2 µg K1F or 2 µg K1H in 100 µL volume either by contralateral IM or IP (intraperitoneal) inoculation.

The leukopenic mouse model. Mice were rendered leukopenic by IP injection of cyclophosphamide (CP) at 150 mg/kg body weight 4 days prior and then at 100 mg/kg body weight 1 day prior to infection. To determine an approximate lethal dose of bacteria, different mice were inoculated across a range of bacterial doses in 10-fold increments spanning 10^3–10^7 CFU (RS218) or 10^4–10^8 CFU (capsule-free RS218 derivative). Simultaneous enzyme (20 µg K1E, K1H) or phage (K1H 10^7 pfu contralateral) treatment was tested in leukopenic mice infected by RS218 at the lethal dose of 2.2–5.1 $\times$ 10^4 CFU. Delayed enzyme (20 µg K1F, K1H) treatment was also tested in leukopenic mice 8 h after infection.

Mouse survival was analyzed by Kaplan-Meier survival curves using SPSS software, where the cumulative survival probability was plotted over the time course of 5 days and statistically evaluated by log rank test [25,38,39].

2.5. Capsule Isolation and Degradation Assay

Capsules were isolated as previously described [11,40] for degradation assays monitored by gel electrophoresis and Alcian Blue staining. To compare activity of purified K1 enzyme to enzyme activity of intact phage, 10–20 µg K1 capsule was mixed with serial dilutions (0.25–8 µg) of K1E, K1F, K1H enzyme or CsCl-purified cognate phage at the calculated amount of enzyme equivalents at 37 °C for 1 h before gel electrophoresis. Enzyme equivalents of phages were calculated from the molecular weight of each enzyme [41,42] and the fact that these phages contain six depolymerase trimers and thus 18 enzyme molecules per phage particle [35]. 10^{10} phage would thus have ~3 $\times$ 10^{-2} µg enzyme (varying somewhat among different enzymes). This calculation assumes that phage stocks plate at a particle:plaque ratio of 1; a reduced efficiency of plating would underestimate the amount of enzyme in the phage sample.

To compare levels of K1 enzyme activity in mouse blood, 100 μL PBS containing 20 μg K1E, K1F or K1H enzyme or no enyzme (control) was injected to mice by IM or IP. Mice were euthanized at 1 h or 24 h after injection and blood was collected to prepare serum. 10–20 μg K1 capsule was incubated with 10 μL K1E serum or 0.3 μL K1F, K1H serum at 37 °C overnight before gel examination. The volumes of serum used in each reaction were determined in preliminary tests to achieve a dynamic range of degradation differentiable by visual inspection. Reactions with control serum were included as negative control, while reactions with serial dilutions of enzymes added immediately to the control serum were included as positive control.

2.6. Resistance Competition Assay (RCA)

The assay measures the in vivo effects of treatment on bacterial numbers. It works by inoculating mice with a mixture of mostly treatment-sensitive bacteria and a small number of treatment-resistant bacteria [36]. Mice are either treated or not (the latter being controls). Resistant bacteria increase in frequency (relative to controls) to the extent that treatment suppresses the population of sensitive bacteria. For this assay, RS218 was mixed with the K5-capsulated ATCC 23506 (ratio approximately 99:1). Six mice were inoculated in the left thigh with the normal bacteria dose (~2.9×10^8 cells) and either treated simultaneously with contralateral IM of 20 μg K1H enzyme (3 mice) or not treated (3 controls). At 4 h post infection, mice were euthanized. The left thigh was removed, homogenized in 10 mL buffer, and the suspension plated at different dilutions on both LB plates and on LB plates saturated with 10^7 pfu K1H phage. The LB plates support growth of both K1 and K5 bacteria, whereas plates with phage grow only the K5-capsulated strain, allowing a determination of K5 (ATCC 23506) frequency. Calculations of the RCA value used the formula in [36]: $RCA = \ln[(p_t(1-p_0))/(p_0(1-p_t))]/t$, with p_0 as the proportion of resistant bacteria in control mice and p_t as the proportion resistant bacteria in treated mice, both measured at time t (hours of treatment). For an RCA value of R, e^{-R} is the per hour growth of bacteria under treatment relative to growth without treatment.

3. Results

3.1. Delayed Treatment Reduces Efficacy

We reported that phage capsule depolymerases were broadly effective in treating lethal *E. coli* infections in mice [11]. That work applied treatment simultaneously with infection. Here we explored the efficacy of delayed treatment, as a delay might better represent clinical therapeutics.

Delayed treatment is better than no treatment only for some enzymes. The delayed treatments of phage capsule depolymerases given 8 h after infection exhibited reduced efficacy (Figure 1). The delayed treatment with K1F or K1H enzyme resulted in 50–60% survival compared to zero survival in control (Figure 1A), while the simultaneous treatment led to 90–100% survival [11]. K1E did not rescue well in simultaneous treatment, thus the delayed treatment was not tested here. Delayed K5 enzyme treatment showed a 60% survival (Figure 1B) compared to 100% survival when treatment was immediate; both were significantly better than no-treatment controls. Delayed treatment with the least active enzyme K30 gp41 did not significantly improve survival (Figure 1C).

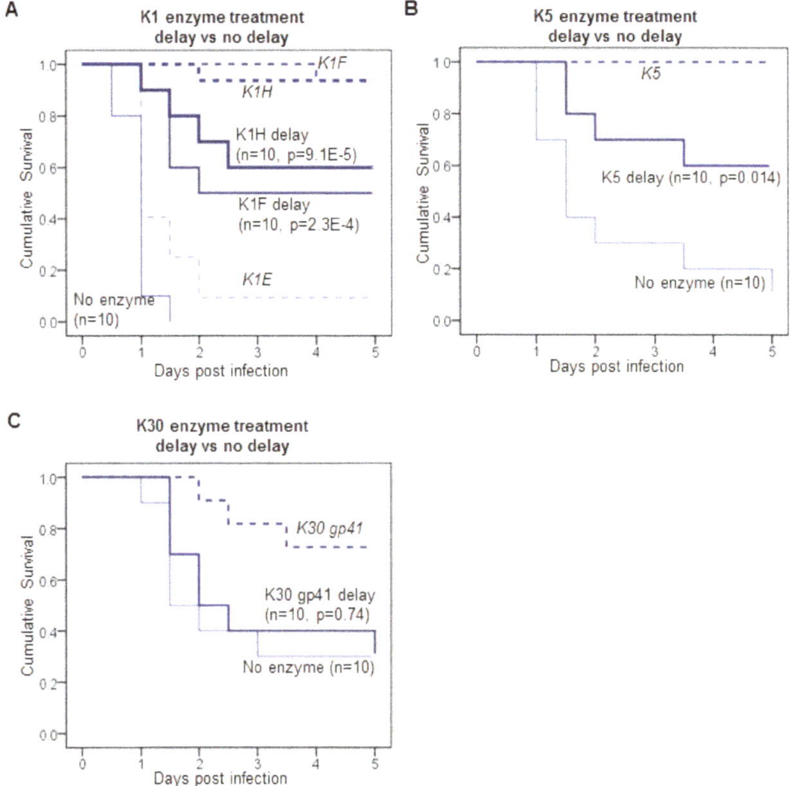

Figure 1. Delayed treatment of infection using capsule depolymerases. (**A**) 1.2–3.4 $\times$ 10^8 CFU of K1-capsulated *E. coli* treated with 20 μg of K1F or K1H enzyme; (**B**) 3.1–6.1 $\times$ 10^8 CFU of K5-capsulated *E. coli* treated with 20 μg of K5 enzyme; (**C**) 1.1–2.8 $\times$ 10^8 CFU of *E. coli* E69 treated with 20 μg of K30 enzyme. All inoculations were IM (intramuscular, thigh); enzyme was administered 8 h after bacteria in the contralateral thigh. Mouse survival was monitored for 5 days and Kaplan-Meier survival curves in solid lines were plotted with the cumulative probability of survival over time for each treatment. Previously reported survival curves of simultaneous enzyme treatments are included for comparison (dashed lines). The mouse number (*n*) of each treatment is given for each curve. Log rank test: *p* values are listed for delayed treatments compared to the no-enzyme control.

Phages outperform enzyme only for K1 bacteria. Phages might be expected to outperform enzymes on the grounds that phage have multiple effects: They amplify within the host, they kill, and cell lysis releases free enzyme as unassembled tailspikes. Yet a superiority of phages under delayed treatment was observed only for K1-capsulated bacteria. K1H or K1E phage yielded higher survival rates than K1 enzymes in delayed treatment (Figure 2A). In contrast, K5 or K30 phages were no better than their enzymes in delayed treatment (Figure 2B,C). As a control for the effects of phage amplification, K1F phage, which does not propagate on RS218, had no effect in mice (data not shown), whereas K1F enzyme efficacy was similar to that of K1H (Figure 1A).

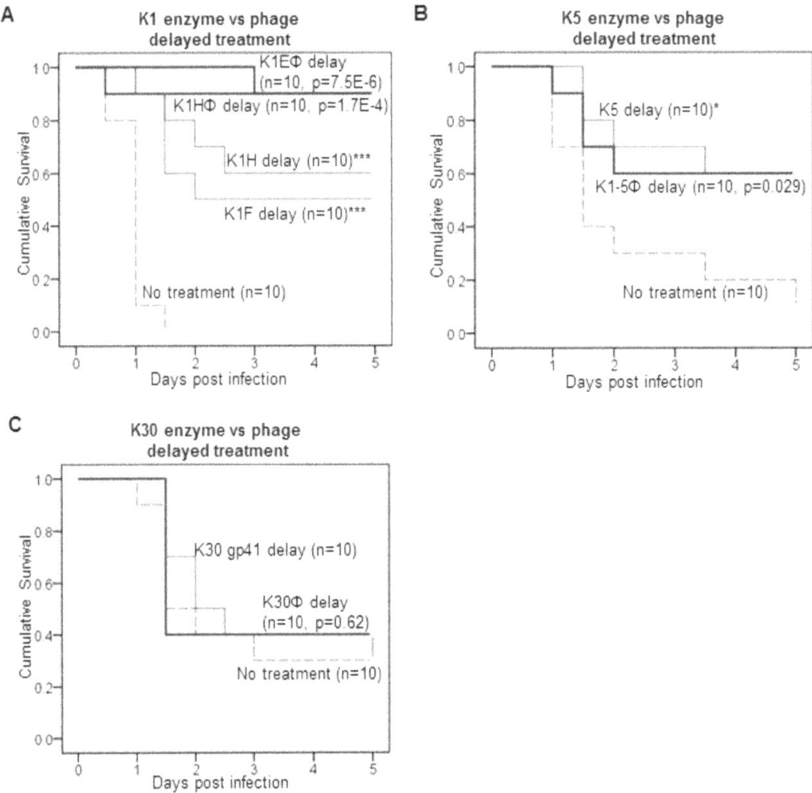

Figure 2. Comparison of depolymerases and cognate phages in treatment. Delayed treatments using phages were carried out in parallel to depolymerase. (**A**) K1E and K1H phage; (**B**) K1-5 phage; (**C**) K30 phage. Depolymerase data are from Figure 1. For phage treatment, mice were infected with intramuscular (IM) inoculations; 8 h later they received ~10^7 pfu phage IM in the contralateral thigh. For comparison to depolymerase treatment, this dose of phage carries about 3×10^{-5} μg of depolymerase (see Methods for calculation). Kaplan-Meier survival curves were plotted as in Figure 1. The mouse number n of each treatment is labeled on each curve. Log rank test: p values for delayed phage treatments compared to the no treatment control are listed, or * $p < 0.05$, *** $p < 0.001$ for delayed enzyme treatment as in Figure 1.

Phages carry active depolymerases as tailspikes that are used in adsorption and penetration of the capsule. It is possible that the activities of the purified enzymes were substantially less than the activities of tailspikes on intact phages. We thus compared the in vitro degrading activities of purified enzyme with activities of intact phages (Figure 3). Degradation by phages was 4–8 fold better than degradation by molar equivalents of purified enzyme. Perhaps surprisingly, in view of the different in vivo performances of the enzymes, in vitro differences among the three K1-specific phages were small.

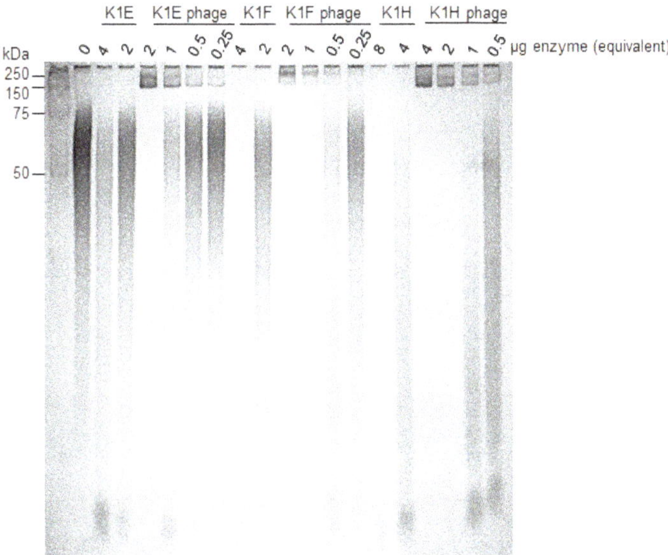

Figure 3. In vitro activity comparison of depolymerases and cognate phages. 10–20 µg of K1 capsule were incubated with serial dilutions of K1E, K1F, K1H depolymerase or phages at the indicated amount of enzyme equivalents (0.25–8 µg). Active enzyme is indicated by loss of signal within the lane; enzyme associated with phages is 4–8 fold more active than free enzyme. Incubation was at 37 °C for 1 h; reactions were fractionated using 12% TBE-PAGE with Alcian Blue staining. Protein standards were loaded in the leftmost lane and their molecular weights indicated.

3.2. *Efficacy Varies with Route of Administration and Enzyme*

K1E enzyme rescued rat pups by IP inoculation [21,22] but was minimally effective in mice by IM inoculation [11]. It does not appear that our purified K1E enzyme is at fault—its in vitro activity approximately matches that of the other K1 enzymes (Figure 3). Motivated by suggestions that administration routes may affect drug bioavailability [43–45], we compared the efficacy of IM versus IP administration routes of the different K1 enzymes; these studies used immediate treatment.

Different enzyme efficacies in mice by different administration routes. Following an IM inoculation of RS218 bacteria, K1E depolymerase efficacy was significantly higher for IP than for IM inoculation at the high dose of 20 µg (Figure 4A). Using only 2 µg for the more active (in vivo) K1F enzyme (Figure 4B) the opposite pattern is suggested, but the small sample sizes provide little power in significance tests. K1H (2 µg dose) yielded similarly low rescue rates for both routes of administration (Figure 4C).

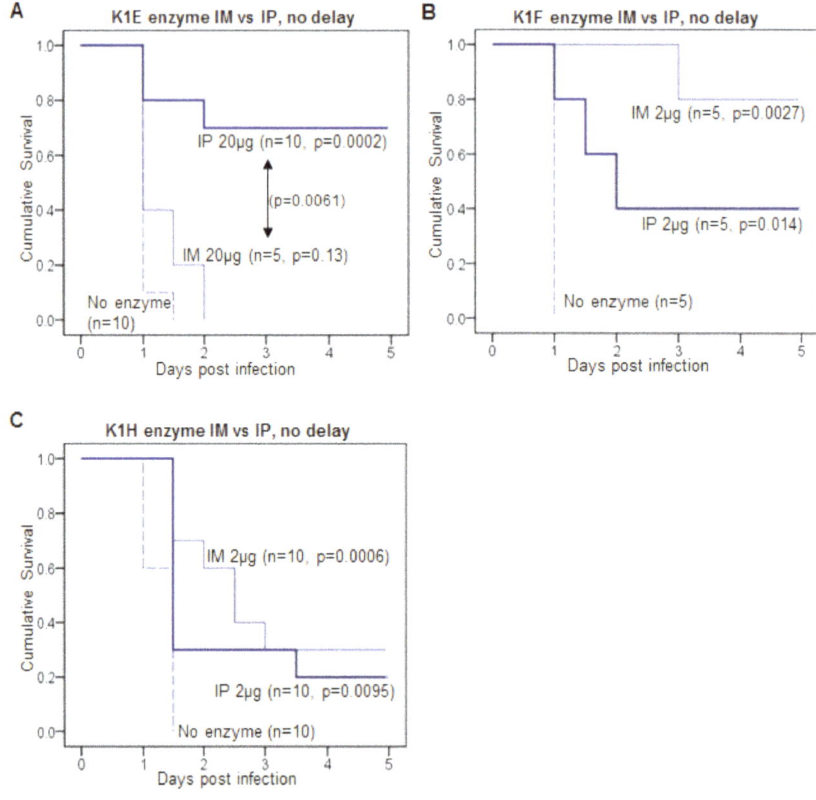

Figure 4. Administration route can affect treatment efficacy. (**A**–**C**) Kaplan-Meier survival curves of depolymerase treatment comparing intramuscular (IM) to intraperitoneal (IP) inoculation in mice. (**A**) Treatment with 20 μg K1E; (**B**) Treatment with 2 μg K1F; (**C**) Treatment with 2 μg K1H. In all mice, 1.2–2.9 × 10^8 CFU of K1-capsulated *E. coli* were injected IM, followed by inoculation of enzyme either IM in the contralateral thigh or IP. Mouse survival was monitored for 5 days and the cumulative probability of survival was plotted for each treatment. The mouse number *n* of each treatment is labeled by each curve. Log rank test: *p* values between enzyme treatment and the no-enzyme control, or significant *p* values between IM and IP treatment are listed.

Basis of the effect of administration route. We tested capsule degradation activity of serum from mice that had been inoculated with enzyme by different routes. If bioavailability of enzyme is affected by route of administration, then serum from mice inoculated by the IP route should have different activity per unit volume than serum from mice inoculated IM. Consistent with treatment efficacy differences, detectable degradation by K1E serum was observed only for the IP route, and then only at 1 h post inoculation (Figure 5A, boxed region). As for K1F and K1H sera, IP delivery resulted in slightly higher capsule degradation activity than IM at 1 h post inoculation, and activity was largely maintained after 24 h exposure (Figure 5B,C). However, both K1F and K1H sera exhibited much higher activities than K1E serum, independent of administration route.

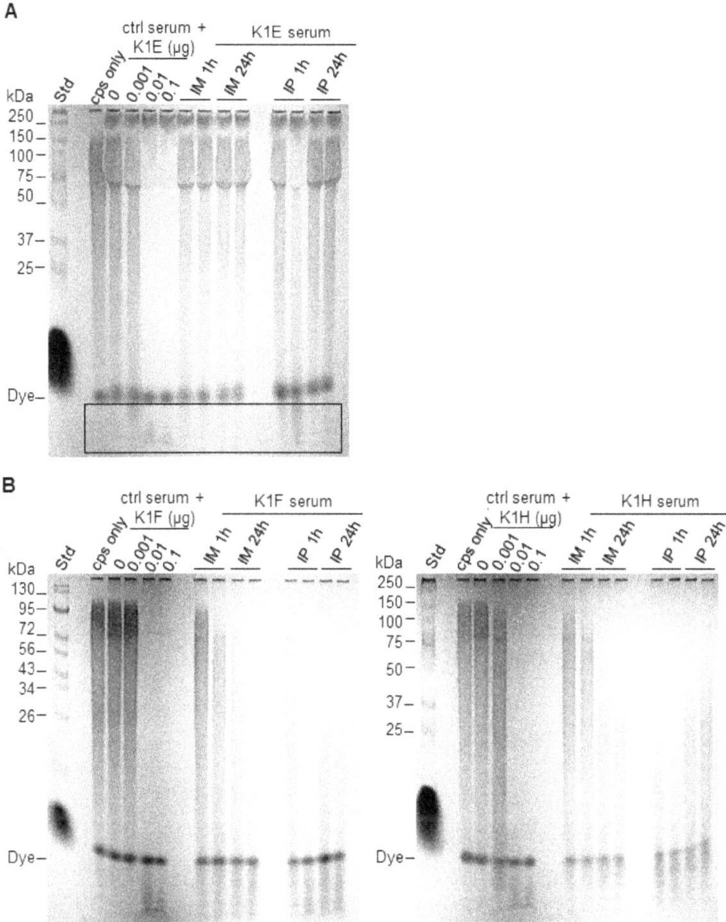

Figure 5. Comparison of depolymerase serum activity using different administration routes. Except for the molecular weight standard ("std") on the left of each panel, each lane fractionates 10–20 ug of K1 capsule ("cps only") or an overnight reaction of capsule incubated with serum and/or enzyme. Lanes with "serum" used serum from mice inoculated with enzyme (**A**) K1E; (**B**) K1F (left) or K1H (right) (IM or IP, 1 mg/kg weight) and sacrificed at 1 or 24 h. Two mice were tested for each route and time point, shown in separate lanes. Lanes with "ctrl serum" had different amounts of free enzyme added to the control serum (from mice not receiving enzyme injection), as the control reactions. Only a slight activity of K1E is evident, and then only for IP 1h (bottom of lane, boxed). In contrast, K1F and K1H both exhibit clear activity by both routes of administration, with a possible effect of route for K1H at 1 h.

3.3. *K1: Treatment Is Successful with Leukopenia*

The above models used immunocompetent mice, requiring high inocula of bacteria to overwhelm innate immunity. To better represent infections that progress from low to high concentrations of bacteria, we tested mice that had been rendered leukopenic, limiting the studies to K1-capsulated bacteria.

Leukopenic mice are far more susceptible to RS218 than are immune-competent mice (Figure 6). A bacterial inoculum slightly exceeding 10^4 was fatal in the leukopenic mice (Figure 6A), compared to the lethal threshold dose of 10^6–10^7 in immunocompetent mice (Figure 6C). In leukopenic mice, doses above 10^4 cfu reduced survival time, to about a half day shorter using 10^7 bacteria (Figure 6A). Phage and enzyme each rescued the leukopenic mice in most cases, whether with immediate (Figure 7A,B) or delayed treatment (Figure 7C).

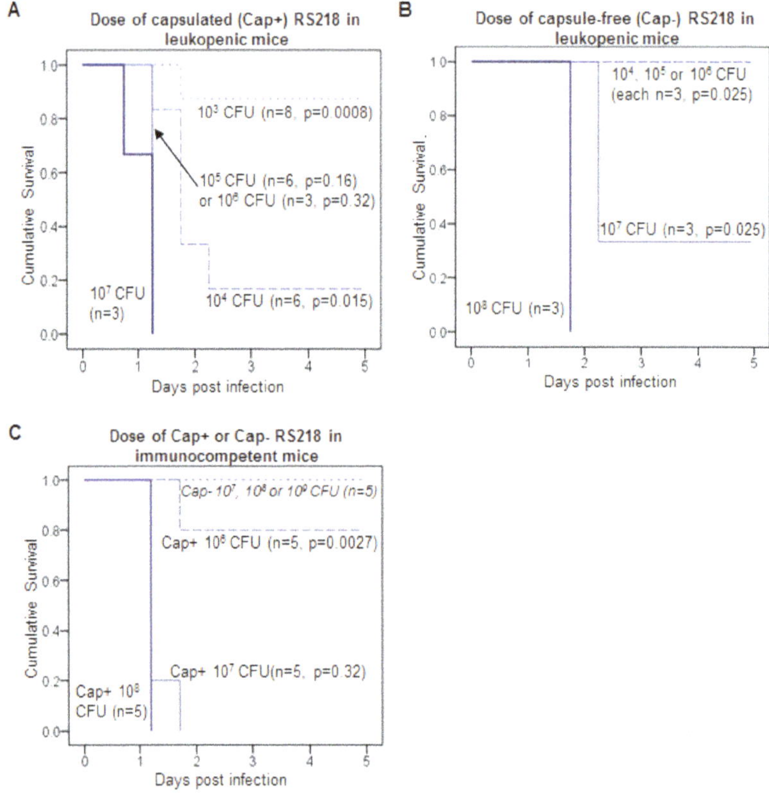

Figure 6. Capsule-free RS218 is avirulent in mice. Kaplan-Meier survival curves of mice per dose of different strains. (**A**) K1-capsulated RS218 (Cap+) in leukopenic mice; (**B**) capsule-free RS218 (Cap−) in leukopenic mice; (**C**) Cap+ and Cap− RS218 in immune competent mice. The median lethal dose is 10^3–10^4 CFU for Cap+ and 10^6–10^7 CFU for Cap- RS218 in leukopenic mice, a difference of 3 orders of magnitude. The median lethal dose in normal mice is 10^6–10^7 CFU for Cap+ RS218, while Cap− RS218 is not lethal at doses as high as 10^9 CFU. Mouse survival was monitored for 5 days and the cumulative probability of survival was plotted for each dose. The mouse number *n* of each treatment is labeled by each curve. Log rank test: *p* values are listed for lower doses compared to the highest dose of each strain.

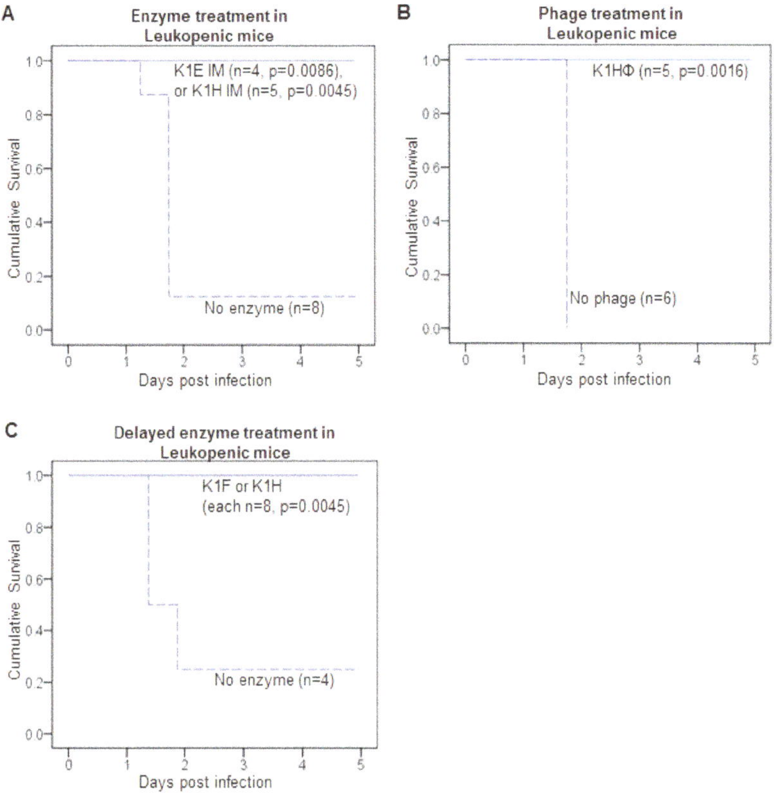

Figure 7. Enzymes and phages rescue infections of leukopenic mice. Total rescue of infections by K1 *E. coli* in leukopenic mice was achieved by simultaneous treatment with (**A**) K1E or K1H depolymerase; (**B**) K1H phage, and by delayed treatment with (**C**) K1F or K1H depolymerase. Each panel shows Kaplan-Meier survival curves with sample sizes given by n. Infection was initiated by intramuscular (IM) inoculations of 2.2–5.1 × 10^4 CFU of K1 *E. coli* in the left thigh. Treatment was an IM injection of 20 μg enzyme or 10^7 phage in the contralateral thigh. Treatment was administered either (**A**,**B**) immediately at the time of infection or (**C**) 8 h after infection. Mouse survival was monitored for 5 days and the cumulative probability of survival was plotted for each treatment. Log rank test: p values are listed for each treatment compared to no-treatment control.

Avirulence of capsule-free bacteria. Our use of leukopenic mice was motivated to assess the potential for treatment failure via the evolution of treatment-resistant bacteria, as observed in a neutropenic mouse-*Pseudomonas* infection model [39]. Resistance is a potential cause of treatment failure with both phages and antibiotics [39,46–50]. Treatment success in our studies indicates that resistance did not ascend, at least not enough to affect survival (true both for immunocompetent and leukopenice mice). We thus addressed the reason evolution of resistance was not a problem in our system despite its cause of treatment failure in others. In the case of enzyme treatment, the relevant "resistance" phenotype is presumably absence of a capsule, since the enzyme has no substrate on those cells; that phenotype is also resistant to phages encoding K1 enzymes [51]. Therefore, the lethal consequences of a capsule-free RS218 derivative in leukopenic mice was evaluated over a series of inoculum sizes (Figure 6B,C). The capsule-free bacteria are profoundly less virulent: The mice could survive an inoculum 10^3–10^4 times larger of capsule-free bacteria than of capsulated bacteria. Thus, resistant (capsule-free) bacteria are more easily controlled by the immune system than are treatment-sensitive, capsulated bacteria and thus are not a reason for treatment failure. (The same argument applies to cells resistant to treatment with phages that require the capsule.) Smith and Huggins (1982) observed reduced virulence of K1 capsule-free bacteria in immunocomptent mice, so these results generalize across both immunocompetent and immune-compromised mice.

Successful, delayed treatment of leukopenic mice. For immunocompetent mice, immediate treatment was superior to delayed treatment for all enzymes. For K1-capsulated bacteria, delayed treatment was still better than the absence of treatment but not as good as immediate treatment. The inferiority of delayed treatment might stem from an artefact of the infection model, specifically that such a high dose of bacteria must be introduced so that the window of opportunity for treatment is short. From the work presented above of lethal inoculum doses in leukopenic vs immunocompetent mice (Figure 6C), leukopenia increases the dynamic range of bacterial densities in which to evaluate treament efficacy—a lower inoculum can be used and the time available to treat increases. To test this latter premise, we attempted delayed treatment of leukopenic mice using K1 enzymes. The bacterial inoculum was approximately 5×10^4 CFU, and enzyme treatment (20 μg K1F or K1H) was given at 8 h. All 16 treated mice survived, whereas only 1 of the 4 controls survived (Figure 7C). This survival rate is significantly higher than that with delayed treatment of immunocompetent mice, but a direct comparison of delayed treatment between leukopenic and immunocompetent mice is not easily interpreted (see Discussion).

3.4. Measuring the Dynamical Impact of Treatment: Resistance Competition Assay

Survival is an endpoint measure of efficacy but gives little insight to the underlying process of bacterial dynamics. Levin [36] developed a metric of the quantitative impact of treatment on bacterial abundances—the resistance competiton assay (RCA). Mice are inoculated with a mix of two bacterial strains, one sensitive and the other resistant to treatment. In untreated controls, the two strains grow in vivo according to their intrinsic abilities, though not necessarily at the same rate. With treatment, the sensitive strain is specifically inhibited. Relative to controls, the resistant strain thus increases its proportion, and the magnitude of this increase depending on how much the sensitive strain is inhibited or killed. An RCA value greater than 0 indicates that the treatment suppresses the sensitive strain, higher values moreso.

Six mice were used for an RCA measure of K1H enzyme, 3 controls and 3 treatments. The resistant strain was the K5-capsulated strain, one that is virulent in vivo (as opposed to a capsule-free RS218). Initial frequency of the resistant strain in the mix was 0.006. At four hours, the frequencies of resistant cells were significantly higher in the enzyme-treated mice than in the controls (Table 2), and the RCA value was 0.30.

Table 2. Frequencies of resistant bacteria in the Resistance Competition Assay.

Control Mice	Treated Mice [1]	Average RCA
0.001, 0.0008, 0.0029	0.0052, 0.0049, 0.0047	0.30

[1] A *t*-test comparing (logged) frequencies of treated mice versus control mice = 3.3 (4 df), $p = 0.015$ (1-tailed). A non-parametric combinations test of the perfect association of high frequencies in treated mice gives $p = 0.05$.

4. Discussion

This study broadly supports a growing realization that phage depolymerases might be useful therapeutics against particular kinds of bacterial infections. The work here also starts to put bounds on the degree of generality in depolymerase utility. Our study used 3 different capsulated *E. coli* strains and tested 3 enzymes against K1 capsules, one enzyme against K5, and one enzyme against K30. The work presented extends tests of therapeutic efficiency to varying conditions and may thus provide insight to efficacy in clinical infections.

K1 depolymerases that performed well in immunocompetent mice also performed well in immune-compromised, leukopenic mice. As expected, leukopenic mice were far less tolerant of bacteria than were immunocompetent mice. Given the artificiality of mouse infections with high doses of bacteria, leukopenic mice may yield a more realistic infection model than do immunocompetent mice, chiefly by increasing the dynamic range of bacterial densities that overwhelm the immune response. In support of this interpretation, leukopenic mice exhibited a modest increase in survival time (when inoculated with low numbers of bacteria), intramuscular treatment with K1E enzyme was improved with leukopenia (Figure 7A), and leukopenic mice were more easily rescued with delayed treatment than were immunocompetent mice. We caution, however, that differences in treatment efficacy between immunocompetent and leukopenic mice are not easily interpreted because of the many differences in the two model infections. Even so, the greater range of useful inoculum sizes afforded by leukopenia increases the latitude of experimental designs.

In contrast to observations with *Pseudomonas* infections of neutropenic mice treated with phages [39], evolution of resistance to K1 capsular depolymerases did not thwart treatment success in leukopenic mice. The absence of resistance evolution likely stems from the avirulence of capsule-free bacteria. The bacterium's only reasonably accessible evolutionary path to avoid the depolymerase is to lose the capsule (it cannot easily generate a new capsule type). This would leave the resistant cell in the same phenotypic state as an otherwise capsulated cell that was stripped of its capsule by enzyme. Thus resistant bacteria and enzyme-treated bacteria may be functionally equivalent in the mouse.

On the negative side for enzyme therapy, delayed treatment of immune competent mice not only reduced efficacy but the reduction was complete for the K30 depolymerase, which lost all efficacy under delayed treatment. However, K30 was also the enzyme showing the weakest effect under simultaneous treatment (e.g., it required the highest dose of all enzymes to rescue mice; Lin et al., 2017 [11]) so there was a strong *a priori* basis for anticipating the large effect of delay with K30 enzyme. In contrast, delayed administration of two K1 enzymes retained efficacy, albeit at a reduced level. Perhaps significantly, with delayed treatment, the two cognate K1-specific phages achieved higher rescue than enzymes. Although no negative consequence of delayed treatment was observed in leukopenic mice (tested only for K1 depolymerases), the outcome with immune competent mice likely reflects a true negative effect of delayed treatment that would be manifested in other contexts.

To gain additional insight to enzyme efficacy, we conducted a resistance competition assay (RCA). RCA values of 2.1 and 1.7 were reported for immediate treatment with streptomycin and a phage requiring the K1-capsule, both highly effective [36], whereas values ranging from 0.2–0.5 were reported for treatments that were only somewhat effective. The observed RCA value for immediate K1H depolymerase treatment determined here is 0.3; over the course of 10 h, this translates into a 95% reduction of bacterial numbers relative to the no treatment control.

The RCA value for K1H, an enzyme that is highly effective in rescuing mice, is thus somewhat lower than expected from previous results. We can suggest several possible reasons for this discrepancy. First, immediate treatment with enzyme may not be quite as effective as with streptomycin or the K1-dependent phage—all treatments yield near 100% survival, but differences could be masked at this upper limit of the dynamic range; indeed, the enzymes do not invariably rescue [11]. Second, treatment spans a time course, and the RCA is measured at a point in time, so 4 h may not be the best time for comparison (e.g., enzyme may be slower to diffuse than antibiotic). Third, assays with phage treatment risk inflating the RCA by allowing bacterial killing to continue during the thigh-processing step. Finally, in comparison to phages, enzymes do not lyse cells and release toxins, so recovery with enzyme treatment may be feasible with a lower dose of killing than is required with phages. Although we cannot yet suggest whether any of these explanations has merit, the comparisons provide interesting avenues for further study.

A greater therapeutic efficacy of phages than of cognate depolymerase enzyme is not surprising, given that phages amplify on the bacteria, whereas enzyme concentration remains static or, more likely, declines due to inactivation and clearance. However, understanding the basis of greater phage efficacy is non-trivial because phage amplification has several components relevant to treatment—direct bacterial killing from phage infection, release of progeny phage particles that contain fresh enzyme molecules, plus the release of additional free enzyme in the form of unassembled tailspikes at the time of cell lysis. A phage inoculum of 10^7 pfu (used here) carries about 2.5–3 $\times$ 10^{-5} μg enzyme, hence ~1.3 $\times$ 10^{-6} less depolymerase delivered in the phage inoculum than delivered in our enzyme treatment. Even with this profound difference and only 4–8 fold greater activity of phage virion-associated enzyme than of the purified form, K1H or K1E phage still gave rise to better survival rates than purified K1 enzymes. This greater efficacy of phages could speak either to a benefit of direct killing or to possibly orders of magnitude greater enzyme produced as the phage population expands. We of course don't know the balance between these two effects. Yet, whatever argument is put forth for the superiority of K1 phages (except K1F, which does not grow on the K1-capsulated host used here) under delayed treatment, K5 phage or K30 phage did not improve survival rate over their pure enzymes at the doses tested.

Reduced efficacy with delayed treatment of any form is not surprising, especially with infections that are rapidly lethal. However, although it is tempting to interpret the effect of a treatment delay as merely giving the bacteria a "head start" to reach a lethal threshold, previous work using K1 infections of mice showed that the bacteria changed state with time to become recalcitrant to treatment [36]. This change in susceptibility may be due to the bacterial environment suddenly changing from laboratory growth media to mouse tissue, as altered bacterial physiology and community lifestyle could affect phage infection [52] and probably phage product treatment. Thus, the effect of therapeutic treatment delay could be one of the bacteria becoming more difficult to treat than of them being more numerous. The use of leukopenic mice as an experimental infection model may help avoid that complication and yield therapeutic regimes more redily translatable to clinical use.

A surprising result was the effect of route of administration on efficacy, at least for enzyme K1E. It is of course easy to argue that each enzyme has its own biochemical properties that may affect pharmacokinetics. For example, unlike the other K1 enzymes used here, K1E purified under our protocol tends to form 18-mers instead of trimers [11]. The larger enzyme complex might affect in vivo distribution and result in poor IM treatment efficiency, but this can only be argued on a *post facto* basis. Whatever the cause, the result indicates that treatment "failure" by one route does not imply failure by other routes.

By applying modifications of animal models, we have obtained more detailed understanding of how phage biology and enzyme biology complicate the pharmacokinetic and pharmacodynamic properties of phage and enzyme in infection treatment. The complex biological and pharmacological properties of phages is one major reason holding back whole phages from drug development and approval [53]. Phage derived enzymes are more similar to conventional antibiotics and thus more suitable than whole phages for the current drug approval processes even if not necessarily more suitable for treatment efficacy. In contrast to phages, depolymerases do not lyse bacteria and thus do not release endotoxin, so there may be circumstances in which enzymes are superior to phages [54,55]. A better understanding of enzyme structure and enzyme kinetics could greatly advance the development and approval of phage-derived enzymes as a novel class of antibiotics.

Author Contributions: Experiment design and Methodology, H.L., J.J.B. and I.J.M.; Investigation, H.L., M.L.P. and J.J.B.; Data Validation and Analysis, H.L., and J.J.B.; Writing, H.L., J.J.B. and I.J.M.

Funding: This research was funded by the NIH grant R21AI 121685-02 to J.J.B. and I.J.M.

Conflicts of Interest: The authors declare no conflict of interest. The funders had no role in the design, execution, interpretation, or writing of the study.

References

1. Lewis, K. Platforms for antibiotic discovery. *Nat. Rev. Drug Discov.* **2013**, *12*, 371–387. [CrossRef] [PubMed]
2. Drulis-Kawa, Z.; Majkowska-Skrobek, G.; Maciejewska, B. Bacteriophages and phage-derived proteins—Application approaches. *Curr. Med. Chem.* **2015**, *22*, 1757–1773. [CrossRef] [PubMed]
3. Cooper, C.J.; Koonjan, S.; Nilsson, A.S. Enhancing whole phage therapy and their derived antimicrobial enzymes through complex formulation. *Pharmaceuticals (Basel)* **2018**, *11*, 34. [CrossRef] [PubMed]

4. Pires, D.P.; Cleto, S.; Sillankorva, S.; Azeredo, J.; Lu, T.K. Genetically Engineered Phages: A Review of Advances over the Last Decade. *Microbiol. Mol. Biol. Rev.* **2016**, *80*, 523–543. [CrossRef] [PubMed]
5. Roach, D.R.; Donovan, D.M. Antimicrobial bacteriophage-derived proteins and therapeutic applications. *Bacteriophage* **2015**, *5*, e1062590. [CrossRef] [PubMed]
6. Abedon, S.T.; García, P.; Mullany, P.; Aminov, R. Editorial: Phage Therapy: Past, Present and Future. *Front. Microbiol.* **2017**, *8*. [CrossRef] [PubMed]
7. Vandenheuvel, D.; Lavigne, R.; Brussow, H. Bacteriophage Therapy: Advances in Formulation Strategies and Human Clinical Trials. *Annu. Rev. Virol.* **2015**, *2*, 599–618. [CrossRef] [PubMed]
8. Rhoads, D.D.; Wolcott, R.D.; Kuskowski, M.A.; Wolcott, B.M.; Ward, L.S.; Sulakvelidze, A. Bacteriophage therapy of venous leg ulcers in humans: Results of a phase I safety trial. *J. Wound Care* **2009**, *18*, 237–238, 240–243. [CrossRef] [PubMed]
9. Sarker, S.A.; Sultana, S.; Reuteler, G.; Moine, D.; Descombes, P.; Charton, F.; Bourdin, G.; McCallin, S.; Ngom-Bru, C.; Neville, T.; et al. Oral Phage Therapy of Acute Bacterial Diarrhea with Two Coliphage Preparations: A Randomized Trial in Children from Bangladesh. *EBioMedicine* **2016**, *4*, 124–137. [CrossRef] [PubMed]
10. Jault, P.; Leclerc, T.; Jennes, S.; Pirnay, J.P.; Que, Y.-A.; Resch, G.; Rousseau, A.F.; Ravat, F.; Carsin, H.; Le Floch, R.; et al. Efficacy and tolerability of a cocktail of bacteriophages to treat burn wounds infected by *Pseudomonas aeruginosa* (PhagoBurn): A randomised, controlled, double-blind phase 1/2 trial. *Lancet Infect. Dis.* **2018**. [CrossRef]
11. Lin, H.; Paff, M.L.; Molineux, I.J.; Bull, J.J. Therapeutic Application of Phage Capsule Depolymerases against K1, K5, and K30 Capsulated *E. coli* in Mice. *Front. Microbiol.* **2017**, *8*, 2257. [CrossRef] [PubMed]
12. Briers, Y.; Lavigne, R. Breaking barriers: Expansion of the use of endolysins as novel antibacterials against Gram-negative bacteria. *Future Microbiol.* **2015**, *10*, 377–390. [CrossRef] [PubMed]
13. Oliveira, H.; Sao-Jose, C.; Azeredo, J. Phage-Derived Peptidoglycan Degrading Enzymes: Challenges and Future Prospects for in vivo Therapy. *Viruses* **2018**, *10*, 292. [CrossRef] [PubMed]
14. Song, J.; Xia, F.; Jiang, H.; Li, X.; Hu, L.; Gong, P.; Lei, L.; Feng, X.; Sun, C.; Gu, J.; et al. Identification and characterization of HolGH15: The holin of Staphylococcus aureus bacteriophage GH15. *J. Gen. Virol.* **2016**, *97*, 1272–1281. [CrossRef] [PubMed]
15. Latka, A.; Maciejewska, B.; Majkowska-Skrobek, G.; Briers, Y.; Drulis-Kawa, Z. Bacteriophage-encoded virion-associated enzymes to overcome the carbohydrate barriers during the infection process. *Appl. Microbiol. Biotechnol.* **2017**, *101*, 3103–3119. [CrossRef] [PubMed]
16. Fischetti, V.A. Development of Phage Lysins as Novel Therapeutics: A Historical Perspective. *Viruses* **2018**, *10*, 310. [CrossRef] [PubMed]
17. Roberts, I.S. The biochemistry and genetics of capsular polysaccharide production in bacteria. *Annu. Rev. Microbiol.* **1996**, *50*, 285–315. [CrossRef] [PubMed]
18. Azeredo, J.; Sutherland, I.W. The use of phages for the removal of infectious biofilms. *Curr. Pharm. Biotechnol.* **2008**, *9*, 261–266. [CrossRef] [PubMed]
19. Fleming, D.; Chahin, L.; Rumbaugh, K. Glycoside Hydrolases Degrade Polymicrobial Bacterial Biofilms in Wounds. *Antimicrob. Agents Chemother.* **2017**, *61*. [CrossRef] [PubMed]
20. Fleming, D.; Rumbaugh, K.P. Approaches to Dispersing Medical Biofilms. *Microorganisms* **2017**, *5*, 15. [CrossRef] [PubMed]
21. Mushtaq, N.; Redpath, M.B.; Luzio, J.P.; Taylor, P.W. Prevention and cure of systemic *Escherichia coli* K1 infection by modification of the bacterial phenotype. *Antimicrob. Agents Chemother.* **2004**, *48*, 1503–1508. [CrossRef] [PubMed]
22. Mushtaq, N.; Redpath, M.B.; Luzio, J.P.; Taylor, P.W. Treatment of experimental *Escherichia coli* infection with recombinant bacteriophage-derived capsule depolymerase. *J. Antimicrob. Chemother.* **2005**, *56*, 160–165. [CrossRef] [PubMed]
23. Scorpio, A.; Tobery, S.A.; Ribot, W.J.; Friedlander, A.M. Treatment of experimental anthrax with recombinant capsule depolymerase. *Antimicrob. Agents Chemother.* **2008**, *52*, 1014–1020. [CrossRef] [PubMed]
24. Negus, D.; Vipond, J.; Hatch, G.J.; Rayner, E.L.; Taylor, P.W. Parenteral Administration of Capsule Depolymerase EnvD Prevents Lethal Inhalation Anthrax Infection. *Antimicrob. Agents Chemother.* **2015**, *59*, 7687–7692. [CrossRef] [PubMed]

25. Lin, T.L.; Hsieh, P.F.; Huang, Y.T.; Lee, W.C.; Tsai, Y.T.; Su, P.A.; Pan, Y.J.; Hsu, C.R.; Wu, M.C.; Wang, J.T. Isolation of a bacteriophage and its depolymerase specific for K1 capsule of Klebsiella pneumoniae: Implication in typing and treatment. *J. Infect. Dis.* **2014**, *210*, 1734–1744. [CrossRef] [PubMed]
26. Pan, Y.J.; Lin, T.L.; Lin, Y.T.; Su, P.A.; Chen, C.T.; Hsieh, P.F.; Hsu, C.R.; Chen, C.C.; Hsieh, Y.C.; Wang, J.T. Identification of capsular types in carbapenem-resistant *Klebsiella pneumoniae* strains by WZC sequencing and implications for capsule depolymerase treatment. *Antimicrob. Agents Chemother.* **2015**, *59*, 1038–1047. [CrossRef] [PubMed]
27. Majkowska-Skrobek, G.; Latka, A.; Berisio, R.; Maciejewska, B.; Squeglia, F.; Romano, M.; Lavigne, R.; Struve, C.; Drulis-Kawa, Z. Capsule-targeting depolymerase, derived from Klebsiella KP36 phage, as a tool for the development of anti-virulent strategy. *Viruses* **2016**, *8*, 324. [CrossRef] [PubMed]
28. Achtman, M.; Mercer, A.; Kusecek, B.; Pohl, A.; Heuzenroeder, M.; Aaronson, W.; Sutton, A.; Silver, R.P. Six widespread bacterial clones among *Escherichia coli* K1 isolates. *Infect. Immun.* **1983**, *39*, 315–335. [PubMed]
29. Orskov, I.; Orskov, F.; Jann, B.; Jann, K. Serology, chemistry, and genetics of O and K antigens of *Escherichia coli*. *Bacteriol. Rev.* **1977**, *41*, 667–710. [PubMed]
30. Vimr, E.R.; Troy, F.A. Regulation of sialic acid metabolism in *Escherichia coli*: Role of N-acylneuraminate pyruvate-lyase. *J. Bacteriol.* **1985**, *164*, 854–860. [PubMed]
31. Smith, H.W.; Huggins, M.B. The association of the O18, K1 and H7 antigens and the ColV plasmid of a strain of *E. coli* with virulence and immunogenicity. *J. Gen. Microbiol.* **1980**, *121*, 387–400. [PubMed]
32. Bull, J.J.; Vimr, E.R.; Molineux, I.J. A tale of tails: Sialidase is key to success in a model of phage therapy against K1-capsulated *Escherichia coli*. *Virology* **2010**, *398*, 79–86. [CrossRef] [PubMed]
33. Scholl, D.; Kieleczawa, J.; Kemp, P.; Rush, J.; Richardson, C.C.; Merril, C.; Adhya, S.; Molineux, I.J. Genomic analysis of bacteriophages SP6 and K1-5, an estranged subgroup of the T7 supergroup. *J. Mol. Biol.* **2004**, *335*, 1151–1171. [CrossRef] [PubMed]
34. Whitfield, C.; Lam, M. Characterisation of coliphage K30, a bacteriophage specific for *Escherichia coli* capsular serotype K30. *FEMS Microbiol. Lett.* **1986**, *37*, 351–355. [CrossRef]
35. Leiman, P.G.; Battisti, A.J.; Bowman, V.D.; Stummeyer, K.; Muhlenhoff, M.; Gerardy-Schahn, R.; Scholl, D.; Molineux, I.J. The structures of bacteriophages K1E and K1-5 explain processive degradation of polysaccharide capsules and evolution of new host specificities. *J. Mol. Biol.* **2007**, *371*, 836–849. [CrossRef] [PubMed]
36. Bull, J.J.; Levin, B.R.; DeRouin, T.; Walker, N.; Bloch, C.A. Dynamics of success and failure in phage and antibiotic therapy in experimental infections. *BMC Microbiol.* **2002**, *2*, 35. [CrossRef]
37. Bull, J.J.; Otto, G.; Molineux, I.J. In vivo growth rates are poorly correlated with phage therapy success in a mouse infection model. *Antimicrob. Agents Chemother.* **2012**, *56*, 949–954. [CrossRef] [PubMed]
38. Rich, J.T.; Neely, J.G.; Paniello, R.C.; Voelker, C.C.; Nussenbaum, B.; Wang, E.W. A practical guide to understanding Kaplan-Meier curves. *Otolaryngol. Head Neck Surg.* **2010**, *143*, 331–336. [CrossRef] [PubMed]
39. Roach, D.R.; Leung, C.Y.; Henry, M.; Morello, E.; Singh, D.; Di Santo, J.P.; Weitz, J.S.; Debarbieux, L. Synergy between the host immune system and bacteriophage is essential for successful phage therapy against an acute respiratory pathogen. *Cell Host Microbe* **2017**, *22*, 38–47. [CrossRef] [PubMed]
40. Pelkonen, S.; Häyrinen, J.; Finne, J. Polyacrylamide gel electrophoresis of the capsular polysaccharides of *Escherichia coli* K1 and other bacteria. *J. Bacteriol.* **1988**, *170*, 2646–2653. [CrossRef] [PubMed]
41. Muhlenhoff, M.; Stummeyer, K.; Grove, M.; Sauerborn, M.; Gerardy-Schahn, R. Proteolytic processing and oligomerization of bacteriophage-derived endosialidases. *J. Biol. Chem.* **2003**, *278*, 12634–12644. [CrossRef] [PubMed]
42. Gerardy-Schahn, R.; Bethe, A.; Brennecke, T.; Mühlenhoff, M.; Eckhardt, M.; Ziesing, S.; Lottspeich, F.; Frosch, M. Molecular cloning and functional expression of bacteriophage PK1E-encoded endoneuraminidase Endo NE. *Mol. Microbiol.* **1995**, *16*, 441–450. [CrossRef] [PubMed]
43. Luke, D.R.; Brunner, L.J.; Vadiei, K. Bioavailability assessment of cyclosporine in the rat. Influence of route of administration. *Drug Metab. Dispos.* **1990**, *18*, 158–162. [PubMed]
44. Yamamura, Y.; Santa, T.; Kotaki, H.; Uchino, K.; Sawada, Y.; Iga, T. Administration-route dependency of absorption of glycyrrhizin in rats: Intraperitoneal administration dramatically enhanced bioavailability. *Biol. Pharm. Bull.* **1995**, *18*, 337–341. [CrossRef] [PubMed]
45. Kijanka, G.; Prokopowicz, M.; Schellekens, H.; Brinks, V. Influence of aggregation and route of injection on the biodistribution of mouse serum albumin. *PLoS ONE* **2014**, *9*, e85281. [CrossRef] [PubMed]

46. Torres-Barcelo, C. Phage Therapy Faces Evolutionary Challenges. *Viruses* **2018**, *10*, 323. [CrossRef] [PubMed]
47. Dennehy, J.J. What Can Phages Tell Us about Host-Pathogen Coevolution? *Int. J. Evol. Biol.* **2012**, *2012*, 396165. [CrossRef] [PubMed]
48. Nilsson, A.S. Phage therapy—Constraints and possibilities. *Ups J. Med. Sci.* **2014**, *119*, 192–198. [CrossRef] [PubMed]
49. Bull, J.J.; Vegge, C.S.; Schmerer, M.; Chaudhry, W.N.; Levin, B.R. Phenotypic resistance and the dynamics of bacterial escape from phage control. *PLoS ONE* **2014**, *9*, e94690. [CrossRef] [PubMed]
50. Levin, B.R.; Bull, J.J. Population and evolutionary dynamics of phage therapy. *Nat. Rev. Microbiol.* **2004**, *2*, 166–173. [CrossRef] [PubMed]
51. Smith, H.W.; Huggins, M.B. Successful treatment of experimental *Escherichia coli* infections in mice using phage: Its general superiority over antibiotics. *J. Gen. Microbiol.* **1982**, *128*, 307–318. [CrossRef] [PubMed]
52. Lourenco, M.; De Sordi, L.; Debarbieux, L. The Diversity of Bacterial Lifestyles Hampers Bacteriophage Tenacity. *Viruses* **2018**, *10*, 327. [CrossRef] [PubMed]
53. Cooper, C.J.; Khan Mirzaei, M.; Nilsson, A.S. Adapting Drug Approval Pathways for Bacteriophage-Based Therapeutics. *Front. Microbiol.* **2016**, *7*, 1209. [CrossRef] [PubMed]
54. Hagens, S.; Habel, A.; von Ahsen, U.; von Gabain, A.; Blasi, U. Therapy of experimental pseudomonas infections with a nonreplicating genetically modified phage. *Antimicrob. Agents Chemother.* **2004**, *48*, 3817–3822. [CrossRef] [PubMed]
55. Matsuda, T.; Freeman, T.A.; Hilbert, D.W.; Duff, M.; Fuortes, M.; Stapleton, P.P.; Daly, J.M. Lysis-deficient bacteriophage therapy decreases endotoxin and inflammatory mediator release and improves survival in a murine peritonitis model. *Surgery* **2005**, *137*, 639–646. [CrossRef] [PubMed]

MDPI

Article

Characterizing Phage Genomes for Therapeutic Applications

Casandra W. Philipson [1,2], Logan J. Voegtly [2,3], Matthew R. Lueder [2,3], Kyle A. Long [2,3], Gregory K. Rice [2,3], Kenneth G. Frey [2], Biswajit Biswas [2], Regina Z. Cer [2,3], Theron Hamilton [2] and Kimberly A. Bishop-Lilly [2,*]

1 Defense Threat Reduction Agency, Fort Belvoir, VA 22060, USA; casandra.w.philipson.civ@mail.mil
2 Biological Defense Research Directorate, Naval Medical Research Center, Fort Detrick, MD 21702, USA; logan.j.voegtly.ctr@mail.mil (L.J.V.); matthew.r.lueder.ctr@mail.mil (M.R.L.); kyle.a.long8.ctr@mail.mil (K.A.L.); gregory.k.rice.ctr@mail.mil (G.K.R.); kenneth.g.frey4.civ@mail.mil (K.G.F.); biswajit.biswas.civ@mail.mil (B.B.); regina.z.cer.ctr@mail.mil (R.Z.C.); theron.c.hamilton.mil@mail.mil (T.H.)
3 Leidos, Reston, VA 20190, USA
* Correspondence: kimberly.a.bishop-lilly.civ@mail.mil; Tel.: +1-301-619-1490

Received: 13 March 2018; Accepted: 9 April 2018; Published: 10 April 2018

Abstract: Multi-drug resistance is increasing at alarming rates. The efficacy of phage therapy, treating bacterial infections with bacteriophages alone or in combination with traditional antibiotics, has been demonstrated in emergency cases in the United States and in other countries, however remains to be approved for wide-spread use in the US. One limiting factor is a lack of guidelines for assessing the genomic safety of phage candidates. We present the phage characterization workflow used by our team to generate data for submitting phages to the Federal Drug Administration (FDA) for authorized use. Essential analysis checkpoints and warnings are detailed for obtaining high-quality genomes, excluding undesirable candidates, rigorously assessing a phage genome for safety and evaluating sequencing contamination. This workflow has been developed in accordance with community standards for high-throughput sequencing of viral genomes as well as principles for ideal phages used for therapy. The feasibility and utility of the pipeline is demonstrated on two new phage genomes that meet all safety criteria. We propose these guidelines as a minimum standard for phages being submitted to the FDA for review as investigational new drug candidates.

Keywords: phage therapy; viral genomes; best practices; IND; high-throughput sequencing

1. Introduction

Phage therapy, the use of bacteriophages to treat bacterial infections, especially in combination with traditional antibiotics, is recognized as a promising strategy to combat multi-drug resistant (MDR) infections [1]. Although phages are generally considered safe [2–4], guidelines for genetic safety assessments of phages prior to clinical use are non-existent. Currently, there is no general approval process for phage therapy in the United States. The Food and Drug Administration (FDA) can grant emergency investigational new drug (eIND) status for phage cocktails in compassionate care cases, however this process requires a request from a medical doctor and protocols remain case-by-case. The lack of guidelines presents one limiting factor for advancing phages as therapeutic agents along the regulatory pipeline. To address this, we present a characterization workflow that implements best-in-field tools to systematically evaluate genetic safety of phage candidates for therapeutic applications. The protocol presented is the minimum standard used by our team to generate IND-enabling data and submit phage therapeutics for FDA approval.

Phage biology is an enormous field with topics ranging from viromes in the sea [5] and human gut [6], to genetic engineering [7], to therapeutic utility and countermeasure development [3]. The number and diversity of discovered bacteriophages is increasing at a rapid rate, especially with respect to viral discovery efforts using high-throughput sequencing. The number of complete phage genomes deposited in the NCBI Genome database has more than doubled in the last three years. As a result of diversity in investigative studies, rigor for sequencing, assembling, finishing and manually polishing phage genomes is reported at varying levels in literature depending on intended use [8–12]. At the assembly stage, algorithmic success often depends on empirically derived heuristics which help overcome complicated repeat patterns in real genomes, random and systematic error in sequencing reads and limitations in computational power. In order to mitigate potential bias introduced by a single assembly algorithm, it is typically necessary to employ a consensus approach utilizing multiple assemblers. This is also true for gene-calling and annotation, which can be performed using fully-automated single-platform tools such as Rapid Annotation using Subsystem Technology RAST [13]. While these platforms democratize genomics and offer efficiencies for first-glance solutions, relying on these tools alone introduces high risk for inaccurate safety assessment. For instance, inaccuracies can arise due to potential misinterpretation of unreliable data that propagate throughout public databases. Despite existing methods to identify start sites and directionality, resolve ends, and predict lifestyle for phages, the criteria employed in these methods are loosely controlled. In this study, we delineate key analytical checkpoints where manual intervention is necessary for achieving standards set forth for genomes used in therapeutic applications. The checkpoints fall within two categories: obtaining a high-quality genome and robust assessment of genetic composition.

Considering genome quality there are standards defined for viral sequences with respect to the level of completeness for desired downstream applications [14]. The recommended category for viruses used in animal models for vaccine development, and by extension phage therapy, is "Finished". Finished status is defined as a single consensus sequence representing 100% of the genome with all open reading frames (ORFs) identified and population diversity, or lack of population diversity as an indicator of purity, of the sequence verified via deep coverage. Phage isolation, sample preparation and rationale for sequencing technology will not be discussed in detail here. However, it is important to note that any contaminants introduced by laboratory protocols, such as bacterial host remnants from phage expansion and those inherent to nucleic acid sample preparation and high-throughput sequencers, can negatively impact the ease and accuracy of obtaining a high-quality genome. As such, contaminant identification is presented as a fundamental step in the safety assessment framework.

In addition to genome quality, properties of "safe" phage therapy candidates have been discussed [15]. Primary determinants for therapeutic selection include: antibacterial virulence, lifestyle, and the absence of deleterious genes. A phage's host-range and antibacterial virulence (efficacy) are evaluated using experimental techniques during initial selection; however, the presence or absence of deleterious genes is analyzed by computational methods. Likewise, phage lifestyle is analyzed first by experimental methods in the laboratory, then additionally evaluated by computational methods. Phage Classification Tool Set (PHACTS) [16] is an example of a tool that can be used for computational analysis of lifestyle. Lifestyle of a phage can be classified by two different states, lytic or lysogenic, with the former being the necessary state for phage therapy candidates. In a lysogenic state, a phage integrates its DNA into the DNA of the host to become a prophage, rendering itself dormant and suppressing the typical anti-bacterial properties exhibited during a lytic state. Additionally, a prophage could provide bacteria with a mutualistic relationship, increasing the fitness of the host and thereby decreasing the effectiveness of phage therapy. Mechanisms underlying lysogeny are largely mediated by the presence of a functional phage-encoded enzyme, integrase. Although temperate phages have been found viable to deliver some non-lytic antimicrobial treatments [17–19], the risks associated with lysogenic phages or prophages are reason enough to avoid using them in phage therapies. Hence, the presence of an integrase gene is undesirable and nullifies a phage's candidacy for therapy unless methods are developed to determine if the integrase is indeed non-functional. Similarly, as some specific bacterial infections can exhibit enhanced virulence mediated by phage-encoded toxins (e.g., Shiga toxin, diphtheria toxin and cholera toxin), toxins must be screened for as well. Detecting any of the following by genetic screening would immediately disqualify a phage for therapeutic use: genes that encode virulence factors, antimicrobial resistance, toxins, or transducable elements.

To date, phage therapy approved under eIND bypasses conventional in vivo safety studies. This necessitates a rigorous safety screening platform to safeguard patients. It remains unclear whether commercialization of phage therapy will facilitate wide-spread use, personalized treatment, or rely on broad-host-range cocktails. For any case, we present a characterization pipeline as a guideline for minimum assessment standards of phages administered to patients (Figure 1). The pipeline, which has been developed and used by our team to prepare phage genomic data in support of IND submission, includes methods to: (i) obtain a high quality whole genome sequence; (ii) identify open reading frames (ORFs); (iii) annotate genes with a consensus function identified across tools; (iv) search for deleterious genetic markers; (v) verify that the sequence is representative of the population; and (vi) perform contaminant analysis. We demonstrate the pipeline's utility by delivering two finished phage genomes that meet safety criteria for phage therapy.

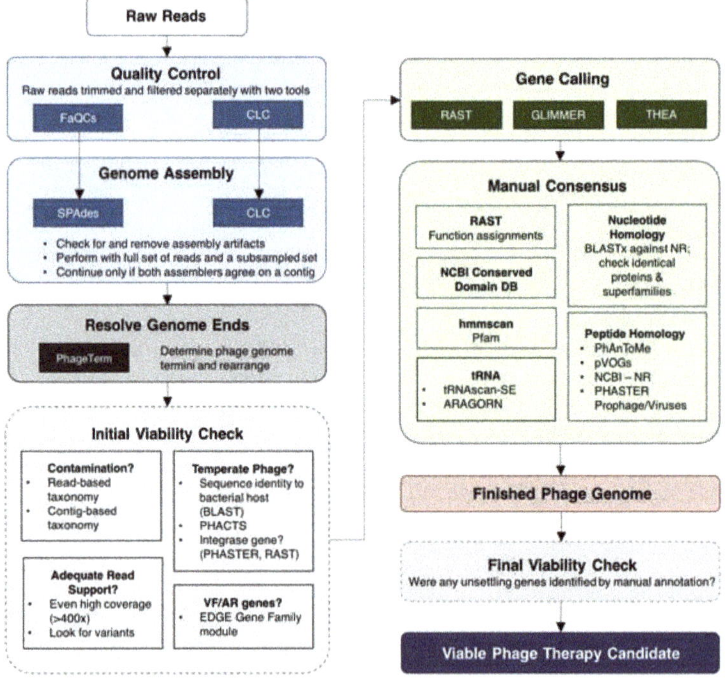

Figure 1. Phage characterization workflow. This pipeline is a simplified representation of tools and methods used to obtain high-quality phage genomes that are deemed viable phage therapy candidates. The pipeline begins with raw reads sequenced on an Illumina machine. To reduce potential bias introduced by bioinformatics tools, quality control and genome assembly are performed using two pipelines in parallel. The final genome sequence is obtained after resolving genome ends. Key viability checkpoints are outlined with dashed borders. In the initial viability check, phages are assessed for problematic genes (antimicrobial resistance (AMR), virulence factors (VF), toxins) and lifestyle. If a candidate passes the initial viability check, a combinatorial approach is applied to identify open reading frames followed by rigorous manual annotation. A final check is performed after completing annotation. Phage candidates that pass the final check point are considered safe for potential use in humans.

2. Materials and Methods

2.1. Phage Isolation and Genomic DNA Extraction

The phages sequenced in this study were isolated from environmental samples by routine isolation techniques [20]. Then they were triple plaque-purified on their respective hosts and inoculated at a multiplicity of infection (MOI) of 0.1 into 100 mL cultures of their respective host bacteria for amplification at 37 °C in preparation for sequencing. Upon lysis of the bacterial cells, the lysate for each phage was filtered through a 0.22 µm filter, DNAse-treated in presence of $MgCl_2$ to degrade DNA that is not protected by viral capsid (e.g., host DNA), Proteinase K- and sodium dodecyl sulfate-treated to inactivate DNAse and disrupt capsid, followed by Phenol-chloroform-isoamyl alcohol extraction, debris removal and polyethylene glycol precipitation of naked DNA in the presence of salt (NaCl) [21]. The nucleic acid pellet was washed with 80% alcohol and dissolved in deionized distilled water before RNAse treatment. The RNAse-treated samples were extracted one more time with Phenol-chloroform isoamyl alcohol and DNA precipitated in presence of absolute alcohol. Finally, the DNA pellet was washed with 70% alcohol before being suspended in deionized distilled water. The resulting phage genomic DNA was subjected to rigorous internal quality control testing, including agarose gel electrophoresis to ensure high molecular weight (indicative of relatively non-sheared, intact genomic DNA), restriction enzyme digests to assess potential genome modifications that prevent manipulability by sequencing library protocols, Qubit measurements (Thermo Fisher Scientific; Waltham, MA, USA) for concentration and Nanodrop measurements (Thermo Fisher Scientific) for purity (optical density 260/230 ratio).

The above protocol was used strictly for preparation of phages for sequencing efforts whereas for clinical use our phage preparations have been conducted via CsCl gradient ultracentrifugation, as in Schooley et al [3], to completely remove any host material such as naked bacterial DNA.

2.2. Library Preparation and Sequencing

Sequencing libraries were constructed using the Accel-NGS® 2S Plus library kit (Swift Biosciences, Ann Arbor, MI, USA) with a slight modification. Briefly, 250 ng of genomic DNA was fragmented using the Covaris M220 (Covaris, Inc, Woburn, MA, USA). Instrument parameters were the factory settings for Illumina TruSeq (350 bp). The sheared gDNA was subjected to a double-sided size selection using AMPure XP beads (Beckman-Coulter, Brea, CA, USA). Selection ratios were 0.75×/0.6×. Size-selected DNA was then used as input for 2S Plus. Library fragments were not amplified using PCR. All libraries were quality checked using the Agilent BioAnalyzer (Agilent Technologies, Santa Clara, CA, USA) and quantitated using the NEBNext® Library Quant kit (New England Biolabs, Inc., Ipswitch, MA, USA). Prior to sequencing, individual libraries were diluted to 2 nM concentration and pooled. Sequencing was performed on a MiSeq (Illumina, Inc., San Diego, CA, USA) using 2 × 300 v3 chemistry. Raw sequencing reads were deposited in NCBI's Sequence Read Archive (SRA) under SRA accession numbers (SRR6764339, SRR6764268).

2.3. Genome Assembly

For each sample, two pipelines were run in parallel for quality control (QC) and assembly (Figure 1). Results from both pipelines were compared to identify a single consensus sequence with high confidence. Specific parameters employed are listed in Supplemental Table S1. If there was not 100% nucleotide identity between the results of the largest contig in both assemblers, the reads were mapped to both contigs in order to manually verify the fidelity of the assembly. Both pipelines implemented the following tasks in order: (1) raw data were processed for QC; (2) all reads that pass QC in Step 1 were assembled de novo; (3) 50,000 to 100,000 quality-controlled reads from Step 1 were subsampled then assembled de novo. The first pipeline combined publicly available tools: FaQCs for QC [22], seqtk for subsampling [23] and SPAdes (version 3.5.0) for assembly [24]. The second pipeline includes NGS Core Tools from CLC Genomics Workbench (version 10, Qiagen,

Redwood City, CA, USA). Both FaQCs and CLC's quality trimmer were set to trim reads to Q30 and remove reads less than 50 bp in length. CLC's quality trimmer was set to remove any reads containing more than two ambiguous bases while FaQCs was set to remove reads with more than two consecutive ambiguous bases. In addition, the way quality trimming is implemented differs between the two pipelines. FaQCs utilizes BWA-style trimming [14] while CLC's quality trimmer utilizes a modified-Mott algorithm. These differences make CLC's quality trimming more stringent. Between the two pipelines, four de novo assemblies were performed for each sample: SPAdes-all reads, SPAdes-subsampled, CLC-all reads, and CLC-subsampled. Subsampling was performed to achieve 80–100× coverage of the genome. If a genome size was unknown, 50,000 reads were subsampled; a maximum of 100,000 reads is recommended for the initial assembly of unknown phage genomes [8]. When sampling paired-end reads, half of the reads should be obtained from each file. SPAdes assembly was performed using default settings whereas CLC assembly parameters were default except for word size, which was set to 64. Assembly artifacts were identified and removed (i.e., 127 bp artificial overlaps at the ends of SPAdes assembled contigs). If the largest contig generated by two different assemblers presented 100% nucleotide identity, regardless of start site or orientation, the resulting sequence proceeded to downstream analysis. Otherwise, all reads were mapped back to the contig to determine genome ends versus sequencing artifacts manually by visualizing read support using CLC. The resultant assembly size is also compared to the range of genome sizes for that particular virus family to make sure that it is relatively consistent with the expected value (virus family being determined either by morphological characteristics, closest sequenced relatives, or both).

2.4. Resolving Genome Ends

The genomic termini and phage packaging strategy were determined using PhageTerm [25]. Briefly, FaQCs processed reads were aligned to the putative phage genome and read build-ups, indicative of over-represented fragment ends, were identified. PhageTerm uses the starting position coverage (SPC) and the coverage in each orientation (COV) to calculate $\tau = \frac{SPC}{COV}$ in each direction for each nucleotide. This metric is used to determine the location of genomic termini as well as classify it as one of the following: 5′ *cos*, 3′ *cos*, direct terminal repeat (DTR) (short), DTR (long), headful (with or without *pac* site detected), Mu-like, or unknown. It automatically rearranges the genome sequence accordingly. All genomes were checked for non-terminal nucleotides with significant *p*-values and aberrant coverage patterns. When a phage genome's termini could not be determined, or in the case where a phage genome had no consistent biological termini (circularly permuted phages), the start site was selected based on the presence and orientation of terminase genes as previously described [26] in order to allow easy comparative analysis of similar phages. If no large terminase gene was found, the start site was selected based on alignment with the closest reference in GenBank.

2.5. High Quality Genome Checkpoint

After ends were resolved and orientation was set, the final genome sequence was validated for quality. To validate the genomic sequence, quality-controlled reads were mapped back to the sequence to ensure it was well-supported by sequencing data. Assembly validation was performed in CLC using clc_mapper with default settings but can be performed at the command line interface using BWA. Three metrics were considered: percentage of reads mapping to the genome, average whole genome coverage and lowest coverage. Assemblies were considered validated if >90% of reads mapped back to the phage genome. Additionally, in order for genomes to proceed from this checkpoint, average whole genome coverage and lowest coverage were at least 100× for complete genomes and ~400× for finished genomes [14]. "Complete" and "finished" genome definitions used here are as defined by Ladner et al.; specifically, complete viral genomes have the whole sequence fully resolved, including ends, whereas finished viral genomes have a complete sequence plus a minimum of 400–1000× coverage depth to resolve population-level variations [14]. Genomes with <400× coverage can proceed through the pipeline, however appropriate detection of single nucleotide polymorphisms (SNPs) in the population requires the deeper coverage metric. Additional sequencing can be performed to obtain more reads for population-level validation of complete phage genomes.

2.6. Phage Lifestyle Checkpoint

Temperate phages do not proceed as viable therapeutic candidates in our pipeline. Therefore, an important checkpoint involves identifying phages that have the potential for temperate behavior. Complete genomes were submitted to RAST [13] and PHAge Search Tool Enhanced Release (PHASTER) [27] for baseline gene calling and functional annotation. Output from these tools was parsed for the presence of "integrase." Additionally, PHACTS was utilized to determine if a phage's overall proteome resembled that of a temperate phage. Briefly, PHACTS is a tool that utilizes a Random Forest classifier to predict phage lifestyle, bacterial host and phage family by comparing proteins from the query to those of phages within the PHACTS database [16]. PHACTS analysis yields a statistically-based score that predicts the likelihood that a phage is prone to a temperate versus lytic lifestyle. Any indicators of temperate behavior result in rejection of the phage candidate.

2.7. Specialty Genes Checkpoint

An initial viability check is performed to identify the presence of toxins, virulence factors, or antimicrobial resistance genes. This step is performed on reads and contigs using the EDGE Bioinformatics Gene Family module (Appendix A). Read-based functional profiling was performed on FaQCs processed reads. ShortBRED (v0.9.4M) searches reads for similarity to antibiotic resistance genes found in three databases: Antibiotic Resistance Genes Database (ARDB) [28], Resfams antibiotic resistance functions [29] and Virulence Factors of Pathogenic Bacteria (VFDB), downloaded December 2015. [30]. Contigs from the SPAdes-all reads assembly were also searched for problematic genes. ORFs were predicted in all contigs >700 base pairs in length using Prodigal [31]. ShortBRED was used to search predicted ORFs against VFDB. Additionally, the Resistance Gene Identifier (RGI), v3.1.1 with database from July 2016, was used to search predicted ORFs against the Comprehensive Antibiotic Resistance Database (CARD) [32]. Databases and sources are listed in Table 1. Default parameters were used for all tools. Detecting any positive hits in this checkpoint renders a phage unfit for therapeutic use. It is important to note that Prodigal is a gene calling algorithm designed to predict ORFs in prokaryotes, thus it is expected to perform well for those organisms. Prodigal is used in this step because it is the gene caller embedded in the EDGE Bioinformatics Gene Family module. We have found this method appropriate for fast candidate viability checks, however none of the gene calls from Prodigal are used for annotation since GLIMMER3 outperforms Prodigal for predicting ORFs in phages. Users are not confined to relying on Prodigal as the method for identifying ORFs or for specialty gene analysis.

Table 1. Databases curated with virulence factor and antimicrobial resistance genes.

Database	# Of Genes in Database	Last Updated [1]	Database Source
ShortBRED VF [2]	26,187	July 2017	https://huttenhower.sph.harvard.edu/shortbred
ShortBRED AR [3]	932	July 2017	https://huttenhower.sph.harvard.edu/shortbred
Virulence Factor DataBase (VFDB)	30,246	February 2018	http://www.mgc.ac.cn/VFs/main.htm
Comprehensive Antibiotic Resistance Database (CARD)	2514	February 2018	https://card.mcmaster.ca/download

[1] Last update available for public download. Database download dates for analyses in this manuscript are described in Materials and Methods. [2] Database built using Victors, VFDB and MvirDB. [3] Database built using CARD.

2.8. Contaminant Analysis

Reads and contigs were analyzed for host and laboratory contamination using EDGE Bioinformatics software [33]. Taxonomy Classification was performed on FaQCs processed reads using four tools: GOTTCHA (version 1.0b) [34], Kraken (version 0.10.4-beta) [35], MetaPhlAn (version 1.7.7) [36] and BWA-mem (version 0.7.9) [37] mapping to RefSeq. Contigs from SPAdes-all reads assembly were classified by aligning contigs to NCBI's RefSeq database using BWA-mem. All programs were run using default parameter settings in EDGE Bioinformatics (Appendix A) [38]. Any contigs >700 bp with >5× coverage obtained in subsampled assemblies were also analyzed by megablast against nr/nt databases. Samples were considered free from contamination if the total assembly size was close to the range of genome sizes within that particular virus family and if less than 5% of reads mapped to host reference genomes.

2.9. Gene Calling and Functional Annotation

Baseline gene predictions and functional annotation were obtained from Classic RAST (Virus domain; genetic code 11; FIGfam version Release70) [13]. Putative ORFs were also predicted using the command line version of GLIMMER-3 (v3.02) [39] and the phage-specific gene caller THEA [40]. Start- and stop-site coordinates were compared for the two approaches and disagreements were considered during manual gene assignments. Annotation was performed manually. The nucleotide sequence for each predicted ORF was queried by BLASTx against NCBI's non-redundant (nr) protein sequences. Peptide sequences for each predicted ORF also underwent homology searches using BLASTp against nr, PhAnTOME [41], pVOGs [42] and the PHASTER Prophage/Virus databases [27]. The following threshold values were applied in general. Putative ORFs with 50–70% sequence identity [43] to a given gene were assigned "putative." When peptide sequences exhibited low identity (less than 50% [44]), protein sequences were also submitted for the analysis of hidden Markov models by hmmscan [45] against the Pfam database [46] and NCBI's Conserved Domain Database. Consensus gene functions were assigned to ORFs manually. tRNAs were identified using tRNAscan-SE [47] and ARAGORN [48]. Specifically, for putative ORFs with identity to potentially harmful gene products, we set a lower threshold (30%) so as to increase sensitivity and err on the side of safety. This threshold was chosen based on the work of Joshi and Xu, in which it is stated that at this threshold the chance for a pair of proteins to share any of the three GO categories at high levels would be 50% or less [43].

3. Results

We present two finished phage genomes that pass all safety checkpoints in the phage characterization workflow and which we have subsequently deposited in GenBank. The two phages, Pseudomonas phage vB_PaeP_130_113 (GenBank accession MH107770) and Staphylococcus phage vB_SauM_0414_108 (GenBank accession MH107769), were selected based on the clinical relevance of their bacterial hosts, *Pseudomonas aeruginosa* and *Staphylococcus aureus*, respectively and demonstrated antibacterial efficacy against clinical isolates in house. *Pseudomonas aeruginosa* and *Staphylococcus aureus* are two of the twelve antibiotic-resistant priority pathogens according to the World Health Organization (WHO); carbapenem-resistant *Pseudomonas aeruginosa* falls within the Priority 1 CRITICAL category. Both pathogens are also categorized as ESKAPE pathogens (*Enterococcus faecium*, *Staphylococcus aureus*, *Klebsiella pneumoniae*, *Acinetobacter baumannii*, *Pseudomonas aeruginosa*, and *Enterobacter species*), microbes responsible for the majority of hospital-acquired antimicrobial-resistant (AMR) infections, by the Infectious Diseases Society of America [49,50]. Results are presented according to key checkpoints in the phage characterization workflow.

3.1. Sequencing and Assembly Statistics

The workflow begins with two pipelines that perform Quality Control and Genome Assembly in parallel, as depicted in Figure 1. There are several QC and assembly checkpoints during this process:

1. Check reads for quality, length, nucleotide composition, ambiguous nucleotides.
2. Do assemblies agree? If no, is there adequate read-support to resolve differences?

3. Is the largest contig a phage sequence? If no, consider contamination analysis.
4. Identify (and later, characterize) all contigs >700 bp with >5× coverage.

Both samples retained 94–98% of reads following quality control (Table 2). For each sample, de novo assemblies were performed by SPAdes and CLC using all quality-controlled reads as well as a subsampled set of quality-controlled reads (four assemblies per sample; see Methods). For the Pseudomonas phage vB_PaeP_130_113 sample, all four de novo assemblies yielded a single contig that was identical among them and therefore proceeded to downstream analysis. The closest relative to Pseudomonas phage vB_PaeP_130_113 is Pseudomonas Phage DL62 (94% query coverage, 94% identity). In contrast, assemblies of randomly subsampled reads, hereafter referred to as "subassemblies", were required to obtain contigs with 100% sequence identity for the Staphylococcus phage vB_SauM_0414_108 sample. These differences highlight the utility of a multipronged approach for obtaining a consensus sequence with high confidence. Reads were mapped back to Staphylococcus phage vB_SauM_0414_108 contigs for assembly validation and the differences observed in assemblies using all reads were deemed artificial overlaps introduced by the algorithms. The validated consensus sequence used for Staphylococcus phage vB_SauM_0414_108 in next steps was the largest contig that presented 100% nucleotide identity from the two subassemblies. The closest relative to Staphylococcus phage vB_SauM_0414_108 is Staphylococcus phage K (95% query coverage, 99% identity). SPAdes all reads assembly contained a 751 bp contig. This contig was disregarded due to low coverage (<1.5×) and because it had no significant sequence similarity when queried (megablast) against the NCBI nr database.

Table 2. Sequencing and Assembly Statistics.

Pipeline Output	Pseudomonas Phage vB_PaeP_130_113	Staphylococcus Phage vB_SauM_0414_108
Total Reads	206,222	347,594
Reads Pass FaQCs (%)	98.82	98.38
Reads Pass CLC (%)	96.97	94.03
Reads sub-sampled (#)	50,000	50,000
SPAdes all reads [1]	1	2
CLC all reads [1]	1	1
SPAdes subsampled [1]	1	1
CLC subsampled [1]	1	1
SPAdes all reads [2]	43,742	141,507
CLC all reads [2]	43,742	141,334
SPAdes subsampled [2]	43,742	141,331
CLC subsampled [2]	43,742	141,330

[1] Number of contigs >700 base pairs long. [2] Length of largest contig (bp), SPAdes assembly artifacts removed.

3.2. High Quality Genomes

The next step in the workflow is the determination of termini position and packaging strategy and involves the following checkpoints:

5. Are genome ends resolved?
6. Is genome supported by adequate even coverage?

Reads that passed FaQCs and the consensus sequence obtained from genome assembly were submitted for analysis using PhageTerm software. Both phages contain direct terminal repeats (DTRs). Pseudomonas phage vB_PaeP_130_113 contains a short 463 bp DTR and Staphylococcus phage vB_SauM_0414_108 has a long DTR spanning 10,296 bp (Table 3). Coverage over the DTR region was approximately twice that of the rest of the genome for both phages. The metric τ, indicative of sudden coverage peaks, also supports the presence of DTRs in both genomes. Finally, read mapping was performed to validate the rearranged genomes using all CLC quality-controlled reads and CLC aligner. The final size of Pseudomonas phage vB_PaeP_130_113 is 44,205 bp with 846.5× average whole genome coverage. The final size of Staphylococcus phage vB_SauM_0414_108 is 151,627 bp with 508.6× average whole genome coverage.

Table 3. Genomic termini statistics.

Phage	Class [1]	DTR Region Length	Start, End τ Metric [2]	Coverage in DTR Region	Coverage Outside of DTR Region
Pseudomonas phage vB_PaeP_130_113	Short DTR	463 bp	0.63, 0.64	1018.0×	634.9×
Staphylococcus phage vB_SauM_0414_108	Long DTR	10,296 bp	0.75, 0.55	753.1×	343.9×

Above metrics are determined by PhageTerm. [1] One of the following: 5′ *cos*, 3′ *cos*, Short DTR, Long DTR, headful (with or without *pac* site detected), Mu-like, or unknown. [2] τ in forward direction for first nucleotide of DTR region, τ in reverse direction for last nucleotide of DTR region.

3.3. Phage Lifestyle

The primary objective of the phage lifestyle step is to identify temperate phages since they are not pursued past this point in our pipeline and involves the following checkpoints:

7. Do any predicted ORFs present sequence identity to known integrase(s)?
8. Do classification algorithms (i.e., PHACTS) bolster confidence?

Genomes were submitted to RAST and PHASTER for rapid preliminary gene calling and annotation. The RAST-generated Genbank file and the PHASTER-generated details.txt file were parsed (using grep) to identify the presence of integrase. In addition, the phage proteomes were analyzed using PHACTS, a computational classification algorithm trained to predict phage lifestyles. Integrase genes were not identified in either genome. In line with this, PHACTS predicted <40% probability that these two phages would exhibit temperate behavior (Table 4). We applied PHACTS analysis to our genomes along with the closest relative for each sample and two phages with integrase. Lytic scores were >0.59 for Pseudomonas phage vB_PaeP_130_113 and its closest relative (Pseudomonas phage DL62, GI:KR054031) as well as Staphylococcus phage vB_SauM_0414_108 and its closest relative (Staphylococcus phage K, GI:KF766114.1). To contrast this, we also present phages with integrases in their genomes (Pseudomonas phage vB_PaeS_PMG1, GI:NC_016765; Staphylococcus phage phiSaus-IPLA88, GI:NC_011614.1) and respective lytic scores of 0.42 or less. These results demonstrate reliable PHACTS predictions for phages with known lifestyles. Taken together, these results strongly suggest that Pseudomonas phage vB_PaeP_130_113 and Staphylococcus phage vB_SauM_0414_108 are likely lytic phages.

Table 4. Phage lifestyle assessment.

Phage	PHACTS Lytic Score	PHACTS Temperate Score	PHACTS Standard Deviation	PHASTER Integrase	RAST Integrase	NCBI Annotated Integrase
Pseudomonas phage vB_PaeP_130_113	0.66	0.34	0.073	No	No	N/A
Pseudomonas phage DL62 (GI:KR054031)	0.73	0.26	0.117	No	No	No
Pseudomonas phage vB_PaeS_PMG1 (GI:NC_016765)	0.42	0.58	0.042	Yes	Yes	Yes
Staphylococcus phage vB_SauM_0414_108	0.60	0.40	0.082	No	No	N/A
Staphylococcus phage K (GI:KF76114)	0.59	0.41	0.107	No	No	No
Staphylococcus phage phiSaus-IPLA88 (GI:NC_011614)	0.28	0.72	0.048	Yes	Yes	Yes

N/A = Not applicable due to in house anntoation.

3.4. Specialty Genes Checkpoint

The next checkpoint involves genes with potential deleterious effects:

9. Are any toxins, virulence factors, or antimicrobial resistance genes detected?

To answer this question, all reads and all annotated coding sequences from contigs were profiled for antimicrobial resistance genes and virulence factors. This analysis was performed using the EDGE Bioinformatics Gene Family module, however the tools and databases are open source and can be run by command line (Table 1). The ShortBRED algorithm was used to perform a targeted search for unique and specific signatures found in four specialty gene databases (Table 1). For read-based analysis ShortBRED searches the Antimicrobial Resistance Database (ARDB), Resfams and the Virulence Factor Database (VFDB). For contig-based analysis, Prodigal performs gene calling, ShortBRED searches coding sequences against VFDB and the Resistance Gene Identifier (RGI) searches the Comprehensive Antibiotic Resistance Database (CARD). By these methods, Pseudomonas phage vB_PaeP_130_113 and Staphylococcus phage vB_SauM_0414_108 were found to encode zero hits to any known problematic specialty gene targets at the read and contig levels.

3.5. Contaminant Analysis

Another key element in the viability check is detection of potential contaminants, or transducing activity which can present similarly as contamination, and this analysis involves the following three checkpoints:

10. What percentage of reads are classified as phage?
11. What is the total assembly size?
12. What percentage of reads map to non-phage contigs from the assembly?

Two approaches were applied to assess contamination or potential transducing ability in a quantitative manner. First, all reads and all assembled contigs were analyzed using the Taxonomy Classification module in EDGE Bioinformatics. Parameters for quality trimming in our local instance of EDGE Bioinformatics software were updated to reflect the FaQCs and SPAdes parameters described in the Methods and Appendix A. The second approach is classification-independent: calculate the percentage of reads that map to non-phage contigs and considered total assembly size. This second measurement is calculated as follows: (1) determine the phage contig(s) versus non-phage contigs through taxonomic classification as described in Section 2.8, *Contaminant Analysis*; (2) calculate the number of reads mapped to the phage contig; (3) subtract the number obtained in Step 2 from the number of reads mapped to whole genome assembly; and (4) calculate the percentage of host sequence (i.e., the number of reads mapped to non-phage contigs divided by the total number of quality-controlled reads multiplied by 100).

Three of four taxonomy tools (GOTTCHA, Kraken and BWA) agreed and identified both reads and contigs from the Pseudomonas phage vB_PaeP_130_113 sample as a Pseudomonas phage (Figure 2A). BWA classification results are presented in detail. We present BWA classification results because this tool performs both read and contig-based classification. Only 46.47% of all reads from Pseudomonas phage vB_PaeP_130_113 were classified using taxonomy-based analysis. However, of the reads that were classified, 99% of classified reads were Viruses, all of which were further classified as Pseudomonas phages (Figure 2B). Of the organisms detected by two or more read-based tools, all have >87% identity to the final phage genome. For contig-based community profiling the top hit was Pseudomonas phage vB_Pae-TbilisiM32 (GI:KX711710), which has 95% query coverage and 94% identity to the final phage contig. The total assembly size for Pseudomonas phage vB_PaeP_130_113 was 44,325 bp. The number of reads mapped to phage contig was 203,003 while the number of reads mapped to the whole assembly was 203,005 and the total trimmed reads is 206,222. This amounts to 0.001% potential host sequence. Taken together, these data indicate that the Pseudomonas phage vB_PaeP_130_113 sample passes this checkpoint.

Likewise, three of four taxonomy tools (GOTTCHA, Kraken and BWA) agreed and identified both reads and contigs from the Staphylococcus phage vB_SauM_0414_108 sample as a Staphylococcus phage (Figure 2C). 79.12% of all reads from Staphylococcus phage vB_SauM_0414_108 were classified by BWA. Of the classified reads, 99.99% were classified as Viruses and all viral reads were classified as viruses of *Staphylococcus* at the species level (Figure 2D). The top five organisms classified by read-based taxonomy have >93% identity to the final phage genome. Only two contigs were >700 bp and only the phage contig had over 5× coverage. The top hit for the largest contig, the unfinished phage contig, was Staphylococcus phage JD007 (GI:JX878671), which has 95% query coverage and 97% identity to the final phage genome. The next largest contig (751 bp with under 1.5× coverage) presents no significant similarity to any known sequences in the nr database (megablast). The total assembly size for the Staphylococcus phage vB_SauM_0414_108 sample was 145,555 bp, with the number of reads mapped to phage contig being 338,323 and the number of reads mapped to assembly being 338,360. The total trimmed reads amount to 341,976. These results indicate 0.011% potential host sequence. This phage also passed criteria for contamination-free sequences and does not, by this analysis, appear to exhibit generalized transduction.

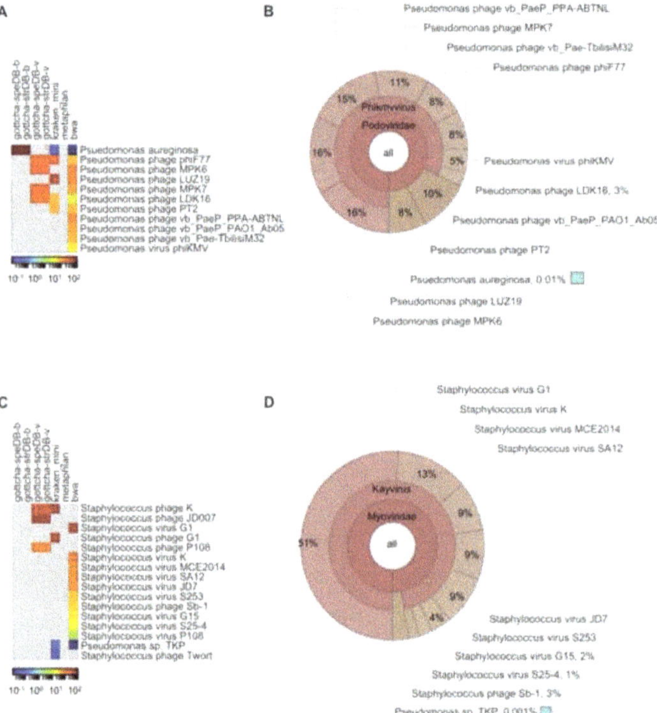

Figure 2. Contaminant analysis using read-based taxonomy classification. Read-based taxonomy results are presented for Pseudomonas phage vB_PaeP_130_113 (**A**,**B**); and Staphylococcus phage vB_SauM_0414_108 (**C**,**D**). Taxonomy results for all classification tools (relative abundance) using all reads that pass QC are presented as heatmaps (**A**,**C**). Reads were classified by GOTTCHA using databases comprised of bacteria (species-level: gottcha-speDB-b; strain-level: gottcha-strDB-b) or viruses (species-level: gottcha-speDB-v; strain-level: gottcha-strDB-v), Kraken (kraken_mini), metaphlan and BWA against RefSeq (BWA-mem). All reads that were classified by BWA are presented as a Krona plots, where percentages are the number of reads that map to each organism divided by the total number of classified reads (**B**,**D**).

3.6. Genome Polishing

After genomes passed all the previous twelve checkpoints, the final phage genomes underwent automated gene-calling followed by manual curation of annotations and the final checkpoint:

13. Were any unsettling genes identified by manual annotation?

Two major considerations at this step include methods to identify genes and the reliability of annotations. We performed gene calling using RAST and two gene prediction algorithms and compared start and stop sites for each ORF. For each putative ORF, the amino acid sequence was searched against a minimum of four databases (see Methods). Consensus annotations were assigned followed by a final check for any notable genes. The finished genome for Pseudomonas phage vB_PaeP_130_113 contains 57 coding sequences (CDS), 35 with assigned functional annotations, 22 hypothetical, and no tRNA (Table 5, Figure 3A). The final genome for Staphylococcus phage vB_SauM_0414_108 contains 241 CDS, 154 with assigned functional annotations, 87 hypothetical, and four tRNA sequences (Table 5, Figure 3B). No notable or problematic genes were found in either genome.

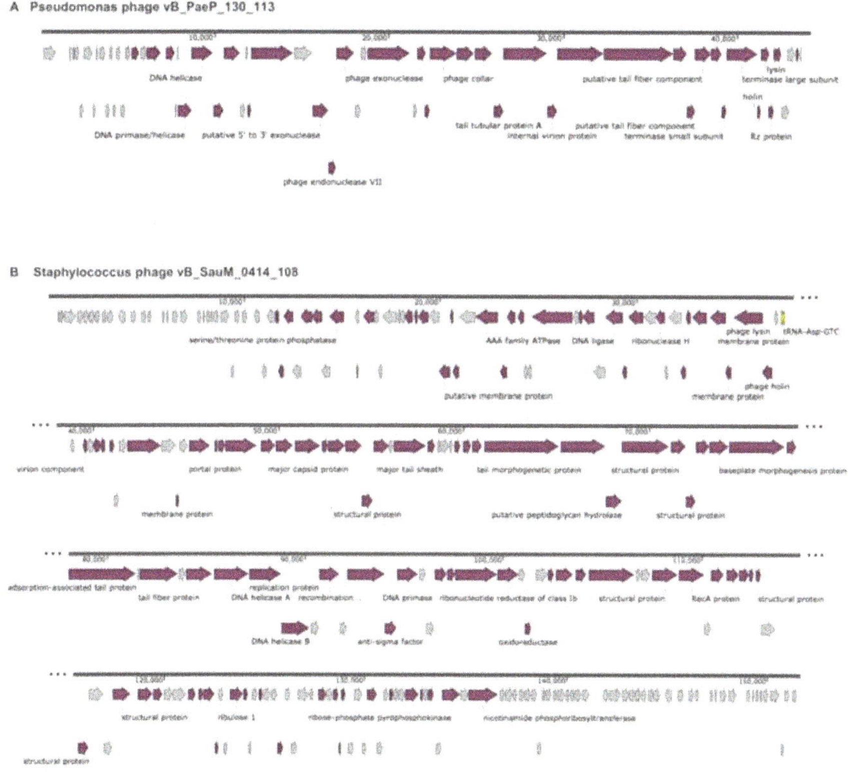

Figure 3. Whole genome maps for finished annotated phage genomes. Annotations for selected predicted open reading frames (ORFs) are presented for Pseudomonas phage vB_PaeP_130_113 (**A**); and Staphylococcus phage vB_SauM_0414_108 (**B**). Mauve colored arrows indicate the ORF has been annotated; grey colored arrows indicate ORFs annotated "hypothetical"; yellow arrows indicate tRNA.

Table 5. Finished genome details.

Phage	Size (bp)	%GC	CDS (#)	Genes with Functional Annotation (#)	Hypothetical Genes (#)	tRNA (#)	Assigned Family
Pseudomonas phage vB_PaeP_130_113	44,205	62.4	57	35	22	0	Podoviridae
Staphylococcus phage vB_SauM_0414_108	151,627	30.4	241	154	87	4	Myoviridae

4. Discussion

The promise of phage therapeutics for devising personalized treatment toward AMR microbes demands a reliable and scalable pipeline to characterize phages. Herein we present an analysis workflow along with recommendations for essential checkpoints to assess the genomic safety of phage candidates. This analysis mirrors efforts carried out by our team for phage genomes submitted to the FDA for IND approval. The two phage genomes delivered here are examples of candidates that would pass all safety criteria. We have made our source code publically available at GitHub (https://github.com/BDRD-Genomics). Below we expand on potential issues that arise at important checkpoints and emphasize the need for manual oversight for genomes being considered for human use.

Phage genomes are hyper-mobile and exhibit high mutation rates, thus a finished genome represents a consensus sequence for the distribution of non-identical related progeny. Detecting minor variants is dependent on sequencing technology and population diversity (i.e., quasi-species). Increasing genomic depth of coverage typically provides additional confidence in overcoming sequencing errors and identifying true single nucleotide polymorphisms (SNPs). Previous studies have determined that 400× coverage is recommended to detect minor variants (present at 1% frequency with 99.999% confidence) in order to accurately describe genetic diversity within a viral population [8]. Although dependent on experimental conditions, there exists an upper range limit in which increasing coverage either produces no additional benefit or has deleterious effects. As noted previously, most genomic assemblers use heuristics in order to solve an NP-hard problem (a class of problems which may take exponential time to solve or are unsolvable) in a reasonable period of time. Generally, these heuristics are geared toward lower coverage levels (<100×). Very high coverage may actually increase assembly fragmentation; with low coverage a given error is typically unique but with high enough coverage the same error may be encountered multiple times and assembled. This creates false branches with a de Bruijn graph-based assembler that may be cut, increasing fragmentation. In order to overcome any potential biases introduced by specific assembler heuristics, we employ multiple assemblers (CLC and SPAdes) at multiple depths of coverage, followed by a consensus based approach to determine the most accurate contig. There are a number of freely available open source assemblers (e.g., Velvet, SOAPdenovo, ABySS), which may be substituted for the proprietary product produced by CLC.

Determining a phage's genomic termini and packaging strategy is necessary for producing the correct nucleotide sequence, selecting the genome start-site, correcting assembly artifacts caused by direct terminal repeats (DTRs), and elucidating the phage's biology. For example, DNA packaging strategy has implications in a large part of a phage's life cycle including: initiation, replication, termination and transcriptional regulation [51]. PhageTerm was selected as the tool of choice for analyzing phage termini because of its automated and user-friendly nature [25]. Like other methods that use high-throughput sequence data to ascertain genomic termini, PhageTerm exploits the random nature of DNA fragmentation during library prep and identifies over-represented fragment ends. Genome orientation cannot be inferred from NGS data for *cos* and DTR phages. PhageTerm will leave the orientation the same as the input reference for these types of phages. In this situation, orientation should be determined based on the orientation of terminase gene(s). Similarly, when a phage's termini cannot be determined, or in the case a phage has no consistent biological termini (circularly permuted phages), it is recommended that the start site be selected relative to the position and orientation of the terminase gene(s). Merril et al. suggest starting circularly permuted phages at, or just upstream from, the large terminase subunit and adjusting the orientation of the genome so it is in the forward direction [26]. However, this often results in the small terminase gene being placed at the opposite end of the genome because it is common for the small terminase subunit to be directly upstream from the gene for the large terminase subunit. Circularly permuted phages put through our workflow start at the small terminase gene so that genes for terminase subunits stay adjacent to each other. The orientation of the genome was adjusted such that it matches the orientation of the small terminase genes. If no small terminase gene can be found, the genome was instead adjusted to the large terminase gene. It is important to note that users should manually inspect the sequence to make sure that the start site does not result in a broken CDS.

Phages that have a high potential to integrate into the chromosomes of their bacterial host are considered undesirable for phage therapy. Our pipeline combines homology-based searches (RAST, PHASTER) and a predictive computational model (PHACTS) to identify risky candidates during the initial viability checkpoint. For annotation-based analysis, integrase, the enzyme that mediates incorporation of phage DNA into bacterial DNA, is the molecular marker used to exclude prophages and phages with temperate potential. Importantly, integrase genes are highly diverse genomic elements, thus relying on sequence similarity to published genomes alone is inadequate for predicting phage lifestyle. PHASTER annotation, for example, only detected 75 of 147 (51%) integrase genes in prophages from *Salmonella enterica* [52]. Moreover, a mutation in a single amino acid residue can render an integrase inactive [53], thus the functional capacity of an integrase is unknown unless it is 100% identical to an experimentally validated annotation. To circumvent the limitations of functional annotation we combine our analysis with PHACTS, a classification algorithm that predicts phage lifestyle based on the entire phage proteome. PHACTS utilizes a novel similarity algorithm and a trained Random Forest classifier to classify phages. These classifications are based on a curated database of phages with annotated lifestyles, which have their proteins aligned against the proteins of the query phage. PHACTS does not always provide clear classifications despite being trained using phages with experimentally proven temperate or lytic lifestyles. It is possible that some discrepancies can be explained by specialized host-phage interactions that govern lifestyle through interference of integrase or other repressor genes (reviewed extensively in [54]). Analysis of repressor genes would be an additional strategy that could add information to phage lifestyle predictions. Another explanation for potentially unclear results from PHACTS could be due to the collection of genomes in the database on which PHACTS is trained. If the query phage has proteins that are similar to those in the database, PHACTS has a higher chance of providing a clear answer. However, if the query phage is fairly unique, it becomes more difficult for the algorithm to make a confident call. The PHACTS training data contains phages that infect *Pseudomonas* and *Staphylococcus* hosts at equal proportions (~50 genomes each), which bolsters confidence in the results presented above. It is important to note that the PHACTS database is limited (zero phages for *Acinetobacter* hosts, for example) and users should be aware that PHACTS predictions are dependent on this database. If users are studying genomes not represented in the database, it is recommended that PHACTS be retrained with a wider variety of phages from a number of different hosts. An additional caveat is that episomal and plasmidial prophages [55,56] might not be detected as such by our analyses unless they exhibit significant nucleotide identity to previously sequenced episomal or plasmidial prophages.

Monitoring the apparent host sequences as a way of assessing potential sample contamination is important for obtaining high quality finished genomes with confidence and ensuring the reproducibility of bioinformatics analysis using raw sequences that are delivered to the FDA as a part of IND-filing. It is also important as a way of monitoring for potential transducing ability. We recommend performing taxonomy classification on all reads and contigs, and, due to the potential uncertainties introduced into the resulting assembly and the qualitative nature of genomic assessments of "safety," disregarding samples with contamination, even if the host sequence is thought to be introduced downstream of phage purification, such as from sequencing run carry-over or bleed-through among samples multiplexed within a sequencing run. EDGE Bioinformatics provides a well-documented web-based interface to perform classification using four different tools that vary in sensitivity and specificity (see Appendix A; [33,38]). This analysis allows users to generate a confident assessment of the proportion of host sequence and/or of potential contaminants from other sources. The tools are dependent on the curation of the respective databases and underlying algorithms. This means that the number of reads mapping to a particular reference will vary among tools, preventing definitive cut-off values for percent contamination. We recommend running taxonomic classification using multiple tools and focusing on calls where agreement is observed across tools. Importantly, the taxonomy tools in EDGE are designed for classifying prokaryotes and viruses; fungal contaminants will not be classified using this method. At a minimum, we recommended samples undergo read and contig-based analysis using BWA-mem mapping to RefSeq. We also suggest calculating the percentage of reads that map to non-phage contigs. This calculation gives an estimation of how pure a phage prep is and will enable a user to discriminate contamination even if the contaminant is not classified by taxonomy tools (i.e., fungal contamination). For instance, assume that a phage sample had 1,263,276 reads mapped to phage contig and the total number of reads mapped to whole assembly was 1,459,830 out of a total of 1,464,496 quality-controlled reads. In this case, the number of reads mapped to non-phage was 196,554 reads or 13.42%. Another factor to consider is the total genome assembly size, in this example, 6,103,974 bp. Since two facts violate the checkpoints (undesirable percentage of reads not mapping to phage contig and total assembly size much larger than any known phage genome), this phage sample would be abandoned and not analyzed further. When considering total genome assembly size, we may deduce that if the total genome assembly size is much too big to be a phage genome alone, it is likely that it includes several contigs resulting from host bacterial sequences. Based on our experience, if the genome assembly size is larger than 1 Mbp or close to the size of host bacterial genomes (i.e., 3 to 6 Mbp) and if the percentage of non-phage reads is greater than 5%, the quality or purity of the phage prep should be questioned, and/or the possibility of transduction considered.

Prior to intensive annotation, our pipeline employs an initial scan against known deleterious genes, using two methods to screen against VFDB. One method queries the phage contig against VFDB using blastn. We have found that some regions in phage genomes share sequence similarity to hits in VFDB, however these hits present low query coverage and can be rapidly excluded as false positives. For this reason, we have selected to use ShortBRED in combination with VFDB as a second method for screening. The ShortBRED algorithm is used to identify unique and distinguishable protein sequences in VFDB and provides highly specific results. If any positive results are identified using ShortBRED, the phage does not pass as a viable candidate. Users should be cautious of VFDB hits when using BLAST; specifically users should investigate whether identity spans the full protein sequence or if functional domains are represented in the sequence.

Even with that robust screening method in place to assist in detection of deleterious genes, thorough annotation as outlined in our pipeline serves the primary role of assessing phage candidates for safety. Identifying and annotating phage genes remains technically difficult due to small size and lack of known homologues. Additionally, in silico gene-prediction and annotation platforms cater to the molecular underpinnings of bacterial genomes. This means that manual refinement is necessary to checking genes and annotations. Relaxing similarity thresholds of search algorithms may be necessary to fully exclude a gene product as hazardous with confidence. With that being said, it is important to note that assigning a gene product "hypothetical protein" is preferred over a more specific annotation with little evidence when depositing final genomes into public repositories. We recommend reviewing guidelines written by Aziz et al. [9], a thorough overview of how to avoid precarious functional assignments. However, we do want to articulate a major warning for identifying and dismissing "red-flag" annotations deposited without experiential evidence. To exemplify this, note the following two genes: toxin TX1 (Pseudomonas phage TH30) (NCBI reference sequence: YP_009226100) and Acriflavin resistance protein (Pseudomonas phage vB_PaeM_C2-10_Ab02) (NCBI reference sequence: CEF89094). Neither of the aforementioned genes have experimental evidence, nor do any of the homologues, all of which are annotated hypothetical. Moreover, the phage containing the gene annotated "toxin TX1" was administered to animals as a therapeutic phage in preclinical experiments, with no adverse effects. These predicted gene products have no similarity to any other toxins. Finally, in some unique cases protein structure modeling can provide clarity. Open-source computational modeling software, like RaptorX [57] and I-TASSER [30], can be used to predict the integrity of enzymes or identify functions based on similarity to known protein structures rather than sequence similarity in order to refine the granularity of annotations. Similarly, publicly available websites like Superfamily [58] provide a platform for structure based searches using hidden Markov models, which can be useful for predicting the function of distantly related proteins. When submitting phage genomes to the FDA for clinical approval, we recommend providing annotation results from all databases and the date databases were accessed. By using orthogonal approaches and combining the results through manual curation we can say with confidence that no known deleterious genes are encoded within candidate phage genomes that pass through our checkpoints, but we cannot exclude the possibility that unknown, previously unsequenced or uncharacterized, deleterious genes could exist that we might not detect through in silico safety analyses.

Another caveat is that in this study we have not addressed the utility of different sequencing chemistries and platforms. In the course of this work we have mainly utilized Illumina short read sequencing technology, although for some genomes we have attempted to resolve potential assembly artifacts through the use of long reads. Short read technologies typically (1) have inherently low error rates and (2) due to the high throughput, resulting deep coverage [59]. On the contrary, long read sequencing platforms such as the Oxford MinION and PacBio hold promise for resolving terminal repeats and other ambiguities in phage genome assemblies, but the combination of higher error rates and less deep coverage [60–62] is likely to make error correction with short reads a necessity in many cases for at least the near future. The pipeline presented in this manuscript would require adjustment to be suitable for long reads, particularly the QC and assembly steps, and subsampling would likely not be required for long reads. However, the downstream processes and checkpoints regarding lifestyle, annotation, and potentially dangerous genes, would remain the same.

In conclusion, we present to the phage therapeutic community a set of guidelines to enable not only the production of high quality phage genomes and but also predictions of phage suitability for therapeutic use. Many of the steps involved are intuitive but some are not. Most of these steps would be employed in genome production and characterization standards for a variety of organisms but in most cases outside of phage therapy, not all of these steps would be required. In many cases there are examples of published genomes that would have benefited from some of these steps. There is currently no single pipeline that is completely automated from end-to-end and would accomplish all the suitability checks and verifications inherent to this type of work but we have made our source code available to the general public and have documented herein the types of human intervention involved and the logic that is applied. We have also made the raw sequence data as well as the polished final products available in public databases for use by others. This work leverages previously developed standards (that were primarily focused on producing genomic data for pathogens or for vaccine candidates [14]) to create the first fully described and published standard for genomes of therapeutic viruses. The approach presented herein should enable researchers to characterize potential therapeutic phages fully prior to IND submission.

5. Conclusions

In conclusion, the pipeline and checkpoints presented here represent a necessary first step toward widespread use of phage therapy in the US. The aim is to produce fully assembled, error-free, well-annotated genomes for lytic phages that do not encode genes likely to promote toxicity or AMR and to do so through a combination of best available tools, well-defined thresholds and necessary human interventions. In support of this movement, we provide the pipeline and our internal thresholds to the scientific community, along with datasets that can be used for training purposes.

Supplementary Materials: The following are available online at http://www.mdpi.com/1999-4915/10/4/188/s1, Table S1: Parameters employed.

Acknowledgments: This work was funded by Naval Medical Research Center's Advanced Medical Development Program, work unit number A1704. Casandra W. Philipson, Kenneth G. Frey, Biswajit Biswas and Kimberly A. Bishop-Lilly are employees of the US government and T.H. is a military service member. This work was prepared as a part of official duties. Title 17 U.S.C. 105 provides that 'Copyright protection under this title is not available for any work of the United States Government.' Title 17 U.S.C. 101 defines a U.S. Government work as a work prepared by a military service member or employee of the U.S. Government as part of a person's official duties. The views expressed in this article are those of the authors and do not necessarily reflect the official policy or position of the Department of the Navy, Defense Threat Reduction Agency, Department of Defense, nor the U.S. Government.

Author Contributions: Kimberly A. Bishop-Lilly and Theron Hamilton conceived and designed the experiments. Casandra W. Philipson, Logan J. Voegtly, Matthew R. Lueder, Kyle A. Long, Gregory K. Rice and Regina Z. Cer designed the phage characterization pipeline and performed bioinformatics analyses. Kenneth G. Frey led sequencing efforts. Biswajit Biswas led phage isolation efforts. Casandra W. Philipson led the manuscript writing effort and all authors contributed to the writing.

Conflicts of Interest: The authors declare no conflict of interest.

Appendix A

EDGE (Enabling the Democratization of Genomics Expertise) is a bioinformatics platform with a user-friendly interface which provides access to cutting edge bioinformatics tools. Of the seven modules available in EDGE, we use four modules to analyze phage sequences. The pre-processing module performs quality control and read trimming; we changed trim quality level to 30, average quality cutoff to 30 and "N" base cutoff to 2. The assembly and annotation module assembles the reads into contigs, performs gene calling using Prodigal and annotation with Prokka; we changed the assembler to SPAdes and Prokka kingdom to Viruses. The taxonomy classification module uses multiple tools to assign taxonomic classifications to reads and contigs; we use default parameters. The gene family module, added in EDGE v1.5, searches specially curated databases for virulence factors and antibiotic resistance genes in our reads and predicted ORFs; we use default parameters.

EDGE code (including the Gene Family analysis) is available at:

- https://github.com/LANL-Bioinformatics/EDGE.

EDGE documentation is available at:

- https://edge.readthedocs.io.

EDGE tutorial is available at:

- https://www.youtube.com/playlist?list=PL7DNo6h5wJsTh2l2GK3N86Imb-9fYQFfH

References

1. Stratton, C.W. Phages, fitness, virulence, and synergy: A novel approach for the therapy of infections caused by pseudomonas aeruginosa. *J. Infect. Dis.* **2017**, *215*, 668–670. [CrossRef] [PubMed]
2. Bogovazova, G.G.; Voroshilova, N.N.; Bondarenko, V.M.; Gorbatkova, G.A.; Afanas'eva, E.V.; Kazakova, T.B.; Smirnov, V.D.; Mamleeva, A.G.; Glukharev Iu, A.; Erastova, E.I.; et al. Immunobiological properties and therapeutic effectiveness of preparations from klebsiella bacteriophages. *Zh. Mikrobiol. Epidemiol. Immunobiol.* **1992**, *3*, 30–33.
3. Schooley, R.T.; Biswas, B.; Gill, J.J.; Hernandez-Morales, A.; Lancaster, J.; Lessor, L.; Barr, J.J.; Reed, S.L.; Rohwer, F.; Benler, S.; et al. Development and use of personalized bacteriophage-based therapeutic cocktails to treat a patient with a disseminated resistant acinetobacter baumannii infection. *Antimicrob. Agents Chemother.* **2017**, *61*, e00954-17. [CrossRef] [PubMed]
4. Zschach, H.; Joensen, K.G.; Lindhard, B.; Lund, O.; Goderdzishvili, M.; Chkonia, I.; Jgenti, G.; Kvatadze, N.; Alavidze, Z.; Kutter, E.M.; et al. What can we learn from a metagenomic analysis of a georgian bacteriophage cocktail? *Viruses* **2015**, *7*, 6570–6589. [CrossRef] [PubMed]

5. Kauffman, K.M.; Hussain, F.A.; Yang, J.; Arevalo, P.; Brown, J.M.; Chang, W.K.; VanInsberghe, D.; Elsherbini, J.; Sharma, R.S.; Cutler, M.B.; et al. A major lineage of non-tailed dsdna viruses as unrecognized killers of marine bacteria. *Nature* **2018**, *554*, 118–122. [CrossRef] [PubMed]
6. Dutilh, B.E.; Cassman, N.; McNair, K.; Sanchez, S.E.; Silva, G.G.; Boling, L.; Barr, J.J.; Speth, D.R.; Seguritan, V.; Aziz, R.K.; et al. A highly abundant bacteriophage discovered in the unknown sequences of human faecal metagenomes. *Nat. Commun.* **2014**, *5*, 4498. [CrossRef] [PubMed]
7. Yosef, I.; Manor, M.; Kiro, R.; Qimron, U. Temperate and lytic bacteriophages programmed to sensitize and kill antibiotic-resistant bacteria. *Proc. Natl. Acad. Sci USA* **2015**, *112*, 7267–7272. [CrossRef] [PubMed]
8. Russell, D.A. Sequencing, assembling, and finishing complete bacteriophage genomes. *Methods Mol. Biol.* **2018**, *1681*, 109–125. [PubMed]
9. Aziz, R.K.; Ackermann, H.W.; Petty, N.K.; Kropinski, A.M. Essential steps in characterizing bacteriophages: Biology, taxonomy, and genome analysis. *Methods Mol. Biol.* **2018**, *1681*, 197–215. [PubMed]
10. Rihtman, B.; Meaden, S.; Clokie, M.R.; Koskella, B.; Millard, A.D. Assessing illumina technology for the high-throughput sequencing of bacteriophage genomes. *PeerJ* **2016**, *4*, e2055. [CrossRef] [PubMed]
11. Ravn, U.; Didelot, G.; Venet, S.; Ng, K.T.; Gueneau, F.; Rousseau, F.; Calloud, S.; Kosco-Vilbois, M.; Fischer, N. Deep sequencing of phage display libraries to support antibody discovery. *Methods* **2013**, *60*, 99–110. [CrossRef] [PubMed]
12. Hayes, S.; Mahony, J.; Nauta, A.; van Sinderen, D. Metagenomic approaches to assess bacteriophages in various environmental niches. *Viruses* **2017**, *9*, 127. [CrossRef] [PubMed]
13. Aziz, R.K.; Bartels, D.; Best, A.A.; DeJongh, M.; Disz, T.; Edwards, R.A.; Formsma, K.; Gerdes, S.; Glass, E.M.; Kubal, M.; et al. The rast server: Rapid annotations using subsystems technology. *BMC Genom.* **2008**, *9*, 75. [CrossRef] [PubMed]
14. Ladner, J.T.; Beitzel, B.; Chain, P.S.; Davenport, M.G.; Donaldson, E.F.; Frieman, M.; Kugelman, J.R.; Kuhn, J.H.; O'Rear, J.; Sabeti, P.C.; et al. Standards for sequencing viral genomes in the era of high-throughput sequencing. *MBio* **2014**, *5*, e01360-14. [CrossRef] [PubMed]
15. Loc-Carrillo, C.; Abedon, S.T. Pros and cons of phage therapy. *Bacteriophage* **2011**, *1*, 111–114. [CrossRef] [PubMed]
16. McNair, K.; Bailey, B.A.; Edwards, R.A. Phacts, a computational approach to classifying the lifestyle of phages. *Bioinformatics* **2012**, *28*, 614–618. [CrossRef] [PubMed]
17. Edgar, R.; Friedman, N.; Molshanski-Mor, S.; Qimron, U. Reversing bacterial resistance to antibiotics by phage-mediated delivery of dominant sensitive genes. *Appl. Environ. Microbiol.* **2012**, *78*, 744–751. [CrossRef] [PubMed]
18. Lu, T.K.; Collins, J.J. Engineered bacteriophage targeting gene networks as adjuvants for antibiotic therapy. *Proc. Natl. Acad. Sci. USA* **2009**, *106*, 4629–4634. [CrossRef] [PubMed]
19. Pei, R.; Lamas-Samanamud, G.R. Inhibition of biofilm formation by T7 bacteriophages producing quorum-quenching enzymes. *Appl. Environ. Microbiol.* **2014**, *80*, 5340–5348. [CrossRef] [PubMed]
20. Regeimbal, J.; Jacobs, A.; Corey, B.; Henry, M.; Thompson, M.; Pavlicek, R.; Quinones, J.; Hannah, R.; Ghebremedhin, M.; Crane, N.; et al. Personalized therapeutic cocktail of wild environmental phages rescues mice from acinetobacter baumannii wound infections. *Antimicrob. Agents Chemother.* **2016**, *60*, 5806–5816. [CrossRef] [PubMed]
21. Paithankar, K.R.; Prasad, K.S.N. Precipitation of DNA by polyethylene glycol and ethanol. *Nucleic Acids Res.* **1991**, *19*, 1346. [CrossRef] [PubMed]
22. Lo, C.C.; Chain, P.S. Rapid evaluation and quality control of next generation sequencing data with FaQCs. *BMC Bioinform.* **2014**, *15*, 366. [CrossRef] [PubMed]
23. Seqtk, Toolkit for Processing Sequences in Fasta/q Formats. Available online: https://github.com/lh3/seqtk (accessed on 14 September 2016).
24. Bankevich, A.; Nurk, S.; Antipov, D.; Gurevich, A.A.; Dvorkin, M.; Kulikov, A.S.; Lesin, V.M.; Nikolenko, S.I.; Pham, S.; Prjibelski, A.D.; et al. Spades: A new genome assembly algorithm and its applications to single-cell sequencing. *J. Comput. Biol.* **2012**, *19*, 455–477. [CrossRef] [PubMed]
25. Garneau, J.R.; Depardieu, F.; Fortier, L.C.; Bikard, D.; Monot, M. Phageterm: A tool for fast and accurate determination of phage termini and packaging mechanism using next-generation sequencing data. *Sci. Rep.* **2017**, *7*, 8292. [CrossRef] [PubMed]

26. Merrill, B.D.; Ward, A.T.; Grose, J.H.; Hope, S. Software-based analysis of bacteriophage genomes, physical ends, and packaging strategies. *BMC Genom.* **2016**, *17*, 679. [CrossRef] [PubMed]
27. Arndt, D.; Grant, J.R.; Marcu, A.; Sajed, T.; Pon, A.; Liang, Y.; Wishart, D.S. Phaster: A better, faster version of the phast phage search tool. *Nucleic Acids Res.* **2016**, *44*, W16–W21. [CrossRef] [PubMed]
28. Liu, B.; Pop, M. ARDB—Antibiotic resistance genes database. *Nucleic Acids Res.* **2009**, *37*, D443–D447. [CrossRef] [PubMed]
29. Gibson, M.K.; Forsberg, K.J.; Dantas, G. Improved annotation of antibiotic resistance determinants reveals microbial resistomes cluster by ecology. *ISME J.* **2015**, *9*, 207–216. [CrossRef] [PubMed]
30. Chen, L.; Zheng, D.; Liu, B.; Yang, J.; Jin, Q. VFDB 2016: Hierarchical and refined dataset for big data analysis—10 years on. *Nucleic Acids Res.* **2016**, *44*, D694–D697. [CrossRef] [PubMed]
31. Hyatt, D.; Chen, G.L.; Locascio, P.F.; Land, M.L.; Larimer, F.W.; Hauser, L.J. Prodigal: Prokaryotic gene recognition and translation initiation site identification. *BMC Bioinform.* **2010**, *11*, 119. [CrossRef] [PubMed]
32. Jia, B.; Raphenya, A.R.; Alcock, B.; Waglechner, N.; Guo, P.; Tsang, K.K.; Lago, B.A.; Dave, B.M.; Pereira, S.; Sharma, A.N.; et al. Card 2017: Expansion and model-centric curation of the comprehensive antibiotic resistance database. *Nucleic Acids Res.* **2017**, *45*, D566–D573. [CrossRef] [PubMed]
33. Li, P.E.; Lo, C.C.; Anderson, J.J.; Davenport, K.W.; Bishop-Lilly, K.A.; Xu, Y.; Ahmed, S.; Feng, S.; Mokashi, V.P.; Chain, P.S. Enabling the democratization of the genomics revolution with a fully integrated web-based bioinformatics platform. *Nucleic Acids Res.* **2017**, *45*, 67–80. [CrossRef] [PubMed]
34. Freitas, T.A.; Li, P.E.; Scholz, M.B.; Chain, P.S. Accurate read-based metagenome characterization using a hierarchical suite of unique signatures. *Nucleic Acids Res.* **2015**, *43*, e69. [CrossRef] [PubMed]
35. Wood, D.E.; Salzberg, S.L. Kraken: Ultrafast metagenomic sequence classification using exact alignments. *Genom. Biol.* **2014**, *15*, R46. [CrossRef] [PubMed]
36. Segata, N.; Waldron, L.; Ballarini, A.; Narasimhan, V.; Jousson, O.; Huttenhower, C. Metagenomic microbial community profiling using unique clade-specific marker genes. *Nat. Methods* **2012**, *9*, 811–814. [CrossRef] [PubMed]
37. Li, H. Aligning sequence reads, clone sequences and assembly contigs with BWA-MEM. *arXiv*, 2013.
38. Philipson, C.W.; Davenport, K.; Voegtly, L.; Lo, C.; Li, P.; Xu, Y.; Shakya, M.; Cer, R.Z.; Bishop-Lilly, K.A.; Hamilton, T.; et al. Brief protocol for EDGE bioinformatics: Analyzing microbial and metagenomic NGS data. *Bio-Protc.* **2017**, *7*. [CrossRef]
39. Delcher, A.L.; Bratke, K.A.; Powers, E.C.; Salzberg, S.L. Identifying bacterial genes and endosymbiont DNA with glimmer. *Bioinformatics* **2007**, *23*, 673–679. [CrossRef] [PubMed]
40. McNair, K.; Zhou, C.; Souza, B.; Edwards, R. Thea: A novel approach to gene identification in phage genomes. *bioRxiv* **2018**. [CrossRef]
41. Hauser, F.; Chen, W.; Deinlein, U.; Chang, K.; Ossowski, S.; Fitz, J.; Hannon, G.J.; Schroeder, J.I. A genomic-scale artificial microrna library as a tool to investigate the functionally redundant gene space in arabidopsis. *Plant Cell* **2013**, *25*, 2848–2863. [CrossRef] [PubMed]
42. Grazziotin, A.L.; Koonin, E.V.; Kristensen, D.M. Prokaryotic virus orthologous groups (PVOGs): A resource for comparative genomics and protein family annotation. *Nucleic Acids Res.* **2017**, *45*, D491–D498. [CrossRef] [PubMed]
43. Joshi, T.; Xu, D. Quantitative assessment of relationship between sequence similarity and function similarity. *BMC Genom.* **2007**, *8*, 222. [CrossRef] [PubMed]
44. Sangar, V.; Blankenberg, D.J.; Altman, N.; Lesk, A.M. Quantitative sequence-function relationships in proteins based on gene ontology. *BMC Bioinform.* **2007**, *8*, 294. [CrossRef] [PubMed]
45. Finn, R.D.; Clements, J.; Arndt, W.; Miller, B.L.; Wheeler, T.J.; Schreiber, F.; Bateman, A.; Eddy, S.R. Hmmer web server: 2015 update. *Nucleic Acids Res.* **2015**, *43*, W30–W38. [CrossRef] [PubMed]
46. Finn, R.D.; Coggill, P.; Eberhardt, R.Y.; Eddy, S.R.; Mistry, J.; Mitchell, A.L.; Potter, S.C.; Punta, M.; Qureshi, M.; Sangrador-Vegas, A.; et al. The pfam protein families database: Towards a more sustainable future. *Nucleic Acids Res.* **2016**, *44*, D279–D285. [CrossRef] [PubMed]
47. Lowe, T.M.; Chan, P.P. Trnascan-se on-line: Integrating search and context for analysis of transfer RNA genes. *Nucleic Acids Res.* **2016**, *44*, W54–W57. [CrossRef] [PubMed]
48. Laslett, D.; Canback, B. Aragorn, a program to detect tRNA genes and tmRNA genes in nucleotide sequences. *Nucleic Acids Res.* **2004**, *32*, 11–16. [CrossRef] [PubMed]

49. Rice, L.B. Federal funding for the study of antimicrobial resistance in nosocomial pathogens: No eskape. *J. Infect. Dis.* **2008**, *197*, 1079–1081. [CrossRef] [PubMed]
50. Boucher, H.W.; Talbot, G.H.; Bradley, J.S.; Edwards, J.E.; Gilbert, D.; Rice, L.B.; Scheld, M.; Spellberg, B.; Bartlett, J. Bad bugs, no drugs: No eskape! An update from the infectious diseases society of america. *Clin. Infect. Dis.* **2009**, *48*, 1–12. [CrossRef] [PubMed]
51. Zhang, X.; Wang, Y.; Tong, Y. Analyzing genome termini of bacteriophage through high-throughput sequencing. *Methods Mol. Biol.* **2018**, *1681*, 139–163. [PubMed]
52. Colavecchio, A.; D'Souza, Y.; Tompkins, E.; Jeukens, J.; Freschi, L.; Emond-Rheault, J.G.; Kukavica-Ibrulj, I.; Boyle, B.; Bekal, S.; Tamber, S.; et al. Prophage integrase typing is a useful indicator of genomic diversity in *Salmonella enterica*. *Front. Microbiol.* **2017**, *8*, 1283. [CrossRef] [PubMed]
53. Bankhead, T.; Segall, A.M. Characterization of a mutation of bacteriophage λ integrase. Putative role in core binding and strand exchange for a conserved residue. *J. Biol. Chem.* **2000**, *275*, 36949–36956. [CrossRef] [PubMed]
54. Feiner, R.; Argov, T.; Rabinovich, L.; Sigal, N.; Borovok, I.; Herskovits, A.A. A new perspective on lysogeny: Prophages as active regulatory switches of bacteria. *Nat. Rev. Microbiol.* **2015**, *13*, 641–650. [CrossRef] [PubMed]
55. Deutsch, D.R.; Utter, B.; Fischetti, V.A. Uncovering novel mobile genetic elements and their dynamics through an extra-chromosomal sequencing approach. *Mob. Genet. Elem.* **2016**, *6*, e100502. [CrossRef] [PubMed]
56. Utter, B.; Deutsch, D.R.; Schuch, R.; Winer, B.Y.; Verratti, K.; Bishop-Lilly, K.A.; Sozhmannan, S.; Fischetti, V.A. Beyond the chromosome: The prevalence of unique extra-chromosomal bacteriophages with integrated virulence genes in pathogenic *Staphylococcus aureus*. *PLoS ONE* **2014**, *9*. [CrossRef] [PubMed]
57. Kallberg, M.; Wang, H.; Wang, S.; Peng, J.; Wang, Z.; Lu, H.; Xu, J. Template-based protein structure modeling using the raptorx web server. *Nat. Protoc.* **2012**, *7*, 1511–1522. [CrossRef] [PubMed]
58. Gough, J.; Karplus, K.; Hughey, R.; Chothia, C. Assignment of homology to genome sequences using a library of hidden markov models that represent all proteins of known structure. *J. Mol. Biol.* **2001**, *313*, 903–919. [CrossRef] [PubMed]
59. Picardi, E.; Horner, D.S.; Chiara, M.; Schiavon, R.; Valle, G.; Pesole, G. Large-scale detection and analysis of RNA editing in grape mtDNA by RNA deep-sequencing. *Nucleic Acids Res.* **2010**, *38*, 4755–4767. [CrossRef] [PubMed]
60. Sauvage, V.; Boizeau, L.; Candotti, D.; Vandenbogaert, M.; Servant-Delmas, A.; Caro, V.; Laperche, S. Early minion™ nanopore single-molecule sequencing technology enables the characterization of hepatitis B virus genetic complexity in clinical samples. *PLoS ONE* **2018**, *13*. [CrossRef] [PubMed]
61. Lu, H.; Giordano, F.; Ning, Z. Oxford nanopore minion sequencing and genome assembly. *Genom. Proteom. Bioinform.* **2016**, *14*, 265–279. [CrossRef] [PubMed]
62. Giordano, F.; Aigrain, L.; Quail, M.A.; Coupland, P.; Bonfield, J.K.; Davies, R.M.; Tischler, G.; Jackson, D.K.; Keane, T.M.; Li, J.; et al. De novo yeast genome assemblies from MinION, PacBio and MISeq platforms. *Sci. Rep.* **2017**, *7*. [CrossRef] [PubMed]

Article

Rapid Identification of Intact Staphylococcal Bacteriophages Using Matrix-Assisted Laser Desorption Ionization-Time-of-Flight Mass Spectrometry

Dana Štveráková [1], Ondrej Šedo [2], Martin Benešík [1], Zbyněk Zdráhal [2,3], Jiří Doškař [1] and Roman Pantůček [1,*]

[1] Department of Experimental Biology, Faculty of Science, Masaryk University, Kotlářská 2, 61137 Brno, Czech Republic; dana.stverak@mail.muni.cz (D.Š.); martin.benesik@mail.muni.cz (M.B.); doskar@sci.muni.cz (J.D.)

[2] Central European Institute of Technology, Masaryk University, Kamenice 5, 62500 Brno, Czech Republic; sedo@post.cz (O.Š.); zbynek.zdrahal@ceitec.muni.cz or zdrahal@sci.muni.cz (Z.Z.)

[3] National Centre for Biomolecular Research, Faculty of Science, Masaryk University, Kamenice 5, 62500 Brno, Czech Republic

* Correspondence: pantucek@sci.muni.cz; Tel.: +420-549-49-6379

Received: 13 March 2018; Accepted: 2 April 2018; Published: 4 April 2018

Abstract: *Staphylococcus aureus* is a major causative agent of infections associated with hospital environments, where antibiotic-resistant strains have emerged as a significant threat. Phage therapy could offer a safe and effective alternative to antibiotics. Phage preparations should comply with quality and safety requirements; therefore, it is important to develop efficient production control technologies. This study was conducted to develop and evaluate a rapid and reliable method for identifying staphylococcal bacteriophages, based on detecting their specific proteins using matrix-assisted laser desorption ionization time-of-flight mass spectrometry (MALDI-TOF MS) profiling that is among the suggested methods for meeting the regulations of pharmaceutical authorities. Five different phage purification techniques were tested in combination with two MALDI-TOF MS matrices. Phages, either purified by CsCl density gradient centrifugation or as resuspended phage pellets, yielded mass spectra with the highest information value if ferulic acid was used as the MALDI matrix. Phage tail and capsid proteins yielded the strongest signals whereas the culture conditions had no effect on mass spectral quality. Thirty-seven phages from *Myoviridae*, *Siphoviridae* or *Podoviridae* families were analysed, including 23 siphophages belonging to the International Typing Set for human strains of *S. aureus*, as well as phages in preparations produced by Microgen, Bohemia Pharmaceuticals and MB Pharma. The data obtained demonstrate that MALDI-TOF MS can be used to effectively distinguish between *Staphylococcus*-specific bacteriophages.

Keywords: MALDI-MS; *Staphylococcus*; bacteriophages; phage therapy; *Kayvirus*; Viral proteins

---

1. Introduction

Due to increasing antibiotic resistance, bacterial infections have become a serious problem in hospital and community environments. A possible approach to combat such infections is phage therapy, either as an alternative to antibiotics or utilizing phage-antibiotic synergy. Experimental phage therapy proved to be successful for the treatment of bacterial infections in animal models and human patients [1]. Interest in phage therapy has therefore increased [2] and applications of phages are currently being investigated extensively [3]. For the rational use of phages, it is necessary to have rapid methods of identification, particularly for new isolates as well as previously characterized phages after passaging for quality control of bacteriophage-based products.

Identification of bacteriophages is usually based on morphological characterization using electron microscopy and by genome sequencing [4]. Multiplex PCR can be used to identify well-characterized phages with defined conserved genes typical for particular groups, for example, phages of dairy bacteria [5] or staphylococcal phages [6]. However, high mosaicism and modular genomic structure of tailed phages complicates their detection [7].

This work focuses on the identification of staphylococcal phages that play an important role in biology, evolution and pathogenicity of staphylococci [8–12]. While lytic phages shape bacterial population dynamics, temperate phages are a driving force in bacterial evolution and can benefit the host bacteria by introducing novel traits such as virulence factors as a consequence of lysogenic conversion [13]. Bacteriophages also facilitate the horizontal transfer of bacterial DNA, including resistance genes, through transduction [14–16].

The taxonomy of staphylococcal phages has recently been updated by the International Committee on Taxonomy of Viruses (ICTV) [17]. All staphylococcal phages belong to the order *Caudovirales*, including *Myoviridae*, *Siphoviridae* and *Podoviridae* families. Staphylococcal phages from the family *Siphoviridae* are temperate and comprise 3 genera that correlate with previously described serological groups [18]. Candidates for combating staphylococcal infections are phages belonging to *Myoviridae* and *Podoviridae* families [8] as they seem unable to lysogenise host cells [11] and recent findings suggest that lytic and polyvalent staphylococcal phages belonging to the *Myoviridae* family are safe for therapeutic applications [19].

Fingerprinting by means of Matrix-Assisted Laser Desorption/Ionization—Time-of-Flight Mass Spectrometry (MALDI-TOF MS) is based on rapid detection of compounds ionized directly or after minimal sample treatment. The greatest success of this method has been achieved in the field of bacterial identification [20]. Protein profiles obtained by MALDI-TOF MS are used nowadays in thousands of clinical laboratories worldwide. Apart from bacteria, several other types of samples, including viruses, can be subjected to MALDI-TOF MS profiling analysis [21]. However, MALDI-TOF MS profiling is not currently used for routine identification of tailed phages.

The first successful attempt at bacteriophage MALDI-TOF MS profiling was documented by Thomas et al. [22]. This study focused on the ssRNA phage MS2 that belongs to the *Leviviridae* family and infects *Escherichia coli*. Treatment of MS2 phage particles, isolated by centrifugation and ultrafiltration, with 50% acetic acid was found to be necessary to obtain protein signals, as an acidic environment causes disassembly of phage protein structures. Acid treatment directly in the MALDI matrix solution was also found to be beneficial, as demonstrated by the successful generation of protein signals after preparing the phage sample in a MALDI matrix comprising 17% formic acid [23,24]. To improve reproducibility and sensitivity, McAlpin et al. [25] suggested a 10 min treatment of *Yersinia pestis* podovirus φA1122 with β-mercaptoethanol to reduce inter-molecular disulphide bonds. This procedure was found to be more efficient than acid treatment.

Despite relatively long-standing knowledge that MALDI-TOF MS fingerprinting of bacteriophages was possible, most studies in the field were aimed at method optimization and involved only a limited set of strains. The largest collection of 12 strains was assessed by Bourdin et al. [26] who used MALDI-TOF MS profiling for phage purification control after a multi-step isolation procedure. These works have demonstrated the usefulness of phage protein profiling for particular analytical purposes, however, comprehensive evaluation of sample preparation methods, MALDI-TOF mass spectral quality, reproducibility and discriminatory power have not yet been assessed critically. For this reason, we conducted a study on a relatively large set involving closely related staphylococcal phage strains, using various combinations of sample preparation techniques to establish the optimum setup and assess its applicability in identifying bacteriophages.

2. Materials and Methods

2.1. Bacterial Strains and Phages

Phages and bacterial strains used in this work are described in Table 1. The bacteriophages from the International Typing Set for human *S. aureus* and their propagation strains (PS) were obtained from Dr. P. Petráš (National Reference Laboratory for *Staphylococci*, National Institute of Public Health, Prague, Czech Republic). Laboratory lysogenised strains PS 47[53$^+$], PS 47[77$^+$] were prepared previously [27]. Phages 11, 80α and K and *S. aureus* RN4220 were kindly provided by Prof. C. Wolz (Interfaculty Inst. of Microbiology and Infection Medicine Tübingen, University of Tübingen, Tübingen, Germany). Propagation strain *S. aureus* RN4220 Δ*tarM* for podoviruses [28] was provided by Prof. A. Peschel (Department of Infection Biology, University of Tübingen, Tübingen, Germany). Phages Twort, 44AHJD and P68 were purchased from Félix d'Hérelle Reference Center for Bacterial Viruses (Université Laval, Québec, QC, Canada) and phage X2 from the National Collection of Type Cultures (Public Health England, Salisbury, UK). Phages B166 and B236 [29], 131 and 812 [30] and phage K1/420 [31] were described previously. Phage PYO was isolated from a phage cocktail containing *Staphylococcus*, *Streptococcus*, *Proteus*, *E. coli* and *Pseudomonas aeruginosa* phages (PYO Bacteriophagum combinierae liquidum, lot no. 000860, BioPharm, Tbilisi, Republic of Georgia). A commercial *Staphylococcus* bacteriophage preparation produced by Microgen (lot no. H141, 0812, 08.14, Moscow, Russia) was purchased in a pharmacy in Samara, Russia. Bacteriophage preparation Stafal® lot no. 140805201 was kindly provided by Bohemia Pharmaceuticals (Prague, Czech Republic). Phage preparation Duofag, staphylococcal phages SAU1 and SAU2 and *P. aeruginosa* phage PAE1 were kindly provided by MB Pharma (Prague, Czech Republic). *S. aureus* strains CCM 4028, CCM 4890 and CCM 8428 were obtained from the Czech Collection of Microorganisms (Masaryk University, Brno, Czech Republic).

Table 1. List of bacteriophages and staphylococcal strains used in this work.

Family	Serogroup/Genus	Phage [1]	Propagation Strains (*S. aureus*)
Siphoviridae	A/*Triavirus*	3A *	PS 3A, CCM 4890
		3C *	PS 3C
		6 *	PS 6
		42E *	PS 42E
		47 *	RN4220, PS 47, PS 47 [53$^+$], PS 47 [77$^+$]
		54 *	PS 54
		75 *	PS 75
		79 *	PS 52A
		81 *	PS 81
		94 *	PS 94
	B/*Phietavirus*	11	CCM 4890
		29 *	RN4220
		52 *	PS 52
		52A *	PS 52A, RN4220
		53 *	CCM 4890
		55 *	RN4220
		71 *	PS 71, CCM 4890
		80 *	PS 80
		80α	RN4220
		83A *	PS 83A
		85 *	PS 85, RN4220
		95 *	PS 95
		96 *	PS 96, RN4220
		B166	CCM 4890
		B236	CCM 4890
		X2	CCM 4890
	L/*Phietavirus*	187	PS 187
	F/*Biseptimavirus*	77 *	PS 77, CCM 4890
		84 *	PS 84

Table 1. *Cont.*

Family	Serogroup/Genus	Phage [1]	Propagation Strains (*S. aureus*)
Myoviridae	D/*Kayvirus*	131	CCM 8428
		812	CCM 4028
		K	RN4220
		K1/420	CCM 8428
		PYO	CCM 8428
	D/*Twortvirus*	Twort	HER 1048 [2]
Podoviridae	G/*P68virus*	44AHJD	RN4220 Δ*tarM*
		P68	RN4220 Δ*tarM*

[1] Phages belonging to the International Typing Set for human *Staphylococcus aureus* are marked with asterisk (*).
[2] *Staphylococcus hyicus*.

2.2. Bacteriophage Propagation and Titration

Phages were propagated on their propagation strains (Table 1) in a liquid medium meat-peptone broth (MPB) prepared from 13 g of nutrient broth (Oxoid, CM0001), 3 g of yeast extract (Oxoid, LP0021) and 5 g of peptone (Oxoid, LP0037) dissolved in distilled water to 1000 mL (pH 7.4). Prophage-less *S. aureus* strains CCM 4890 [32], RN4220 [33] and CCM 8428 [34] were used for propagation of phages with known genomic sequences to eliminate the possibility of false positive signals in MALDI-TOF MS spectra caused by induced prophages. Phages 29, 42E and 79 were propagated in a soft agar layer with 0.7% (w/v) of bacteriological agar (Oxoid, LP0011) on top of solid meat-peptone agar with 1.5% (w/v) technical agar (Oxoid, LP0012). Two other types of growth media were used to test the influence of medium on mass spectral quality: brain heart infusion (BHI) broth (Oxoid, CM1135; pH 7.4) and 2× yeast-tryptone broth (2YT) consisting of 16 g of tryptone (Oxoid, LP0042), 10 g of yeast extract (Oxoid, LP0021) and 5 g of NaCl dissolved in distilled water to 1000 mL (pH 7.4).

For phage titration, ten-fold serial dilutions of phage in MPB were prepared. An overnight culture of propagation strain in MPB was mixed with 1/10 volume of 0.02 M $CaCl_2$. Bacterial suspension (0.1 mL) was added to 3 mL of 0.7% meat-peptone soft agar cooled to 45 °C and overlaid on 1.5% meat peptone agar plates. The plates were left to dry for 10 min. Phage dilutions were dropped onto the top agar and left to dry. Plates were incubated overnight at 37 °C.

2.3. Concentrating Bacteriophage to Pellets

Phage lysate, obtained after complete lysis of the bacteria, was centrifuged at 4500× *g* for 30 min at 4 °C and filtered through 0.45 µm pore-sized polyethersulfone syringe filters (Techno Plastic Products, Trasadingen, Switzerland) to remove bacterial debris. Phages were pelleted by centrifugation at 54,000× *g* for 2.5 h at 4 °C in a JA-30.50 Ti rotor (Beckman, Brea, CA, USA). For preparation of pellets from 3 mL of phage lysates, conical tubes (part no. 358119, Beckman) and adapters (part no. 358153, Beckman) with an SW 55 Ti rotor (Beckman) were used. The resulting pellet was resuspended in 350 µL of phage buffer (5×10^{-2} mol/L Tris pH 8.0, 10^{-2} mol/L $CaCl_2$, 10^{-2} mol/L NaCl) overnight at 4 °C. For resuspending the pellets from 3 mL samples, 30 µL of phage buffer was used.

2.4. Purification of Phage Particles by CsCl Density Gradient Centrifugation

This procedure was carried out according to the description by Nováček et al. [31]. Phage pellets resuspended in phage buffer were used as an input material for CsCl density gradient centrifugation. Soluble proteins from resuspended pellets were removed by extraction with an equal volume of chloroform. The resulting aqueous fraction (approximately 1.2 mL) was overlaid onto a preformed CsCl (Sigma-Aldrich, St. Louis, MO, USA) density gradient (1 mL of each 1.45 g/mL 1.50 g/mL, 1.70 g/mL of CsCl in phage buffer) and centrifuged at 194,000× *g* for 4 h at 12 °C using a SW 55 Ti rotor (Beckman). Phage particles forming a visible zone were collected by puncturing the tube with an 0.8 mm gauge needle and syringe. Caesium chloride was removed from the phage-containing fraction by dialysis against an excess of phage buffer at 4 °C overnight using Visking dialysis tubing type 8/32", 0.05 mm thick (part no. 1780.1, Carl Roth, Karlsruhe, Germany).

2.5. Fast Protein Liquid Chromatography (FPLC)

Bacteriophage lysates were purified using a monolithic column, CIMmultus™ QA 1 mL (Bia separations, Ajdovščina, Slovenia) and FPLC NGC (Bio-Rad, Hercules, CA, USA). Phages were purified according to column manufacturer's recommendation with minor modifications [35]. After removing bacterial debris by centrifugation and filtration, 25 mL of phage lysate (10^{8-9} PFU/mL) were mixed with 100 mM sodium phosphate buffer pH 7 at a ratio of 1:1. The column was equilibrated with 100 mM sodium phosphate buffer pH 7. The phage suspension was loaded onto the column and a linear gradient of 100 mM sodium phosphate, 2 M NaCl, pH 7 buffer was applied for elution of phages. NaCl was removed from the phage-containing fraction by dialysis as described following CsCl density gradient centrifugation.

2.6. Phage Purification by Ultrafiltration

Filtered phage lysate was used for tangential flow filtration using Pellicon XL 50 Ultrafiltration Cassettes (Millipore, Burlington, MA, USA) Biomax 300 (for *Podoviridae* and *Myoviridae* phages) and Biomax 500 (for *Siphoviridae* phages) according to the manufacturer's instructions. All fluids were loaded onto the cassette with a peristaltic pump set at 40–50 mL per minute. The cassette was rinsed with 300 mL of sterile distilled water and then 300 mL of phage buffer. 350 mL of phage lysate was loaded onto the cassette and after concentrating/thickening of phages to 10–50 mL, phages were mixed with 1000 mL of phage buffer and thickened again to 10–50 mL.

2.7. Phage Precipitation by Polyethylene Glycol (PEG)

Phages were purified with PEG 8000 (Sigma-Aldrich) according to Sambrook et al. [36] with minor modifications. NaCl and PEG 8000 were dissolved in 30 mL of filtered phage lysate to final concentrations of 0.5 M and 10 % (w/v), respectively, by brief stirring. Phages were precipitated overnight at 4 °C, pelleted by centrifugation at 10,000× *g* for 15 min at 4 °C and the supernatant was discarded. Pellets were dissolved in 0.5 mL of phage buffer overnight at 4 °C. Phage particles were then separated from co-precipitated bacterial debris by centrifugation at 5000× *g* for 10 min at 4 °C. The residual PEG was removed by gentle extraction for 1 min with an equal volume of chloroform. The phage-containing aqueous phase was separated by centrifugation at 5000× *g* for 15 min, collected and filtered through 0.22 µm pore-sized polyethersulfone syringe filters (Techno Plastic Products, Trasadingen, Switzerland).

2.8. Sample Preparation for MALDI-MS

The phage samples were mixed with a MALDI matrix solution in a 1:4 v/v ratio, the resulting mixtures were applied to three positions on the stainless steel MALDI sample plate in a volume of 0.6 µL and were allowed to dry at room temperature. Ferulic acid (FerA, 12.5 mg/mL in a water:acetonitrile:formic acid, 50:33:17, v/v mixture), or, alternatively, alpha-cyano-4-hydroxycinnamic acid (HCCA, saturated solution in water:acetonitrile:trifluoroacetic acid, 47.5:50:2.5, v/v mixture) were used as the MALDI matrix solutions.

For peptide mass fingerprinting and MS/MS analyses, the phage preparations in a volume of 5 µL were incubated with 50 ng of trypsin in 25 mM ammonium bicarbonate for two h at 40 °C. The proteolytic mixture in a volume of 1 µL was applied to an AnchorChip MALDI sample plate, mixed with 0.6 µL of the MALDI matrix (alpha-cyano-4-hydroxycinnamic acid, 2 mg/mL in a water:acetonitrile:trifluoroacetic acid, 32:66:2, v/v mixture) and allowed to dry at room temperature.

2.9. MALDI-MS Profiling Analysis

MALDI-TOF mass spectral fingerprints were obtained using an Ultraflextreme instrument (Bruker Daltonik, Bremen, Germany) operated in the linear positive mode under FlexControl 3.4 software. External calibration of the mass spectra in the linear positive mode was performed using lysozyme (its monomer, dimer and multiply protonated ions). Laser power was set to 120% of the threshold laser power for a particular type of sample. Three independent spectra comprising 1000 laser shots each were acquired from each of the wells. Within an individual well, a minimum of 200 and a maximum of 400 shots were obtained from one position. The mass spectra were recorded within the *m/z* range of 2–100 kDa. Mass spectra were processed using Flex Analysis (version 3.4; Bruker Daltonik). The MALDI-TOF mass spectra-based dendrogram was constructed using the Pearson's product moment coefficient as a measure of similarity and the unweighted pair group average linked method (UPGMA) as a grouping method using the Biotyper 3.1 software (Bruker Daltonik). The protein fingerprints of phages with known genomic sequences were closely examined by comparing the *m/z* values with the predicted molecular weight (Mw) of phage structural proteins from the NCBI Protein database or from custom RAST annotations [37]. For matching the *m/z* values to Mw of phage structural proteins, a 500 ppm tolerance was taken into account. The first amino acid of a protein was identified using TermiNator [38,39] and the Mw was calculated using ExPASy ProtParam [40].

2.10. MALDI-MS/MS Analysis

Analyses of digested samples were carried out using the same instrument operated in the reflectron positive mode. Seven peptide standards (Bruker Daltonik) covering the mass range of 700–3100 Da were used for external mass calibration. Peptide maps were acquired with 800 laser shots. Peaks with minimum S/N = 10 were picked out for MS/MS analysis employing the LIFT arrangement with 600 laser shots for each peptide. The MASCOT 2.2 (MatrixScience, London, UK) search engine was used for processing the MS and MS/MS data. Database searches were carried out on the NCBIprot database (non-redundant, taxonomy All Entries; downloaded from ftp://ftp.ncbi.nih.gov/blast/db/FASTA/). For MALDI-MS data, a mass tolerance of 30 ppm was allowed for peptide mapping and 0.5 Da for MS/MS ion searches. Oxidation of methionine as an optional modification and two enzyme miscleavages were set for all searches. Peptides with statistically significant peptide scores ($p < 0.05$) were considered. Manual MS/MS spectral assignment validation was carried out.

3. Results

3.1. Comparison of Phage Purification Techniques for Sample Preparation and MALDI-TOF MS Method Development

Five different techniques that are commonly used for phage purification and concentration were tested together with two MALDI matrices for developing a rapid and reliable MALDI-TOF MS-based method for identification of *S. aureus* bacteriophages. Traditional techniques used for the concentration and purification of phages involve centrifugation, filtration, precipitation and zonal ultracentrifugation [41]. In addition to these traditional techniques, FPLC has been recently proved to be efficient for phage purification [42]. The following types of phage samples were analysed by MALDI-TOF MS: (i) phage pellets dissolved in phage buffer; (ii) phages purified by CsCl density gradient centrifugation; (iii) PEG precipitated phages; (iv) phages purified using FPLC and a poly(glycidyl methacrylate-co-ethylene dimethacrylate) monolith column modified with quaternary amine groups; and (v) phages concentrated and purified with Pellicon XL 50 ultrafiltration cassettes. As a measure of the feasibility of the sample preparation method, the mass spectral quality, in terms of the number of peaks detected, the signal-to-noise ratio and reproducibility based on three independent experiments separated by a minimum of weekly intervals, were assessed on five phage specimens belonging to five different genera from three families: phage K1/420 belonging to the *Myoviridae* family and *Kayvirus* genus; phage P68 belonging to the *Podoviridae* family and the *P68virus* genus; phages 3A, 71 and 77 belonging to the *Siphoviridae* family and *Triavirus*, *Phietavirus* and *Biseptimavirus* genera, respectively. The results are summarized in Table 2 and depicted in Figure 1.

Table 2. Evaluation of the mass spectral quality in terms of signal reproducibility (peaks detected in 100% or 70–99% of mass spectra acquired for each sample) and signal-to-noise ratio of the most intense peak when using five different isolation methods and two matrix-assisted laser desorption ionization (MALDI) matrices.

Phage	Isolation Method	MALDI Matrix	One Term			Three Terms	
			100% Peaks	70–99% Peaks	Maximal Signal-to-Noise	100% Peaks	70–99% Peaks
3A	CsCl gradient	FerA	20	22	504 ± 141	12	18
	CsCl gradient	HCCA	3	2	107 ± 41	-	-
	FPLC	FerA	4	5	34 ± 11	-	-
	Ultrafiltration	FerA	1	8	16 ± 4	-	-
	Pellet	FerA	23	9	517 ± 219	8	8
	Pellet	HCCA	15	4	44 ± 4	-	
71	CsCl gradient	FerA	8	13	397 ± 307	3	15
	CsCl gradient	HCCA	2	2	5 ± 1	-	-
	FPLC	FerA	5	3	157 ± 72	-	-
	Ultrafiltration	FerA	3	6	17 ± 1	-	-
	Pellet	FerA	11	11	557 ± 108	9	4
	Pellet	HCCA	14	9	10 ± 1	-	-
77	CsCl gradient	FerA	10	21	591 ± 96	6	20
	CsCl gradient	HCCA	11	3	64 ± 11	-	-
	FPLC	FerA	0	0	N/A	-	-
	Ultrafiltration	FerA	6	3	19 ± 7	-	-
	Pellet	FerA	12	9	247 ± 36	8	11
	Pellet	HCCA	9	6	15 ± 7	-	-
K1/420	CsCl gradient	FerA	41	13	398 ± 122	21	12
	CsCl gradient	HCCA	17	3	429 ± 30	-	-
	FPLC	FerA	0	0	N/A	-	-
	Ultrafiltration	FerA	0	0	N/A	-	-
	Pellet	FerA	17	7	125 ± 23	8	7
	Pellet	HCCA	5	5	29 ± 3	-	-
P68	CsCl gradient	FerA	13	14	452 ± 122	7	9
	CsCl gradient	HCCA	4	5	281 ± 18	-	-
	FPLC	FerA	19	11	227 ± 23	-	-
	Ultrafiltration	FerA	5	1	41 ± 15	-	-
	Pellet	FerA	9	9	248 ± 19	5	4
	Pellet	HCCA	4	4	399 ± 166	-	-

Legend: N/A, not applicable; -, not analysed.

Caesium chloride density gradient centrifugation provided reproducible MALDI-TOF mass spectra with the highest number of peaks and signal-to-noise ratios for most of the five samples assessed. This method is the most precise for phage purification and is routinely used in structural studies of phages. On the other hand, CsCl density gradient centrifugation is an expensive and time-consuming method that is inaccessible to many users and inapplicable for large scale production of phages. Therefore, MALDI-TOF MS phage identification using other affordable phage purification methods was investigated. PEG precipitation is an easy method for small-scale phage purification but the residues of PEG 8000 in samples interfered with the ionization of proteins, so this method was determined to be unsuitable for subsequent MALDI-TOF MS profiling analysis (Figure 1). When FPLC was used for phage purification, the mass spectra showed approximately 50% of signals compared to spectra of the CsCl purified phages. Similarly, ultrafiltration was not determined to be a suitable method, yielding a considerably lower number of signals (Table 2). To improve the mass spectral quality, treatment of the phages with β-mercaptoethanol for disassembling the phage particles was assessed, as described by McAlpin et al. [25]. Apart from the recommended protocol (10 min treatment at a 1:10 v/v ratio), a treatment of phage 3A purified by 500 kDa Pellicon XL 50 ultrafiltration cassette was modified to 30 or 60 min at v/v ratio of 1:5 or 1:30. However, no improvement in mass spectral quality was achieved by any of the β-mercaptoethanol treatment conditions tested (Figure S1).

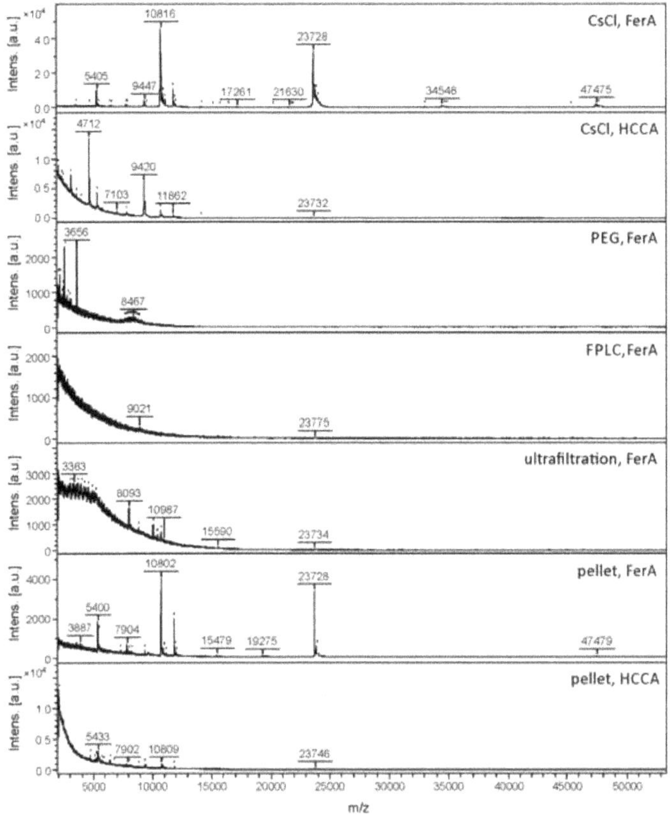

Figure 1. Comparison of the mass spectra of phage 77 prepared by five different sample preparation methods and by using two different matrix-assisted laser desorption ionization matrices, ferulic acid (FerA) and alpha-cyano-4-hydroxycinnamic acid (HCCA) shown for samples prepared by CsCl density gradient centrifugation and pellet dissolution.

The simplest setup consisting of phage pellet dissolution in phage buffer yielded surprisingly high-quality spectra comparable to those acquired from phages purified by CsCl density gradient centrifugation. Although in this case, the phage was not completely separated from bacterial cell debris, the quantity of phage particles was sufficient to provide mass spectra containing the majority of peaks (80–90%) corresponding to those detected after CsCl purification (Figure 1). Other components of the pellet, such as remaining bacterial cell debris evidently do not cause interference by suppressing ionization of the bacteriophage proteins nor by the appearance of superfluous signals unrelated to bacteriophages.

The influence of sample preparation technique on mass spectral quality was assessed in more depth by cluster analysis involving mass spectra obtained by three repeated analyses, using both CsCl density gradient centrifugation and pellet dissolution (Figure 2). The distances corresponding to variability among three repeated experiments were greater than those arising from different sample preparation methods. The influence of sample preparation method on analytical output was found to be insignificant. The simpler sample preparation procedure thus yielded satisfactory mass spectral outputs; however, to accurately determine the relationship between mass spectra and phage protein components, CsCl density gradient centrifugation was the most appropriate.

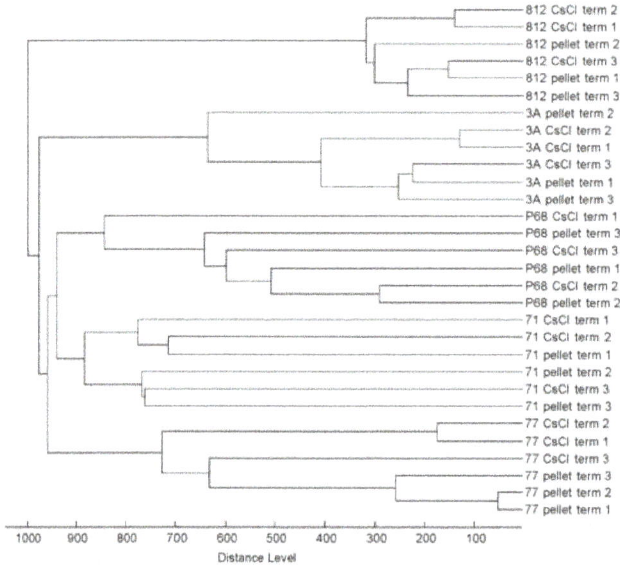

Figure 2. Cluster analysis based on matrix-assisted laser desorption ionization-time-of-flight (MALDI-TOF) mass spectra of three independent analyses of five phage strains prepared by CsCl density gradient centrifugation and pellet dissolution.

A disadvantage of the MALDI-MS profiling method using ferulic acid as the MALDI matrix was the inability to achieve automatic mass spectral acquisition due to non-homogeneous crystallization on the MALDI target. For that reason, another matrix solution (HCCA) was tested, selected on the basis of results of Bourdin et al. [26]. This matrix solution is also used for routine bacterial identification; its use in phage profiling would thus be more compatible with a microbiological laboratory workflow. The mass spectra obtained using HCCA as a matrix revealed signals in a narrower mass range (Figure 1). In addition, in some cases, almost all peaks detected corresponded to different ionization forms of the same protein caused by multiple protonation or gas-phase oligomer formation. The information content of the mass spectra was therefore limited, negatively influencing the discriminatory power of the method. In addition, the use of HCCA as a matrix was not advisable in combination with samples prepared by pellet dissolution, as all mass spectral quality parameters were significantly deteriorated in that case (see Figure 1 and Table 2). For that reason, ferulic acid as a MALDI matrix, in combination with phage pellet dissolution, was used in the remaining profiling experiments.

3.2. *Effect of Culture Conditions on the Quality of Phage MALDI-TOF Mass Spectra*

Culture conditions can affect bacterial MALDI-MS protein fingerprints. As published previously [43], there were small differences in the MALDI-TOF mass spectra of bacteria cultured under different conditions. Nevertheless, despite those differences, whole bacterial cells were identifiable by MALDI-TOF MS regardless of the medium used. However, it is not known whether phage proteomes are influenced by different phage propagation conditions. As a parameter of sample preparation, the influence of culture medium on MALDI-TOF mass spectral quality was tested. For that purpose, the five phage strains 3A, 71, 77, K1/420 and P68 used for method development were grown on prophage-less *S. aureus* strains in three different growth media (MPB, BHI and 2YT). As an example, the mass spectra of phage 3A are compared in Figure 3a. The main differences between the mass spectra were the relative intensities of signals in the high-mass range. Differences were also visible in the low-mass range (especially at *m/z* lower than 3.5 kDa). These probably resulted in more distant positions of the analyses in the dendrogram (Figure 4), equivalent to mass spectral differences due to the accuracy of the method itself (distance level equal to 1000). Similar effect of the culture conditions on the mass spectral quality was determined also for the remaining four strains 71, 77, K1/420 and P68 showing only minor qualitative differences in their mass spectra. The proportion of peaks conserved regardless of the cultivation conditions was greater than 80% in all cases including nine replicate analyses per sample.

The influence of different *S. aureus* strains used for phage propagation was also tested, especially for the possibility of induced prophages that could affect the protein fingerprint obtained by MALDI-TOF MS. Bacteriophage 47 was grown on prophage-less strain RN4220, on strain PS 47 (=RN1) known to carry three prophages 11, 12 and 13 and on two quadruple lysogenic strains derived from PS 47 harbouring additional prophages 53 or 77. The MALDI-TOF mass spectra shown in Figure 3b reveal that the vast majority of peaks were conserved, regardless of the propagation strain. Qualitative mass spectral differences were associated with peaks of low intensity in the low-mass range (lower than 7 kDa), representing less than 15% of all peaks observed in the individual mass spectra. The most significant differences were of a quantitative nature, where the relative intensities of peaks at *m/z* greater than 20 kDa and their signal-to-noise ratios, were influenced by the propagation strain.

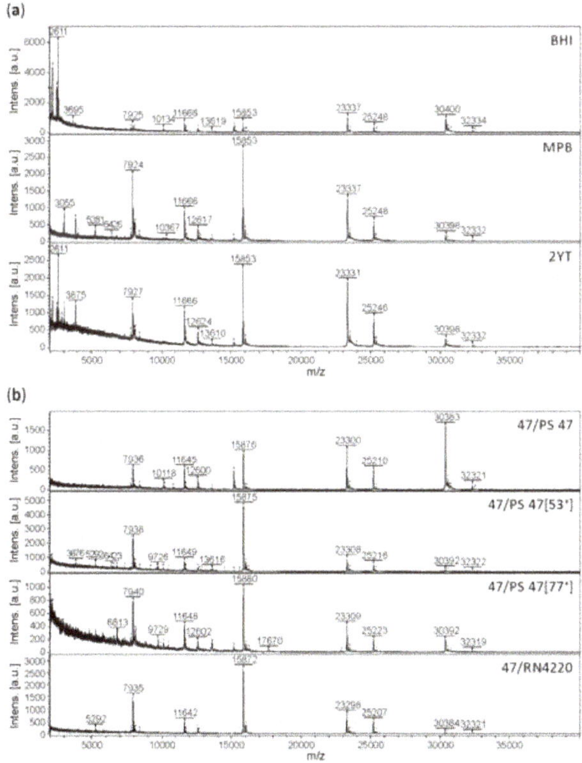

Figure 3. Comparison of the mass spectra of phage 3A using three different cultivation media (**a**) and phage 47 grown on four different propagation strains (**b**).

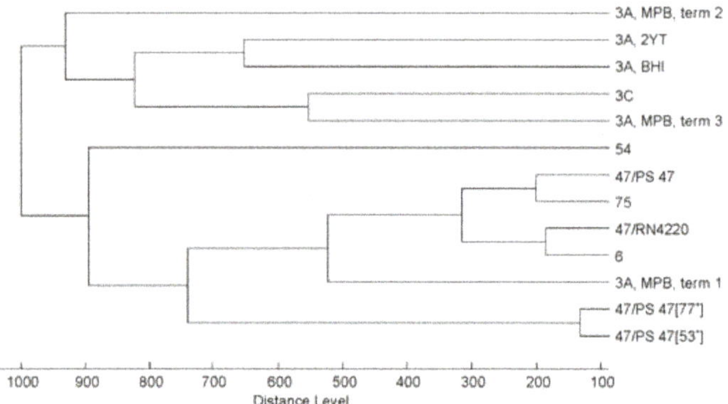

Figure 4. Cluster analysis of *Triavirus* mass spectra demonstrating the influence of the mass spectral variability of the method itself (phage 3A, term 1, term 2, term 3), variability resulting from different propagation strains (phage 47 propagated on *S. aureus* RN4220, PS 47, PS 47 [53^+] and PS 47 [77^+]) and different cultivation media (phage 3A propagated in meat-peptone broth (MPB), brain heart infusion (BHI) and 2× yeast-tryptone broth (2YT).

The impact of propagation strain on the discriminatory power of the method was tested by means of cluster analysis based on MALDI-TOF mass spectra of six *Triavirus* strains of the *Siphoviridae* family that have very similar MALDI-TOF mass spectra (Figure 4). While the accuracy of the method, as reflected by the distance between three independent analyses of phage strain 3A, reached the maximum dendrogram distance (1000), mass spectral differences of phage 47 arising from different propagation strains resulted in a distance of 750. Note that due to the clustering method used (see Section 2.9), these threshold values are dependent of the collection of mass spectra used for the generation of individual dendrograms and the distance level representing the minimal similarity is always considered to be the highest distance (1000). The influence of propagation strain was thus lower than the overall accuracy of the method and its impact on the discriminatory power of the method was practically insignificant.

3.3. *Profiling of 37 Phages by MALDI-TOF MS*

Under the sample preparation arrangement proposed, 37 phages were purified and analysed. Mass spectra of *Myoviridae* and *Siphoviridae* were found to be richer than the mass spectra of *Podoviridae*. This could be explained by the fact that *Myoviridae* and *Siphoviridae* include more complex phages with a higher number of structural proteins of molecular weights detectable by MALDI-TOF MS. The dendrogram based on cluster analysis of the mass spectral signals (Figure 5) revealed four significant groups. The first one corresponded to phages of the *Kayvirus* genus (K, K1/420, 812, 131 and PYO). This grouping was supported by visual inspection of the mass spectra where all phage strains shared distinctive peaks at m/z = 10,204, 11,533, 15,796, 17,745, 19,112 and 23,067. This mass spectral profile allows unambiguous identification of kayviruses that are frequently used in phage therapy.

Six phages of the second well separated cluster (3A, 3C, 6, 47, 54 and 75) belonged to the *Triavirus* genus and shared several common signals in their MALDI-MS profiles (m/z = 7936, 11,645, 12,598, 15,871, 23,295, 25,208 and 30,373). The third well-defined group consisted of transducing phages from the genus *Phietavirus* where phages 11, 53 and 80α are closely related at the genomic level. Mass spectra of other phages from this branch did not show major similarities except phages 29 and 55, that shared the peak at m/z = 20,786, corresponding to the major tail protein. The last group consisted of the only two podoviruses P68 and 44AHJD present in this analysis. They shared peaks at m/z= 6916, 15,112 and 46,769. The profiles of other phages were not clustered into groups corresponding to their genomic relatedness. MALDI-TOF mass spectra of *Myoviridae* phage Twort (*Twortvirus* genus) and

Siphoviridae phages 187 and B166 (supposed phietaviruses) showed no similarities with MALDI-TOF mass spectra of any other phages included in the clustered analysis and their genomes and proteomes also differed [29,44].

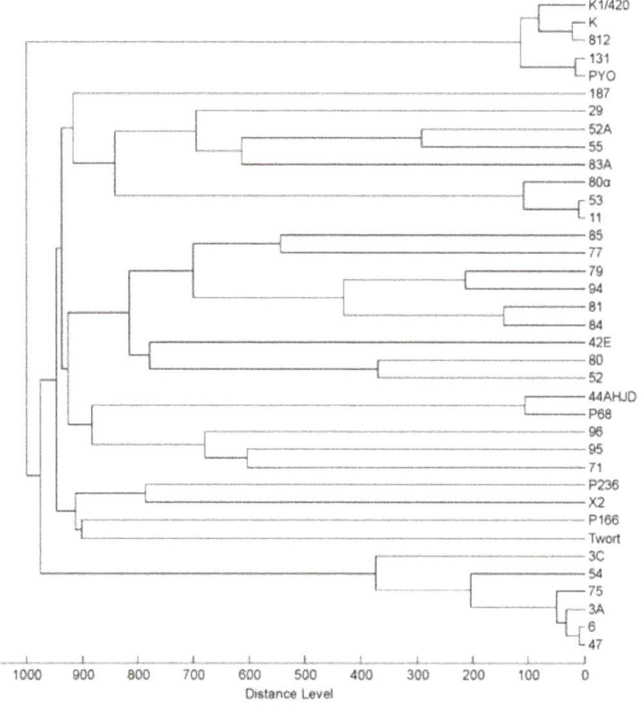

Figure 5. Cluster analysis of mass spectra of all 37 phage strains involved in the study.

On the basis of *m/z* values, peaks from mass spectra were assigned to corresponding phage virion proteins annotated in databases (Table 3). The most frequently detected proteins were the major tail protein, Ig-like domain and major capsid protein. Head and tail connecting protein, baseplate protein and various hypothetical proteins were also assigned. In phage K1/420, representing *Kayvirus*, the identity of proteins was confirmed by MALDI-MS/MS analysis of peptides obtained by tryptic digestion of phage K1/420. The masses of proteins listed in Table 4 correspond to those observed in the MALDI-MS profiles.

Table 3. List of proteins identified in MALDI-TOF MS spectra.

Phage	NCBI Genome Accession No.	Protein Function	NCBI Protein Accession no.	Mw Theoretical	Mw Experimental
3A	NC_007053	Major tail protein	YP_239944	23,335	23,336
		Ig-like domain	YP_239945	15,855	15,857
		Unknown	YP_239952	10,372	10,373
42E	NC_007052	Major tail protein	YP_239866	23,295	23,290
		Ig-like domain	YP_239868	15,871	15,873
47	NC_007054	Major tail protein	YP_240012	23,295	23,298
		Ig-like domain	YP_240013	15,871	15,875
11	NC_004615	Major tail protein	NP_803292	21,382	21,376
		Head-tail connector protein	NP_803289	12,660	12,658

Table 3. *Cont.*

Phage	NCBI Genome Accession No.	Protein Function	NCBI Protein Accession no.	Mw Theoretical	Mw Experimental
80α	NC_009526	Major tail protein	YP_001285367	21,395	21,406
		Unknown	YP_001285362	10,790	10,797
53	NC_007049	Major tail protein	YP_239653	21,395	21,405
		Head-tail connector protein	YP_239650	12,660	12,667
55	NC_007060	Major capsid protein	YP_240459	29,487	29,497
71	NC_007059	Head-tail connector protein	YP_240387	11,759	11,758
		Unknown	YP_240388	12,857	12,857
B166	NC_028859	Major tail protein	AKC04659	20,406	20,413
B236	NC_028915	Major capsid protein	YP_009209168	29,489	29,493
187	NC_007047	Major capsid protein	YP_239493	32,997	32,987
77	NC_005356	Major tail protein	NP_958612	23,730	23,731
		Unknown	NP_958619	10,805	10,801
K1/420	KJ206563	Tail tube protein	AHY26502	15,794	15,796
		Baseplate protein	AHY26518	19,109	19,111
		Putative baseplate component	AHY26523	14,480	14,482
		Ig-like protein	AHY26552	23,069	23,070
		Tail morphogenetic protein	AHY26553	17,718	17,718
		Major tail protein	AHY26554	7818	7819
812	KJ206559	Tail tube protein	*	15,794	15,800
		Baseplate protein	AHY25649	19,109	19,114
		Putative baseplate component	AHY25654	14,480	14,479
		Ig-like protein	AHY25683	23,069	23,073
		Tail morphogenetic protein	*	17,718	17,719
		Major tail protein	AHY25685	7818	7817
131	RAST annotations were used	Tail tube protein	*	15,794	15,792
		Baseplate protein	*	19,109	19,109
		Putative baseplate component	*	14,480	14,484
		Ig-like protein	*	23,069	23,069
		Tail morphogenetic protein	*	17,718	17,719
K	NC_005880.2	Tail tube protein	YP_009041323	15,794	15,802
		Baseplate protein	YP_009041339	19,109	19,104
		Putative baseplate component	YP_009041343	14,480	14,488
		Ig-like protein	YP_009041372	23,069	23,071
		Tail morphogenetic protein	YP_009041373	17,718	17,711
Twort	NC_007021	Structural protein	YP_238577	18,759	18,763
		Unknown	YP_238687	3923.5	3920
		Unknown	YP_238618	7638	7639
P68	NC_004679	Major capsid protein	NP_817336	46,769	46,764
		Unknown	NP_817338	15,112	15,104
		Unknown	NP_817337	6916	6915
44AHJD	NC_004678	Major capsid protein	NP_817314	46,769	46,770
		Unknown	NP_817316	15,112	15,110
		Unknown	NP_817315	6916	6917

* Newly sequenced phage genomes annotated using RAST.

Table 4. List of proteins identified by MALDI-MS/MS after tryptic digestion of phage K1/420.

NCBI Accession No.	Protein Name	Predicted Protein Function	Mw	No. of Matched Peptides	Mascot Score	Sequence Coverage
YP_240967	ORF189	Major tail protein	7818	8	1276	96%
YP_240933	ORF117	Unknown	14,480	5	291	51%
YP_007112937	F867_gp192	Ig-like protein	23,069	3	188	26%

3.4. Practicability of the Method

MALDI-TOF MS is known for its high sensitivity; the detection limit for peptides is reaching, in some cases, attomolar levels. To estimate the detection limit for phages, a series of 10-fold dilutions of CsCl purified phage K1/420 with an initial titre 1×10^9 PFU/mL were prepared. Signals enabling phage identification were detected only in the 10-fold diluted sample. The 100-fold diluted sample yielded only one signal, corresponding to one of the dominant signals of the undiluted phage sample. Further dilution resulted in the absence of any signals (Figure S2). The detection limit of the method is therefore 1×10^7 PFU/mL but practically, 1×10^8 PFU/mL are needed for reliable identification.

To examine the applicability of the method, mass spectra were obtained from pellets formed from 3 mL volumes of lysates of five phages from the initial method evaluation set. Most significantly, phage titre had an impact on mass spectral quality; while no significant decrease in the number of signals compared to the standard procedure was observed for phage strains K1/420 and 3A (initial titre was 7×10^9 and 1×10^{10} PFU/mL), the number of signals decreased to approximately 50% for phages 71 and 77 (initial titre was 1×10^9 PFU/mL for both phages) and to less than 30% in the case of phage P68 with an initial titre of 1×10^8 PFU/mL.

To examine the possibility of identifying phages in commercial phage preparations, *Staphylococcus* bacteriophage produced by Microgen, Stafal®, Duofag and three individual phages designated SAU1, SAU2 and PAE1 that are components of Duofag, were pelleted from a 20 mL volume and analysed by MALDI-TOF MS. The peak lists obtained were compared to a custom database containing 37 bacteriophage profiles (generated in Section 3.3) using a data treatment approach common to routine bacterial identification using Biotyper software. Similarities in the experimental mass spectra to the individual database entries were expressed by log(scores) that indicate the confidence of identification. For bacterial identification, the log(score) thresholds of 1.700 and 2.000 indicate species identification at lower and higher confidence levels, respectively. From the six phage preparations tested in nine replicates, four samples were assigned to database entries of strains of the *Kayvirus* genus with high confidence (the log(scores) from nine analyses were the following: SAU1: 2.06 ± 0.06, SAU2 2.02 ± 0.05, Stafal®: 2.22 ± 0.03 and Duofag: 2.01 ± 0.07) and the remaining sample from the Microgen preparation to the same database entries at a lower confidence level (log(score) = 1.87 ± 0.08). Mass spectra of phage PAE1 did not match any of the database entries (log(score) = 1.07 ± 0.11), which is a true negative result as this phage infects *P. aeruginosa* and differs from staphylococcal phages in our database. It should also be noted that due to the increased number of peaks in the low-mass range, most probably related to a higher complexity of the sample matrix, the identification was based only on signals at $m/z > 3.5$ kDa. Importantly, the score difference between the identification hits of kayviruses and the remaining hits for all samples was always greater than 0.3, which assured no ambiguity in the identification outputs.

4. Discussion

Although phage research started over 100 years ago and their therapeutic applications have been studied from the beginning [45], necessary safety requirements for phage preparations are still being discussed [2,46]. Different phage preparations are currently available on the market or under specific experimental treatment programs in Georgia, Poland, Russia, Slovakia and several clinical trials related to phage therapy have been started in the Western World. The production of bacteriophage preparations should comply with recently established requirements including phage identification as a component of quality control [2]. MALDI-TOF MS is among the recommended methods for phage identification in master and working seed lots [46].

Bacteriophage profiling by MALDI-TOF MS described in the literature has mainly been employed in bacterial identification based on phage amplification and detection of specific protein signals. Bacteriophage amplification was also found to be feasible for the identification of components in a bacterial mixture, which is difficult in direct bacterial profiling by MALDI-TOF MS. This was demonstrated by Rees and Voorhees [24] who were able to determine both components of an *E. coli*—*Salmonella* ssp. mixture by detection of proteins from phages MS2 and MPSS-1 inoculated with the bacteria. Cox et al. [47] employed modelling of *Y. pestis* phage φA1122 and *E. coli* MS2 phage amplification to predict optimum growth conditions and thus to simplify the analysis workflow, which normally includes monitoring phage growth by repeated analyses at certain times. To avoid false positive identifications, undetectably low phage titres must be introduced. Therefore, Pierce et al. [48] proposed the use of isotopically labelled 15N *Siphoviridae* bacteriophage 53, whereas the presence of *S. aureus* was based on detection of an unlabelled form of phage 53 capsid protein.

Over the past decade, several variants of the method based on proteomics approaches have been described: MALDI-TOF MS analysis after microwave-assisted acid digestion of MS2 phages isolated by centrifugation and ultrafiltration represents a simple and rapid method [49]. In combination with subsequent digestion by trypsin, MALDI-TOF MS was used for detection of the amplified staphylococcal phage K, in the presence of an antibiotic, where the phage protein tryptic peptides were detected only for antibiotic resistant bacterial strains [50]. Instead of MALDI, electrospray (ESI) has also been employed. After centrifugation, ultrafiltration and acidification, the diluted solution of MS2 protein extract yielded coat protein signals after direct injection to ESI-MS. The identity of the proteins was verified by ESI-MS/MS (Top-down) by Cargile et al. [51] and Wick et al. [52]. An even more sophisticated approach is represented by LC-ESI-MS/MS of phage proteins digested by trypsin [53].

We have shown that pelleting of phages is time- and cost-effective and this technique can be used for sample preparation for MALDI-TOF MS even from a volume of 3 mL of phage lysate containing a sufficiently high titre (approximately 10^9 PFU/mL). The risk of obtaining limited analytical output when dealing with small sample volumes at lower titres should always be considered, especially when CsCl purification is not involved. The mass spectral quality should be visually monitored to avoid possible ambiguities arising from limited information gained from lower numbers of protein components detected. Similar detection limits were shown by Rees and Barr [50]. Although pelleted phages lack the purity of phages isolated by CsCl density gradient centrifugation, the quality of the MALDI-TOF mass spectra acquired from pellets is comparable to those acquired from CsCl purified phages. At the same time, almost no differences between phages propagated in different types of media or on different propagation strains were found even when prophages were present in the genome of a propagation strain. Prophages are induced spontaneously under various stress conditions [54–56]. In our previous work [57] we showed that it was possible to detect spontaneously induced phages by PCR as they can contaminate the lysates of phages propagated on lysogenic strains. The frequency of spontaneous phage induction from lysogenic bacterial strains, as demonstrated by the appearance of infecting phage particles, has been shown to be low—around 10^{-8} to 10^{-5} PFU per bacterial cell [58]. Under experimental conditions, the cell count in cultivation reaches a maximum of 10^8 cells per mL, therefore 10^3 induced phages per mL may be present. According to the experimentally set phage detection limit of MALDI-TOF MS (10^7 PFU/mL), induced phages cannot be detected, particularly in the presence of an excess of propagated phages. Due to ionization suppression effects, MALDI-TOF MS profiling analysis principally determinates only the majority component of the sample. Without employing other separation steps, the method cannot be used for detection of minority components, such as traces of unwanted contaminants of commercial preparations.

The results of cluster analysis of MALDI-TOF mass spectra obtained from 37 phages including 29 *Siphoviridae*, 6 *Myoviridae* and 2 *Podoviridae* did not always correlate with their genome-based taxonomic status. Therefore, the method is not a suitable tool for classification of unknown phage strains to established phage genera. However, due to the satisfactory degree of repeatability of the mass spectra acquired from phages propagated under different conditions, the method can be used for direct phage strain identification or classification to a group of closely related ones. The possibility of reliable phage strain identification will always rely on the uniqueness of signals observed. Strain typing would be practically possible only after comprehensive examination of a wide of range of strains to affirm the specificity of the signals. It is evident from cluster analysis involving three independent cultivations of selected phages (Figure 4), that some strains would not be distinguishable, as was demonstrated on 3A-related phages from *Triavirus* genus. This is due to the fact that the variability of their mass spectra induced by the analytical method itself was greater than differences between mass spectra among different phages within a particular cluster.

Cluster analysis based on MALDI-TOF mass spectra showed that phages from the *Kayvirus* genus were clustered separately from other phages. Distinguishing kayviruses from other phages is important as these phages have been successfully used in the treatment of staphylococcal infections in humans and animals [19,30,59–63]. For most of the tested phages, the MALDI-TOF MS output was strain specific but kayviruses share 99% amino acid identity in most structural virion proteins, therefore they can be considered as variants of one phage strain, where individual mutant variants differ in genes encoding non-structural proteins. It is thus logical that on the basis of their MALDI-TOF MS protein profiles, these strains remain indistinguishable. MALDI-TOF mass spectra with peaks at *m/z* values typical for *Kayvirus* genus were obtained also from pelleted phage preparations. These signals then permitted the identification of strains using a workflow that is familiar to users of MALDI-MS systems in microbial diagnostics. Interpretation of the scoring outputs that was adopted from a routine setup used in bacterial identification represents a field for specific evaluation and probably, different score thresholds should be applied for phage analysis. *Kayvirus* characteristic peaks were found even in the Duofag phage cocktail that consists of two staphylococcal phages and one *P. aeruginosa* phage.

Our findings suggest that MALDI-TOF MS could be used not only for identification of laboratory cultured phages but also in the verification of phages in ready-to-use preparations. The identification success rate is dependent mostly on phage titres, while the cultivation conditions and sample purity do not play a key role. The reliability of identification of individual strains should be evaluated in comparison with other closely related strains.

Supplementary Materials: The following are available online at http://www.mdpi.com/1999-4915/10/4/176/s1, Figure S1: Comparison of MALDI-TOF mass spectra obtained from phage 3A purified by 500 kDa Pellicon XL 50 ultrafiltration cassette before and after 10 min treatment with β-mercaptoethanol in 1:10 *v*/*v* ratio, Figure S2: MALDI-TOF mass spectra obtained from series of 10-fold dilutions of CsCl purified *Kayvirus* K1/420 with an initial titre 1×10^9 PFU/mL.

Acknowledgments: This work was supported by the Ministry of Health of the Czech Republic (grant number NT16-29916A). The work was also supported from European Regional Development Fund-Project "CIISB4HEALTH" (No. CZ.02.1.01/0.0/0.0/16_013/0001776). Ondrej Šedo and Zbyněk Zdráhal are grateful for support from the Czech Science Foundation (project no. P206/12/G151). CIISB research infrastructure project LM2015043 funded by MEYS CR is gratefully acknowledged for the financial support of the MALDI-TOF MS measurements at the Proteomics Core Facility.

Author Contributions: Roman Pantůček and Dana Štveráková designed the study. Roman Pantůček, Ondrej Šedo, Dana Štveráková and Martin Benešík conceived and planned the experiments; Dana Štveráková, Ondrej Šedo and Martin Benešík performed the experiments; Ondrej Šedo and Dana Štveráková analysed the data; Zbyněk Zdráhal and Jiří Doškař consulted the theory and experimental design; Dana Štveráková, Ondrej Šedo and Roman Pantůček wrote the paper.

Conflicts of Interest: The authors declare no conflict of interest.

References

1. Melo, L.D.R.; Oliveira, H.; Santos, S.B.; Sillankorva, S.; Azeredo, J. Phages against infectious diseases. In *Bioprospecting*; Topics in Biodiversity and Conservation; Springer: Cham, Switzerland, 2017; Volume 16, pp. 269–294. ISBN 978-3-319-47933-0.
2. Sybesma, P.; Pirnay, J.-P. Expert round table on acceptance and re-implementation of bacteriophage therapy Silk route to the acceptance and re-implementation of bacteriophage therapy. *Biotechnol. J.* **2016**, *11*, 595–600. [CrossRef]
3. Bárdy, P.; Pantůček, R.; Benešík, M.; Doškař, J. Genetically modified bacteriophages in applied microbiology. *J. Appl. Microbiol.* **2016**, *121*, 618–633. [CrossRef] [PubMed]
4. Ackermann, H.-W. Phage classification and characterization. *Methods Mol. Biol.* **2009**, *501*, 127–140. [CrossRef] [PubMed]
5. Del Rio, B.; Binetti, A.G.; Martín, M.C.; Fernández, M.; Magadán, A.H.; Alvarez, M.A. Multiplex PCR for the detection and identification of dairy bacteriophages in milk. *Food Microbiol.* **2007**, *24*, 75–81. [CrossRef] [PubMed]

6. Kahánková, J.; Pantůček, R.; Goerke, C.; Růžičková, V.; Holochová, P.; Doškař, J. Multilocus PCR typing strategy for differentiation of *Staphylococcus aureus* siphoviruses reflecting their modular genome structure. *Environ. Microbiol.* **2010**, *12*, 2527–2538. [CrossRef] [PubMed]
7. Brüssow, H.; Hendrix, R.W. Phage genomics: Small is beautiful. *Cell* **2002**, *108*, 13–16. [CrossRef]
8. Deghorain, M.; Van Melderen, L. The staphylococci phages family: An overview. *Viruses* **2012**, *4*, 3316–3335. [CrossRef] [PubMed]
9. Howard-Varona, C.; Hargreaves, K.R.; Abedon, S.T.; Sullivan, M.B. Lysogeny in nature: Mechanisms, impact and ecology of temperate phages. *ISME J.* **2017**, *11*, 1511–1520. [CrossRef] [PubMed]
10. Zeman, M.; Mašlaňová, I.; Indráková, A.; Šiborová, M.; Mikulášek, K.; Bendíčková, K.; Plevka, P.; Vrbovská, V.; Zdráhal, Z.; Doškař, J.; et al. *Staphylococcus sciuri* bacteriophages double-convert for staphylokinase and phospholipase, mediate interspecies plasmid transduction and package *mecA* gene. *Sci. Rep.* **2017**, *7*, 46319. [CrossRef] [PubMed]
11. Xia, G.; Wolz, C. Phages of *Staphylococcus aureus* and their impact on host evolution. *Infect. Genet. Evol.* **2014**, *21*, 593–601. [CrossRef] [PubMed]
12. Goerke, C.; Pantůček, R.; Holtfreter, S.; Schulte, B.; Zink, M.; Grumann, D.; Bröker, B.M.; Doškař, J.; Wolz, C. Diversity of prophages in dominant *Staphylococcus aureus* clonal lineages. *J. Bacteriol.* **2009**, *191*, 3462–3468. [CrossRef] [PubMed]
13. Brüssow, H.; Canchaya, C.; Hardt, W.-D. Phages and the evolution of bacterial pathogens: From genomic rearrangements to lysogenic conversion. *Microbiol. Mol. Biol. Rev.* **2004**, *68*, 560–602. [CrossRef] [PubMed]
14. Haaber, J.; Leisner, J.J.; Cohn, M.T.; Catalan-Moreno, A.; Nielsen, J.B.; Westh, H.; Penadés, J.R.; Ingmer, H. Bacterial viruses enable their host to acquire antibiotic resistance genes from neighbouring cells. *Nat. Commun.* **2016**, *7*, 13333. [CrossRef] [PubMed]
15. Mašlaňová, I.; Doškař, J.; Varga, M.; Kuntová, L.; Mužík, J.; Malúšková, D.; Růžičková, V.; Pantůček, R. Bacteriophages of *Staphylococcus aureus* efficiently package various bacterial genes and mobile genetic elements including SCCmec with different frequencies. *Environ. Microbiol. Rep.* **2013**, *5*, 66–73. [CrossRef] [PubMed]
16. Mašlaňová, I.; Stříbná, S.; Doškař, J.; Pantůček, R. Efficient plasmid transduction to *Staphylococcus aureus* strains insensitive to the lytic action of transducing phage. *FEMS Microbiol. Lett.* **2016**, *363*, fnw211. [CrossRef] [PubMed]
17. Adams, M.J.; Lefkowitz, E.J.; King, A.M.Q.; Harrach, B.; Harrison, R.L.; Knowles, N.J.; Kropinski, A.M.; Krupovic, M.; Kuhn, J.H.; Mushegian, A.R.; et al. Ratification vote on taxonomic proposals to the International Committee on Taxonomy of Viruses (2016). *Arch. Virol.* **2016**, *161*, 2921–2949. [CrossRef] [PubMed]
18. Doškař, J.; Pallová, P.; Pantůček, R.; Rosypal, S.; Růžičková, V.; Pantůčková, P.; Kailerová, J.; Klepárník, K.; Malá, Z.; Boček, P. Genomic relatedness of phages of the International Typing Set and detection of serogroup A, B and F prophages in lysogenic strains. *Can. J. Microbiol.* **2000**, *46*, 1066–1076. [CrossRef] [PubMed]
19. Cui, Z.; Guo, X.; Dong, K.; Zhang, Y.; Li, Q.; Zhu, Y.; Zeng, L.; Tang, R.; Li, L. Safety assessment of *Staphylococcus* phages of the family *Myoviridae* based on complete genome sequences. *Sci. Rep.* **2017**, *7*, 41259. [CrossRef] [PubMed]
20. Bizzini, A.; Greub, G. Matrix-assisted laser desorption ionization time-of-flight mass spectrometry, a revolution in clinical microbial identification. *Clin. Microbiol. Infect.* **2010**, *16*, 1614–1619. [CrossRef] [PubMed]
21. Fenselau, C.; Demirev, P.A. Characterization of intact microorganisms by MALDI mass spectrometry. *Mass Spectrom. Rev.* **2001**, *20*, 157–171. [CrossRef] [PubMed]
22. Thomas, J.J.; Falk, B.; Fenselau, C.; Jackman, J.; Ezzell, J. Viral characterization by direct analysis of capsid proteins. *Anal. Chem.* **1998**, *70*, 3863–3867. [CrossRef] [PubMed]
23. Madonna, A.J.; Van Cuyk, S.; Voorhees, K.J. Detection of *Escherichia coli* using immunomagnetic separation and bacteriophage amplification coupled with matrix-assisted laser desorption/ionization time-of-flight mass spectrometry. *Rapid Commun. Mass Spectrom.* **2003**, *17*, 257–263. [CrossRef] [PubMed]
24. Rees, J.C.; Voorhees, K.J. Simultaneous detection of two bacterial pathogens using bacteriophage amplification coupled with matrix-assisted laser desorption/ionization time-of-flight mass spectrometry. *Rapid Commun. Mass Spectrom.* **2005**, *19*, 2757–2761. [CrossRef] [PubMed]
25. McAlpin, C.R.; Cox, C.R.; Matyi, S.A.; Voorhees, K.J. Enhanced matrix-assisted laser desorption/ionization time-of-flight mass spectrometric analysis of bacteriophage major capsid proteins with β-mercaptoethanol pretreatment. *Rapid Commun. Mass Spectrom.* **2010**, *24*, 11–14. [CrossRef] [PubMed]

26. Bourdin, G.; Schmitt, B.; Guy, L.M.; Germond, J.-E.; Zuber, S.; Michot, L.; Reuteler, G.; Brüssow, H. Amplification and purification of T4-like *Escherichia coli* phages for phage therapy: From laboratory to pilot scale. *Appl. Environ. Microbiol.* **2014**, *80*, 1469–1476. [CrossRef] [PubMed]
27. Borecká, P.; Rosypal, S.; Pantůček, R.; Doškař, J. Localization of prophages of serological group B and F on restriction fragments defined in the restriction map of *Staphylococcus aureus* NCTC 8325. *FEMS Microbiol. Lett.* **1996**, *143*, 203–210. [CrossRef] [PubMed]
28. Li, X.; Gerlach, D.; Du, X.; Larsen, J.; Stegger, M.; Kühner, P.; Peschel, A.; Xia, G.; Winstel, V. An accessory wall teichoic acid glycosyltransferase protects *Staphylococcus aureus* from the lytic activity of *Podoviridae*. *Sci. Rep.* **2015**, *5*, 17219. [CrossRef] [PubMed]
29. Botka, T.; Růžičková, V.; Konečná, H.; Pantůček, R.; Rychlík, I.; Zdráhal, Z.; Petráš, P.; Doškař, J. Complete genome analysis of two new bacteriophages isolated from impetigo strains of *Staphylococcus aureus*. *Virus Genes* **2015**, *51*, 122–131. [CrossRef] [PubMed]
30. Pantůček, R.; Rosypalová, A.; Doškař, J.; Kailerová, J.; Růžičková, V.; Borecká, P.; Snopková, Š.; Horváth, R.; Götz, F.; Rosypal, S. The polyvalent staphylococcal phage φ812: Its host-range mutants and related phages. *Virology* **1998**, *246*, 241–252. [CrossRef] [PubMed]
31. Nováček, J.; Šiborová, M.; Benešík, M.; Pantůček, R.; Doškař, J.; Plevka, P. Structure and genome release of Twort-like *Myoviridae* phage with a double-layered baseplate. *Proc. Natl. Acad. Sci. USA* **2016**, *113*, 9351–9356. [CrossRef] [PubMed]
32. Yoshizawa, Y. Isolation and characterization of restriction negative mutants of *Staphylococcus aureus*. *Jikeikai Med. J.* **1985**, *32*, 415–421.
33. Novick, R. Properties of a cryptic high-frequency transducing phage in *Staphylococcus aureus*. *Virology* **1967**, *33*, 155–166. [CrossRef]
34. Moša, M.; Boštík, J.; Pantůček, R.; Doškař, J. Medicament in the Form of Anti-Staphylococcus Phage Lysate, Process of Its Preparation and Use. Patent Application CZ201200668-A3, 27 September 2012.
35. Kramberger, P.; Honour, R.C.; Herman, R.E.; Smrekar, F.; Peterka, M. Purification of the *Staphylococcus aureus* bacteriophages VDX-10 on methacrylate monoliths. *J. Virol. Methods* **2010**, *166*, 60–64. [CrossRef] [PubMed]
36. Sambrook, J.; Maniatis, T.; Fritsch, E.F. *Molecular Cloning: A Laboratory Manual*, 2nd ed.; Cold Spring Harbor Laboratory Press: Cold Spring Harbor, NY, USA, 1987; ISBN 978-0-87969-309-1.
37. Aziz, R.K.; Bartels, D.; Best, A.A.; DeJongh, M.; Disz, T.; Edwards, R.A.; Formsma, K.; Gerdes, S.; Glass, E.M.; Kubal, M.; et al. The RAST server: Rapid annotations using subsystems technology. *BMC Genom.* **2008**, *9*, 75. [CrossRef] [PubMed]
38. Frottin, F.; Martinez, A.; Peynot, P.; Mitra, S.; Holz, R.C.; Giglione, C.; Meinnel, T. The proteomics of N-terminal methionine cleavage. *Mol. Cell. Proteom.* **2006**, *5*, 2336–2349. [CrossRef] [PubMed]
39. Martinez, A.; Traverso, J.A.; Valot, B.; Ferro, M.; Espagne, C.; Ephritikhine, G.; Zivy, M.; Giglione, C.; Meinnel, T. Extent of N-terminal modifications in cytosolic proteins from eukaryotes. *Proteomics* **2008**, *8*, 2809–2831. [CrossRef] [PubMed]
40. Gasteiger, E.; Hoogland, C.; Gattiker, A.; Duvaud, S.; Wilkins, M.R.; Appel, R.D.; Bairoch, A. Protein identification and analysis tools on the ExPASy server. In *The Proteomics Protocols Handbook*; Humana Press: New York, NY, USA, 2005; pp. 571–607. ISBN 978-1-59259-890-8.
41. Bonilla, N.; Rojas, M.I.; Netto Flores Cruz, G.; Hung, S.-H.; Rohwer, F.; Barr, J.J. Phage on tap—A quick and efficient protocol for the preparation of bacteriophage laboratory stocks. *PeerJ* **2016**, *4*, e2261. [CrossRef] [PubMed]
42. Adriaenssens, E.M.; Lehman, S.M.; Vandersteegen, K.; Vandenheuvel, D.; Philippe, D.L.; Cornelissen, A.; Clokie, M.R.J.; Garría, A.J.; De Proft, M.; Maes, M.; et al. CIM®monolithic anion-exchange chromatography as a useful alternative to CsCl gradient purification of bacteriophage particles. *Virology* **2012**, *434*, 265–270. [CrossRef] [PubMed]
43. Valentine, N.; Wunschel, S.; Wunschel, D.; Petersen, C.; Wahl, K. Effect of culture conditions on microorganism identification by matrix-assisted laser desorption ionization mass spectrometry. *Appl. Environ. Microbiol.* **2005**, *71*, 58–64. [CrossRef] [PubMed]
44. Kwan, T.; Liu, J.; DuBow, M.; Gros, P.; Pelletier, J. The complete genomes and proteomes of 27 *Staphylococcus aureus* bacteriophages. *Proc. Natl. Acad. Sci. USA* **2005**, *102*, 5174–5179. [CrossRef] [PubMed]
45. Salmond, G.P.C.; Fineran, P.C. A century of the phage: Past, present and future. *Nat. Rev. Microbiol.* **2015**, *13*, 777–786. [CrossRef] [PubMed]

46. Pirnay, J.-P.; Merabishvili, M.; Raemdonck, H.V.; Vos, D.D.; Verbeken, G. Bacteriophage production in compliance with regulatory requirements. *Methods Mol. Biol.* **2018**, *1693*, 233–252. [CrossRef] [PubMed]
47. Cox, C.R.; Rees, J.C.; Voorhees, K.J. Modeling bacteriophage amplification as a predictive tool for optimized MALDI-TOF MS-based bacterial detection. *J. Mass Spectrom.* **2012**, *47*, 1435–1441. [CrossRef] [PubMed]
48. Pierce, C.L.; Rees, J.C.; Fernández, F.M.; Barr, J.R. Viable *Staphylococcus aureus* quantitation using ^{15}N metabolically labeled bacteriophage amplification coupled with a multiple reaction monitoring proteomic workflow. *Mol. Cell. Proteom.* **2012**, *11*, M111.012849. [CrossRef] [PubMed]
49. Swatkoski, S.; Russell, S.; Edwards, N.; Fenselau, C. Analysis of a model virus using residue-specific chemical cleavage and MALDI-TOF mass spectrometry. *Anal. Chem.* **2007**, *79*, 654–658. [CrossRef] [PubMed]
50. Rees, J.C.; Barr, J.R. Detection of methicillin-resistant *Staphylococcus aureus* using phage amplification combined with matrix-assisted laser desorption/ionization mass spectrometry. *Anal. Bioanal. Chem.* **2017**, *409*, 1379–1386. [CrossRef] [PubMed]
51. Cargile, B.J.; McLuckey, S.A.; Stephenson, J.L. Identification of bacteriophage MS2 coat protein from *E. coli* lysates via ion trap collisional activation of intact protein ions. *Anal. Chem.* **2001**, *73*, 1277–1285. [CrossRef] [PubMed]
52. Wick, C.H.; Elashvili, I.; Stanford, M.F.; McCubbin, P.E.; Deshpande, S.V.; Kuzmanovic, D.; Jabbour, R.E. Mass spectrometry and integrated virus detection system characterization of MS2 bacteriophage. *Toxicol. Mech. Methods* **2007**, *17*, 241–254. [CrossRef] [PubMed]
53. Serafim, V.; Pantoja, L.; Ring, C.J.; Shah, H.; Shah, A.J. Rapid identification of *E. coli* bacteriophages using mass spectrometry. *J. Proteom. Enzymol.* **2017**, *6*, 1000130. [CrossRef]
54. Goerke, C.; Köller, J.; Wolz, C. Ciprofloxacin and trimethoprim cause phage induction and virulence modulation in *Staphylococcus aureus*. *Antimicrob. Agents Chemother.* **2006**, *50*, 171–177. [CrossRef] [PubMed]
55. Selva, L.; Viana, D.; Regev-Yochay, G.; Trzcinski, K.; Corpa, J.M.; Lasa, Í.; Novick, R.P.; Penadés, J.R. Killing niche competitors by remote-control bacteriophage induction. *Proc. Natl. Acad. Sci. USA* **2009**, *106*, 1234–1238. [CrossRef] [PubMed]
56. Tang, Y.; Nielsen, L.N.; Hvitved, A.; Haaber, J.K.; Wirtz, C.; Andersen, P.S.; Larsen, J.; Wolz, C.; Ingmer, H. Commercial biocides induce transfer of prophage Φ13 from human strains of *Staphylococcus aureus* to livestock CC398. *Front. Microbiol.* **2017**, *8*, 2418. [CrossRef] [PubMed]
57. Kahánková, J.; Španová, A.; Pantůček, R.; Horák, D.; Doškař, J.; Rittich, B. Extraction of PCR-ready DNA from *Staphylococcus aureus* bacteriophages using carboxyl functionalized magnetic nonporous microspheres. *J. Chromatogr. B* **2009**, *877*, 599–602. [CrossRef] [PubMed]
58. Czyz, A.; Los, M.; Wrobel, B.; Wegrzyn, G. Inhibition of spontaneous induction of lambdoid prophages in *Escherichia coli* cultures: Simple procedures with possible biotechnological applications. *BMC Biotechnol.* **2001**, *1*, 1. [CrossRef]
59. O'Flaherty, S.; Ross, R.P.; Meaney, W.; Fitzgerald, G.F.; Elbreki, M.F.; Coffey, A. Potential of the polyvalent anti-*Staphylococcus* bacteriophage K for control of antibiotic-resistant staphylococci from hospitals. *Appl. Environ. Microbiol.* **2005**, *71*, 1836–1842. [CrossRef] [PubMed]
60. Kvachadze, L.; Balarjishvili, N.; Meskhi, T.; Tevdoradze, E.; Skhirtladze, N.; Pataridze, T.; Adamia, R.; Topuria, T.; Kutter, E.; Rohde, C.; et al. Evaluation of lytic activity of staphylococcal bacteriophage Sb-1 against freshly isolated clinical pathogens. *Microb. Biotechnol.* **2011**, *4*, 643–650. [CrossRef] [PubMed]
61. Vandersteegen, K.; Mattheus, W.; Ceyssens, P.-J.; Bilocq, F.; De Vos, D.; Pirnay, J.-P.; Noben, J.-P.; Merabishvili, M.; Lipinska, U.; Hermans, K.; et al. Microbiological and molecular assessment of bacteriophage ISP for the control of *Staphylococcus aureus*. *PLoS ONE* **2011**, *6*, e24418. [CrossRef] [PubMed]
62. Łobocka, M.; Hejnowicz, M.S.; Dąbrowski, K.; Gozdek, A.; Kosakowski, J.; Witkowska, M.; Ulatowska, M.I.; Weber-Dąbrowska, B.; Kwiatek, M.; Parasion, S.; et al. Genomics of staphylococcal Twort-like phages—Potential therapeutics of the post-antibiotic era. *Adv. Virus Res.* **2012**, *83*, 143–216. [CrossRef] [PubMed]
63. Międzybrodzki, R.; Kłak, M.; Jończyk-Matysiak, E.; Bubak, B.; Wójcik, A.; Kaszowska, M.; Weber-Dąbrowska, B.; Łobocka, M.; Górski, A. Means to facilitate the overcoming of gastric juice barrier by a therapeutic staphylococcal bacteriophage A5/80. *Front. Microbiol.* **2017**, *8*. [CrossRef] [PubMed]

Communication

The Magistral Phage

Jean-Paul Pirnay [1,*], Gilbert Verbeken [1], Pieter-Jan Ceyssens [2], Isabelle Huys [3], Daniel De Vos [1], Charlotte Ameloot [4] and Alan Fauconnier [4,5]

1 Laboratory for Molecular and Cellular Technology, Queen Astrid Military Hospital, Bruynstraat 1, 1120 Brussel, Belgium; gilbert.verbeken@mil.be (G.V.); danielmarie.devos@mil.be (D.D.V.)
2 Bacterial Diseases, Unit Antibiotic Resistance, Scientific Institute of Public Health, Rue Engelandstraat 642, 1180 Brussel, Belgium; pieterjan.ceyssens@wiv-isp.be
3 Department of Pharmaceutical and Pharmacological Sciences, KU Leuven, O&N2, Herestraat 49, Box 521, 3000 Leuven, Belgium; isabelle.huys@pharm.kuleuven.be
4 Federal Agency for Medicines and Health Products, Place Victor Horta 40/40, 1060 Brussels, Belgium; charlotte.ameloot@fagg-afmps.be (C.A.); alan.fauconnier@invivo.be (A.F.)
5 Culture In Vivo ASBL, rue du Progrès, 4, boîte 7, 1400 Nivelles, Belgium
* Correspondence: jean-paul.pirnay@mil.be; Tel.: +32-2-264-4844

Received: 15 January 2018; Accepted: 3 February 2018; Published: 6 February 2018

Abstract: Since time immemorial, phages—the viral parasites of bacteria—have been protecting Earth's biosphere against bacterial overgrowth. Today, phages could help address the antibiotic resistance crisis that affects all of society. The greatest hurdle to the introduction of phage therapy in Western medicine is the lack of an appropriate legal and regulatory framework. Belgium is now implementing a pragmatic phage therapy framework that centers on the magistral preparation (compounding pharmacy in the US) of tailor-made phage medicines.

Keywords: antibiotic; antimicrobial resistance; magistral preparation; compounding pharmacy; phage therapy; regulatory framework; personalized medicine

---

1. The Age of the Superbug

On 21 September 2016, the UN General Assembly convened a meeting on antimicrobial resistance (AMR) at the UN headquarters in New York. It was only the fourth time the General Assembly addressed a health emergency. This high-level meeting resulted in a UN resolution focused on combatting the AMR health threat. World leaders acknowledged that global AMR poses a fundamental long-term threat to human health, the production of food, and sustainable development. Based on scenarios of rising drug resistance for six pathogens, experts estimated that by 2050 the burden of AMR could rise to 10 million people dying every year and an economic cost of $100 trillion [1].

Commercial antibiotics that are currently used in public health, animal, food, agriculture and aquaculture sectors, are immutable chemicals that are based on natural antibiotics produced by soil bacteria or fungi to—depending on their concentration—either combat competitors, communicate with other organisms, or act as pleiotropic effectors of metabolic pathways. It was, therefore, to be expected that bacteria would be extremely proficient at evolving resistance to such antibiotics, especially when these are excessively and often unnecessarily used. Selective pressures imposed by humans have resulted in the emergence of "superbugs", or bacteria that are resistant to virtually all commercial antibiotics. Experts fear that society could return to a pre-antibiotic era, when simple infections could wipe out entire populations and surgical interventions were life threatening. Today, it seems that all "easy" antibiotics have been exploited and industry has been reluctant to put new efforts into the discovery and development of new classes of antibiotics. These are expensive to develop and are bound to offer a poor return on investment as they are only taken for a short period and their use is likely to be restricted in the future.

Therefore, the UN committed to work at national, regional and global levels to support the development of new antimicrobial agents and therapies [2].

2. Phage Therapy

One of the promising "new" treatments that is increasingly highlighted—*inter alia* during the recent UN General Assembly—is phage therapy, the therapeutic use of bacteriophages (phages in short)—the viruses of bacteria—to treat bacterial infections [3]. Since time immemorial, phages control their hosts, the bacteria, on our planet. When discovered in the early twentieth century, they were immediately applied in medicine. It soon appeared that phages are exquisitely host-specific. Most phages can only lyse a subset of a bacterial species. Physicians must thus first know which bacteria cause the infections before they can treat the patients.

As could be expected, it was shown that bacteria could also evolve to evade phage infection, even when potent phages are applied simultaneously [4]. However, the main advantage of phages over antibiotics is their ability to mutate at least as fast as their hosts, enabling them to evolve new infectivity and thus regain the "upper hand" over bacteria. Bacteria and phage are thus involved in a continuous arms race of co-evolving infectivity and defense mechanisms.

The advent of broad-spectrum antibiotics, which target a wide range of bacterial infections and could thus be used empirically, heralded the decline of phage therapy in the Western world. The success stories of the many phage applications in the past, mainly on the east side of the Iron Curtain, where phage therapy remained an established treatment, together with the increasing number of virtually untreatable bacterial infections, has created a growing demand for phage therapy. Some successful intravenous applications of phages to treat terminally ill patients in the Western world have recently been published in the scientific literature [5,6].

The Promise of the Phage Therapy Medicinal Product

At their reintroduction in the Western world, phage preparations were classified as medicinal products (European Union) or drugs (US), based on the literal implementation of definitions. Namely, any substance presented as having properties for treating or preventing disease in human beings is considered to be a medicinal product or a drug. As a result, a large body of costly and time-consuming requirements and procedures for manufacturing and for obtaining marketing authorization for medicinal products (drugs in the US) for human use were imposed on phage therapy medicinal products (PTMPs).

On the one hand, it turns out that the established pharmaceutical industry is not interested in PTMPs, mainly because of limitations in intellectual property protection of natural entities such as genes or phages and because of phage specificity and bacterial resistance issues, which compromise widespread and long-term use of immutable pre-defined PTMPs. On the other hand, it is becoming clear that medicinal product provisions, which were originally developed to cater for widely used and mass-produced chemical molecules such as aspirin and antibiotics, are not compatible with sustainable (non-empirical) or customized phage therapy approaches in which phages need to be selected and produced ad hoc [7]. Pre-defined PTMPs, could make it through the medicinal product funnel, but such preparations are less flexible to deal with changes in the incidences of infecting bacterial species in certain settings or geographical areas, or with the emergence of mutated bacterial strains. The long-term use of immutable PTMPs is also bound to elicit considerable bacterial phage resistance, although not much is known about the rate at which this would occur in clinical settings. Overall, the efficacy of PTMPs is likely to decrease over time and they would need to be regularly adapted and re-approved for use.

Some of these issues crystallized during PhagoBurn (www.phagoburn.eu), the first major trial under modern medicinal product regulatory standards in the European Union [8]. Cocktails of 12 and 13 phages were needed to ensure a certain activity against a collection of *Pseudomonas aeruginosa* and *E. coli* isolates, respectively. Manufacturing of one batch of the investigational products ended up taking

20 months and the largest part of the study budget. In addition, phage specificity issues hampered the recruitment of patients. Because each of the two study products, which couldn't be applied simultaneously, targeted only one of the multiple bacterial species that are known to (simultaneously) infect or colonize burn wounds, physicians were reluctant to include patients [8]. Regardless of the final clinical outcome of PhagoBurn, the preliminary phase of the study showed at least that dedicated and realistic production and documentation requirements are urgently needed to enable the timely supply of secure phage preparations. This would enable clinicians to conduct the desperately needed safety and efficacy studies and to deal with urgent individual or local infection issues or public health threats (e.g., the 2011 *E. coli* O104:H4 outbreak in Germany).

Meanwhile, sporadic phage applications are carried out in the West, often under the umbrella of Article 37 (Unproven Interventions in Clinical Practice) of the Declaration of Helsinki (www.wma.net). In addition, several European and US patients suffering from chronic, extremely resistant or difficult to treat bacterial infections are known to have travelled to a phage therapy center in Tbilisi, Georgia (www.eliavaphagetherapy.com, www.phagetherapycenter.com), for treatment.

3. Enter the Magistral Phage

On 5 July 2016, during a meeting of the Belgian Chamber of Representatives and in response to two parliamentary questions related to the implementation of phage therapy [9], the Belgian Minister of Social Affairs and Public Health acknowledged that phage therapy has no specific regulation in Europe and that there is a consensus that phage preparations are medicinal products. However, according to the Minister it is difficult to determine whether we should deal with industrially-prepared medicinal products or rather with magistral preparations, the former being subject to constraints related to their production and marketing authorization, unlike the latter.

3.1. Magistral Preparations

In European and Belgian law, the notion of a magistral preparation (compounded prescription drug product in the US) is defined as "any medicinal product prepared in a pharmacy in accordance with a medical prescription for an individual patient" (Article 3 of Directive 2001/83 and Article 6 quater, § 3 of the Law of 25 March 1964). Magistral preparations are mixed from their constituent ingredients by a pharmacist (or at least under his/her supervision), for a given patient according to a prescription by a physician and following the technical and scientific standards of the pharmaceutical art. The magistral formula is a practical way for a medical doctor to personalize patient treatments to specific needs and to make medications available that do not exist commercially. Some medicines, such as natural hormone combination products and allergens, are not produced by commercial manufacturers because they lack patent protection and hence return on investment for pharmaceutical companies, but are actually delivered as magistral preparations. Owing to the emergence of innovative medicines for rare diseases or for personalized therapies, magistral preparations are increasingly in demand.

3.2. The Belgian Magistral Phage Medicine Strategy

The Community code leaves the door open for some flexibility to implement certain national solutions relating to medicines for human use [10]. As such, the Belgian Minister of Public Health asked the Federal Agency for Medicines and Health Products (FAMHP, the Belgian competent authority for medicines) to help set up a national strategy for magistral phage medicines. In general, active ingredients of magistral preparations must meet the requirements of the European Pharmacopoeia, of the Belgian Pharmacopoeia or of an official pharmacopoeia [10]. If no such document exists, then the active ingredients must be authorized by the Minister of Public Health, following a favorable opinion of the national Pharmacopoeia Commission [10]. In addition, non-authorized ingredients may also be used in magistral preparations, providing that they are accompanied by a certificate of analysis issued by a Belgian Approved Laboratory [10]. The so-called "Belgian Approved Laboratories" are quality control laboratories which are granted an accreditation by the Belgian regulatory authorities. This status allows them to perform the batch release testing of medicinal products. This national accreditation is equivalent to—and gradually replaced by—the GMP certification for the batch release testing of medicinal products. Belgian Approved Laboratories can be either private (e.g., subcontractor of the pharmaceutical industry) or partially or entirely public (e.g., academic laboratories and scientific institutes). Some of them belong to the European Official Medicines Control Laboratories (OMCL) network, which is made up of independent public laboratories that have been appointed by their respective national authority.

The option of the "non-authorized ingredient" was chosen in this case because of the enormous variety of phages that could qualify as active ingredients and should then, each individually, obtain an authorization issued by the Minister of Public Health [9]. The Scientific Institute of Public Health was identified as a suitable Belgian Approved Laboratory for issuing valid certificates of analysis for batches of phage active ingredients. Although the standard procedure for unauthorized active ingredients only involves the medical doctor, his patient, the manufacturer of the active substances, the approved laboratory and the pharmacist, it was decided—in joint consultation and because of the innovative and very specific character of phage therapy—to involve the FAMHP in the elaboration of a Belgian magistral phage medicine procedure.

In practice, and to consolidate the opening left by the Minister of Public Health, a formal question and answer session was initiated between the military hospital and the FAMHP within the context of the existing national Scientific-Technical Advice (STA) procedure. On 26 October 2016, it was formally agreed that natural phages whose derivative finished products are not fully compliant with the requirements relating to medicinal products for human use (Directive 2001/83), and for which there is no monograph in an official pharmacopoeia, can be processed by a pharmacist as active pharmaceutical ingredients (APIs) in magistral preparations, providing compliance to a number of logical provisions:

- Phages should be delivered in the form of a magistral preparation to a specific (nominal) patient.
- Magistral preparations should always be delivered under the direct responsibility of a medical doctor and a pharmacist.
- The relevant characteristics and qualities of the phage APIs should be defined in an internal monograph (prepared by the supplier).
- Before the pharmacist can use the unlicensed material, he/she must ascertain—based on certificates of analysis issued by a Belgian Approved Laboratory—that the raw materials conform to the provisions of the internal monograph.
- Even if not legally required, it is recommended that the supplier submits the monograph for assessment by the FAMHP.

The general concept of the Belgian magistral phage medicine strategy is depicted in Figure 1. A single characterized phage seed lot is selected from a phage bank. To prevent the unwanted drift of properties resulting from repeated subcultures, the production of medicines obtained by microbial culture is best based on a system of banked master and working seed lots. From this phage seed lot, a phage API is produced according to a monograph. A Belgian Approved Laboratory performs External Quality Assessments to evaluate the API's properties and quality. Each batch of these phage APIs will have a batch record, which describes the production process for that batch in detail. Phage APIs can be produced by both private companies and public institutions. The phage API, accompanied by its batch record protocol and the results of the External Quality Assessments, is then transferred to the hospital pharmacy for possible incorporation in magistral formulas. Ideally, active phage APIs are selected against the target bacteria. In comparison to an antibiogram (to test antibiotic sensitivity), as it were, a "phagogram" is performed. Today, no formal guidelines exist with regard to the clinical use (e.g., medical indications, formulations and posology) of magistral phage medicines. However, it is the intention to draft these guidelines as quickly as possible, at the Belgian level and possibly at the European level.

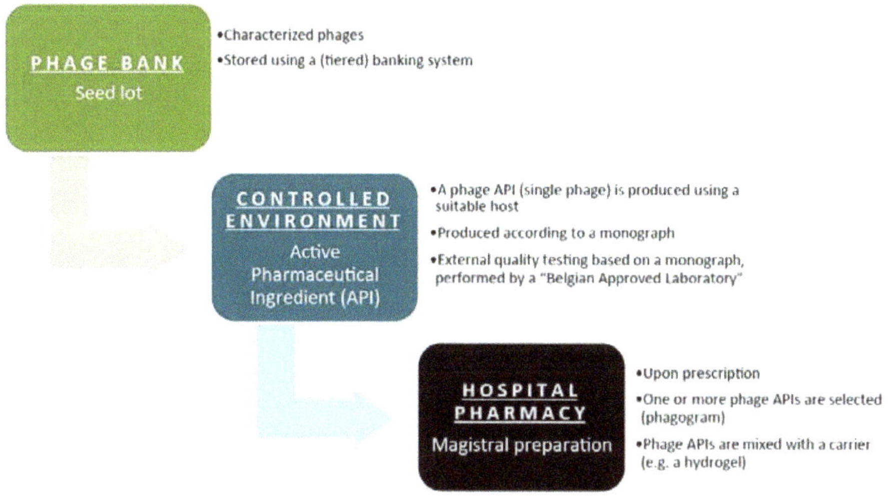

Figure 1. General flowchart of the magistral phage medicine process.

3.3. *Phage API Monograph*

Next, experts of the Queen Astrid military hospital in Brussels, the FAMHP and the Belgian Scientific Institute of Public Health elaborated a pragmatic supplier monograph for phage APIs with a limited use (to hospital pharmacies) status. This document was conceived as a general (applicable to most phages) and evolving document. On 10 January 2018, version 1.0 of the monograph (Supplementary Document 1) received a formal positive advice by the FAMHP.

3.4. *Pricing and Reimbursement*

In terms of pricing, the total cost of a magistral preparation is a reflection of the costs for the products in the preparation, eventually the costs of the prescribed excipients or recipients and an honorarium for the pharmacist for the magistral preparation. Reimbursement of a magistral preparation in Belgium is subject to several criteria: (1) the pharmacist receives a prescription from a physician; (2) this pharmacist makes the magistral preparation and delivers it; (3) products in the magistral preparation are listed on a predefined list of products eligible for reimbursement; and (4) the conditions for reimbursement need to be respected. Bacteriophages are at the moment not listed as products eligible for reimbursement. Therefore, depending on the ultimate price set for a phage magistral preparation, this might (or not) influence the access of phage therapy to patients.

4. Conclusions

It seems to be a matter of time before phage therapy regains its status as an established antibacterial tool. However, this will not only depend on the credibility of "phage researchers", but also on the political context in which they are working. Phage therapy is not sustainable without reimbursement of the researchers providing the therapeutic phages, and so far, phage research is underfunded. Just as drug companies are allowed to profit for some time after developing a drug, there must be some form of compensation for the investigators isolating, characterizing, and optimizing the phages that will be included in future therapeutic phage banks, all before a pharmacist gains access to them for combination. Phage researchers should not be expected to automatically be "altruists", and compensation must be given for their efforts at developing phage therapy as a medicine.

Believing that Belgium could do some pioneering work in the phage therapy field, the Belgian Minister of Public Health and the FAMHP opened the door to phage medicines that take into account the unique characteristics of phages and the need for personalized "sur-mesure" and sustainable phage therapy approaches [7]. There is every reason to believe that the resulting Belgian "magistral phage medicine" framework will be flexible enough to exploit and further explore the specific nature of phages as co-evolving antibacterials whilst giving precedence to patients' safety. Importantly, this Belgian solution avoids the application of certain medicinal product requirements that restrain flexible phage therapy approaches, such as compliance to Good Manufacturing Practice (GMP). There are indications that other (EU) countries might also adopt this phage therapy framework in the near future, in anticipation of a European solution. Recently, the biological master file concept was put forward as a European solution to overcome the regulatory challenges of personalized medicines in general and phage medicines more specifically [11].

Supplementary Materials: The following are available online at www.mdpi.com/xxx/s1, Document 1. Phage API monograph (version 1.0).

Author Contributions: Jean-Paul Pirnay, Gilbert Verbeken, Pieter-Jan Ceyssens, Isabelle Huys, Daniel De Vos, Charlotte Ameloot and Alan Fauconnier contributed to the conception and writing of this paper.

Conflicts of Interest: The authors declare no conflict of interest.

References

1. Tackling Drug-Resistant Infections Globally: Final Report and Recommendations. The Review on Antimicrobial Resistance. 2016 Release. Available online: https://amr-review.org/sites/default/files/160525_Final%20paper_with%20cover.pdf (accessed on 4 February 2018).
2. United Nations. Draft Political Declaration of the High-Level Meeting of the General Assembly on Antimicrobial Resistance (16-16108 (E)). 2016 Release. Available online: http://www.un.org/pga/71/wp-content/uploads/sites/40/2016/09/DGACM_GAEAD_ESCAB-AMR-Draft-Political-Declaration-1616108E.pdf (accessed on 17 December 2017).
3. Thiel, K. Old dogma, new tricks—21st Century phage therapy. *Nat. Biotechnol.* **2004**, *22*, 31–36. [CrossRef] [PubMed]
4. Hall, A.R.; De Vos, D.; Friman, V.P.; Pirnay, J.P.; Buckling, A. Effects of sequential and simultaneous applications of bacteriophages on populations of *Pseudomonas aeruginosa* in vitro and in wax moth larvae. *Appl. Environ. Microbiol.* **2012**, *78*, 5646–5652. [CrossRef] [PubMed]
5. Jennes, S.; Merabishvili, M.; Soentjens, P.; Pang, K.W.; Rose, T.; Keersebilck, E.; Soete, O.; François, P.M.; Teodorescu, S.; Verween, G.; et al. Use of bacteriophages in the treatment of colistin-only-sensitive *Pseudomonas aeruginosa* septicaemia in a patient with acute kidney injury—A case report. *Crit. Care* **2017**, *21*, 129. [CrossRef] [PubMed]
6. Schooley, R.T.; Biswas, B.; Gill, J.J.; Hernandez-Morales, A.; Lancaster, J.; Lessor, L.; Barr, J.J.; Reed, S.L.; Rohwer, F.; Benler, S.; et al. Development and use of personalized bacteriophage-based therapeutic cocktails to treat a patient with a disseminated resistant *Acinetobacter baumannii* Infection. *Antimicrob. Agents Chemother.* **2017**, *61*, e00954-17. [CrossRef] [PubMed]
7. Pirnay, J.P.; De Vos, D.; Verbeken, G.; Merabishvili, M.; Chanishvili, N.; Vaneechoutte, M.; Zizi, M.; Laire, G.; Lavigne, R.; Huys, I.; et al. The phage therapy paradigm: Prêt-à-porter or sur-mesure? *Pharm. Res.* **2011**, *28*, 934–937. [CrossRef] [PubMed]
8. Servick, K. Beleaguered phage therapy trial presses on. *Science* **2016**, *352*, 1506. [CrossRef] [PubMed]
9. Commission De La Santé Publique, De L'environnement Et Du Renouveau De La Société. Questions Jointes De Mme Muriel Gerkens Et M. Philippe Blanchart À La Ministre Des Affaires Sociales Et De La Santé Publique Sur "La Phagothérapie" À La Ministre Des Affaires Sociales Et De La Santé Publique" (N° 11955 and N° 12911). 2016 Release. Available online: https://www.dekamer.be/doc/CCRA/pdf/54/ac464.pdf (accessed on 17 December 2017).

10. Fauconnier, A. Guidelines for Bacteriophage Product Certification. *Methods Mol. Biol.* **2018**, *1693*, 253–268. [PubMed]
11. Fauconnier, A. Regulating phage therapy: The biological master file concept could help to overcome regulatory challenge of personalized medicines. *EMBO Rep.* **2017**, *18*, 198–200. [CrossRef] [PubMed]

Article

Characterization of vB_SauM-fRuSau02, a Twort-Like Bacteriophage Isolated from a Therapeutic Phage Cocktail

Katarzyna Leskinen [1], Henni Tuomala [1,2], Anu Wicklund [1,2], Jenni Horsma-Heikkinen [1], Pentti Kuusela [2], Mikael Skurnik [1,2] and Saija Kiljunen [1,2,*]

1 Department of Bacteriology and Immunology, Medicum, Research Programs Unit, Immunobiology Research Program, University of Helsinki, Helsinki 00290, Finland; katarzyna.leskinen@helsinki.fi (K.L.); henni.tuomala@helsinki.fi (H.T.); anumaria.wicklund@gmail.com (A.W.); jenni.horsma@helsinki.fi (J.H.-H.); mikael.skurnik@helsinki.fi (M.S.)

2 Division of Clinical Microbiology, HUSLAB, University of Helsinki and Helsinki University Hospital, Helsinki 00290, Finland; pentti.kuusela@hus.fi

* Correspondence: saija.kiljunen@helsinki.fi; Tel.: +358-2941-2673

Academic Editors: Harald Brüssow and Eric Freed
Received: 29 June 2017; Accepted: 11 September 2017; Published: 14 September 2017

Abstract: *Staphylococcus aureus* is a commensal and pathogenic bacterium that causes infections in humans and animals. It is a major cause of nosocomial infections worldwide. Due to increasing prevalence of multidrug resistance, alternative methods to eradicate the pathogen are necessary. In this respect, polyvalent staphylococcal myoviruses have been demonstrated to be excellent candidates for phage therapy. Here we present the characterization of the bacteriophage vB_SauM-fRuSau02 (fRuSau02) that was isolated from a commercial *Staphylococcus* bacteriophage cocktail produced by Microgen (Moscow, Russia). The genomic analysis revealed that fRuSau02 is very closely related to the phage MSA6, and possesses a large genome (148,464 bp), with typical modular organization and a low G+C (30.22%) content. It can therefore be classified as a new virus among the genus *Twortlikevirus*. The genome contains 236 predicted genes, 4 of which were interrupted by insertion sequences. Altogether, 78 different structural and virion-associated proteins were identified from purified phage particles by liquid chromatography-tandem mass spectrometry (LC-MS/MS). The host range of fRuSau02 was tested with 135 strains, including 51 and 54 *Staphylococcus aureus* isolates from humans and pigs, respectively, and 30 coagulase-negative *Staphylococcus* strains of human origin. All clinical *S. aureus* strains were at least moderately sensitive to the phage, while only 39% of the pig strains were infected. Also, some strains of *Staphylococcus intermedius*, *Staphylococcus lugdunensis*, *Staphylococcus epidermidis*, *Staphylococcus haemolyticus*, *Staphylococcus saprophyticus* and *Staphylococcus pseudointer* were sensitive. We conclude that fRuSau02, a phage therapy agent in Russia, can serve as an alternative to antibiotic therapy against *S. aureus*.

Keywords: *Staphylococcus aureus*; bacteriophage; phage therapy; vB_SauM-fRuSau02; *Twortlikevirus*

1. Introduction

Staphylococcus aureus is a commensal and pathogenic bacterium that causes opportunistic infections in humans and animals. Approximately 20% of humans have persistent and 30% sporadic nasal *S. aureus* colonization [1]. As a pathogen, *S. aureus* causes a broad spectrum of infections in humans ranging from simple abscesses to fatal sepsis, including pneumonia, endocarditis, meningitis, mastitis, food poisoning, and toxic shock syndrome [2]. Currently the antibiotic resistance of this species poses a threat to public health. Even though the incidence of severe infections caused by methicillin-resistant *S. aureus* (MRSA) is decreasing [3], MRSA still is an important cause of nosocomial infections worldwide [4,5]. The emergence of multidrug resistance results in difficulties in eradication of the pathogen with the use of conventional therapies and thus requires development of alternatives to antibiotic-based therapies.

One promising alternative to treat infections caused by antibiotic resistant bacteria is phage therapy, where the natural predators of bacteria (bacteriophages, phages) are used to kill the pathogens [6–8]. The history of phage therapy has been extensively reviewed elsewhere [9,10] and is not discussed here. To be considered safe, phage therapy has to meet a number of criteria: phages used for therapeutic purposes need to be strictly lytic and they should not carry known genes coding for toxins or other harmful substances [11]. Furthermore, the host bacteria used for phage production should have as few prophages as possible and the therapeutic phage preparation should not contain high concentration of bacterial toxins.

All known *S. aureus* phages belong to order *Caudovirales*, i.e., they are tailed phages with an icosahedral capsid that surrounds the double-stranded DNA genome [12,13]. Staphylococcal phages can be classified into three categories: (1) podoviruses with <20 kb genomes; (2) siphoviruses with ~40 kb genomes; and (3) myoviruses with >125 kb genomes [12]. Of these phage groups, staphylococcal siphoviruses are generally temperate and often carry genes promoting bacterial virulence [13], which makes them inappropriate for therapeutic applications. Staphylococcal podoviruses, on the other hand, are strictly lytic but extremely rare and difficult to find [14]. From therapeutic point of view, myoviruses are considered the most interesting staphylococcal phages [14,15].

Many of the staphylococcal myoviruses are classified into the genus *Twortlikevirus* of the *Spounavirinae* subfamily and are related at genetic and proteomic level [16]. The *Twortlikevirus* genus consists of phages with genomes of 127–141 kb, low G+C content (30–31%), and 183 to 217 open reading frames (ORFs) [17]. Currently, this genus contains over 25 members, including phage Twort, G1 [18], K [19], MSA6 [20], GH15 [21], Romulus, and Remus [17]. A typical feature for Twort-like viruses is the presence of long terminal repeats (LTRs), several thousand base pair-long direct repeats at the ends of the genome. The nucleotide sequence and length of LTR regions differ among the representatives of the genus and may influence the host range [15]. Twort-like viruses are also known for their broad host range. This phenomenon is mainly accounted to the presence of multiple receptor binding proteins in the viral capsid that allow them to utilize at least two adsorption apparatuses and recognize different structures [22]. This feature, together with their strictly lytic life cycle, makes Twort-like viruses particularly suitable for clinical applications [15].

In this paper, we report the isolation and analysis of a Twort-like *S. aureus* phage, vB_SauM-fRuSau02 (fRuSau02). The phage was isolated from a therapeutic bacteriophage product from Microgen Company (Moscow, Russia). The product was purchased in a pharmacy in Saint Petersburg, Russia, and was meant to treat infections typically caused by *S. aureus*. However, no information about the phage composition or the efficacy of the phage cocktail was available. Phage fRuSau02 was the only phage we were able to isolate from this product. Here, we show the analysis of fRuSau02 at a genetic and proteome level, the latter of which allowed us to identify the majority of the phage structural proteins. Additionally, we provide an insight into the reasons why this phage might be well-suited for clinical applications by testing its growth efficiency and host range with a broad range of human and porcine isolates. We also present an evaluation of the fRuSau02 production in different host strains, with intention to select an optimal producer strain for clinical applications.

2. Materials and Methods

2.1. Bacterial Strains, Phages and Media

The bacterial strains used in this work are described in Table S1. The collection of human isolates used in this study was provided by The Hospital District of Helsinki and Uusimaa Laboratories

(HUSLAB), Finland. All staphylococcal and phage incubations were done at 37 °C using Luria Broth (LB) [23] medium. Soft agar medium included additionally 0.35 or 0.4% (*w*/*v*) agar (Becton Dickinson, Franklin Lakes, NJ, USA), and LB agar plates were solidified with 1.5% (*w*/*v*) of agar. fRuSau02 was isolated using a clinical *S. aureus* strain 13KP (Table S1) as a host, and the same strain was then used as a standard host strain for the phage. The phage lysates were produced from semiconfluent soft-agar plates as described elsewhere [23].

2.2. Phage Purification

The fRuSau02 lysate (5×10^{10} plaque-forming units (PFU)/mL) was ultrafiltrated with Amicon Ultra-4 (100 kDa) Centrifugal Filter Units (Merck Millipore, Billerica, MA, USA) to one quarter of the initial volume. Three volumes of chromatography buffer A (20 mM Tris-Cl, pH 7.5) were added and the ultrafiltration was repeated. The volume was adjusted with buffer A. The ultrafiltrated phage sample was then purified with ion exchange chromatography (IEX) using Äkta Purifier (GE Healthcare, Chicago, IL, USA) and a CIM QA-1 tube monolithic column with a 6-µm pore size (BIA Separations, Ajdovščina, Slovenia). The sample was injected to the column in buffer A, washed with buffer A containing 350 mM NaCl and eluted with buffer A with 450 mM NaCl. The phage-containing fractions of two purification batches were pooled, and an Amicon Ultra was used to concentrate the product and to change the buffer to TM (50 mM Tris, pH 7.5–10 mM Mg_2SO_4). Purified phage samples were stored at 4 °C.

2.3. Electron Microscopy

IEX-purified phage lysate was pelleted by centrifugation at 16,000× *g*, 4 °C, for 2 h and resuspended into 0.1 M ammonium acetate. Subsequently, the phage particles were allowed to sediment on 200 mesh pioloform-coated copper grids for 1 min and stained negatively using 3% uranyl acetate. Samples were examined with a JEOL JEM-1400 transmission electron microscope JEOL Ltd., Tokyo, Japan) under 80 kV at the Electron Microscopy Unit (Institute of Biotechnology, University of Helsinki, Helsinki, Finland). Pictures were taken using Gatan Orius SC 1000B bottom-mounted Charged Coupled Device (CCD)-camera (Gatan Inc., Pleasanton, CA, USA). Ten virions were measured and data were used to calculate mean values and standard deviations.

2.4. Infection Growth Curves

Overnight bacterial cultures of *S. aureus* 13KP were diluted to a ratio of 1:100 in fresh LB medium, and 180-µL aliquots were distributed into honeycomb plate wells (Growth Curves Ab Ltd., Helsinki, Finland), where they were mixed with 20-µL aliquots of different fRuSau02 phage stock dilutions. The phage stock and bacterial culture were mixed to achieve multiplicity of infection (MOI) values ranging between 5×10^{-7} and 500. A negative control was obtained by mixing 20 µL of phage stock with 180 µL of LB, whereas the positive control consisted of 180 µL of bacterial culture and 20 µL of fresh LB medium. The growth experiment was carried out at 37 °C using a Bioscreen C incubator (Growth Curves Ab Ltd.) with continuous shaking. The optical density at 600 nm (OD_{600}) of the cultures was measured every 1 h. The averages were calculated from values obtained for the bacteria grown in five parallel wells.

2.5. DNA Isolation and Phage Genome Sequencing

fRuSau02 DNA was isolated from crude phage lysate with Invisorb Spin Virus DNA Mini Kit (Stratec Biomedical, Birkenfeld, Germany). Sequencing was performed at the Institute for Molecular Medicine Finland (FIMM) Technology Centre Sequencing Unit [24]. For next-generation sequencing, the DNA library was constructed with Nextera sample prep kit (Illumina, San Diego, CA, USA). Paired-end sequencing was done using Illumina MiSeq PE300 sequencer (Illumina, San Diego, CA, USA) with the read length of 300 nucleotides. TheA5 (Andrew And Aaron's Awesome Assembly)-miseq integrated pipeline for de novo assembly of microbial genomes was used to obtain the genome sequence [25]. fRuSau02 sequence was submitted to GenBank with accession number MF398190.

2.6. Determination of Physical Ends of the Phage Genome

To determine the physical ends of the phage genome, the approximate positions of the terminal repeats were estimated based on the sequence read numbers using the Integrative Genomics Viewer (IGV) [26,27]. The genome was manually edited according to the estimated physical ends and subjected to virtual digestions with several restriction endonucleases. Two enzymes (NheI and PstI), yielding identifiable end fragments, were used to digest the fRuSau02 DNA, and the resulting fragment distributions were compared to the virtual digestions. The NheI fragments corresponding to the physical ends of the phage genome were isolated from a preparative agarose gel and ligated to pUC19 digested with XbaI and SmaI. The ligation mixtures were used as a template for PCR reaction with pUC19—specific primers Puc19-F (CCTCTTCGCTATTACGCCAG) and pUC19-R (CAACGCAATTAATGTGAGTTAGCT). The PCR products corresponding to the sizes of the genome end fragments (2.7 kb and 4.5 kb for left and right ends, respectively) were isolated from a preparative agarose gel and sequenced with ABI3730XL DNA Analyzer (Applied Biosystems, Foster City, CA, USA) capillary sequencer with primers Puc19-F and Puc19-R at FIMM [24]. The sequence information was used to deduce the actual sequence of the genome ends and terminal repeats.

2.7. In Silico Analysis of Phage Genome

The phage genome was autoannotated using Rapid Annotation Using Subsystem Technology (RAST [28] and proofread manually. Promoters and terminators were predicted using PePPER [29] and ARNold [30,31], respectively, with subsequent manual verification. The promoter consensus sequence was analyzed using MEME [32]. A comparative genome figure was generated using CGView [33]. The genome-wide comparison of bacteriophages was conducted with EMBOSS Stretcher [34].

Phylogeny analysis was conducted with the VICTOR Virus Classification and Tree Building Online Resource [35] using the Genome-BLAST Distance Phylogeny (GBDP) method [36] under settings recommended for prokaryotic viruses [35]. The resulting intergenomic distances (including 100 replicates each) were used to infer a balanced minimum evolution tree with branch support via FASTME including Subtree Pruning and Regrafting (SPR) postprocessing [37] for the formula D0. The tree was rooted at the midpoint [38] and visualized with FigTree [39]. Taxon boundaries at the species, genus and family level were estimated with the OPTSIL program [40], the recommended clustering thresholds [35] and an F value (fraction of links required for cluster fusion) of 0.5 [41].

2.8. Proteome Analysis

IEX-purified phages were concentrated by centrifugation for 2 h at 4 °C and 16,000× *g*. Prior to digestion of proteins to peptides with trypsin, the proteins in the samples were reduced with tris(2-carboxyethyl)phosphine (TCEP) and alkylated with iodoacetamide. Tryptic peptide digests were purified by C18 reversed-phase chromatography columns [42] and the mass spectrometry (MS) analysis was performed on an Orbitrap Elite Electron-Transfer Dissociation (ETD) mass spectrometer (Thermo Scientific, Waltham, MA, USA), using Xcalibur version 2.7.1, coupled to an Thermo Scientific nLCII nanoflow High Pressure Liquid Chromatography (HPLC) system. Peak extraction and subsequent protein identification was achieved using Proteome Discoverer 1.4 software (Thermo Scientific). Calibrated peak files were searched against the fRuSau02 and *Staphylococcus aureus* subsp. *aureus* ST398 proteins (ASM188707v1, NCBI) by a SEQUEST search engine. Error tolerances on the precursor and fragment ions were ±15 ppm and ±0.6 Da, respectively. For peptide identification, a stringent cut-off (0.5% false discovery rate) was used. The LC-MS/MS was performed at the Proteomics Unit, Institute of Biotechnology, University of Helsinki.

2.9. Host Range Screening

The fRuSau02 host range was analyzed by spot assay for most of the bacterial stains studied. Some pig isolates failed to grow on soft agar, and their sensitivity was studied by a liquid culture method. For this, bacteria were cultured overnight in Brain Heart Infusion (BHI) medium (Becton, Dickinson and Company, Franklin Lakes, NJ, USA). Cultures were diluted 1:100 in BHI and aliquoted into 200 µL aliquots to 96-well plates. To these, 10 µL of phage fRuSau02 (6.8×10^6 PFU) was added and the plate was incubated at 37 °C with moderate shaking. For non-infected controls, 10 µL of LB was added instead of the phage. Each strain was studied in triplicate wells. Bacterial growth was monitored by measuring OD_{600} with FLUOstar OPTIMA plate reader (BMG LABTECH GmbH, Ortenberg, Germany) at 60 min intervals for 4 h, and the inhibition of growth in phage-infected wells compared to non-infected control wells indicated a sensitive strain.

2.10. Efficiency of Plating and Adsorption Assay

Bacterial strains were checked by the efficiency of plating (EOP), as described earlier [43]. In brief, *S. aureus* strains 13KP, Newman, TB4, and tagO were pre-grown for 2–3 h at 37 °C. Subsequently 300 PFU of phage was mixed with 90 µL/OD_{600} of bacterial culture and 3 mL of soft agar (0.35% LB agar) and poured over an LB plate. Following 24 h incubation at 37 °C, PFUs were counted and the size and morphology of the plaques was evaluated. The experiment was performed in triplicates, and negative controls without the bacteriophages were prepared. To estimate the adsorption of phage particles on the surface of different *S. aureus* strains, a phage adsorption assay was conducted as described earlier [44]. Briefly, approximately 2.5×10^6 PFU of fRuSau02 was mixed with 500 µL of bacterial overnight cultures (OD_{600} = 3.3). The suspension was incubated at room temperature for 5 min, centrifuged at 16,000× *g* for 3 min, and the phage titer remaining in the supernatant was determined. The phage titer in the control supernatant was set to 100%. LB was used as a non-adsorbing control. Each assay was performed in triplicates.

2.11. Staphylococcal Enterotoxin Measurement

Staphylococcal enterotoxins were measured from the phage lysates with the Transia Plate Staphylococcal Enterotoxins assay (BioControl Systems, Inc., Bellevue, WA, USA) using staphylococcal enterotoxin A (Sigma-Aldrich, St. Louis, MO, USA) as standard. Absorbance at 450 nm was recorded with Hidex Sense Microplate Reader (Hidex, Turku, Finland).

3. Results

3.1. Isolation and Morphology

Phage fRuSau02 was isolated from the *Staphylococcus* bacteriophage cocktail produced by the Microgen Company (Moscow, Russia; series: H52, 0813, PN001973/01). Electron microscopy of the negatively-stained fRuSau02 particles revealed that the phage had an icosahedral head with a contractile tail and a basal tuft attached to the tail (Figure 1). The dimensions of the head were 86 nm (vertical) and 83 nm (horizontal), and the tail length without the base plate was 192 nm. Standard deviations were 3.1, 2.8, and 5.3 nm respectively. Only one particle with contracted tail was found (Figure 1b), and the contracted part was 96 nm. Based on the morphological characteristics, phage fRuSau02 belongs to the order *Caudovirales* and the family *Myoviridae* [45,46].

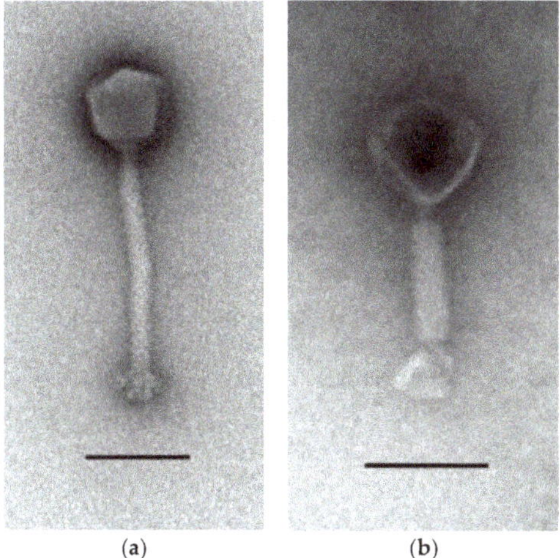

Figure 1. Electron micrographs of negatively stained vB_SauM-fRuSau02 (fRuSau02) particles. Phage particles with non-contracted (**a**) and contracted (**b**) tails are shown. Bars represent 100 nm.

3.2. The Efficiency of Infection

To examine the efficiency of fRuSau02 infection, *S. aureus* 13KP was infected in liquid culture at different MOI values and the bacterial growth was assessed by following the optical density of the culture. The study showed that fRuSau02 is able to efficiently lyse the culture at MOIs above 5×10^{-5} (Figure 2), whereas at MOIs values below this limit there was no lysis observed (data not shown). Additionally, there was no re-growth of the bacterial culture observed within the 24 h time period of the experiment. In coherence, the prolonged incubation of the infected bacteria in the soft agar did not result in emergence of resistance within the first 7 days of incubation indicating low rate of phage-resistance development among the bacteria. Efficient infection and complete lysis of bacterial culture with very low MOI values may be a common feature of twort-like phages, as MOI 1×10^{-4} was earlier shown to be optimal for the production of high-titer lysate of phage MSA6 [20].

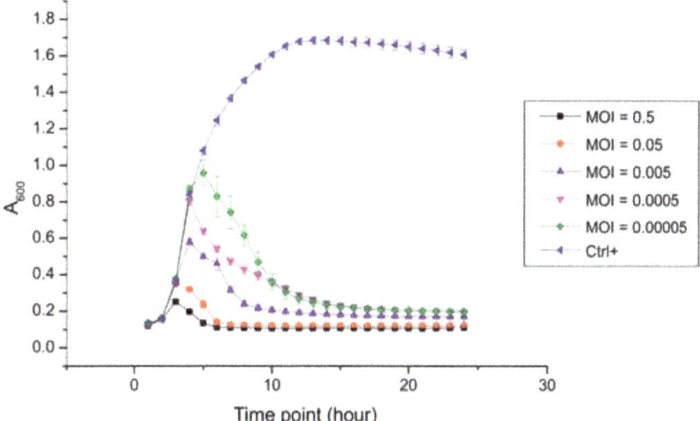

Figure 2. Growth curves of *Staphylococcus aureus* of 13KP infected with fRuSau02. Bacteria were cultured with different concentrations of phage virions in Luria Broth (LB) at 37 °C. Each curve represents the average results for five replicates, error bars represent standard deviation (SD). MOI: multiplicity of infection.

3.3. General Genome Analysis

The linear double-stranded DNA of fRuSau02 comprises 148,464 bp encoding 236 putative ORFs (Figure 3, Table 1). The two terminally redundant 8076 bp long ends encode 20 putative terminal repeat proteins. Sixty-five of the predicted genes are transcribed from the minus strand, including the genes likely to be involved in bacterial cell lysis (holin and numerous putative membrane proteins). Additionally, two single genes from the terminally redundant region, *treI* and *treM*, are also encoded on the minus strand. *In silico* analysis predicted the presence of 43 bacterial promoters and 32 terminators. The analysis failed to identify promoters within the 39,000–66,500 bp range in the fRuSau02 genome where the structural proteins are encoded. The consensus sequence of the promoters was identified (Figure 4 and Table S2). Interestingly, three predicted promoters located in front of spliced or intron encoded genes (*lysK.1, ksaI, I-MsaI*) presented a distinctive promoter sequence that did not follow the consensus sequence (Table S2). In addition to protein-coding sequences, three functional transfer RNA(tRNA) genes encoding $tRNA^{Met}$, $tRNA^{Phe}$, and $tRNA^{Asp}$ were detected. Additionally, no known genes encoding integrases, lysogeny- or virulence-associated or toxic proteins were identified and therefore this bacteriophage can be considered as virulent and potentially safe for phage therapy. Similarly to phage K, the genome of fRuSau02 completely lacks GATC sites that could be recognized by host-encoded restriction endonucleases [19].

Table 1. The structural proteins of phage fRuSau02 identified using liquid chromatography tandem mass spectrometry (LC-MS/MS).

Locus	Name	No of AA	M_W * [kDa]	pI * (calc.)
RS_018	TreR, terminal repeat encoded protein R	156	17.8	3.78
RS_033	phage structural protein	105	11.8	6.76
RS_036	phage structural protein	64	7.6	4.65
RS_037	phage structural protein	245	28.6	6.58
RS_041	phage structural protein	57	6.8	5.26
RS_042	phage structural protein	160	18.8	4.64
RS_046	putative membrane protein MbpR	91	10.9	5.01
RS_048	phage structural protein	372	42.2	4.84
RS_050	phage structural protein	138	16.0	5.22
RS_051	HmzG, DNA-binding protein	100	11.3	4.91
RS_055	phage structural protein	87	10.1	5.91
RS_059	Lig, putative DNA or RNA ligase	298	35.0	5.57
RS_061	Phr, putative PhoH-related protein	246	28.6	5.29
RS_063	Rbn, phage ribonuclease H	141	15.8	7.27
RS_067	phage structural protein	75	9.2	9.95
RS_070	putative membrane protein MbpS	263	29.3	8.82
RS_072	LysK.1, phage lysin	209	23.1	9.66
RS_074	LysK.2, phage lysin	267	29.8	9.45
RS_075	HolA, phage holin	167	18.1	4.25
RS_078	DmcB	69	8.0	5.97
RS_080	putative membrane protein MbpC	108	13.0	5.54
RS_082	putative membrane protein MbpD	88	10.3	8.31
RS_085	Ter.1, phage terminase	65	7.7	9.60
RS_087	Ter.2, phage terminase	515	59.7	6.10
RS_088	phage structural protein	266	29.8	5.30
RS_094	Prt, portal protein	563	64.0	6.42
RS_095	Pro, prohead protease	257	28.6	5.01
RS_096	phage structural protein	318	35.9	4.46
RS_097	Mcp, major capsid protein	463	51.2	5.24

Table 1. *Cont.*

Locus	Name	No of AA	M_W * [kDa]	pI * (calc.)
RS_098	phage structural protein	98	11.3	9.42
RS_099	phage structural protein	302	34.1	5.24
RS_100	phage structural protein	292	33.7	5.82
RS_101	phage structural protein	206	23.7	10.32
RS_102	phage structural protein	278	31.7	4.79
RS_104	Tsp, major tail sheath protein	587	64.5	4.98
RS_105	TmpA, tail tube protein	142	15.9	5.54
RS_109	phage structural protein	103	12.2	6.13
RS_110	phage structural protein	152	18.1	4.79
RS_111	TmpB, tail morphogenic protein	178	20.9	4.40
RS_112	TmpC, phage DNA transfer protein	1351	143.7	9.11
RS_113	TmpD, tail murein hydrolase	808	91.2	6.74
RS_114	TmpE, putative peptidoglycan hydrolase	295	34.6	4.60
RS_115	Glycerophosphoryl diester phosphodiesterase	848	96.0	4.96
RS_116	phage structural protein	263	29.3	8.19
RS_117	phage structural protein	174	19.9	4.61
RS_118	BmpA, baseplate morphogenetic protein	234	26.6	4.77
RS_119	BmpB, baseplate morphogenetic protein	348	39.2	4.86
RS_120	TmpF, tail morphogenetic protein	1019	116.2	5.08
RS_121	BmpC, baseplate morphogenetic protein	173	19.2	5.39
RS_122	TmpG, tail morphogenetic protein	1152	129.0	5.19
RS_124	receptor binding protein	640	72.6	7.39
RS_126	receptor binding protein	458	50.3	6.27
RS_127	DhlA, DNA helicase	582	67.2	5.85
RS_129	DhlB, DNA helicase	480	54.5	5.72
RS_132	RncB, recombination nuclease B	639	73.4	5.19
RS_133	Asf, anti-sigma factor	198	23.2	6.81
RS_137	phage structural protein	202	23.6	5.72
RS_139	NrdE, ribonucleotide reductase	704	80.1	5.64
RS_140	NrdF, ribonucleotide reductase	349	40.4	4.78
RS_141	phage structural protein	109	12.4	4.68
RS_143	phage structural protein	179	21.1	6.95
RS_152	phage structural protein	423	46.8	4.75
RS_153	Rec.1, phage recombinase	74	7.9	6.61
RS_155	Rec.2, phage recombinase	315	35.7	5.16
RS_157	Sig, sigma factor	220	26.6	5.36
RS_158	phage structural protein	210	23.2	4.84
RS_159	TmpH, phage major tail protein	73	7.9	4.54
RS_160	phage structural protein	86	10.3	5.91
RS_163	putative membrane protein MbpG	122	14.0	5.95
RS_165	phage structural protein	178	20.8	7.47
RS_168	phage structural protein	287	32.3	5.76
RS_169	phage structural protein	243	28.3	5.34
RS_170	phage structural protein	152	17.8	4.98
RS_173	putative membrane protein MbpH	132	15.4	8.94
RS_175	phage structural protein	80	9.4	9.31
RS_181	phage structural protein	98	11.3	7.24
RS_196	phage structural protein	87	9.9	10.05
RS_206	NadV, nicotinamide phosphoribosyltransferase	489	56.1	5.44

* M_W: Molecular weight, pI: Isoelectric point.

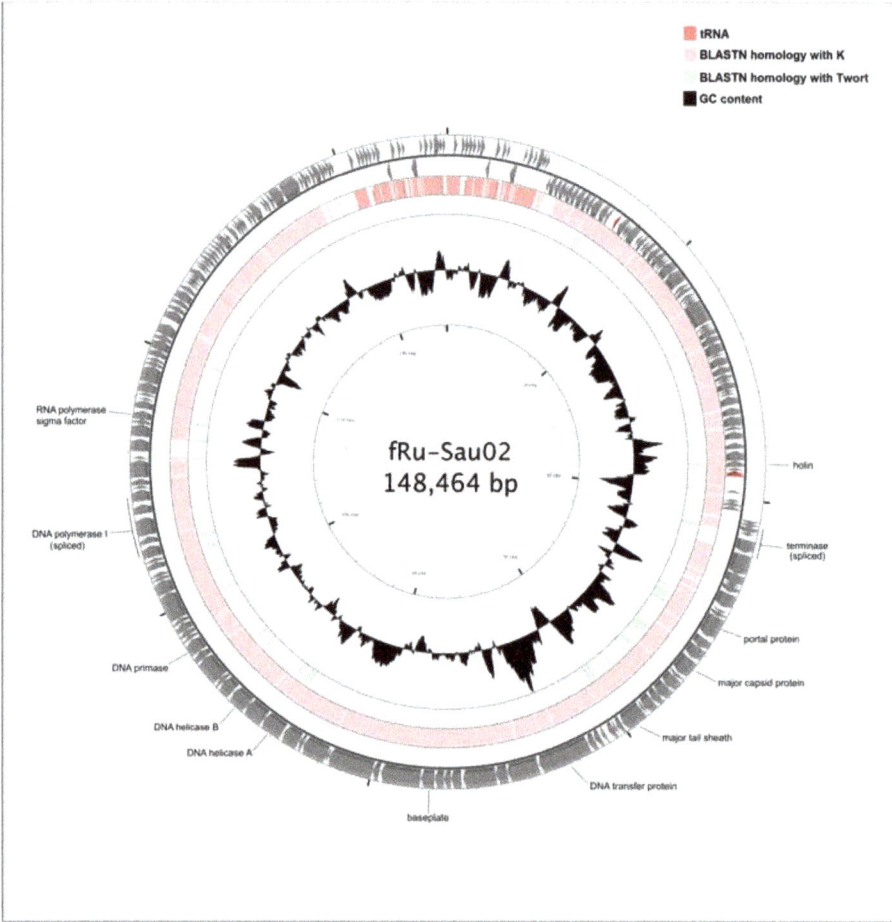

Figure 3. Genome comparison of three Twort-like phages. The outer ring represents the open reading frames (ORFs) of the circularized fRuSau02 phage. The two other rings display the identity between fRuSau02 and K (lavender) and between fRuSau02 and Twort (green). The inner ring shows the GC content of the fRuSau02 genome (black). Selected gene functions are indicated. The figure was generated with CGView [33].

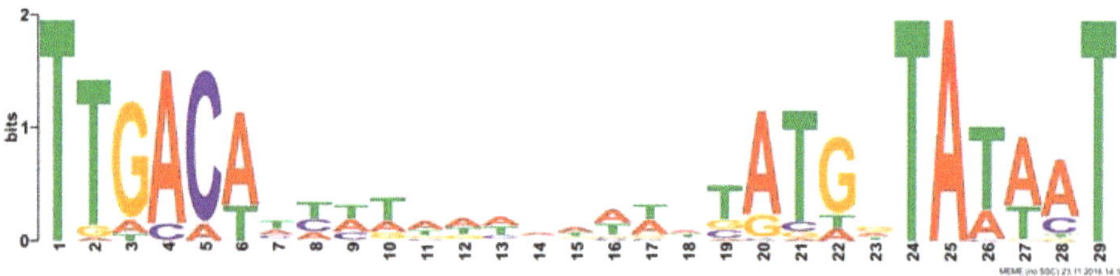

Figure 4. Consensus sequence of the phage fRuSau02 putative promoters. The promoter sequences are listed in supplementary Table S2.

3.4. Comparative Genome Analysis

Bioinformatic analysis of the fRuSau02 genome revealed that the phage has a genome size and organization typical for Twort-like viruses [15]. It is most closely related to phage MSA6 (JX080304), the two viruses showing 99.6% identity at the nucleotide level. The DNA sequence comparison with other Twort-like viruses revealed identity in the range of 39.0–96.0% (Table S3). The highest identity was observed with phages A5W (EU418428)—96.0%, Staph1N (JX080300)—96.0%, and Fi200W (JX080303)—95.1%. The genomic comparison of fRuSau02 with phages K and Twort showed identity rates of 93.5% and 46.5%, respectively. Figure 3 shows a BLASTN comparison of the genomes of phages fRuSau02, K and Twort. The whole-genome level nucleotide phylogeny analysis of fRuSau02 and the 34 phage genomes described in Table S3 showed that fRuSau02 clusters in the same species with A5W, Staph1N, MSA6, Fi200W, and 676Z (Figure 5). The analysis yielded average support of 16% and the OPTSIL clustering resulted altogether to 22, 2, and 1 clusters at species, genus, and family levels, respectively.

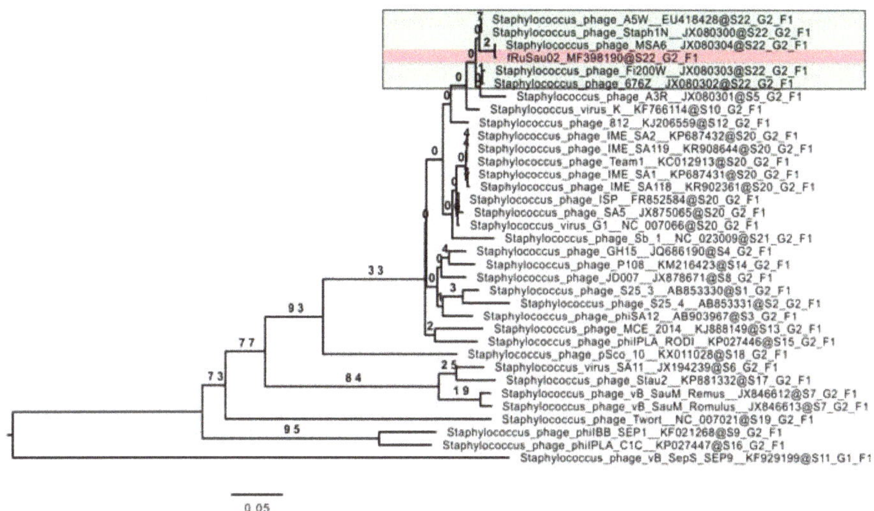

Figure 5. Genome-wide nucleotide phylogeny of 35 Twort-like viruses. The analysis was conducted with VICTOR Virus Classification and Tree Building Online Resource [35] with settings recommended for prokaryotic viruses. The tree was visualized with FigTree [39]. The analysis yielded 22 clusters at species (S1–S22) and two at genus (G1–G2) level, respectively. All the phages clustered to the same family (F1). Phage fRuSau02 is indicated with red box and the phages belonging to the same species with it in the green box.

Most fRuSau02 nucleotide differences to MSA6 were single base pair substitutions or small indels in intergenic regions. There were 25 coding regions having differences to MSA6, eight of which had silent mutations, leading to proteins with 100% amino acid identity with their MSA6 counterparts (terminal repeat encoded protein TreP, phage terminase Ter.2, tail morphogenetic proteins TmpB and TmpG, DNA helicase DhlA, DNA primase/DNA helicase Pri, ribonucleotide reductase NrdE, and intron encoded endonuclease I-MsaI). Ten proteins had difference(s) to the corresponding proteins in MSA6 but showed 100% amino acid identities to their homologs in other Twort-like viruses: terminal repeat encoded protein TreK was identical to phage G1 ORF159 but showed only 87.4% identity to MSA6 TreK. Terminal repeat encoded protein TreG was identical to phage K Gp007, putative membrane protein MbpP to G1 ORF007, hypothetical protein DmcA to Gp122 of phage JD007, putative membrane protein MbpE to phage G1 ORF120, major tail sheath Tsp to phage K Gp166, hypothetical protein RS_209 to Team1 Gp041, putative receptor binding protein RS_126 to SA5 ORF40, terminal repeat encoded protein TreB to A5W TreB and to G1 ORF231, and putative portal protein Prt to G1 ORF014. Each of these proteins had one to two amino acid difference(s) to their MSA6 homologs. Seven proteins had at least one amino acid (according to present knowledge) unique to fRuSau02, i.e., have not been observed in homologous proteins of any other Twort-like virus: putative membrane protein MbpC, tail morphogenetic protein TmpC, putative receptor binding protein RS_124, putative polymerase-associated exonuclease PolA.2, putative RNA polymerase sigma factor Sig, hypothetical protein RS_200, and putative membrane protein MbpI.

Of the "unique" fRuSau02 proteins, RS_124 is of outmost interest. It is 98.1% identical at the amino acid level to the *orf103* gene product of phage ΦSA012, shown to be one out of two receptor binding proteins (RBPs) of this phage [22]. Phage fRuSau02 RS_124 has histidine in position 306, where all the other Twort-like viruses sequenced so far have proline. The residue 306 is part of a carbohydrate binding domain, which is formed by amino acids 213–336. Preliminary structural modelling of this region showed that H306 fits nicely into an anti-parallel β-sheet structure, which is completely distorted by H306P change (not shown). This suggests that the structure of the receptor binding protein of fRuSau02 may be different from all the other Twort-like viruses characterized so far.

3.5. Genes Interrupted by Self-Splicing Elements

The presence of mobile splicing elements in the genomes of Twort-like viruses is a characteristic feature of staphylococcal myoviruses [17–19]. In the case of fRuSau02 phage, four protein-encoding genes were found to be interrupted by different insertion sequences (Figure S1): (1) The gene encoding phage lysin (Lys) is fragmented into two by an intron-encoded HNH homing endonuclease gene (*I-KsaI*). (2) The terminase large subunit gene is divided by the intron, encoding I-MsaI, a protein of unknown function. (3) The gene encoding the putative DNA polymerase-associated exonuclease (PolA) contains two introns encoding proteins I-KsaII and I-KsaIII. (4) The gene encoding phage RecA-like recombinase (Rec) contains the intron-encoded endonuclease gene *I-MsaII*. All of the intervening sequences are predicted to be group I introns encoding putative endonucleases. The functionality of the spliced genes was shown by the fact that the presence of full proteins encoded by three out of four spliced genes was observed in the LC-MS/MS analysis. Namely, lysin, terminase large subunit, and phage recombinase were among the proteins identified in the LC-MS/MS analysis of the purified phage particles (Table 1), with peptides present in both the C- and N-terminal parts of the proteins (see below). While this does not conclusively prove that the polypeptides are continuous, this is the most likely option. The gene splicing pattern of lysin and polymerase encoding genes is identical to the one presented by staphylococcal phages G1, K and ISP [19,47,48]. However, fRuSau02 has additional insertions in the terminase large subunit and recombinase genes that were also present in phage Team1. Unlike in the more distant phages Remus/Romulus and Twort, there were no intein domains identified in the fRuSau02 genome [17,49,50].

3.6. Proteomic Analysis of the Phage Structural Proteins

To confirm the identification and expression of the phage structural proteins, a proteomic analysis of the purified phage particles using LC-MS/MS was performed. The comparative analysis of obtained peptide sequences with the sequences of predicted phage proteins allowed for the identification of these structural proteins. To exclude the possibility of obtaining false positive results due to similarity with bacterial proteins that could be carried over from the lysate during the sample preparation, the obtained peptide sequences were compared simultaneously against the phage and bacterial protein sequences. Altogether, 81 phage proteins were identified in the LC-MS/MS analysis, of which 78 fulfilled the inclusion criteria (>2 unique peptides and/or >5% coverage) (Table 1). The analysis of the structural proteome identified the capsid (Mcp) and the tail (Tsp, TmpA, TmpB, TmpC, TmpD, TmpE, TmpF, TmpG, TmpH, BmpA, BmpB, BmpC) proteins, receptor binding proteins (RS_124 and RS_126), the portal protein (Prt), and putative membrane proteins (MbpC, MbpD, MbpG, MbpH, MbpR,

MbpS), as well as phage holin (HolA). Additionally, this study showed the presence of ribonucleotide reductases (NrdE and NrdF), DNA-binding protein (HmzG), putative ligase (Lig), recombination nuclease B (RncB), ribonuclease H (Rbn), and DNA helicases (DhlA and DhlB) as well as sigma and anti-sigma factors (Sig, Asf). Altogether, 33 of the identified proteins were annotated as novel phage structural proteins. As already described above, the LC-MS/MS analysis showed also the presence of three proteins: the phage lysin (LysK), terminase (Ter) and recombinase (Rec) encoded by the genes interrupted by the intervening sequences. The prohead protease Pro was also identified in the LC-MS/MS analysis. The presence of the protease among the structural proteins was described previously for *Lactobacillus delbrueckii*-specific phages [51,52] but may be the result of lack of dissociation from the head after the assembly.

3.7. *fRuSau02 Host Range*

A collection of 135 *Staphylococcus* strains, including 51 human and 54 porcine *S. aureus* isolates and 30 coagulase-negative *Staphylococcus* strains of human origin were used to assess the host range (Table 2). Of the 50 clinical *S. aureus* strains collected for this study, 35 were methicillin-sensitive and 15 methicillin-resistant (Table S1). Phage fRuSau02 was able to infect 49 (96%) coagulase-positive and 15 (50%) coagulase-negative strains of human origin, whereas the rate of infection of pig isolates was much lower with only 18 (33%) strains infected. The infectivity of *S. aureus* strains did not depend on their response to methicillin. Some of the bacterial strains instead of clear lysis displayed turbid lysis or only slower growth rate. Counting together the clear and turbid lysis, 33 (61%) of pig isolates and 5 (17%) of coagulase-negative strains were resistant to phage infection. Importantly, all *S. aureus* human isolates (including MRSA strains) were at least moderately sensitive to fRuSau02. Further, patient isolates of coagulase-negative *Staphylococcus* strains, including *S. intermedius*, *S. lugdunensis*, *S. epidermidis*, *S. haemolyticus*, *S. saprophyticus* and *S. pseudointer*, showed different rates of infection depending on the strain, however, at least one strain of each species was susceptible to the phage. Further, the coagulase-negative strains and several pig isolates displayed lower efficiency of infection and turbidity of the plaques, while human *S. aureus* isolates were characterized by big (1–3 mm) clear plaques (data not shown).

Table 2. The infectivity of fRuSau02 for different staphylococcal isolates. The details and strain references are listed in supplementary material Table S1.

	fRuSau02 Infectivity					
Bacterial Hosts	**Infected ***		**Intermediate ***		**Resistant ***	
Coagulase-Positive Human Isolates (*n* = 51)						
S. aureus	49	(96%)	2	(4%)	0	(0%)
Coagulase-Negative Human Isolates (*n* = 30)						
S. intermedius	0		3		2	
S. lugdunensis	1		4		0	
S. epidermidis	0		1		4	
S. haemolyticus	0		2		3	
S. saprophyticus	1		2		2	
S. pseudointer	0		4		1	
ALL	2	(7%)	16	(53%)	12	(40%)
Coagulase-Positive Porcine Isolates (*n* = 54)						
S. aureus	18	(33%)	3	(6%)	33	(61%)

* Infected indicates clear lysis or growth inhibition for spot and liquid assays, respectively, intermediate turbid lysis or slower growth rate, and resistant no infection.

3.8. fRuSau02 Receptor

Staphylococcal Twort-like phages have been shown to utilize cell wall teichoic acids (WTAs) as their receptors [22]. To test whether this is also the case with fRuSau02, we analyzed the infectivity of fRuSau02 in *S. aureus* strain tagO. This strain carries a mutation in the gene encoding TagO, an enzyme that catalyzes the transfer of *N*-acetylglucosamine to bactoprenol in the first step of teichoic acid biosynthesis [53,54]. As shown in Figure 6A, the phage failed to reproduce in the tagO strain. Furthermore, the adsorption assay revealed that fRuSau02 was not able to adsorb to this strain (Figure 6B). It thus seems that like for other staphylococcal Twort-like phages, WTAs serve as receptors for fRuSau02.

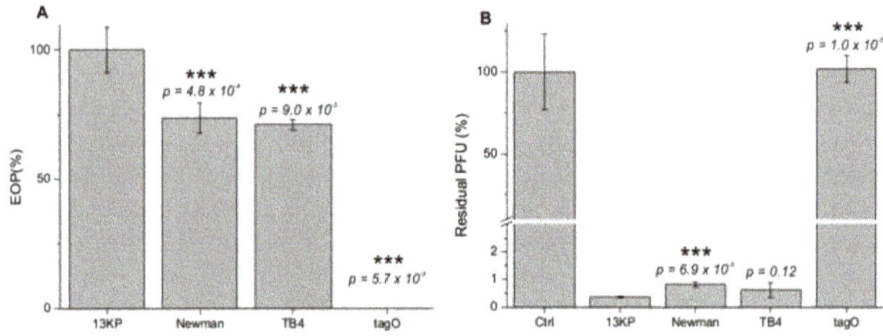

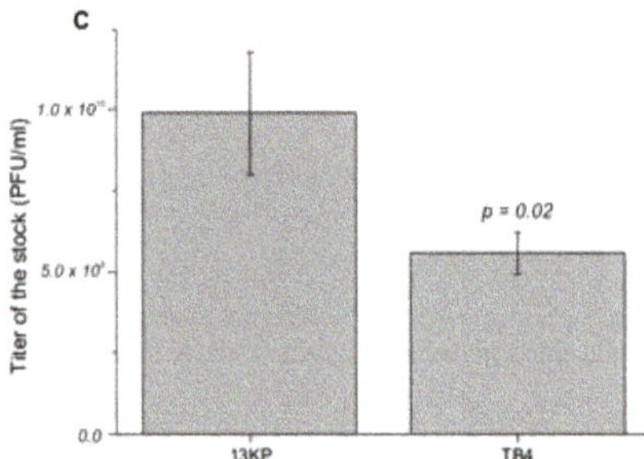

Figure 6. Suitability of host strains for production of fRuSau02. (**A**) The efficiency of plating (EOP) counted as the number of plaque-forming units (PFU) obtained from the same amount of phage lysate for different bacterial strains. The result obtained for the reference strain 13KP was set as 100%; (**B**) Adsorption of fRuSau02 to bacterial surface. Ctrl represents LB as negative control, in which the residual PFU was set to 100%; (**C**) The titer of fRuSau02 lysate produced in strains 13KP and TB4. Error bars indicate SD, *p*-values the level of significance between 13KP and other strains, *** indicates that the difference is statistically significant at the *p*-value < 0.001 level.

3.9. The Choice of Optimal Host Strain for Therapeutic Phage Production

In an ideal situation, host strains used for the production of therapeutic phages should be free of prophages. This is because prophages encode virulence factors, such as staphylococcal enterotoxins [55]. In addition, they can be induced from cells during the infection by therapeutic phage and cause genome variations [56]. To study the possibility to produce fRuSau02 in the prophage-free *S. aureus* strain, we compared the efficiency of plating (EOP) in strains 13KP, Newman, and TB4. Of these, 13KP is the strain that was used as a host strain during the phage isolation, and it was used as a control strain with 100% EOP. TB4 is a prophage-free strain [53] and Newman the parental strain for TB4 [57].

The EOP assay showed that both TB4 and Newman had reduced infectivity compared to the reference strain 13KP (Figure 6A). This was not due to the lowered adsorption efficiency of fRuSau2 to these strains, as the adsorption assay did not reveal significant difference between 13KP and TB4. The phage adsorption to Newman was significantly reduced from 13KP, even though the residual PFU even in this strain was only 0.8% (Figure 6B). The potential of TB4 strain for phage production was further studied by preparing phage stocks. Semi-confluent soft-agar overlay plates were prepared using host-strain adjusted amount of fRuSau02 bacteriophage. Phage stocks were prepared in three parallels and titrated (Figure 6C). The results showed that the difference in the amount of phage obtained using 13KP and TB4 as host was statistically significant ($p = 0.020$), however even the titer obtained in TB4 was sufficient for phage production purposes.

As genes for bacterial toxins often reside in prophage genomes, we wanted to analyze whether fRuSau02 lysate produced in TB4 strain contains less toxins than the lysate produced in 13KP or Newman strains. To this end, the staphylococcal enterotoxins were measured from the phage lysates with the Transia Plate Staphylococcal Enterotoxins assay that detects enterotoxins A, B, C, D, and E. The phage lysate produced in 13KP was clearly positive for enterotoxins, with a concentration that approximately corresponded to 320 ng/mL of staphylococcal enterotoxin A. The phage lysate produced in Newman strain had clearly less toxins (~6 ng/mL) and the lysate produced in TB4 remained negative, indicating that the enterotoxin concentration was lower than the detection limit of the assay (~1 ng/mL of enterotoxin A). It should be noted here that the assay is not validated for quantitative analysis, thus the concentrations need to be considered approximates.

To conclude, strain TB4 should be considered as potential bacterial host for the production of bacteriophage preparations for phage therapy, as it does not possess the risk of temperate phage or enterotoxin carry-over.

4. Discussion

This study reports a new bacteriophage, fRuSau02, isolated from a commercial *Staphylococcus* bacteriophage cocktail produced by Microgen. The genomic analysis revealed that fRuSau02 is very closely related to phage MSA6 and many other Twort-like viruses. Bacteriophage fRuSau02 possesses a large genome (148,464 bp) with typical modular organization and a low G+C content and therefore can be classified as a member of the genus *Twortlikevirus*. In coherence, the morphology of the fRuSau02 phage is similar to MSA6 [20], as well as the staphylococcal phage K [19] and *Listeria* phage A511 [58]. The phylogeny analysis of 35 Twort-like phages clustered fRuSau02 and MSA6 in a same species together with A5W, Staph1N, Fi200W, and 676Z. Phage Twort, the type representative of this genus, is more distantly related to fRuSau02, the two phages displaying only 46.5% identity at the nucleotide level. Perhaps the most significant difference between fRuSau02 and MSA6 was the H306P change in the putative receptor binding protein (RS_124 and ORF094 in fRuSau02 and MSA6, respectively). Histidine in position 306 seems unique for fRuSau02, as homologous proteins of other Twort-like viruses analyzed so far, for example G1ORF008 of phage G1, ORF107 of Sb-1, and ORF125 of Team1, all have P306. The preliminary structural modelling indicated the H306P change alters the structure of the carbohydrate-binding region of the RBP, which may have a profound effect on the phage host range.

The in silico analysis revealed the presence of bacterial promoters in the genome of fRuSau02. However, both the presence of genes encoding for the phage sigma and anti-sigma factors and the genomic region of 27.5 kb that does not contain any promoters suggest the existence of phage promoters. Although the performed bioinformatics study failed to reveal any conserved sequences present upstream of the genes of this module, we believe that fRuSau02 possesses two types of promoters. Most likely, in the beginning of the infection process viral genes are transcribed by the bacterial sigma factor. During this step, the sigma and anti-sigma factors encoded by the phage genome are also transcribed. In the later stages of infection, the anti-sigma factor inhibits the activity of the bacterial factor and allows the phage sigma factor to lead the transcription from its own unique promoters. Such a process would allow the bacteriophage to have a high and uniform rate of transcription of late structural genes with a minimal transcription of bacterial genes. Further studies aiming at the recognition of transcriptional starting sites are needed to indicate the possible viral promoters and to validate the annotated bacterial promoters.

The LC-MS/MS analysis revealed 78 phage structural proteins. Due to the high sensitivity of the method, we have to take into consideration the possibility that some of the identified proteins are carried over from the lysate and co-isolated with the phage particles. On the other hand, previous studies showed that some proteins are commonly packed together with the DNA due to their association with nucleic acids [59,60]. For example, the phage sigma and anti-sigma factors (RS_157 and RS_133, respectively) were among the proteins identified, however, it is not likely that they are structural proteins of the phage particles. Earlier studies showed that the primary staphylococcal polymerase σ^{SA}, directs transcription of early genes in Twort-like viruses [47]. In addition, both bacterial RNA polymeras (RNAP) subunits and the sigma factor were among the bacterial proteins identified in the LC-MS/MS study (data not shown) suggesting that these proteins were co-isolated together with the phage particles. It is possible that these proteins show physical properties that make them either more prone to be co-isolated with phage particles during the purification process or they display unspecific binding to the capsid proteins. Similarly, the proteome analysis showed the presence of putative membrane proteins (MbpC, MbpD, MbpG, MbpH, MbpR, MbpS) that can be a part of the structural proteome used during the infection step or during the assembly and lysis. However, due to the fact that they bind to the membranes, it is also possible that they were carryover from the phage lysate.

Phage fRuSau02 was shown to infect a considerable number of human *S. aureus* isolates, however, the rates of infectivity were much lower among the animal isolates. Similar host range pattern has earlier been observed with phage ISP, which also infects efficiently human *S. aureus* isolates but is unable to infect *S. aureus* strains isolated from pigs [48]. The resistance of the pig strains may be due to minor structural differences of WTA between different *S. aureus* strains. It is also plausible that some strains developed phage resistance without the introduction of modifications in the phage receptor structures. For example, the presence of CRISPR sequences in the genome of *S. aureus* allows the bacteria to acquire the immunity against encountered phages [61]. The pig MRSA strains often belong to only few clonal complexes [62], which may explain their different phage profile compared to *S. aureus* stains isolated from other sources. Interestingly, the host profile of fRuSau02 may be somewhat different from ISP, as fRuSau02 was also able to infect some coagulase-negative staphylococcal strains, including *S. haemolyticus* earlier shown to be resistant for ISP [48].

Phages have important potential as antimicrobial agents and may serve as an alternative to antibiotics, especially in case of multi-drug resistant pathogens. Phage therapy is a possible cure for community-acquired and nosocomial infections caused by drug-resistant *Staphylococcus*, as well as a good candidate for prevention of bacterial contamination in industry and animal husbandry [53–67]. Twort-like phages are perhaps the most studied *S. aureus* phage group for clinical applications [15]. For example, phage ISP is one component of a phage therapy cocktail BFC-1, developed for the treatment of burn wound infections [68,69].

To conclude, both our analyses and the fact that fRuSau02 was isolated from a commercial therapeutic phage cocktail suggest that it should be considered as well suited for human phage therapy against coagulase-positive and to some extent also coagulase-negative staphylococcal strains. However, its capacity for the prevention and control of MRSA carriage and/or contamination in the animal husbandry and food industry may be more limited. The efficacy and safety of fRuSau02 as the therapeutic tool is still to be elucidated. Further research that includes pharmacological trials is essential to confirm the possible role of fRuSau02 in the treatment of different forms of MRSA infections in humans.

Supplementary Materials: The following are available online at www.mdpi.com/1999-4915/9/9/258/s1. Table S1: Host range analysis of fRuSau02; Table S2: Putative promoter sequences identified in the fRuSau02 genome; Table S3: Comparative nucleotide analysis between the genomes of fRuSau02 and selected *Staphylococcus* phages; Figure S1: Split genes in the genome of fRuSau02.

Acknowledgments: Helsinki University Hospital special state subsidy for health science research grants, the Academy of Finland (project 1288701) and Jane & Aatos Erkko Foundation are acknowledged for funding (to MS). KL was supported by the Emil Aaltonen Foundation. Annamari Heikinheimo is thanked for providing the porcine *S. aureus* strain collection and Taeok Bae for sharing the TB4 and *tagO* strains. Arnab Bhattacharjee is acknowledged for his help with structural modelling and Joseph Michael Ochieng' Oduor for critically reading the manuscript.

Author Contributions: K.L., A.W., M.S. and S.K. conceived and designed the experiments; K.L., H.T., A.W., J.H.H. and S.K. performed the experiments; K.L., A.W., M.S. and S.K. analyzed the data; P.K. contributed clinical bacterial strains; K.L. and S.K. wrote the paper.

Conflicts of Interest: The authors declare no conflict of interest. The founding sponsors had no role in the design of the study; in the collection, analyses, or interpretation of data; in the writing of the manuscript, and in the decision to publish the results.

References

1. Gordon, R.J.; Lowy, F.D. Pathogenesis of methicillin-resistant *Staphylococcus aureus* infection. *Clin. Infect. Dis.* **2008**, *46*, S350–S359. [CrossRef] [PubMed]
2. Tong, S.Y.; Davis, J.S.; Eichenberger, E.; Holland, T.L.; Fowler, V.G., Jr. *Staphylococcus aureus* infections: Epidemiology, pathophysiology, clinical manifestations, and management. *Clin. Microbiol. Rev.* **2015**, *28*, 603–661. [CrossRef] [PubMed]
3. Dantes, R.; Mu, Y.; Belflower, R.; Aragon, D.; Dumyati, G.; Harrison, L.H.; Lessa, F.C.; Lynfield, R.; Nadle, J.; Petit, S.; et al. National burden of invasive methicillin-resistant *Staphylococcus aureus* infections, united states, 2011. *JAMA Intern. Med.* **2013**, *173*, 1970–1978. [PubMed]
4. Bal, A.M.; Coombs, G.W.; Holden, M.T.G.; Lindsay, J.A.; Nimmo, G.R.; Tattevin, P.; Skov, R.L. Genomic insights into the emergence and spread of international clones of healthcare-, community- and livestock-associated meticillin-resistant *Staphylococcus aureus*: Blurring of the traditional definitions. *J. Glob. Antimicrob. Resist.* **2016**, *6*, 95–101. [CrossRef] [PubMed]
5. Enright, M.C.; Robinson, D.A.; Randle, G.; Feil, E.J.; Grundmann, H.; Spratt, B.G. The evolutionary history of methicillin-resistant *Staphylococcus aureus* (MRSA). *Proc. Natl. Acad. Sci. USA* **2002**, *99*, 7687–7692. [CrossRef] [PubMed]
6. Chan, B.K.; Abedon, S.T.; Loc-Carrillo, C. Phage cocktails and the future of phage therapy. *Future Microbiol.* **2013**, *8*, 769–783. [CrossRef] [PubMed]
7. Ryan, E.M.; Gorman, S.P.; Donnelly, R.F.; Gilmore, B.F. Recent advances in bacteriophage therapy: How delivery routes, formulation, concentration and timing influence the success of phage therapy. *J. Pharm. Pharmacol.* **2011**, *63*, 1253–1264. [CrossRef] [PubMed]
8. Skurnik, M.; Strauch, E. Phage therapy: Facts and fiction. *Int. J. Med. Microbiol.* **2006**, *296*, 5–14. [CrossRef] [PubMed]
9. Abedon, S.T.; Kuhl, S.J.; Blasdel, B.G.; Kutter, E.M. Phage treatment of human infections. *Bacteriophage* **2011**, *1*, 66–85. [CrossRef] [PubMed]
10. Wittebole, X.; De Roock, S.; Opal, S.M. A historical overview of bacteriophage therapy as an alternative to antibiotics for the treatment of bacterial pathogens. *Virulence* **2014**, *5*, 226–235. [CrossRef] [PubMed]
11. Pirnay, J.P.; Blasdel, B.G.; Bretaudeau, L.; Buckling, A.; Chanishvili, N.; Clark, J.R.; Corte-Real, S.; Debarbieux, L.; Dublanchet, A.; De Vos, D.; et al. Quality and safety requirements for sustainable phage therapy products. *Pharm. Res.* **2015**, *32*, 2173–2179. [CrossRef] [PubMed]
12. Deghorain, M.; Van Melderen, L. The staphylococci phages family: An overview. *Viruses* **2012**, *4*, 3316–3335. [CrossRef] [PubMed]
13. Xia, G.; Wolz, C. Phages of *Staphylococcus aureus* and their impact on host evolution. *Infect. Genet. Evol.* **2014**, *21*, 593–601. [CrossRef] [PubMed]
14. Kazmierczak, Z.; Gorski, A.; Dabrowska, K. Facing antibiotic resistance: *Staphylococcus aureus* phages as a medical tool. *Viruses* **2014**, *6*, 2551–2570. [CrossRef] [PubMed]

15. Lobocka, M.; Hejnowicz, M.S.; Dabrowski, K.; Gozdek, A.; Kosakowski, J.; Witkowska, M.; Ulatowska, M.I.; Weber-Dabrowska, B.; Kwiatek, M.; Parasion, S.; et al. Genomics of staphylococcal Twort-like phages—Potential therapeutics of the post-antibiotic era. *Adv. Virus Res.* **2012**, *83*, 143–216. [PubMed]
16. Lavigne, R.; Darius, P.; Summer, E.J.; Seto, D.; Mahadevan, P.; Nilsson, A.S.; Ackermann, H.W.; Kropinski, A.M. Classification of *Myoviridae* bacteriophages using protein sequence similarity. *BMC Microb.* **2009**, *9*, 224. [CrossRef] [PubMed]
17. Vandersteegen, K.; Kropinski, A.M.; Nash, J.H.; Noben, J.P.; Hermans, K.; Lavigne, R. Romulus and Remus, two phage isolates representing a distinct clade within the *Twortlikevirus* genus, display suitable properties for phage therapy applications. *J. Virol.* **2013**, *87*, 3237–3247. [CrossRef] [PubMed]
18. Kwan, T.; Liu, J.; DuBow, M.; Gros, P.; Pelletier, J. The complete genomes and proteomes of 27 *Staphylococcus aureus* bacteriophages. *Proc. Natl. Acad. Sci. USA* **2005**, *102*, 5174–5179. [CrossRef] [PubMed]
19. O'Flaherty, S.; Coffey, A.; Edwards, R.; Meaney, W.; Fitzgerald, G.F.; Ross, R.P. Genome of staphylococcal phage K: A new lineage of myoviridae infecting gram-positive bacteria with a low G+C content. *J. Bacteriol.* **2004**, *186*, 2862–2871. [CrossRef] [PubMed]
20. Kwiatek, M.; Parasion, S.; Mizak, L.; Gryko, R.; Bartoszcze, M.; Kocik, J. Characterization of a bacteriophage, isolated from a cow with mastitis, that is lytic against *Staphylococcus aureus* strains. *Arch. Virol.* **2012**, *157*, 225–234. [CrossRef] [PubMed]
21. Gu, J.; Liu, X.; Lu, R.; Li, Y.; Song, J.; Lei, L.; Sun, C.; Feng, X.; Du, C.; Yu, H.; et al. Complete genome sequence of *Staphylococcus aureus* bacteriophage GH15. *J. Virol.* **2012**, *86*, 8914–8915. [CrossRef] [PubMed]
22. Takeuchi, I.; Osada, K.; Azam, A.H.; Asakawa, H.; Miyanaga, K.; Tanji, Y. The presence of two receptor-binding proteins contributes to the wide host range of staphylococcal Twort-like phages. *Appl. Environ. Microbiol.* **2016**, *82*, 5763–5774. [CrossRef] [PubMed]
23. Sambrook, J.; Russell, D.W. *Molecular Cloning, a Laboratory Manual*, 3rd ed.; Cold Spring Harbor Laboratory Press: New York, NY, USA, 2001.
24. FIMM Sequencing Unit. Available online: https://www.fimm.fi/en/services/technology-centre/sequencing/ (accessed on 13 September 2017).
25. Coil, D.; Jospin, G.; Darling, A.E. A5-miseq: An updated pipeline to assemble microbial genomes from Illumina MiSeq data. *Bioinformatics* **2015**, *31*, 587–589. [CrossRef] [PubMed]
26. Robinson, J.T.; Thorvaldsdottir, H.; Winckler, W.; Guttman, M.; Lander, E.S.; Getz, G.; Mesirov, J.P. Integrative Genomics Viewer. *Nat. Biotechnol.* **2011**, *29*, 24–26. [CrossRef] [PubMed]
27. Thorvaldsdottir, H.; Robinson, J.T.; Mesirov, J.P. Integrative Genomics Viewer (IGV): High-performance genomics data visualization and exploration. *Brief. Bioinform.* **2013**, *14*, 178–192. [CrossRef] [PubMed]
28. RAST (Rapid Annotation Using Subsystem Technology). Available online: http://rast.nmpdr.org/ (accessed on 13 September 2017).
29. De Jong, A.; Pietersma, H.; Cordes, M.; Kuipers, O.P.; Kok, J. PePPER: A webserver for prediction of prokaryote promoter elements and regulons. *BMC Genom.* **2012**, *13*, 299. [CrossRef] [PubMed]
30. Gautheret, D.; Lambert, A. Direct RNA motif definition and identification from multiple sequence alignments using secondary structure profiles. *J. Mol. Biol.* **2001**, *313*, 1003–1011. [CrossRef] [PubMed]
31. Macke, T.J.; Ecker, D.J.; Gutell, R.R.; Gautheret, D.; Case, D.A.; Sampath, R. RNAmotif, an RNA secondary structure definition and search algorithm. *Nucl. Acids Res.* **2001**, *29*, 4724–4735. [CrossRef] [PubMed]
32. Bailey, T.L.; Boden, M.; Buske, F.A.; Frith, M.; Grant, C.E.; Clementi, L.; Ren, J.Y.; Li, W.W.; Noble, W.S. MEME suite: Tools for motif discovery and searching. *Nucl. Acids Res.* **2009**, *37*, W202–W208. [CrossRef] [PubMed]
33. Stothard, P.; Wishart, D.S. Circular genome visualization and exploration using GCView. *Bioinformatics* **2005**, *21*, 537–539. [CrossRef] [PubMed]
34. EMBOSS Stretcher. Available online: http://www.ebi.ac.uk/Tools/psa/emboss_stretcher (accessed on 13 September 2017).
35. Meier-Kolthoff, J.P.; Göker, M. Victor: Genome-based phylogeny and classification of prokaryotic viruses. *Bioinformatics* **2017**, 1–9. [CrossRef]
36. Meier-Kolthoff, J.P.; Auch, A.F.; Klenk, H.P.; Goker, M. Genome sequence-based species delimitation with confidence intervals and improved distance functions. *BMC Bioinform.* **2013**, *14*, 60. [CrossRef] [PubMed]
37. Lefort, V.; Desper, R.; Gascuel, O. Fastme 2.0: A comprehensive, accurate, and fast distance-based phylogeny inference program. *Mol. Biol. Evol.* **2015**, *32*, 2798–2800. [CrossRef] [PubMed]
38. Farris, J.S. Estimating phylogenetic trees from distance matrices. *Am. Nat.* **1972**, *106*, 645–668. [CrossRef]

39. FigTree. Available online: http://tree.bio.ed.ac.uk/software/figtree/ (accessed on 13 September 2017).
40. Göker, M.; Garcia-Blazquez, G.; Voglmayr, H.; Telleria, M.T.; Martin, M.P. Molecular taxonomy of phytopathogenic fungi: A case study in *Peronospora*. *PLoS ONE* **2009**, *4*, e6319. [CrossRef] [PubMed]
41. Meier-Kolthoff, J.P.; Hahnke, R.L.; Petersen, J.; Scheuner, C.; Michael, V.; Fiebig, A.; Rohde, C.; Rohde, M.; Fartmann, B.; Goodwin, L.A.; et al. Complete genome sequence of DSM 30083(T), the type strain (U5/41(T)) of *Escherichia coli*, and a proposal for delineating subspecies in microbial taxonomy. *Stand. Genom.Sci.* **2014**, *9*, 2. [CrossRef] [PubMed]
42. Varjosalo, M.; Keskitalo, S.; Van Drogen, A.; Nurkkala, H.; Vichalkovski, A.; Aebersold, R.; Gstaiger, M. The protein interaction landscape of the human CMGC kinase group. *Cell Rep.* **2013**, *3*, 1306–1320. [CrossRef] [PubMed]
43. Kutter, E. Phage host range and efficiency of plating. *Methods Mol. Biol.* **2009**, *501*, 141–149. [PubMed]
44. Leon-Velarde, C.G.; Happonen, L.; Pajunen, M.; Leskinen, K.; Kropinski, A.M.; Mattinen, L.; Rajtor, M.; Zur, J.; Smith, D.; Chen, S.; et al. *Yersinia enterocolitica*-specific infection by bacteriophages TG1 and varphiR1-RT is dependent on temperature-regulated expression of the phage host receptor OmpF. *Appl. Environ. Microbiol.* **2016**, *82*, 5340–5353. [CrossRef] [PubMed]
45. Ackermann, H.W. Bacteriophage observations and evolution. *Res. Microbiol.* **2003**, *154*, 245–251. [CrossRef]
46. Maniloff, J.; Ackermann, H.W. Taxonomy of bacterial viruses: Establishment of tailed virus genera and the order *Caudovirales*. *Arch. Virol.* **1998**, *143*, 2051–2063. [CrossRef] [PubMed]
47. Dehbi, M.; Moeck, G.; Arhin, F.F.; Bauda, P.; Bergeron, D.; Kwan, T.; Liu, J.; McCarty, J.; Dubow, M.; Pelletier, J. Inhibition of transcription in *Staphylococcus aureus* by a primary sigma factor-binding polypeptide from phage G1. *J. Bacteriol.* **2009**, *191*, 3763–3771. [CrossRef] [PubMed]
48. Vandersteegen, K.; Mattheus, W.; Ceyssens, P.J.; Bilocq, F.; De Vos, D.; Pirnay, J.P.; Noben, J.P.; Merabishvili, M.; Lipinska, U.; Hermans, K.; et al. Microbiological and molecular assessment of bacteriophage ISP for the control of *Staphylococcus aureus*. *PLoS ONE* **2011**, *6*, e24418. [CrossRef] [PubMed]
49. Landthaler, M.; Begley, U.; Lau, N.C.; Shub, D.A. Two self-splicing group i introns in the ribonucleotide reductase large subunit gene of *Staphylococcus aureus* phage Twort. *Nucl. Acids Res.* **2002**, *30*, 1935–1943. [CrossRef] [PubMed]
50. Landthaler, M.; Shub, D.A. Unexpected abundance of self-splicing introns in the genome of bacteriophage Twort: Introns in multiple genes, a single gene with three introns, and exon skipping by group I ribozymes. *Proc. Natl. Acad. Sci. USA* **1999**, *96*, 7005–7010. [CrossRef] [PubMed]
51. Casey, E.; Mahony, J.; Neve, H.; Noben, J.P.; Dal Bello, F.; van Sinderen, D. Novel phage group infecting *Lactobacillus delbrueckii* subsp. *Lactis*, as revealed by genomic and proteomic analysis of bacteriophage Ldl1. *Appl. Environ. Microbiol.* **2015**, *81*, 1319–1326. [PubMed]
52. Casey, E.; Mahony, J.; O'Connell-Motherway, M.; Bottacini, F.; Cornelissen, A.; Neve, H.; Heller, K.J.; Noben, J.P.; Dal Bello, F.; van Sinderen, D. Molecular characterization of three *Lactobacillus delbrueckii* subsp. *Bulgaricus* phages. *Appl. Environ. Microbiol.* **2014**, *80*, 5623–5635. [CrossRef] [PubMed]
53. Bae, T.; Baba, T.; Hiramatsu, K.; Schneewind, O. Prophages of *Staphylococcus aureus* Newman and their contribution to virulence. *Mol. Microbiol.* **2006**, *62*, 1035–1047. [CrossRef] [PubMed]
54. Soldo, B.; Lazarevic, V.; Karamata, D. TagO is involved in the synthesis of all anionic cell-wall polymers in *Bacillus subtilis* 168. *Microbiology* **2002**, *148*, 2079–2087. [CrossRef] [PubMed]
55. Betley, M.J.; Mekalanos, J.J. Staphylococcal enterotoxin a is encoded by phage. *Science* **1985**, *229*, 185–187. [CrossRef] [PubMed]
56. Goerke, C.; Wirtz, C.; Fluckiger, U.; Wolz, C. Extensive phage dynamics in *Staphylococcus aureus* contributes to adaptation to the human host during infection. *Mol. Microbiol.* **2006**, *61*, 1673–1685. [CrossRef] [PubMed]
57. Miller, K.D.; Hetrick, D.L.; Bielefeldt, D.J. Production and properties of *Staphylococcus aureus* (strain Newman D2C) with uniform clumping factor activity. *Thromb. Res.* **1977**, *10*, 203–211. [CrossRef]
58. Klumpp, J.; Dorscht, J.; Lurz, R.; Bielmann, R.; Wieland, M.; Zimmer, M.; Calendar, R.; Loessner, M.J. The terminally redundant, nonpermuted genome of *Listeria* bacteriophage A511: A model for the SPO1-like myoviruses of gram-positive bacteria. *J. Bacteriol.* **2008**, *190*, 5753–5765. [CrossRef] [PubMed]
59. Thomas, J.A.; Benitez Quintana, A.D.; Bosch, M.A.; Coll De Pena, A.; Aguilera, E.; Coulibaly, A.; Wu, W.; Osier, M.V.; Hudson, A.O.; Weintraub, S.T.; et al. Identification of essential genes in the *Salmonella* phage SPN3US reveals novel insights into giant phage head structure and assembly. *J. Virol.* **2016**, *90*, 10284–10298. [CrossRef] [PubMed]

60. Thomas, J.A.; Weintraub, S.T.; Wu, W.; Winkler, D.C.; Cheng, N.; Steven, A.C.; Black, L.W. Extensive proteolysis of head and inner body proteins by a morphogenetic protease in the giant *Pseudomonas aeruginosa* phage phiKZ. *Mol. Microbiol.* **2012**, *84*, 324–339. [CrossRef] [PubMed]
61. Yang, S.; Liu, J.; Shao, F.; Wang, P.; Duan, G.; Yang, H. Analysis of the features of 45 identified CRISPR loci in 32 *Staphylococcus aureus*. *Biochem. Biophys. Res. Commun.* **2015**, *464*, 894–900. [CrossRef] [PubMed]
62. Heikinheimo, A.; Johler, S.; Karvonen, L.; Julmi, J.; Fredriksson-Ahomaa, M.; Stephan, R. New dominant spa type t2741 in livestock-associated MRSA (CC398-MRSA-V) in finnish fattening pigs at slaughter. *Antimicrob. Resist. Infect. Control* **2016**, *5*, 6. [CrossRef] [PubMed]
63. Cisek, A.A.; Dabrowska, I.; Gregorczyk, K.P.; Wyzewski, Z. Phage therapy in bacterial infections treatment: One hundred years after the discovery of bacteriophages. *Curr. Microbiol.* **2017**, *74*, 277–283. [CrossRef] [PubMed]
64. Endersen, L.; O'Mahony, J.; Hill, C.; Ross, R.P.; McAuliffe, O.; Coffey, A. Phage therapy in the food industry. *Annu. Rev. Food Sci. Technol.* **2014**, *5*, 327–349. [CrossRef] [PubMed]
65. Gutierrez, D.; Rodriguez-Rubio, L.; Martinez, B.; Rodriguez, A.; Garcia, P. Bacteriophages as weapons against bacterial biofilms in the food industry. *Front. Microbiol.* **2016**, *7*, 825. [CrossRef] [PubMed]
66. Kazi, M.; Annapure, U.S. Bacteriophage biocontrol of foodborne pathogens. *J. Food Sci. Technol.* **2016**, *53*, 1355–1362. [CrossRef] [PubMed]
67. Kutter, E.M.; Kuhl, S.J.; Abedon, S.T. Re-establishing a place for phage therapy in western medicine. *Future Microbiol.* **2015**, *10*, 685–688. [CrossRef] [PubMed]
68. Merabishvili, M.; Pirnay, J.P.; Verbeken, G.; Chanishvili, N.; Tediashvili, M.; Lashkhi, N.; Glonti, T.; Krylov, V.; Mast, J.; Van Parys, L.; et al. Quality-controlled small-scale production of a well-defined bacteriophage cocktail for use in human clinical trials. *PLoS ONE* **2009**, *4*, e4944. [CrossRef] [PubMed]
69. Rose, T.; Verbeken, G.; Vos, D.D.; Merabishvili, M.; Vaneechoutte, M.; Lavigne, R.; Jennes, S.; Zizi, M.; Pirnay, J.P. Experimental phage therapy of burn wound infection: Difficult first steps. *Int. J. Burns Trauma* **2014**, *4*, 66–73. [PubMed]

Review

Evaluation of Phage Therapy in the Context of *Enterococcus faecalis* and Its Associated Diseases

Andrei S. Bolocan [1,2,†], Aditya Upadrasta [1,2,†], Pedro H. de Almeida Bettio [1,2], Adam G. Clooney [1,2], Lorraine A. Draper [1,2], R. Paul Ross [1,2,3] and Colin Hill [1,2,*]

1 APC Microbiome Ireland, University College Cork, Cork T12 YT20, Ireland; andrei.s.bolocan@gmail.com (A.S.B.); aupadrasta@gmail.com (A.U.); pedro.almeida.bettio@gmail.com (P.H.d.A.B.); adam.clooney@ucc.ie (A.G.C.); L.Draper@ucc.ie (L.A.D.); p.ross@ucc.ie (R.P.R.)

2 School of Microbiology, University College Cork, Cork T12 YN60, Ireland

3 Teagasc Food Research Centre, Moorepark, Fermoy, Cork P61 C996, Ireland

* Correspondence: c.hill@ucc.ie

† These authors contributed equally to this work.

Received: 19 March 2019; Accepted: 17 April 2019; Published: 20 April 2019

Abstract: Bacteriophages (phages) or bacterial viruses have been proposed as natural antimicrobial agents to fight against antibiotic-resistant bacteria associated with human infections. *Enterococcus faecalis* is a gut commensal, which is occasionally found in the mouth and vaginal tract, and does not usually cause clinical problems. However, it can spread to other areas of the body and cause life-threatening infections, such as septicemia, endocarditis, or meningitis, in immunocompromised hosts. Although *E. faecalis* phage cocktails are not commercially available within the EU or USA, there is an accumulated evidence from in vitro and in vivo studies that have shown phage efficacy, which supports the idea of applying phage therapy to overcome infections associated with *E. faecalis*. In this review, we discuss the potency of bacteriophages in controlling *E. faecalis*, in both in vitro and in vivo scenarios. *E. faecalis* associated bacteriophages were compared at the genome level and an attempt was made to categorize phages with respect to their suitability for therapeutic application, using orthocluster analysis. In addition, *E. faecalis* phages have been examined for the presence of antibiotic-resistant genes, to ensure their safe use in clinical conditions. Finally, the domain architecture of *E. faecalis* phage-encoded endolysins are discussed.

Keywords: phage therapy; *E. faecalis*; OrthoMCL

---

1. Introduction

Enterococcus is a genus of gram-positive non-spore-forming bacteria that typically inhabit the gastrointestinal tract (GIT), which currently contains thirty five well-recognized species [1], including *Enterococcus faecalis*. The enterococci possess a remarkable ability to adapt to different environments and have a propensity to acquire antibiotic resistance, which has led to the emergence of multi-drug resistant variants, across the genus [1]. *E. faecalis* is mainly described as a core commensal member of the human gut, but it can also act as an opportunistic pathogen and translocate across the mucosal barrier to cause systemic infections [2,3]. More than 90% of the bacterial isolates frequently recovered from clinical specimens (blood, and other infectious site samples) are *E. faecalis* and *E. faecium* [4,5]. Life-threatening infections generally linked to *E. faecalis* include endocarditis, bacteremia, urinary tract infections, meningitis, and root canal infections. In contrast, *E. faecalis* Symbioflor 1 strain (Symbiopharm, Herborn, Germany) has been demonstrated to be a safe and effective probiotic and a few other enterococcal strains have been used as starter cultures in the cheese industry [6]. However, the genus *Enterococcus* is not listed in the Qualified Presumption of Safety (QPS) of the European Food

Safety Authority, nor does it have a generally regarded as safe (GRAS) status [6]. Hence the continued use of enterococci in traditional fermented foods and as probiotics, is controversial, because of their association with human infections [7].

Antimicrobial resistance (AMR) causes 700,000 global deaths each year, and it is estimated that it will rise to 10 million deaths by 2050 [7,8]. The high prevalence of Multi-Drug Resistant (MDR) bacteria and inefficiency of available antibiotics to overcome infectious diseases, has inspired a search for viable alternatives. Bacteriophages, also known as phages, and their associated cell wall lysing enzymes (endolysins), have the potential to be useful tools to combat MDR pathogens [9–11].

Phages are prokaryotic viruses that have the ability to infect and replicate within their host bacterial cell, and to subsequently lyse the cell, to release their progeny. Based on their replication strategy, phages can undergo two different life cycles; the lytic (virulent) and the lysogenic (temperate). Naturally virulent phages are suitable candidates for phage therapy, but temperate phages are not as useful. However, genome engineering strategies can be applied to convert temperate phages to virulent, for their effective use in phage therapy [12]. Phage therapy is described as the application of phages to treat bacterial infections [13,14]. There are some indications that phages could be suitable alternatives to combat *Enterococcus*-associated infections [2,15–18]. In this review, we focus on (i) phage therapy to treat *E. faecalis* infections using in vitro and in vivo models; (ii) the genetic relationships between currently isolated *E. faecalis* bacteriophages; (iii) identification of candidates suitable for phage therapy; (iv) *E. faecalis* phages endolysins as alternative to phage therapy; and (v) conclusions and recommendations for further development of *E. faecalis* phage therapy.

2. The Necessity of *E. faecalis* Phage Therapy

E. faecalis is one of the first colonizers of the human GIT and it plays a role in intestinal immune development at the very early stages of life [19]. *E. faecalis* is a ubiquitous microorganism that possesses the ability to survive and persist in a broad range of environments. In susceptible hosts, *E. faecalis* can act as an opportunistic pathogen, causing severe infections, including urinary tract infections (UTIs), endocarditis, bacteremia, catheter-related infections, wound infections, and intra-abdominal and pelvic infections [1].

An important question is, what makes this bacterium an opportunistic pathogen and under what circumstances? The key factors linked to the pathogenic role of *E. faecalis* in the GIT is its ability to generate reactive oxygen species (ROS) and extracellular superoxide, which can cause genomic instability and damage to the colonic DNA [20]. Opportunistic infection has been associated with the production of virulence factors, adherence to Caco-2 and HEP-2 cells, capacity for biofilm formation and resistance to antimicrobials [21–23]. Numerous virulence factors have been identified that are associated with a wide range of *E. faecalis* infections; namely, aggregation substance (AS), adhesion to collagen of *E. faecalis* (Ace), cell wall glycopeptides, gelatinase (GelE) and biofilm-associated Pili (Ebp), Enterococcal fibronectin-binding protein A (EfbA), membrane metalloprotease (Eep), and biofilm formation. AS is a pheromone-inducible plasmid-encoded cell surface protein, involved in bacterial aggregation during conjugation, via binding to the enterococcal binding substance (EBS) [22–26]. There are three AS proteins (Asa1, Asc10, and Asp1), which belong to a family of surface adhesions and are highly similar to each other. These factors are responsible for the initial adherence and biofilm formation at infected sites [25,27]. Other important cell wall-associated virulence factors are pili and fimbriae, which are anchored to the outer cell surface of the bacterium and aid the bacterium to adhere to host cells. In *E. faecalis*, these are encoded by a three-gene locus (ebpABC), with an associated enzyme sortase, srtC. This ebpABC locus has also been shown to encode proteins involved in biofilm formation [24,28].

Other virulence factors such as Ace, a cell-wall anchored adhesion, plays a pivotal role in in vitro adherence [27,29]. Similarly, EfbA, located on the outer cell membrane, confers adhesion to the host glycoprotein fibronectin [30]. One more critical virulence factor is GelE, an extracellular zinc-metallo protease that contributes to the degradation of various host proteins, such as collagen, fibrinogen, fibrin, and immune complement components C3 and C3a. Many of these factors associated with virulence are also known to promote biofilm formation in *E. faecalis*, suggesting that biofilms are crucial to development of severe infections [31].

In addition, *E. faecalis* is intrinsically resistant to numerous antibiotics, such as penicillin, ampicillin, piperacillin, imipenem, and vancomycin—which have only bacteriostatic rather than bactericidal effects [32]. Over the last decade vancomycin-resistant *E. faecalis* (VREF), together with the other vancomycin-resistant enterococci (VRE), have generated much concern. In the context of a cumulative mortality rate of 20–40% for infective endocarditis, generated by *E. faecalis* and *E. faecium*, *E. faecalis* accounts for approximately 97% of cases [33]. In contrast to that, in leukemia patients, the VR *E. faecium* is more prevalent, accounting for 84%, followed by *E. faecalis* accounting for 6% and the rest 10% was occupied by all other *Enterococcus* sp. [34] and these percentages slightly varied in different studies [35]. In addition, it has been reported that VR *E. faecium* was the leading cause of early infection-related mortality in older (≥60 years) acute leukemia patients, who were receiving induction chemotherapy [36]. Moreover, enterococcal bloodstream infections occurs frequently in patients with acute leukemia, and causes significant morbidity and mortality (87% due to *E. faecium*, while only 13% due to *E. faecalis*) [37]. However, the role of *E. faecalis* and *E. faecium* in colorectal cancer and other diseases such as inflammatory bowel disease (IBD), remains unclear, and their involvement in colorectal cancer is still under investigation [38]. It is presumed that it is the inefficient activity of β-lactams, as well as the biofilm-forming ability of *E. faecalis* which makes these infections difficult to treat. Often, combinations of antibiotic therapies are required for treatment of severe infections associated with *E. faecalis*. However, even these antibiotic treatment options are limited, considering that 50% of isolates exhibit a high-level of aminoglycoside resistance, mediated by aminoglycoside-modifying enzymes, which eliminate the synergistic bactericidal effect, usually seen when a cell wall-active agent is combined with an aminoglycoside [33,39].

3. Strategies for Obtaining *E. faecalis* Phages for Phage Therapy

There are several advantages associated with bacteriophages over antibiotics to treat bacterial infections. For example, unlike antibiotics, bacteriophages are highly specific to their corresponding target and, thus, do not perturb indigenous microbial communities [13,40–42]. Phages targeting *Enterococcus* spp. have been isolated from various sources, like sewage, animal yard effluents, human feces, urogenital secretions or by inducing chromosomally integrated prophages [17,43–46].

In general, plaque and spot assays are the methods applied by researchers to isolate phages, using bacterial hosts of interest. In an attempt to increase the recovery of phages from environments where they are scarce, a pre-enrichment step has been widely used, prior to plaque/spot assay. In the case of *E. faecalis*, typically, vancomycin-resistant strains or other clinical isolates have been used for screening, in order to realize the potential of phages as novel therapeutics [38,41].

Many factors can affect the process of phage isolation. For example, poor or invisible plaque morphology, difficulty in obtaining confluency of bacterial lawns, poor enrichment of samples containing very low numbers of phages, or sample availability [47]. Furthermore, bacterial host strains might adapt to routine laboratory culturing practices resulting in changes to their cell physiology. Such genotypic and phenotypic changes which occur during sub-culturing, can reduce the chances for the discovery new phages. To overcome such hurdles, Purnell et al. [37], suggest the isolation of target bacterial hosts, and their cognate bacteriophages, from the same sample, to achieve a higher success rate. Therefore, it is advisable to obtain a fresh culture from the glycerol stock and avoid multiple sub-culturing and serial broth-to-broth transfers, prior to phage isolation. In addition, bacteria can rapidly evolve to overcome phage infection by means of spontaneous mutation, or by acquiring CRISPR-*cas* mediated adaptive immunity, resulting in bacteriophage-insensitive mutants (BIMs) [48–50]. In addition, since multiple bacterial strains can be involved in diseases, the application of phage cocktails are deemed to be more appropriate over single-phage preparations, in therapeutic interventions [16].

4. Orthocluster Analysis of *E. faecalis* Phages

On the 30 December 2018, fifty-four *Enterococcus* phage genome sequences were available (http://millardlab.org/bioinformatics/bacteriophage-genomes/), of which 89% had *E. faecalis* and 11% had *E. faecium* as a target (Table S1). Usually, these phages infect both species at varying efficiencies [16,17,51–54].

To determine the gene content relationship between these bacteriophages, a cluster analysis was performed on the basis of the percentage of shared orthologous genes. For the orthocluster analysis, the phage genomes were downloaded from the NCBI database, and potential Open Reading Frames (ORFs) were predicted by Prodigal [55]. Identification of the bacteriophage protein Orthologous Groups (OG, cluster of proteins from at least two phages) was performed, using orthoMCL [56]. OrthoMCL phage clusters identified from this analysis were defined as "orthoclusters". This analysis allowed the identification of ten distinct and well-supported (100% bootstrap support) clusters of *Enterococcus* phage genomes. Of the fifty-four *Enterococcus* phage genomes, fifty-two fell into one of the ten distinct clusters, designated as orthoclusters I–V, VII, IX–X, as depicted in Figure 1. The remaining two phages used in this analysis, did not cluster with any other phages. Therefore, we hypothesize that the phages EF62phi and phiFL4, formed two different orthoclusters, V and VII, respectively. The distinct orthoclusters, typically contain phages of the same family, with similar genome size, GC content and morphology. The clustering was in good agreement with classical taxonomical phage families, as determined by the morphology and genome analysis—virulent *Myoviridae* family—orthocluster II, virulent *Siphoviridae* family—orthoclusters I, III, V, VII, IX, and X, temperate *Siphoviridae* family—orthoclusters IV and VIII, and temperate *Podoviridae* family—orthocluster V, and virulent *Podoviridae* family—orthocluster VI.

With respect to phage therapy, orthoclusters comprising native virulent phages, are of immense interest. Of the *Enterococcus* phages characterized to date, 77% are known to be virulent, and belong to the orthoclusters I, II, III, IV, VI, IX, and X. Although temperate phages have less obvious usefulness with respect to phage therapy, molecular mechanisms of phage conversion from temperate to virulent, might make this possible.

Orthocluster I, which is supported by a bootstrap value of 1000, contains 19 phages belonging to the *Siphoviridae* family. This orthocluster is particularly interesting as the phages differ significantly from each other, in terms of their genome length and mean GC content, features which are conserved among the other orthoclusters. The genome sizes range from ~17 kb to ~42 kb, and the mean GC content varies from 17.35% to 36.7%. The suitability of these phages for phage therapy is questionable, as the orthologous group 32, which belongs to orthocluster I, contains the putative metallo-beta-lactamase gene, a gene related to antibiotic resistance (Figure 2) [57,58]. All phages harbor this gene, except for EFRM31 and EFAP_1, within the orthocluster I. However, the functionality of this gene is currently unknown. Further studies are warranted to evaluate these phages and their involvement in antibiotic gene dissemination in the gut. In addition, gene editing tools could be applied to either delete or inactivate the metallo-beta-lactamase gene, before considering therapeutic applications. A study by Nezhad Fard et al. [59], demonstrated that the phage EFRM31 was efficient at transducing gentamicin resistance to multiple enterococcal species. In fact, this was the first example of inter-species host range generalized transduction, and thus, it did not support a role for such phages in therapeutic applications.

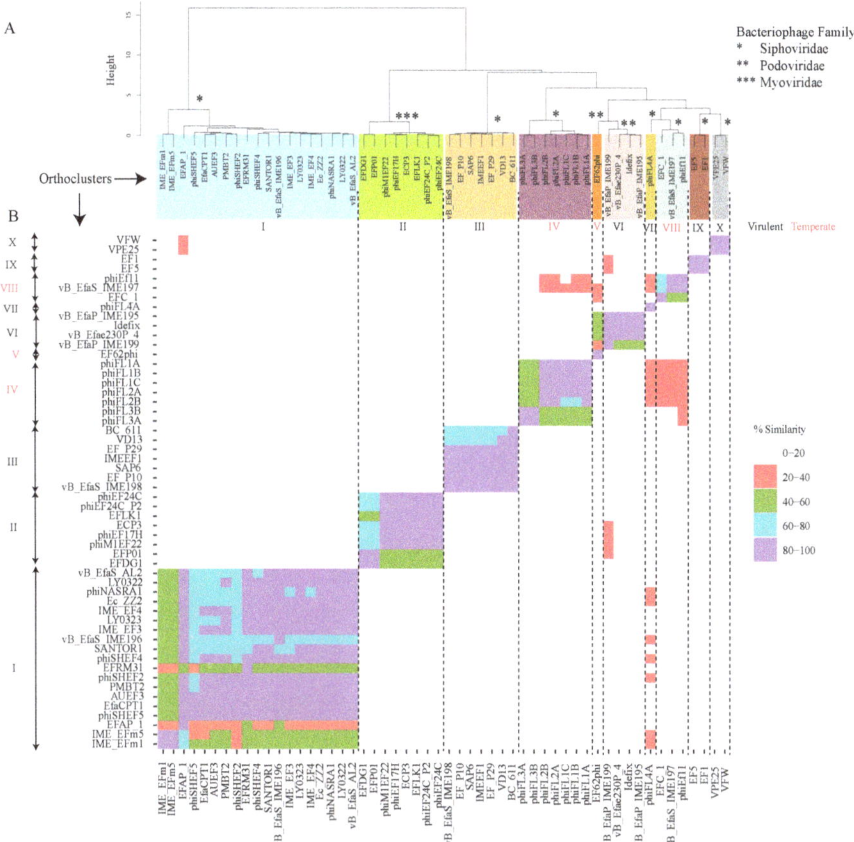

Figure 1. Genomic comparison of *Enterococcus* phages. (**A**) Neighbor-joining tree based on the percentage of shared orthologous genes (1000 bootstrap replicates); squares indicate the 10 phage putative orthoclusters. (**B**) Dot plot comparison of amino acids identity among the 10 orthoclusters; genes that share more than 40% homology were considered as being part of the same orthologous group. The vertical axis shows phage clusters and phage IDs.

Interestingly, Orthocluster II incorporates all the *Myoviridae* phages described so far, which infect *E. faecalis*. These phages can infect and proliferate in multiple strains of *E. faecalis* and *E. faecium* strains. The size of the genomes ranged between ~130 kb to ~150 kb, and the mean GC content was estimated to be 35.3% to 37.2%. These phages were related to SPO1-like viruses, such as the *Staphylococcus* phage K, *Listeria* phage P100, and *Lactobacillus* phage LP65. Interestingly, no *E. faecalis* temperate phages belonging to the *Myoviridae* family have ever been described [16,60].

Orthocluster III contains the most studied *E. faecalis* virulent phages from the *Siphoviridae* family (genus *Sap6virus*). The size of the genomes ranged between ~53 kb to ~59 kb, and the mean GC content was estimated to be 39% to 40%. These phages exhibited a broad host range and a high level of efficiency in in vitro and in vivo studies, which have been discussed in more detail, later on. Genome analysis did not reveal any putative virulence factors or antibiotic-resistant genes, and to date no transduction potential has been described. Members of this orthocluster should, therefore, be considered and studied with respect to their therapeutic potential [61,62].

The phages from Orthocluster IV were induced using norfloxacin and UV from bacteremia isolates of the *Enterococcus* sp. These temperate phages belonged to the *Siphoviridae* family, with a genome size of 30–40 kb, and a mean GC content of 30%–40%. Currently, only virulent phages have been considered as suitable candidates for phage therapy, but there is a possibility to convert these lysogenic phages to virulent entities, which would allow us to investigate these phages in the context of phage therapy. However, the use of genetically-modified phages, is not acceptable, for now [12]. Further inspection of the orthocluster IV harboring temperate phages, revealed their ability to pack its bacterial host DNA, a generalized transduction potential event observed in some other temperate phages, as well. As a result, these phages are not suitable for phage therapy. It is unfortunate that on rare occasions generalized transduction events have also been observed in some virulent phages [43].

The *Podoviridae* phage, EF62phi (~30 kb, mean GC content 32.7%) which forms the putative orthocluster V, is a pseudotemperate linear bacteriophage identified in the genome of *E. faecalis* strain 62, isolated from a healthy Norwegian infant. EF62ph is the only pseudotemperate enterococcal phage described to date. EF62ph is maintained in the bacterial genome by means of RepB and a toxin–antitoxin system [63]. There have been no studies, so far, on pseudotemperate enterococcal phages and their involvement in phage therapy.

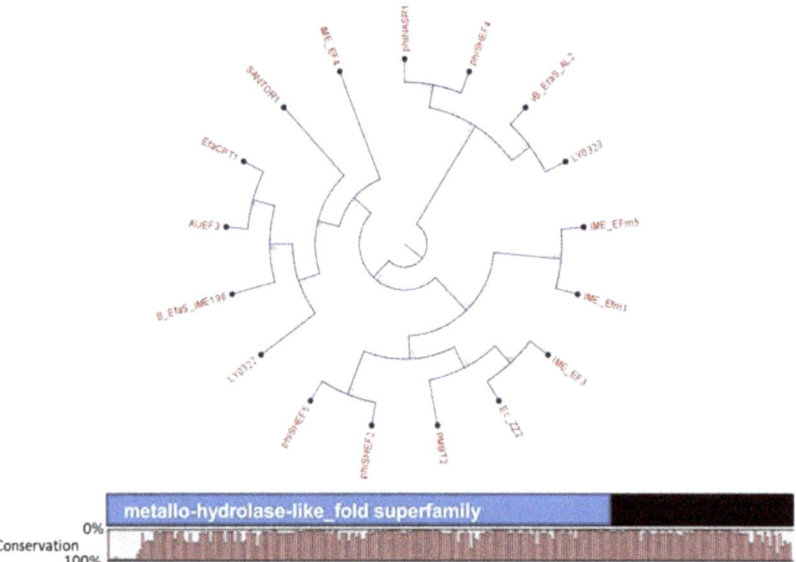

Figure 2. Maximum likelihood phylogenetic analysis sequence relatedness of the *Enterococcus faecalis* phage putative metallo-beta-lactamase gene (orthologous group 32); tree node labels represent bootstrap values.

Orthocluster VI is comprised of the *Podoviridae* phages, of the genus *Ahjdlikevirus*. These phages have been isolated from sewage, and infect both *E. faecalis* and *E. faecium* strains. The size of the genomes range from ~17 kb to ~18 kb, and have a mean GC content of 33.2% to 34.6%. With no evidence of antibiotic-resistance-associated genes or transduction potential, these phages should be explored further for potential therapeutic applications [17,54,64].

The phage phiFL4A, which forms the putative orthocluster VII (*Siphoviridae* family, *Phifelvirus* genus, 37 kb, mean GC content 37.8%) was induced from bacteremia isolates, using mitomycin C, in the same study as that of the phages of orthocluster IV. This phage is also temperate and has the ability of generalized transduction and, therefore, is not eligible for phage therapy [43].

Orhocluster VIII contains three temperate prophages and is part of the *Siphoviridae* family. phiEf11 was induced with mitomycin C from the root isolate *E. faecalis* TUSoD11 [65], EFC1 was induced with mitomycin C from the raw milk isolate *E. faecalis* KBL101 [66] and vB_EfaS_IME197 was isolated from sewage. The size of the genomes range from ~40 kb to ~42 kb and the mean GC content from 34% to 35%. This group is particularly interesting from the point of view of phage therapy, as Ef11 phage have been converted from temperate to virulent, followed by successful testing against *E. faecalis*. Therefore, the temperate phages from this orthocluster opens the direction for a new type of *E. faecalis* phage therapy, based on genetically engineered phages [67–69]

The phages that form orthocluster IX, EF1, and EF5, were previously annotated as part of the *Myoviridae* family. However, our genome annotation using RASTtk and BLAST suggest that these two virulent phages are part of the *Siphoviridae* family. By comparison with the other *Siphoviridae* virulent *E. faecalis* phages, these two phages have a large genome of 141.996 kb, with a mean content GC of 31.9%. Larger genomes are typical for *Myoviridae* family, which may be the reason for their previous attribution in the database. No therapeutic studies have been performed using these phages and, therefore, their potential role in phage therapy could not be predicted. Despite this, our genome analysis did not reveal any genes that would hinder further research of these phages for therapeutic potential.

Phages VPE25 and VFW formed orthocluster X. They were isolated from sewage and shared 95% homology at the nucleotide level. The size of the genomes of both phages was ~86 kb, with a mean GC of 33.2%. VPE25 and VFW were obligate lytic and their isolation, using VR *E. faecalis* V583 as a host, suggested them to be putative candidates for therapy [70].

Phages from each of the described orthocluster are now discussed in more details, with respect to the published in vitro and in vivo phage therapy studies.

5. *E. faecalis* Phage Therapy in In Vitro Models

5.1. Biofilm Eradication

Various studies describe the ability of single phage or phage cocktails in the treatment of bacterial biofilms. For example, biofilms formed by pathogenic bacteria *Streptococcus mutants* [71], *E. coli* [72], *Pseudomonas aeruginosa* [73], *Staphylococcus aureus* [74], and *E. faecalis* [75], can be disrupted by phages. Phage treatment is more efficient against biofilms, compared to conventional antibiotics, since, as the phages infect the bacteria from the upper layer, upon replication they release a new virion progeny, which subsequently attacks the bottom layer(s). As a result of this layer-by-layer mode of action, the biofilms are effectively eradicated [75,76]. Microtiter plates are the most commonly used method for studying biofilm formation, and to test the activity of antimicrobial compounds. More advanced techniques like confocal microscopy can also be applied for the visualization of biofilm matrices, before and after phage treatment [77]. Using this method, the efficiency of phage EFDG1 (orthocluster II) to reduce two-week-old biofilms of *E. faecalis* V583 has been described [18]. The genetically-engineered orthocluster VIII phage phiEf11 (phiEf11/phiFL1C(Δ36)PnisA [67]), reduced the static biofilm of *E. faecalis* strains JH2-2 (pMSP3535 nisR/K) and V583 (pMSP3535nisR/K), which had formed on coverslips. After 24 and 48 h of incubation, a 10–100-fold decrease in viable cells (CFU/biofilm) was observed [69].

5.2. Human Root Canal Model (In Vitro/Ex Vivo)

E. faecalis has been found, over time, to be more prevalent (24% to 77% of cases) in asymptomatic and persistent endodontic infections [78,79]. The extreme survival ability and highly adaptive nature of *E. faecalis* in harsh environments, allows the bacterium to cause persistent infections in root canals. Furthermore, it can resist nutritional deprivation and invade dental tubules to form endodontic biofilms. In this scenario, treatment with 2% chlorhexidine, combined with sodium hypochlorite, is generally effective. However, a number of failures have been recorded in endodontic treatment, due to technical difficulties associated with dental practices [78,80]. Therefore, the development of alternative strategies are necessary to prevent such situations. In this regard, the efficacy of phage treatment has been evaluated using an ex vivo two chamber bacterial leakage model of human teeth [18]. No turbidity was observed in the obturated root canals, which were subjected to 10^8 PFU/mL of EFDG1 phage (orthocluster II) irrigation and the results also indicated a 7-log reduction of bacterial leakage, from the root apex, when compared to the control. In a similar study, Paisano et al. [81] showed that a phage lysate of 2×10^8 PFU/mL was able to significantly inhibit *E. faecalis* in human dental roots inoculated for 6 days with a suspension of *E. faecalis* ATCC 29212 at the three different multiplicities of infection; 0.1, 1.0, and 10.0. Moreover, in the study of Tinoco et al [12]. extracted human dentin root segments were cemented into a sealable double-chamber and inoculated for 7 days, with an overnight suspension of either VR *E. faecalis* V583, or *E. faecalis* JH2-2, which is vancomycin sensitive, but resistant to fusidic acid and rifampin. The treatment with genetically-engineered phage, phiEf11/phiFL1C (Δ36)PnisA, generated a reduction of 18% for the JH2-2-infected models, and by 99% for the V583-infected models. These examples certainly strengthen the efficacy of phage therapy in the treatment of *E. faecalis* root canal infections.

5.3. Fibrin Clot Model

Clots are gel-like clumps of blood that occurs when thrombin converts fibrinogen to fibrin, a structural protein that assembles into a polymer [82]. An in vitro fibrin clot model has been successfully used to test the role of antibiotics in the treatment of bacterial endocarditis [83], demonstrating the in vitro clotting ability of bacterial strains *Bacillus cereus* [84], *Staphylococcus aureus* [85], *E. faecalis* [86], and *E. faecium* [83,84]. Recently, the in vitro fibrin clot model has been used to demonstrate the efficacy of individual phages and phage cocktails [16]. The authors spiked the plasma with vancomycin-resistant and sensitive *E. faecalis* strains, and triggered the plasma coagulation with the addition of bovine thrombin and $CaCl_2$. The resultant clots were subjected to a 10^8 PFU/mL bacteriophage treatment. Bacterial counts were significantly reduced by 3–6 logs, after treatment with phage(s) EFDG1 and EFLK1 (orthocluster II).

5.4. *E. faecalis* Phages as Biocontrol Agents

Bacteriophages have long been recognized as effective biological entities in the control of undesired foodborne bacteria. In 2007, a *Listeria*-specific bacteriophage preparation, Listex P100, obtained U.S. FDA approval for use as a biopreservative, in ready-to-eat meat products (U.S. Food and Drug Administration, 2007). In a recent study, phage Q69 has been shown to be effective against *E. faecalis*, in a cheese model system. This phage significantly reduced *E. faecalis* numbers and subsequently eliminated the accumulation of toxic biogenic amine tyramine, during cheese ripening [87].

6. *E. faecalis* Phage Therapy in In Vivo Models

To date, we are only aware of a single human study describing the phage treatment *of E. faecalis* associated chronic prostatitis (Table 1). Three subjects were selected for phage therapy who had failed to respond to antibiotic, auto-vaccine, and laser bio-stimulation treatments. During phage treatment, 10 mL of bacterial phage lysate was rectally applied, twice daily, for 30 days. In all three cases, the pathogen was eradicated, clinical symptoms abated, and early disease recurrence was not observed [88].

Table 1. Target infections, phage dosage, and outcomes in *Enterococcus faecalis* phage therapy in vivo models.

Disease (Target Strain)	No (n) and Type of Subjects	Form and Dosage	Application Route and Clinical Outcome	Reference
Chronic bacterial prostatitis	$n = 3$ human male	Phage lysate ~10^7–10^9 PFU/mL	Rectal Pathogen eradication, Abatement of clinical symptoms Lack of early disease recurrence	[88]
Infection (EF14 VRE2)	$n = 20$; BALB/c mice female 6 to 8 week old	CsCl; 1×10^{12} PFU/mL;	Intraperitoneal; Significantly effective, Efficiently rescued mice;	[89]
Bacteremia (VAN)	$n = 5$ BALB/c mice 1 month old	CsCl 3×10^8 PFU/mL	Intraperitoneal 100% survival 45 min after bacterial challenge 50% of moribund mice rescued after delayed phage administration	[90]
Sepsis 002	$n = 8$ 7 different dosage groups BALB/c female mice 6 to 8 weeks old	PEG 3.9×10^9 PFU/mL or 0.2 mg endolysin	Intraperitoneal 60% survival at 30 min post bacterial inoculation 40% survival at 4 h post bacterial administration	[52]
E. faecalis challenge	$n = 10$ 5 different dosage groups BALB/c F 6 to 8 weeks old	CsCl 4×10^3, 4×10^4, 4×10^5, 4×10^6, 4×10^7 PFU/mouse	Intraperitoneal Mice were protected from the infection	[91]
Septic peritonitis	$n = 15$ 4 groups ICR(CD-1C)	Dialyzed phage lysate 2×10^8	Intraperitoneal 100% survival No harmful effect on the microbiome	[60]
E. faecalis challenge (VAN)	$n = 5$ 4 different groups BALB/c n female mice 6 to 8 weeks old	LysEF-P10 endolysin 1 μg, 5 μg, 10 μg	Intraperitoneal Reduced *E. faecalis* colonization Alleviated the gut microbiota imbalance caused by VRE	[92]

VAN- experiment performed using vancomycin resistant *E. faecalis*; CsCl- Cesium chloride gradient purified phages; PEG- phage prepared by PEG precipitation.

Other positive results obtained on treating infectious disease unresponsive to antibiotics, caused by other bacteria, such as *S. aureus*, *E. coli*, *Klebsiella*, *Proteus*, *Pseudomonas*, and *Enterobacter*, support the idea of using phage therapy against antibiotic-resistant *E. faecalis* [93]. All highlight the efficiency of phages in disease resolution, and as future options for treating multi-drug-resistant bacterial infections. Another example describes a life-threatening multi-drug-resistant pathogen *Acinetobacter baumannii* infection, which was treated with an intravenous bacteriophage cocktail. This reversed the patient's clinical trajectory, cleared the *A. baumannii* infection, and restored the individual from a state of coma to complete health [94]. More clinical scenarios like these will undoubtedly open new avenues for phages or phage-derived enzybiotics as biotherapeutics, to combat situations where antibiotic treatments are no longer viable.

6.1. Vertebrate Models

Meanwhile, some studies have shown the efficacy of phages, in vivo, against *E. faecalis*, using mouse models (Table 1). An intraperitoneal application of phages, significantly rescued mice, when deliberately challenged with the *E. faecalis* EF14 and *E. faecalis* VRE2 strains [95]. Similarly, another study has showed that mice treated with different phage doses were protected from the VREF systemic infection, and alleviated the gut microbial imbalance that occurred as a result of infection [91]. In another study, a single dose of the lytic phage cocktail was effective in completely reversing a 100% mortality in a septic peritonitis mouse model caused by VREF, and without causing any collateral damage to the gut microbiome [60]. Furthermore, phage therapy has proven to be safe and effective in treating *E. faecalis*-induced bacteremia [90] and sepsis [52], in mouse models.

6.2. Invertebrate Models

The larvae of wax moth *Galleria mellonella* has been used as a model system to examine pathogenesis of many bacteria, such as *S. aureus*, *P. aeruginosa*, *L. monocytogenes*, *Klebsiella pneumoniae*, *E. faecalis*, and *E. faecium*, and the fungi *Candida albicans* and *Aspergillus fumigatus* [96–100]. This model involves monitoring *G. mellonella* caterpillars infected with bacterial culture, followed by the administration of a test drug or saline solution as a negative control. A number of *E. faecalis* virulence gene factors have been associated with larval mortality [101]. This method has been demonstrated as a suitable model for studying *E. faecalis*-drug interaction, for example, studies have used distamycin, linezolid, rifampicin, and extracts of *Zingiber officinale* [101–103]. The most significant advantage of this model is that it allows a precise measurement of the inoculum and the quantity of the administrated drug, over time. Not only are promising results obtained using this larval model, but it involves simple methodological approaches. To date, there are no reports of phages treatment of *E. faecalis* in *G. mellonella*. However, Yasmin et al. [43] infected *G. mellonella* with *E. faecalis* JH2-2 lysogenized by phiFL3A and phiFL3B (orthocluster IV), and found that it increased the mortality of caterpillars. Conversely, some of the other lysogens obtained in the same study, but with different phages, such as phiFL1B and phiFL2B (orthocluster IV), and phiFL4A (putative orthocluster VII), did not show any death in the caterpillars, when compared to the JH2-2 generic strain group. This *G. mellonella* model could be a valuable tool to pre-screen the ability of phages in an in vivo scenario, before performing large scale animal trials. In fact, the *Galleria* larval model has been used to examine the therapeutic potential of bacteriophages against other bacterial pathogens, such as *C. difficile* [104], *Burkholderia cepacia* [105], *Pseudomonas aeruginosa* [106], *Escherichia coli*, *K. pneumoniae*, *Enterobacter cloacae* [100], and *Cronobacter sakazakii* [107].

7. *E. faecalis* Phage Endolysins as Viable Alternatives for Phage Therapy

Endolysins, also termed phage lysins, have the ability to degrade the peptidoglycan layer of bacterial cell walls, leading to cell death. These phage-derived enzymes allow the release of nascent virions, following intracellular replication [108]. Endolysins possess a wide degree of killing activity, which also makes them potential therapeutic agents. Considering the bottlenecks associated with the production and purification of phages, to ensure the removal of host-derived endotoxins for therapeutic use, endolysin manufacture is a less arduous process, with a potentially similar outcome. Moreover, with the advent of mass sequencing technologies and the availability of curated gene functional databases, it is now possible to access the genomes of uncultured phages and their enigmatic gene content, to develop potential lytic enzymes, without the necessity for phage isolation. In fact, an in silico examination of uncultured phage genomes, revealed enormous diversity among endolysins [109]. With a varied host specificity and domain architecture, the development of robust novel antimicrobials for future application are within our reach.

7.1. Domain Architecture of E. faecalis Phage Endolysins

Based on their muralytic activity, four types of phage endolysins have already been identified; type I (lysozymes) and type II (transglycosidases); both of which act on the glycosidic bond linking the amino sugars in the cell wall. Type III (amidases) and type IV (endopeptidases), both act on the amide and peptide bonds of the oligopeptide cross-linking stems [110]. Endolysins typically consist of an N-terminal catalytic domain targeting the peptidoglycan network, and a C-terminal cell wall binding domain (termed as carbohydrate binding domain, CBD), which initializes the binding for corresponding enzymatic action, against the specific substrate (Loessner, 2005). A comprehensive in silico analysis on endolysin classes revealed that most (more than 74%) of the *E. faecalis* phage endolysins have an LysM module as a part of their Cell Binding Domain (CBD), whereas the Enzyme Catalytic Domain (ECD) consists of a glycosidase hydrolase (GH) module GH25 (the predominant one 50–74%) and cysteine, and hsitidine-dependent amidohydrolase/peptidase (CHAP) (accounting for less than 25%) (Oliveira et al. [111]). We identified a total of 54 putative and reference endolysin sequences in *E. faecalis* phages (Figure 3). They were clustered into orthologous groups (OGs) using OrthoMCL with default settings (Li et al. [57]). All but one (an endolysin associated with the phage EF62phi) clustered into one of the four distinct orthologous groups (OG 22, OG 28, OG 78, and OG 236), which mirrored the orthologous groups of their parental phages (Figures 2 and 3).

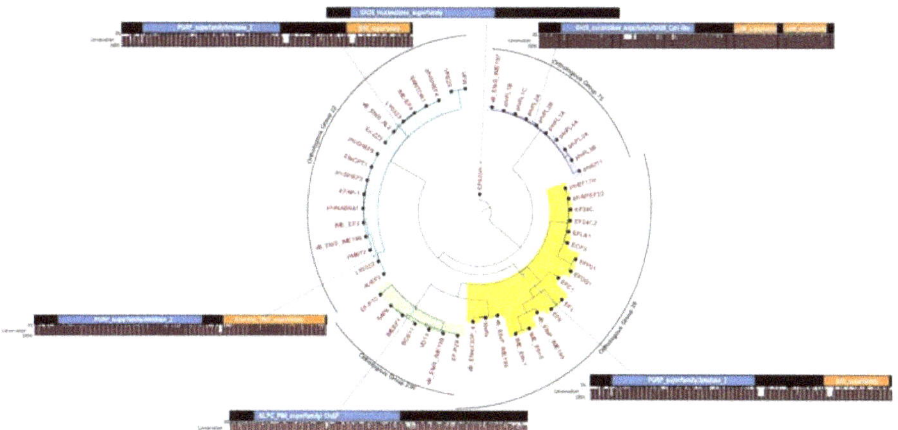

Figure 3. Maximum likelihood phylogenetic analysis sequence relatedness of *E. faecalis* phage endolysin functional domains; tree node labels represent the bootstrap values; the sequence similarity between functional domains is evidenced by using identical filling patterns; in blue—active domain; in orange—biding in domain; each of the four orthologous group is represented by a different color; Ef62phi could not be associated to any orthologous group.

One representative sequence was selected from each OG and subjected to HHMER [112] or HHPRED [113] analysis, to determine the protein domain architecture. Proteins assigned to the same OG often displayed the identical domain architectures, although a few exceptions were observed.

In the case of ECD, three major domains—GH25, Amidase_2, and CHAP—were observed across the four OGs, whereas in CBD, three domains—LysM, SH3, and PET-M23 (ZoocinA)—were identified (Figure 3). This observation was consistent with the findings of Oliveira et al. [114].

7.2. Applications of E. faecalis Phage Endolysins

Of note, endolysins could also be used in combination with traditional antibiotics to treat polyantibiotic-resistant bacterial pathogens. Many studies have shown the successful application of phage endolysins, in treating multi-drug resistant bacterial infections caused by *A. baumannii*, *S. aureus*, Methicillin resistance *S. aureus* (MRSA), *E. coli*, *Proteus mirabilis*, *Klebsiella*, *Pseudomonas*, *Morganella*, *Enterobacter*, *Enterococcus*, and *Salmonella* [111]. A small number of studies have demonstrated the in vivo efficacy of *E. faecalis* specific endolysins. One recent study evaluated endolysin LysEF-P10 to treat multi-drug resistant *E. faecalis* in a mouse model [92]. Here, a single intraperitoneal dose of 5 µg LysEF-P10 endolysin, was sufficient to eliminate the vancomycin resistant strain from the gut, without causing any collateral damage to the gut communities. Another study described the use of the endolysin IME-EF1, which protected 80% of mice challenged with a lethal dose of *E. faecalis* 002, and significantly reduced bacterial proliferation in the blood [52]. Several studies have described the in vitro antimicrobial action of *E. faecalis* endolysins. Heterologous expression of two endolysins Lys168 and Lys170 derived from *E. faecalis*, displayed a promising activity against clinical isolates of exponentially growing vancomycin-resistant and sensitive *E. faecalis* cultures, but failed to display a similar activity against log phase cultures [62]. Lys170 contains a catalytic domain of the amidase-2 family, which has an N-acetlymuramoyl-L-alanine amidase activity, while Lys168 was identified as being unique among the enterococcal phage endolysins, and highly similar to the endolysin of *S. aureus* phage SAP6, therefore, distantly related to all CHAP domain containing enterococcal endolysins [62]. In a follow-up study, these authors used a domain shuffling approach, by fusing a peptidase M23 catalytic domain to a cell-wall-binding domain of the native endolysin Lys170, to generate a bacteriolysin-like chimera, designated as EC300, to improve its anti- *E. faecalis* activity [115]. A recent study highlighted the advantage of using the phage endolysin IME-EFm5, over a narrow host range *E. faecalis* phage. Interestingly, the endolysin of phage IME-EFm5, displayed lytic activity against almost all tested strains [15]. Similarly, an expanded lytic activity of the *E. faecalis* bacteriophage ϕEF24C endolysin, ORF9 has been observed when heterologously expressed in *E. coli*. Further analysis has revealed that ORF9 belongs to the family of N-acetlymuramoyl-L-alanine amidases [44,116].

Antibacterial activity of a thermostable endolysin VD13 with an N-terminal CHAP domain has been demonstrated in vitro, against *E. faecalis*, with no activity observed against *E. faecium* or any other non-enterococcal strains tested [51]. In general, phage endolysins display a wider spectrum of activity than their parental phage counterparts.

8. Conclusions

We conclude that phages could provide a viable alternative therapy to antibiotics in the fight against *E. faecalis* infections. To date, only one clinical study has demonstrated the efficiency of *E. faecalis* phages in a clinical setting. However, there are increased chances of developing a successful phage therapy approach to an *E. faecalis* control, based on the in vitro and in vivo studies described in this review. As far as we are aware, no current phage clinical trials are focused on *E. faecalis*, but the outcomes of trials targeting other pathogens might be useful for the design of future *E. faecalis* phage therapy.

One of the issues of phage therapy is the narrow host range of the phages. In the case of *E. faecalis*, the diversity of phages showed in this review, based on the orthocluster identification, support the idea of expanding the phage host range by creating phage cocktails with a broader host range. It is unlikely that resistance will simultaneously occur for all virulent phages.

If this approach fails, there is the possibility of engineering temperate phages, as was done successfully for the *E. faecalis* phage phiEf11. Moreover, even if phages fail in providing a therapy for *E. faecalis*, their endolysins might prove to be a suitable alternative in the fight against *E. faecalis*-associated disease.

Supplementary Materials: The following are available online at http://www.mdpi.com/1999-4915/11/4/366/s1. Table S1: *E. faecalis* phages selected for the OrthoMCL analyses.

Author Contributions: Bioinformatics, writing, and editing, A.S.B.; writing and editing, A.U.; P.H.d.A.B.; writing, bioinformatics, and editing A.G.C.; writing and editing, L.A.D.; supervision and editing R.P.R. and C.H.

Funding: This work was conducted with the financial support of Science Foundation Ireland (SFI) under Grant Number SFI/12/RC/2273 a Science Foundation of the Ireland's Spokes Programme, which is co-funded under the European Regional Development Fund under Grant Number SFI/14/SP APC/B3032, and a research grant from Janssen Biotech, Inc.

Conflicts of Interest: The authors declare no conflict of interest.

References

1. Arias, C.A.; Contreras, G.A.; Murray, B.E. Management of multidrug-resistant enterococcal infections. *Clin. Microbiol. Infect.* **2010**, *16*, 555–562. [CrossRef] [PubMed]
2. Sava, I.G.; Heikens, E.; Huebner, J. Pathogenesis and immunity in enterococcal infections. *Clin. Microbiol. Infect.* **2010**, *16*, 533–540. [CrossRef]
3. Berg, R.D. The indigenous gastrointestinal microflora. *Trends Microbiol.* **1996**, *4*, 430–435. [CrossRef]
4. Facklam, R.R.; da Carvalho, M.G.S.; Teixeira, L.M. History, Taxonomy, Biochemical Characteristics, and Antibiotic Susceptibility Testing of Enterococci. In *The Enterococci*; American Society of Microbiology: Washington, DC, USA, 2002; pp. 1–54.
5. Simonsen, G.S.; Småbrekke, L.; Monnet, D.L.; Sørensen, T.L.; Møller, J.K.; Kristinsson, K.G.; Lagerqvist-Widh, A.; Torell, E.; Digranes, A.; Harthug, S.; et al. Prevalence of resistance to ampicillin, gentamicin and vancomycin in Enterococcus faecalis and Enterococcus faecium isolates from clinical specimens and use of antimicrobials in five Nordic hospitals. *J. Antimicrob. Chemother.* **2003**, *51*, 323–331. [CrossRef] [PubMed]
6. Hanchi, H.; Mottawea, W.; Sebei, K.; Hammami, R. The Genus Enterococcus: Between Probiotic Potential and Safety Concerns—An Update. *Front. Microbiol.* **2018**, *9*, 1791. [CrossRef] [PubMed]
7. O'Neill, J. AMR Review Paper-Tackling a Crisis for the Health and Wealth of Nations. Available online: http://www.jpiamr.eu/wp-content/uploads/2014/12/AMR-Review-Paper-Tackling-a-crisis-for-the-health-and-wealth-of-nations_1-2.pdf (accessed on 19 April 2019).
8. Tagliabue, A.; Rappuoli, R. Changing Priorities in Vaccinology: Antibiotic Resistance Moving to the Top. *Front. Immunol.* **2018**, *9*, 1068. [CrossRef] [PubMed]
9. Bolocan, A.S.; Callanan, J.; Forde, A.; Ross, P.; Hill, C. Phage therapy targeting Escherichia coli-a story with no end? *FEMS Microbiol. Lett.* **2016**, *363*. [CrossRef] [PubMed]
10. Kortright, K.E.; Chan, B.K.; Koff, J.L.; Turner, P.E. Phage Therapy: A Renewed Approach to Combat Antibiotic-Resistant Bacteria. *Cell Host Microbe* **2019**, *25*, 219–232. [CrossRef] [PubMed]
11. Brüssow, H. Phage therapy: The Escherichia coli experience. *Microbiology* **2005**, *151*, 2133–2140. [CrossRef]
12. Pires, D.P.; Cleto, S.; Sillankorva, S.; Azeredo, J.; Lu, T.K. Genetically Engineered Phages: A Review of Advances over the Last Decade. *Microbiol. Mol. Biol. Rev.* **2016**, *80*, 523–543.
13. Moelling, K.; Broecker, F.; Willy, C. A Wake-Up Call: We Need Phage Therapy Now. *Viruses* **2018**, *10*, 688. [CrossRef] [PubMed]
14. Patey, O.; McCallin, S.; Mazure, H.; Liddle, M.; Smithyman, A.; Dublanchet, A. Clinical Indications and Compassionate Use of Phage Therapy: Personal Experience and Literature Review with a Focus on Osteoarticular Infections. *Viruses* **2018**, *11*, 18. [CrossRef] [PubMed]
15. Gong, P.; Cheng, M.; Li, X.; Jiang, H.; Yu, C.; Kahaer, N.; Li, J.; Zhang, L.; Xia, F.; Hu, L.; et al. Characterization of Enterococcus faecium bacteriophage IME-EFm5 and its endolysin LysEFm5. *Virology* **2016**, *492*, 11–20. [CrossRef]
16. Khalifa, L.; Gelman, D.; Shlezinger, M.; Dessal, A.L.; Coppenhagen-Glazer, S.; Beyth, N.; Hazan, R. Defeating Antibiotic- and Phage-Resistant Enterococcus faecalis Using a Phage Cocktail in Vitro and in a Clot Model. *Front. Microbiol.* **2018**, *9*, 326. [CrossRef]

17. Lossouarn, J.; Briet, A.; Moncaut, E.; Furlan, S.; Bouteau, A.; Son, O.; Leroy, M.; DuBow, M.; Lecointe, F.; Serror, P.; et al. Enterococcus faecalis Countermeasures Defeat a Virulent Picovirinae Bacteriophage. *Viruses* **2019**, *11*, 48. [CrossRef] [PubMed]
18. Khalifa, L.; Brosh, Y.; Gelman, D.; Coppenhagen-Glazer, S.; Beyth, S.; Poradosu-Cohen, R.; Que, Y.A.; Beyth, N.; Hazan, R. Targeting Enterococcus faecalis biofilms with phage therapy. *Appl. Environ. Microbiol.* **2015**, *81*, 2696–2705. [CrossRef] [PubMed]
19. Fanaro, S.; Chierici, R.; Guerrini, P.; Vigi, V. Intestinal microflora in early infancy: Composition and development. *Acta Paediatr.* **2007**, *92*, 48–55. [CrossRef]
20. Huycke, M.M.; Abrams, V.; Moore, D.R. Enterococcus faecalis produces extracellular superoxide and hydrogen peroxide that damages colonic epithelial cell DNA. *Carcinogenesis* **2002**, *23*, 529–536. [CrossRef]
21. Bhatty, M.; Cruz, M.R.; Frank, K.L.; Laverde Gomez, J.A.; Andrade, F.; Garsin, D.A.; Dunny, G.M.; Kaplan, H.B.; Christie, P.J. *Enterococcus faecalis* pCF10-encoded surface proteins PrgA, PrgB (aggregation substance) and PrgC contribute to plasmid transfer, biofilm formation and virulence. *Mol. Microbiol.* **2015**, *95*, 660–677. [CrossRef] [PubMed]
22. Kayaoglu, G.; Ørstavik, D. Virulence factors of Enterococcus faecalis: Relationship to endodontic disease. *Crit. Rev. Oral Biol. Med.* **2004**, *15*, 308–320. [CrossRef] [PubMed]
23. Upadhyaya, P.; Ravikumar, K.; Umapathy, B. Review of virulence factors of enterococcus: An emerging nosocomial pathogen. *Indian J. Med. Microbiol.* **2009**, *27*, 301. [CrossRef]
24. Singh, K.V.; Nallapareddy, S.R.; Murray, B.E. Importance of the *ebp* (Endocarditis- and Biofilm-Associated Pilus) Locus in the Pathogenesis of *Enterococcus faecalis* Ascending Urinary Tract Infection. *J. Infect. Dis.* **2007**, *195*, 1671–1677. [CrossRef]
25. Singh, K.V.; Nallapareddy, S.R.; Sillanpää, J.; Murray, B.E. Importance of the Collagen Adhesin Ace in Pathogenesis and Protection against Enterococcus faecalis Experimental Endocarditis. *PLoS Pathol.* **2010**, *6*, e1000716. [CrossRef]
26. Toledo-Arana, A.; Valle, J.; Solano, C.; Arrizubieta, M.J.; Cucarella, C.; Lamata, M.; Amorena, B.; Leiva, J.; Penadés, J.R.; Lasa, I. The enterococcal surface protein, Esp, is involved in Enterococcus faecalis biofilm formation. *Appl. Environ. Microbiol.* **2001**, *67*, 4538–4545. [CrossRef] [PubMed]
27. Nallapareddy, S.R.; Singh, K.V.; Duh, R.W.; Weinstock, G.M.; Murray, B.E. Diversity of ace, a gene encoding a microbial surface component recognizing adhesive matrix molecules, from different strains of Enterococcus faecalis and evidence for production of ace during human infections. *Infect. Immun.* **2000**, *68*, 5210–5217. [CrossRef] [PubMed]
28. Montealegre, M.C.; La Rosa, S.L.; Roh, J.H.; Harvey, B.R.; Murray, B.E. The Enterococcus faecalis EbpA Pilus Protein: Attenuation of Expression, Biofilm Formation, and Adherence to Fibrinogen Start with the Rare Initiation Codon ATT. *MBio* **2015**, *6*, e00467-15. [CrossRef] [PubMed]
29. Hubble, T.S.; Hatton, J.F.; Nallapareddy, S.R.; Murray, B.E.; Gillespie, M.J. Influence of Enterococcus faecalis proteases and the collagen-binding protein, Ace, on adhesion to dentin. *Oral Microbiol. Immunol.* **2003**, *18*, 121–126. [CrossRef] [PubMed]
30. Singh, K.V.; La Rosa, S.L.; Somarajan, S.R.; Roh, J.H.; Murray, B.E. The fibronectin-binding protein EfbA contributes to pathogenesis and protects against infective endocarditis caused by Enterococcus faecalis. *Infect. Immun.* **2015**, *83*, 4487–4494. [CrossRef] [PubMed]
31. Nallapareddy, S.R.; Qin, X.; Weinstock, G.M.; Höök, M.; Murray, B.E. Enterococcus faecalis adhesin, ace, mediates attachment to extracellular matrix proteins collagen type IV and laminin as well as collagen type I. *Infect. Immun.* **2000**, *68*, 5218–5224. [CrossRef]
32. Kristich, C.J.; Rice, L.B.; Arias, C.A. *Enterococcal Infection—Treatment and Antibiotic Resistance*; Massachusetts Eye and Ear Infirmary: Boston, MA, USA, 2014.
33. Beganovic, M.; Luther, M.K.; Rice, L.B.; Arias, C.A.; Rybak, M.J.; Laplante, K.L. A Review of Combination Antimicrobial Therapy for Enterococcus Faecalis Bloodstream Infections and Infective Endocarditis Citation/Publisher Attribution. *Clin. Infect. Dis.* **2018**, *2*, 303–309. [CrossRef]
34. Matar, M.J.; Safdar, A.; Rolston, K.V.I. Relationship of colonization with vancomycin-resistant enterococci and risk of systemic infection in patients with cancer. *Clin. Infect. Dis.* **2006**, *42*, 1506–1507. [CrossRef] [PubMed]

35. Gedik, H.; Şimşek, F.; Kantürk, A.; Yıldırmak, T.; Arıca, D.; Aydın, D.; Yokuş, O.; Demirel, N. Vancomycin-resistant enterococci colonization in patients with hematological malignancies: Screening and its cost-effectiveness. *Afr. Health Sci.* **2014**, *14*, 899. [PubMed]
36. Hicks, K.L.; Breto, L.; Halbur, L. *Vancomycin Resistant Enterococcus as a Leading Cause of Early Infection-Related Mortality in Older (≥60 Years) AML Patients Admitted to a Community Hospital for Standard Induction Chemotherapy*; American Society of Hematology: Washington, DC, USA, 2006.
37. Messina, J.A.; Sung, A.D.; Chao, N.J.; Alexander, B.D. *The Timing and Epidemiology of Enterococcus Faecium and E. Faecalis Bloodstream Infections (BSI) in Patients with Acute Leukemia Receiving Chemotherapy*; American Society of Hematology: Washington, DC, USA, 2017.
38. De Almeida, C.V.; Taddei, A.; Amedei, A. The controversial role of *Enterococcus faecalis* in colorectal cancer. *Ther. Adv. Gastroenterol.* **2018**, *11*, 175628481878360. [CrossRef] [PubMed]
39. Koehler, P.; Jung, N.; Cornely, O.A.; Rybniker, J.; Fätkenheuer, G. Combination Antimicrobial Therapy for *Enterococcus faecalis* Infective Endocarditis. *Clin. Infect. Dis.* **2019**. [CrossRef] [PubMed]
40. Abdelkader, K.; Gerstmans, H.; Saafan, A.; Dishisha, T.; Briers, Y. The Preclinical and Clinical Progress of Bacteriophages and Their Lytic Enzymes: The Parts are Easier than the Whole. *Viruses* **2019**, *11*, 96. [CrossRef]
41. Elbreki, M.; Ross, R.P.; Hill, C.; O'Mahony, J.; McAuliffe, O.; Coffey, A. Bacteriophages and Their Derivatives as Biotherapeutic Agents in Disease Prevention and Treatment. *J. Viruses* **2014**, *2014*, 1–20. [CrossRef]
42. Purnell, S.E.; Ebdon, J.E.; Taylor, H.D. Bacteriophage lysis of enterococcus host strains: A tool for microbial source tracking? *Environ. Sci. Technol.* **2011**, *45*, 10699–10705. [CrossRef] [PubMed]
43. Yasmin, A.; Kenny, J.G.; Shankar, J.; Darby, A.C.; Hall, N.; Edwards, C.; Horsburgh, M.J. Comparative genomics and transduction potential of Enterococcus faecalis temperate bacteriophages. *J. Bacteriol.* **2010**, *192*, 1122–1130. [CrossRef]
44. Uchiyama, J.; Rashel, M.; Maeda, Y.; Takemura, I.; Sugihara, S.; Akechi, K.; Muraoka, A.; Wakiguchi, H.; Matsuzaki, S. Isolation and characterization of a novel Enterococcus faecalis bacteriophage ϕEF24C as a therapeutic candidate. *FEMS Microbiol. Lett.* **2008**, *278*, 200–206. [CrossRef] [PubMed]
45. Santiago-Rodriguez, T.M.; De vila, C.; Gonzalez, J.; Bonilla, N.; Marcos, P.; Urdaneta, M.; Cadete, M.; Monteiro, S.; Santos, R.; Domingo, J.S.; et al. Characterization of Enterococcus faecalis-infecting phages (enterophages) as markers of human fecal pollution in recreational waters. *Water Res.* **2010**, *44*, 4716–4725. [CrossRef] [PubMed]
46. Bonilla, N.; Santiago, T.; Marcos, P.; Urdaneta, M.; Santo Domingo, J.; Toranzos, G.A. Enterophages, a group of phages infecting Enterococcus faecalis, and their potential as alternate indicators of human faecal contamination. *Water Sci. Technol.* **2010**, *61*, 293–300. [CrossRef] [PubMed]
47. Mullan, M. Factors Affecting Plaque Formation by Bacteriophages. *Dairy Sci.* **2002**, *5*, 1–12.
48. Örmälä, A.-M.; Jalasvuori, M. Phage therapy. *Bacteriophage* **2013**, *3*, e24219. [CrossRef]
49. Rostøl, J.T.; Marraffini, L. (Ph)ighting Phages: How Bacteria Resist Their Parasites. *Cell Host Microbe* **2019**, *25*, 184–194. [CrossRef] [PubMed]
50. Duerkop, B.A.; Palmer, K.L.; Horsburgh, M.J. Enterococcal Bacteriophages and Genome Defense. In *Enterococci: From Commensals to Leading Causes of Drug Resistant Infection*; Eye and Ear Infirmary: Boston, MA, USA, 2014.
51. Swift, S.M.; Rowley, D.T.; Young, C.; Franks, A.; Hyman, P.; Donovan, D.M. The endolysin from the *Enterococcus faecalis* bacteriophage VD13 and conditions stimulating its lytic activity. *FEMS Microbiol. Lett.* **2016**, *363*, fnw216. [CrossRef]
52. Zhang, W.; Mi, Z.; Yin, X.; Fan, H.; An, X.; Zhang, Z.; Chen, J.; Tong, Y. Characterization of Enterococcus faecalis phage IME-EF1 and its endolysin. *PLoS ONE* **2013**, *8*, e80435. [CrossRef] [PubMed]
53. Wang, Y.; Wang, W.; Lv, Y.; Zheng, W.; Mi, Z.; Pei, G.; An, X.; Xu, X.; Han, C.; Liu, J.; et al. Characterization and complete genome sequence analysis of novel bacteriophage IME-EFm1 infecting Enterococcus faecium. *J. Gen. Virol.* **2014**, *95*, 2565–2575. [CrossRef] [PubMed]
54. Wang, R.; Xing, S.; Zhao, F.; Li, P.; Mi, Z.; Shi, T.; Liu, H.; Tong, Y. Characterization and genome analysis of novel phage vB_EfaP_IME195 infecting Enterococcus faecalis. *Virus Genes* **2018**, *54*, 804–811. [CrossRef]
55. Hyatt, D.; Chen, G.-L.; LoCascio, P.F.; Land, M.L.; Larimer, F.W.; Hauser, L.J. Prodigal: Prokaryotic gene recognition and translation initiation site identification. *BMC Bioinform.* **2010**, *11*, 119. [CrossRef]
56. Li, L.; Stoeckert, C.J.; Roos, D.S.; Roos, D.S. OrthoMCL: Identification of ortholog groups for eukaryotic genomes. *Genome Res.* **2003**, *13*, 2178–2189. [CrossRef] [PubMed]

57. Li, X.; Ding, P.; Han, C.; Fan, H.; Wang, Y.; Mi, Z.; Feng, F.; Tong, Y. Genome analysis of Enterococcus faecalis bacteriophage IME-EF3 harboring a putative metallo-beta-lactamase gene. *Virus Genes* **2014**, *49*, 145–151. [CrossRef]
58. Son, J.S.; Jun, S.Y.; Kim, E.B.; Park, J.E.; Paik, H.R.; Yoon, S.J.; Kang, S.H.; Choi, Y.-J. Complete genome sequence of a newly isolated lytic bacteriophage, EFAP-1 of *Enterococcus faecalis*, and antibacterial activity of its endolysin EFAL-1. *J. Appl. Microbiol.* **2010**, *108*, 1769–1779. [CrossRef] [PubMed]
59. Mazaheri Nezhad Fard, R.; Barton, M.D.; Heuzenroeder, M.W. Bacteriophage-mediated transduction of antibiotic resistance in enterococci. *Lett. Appl. Microbiol.* **2011**, *52*, 559–564. [CrossRef]
60. Gelman, D.; Beyth, S.; Lerer, V.; Adler, K.; Poradosu-Cohen, R.; Coppenhagen-Glazer, S.; Hazan, R. Combined bacteriophages and antibiotics as an efficient therapy against VRE Enterococcus faecalis in a mouse model. *Res. Microbiol.* **2018**, *169*, 531–539. [CrossRef] [PubMed]
61. Lee, Y.-D.; Park, J.-H. Complete genome sequence of enterococcal bacteriophage SAP6. *J. Virol.* **2012**, *86*, 5402–5403. [CrossRef]
62. São-José, C.; Proença, D.; Leandro, C.; Fernandes, S.; Pimentel, M.; Mato, R.; Lopes, F.; Santos, S.; Cavaco-Silva, P.; Silva, F.A. Phage Endolysins with Broad Antimicrobial Activity Against Enterococcus faecalis Clinical Strains. *Microb. Drug Resist.* **2012**, *18*, 322–332.
63. Brede, D.A.; Snipen, L.G.; Ussery, D.W.; Nederbragt, A.J.; Nes, I.F. Complete genome sequence of the commensal Enterococcus faecalis 62, isolated from a healthy Norwegian infant. *J. Bacteriol.* **2011**, *193*, 2377–2378. [CrossRef] [PubMed]
64. Xing, S.; Zhang, X.; Sun, Q.; Wang, J.; Mi, Z.; Pei, G.; Huang, Y.; An, X.; Fu, K.; Zhou, L.; et al. Complete genome sequence of a novel, virulent Ahjdlikevirus bacteriophage that infects Enterococcus faecium. *Arch. Virol.* **2017**, *162*, 3843–3847. [CrossRef] [PubMed]
65. Stevens, R.H.; Porras, O.D.; Delisle, A.L. Bacteriophages induced from lysogenic root canal isolates of *Enterococcus faecalis*. *Oral Microbiol. Immunol.* **2009**, *24*, 278–284. [CrossRef] [PubMed]
66. Yoon, B.H.; Chang, H.-I. Genomic annotation for the temperate phage EFC-1, isolated from Enterococcus faecalis KBL101. *Arch. Virol.* **2015**, *160*, 601–604. [CrossRef]
67. Zhang, H.; Fouts, D.E.; DePew, J.; Stevens, R.H. Genetic modifications to temperate Enterococcus faecalis phage Ef11 that abolish the establishment of lysogeny and sensitivity to repressor, and increase host range and productivity of lytic infection. *Microbiology* **2013**, *159*, 1023–1035. [CrossRef] [PubMed]
68. Stevens, R.H.; Zhang, H.; Hsiao, C.; Kachlany, S.; Tinoco, E.M.B.; Depew, J.; Fouts, D.E.; Roy, F.; Stevens, H. Structural proteins of Enterococcus faecalis bacteriophage ϕEf11. *Bacteriophage* **2016**, *6*, e1251381. [CrossRef] [PubMed]
69. Tinoco, J.M.; Buttaro, B.; Zhang, H.; Liss, N.; Sassone, L.; Stevens, R. Effect of a genetically engineered bacteriophage on Enterococcus faecalis biofilms. *Arch. Oral Biol.* **2016**, *71*, 80–86. [CrossRef]
70. Duerkop, C.A.; Huo, B.A.; Bhardwaj, W.; Palmer, P.L.; Hooper, K.L. Molecular basis for lytic bacteriophage resistance in enterococci. *MBio* **2016**, *7*, 1304–1320. [CrossRef] [PubMed]
71. Dalmasso, M.; de Haas, E.; Neve, H.; Strain, R.; Cousin, F.J.; Stockdale, S.R.; Ross, R.P.; Hill, C. Isolation of a Novel Phage with Activity against Streptococcus mutans Biofilms. *PLoS ONE* **2015**, *10*, e0138651. [CrossRef]
72. Nazareth, N.; Magro, F.; Machado, E.; Ribeiro, T.G.; Martinho, A.; Rodrigues, P.; Alves, R.; Macedo, G.N.; Gracio, D.; Coelho, R.; et al. Prevalence of Mycobacterium avium subsp. paratuberculosis and Escherichia coli in blood samples from patients with inflammatory bowel disease. *Med. Microbiol. Immunol.* **2015**, *204*, 681–692. [CrossRef]
73. Fong, S.A.; Drilling, A.; Morales, S.; Cornet, M.E.; Woodworth, B.A.; Fokkens, W.J.; Psaltis, A.J.; Vreugde, S.; Wormald, P.-J. Activity of Bacteriophages in Removing Biofilms of Pseudomonas aeruginosa Isolates from Chronic Rhinosinusitis Patients. *Front. Cell. Infect. Microbiol.* **2017**, *7*, 418. [CrossRef]
74. Abdulamir, A.S.; Jassim, S.A.A.; Hafidh, R.R.; Bakar, F.A. The potential of bacteriophage cocktail in eliminating Methicillin-resistant Staphylococcus aureus biofilms in terms of different extracellular matrices expressed by PIA, ciaA-D and FnBPA genes. *Ann. Clin. Microbiol. Antimicrob.* **2015**, *14*, 49. [CrossRef]
75. Khalifa, L.; Shlezinger, M.; Beyth, S.; Houri-Haddad, Y.; Coppenhagen-Glazer, S.; Beyth, N.; Hazan, R. Phage therapy against Enterococcus faecalis in dental root canals. *J. Oral Microbiol.* **2016**, *8*, 32157. [CrossRef] [PubMed]
76. Szafrański, S.P.; Winkel, A.; Stiesch, M. The use of bacteriophages to biocontrol oral biofilms. *J. Biotechnol.* **2017**, *250*, 29–44. [CrossRef] [PubMed]

77. Sutherland, I.W.; Hughes, K.A.; Skillman, L.C.; Tait, K. The interaction of phage and biofilms. *FEMS Microbiol. Lett.* **2004**, *232*, 1–6. [CrossRef]
78. Stuart, C.H.; Schwartz, S.A.; Beeson, T.J.; Owatz, C.B. Enterococcus faecalis: Its role in root canal treatment failure and current concepts in retreatment. *J. Endod.* **2006**, *32*, 93–98. [CrossRef] [PubMed]
79. Wang, Q.Q.; Zhang, C.F.; Chu, C.H.; Zhu, X.F. Prevalence of Enterococcus faecalis in saliva and filled root canals of teeth associated with apical periodontitis. *Int. J. Oral Sci.* **2012**, *4*, 19–23. [CrossRef]
80. Sundqvist, G.; Figdor, D.; Persson, S.; Sjögren, U. Microbiologic analysis of teeth with failed endodontic treatment and the outcome of conservative re-treatment. *Oral Surg. Oral Med. Oral Pathol. Oral Radiol. Endodontol.* **1998**, *85*, 86–93. [CrossRef]
81. Paisano, A.F.; Spira, B.; Cai, S.; Bombana, A.C. In vitro antimicrobial effect of bacteriophages on human dentin infected with Enterococcus faecalis ATCC 29212. *Oral Microbiol. Immunol.* **2004**, *19*, 327–330. [CrossRef] [PubMed]
82. Furie, B.; Furie, B.C. The molecular basis of blood coagulation. *Cell* **1988**, *53*, 505–518. [CrossRef]
83. Hershberger, E.; Coyle, E.A.; Kaatz, G.W.; Zervos, M.J.; Rybak, M.J. Comparison of a Rabbit Model of Bacterial Endocarditis and an In Vitro Infection Model with Simulated Endocardial Vegetations. *Antimicrob. Agents Chemother.* **2000**, *44*, 1921–1924. [CrossRef]
84. Kastrup, C.J.; Boedicker, J.Q.; Pomerantsev, A.P.; Moayeri, M.; Bian, Y.; Pompano, R.R.; Kline, T.R.; Sylvestre, P.; Shen, F.; Leppla, S.H.; et al. Spatial localization of bacteria controls coagulation of human blood by "quorum acting". *Nat. Chem. Biol.* **2008**, *4*, 742–750. [CrossRef]
85. McGrath, B.J.; Kang, S.L.; Kaatz, G.W.; Rybak, M.J. Bactericidal activities of teicoplanin, vancomycin, and gentamicin alone and in combination against Staphylococcus aureus in an in vitro pharmacodynamic model of endocarditis. *Antimicrob. Agents Chemother.* **1994**, *38*, 2034–2040. [CrossRef] [PubMed]
86. Houlihan, H.H.; Stokes, D.P.; Rybak, M.J. Pharmacodynamics of vancomycin and ampicillin alone and in combination with gentamicin once daily or thrice daily against Enterococcus faecalis in an in vitro infection model. *J. Antimicrob. Chemother.* **2000**, *46*, 79–86. [CrossRef]
87. Ladero, V.; Gómez-Sordo, C.; Sánchez-Llana, E.; del Rio, B.; Redruello, B.; Fernández, M.; Martín, M.C.; Alvarez, M.A. Q69 (an E. faecalis-Infecting Bacteriophage) As a Biocontrol Agent for Reducing Tyramine in Dairy Products. *Front. Microbiol.* **2016**, *7*, 445. [CrossRef]
88. Letkiewicz, S.; Międzybrodzki, R.; Fortuna, W.; Weber-Dąbrowska, B.; Górski, A. Eradication of Enterococcus faecalis by phage therapy in chronic bacterial prostatitis—Case report. *Folia Microbiol.* **2009**, *54*, 457–461. [CrossRef] [PubMed]
89. Uchiyama, J.; Rashel, M.; Takemura, I.; Wakiguchi, H.; Matsuzaki, S. In silico and in vivo evaluation of bacteriophage ϕEF24C, a candidate for treatment of Enterococcus faecalis infections. *Appl. Environ. Microbiol.* **2008**, *74*, 4149–4163. [CrossRef]
90. Biswas, B.; Adhya, S.; Washart, P.; Paul, B.; Trostel, A.N.; Powell, B.; Carlton, R.; Merril, R.; Merril, C.R. Bacteriophage Therapy Rescues Mice Bacteremic from a Clinical Isolate of Vancomycin-Resistant Enterococcus Bacteriophage Therapy Rescues Mice Bacteremic from a Clinical Isolate of Vancomycin-Resistant Enterococcus faecium. *Infect. Immun.* **2002**, *70*, 204–210. [CrossRef]
91. Cheng, M.; Liang, J.; Zhang, Y.; Hu, L.; Gong, P.; Cai, R.; Zhang, L.; Zhang, H.; Ge, J.; Ji, Y. The bacteriophage EF-P29 efficiently protects against lethal vancomycin-resistant Enterococcus faecalis and alleviates gut microbiota imbalance in a murine bacteremia model. *Front. Microbiol.* **2017**, *8*, 837. [CrossRef]
92. Cheng, M.; Zhang, Y.; Li, X.; Liang, J.; Hu, L.; Gong, P.; Zhang, L.; Cai, R.; Zhang, H.; Ge, J.; et al. Endolysin LysEF-P10 shows potential as an alternative treatment strategy for multidrug-resistant Enterococcus faecalis infections. *Sci. Rep.* **2017**, *7*, 10164. [CrossRef]
93. Weber-Dąbrowska, B.; Mulczyk, M.; Górski, A. Bacteriophage Therapy of Bacterial Infections: An Update of our Institute's Experience. In *Inflammation*; Springer: Dordrecht, The Netherlands, 2001; pp. 201–209.
94. Schooley, R.T.; Biswas, B.; Gill, J.J.; Hernandez-Morales, A.; Lancaster, J.; Lessor, L.; Barr, J.J.; Reed, S.L.; Rohwer, F.; Benler, S.; et al. Development and Use of Personalized Bacteriophage-Based Therapeutic Cocktails to Treat a Patient with a Disseminated Resistant Acinetobacter baumannii Infection. *Antimicrob. Agents Chemother.* **2017**, *61*, e00954-17. [CrossRef] [PubMed]
95. Uchiyama, J.; Matsui, H.; Murakami, H.; Kato, S.; Watanabe, N.; Nasukawa, T.; Mizukami, K.; Ogata, M.; Sakaguchi, M.; Matsuzaki, S.; et al. Potential Application of Bacteriophages in Enrichment Culture for Improved Prenatal Streptococcus agalactiae Screening. *Viruses* **2018**, *10*, 552. [CrossRef] [PubMed]

96. Kavanagh, K.; Fallon, J.P. Galleria mellonella larvae as models for studying fungal virulence. *Fungal Biol. Rev.* **2010**, *24*, 79–83. [CrossRef]
97. Mukherjee, K.; Altincicek, B.; Hain, T.; Domann, E.; Vilcinskas, A.; Chakraborty, T. Galleria mellonella as a model system for studying Listeria pathogenesis. *Appl. Environ. Microbiol.* **2010**, *76*, 310–317. [CrossRef]
98. Ramarao, N.; Nielsen-Leroux, C.; Lereclus, D. The insect Galleria mellonella as a powerful infection model to investigate bacterial pathogenesis. *J. Vis. Exp.* **2012**, *70*, e4392. [CrossRef] [PubMed]
99. Kamal, F.; Dennis, J.J. Burkholderia cepacia complex phage-antibiotic synergy (PAS): Antibiotics stimulate lytic phage activity. *Appl. Environ. Microbiol.* **2015**, *81*, 1132–1138. [CrossRef] [PubMed]
100. Manohar, P.; Nachimuthu, R.; Lopes, B.S. The therapeutic potential of bacteriophages targeting gram-negative bacteria using Galleria mellonella infection model. *BMC Microbiol.* **2018**, *18*, 97. [CrossRef] [PubMed]
101. Goh, H.M.S.; Yong, M.H.A.; Chong, K.K.L.; Kline, K.A. Model systems for the study of Enterococcal colonization and infection. *Virulence* **2017**, *8*, 1525–1562. [CrossRef]
102. Eiko Maekawa, L.; Dennis Rossoni, R.; Oliveira Barbosa, J.; Olavo Cardoso Jorge, A.; Campos Junqueira, J.; Carneiro Valera, M.; José Longo, F. Different Extracts of Zingiber Officinale Decrease Enterococcus Faecalis Infection in Galleria Mellonella. *Braz. Dent. J.* **2015**, *26*, 105–109. [CrossRef]
103. Luther, M.K.; Arvanitis, M.; Mylonakis, E.; LaPlante, K.L. Activity of daptomycin or linezolid in combination with rifampin or gentamicin against biofilm-forming Enterococcus faecalis or E. faecium in an in vitro pharmacodynamic model using simulated endocardial vegetations and an in vivo survival assay using Galleria mellonella larvae. *Antimicrob. Agents Chemother.* **2014**, *58*, 4612–4620. [PubMed]
104. Nale, J.Y.; Chutia, M.; Carr, P.; Hickenbotham, P.T.; Clokie, M.R.J. "Get in early"; Biofilm and wax moth (Galleria mellonella) models reveal new insights into the therapeutic potential of Clostridium difficile bacteriophages. *Front. Microbiol.* **2016**, *7*, 1383. [CrossRef]
105. Seed, K.D.; Dennis, J.J. Experimental bacteriophage therapy increases survival of Galleria mellonella larvae infected with clinically relevant strains of the Burkholderia cepacia complex. *Antimicrob. Agents Chemother.* **2009**, *53*, 2205–2208. [CrossRef]
106. Beeton, M.L.; Alves, D.R.; Enright, M.C.; Jenkins, A.T.A. Assessing phage therapy against Pseudomonas aeruginosa using a Galleria mellonella infection model. *Int. J. Antimicrob. Agents* **2015**, *46*, 196–200. [CrossRef]
107. Abbasifar, R.; Kropinski, A.M.; Sabour, P.M.; Chambers, J.R.; MacKinnon, J.; Malig, T.; Griffiths, M.W. Efficiency of bacteriophage therapy against Cronobacter sakazakii in Galleria mellonella (greater wax moth) larvae. *Arch. Virol.* **2014**, *159*, 2253–2261. [CrossRef] [PubMed]
108. Schmelcher, M.; Donovan, D.M.; Loessner, M.J. Bacteriophage endolysins as novel antimicrobials. *Future Microbiol.* **2012**, *7*, 1147–1171. [CrossRef]
109. Fernández-Ruiz, I.; Coutinho, F.H.; Rodriguez-Valera, F. Thousands of Novel Endolysins Discovered in Uncultured Phage Genomes. *Front. Microbiol.* **2018**, *9*, 1033. [CrossRef]
110. Young, R.Y. Bacteriophage holins: Deadly diversity. *J. Mol. Microbiol. Biotechnol.* **2002**, *4*, 21–36.
111. Hill, C.; Mills, S.; Ross, R.P. Phages & antibiotic resistance: Are the most abundant entities on earth ready for a comeback? *Future Microbiol.* **2018**, *13*, 711–726.
112. Finn, R.D.; Clements, J.; Eddy, S.R. HMMER web server: Interactive sequence similarity searching. *Nucleic Acids Res.* **2011**, *39*, W29–W37. [CrossRef] [PubMed]
113. Söding, J.; Biegert, A.; Lupas, A.N. The HHpred interactive server for protein homology detection and structure prediction. *Nucleic Acids Res.* **2005**, *33*, W244–W248. [CrossRef]
114. Oliveira, H.; Melo, L.D.R.; Santos, S.B.; Nóbrega, F.L.; Ferreira, E.C.; Cerca, N.; Azeredo, J.; Kluskens, L.D. Molecular aspects and comparative genomics of bacteriophage endolysins. *J. Virol.* **2013**, *87*, 4558–4570. [CrossRef]
115. Proença, D.; Leandro, C.; Garcia, M.; Pimentel, M.; São-José, C. EC300: A phage-based, bacteriolysin-like protein with enhanced antibacterial activity against Enterococcus faecalis. *Appl. Microbiol. Biotechnol.* **2015**, *99*, 5137–5149. [CrossRef]
116. Uchiyama, J.; Takemura, I.; Hayashi, I.; Matsuzaki, S.; Satoh, M.; Ujihara, T.; Murakami, M.; Imajoh, M.; Sugai, M.; Daibata, M. Characterization of lytic enzyme open reading frame 9 (ORF9) derived from Enterococcus faecalis bacteriophage ϕEF24C. *Appl. Environ. Microbiol.* **2011**, *77*, 580–585. [CrossRef]

Review

Current State of Compassionate Phage Therapy

Shawna McCallin [1,2,*], Jessica C. Sacher [3], Jan Zheng [3] and Benjamin K. Chan [4]

1 Unit of Regenerative Medicine, Department of Musculoskeletal Medicine, Service of Plastic, Reconstructive, & Hand Surgery, University Hospital of Lausanne (CHUV), 1066 Epalignes, Switzerland
2 Swiss Federal Institute of Technology Lausanne (EPFL), 1015 Lausanne, Switzerland
3 Phage Directory, Atlanta, GA 30303, USA; jessica@phage.directory (J.C.S.); jan@phage.directory (J.Z.)
4 Yale University, New Haven, CT 06520, USA; b.chan@yale.edu
* Correspondence: Shawna.mccallin@gmail.com

Received: 1 March 2019; Accepted: 6 April 2019; Published: 12 April 2019

Abstract: There is a current unmet medical need for the treatment of antibiotic-resistant infections, and in the absence of approved alternatives, some clinicians are turning to empirical ones, such as phage therapy, for compassionate treatment. Phage therapy is ideal for compassionate use due to its long-standing historical use and publications, apparent lack of adverse effects, and solid support by fundamental research. Increased media coverage and peer-reviewed articles have given rise to a more widespread familiarity with its therapeutic potential. However, compassionate phage therapy (cPT) remains limited to a small number of experimental treatment centers or associated with individual physicians and researchers. It is possible, with the creation of guidelines and a greater central coordination, that cPT could reach more of those in need, starting by increasing the availability of phages. Subsequent steps, particularly production and purification, are difficult to scale, and treatment paradigms stand highly variable between cases, or are frequently not reported. This article serves both to synopsize cPT publications to date and to discuss currently available phage sources for cPT. As the antibiotic resistance crisis continues to grow and the future of phage therapy clinical trials remains undetermined, cPT represents a possibility for bridging the gap between current treatment failures and future approved alternatives. Streamlining the process of cPT will help to ensure high quality, therapeutically-beneficial, and safe treatment.

Keywords: bacteriophage therapy; compassionate use; antibiotic resistance

---

1. Introduction

The first documented therapeutic case of harnessing the natural antibacterial mechanism of bacteriophages, or phages, for the treatment of a human bacterial infection predates the discovery of antibiotics by two decades [1]. Phages were used experimentally for the treatment of various bacterial infections throughout the 1920s, including cholera (reviewed in [2]), dysentery [3], and staphylococcal infections [4] to varying degrees of success [5,6]. For these early applications, phages needed to be isolated from environmental sources, cultivated on bacterial hosts, and purified in line with technology at the time. The deemed founder of phage therapy, F. d'Hérelle, had a heavy hand in the spread of phage therapy during these early years, which he encouraged by traveling to different countries, such as the Soviet Union, India, Egypt, and others, bringing with him phages and the knowledge of how to use them against human bacterial infections [2,7,8].

As phages fell to the wayside with the pursuit of antibiotics in Western medicine in the 1940s, Soviet researchers continued phage development at the G. Eliava Institute of Bacteriophages, Microbiology, and Virology in Tbilisi, Georgia [7,9]. There, phages were isolated from environmental sources and accumulated into a phage bank that exists to this day. This collection provides a large repertoire from which phages can either be incorporated into pre-formulated products or selectively matched against bacterial isolates for personalized therapies. As a result of historical clinical trials and experience accrued during the twentieth century, phages exist alongside antibiotics as approved medicines in some former Soviet Union countries. However, historical data from one country holds little scientific weight in present day evaluations of unapproved medicines in others.

Now, the rest of the world has a re-found interest in revitalizing phage therapy that has paralleled the rise of antibiotic resistance [10–14]. For phage therapy to be recognized as an effective alternative to antibiotics, it will require efficacy data from randomized, controlled clinical trials (RCTs). The three phage RCTs completed to date have failed to produce robust conclusions on efficacy, therefore leaving phage therapy in limbo in the approval process until future trials are conducted [15–17]. Only one RCT for phage products is currently open for enrollment (ClinicalTrials.gov Identifier: NCT03808103), although several are scheduled for patient enrollment in the near future. In the interim, several competency centers, physicians, and researchers are invoking phage therapy for compassionate means in order to respond to the current clinical needs of patients suffering from antibiotic failure.

2. Compassionate Use

Compassionate treatment denotes the use of unapproved medicines outside of clinical trials for the treatment of patients for which approved therapeutic options have been exhausted. The principle of compassionate use is codified in the "Helsinki Declaration of Ethical Principles for Medical Research Involving Human Subjects", which is an international agreement on facets of clinical research, such as patient consent and placebo control [18]. Article 37 specifically asserts a physician's authority to act in the best interest of their patient by using experimental treatments in the absence of approved options, although the support of using unproven treatments was not stipulated by the Declaration until its amendment in 2000 (v2000, Article 32) [19]. In its current state, it reads in its entirety, "*In the treatment of an individual patient, where proven interventions do not exist or other known interventions have been ineffective, the physician, after seeking expert advice, with informed consent from the patient or a legally authorised representative, may use an unproven intervention if in the physician's judgement it offers hope of saving life, re-establishing health or alleviating suffering. This intervention should subsequently be made the object of research, designed to evaluate its safety and efficacy. In all cases, new information must be recorded and, where appropriate, made publicly available*" [18].

The term "compassionate use" can therefore be referred to both vernacularly in this general sense, as well as formally as a regulatory pathway (also referred to as "expanded access" or "special access"). The process and conditions for compassionate use are stipulated by regulatory agencies, such as the Food and Drug Administration (FDA) in the United States, the Therapeutic Goods Administration (TGA) in Australia, or the European Medicines Agency (EMA) in the European Union (EU), although EU member states apply EMA directives independently and may be governed by additional national regulation [20–22]. The objective of compassionate treatment differs from RCTs in that its primary aim is to provide therapeutic benefit to the patient, rather than to evaluate the efficacy of the experimental treatment (although safety may be evaluated). While the term "compassionate" is frequently associated with case reports of phage therapy, it does not inherently signify regulatory adherence, and legal processes that are required for compassionate treatment vary from country to country [23].

The general prevalence and importance of compassionate use is changing, with an increase in access requests and legal support [24–26]. Instigation of compassionate treatment also increasingly arises from patient advocacy groups or patients via social media platforms, to bring attention to, put pressure on, and finance access to unapproved therapies [27]. "Right-to-try" legislation in the US aims to expedite treatment of severely ill patients with unapproved medicines, albeit with lower regulatory and safety oversight [25]. While the intention is to increase therapeutic options for patients and highlight the inability of current pathways to respond punctually to medical needs, it is not without consequence for ethical considerations, such as equal access, unfulfilled expectations, data collection/usage, or financial responsibility [23–25,28].

3. Compassionate Phage Therapy (cPT)

The potential utility of cPT is considered after antibiotic failure is clearly documented, attempts to use conventional treatment have been exhausted, and there are no active clinical trials suitable for enrolment (Figure 1). The possibility of using phages may be suggested by the physician, medical entourage, or the patient themselves. Both the consent of the physician and the patient or guardian are essential for continuing the process of cPT, which may or may not be subject to additional institutional or national regulation on the use of unapproved or experimental therapies. cPT has been approved under emergency investigational new drug (eIND) and expanded access schemes by the FDA, a temporary use authorization (ATU) by the French National Agency for Medicines and Health Products Safety (ANSM) in France, by special access schemes by the TGA in Australia, and by national regulation in Poland. Expanded/special access schemes facilitate access to products in clinical development for compassionate treatments and several phage products have fallen under such schemes in the US and Australia [29–31]. Without local support, physically- and financially-able patients have the option of traveling to receive phage therapy in countries where it is an approved medicine: For instance, the Eliava Institute in Tbilisi, Georgia has provided treatment to a number of international patients on-site [32–34]. The exact process for organizing cPT remains highly variable at present due to its compassionate nature. It can represent a time-consuming endeavor for new cases, to the extent that it may deter motivation to pursue cPT as an option or delay the initiation of treatment, which may influence therapeutic outcomes. Competency centers or individuals experienced with cPT have the advantage of activating familiar pathways for subsequent treatments, and it is the experience of the authors that these centers and individuals are generally willing to be consulted for information on how to best initiate and follow through with cPT. The cost of providing a phage suitable for human application is currently high, with the financial burden falling on the phage provider for most cPT cases, although this may vary between countries and the regulatory status of phage therapy.

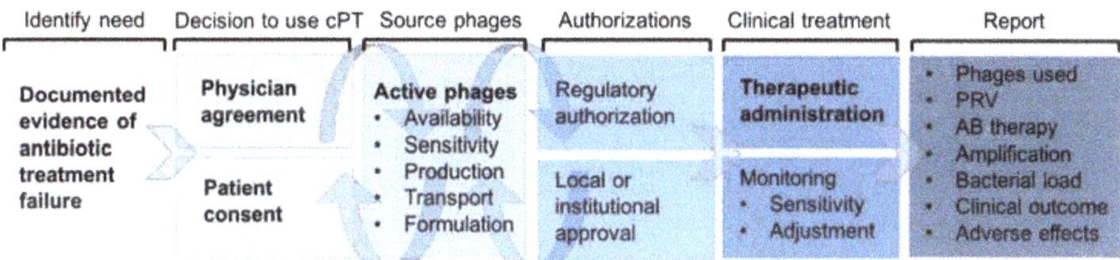

Figure 1. General process and considerations for compassionate phage therapy (cPT). Required steps are shown in bold. Circular arrows indicate processes that are dynamic and do not occur necessarily in a chronological order. PRV: Phage-resistant variant. AB: Antibiotic.

Some countries have established legislation for phage therapy without marketing approvals for phage products, such as the Ludwik Hirszfeld Institute of Immunology and Experimental Therapy in Poland, which has been treating patients with phages experimentally with outpatient care since the 1970s. The Phage Therapy Unit (PTU) was opened there in Poland in 2005, which operates phage treatment under a national regulation scheme, and researchers have published summaries and case reports on nearly 1500 patients since 2000 [35–40]. Costs for cPT are more realistically managed in Poland, where research institutions, such as the PTU, are not permitted to cover healthcare-related costs, leaving payment to the patients, insurance companies, or sponsors. The Center for Innovative Phage Applications and Therapeutics (IPATH) at the University of California San Diego School of Medicine opened mid-2018 as the only present-day phage center in North America with a clear intention of using phages for compassionate needs and for the eventual elaboration of clinical trials [41]. Experience with several cPT treatments in Belgium led to a recently orchestrated permission to use phages as active ingredients of magistral preparations (known as compounded prescription drugs in the US) [42].

This framework allows phages to be prescribed for individual patients as long as they are produced according to an internal monograph. Phages are still considered "non-authorized" components of the preparation, however, and the availability of magistral phage preparations is still limited, even within Belgium. While this model is distinct from compassionate use, it illustrates how compassionate use can lead to the elaboration of alternative approval pathways with clearly-defined guidelines, even if they are unlikely to be replicated in countries, such as the US, where compounded components require authorizations. Beyond such phage competency centers, unassociated physicians have occasionally independently administered phages from academic, biotech, and commercial sources for the treatment of antibiotic resistant infections [32,43–47].

There are more than 25 reports of cPT since 2000, half of which have been published in the past two years and represent different infections, phages, pathogens, and administration routes that collectively represent the application of phages to nearly 2000 people (Table 1). These case studies are published either as periodic updates on the experiences of competency centers or zealous physicians or researchers, or as isolated one-off applications. They vary widely in the information included within the publication, concerning treatment outcomes, concomitant antibiotic use, and microbiological assessment. Instances of cPT usually incur a lag time to publication or are presented at conferences or published as press releases rather than peer-reviewed publications, meaning that there are more cPT cases occurring than published through scientific channels. Indeed, Ampliphi Biosciences have announced via press-release an 84% clinical success rate through their expanded access programs for the treatment of *Staphylococcus aureus* or *Pseudomonas aeruginosa* infections [30,48].

From published cases, treatment with cPT for *S. aureus* infections has been reported the most frequently, followed by *P. aeruginosa* and *Escherichia coli*, and to a lesser extent, *Enterococcus* sp., and *Acinetobacter baumannii* (Table 1). Cases include the treatment of a myriad of different indications for both chronic and acute conditions, including bone-and-joint, urogenital, respiratory, wound, cardiac, and systemic infections, via various administration routes. Positive treatment outcomes range from 40 to 100% of patients included in reports of more than one participant, depending on the size of the study and heterogeneity of treatment strategies (monotherapy versus cocktail; phage substitution; combination with antibiotics). The development of resistance to applied phages was microbiologically documented in only five reports and largely uninvestigated or unreported in most studies, even in the event of unsatisfactory clinical outcomes. Larger reports show treatment failure rates between 4% and 60%, again with differing methodology between studies with little analytical explanation as to how or why failure occurred. Even definitions of clinical "success" or "failure" may vary, therefore cautioning against the over-interpretation of some cPT results. While publishing cases of cPT helps foster familiarity with phage therapy and support claims of safety, it is not possible to draw conclusions on broader efficacy or to use compassionate treatments in lieu of clinical trials. More standardized reporting guidelines would, however, be useful in order to make comparisons between treatments, particularly in terms for the development of phage resistance (Oechslin and McCallin, submitted).

Table 1. Summary of 29 publicly-available, published cases of cPT as of April 2019 in chronological order of most recent publication. Causative pathogens, types of infections (mono/polymicrobial; clinical indication), and administration routes vary between studies. The definition of success may be specific to authors, but overall indicates observed clinical amelioration and/or pathogen clearance. Concomitant antibiotic therapy is indicated for the number of patients per study if ≥1. Plausible reasons for cPT failure are listed when available, as well as the investigation into bacterial development of resistance to applied phages. Phage sources used for treatment are listed and further information can be found in cited references.

Pathogen	Infection	Admin Route	N*	Clinical Outcome	AB (N*)	Failure /PRV+	Phage Source	Ref.
A. baumannii, K. pneumoniae	Bone	iv	1	Success	Yes	na/no	Military	[49]
S. aureus; P. aeruginosa; E. coli; Proteus[PM]	Bone; GI; ENT; urogenital	Local; oral; rectal; joint injection	15	High success rate (12/15); all cases improved	Yes	2° pathogen for 1 patient; unclear results for 2 patients	Mostly commercial	[33]
S. aureus	Bone	Soft-tissue injection	1	Success	Int.	na/nr	Commercial (Eliava)	[45]
S. aureus; E. coli; Proteus; Streptococcus; P. aeruginosa	UTI	Local via catheter	9	Bacterial load decrease in 67% (6/9); pathogen clearance for 3 patients	Yes (1)	No decrease for 1 patient; 2° infection for 1 patient/ nr	Commercial (Eliava); adapted to strains	[50]
Achromabacter xylosoxidans	Cystic Fibrosis infection	Inhaled; oral	1	Improved lung function and general condition	Yes, post	na/nr	Environ.	[51]
P. aeruginosa	Recurrent pneumonia	Inhaled; iv	1	Success	Yes	na/ Yes (PS)	Environ., biotech; military	[52]
S. aureus	Bone	Local	1	Success	Yes	na/nr	Biotech	
S. aureus, P. aeruginosa[PM]	Bone	Local	3	nr	nr	nr/nr	Biotech	[53]
P. aeruginosa	Bone	Local	1	Success for bacterial clearance†	Yes	na/nr	Biotech	[54]
E. coli; Proteus; S. aureus; P. aeruginosa; Streptococcus; Enterococcus	Burns, ulcers, wounds	Topical; sc	234: (27; 90; 94; 23)	Overall high success rate; varied by study	Varied with study	Varied with study/nr	Commercial; unspecified	Review of 4 cases in Russian [55]
P. aeruginosa	Aortic valve graft	Direct via fistula	1	Success	Yes	na/nr	Academic	[43]
A. baumanii	Post-operative cranial infection	iv	1	Infection site cleared; blood cultures negative†	No	Treatment discontinued/nr	Military	[56]
S. aureus	Chronic skin infection	Topical; oral	1	Decreased bacterial load; improved clinical condition	No	Prolonged treatment/Yes (PS)	Commercial (Eliava)	[34]
A. baumanii	Necrotizing pancreatitis	iv; local	1	Success	Yes	na/Yes (PS)	Environ., military; biotech; phage bank	[57]
P. aeruginosa	Infected wound/ septicemia	iv; local	1	Wounds remained colonized, blood cultures were negative†	Int	Bacteremia resolved, but local infection persisted/nr	Military	[58]
P. aeruginosa	Bacteremia	iv	1	Bacteremia eradicated twice; subsequent regrowth †	Yes	Slow bacterial regrowth/PRV likely	Military	[44]
S. aureus	Diabetic toe ulcer infection	Topical	6	Success; avoided amputation	nr	na/nr	Commercial (Eliava)	[59]

Table 1. *Cont.*

Pathogen	Infection	Admin Route	N*	Clinical Outcome	AB (N*)	Failure /PRV+	Phage Source	Ref.
S. aureus	Corneal abscess	Topical, nasal, iv	1	Success	nr	na/nr	Commercial (Eliava)	[32]
P. aeruginosa; *S. aureus*PM	Burn wound infections	Topical	9	Modest reduction in bacterial load for 8 patients	Just prior	nr/nr	Military	[60]
Staphylococcus; *Enterococcus*; *Pseudomonas*; *E. coli*; *Proteus*; *Enterococcus*; etcPM	UTI; urogenital; soft tissue; skin; orthopedic; respiratory; bacteremia; etc.	Topical, oral, rectal, vaginal, inhaled	157	Good clinical outcomes for 44% of patients (success for 18%)	Yes (29%)	Inadequate response for 60% of patients/Yes	In-house	[40]
P. aeruginosa	UTI	Local in bladder	1	Success	Yes	na/No	Commercial (Eliava)	[47]
Enterococcus faecalis	Prostatitis	Rectal	3	Success	No	na/nr	In-house	[36]
S. aureus	GI Carrier status	Oral	1	Success	No	na/nr	In-house	[35]
P. aeruginosa	Burn wound	Topical	1	Successful grafting	Yes	na/nr	Academic	[61]
S. aureus	Wounds	Topical	2	Success	Yes	na/nr	Commercial	[62]
S. aureus; *E. coli*; *P. aeruginosa*; *Klebsiella*; etcPM	Septicemia	Oral	94	85% success rate	Yes (n = 71)	Phage ineffective for 15% of patients/nr	In-house	[39]
Staphylococcus; *E. coli*; *Proteus*; *Streptococcus*; *P. aeruginosa* PM	Venous ulcers and wounds	Topical	96	70% healing	Yes	No clinical improvement for 5 patients	Commercial	[63]
S. aureus; *E. coli*; *P. aeruginosa*; *Klebsiella* PM	Various infections in cancer patients	Oral, local	20	Healing in all patients	nr	na/nr	In-house	[38]
S. aureus; *E. coli*; *Proteus*; *P. aeruginosa*; *Klebsiella*; *Enterobacter*PM	Septicemia; ENT; UTI; meningitis; respiratory; wounds; bone; etc.	Oral; topical; local	1307	Full recovery 86%; 11% transient improvement	nr	No effect in 3.8% of study population (n = 50)	In-house	[37]

* Number patients in study; AB: Concomitant antibiotic treatment with number of patients in (); PRV: Phage-resistant variants reported; PM includes polymicrobial infections; GI: Gastrointestinal; ENT: Ear Nose Throat; 2° Secondary; iv: intravenous; Int: intermittent; † Deceased; PS: Phage Substitution; sc: subcutaneous; na: not applicable; nr: not reported; Environ: Environmental.

4. Sources and Availability of Phages for cPT

An essential prerequisite for cPT is the availability of phages active against the patient isolate that can then be sufficiently purified to support clinical application. While evident, this can be a limiting factor for cPT considering both the high level of specificity of phage–bacterial interactions and time-to-treatment constraints for acute infections. Possible sources of phages for cPT are summarized below, all of which have contributed by varying extents to cPT efforts.

4.1. Environmentally-Sourced Phages

Phages are naturally present in abundance from environmental samples, particularly in bacteria-rich environments, such as sewage or from infections themselves, and natural environments have been the primary source for all phages used in cPT to date [64]. However, starting from this point entails phage isolation, propagation, and characterization that can delay treatment considerably, and requires research infrastructure and expertise. Rare or less-studied pathogens may necessitate environmental isolation of new phages, whereas phages against well-known pathogens (e.g., *S. aureus*, *P. aeruginosa*, *E. coli*) are already widely available.

4.2. Academically-Sourced Phages

Phages are the subject of fundamental and translational research in numerous academic laboratories around the world. As such, phages sourced from academic labs often offer the benefit of additional characterization, such as genome sequencing, host range analysis, and in vitro/in vivo studies that can provide further information to support their use for cPT. Examples of cPT cases that used phages sourced from academic labs include Schooley et al. [57] and Chan et al. [43]. In addition to academic labs, phages can also be sourced from established phage banks or repositories, some of which provide phages across international borders. Examples of phage banks include the Félix d'Hérelle Reference Center for Bacterial Viruses at the University of Laval [65], the Leibniz Institute DSMZ-German Collection of Microorganisms and Cell Cultures [66] and the Bacteriophage Bank of Korea [67]. Phages sourced from such banks are also often well-characterized, but may incur standard purchasing costs, while academically-sourced phages tend to be supplied pro bono. Large phage banks can provide the benefit of wider pathogen coverage, while some academic phage collections only include phages against one or a select few pathogens. In addition to large phage banks that serve the international community, other phage banks are intended to supply phages for in-house or local cases. For instance, the collection at the PTU contains over 500 phages that cover 15 bacterial pathogens; however, their phages have not been reported for cPT outside of Poland [40]. Both academically- and bank-sourced phages may be liable to intellectual property (IP) constraints, though to different degrees, or require a material transfer agreement (MTA) that limits and delineates the use of the phage(s) supplied.

4.3. Phage Products in Clinical Development

Phages are progressively being developed for clinical use by biotech companies. Such companies as Pherecydes Pharma (France), Ampliphi Biosciences (US, Australia), and Adaptive Phage Therapeutics (US) have participated in the supply and preparation of phages for cPT patients [52–54,57]. Phages from clinical developers are well-suited for cPT, but phage biotechs understandably retain the right to decline phage supply in consideration of their capacity and business interests.

4.4. Eastern European Phage Products

Commercially-available phages and phage preparations from Eastern European countries are an additional phage source that have been used in clinical trials [16] and in compassionate treatments, both within countries where they exist as registered products and in Western countries [32–34,45–47,51,55]. While standard commercial preparations have a predefined composition of phages, the Eliava Institute offers personalized [34,47] or adapted phage compositions [50] that have been used in Tbilisi or sent to other countries such as France, the US, or Australia. However, the use of commercial preparations from Eastern Europe for cPT in countries where phage therapy is not approved may, or may not, lead to importation or approval difficulties depending on regulatory adherence and requirements.

4.5. Crowd-Sourcing Phages

The importance of phages, whatever the source, for cPT is that they have activity against the patient isolate, can be purified and formulated for safe administration, and are readily available to conduce punctual treatment. The need for coordinated phage sharing was documented within a cPT case for the treatment of a multidrug-resistant *A. baumannii* infection with phages [57]. In this case, a total of nine phages from three different sources were required, and the effort was largely coordinated by the patient's wife via email and social media outlets due to the absence of established or official channels. Following this case, in 2017, an initiative to organize such sharing was founded called Phage Directory (https://phage.directory) [68]. One focus of this initiative is to keep a register of academic phage researchers, phage banks, and phage companies that are willing to contribute phages for cPT in order to locate active phages in the most time-efficient manner. For example, in late 2018, Phage Directory helped coordinate the sourcing of *Klebsiella pneumoniae* phages for a patient in Helsinki,

Finland by sending an electronic alert to its network of registered labs and phage collections. This effort resulted in >175 phages being contributed by ten different groups over the span of three weeks, all of which were tested against the patient's isolate [31]. As of January 2019, there were 36 academic phage laboratories and one phage bank registered on Phage Directory, representing more than 20 different countries with phages covering more than 32 host genera (Figure 2). While this sharing network may be less important for established centers or for those with direct access to large phage collections, it certainly facilitates access for geographically-removed patients or physicians without phage research support or established connections.

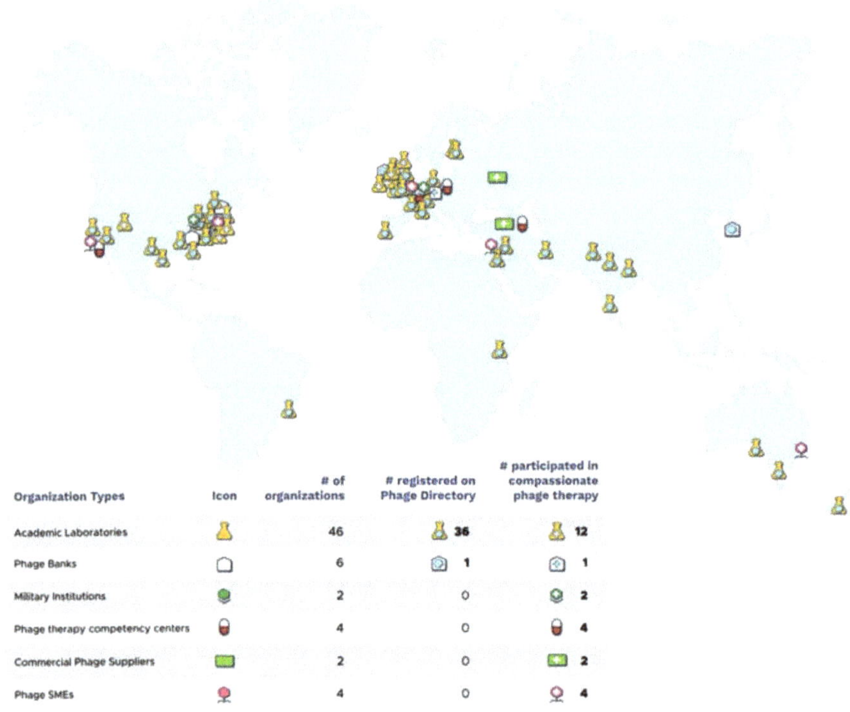

Organization Types	Icon	# of organizations	# registered on Phage Directory	# participated in compassionate phage therapy
Academic Laboratories		46	36	12
Phage Banks		6	1	1
Military Institutions		2	0	2
Phage therapy competency centers		4	0	4
Commercial Phage Suppliers		2	0	2
Phage SMEs		4	0	4

Figure 2. Geographic distribution of organizations (grouped by type) that have either previously participated in cPT cases or demonstrated intent to do so in the future through registration with Phage Directory (numbers current as of January 2019). Phage organizations not having yet contributed to cPT are not listed here. SME: Small- and medium-sized enterprises.

4.6. *Logistical Constraints*

Phage sharing still requires the shipping of bacterial strains and/or phages to various locations across the world; often phage biotechs or phage banks prefer bacterial strains to be shipped to them for sensitivity testing, whereas academics have been more willing to send phages directly to other researchers. From a regulatory point of view, shipping phages does not raise biosafety concerns. However, shipping pathogenic bacteria does and is subject to pathogen transport regulations regarding labeling, packaging, and documentation. In either situation, this step for cPT is time-consuming and expensive, and represents a point of intervention for simplifying cPT. The ability to centralize stock of phages available for cPT or to perform on-site susceptibility testing would reduce costs, standardize susceptibility testing, and reduce time-to-treatment for cPT and clinical trials alike.

5. Beyond Availability

Active phages are indeed indispensable for cPT, but several subsequent considerations need to be addressed in order to assure a sound therapy. How phages are transported, amplified, purified, and formulated into clinically-applicable formulations remains variable between cPT cases. These processes require the oversight of, and close collaboration between, competent phage scientists and physicians. Disorganization is a risk factor for errors to arise throughout this process, and measures must be taken so as not to compromise to the integrity of the phage product and subsequent therapy. Verifying phage viability, compatibility with medical devices (such as tubing or nebulizers), and sustained activity against a patient's infection throughout treatment are not systematically included for cPT, although they are important factors for achieving intended therapeutic benefits. Compassionate use is not subject to consistent procedures, and the employment of non-standardized methods for sensitivity testing, purification, or formulation could contribute to variable treatment outcomes that may be difficult to explain without thorough analysis. Phage therapy walks a thin wire for retaining support to becoming part of modern medicine due to the historical hangover of inconsistent observations in early trials, which continues to cast doubt on the potential of phage therapy today (reviewed in [69–71]). Another risk for cPT treatment is that candidate patients often have confounding medical conditions that complicate prognosis, although phage administration has never been linked to cause of mortality [44,56]. These considerations make the effort to ensure that cPT is consistent and cooperative all the more important.

However well-coordinated these processes become, the cost of providing cPT treatment is a constraint on its scalability. Financial estimates for production costs and manpower needed on a per-case basis are difficult to come by, but have ranged in the tens of thousands of US dollars in countries where phage therapy does not have a legal framework (personal communications). cPT is currently provided at no cost to the patient or treating institution in cases of cPT in the US, France, or Australia; a model with little financial viability for either small biotechs or research labs. However, cPT does not represent an avenue for commercialization. As clinical trials open, it is thought that more patients will be able to access treatments through expanded access schemes or even through participation in ongoing trials. The most scalable option is indeed to obtain marketing authorizations for phage products, which, in a catch-22 situation, does little to address the issue of the current medical need for cPT now.

On a final note, inconsistent, incomplete, or a lack of cPT reporting altogether is a missed opportunity for gaining a better understanding of the antibacterial activity of phages in humans and for further developing human phage therapy. The last phrase of Article 37 iterates the importance of recording information gleaned from compassionate use cases and making it publicly available [18]. However, cPT reporting is frequently neglected or delayed for long periods of time following treatment, with traditional news and social media-based reporting often outpacing peer-reviewed publications. Data gathering has been identified as a problem with compassionate programs [72], which is further complicated when compassionate treatment is provided by multiple sources, as in the case with cPT, instead of a singular manufacturer. A better-structured, data-supported coordination of cPT would enable this treatment option to not only become more widespread and ensure safer practices for patients, but also to provide invaluable information to help refine future phage treatments. The focus of compassionate treatment is unquestionably to provide benefit to the patient, but in consideration of the higher success rate with cPT compared to meager RCT results, it is both wasteful and borderline unethical to not thoroughly record and analyze non-efficacy data from cPT cases, such as doses, frequency, or changes to phage sensitivity profiles. Information including pharmacokinetics, concomitant treatment with antibiotics, and the apparition of phage-resistant variants from cPT would be extremely useful in shaping future phage therapy endeavors and avoiding the clinical futility that has been associated with recent phage RCTs. A detailed set of suggested criteria that phage research and therapy should report has been proposed by Abedon [73]. Here we have presented several generalities that should be addressed for cPT, which then next requires a practical proposal to be formulated, supported, and adhered to by multiple stakeholders for the creation of clear policy and actual implementation.

6. Conclusions

The duration of time until approved alternatives to antibiotics become available is unreassuringly unknown. Traditional drug development pipelines estimate four to ten years for widespread marketing and distribution of any new medicine or therapy, leaving approved phage products something for the future. This substantial lag time between current need and the earliest foreseeable approvals for new antibacterials leaves a considerable number of patients in a highly precarious situation: Reports estimate that approximately 700,000 deaths are caused by antibiotic resistance each year already [74], and claim an even higher number of disability-adjusted life-years and financial burden [75]. The success rates of the cPT cases that have been reported on to date, as well as the willingness of the phage community to participate in cPT efforts for critically-ill patients, emphasizes the potential role that cPT could play in filling this gap between faltering antibiotics and the development of viable alternatives. However, the case reports of cPT over the past decade have addressed only a negligible proportion of antibiotic-resistant cases and remain geographically concentrated around experimental centers or related to a small number of physicians and researchers with the required know-how. The most impactful way to address antibiotic resistance would be to generate efficacy data through clinical trials that would lead to marketing approvals. In the meantime, with better organizing of cPT in terms of phage availability, logistics, and data reporting, progress can be made in the here and now toward alleviating clinical failures due to antibiotic resistance.

Author Contributions: All authors significantly contributed to the formulation, writing, and editing of this manuscript. J.Z. was responsible for the graphical representation of the Phage Directory network.

Acknowledgments: We would like to thank the Cystic Fibrosis Foundation for their support.

Conflicts of Interest: Two authors, Jessica C. Sacher and Jan Zheng, are co-founders of Phage Directory.

References

1. D'Herelle, F. Bacteriophage as a Treatment in Acute Medical and Surgical Infections. *Bull. N. Y. Acad. Med.* **1931**, *7*, 329–348. [PubMed]
2. Summers, W.C. Cholera and plague in India: The bacteriophage inquiry of 1927–1936. *J. Hist. Med. Allied Sci.* **1993**, *48*, 275–301. [CrossRef] [PubMed]
3. D'Hérelle, F. Sur un microbe invisible antagoniste des bacilles dysentériques. *Acad. Sci. Paris* **1917**, *165*, 373–375.
4. Bruynoghe, R.; Maisin, J. Essais de thérapeutique au moyen du bactériophage. *CR Soc. Biol.* **1922**, *85*, 1120–1121.
5. Krueger, A.P.; Scribner, E.J. The Bacteriophage: Its Nature and Therapeutic Use. *JAMA* **1941**, *116*, 2269–2277. [CrossRef]
6. Eaton, M.D.; Bayne-Jones, S. Bacteriophage Therapy: Review of the Principles and Results of the use of Bacteriophages in the Treatment of Infections. *JAMA* **1934**, *103*, 1769–1776. [CrossRef]
7. Parfitt, T. Georgia: An unlikely stronghold for bacteriophage therapy. *Lancet* **2005**, *365*, 2166–2167. [CrossRef]
8. Dublanchet, A. *Autobiographie de Félix d'Hérelle. Les pérégrinations d'un bactériologiste*; Pech, K., Ed.; Lavoisier: Paris, France, 2017; p. 347.
9. Kutateladze, M. Experience of the Eliava Institute in bacteriophage therapy. *Virol. Sin.* **2015**, *30*, 80–81. [CrossRef] [PubMed]
10. Lin, D.M.; Koskella, B.; Lin, H.C. Phage therapy: An alternative to antibiotics in the age of multi-drug resistance. *World J. Gastrointest. Pharmacol. Ther.* **2017**, *8*, 162–173. [CrossRef] [PubMed]
11. McCallin, S.; Brüssow, H. Phage therapy: An alternative or adjunct to antibiotics? *Emerg. Top. Life Sci.* **2017**. [CrossRef]
12. Nobrega, F.L.; Costa, A.R.; Kluskens, L.D.; Azeredo, J. Revisiting phage therapy: new applications for old resources. *Trends Microbiol.* **2015**, *23*, 185–191. [CrossRef] [PubMed]

13. Reardon, S. Phage therapy gets revitalized. *Nature* **2014**, *510*, 15–16. [CrossRef]
14. Kortright, K.E.; Chan, B.K.; Koff, J.L.; Turner, P.E. Phage Therapy: A Renewed Approach to Combat Antibiotic-Resistant Bacteria. *Cell Host Microbe.* **2019**, *25*, 219–232. [CrossRef]
15. Jault, P.; Leclerc, T.; Jennes, S.; Pirnay, J.P.; Que, Y.A.; Resch, G.; Rousseau, A.F.; Ravat, F.; Carsin, H.; Le Floch, R.; et al. Efficacy and tolerability of a cocktail of bacteriophages to treat burn wounds infected by Pseudomonas aeruginosa (PhagoBurn): A randomised, controlled, double-blind phase 1/2 trial. *Lancet Infect. Dis.* **2019**, *19*, 35–45. [CrossRef]
16. Sarker, S.A.; Sultana, S.; Reuteler, G.; Moine, D.; Descombes, P.; Charton, F.; Bourdin, G.; McCallin, S.; Ngom-Bru, C.; Neville, T.; et al. Oral Phage Therapy of Acute Bacterial Diarrhea With Two Coliphage Preparations: A Randomized Trial in Children From Bangladesh. *EBioMedicine* **2016**, *4*, 124–137. [CrossRef]
17. Wright, A.; Hawkins, C.H.; Anggard, E.E.; Harper, D.R. A controlled clinical trial of a therapeutic bacteriophage preparation in chronic otitis due to antibiotic-resistant Pseudomonas aeruginosa; a preliminary report of efficacy. *Clin. Otolaryngol.* **2009**, *34*, 349–357. [CrossRef]
18. World Medical, A. World Medical Association Declaration of Helsinki: ethical principles for medical research involving human subjects. *JAMA* **2013**, *310*, 2191–2194. [CrossRef]
19. Carlson, R.V.; Boyd, K.M.; Webb, D.J. The revision of the Declaration of Helsinki: past, present and future. *Br. J. Clin. Pharmacol.* **2004**, *57*, 695–713. [CrossRef] [PubMed]
20. Balasubramanian, G.; Morampudi, S.; Chhabra, P.; Gowda, A.; Zomorodi, B. An overview of Compassionate Use Programs in the European Union member states. *Intractable Rare Dis. Res.* **2016**, *5*, 244–254. [CrossRef]
21. Jarow, J.P.; Lurie, P.; Ikenberry, S.C.; Lemery, S. Overview of FDA's Expanded Access Program for Investigational Drugs. *Ther. Innov. Regul. Sci.* **2017**, *51*, 177–179. [CrossRef]
22. Donovan, P. Access to unregistered drugs in Australia. *Aust. Prescr.* **2017**, *40*, 194–196. [CrossRef]
23. Borysowski, J.; Ehni, H.J.; Gorski, A. Ethics review in compassionate use. *BMC Med.* **2017**, *15*, 136. [CrossRef]
24. Bunnik, E.M.; Aarts, N.; van de Vathorst, S. The changing landscape of expanded access to investigational drugs for patients with unmet medical needs: Ethical implications. *J. Pharm. Policy Pract.* **2017**, *10*, 10. [CrossRef]
25. Holbein, M.E.; Berglund, J.P.; Weatherwax, K.; Gerber, D.E.; Adamo, J.E. Access to Investigational Drugs: FDA Expanded Access Programs or "Right-to-Try" Legislation? *Clin. Transl. Sci.* **2015**, *8*, 526–532. [CrossRef] [PubMed]
26. Gaffney, A. FDA Sees 92% Increase in Requests for Compassionate Access to Medicines in 2014. Regulatory Focus 2015. Available online: http://www.raps.org/Regulatory-Focus/News/2015/01/23/21151/Compassionate-Use-Requests-Increase-92-Percent/# (accessed on 5 January 2019).
27. Mackey, T.K.; Schoenfeld, V.J. Going "social" to access experimental and potentially life-saving treatment: an assessment of the policy and online patient advocacy environment for expanded access. *BMC Med.* **2016**, *14*, 17. [CrossRef] [PubMed]
28. Joffe, S.; Lynch, H.F. Federal Right-to-Try Legislation—Threatening the FDA's Public Health Mission. *N. Engl. J. Med.* **2018**, *378*, 695–697. [CrossRef]
29. Corporation, A.B. AmpliPhi to Collaborate with Western Sydney Local Health District and Westmead Institute for Medical Research on Expanded Access for Investigational Bacteriophage Therapeutics AB-SA01 and AB-PA01. San Diego, CA, 2018. Available online: https://investor.ampliphibio.com/press-release/featured/ampliphi-collaborate-western-sydney-local-health-district-and-westmead (accessed on 5 January 2019).
30. Corporation, A.B. AmpliPhi Biosciences Announces Presentation of Positive Clinical Data from Its Expanded Access Program for Serious Saureus Infections at IDWeek 2018 Conference. San Diego, CA, 2018. Available online: https://investor.ampliphibio.com/press-release/featured/ampliphi-biosciences-announces-presentation-positive-clinical-data-its (accessed on 5 January 2019).
31. Sacher, J.; Zheng, J.; McCallin, S. Sourcing phages for compassionate use. *Microbiol. Aust.* **2019**. [CrossRef]
32. Fadlallah, A.; Chelala, E.; Legeais, J.-M.M. Corneal Infection Therapy with Topical Bacteriophage Administration. *Open Ophthalmol. J.* **2015**, *9*, 167–168. [CrossRef]
33. Patey, O.; McCallin, S.; Mazure, H.; Liddle, M.; Smithyman, A.; Dublanchet, A. Clinical Indications and Compassionate Use of Phage Therapy: Personal Experience and Literature Review with a Focus on Osteoarticular Infections. *Viruses* **2018**, *11*. [CrossRef] [PubMed]
34. Zhvania, P.; Hoyle, N.S.; Nadareishvili, L.; Nizharadze, D.; Kutateladze, M. Phage Therapy in a 16-Year-Old Boy with Netherton Syndrome. *Front. Med. (Lausanne)* **2017**, *4*, 94. [CrossRef] [PubMed]

35. Leszczyński, P.; Weber-Dabrowska, B.; Kohutnicka, K.M.; uzcak, M.; Górecki, A.; Górski, A. Successful eradication of methicillin-resistant Staphylococcus aureus (MRSA) intestinal carrier status in a healthcare worker—Case report. *Folia. Microbiol.* **2006**, *51*, 336–338. [CrossRef]
36. Letkiewicz, S.; Międzybrodzki, R.; Fortuna, W.; Weber-Dąbrowska, B.; Górski, A. Eradication of Enterococcus faecalis by phage therapy in chronic bacterial prostatitis—case report. *FEMS Immunol. Med. Microbiol.* **2010**, *54*, 457–461. [CrossRef]
37. Weber-Dabrowska, B.; Mulczyk, M.; Gorski, A. Bacteriophage therapy of bacterial infections: an update of our institute's experience. *Arch. Immunol. Ther. Exp. (Warsz)* **2000**, *48*, 547–551. [PubMed]
38. Weber-Dabrowska, B.; Mulczyk, M.; Gorski, A. Bacteriophage therapy for infections in cancer patients. *Clin. Appl. Immunol. Rev.* **2001**, *1*, 4. [CrossRef]
39. Weber-Dabrowska, B.; Mulczyk, M.; Gorski, A. Bacteriophages as an efficient therapy for antibiotic-resistant septicemia in man. *Transplant. Proc.* **2003**, *35*, 1385–1386. [CrossRef]
40. Miedzybrodzki, R.; Borysowski, J.; Weber-Dabrowska, B.; Fortuna, W.; Letkiewicz, S.; Szufnarowski, K.; Pawelczyk, Z.; Rogoz, P.; Klak, M.; Wojtasik, E.; et al. Clinical aspects of phage therapy. *Adv. Virus Res.* **2012**, *83*, 73–121. [CrossRef]
41. Center for Innovative Phage Applications and Therapeutics. Available online: http://ipath.ucsd.edu/ (accessed on 5 January 2019).
42. Pirnay, J.P.; Verbeken, G.; Ceyssens, P.J.; Huys, I.; De Vos, D.; Ameloot, C.; Fauconnier, A. The Magistral Phage. *Viruses* **2018**, *10*. [CrossRef] [PubMed]
43. Chan, B.K.; Turner, P.E.; Kim, S.; Mojibian, H.R.; Elefteriades, J.A.; Narayan, D. Phage treatment of an aortic graft infected with Pseudomonas aeruginosa. *Evol. Med. Public Health* **2018**, *2018*, 60–66. [CrossRef]
44. Duplessis, C.; Biswas, B.; Hanisch, B.; Perkins, M.; Henry, M.; Quinones, J.; Wolfe, D.; Estrella, L.; Hamilton, T. Refractory Pseudomonas Bacteremia in a 2-Year-Old Sterilized by Bacteriophage Therapy. *J. Pediatric. Infect. Dis. Soc.* **2017**, *7*, 253–256. [CrossRef] [PubMed]
45. Fish, R.; Kutter, E.; Bryan, D.; Wheat, G.; Kuhl, S. Resolving Digital Staphylococcal Osteomyelitis Using Bacteriophage—A Case Report. *Antibiotics (Basel)* **2018**, *7*. [CrossRef] [PubMed]
46. Fish, R.; Kutter, E.; Wheat, G.; Blasdel, B.; Kutateladze, M.; Kuhl, S. Compassionate Use of Bacteriophage Therapy for Foot Ulcer Treatment as an Effective Step for Moving Toward Clinical Trials. *Methods Mol. Biol.* **2018**, *1693*, 159–170. [CrossRef] [PubMed]
47. Khawaldeh, A.; Morales, S.; Dillon, B.; Alavidze, Z.; Ginn, A.N.; Thomas, L.; Chapman, S.J.; Dublanchet, A.; Smithyman, A.; Iredell, J.R. Bacteriophage therapy for refractory Pseudomonas aeruginosa urinary tract infection. *J. Med. Microbiol.* **2011**, *60*, 1697–1700. [CrossRef]
48. Corporation, A.B. AmpliPhi Biosciences Announces First Intravenous Treatment of a Patient with AB-SA01 Targeting Staphylococcus aureus. San Diego, USA. 2017. Available online: https://www.ampliphibio.com/ampliphi-biosciences-announces-first-intravenous-treatment-of-a-patient-with-ab-sa01-targeting-staphylococcus-aureus/ (accessed on 5 January 2019).
49. Nir-Paz, R.; Gelman, D.; Khouri, A.; Sisson, B.M.; Fackler, J.; Alkalay-Oren, S.; Khalifa, L.; Rimon, A.; Yerushalmy, O.; Bader, R.; et al. Successful treatment of antibiotic resistant poly-microbial bone infection with bacteriophages and antibiotics combination. *Clin. Infect. Dis.* **2019**. [CrossRef]
50. Ujmajuridze, A.; Chanishvili, N.; Goderdzishvili, M.; Leitner, L.; Mehnert, U.; Chkhotua, A.; Kessler, T.M.; Sybesma, W. Adapted Bacteriophages for Treating Urinary Tract Infections. *Front. Microbiol.* **2018**, *9*, 1832. [CrossRef] [PubMed]
51. Hoyle, N.; Zhvaniya, P.; Balarjishvili, N.; Bolkvadze, D.; Nadareishvili, L.; Nizharadze, D.; Wittmann, J.; Rohde, C.; Kutateladze, M. Phage therapy against Achromobacter xylosoxidans lung infection in a patient with cystic fibrosis: a case report. *Res. Microbiol.* **2018**, *169*, 540–542. [CrossRef] [PubMed]
52. Aslam, S.; Yung, J.; Dan, S.; Reed, S.; LeFebvre, M.; Logan, C.; Taplitz, R.; Law, N.; Golts, E.; Afshar, S.; et al. Bacteriophage Treatment in a Lung Transplant Recipient. *J. Heart Lung Transplant.* **2018**, *37*, S155–S156. [CrossRef]
53. Ferry, T.; Boucher, F.; Fevre, C.; Perpoint, T.; Chateau, J.; Petitjean, C.; Josse, J.; Chidiac, C.; L'Hostis, G.; Leboucher, G.; et al. Innovations for the treatment of a complex bone and joint infection due to XDR Pseudomonas aeruginosa including local application of a selected cocktail of bacteriophages. *J. Antimicrob. Chemother.* **2018**, *73*, 2901–2903. [CrossRef] [PubMed]

54. Ferry, T.; Leboucher, G.; Fevre, C.; Herry, Y.; Conrad, A.; Josse, J.; Batailler, C.; Chidiac, C.; Medina, M.; Lustig, S.; et al. Salvage Debridement, Antibiotics and Implant Retention ("DAIR") With Local Injection of a Selected Cocktail of Bacteriophages: Is It an Option for an Elderly Patient With Relapsing Staphylococcus aureus Prosthetic-Joint Infection? *Open Forum. Infect. Dis.* **2018**, *5*, ofy269. [CrossRef]
55. Morozova, V.V.; Vlassov, V.V.; Tikunova, N.V. Applications of Bacteriophages in the Treatment of Localized Infections in Humans. *Front. Microbiol.* **2018**, *9*, 1696. [CrossRef]
56. LaVergne, S.; Hamilton, T.; Biswas, B.; Kumaraswamy, M.; Schooley, R.T.; Wooten, D. Phage Therapy for a Multidrug-Resistant Acinetobacter baumannii Craniectomy Site Infection. *Open Forum. Infect. Dis.* **2018**, *5*, ofy064. [CrossRef]
57. Schooley, R.T.; Biswas, B.; Gill, J.J.; Hernandez-Morales, A.; Lancaster, J.; Lessor, L.; Barr, J.J.; Reed, S.L.; Rohwer, F.; Benler, S.; et al. Development and Use of Personalized Bacteriophage-Based Therapeutic Cocktails To Treat a Patient with a Disseminated Resistant Acinetobacter baumannii Infection. *Antimicrob. Agents Chemother.* **2017**, *61*, 17. [CrossRef] [PubMed]
58. Jennes, S.; Merabishvili, M.; Soentjens, P.; Pang, K.; Rose, T.; Keersebilck, E.; Soete, O.; François, P.-M.; Teodorescu, S.; Verween, G.; et al. Use of bacteriophages in the treatment of colistin-only-sensitive Pseudomonas aeruginosa septicaemia in a patient with acute kidney injury—a case report. *Critical. Care* **2017**, *21*, 129. [CrossRef]
59. Fish, R.; Kutter, E.; Wheat, G.; Blasdel, B.; Kutateladze, M.; Kuhl, S. Bacteriophage treatment of intransigent diabetic toe ulcers: A case series. *J. Wound Care* **2016**, *25* (Suppl. 7), S27–33. [CrossRef]
60. Rose, T.; Verbeken, G.; Vos, D.D.; Merabishvili, M.; Vaneechoutte, M.; Lavigne, R.; Jennes, S.; Zizi, M.; Pirnay, J.P. Experimental phage therapy of burn wound infection: difficult first steps. *Int. J. Burns Trauma* **2014**, *4*, 66–73. [PubMed]
61. Marza, J.A.; Soothill, J.S.; Boydell, P.; Collyns, T.A. Multiplication of therapeutically administered bacteriophages in Pseudomonas aeruginosa infected patients. *Burns* **2006**, *32*, 644–646. [CrossRef]
62. Jikia, D.; Chkhaidze, N.; Imedashvili, E.; Mgaloblishvili, I.; Tsitlanadze, G.; Katsarava, R.; Morris, G.J.; Sulakvelidze, A. The use of a novel biodegradable preparation capable of the sustained release of bacteriophages and ciprofloxacin, in the complex treatment of multidrug-resistant Staphylococcus aureus-infected local radiation injuries caused by exposure to Sr90. *Clin. Exp. Dermatol.* **2005**, *30*, 23–26. [CrossRef] [PubMed]
63. Markoishvili, K.; Tsitlanadze, G.; Katsarava, R.; Morris, J.G., Jr.; Sulakvelidze, A. A novel sustained-release matrix based on biodegradable poly(ester amide)s and impregnated with bacteriophages and an antibiotic shows promise in management of infected venous stasis ulcers and other poorly healing wounds. *Int. J. Dermatol.* **2002**, *41*, 453–458. [CrossRef]
64. Weber-Dąbrowska, B.; Jończyk-Matysiak, E.; Żaczek, M.; obocka, M.; usiak-Szelachowska, M.; Górski, A. Bacteriophage Procurement for Therapeutic Purposes. *Front. Microbiol.* **2016**, *7*, 1177. [CrossRef]
65. Félix d'Hérelle Reference Center for Bacterial Viruses. Available online: https://www.phage.ulaval.ca/en/home/ (accessed on 5 January 2019).
66. Phages. Available online: https://www.dsmz.de/catalogues/catalogue-microorganisms/groups-of-organisms-and-their-applications/phages.html (accessed on 5 January 2019).
67. The Bacteriophage Bank of Korea. Available online: http://www.phagebank.or.kr/intro/eng_intro.jsp (accessed on 5 January 2019).
68. Phage Directory. Available online: https://phage.directory/ (accessed on 5 January 2019).
69. Fruciano, D.E.; Bourne, S. Phage as an antimicrobial agent: D'Herelle's heretical theories and their role in the decline of phage prophylaxis in the West. *Can. J. Infect. Dis. Med. Microbiol.* **2007**, *18*, 19–26. [CrossRef]
70. Sulakvelidze, A.; Alavidze, Z.; Morris, J.G., Jr. Bacteriophage therapy. *Antimicrob. Agents Chemother.* **2001**, *45*, 649–659. [CrossRef]
71. Summers, W.C. The strange history of phage therapy. *Bacteriophage* **2012**, *2*, 130–133. [CrossRef] [PubMed]
72. Calandra, G.B.; Garelik, J.P.; Kohler, P.T.; Brown, K.R. Problems and benefits of an antibiotic compassionate therapy program. *Rev. Infect. Dis.* **1987**, *9*, 1095–1101. [CrossRef] [PubMed]
73. Abedon, S.T. Information Phage Therapy Research Should Report. *Pharmaceuticals (Basel)* **2017**, *10*. [CrossRef]

74. O'Neill, J. *Tackling Drug-Resistant Infections Globally: Final Report and Recommendations;* Review on Antimicrobial Resistance: London, UK, 2016.
75. Cassini, A.; Högberg, L.D.; Plachouras, D.; Quattrocchi, A.; Hoxha, A.; Simonsen, G.S.; Colomb-Cotinat, M.; Kretzschmar, M.E.; Devleesschauwer, B.; Cecchini, M.; et al. Attributable deaths and disability-adjusted life-years caused by infections with antibiotic-resistant bacteria in the EU and the European Economic Area in 2015: A population-level modelling analysis. *Lancet Infect Dis. 19*, 56–66. [CrossRef]

Review

Towards Inhaled Phage Therapy in Western Europe

Sandra-Maria Wienhold [1], Jasmin Lienau [1] and Martin Witzenrath [1,2,*]

[1] Division of Pulmonary Inflammation, Charité–Universitätsmedizin Berlin, Freie Universität Berlin, Humboldt-Universität zu Berlin, and Berlin Institute of Health, 10117 Berlin, Germany; sandra.wienhold@charite.de (S.-M.W.); jasmin.lienau@charite.de (J.L.)

[2] Department of Infectious Diseases and Respiratory Medicine, Charité–Universitätsmedizin Berlin, Freie Universität Berlin, Humboldt-Universität zu Berlin, and Berlin Institute of Health, 10117 Berlin, Germany

* Correspondence: martin.witzenrath@charite.de; Tel.: +49-30-450-553-122

Received: 15 February 2019; Accepted: 20 March 2019; Published: 23 March 2019

Abstract: The emergence of multidrug-resistant bacteria constitutes a great challenge for modern medicine, recognized by leading medical experts and politicians worldwide. Rediscovery and implementation of bacteriophage therapy by Western medicine might be one solution to the problem of increasing antibiotic failure. In some Eastern European countries phage therapy is used for treating infectious diseases. However, while the European Medicines Agency (EMA) advised that the development of bacteriophage-based therapies should be expedited due to its significant potential, EMA emphasized that phages cannot be recommended for approval before efficacy and safety have been proven by appropriately designed preclinical and clinical trials. More evidence-based data is required, particularly in the areas of pharmacokinetics, repeat applications, immunological reactions to the application of phages as well as the interactions and effects on bacterial biofilms and organ-specific environments. In this brief review we summarize advantages and disadvantages of phage therapy and discuss challenges to the establishment of phage therapy as approved treatment for multidrug-resistant bacteria.

Keywords: bacteriophage; phage therapy; multidrug-resistant bacteria; antimicrobial resistance

---

1. Introduction

Antimicrobial drug resistance (AMR) is a growing challenge worldwide. The emergence of new resistance mechanisms and their broad distribution through vertical and horizontal gene transfer is alarming. Consequently, multidrug-resistant (MDR) bacteria are spreading globally [1]. Due to the lack of a global tracking system, the full impact of infections with MDR bacteria is still unknown. A recent study estimated that approximately 33,000 people died in 2015 in the European Union as consequence of an infection with a resistant pathogen [2]. In the U.S. about 23,000 people die each year due to infections with resistant bacteria and far more people are infected [3]. Besides the medical aspect the socio-economic burden for health care systems is enormous [3]. Previously, the focus was mainly on gram-positive bacteria such as methicillin-resistant *Staphylococcus aureus* (MRSA) or vancomycin-resistant enterococci (VRE) but in recent years gram-negative bacteria resistant against 3 or 4 classes of antimicrobial drugs or even pan-resistant bacteria are rapidly gaining importance. In this respect, particularly noteworthy are *Pseudomonas aeruginosa* and *Acinetobacter baumannii* [4]. Many advanced therapies for cancer or autoimmune diseases, as well as transplantations are no longer effective when patients suffer from untreatable nosocomial infections [5]. Even commensal and opportunistic bacteria could then become problematic and jeopardize medical progress [6]. Moreover, novel antibiotics are rare as the pharmaceutical industry has minimized research and development programs in infectious diseases for different reasons [7]. Meanwhile, as AMR poses a major public health concern, this issue has been discussed at the highest political levels (from the United Nations and the WHO to local authorities). The "Leaders´ Declaration G7 Summit, 7–8 June 2015" (held in Elmau, Germany) stated: "We will foster the prudent use of antibiotics and will engage in stimulating basic research, research on epidemiology, infection prevention and control, and the development of new antibiotics, alternative therapies, vaccines and rapid point-of-care diagnostics" [8].

Already 100 years ago, a decade before the discovery of penicillin, bacteriophages (phages) were considered for clinical use [9]. However, driven by the easy use and broader antibacterial spectrum of antibiotics, phage therapy was seldom used during the last few decades, especially in Western Countries [10]. Only in some countries of the former Soviet Union, such as Georgia and Russia but also in Poland, have physicians continued to use phages and generated valuable practical experience [11]. The recent rediscovery and reintroduction of bacteriophage therapy in the Western World may possibly provide an attractive solution to the increasing failure of antibiotics. Since then, phages have been shown to be effective in treating bacterial infections in several experimental animal studies, as well as in case reports and clinical trials in humans [12]. In *Staphylococcus aureus* induced sepsis in mice for example, intraperitoneal (i.p.) application of phages 6 h after infection resulted in survival rates of 67%, whereas only 10% of control mice survived [13]. Systemic phage lysin application increased survival of mice with severe pneumonia due to *S. pneumoniae* from 0 to 100% [14]. Furthermore, it was also shown that inhaled application of the bacteriophage endolysin Cpl-1 is a safe and efficient therapy in severe pneumococcal pneumonia in mice [15]. So far, lysins seem to be more effective for treating gram-positive bacteria and are currently being tested in clinical trials [16]. Improvement of the enzymes' penetrative abilities through the outer membranes is necessary for their efficient use in gram-negative bacteria [6,17].

In *P. aeruginosa* lung infection, intranasal application of phages, given 24 h prior or 2 h after infection, protected all mice from lethal infection [18]. In diabetic and nondiabetic mice with severe bacteraemia due to i.p. injection of MDR *P. aeruginosa*, a single i.p. injection of phages 20 min after bacterial injection increased survival [19]. The authors reported a survival rate of 90% in diabetic and 100% in nondiabetic mice even when treatment started 4 h after bacterial challenge. Treatment started 6 h after infection resulted in lower survival rates among diabetic mice. Further delay of treatment (12 h) also reduced the effectiveness of phage therapy in nondiabetic mice [19]. This suggests, phage therapy is effective in both immunocompetent and -incompetent mice. In the UK a clinical trial (double-blind placebo-controlled, randomized phase I/II) for treatment of chronic otitis media investigated the effect of a phage-cocktail of 6 phages against MDR *P. aeruginosa*. The study demonstrated physical improvement of the patients and distinctly lower *P. aeruginosa* counts compared to the placebo treated group after a single aural application. Notably, no side effects were reported [20]. These studies along with others indicate phage therapy could be a promising prospect for the treatment of MDR infectious diseases [21]. In laboratory animals, phages can generally be administered via different routes for example, i.p., subcutaneous (s.c.), intramuscular (i.m.), intravenous (i.v.), oral, inhaled or topical [22,23], with the success of phage therapy depending on both the application route and the target organ. After parenteral application, phages are quickly distributed in the systemic circulation [23,24]. McVay et al. [24] investigated different application routes for phage therapy in mice subjected to burn injury and subsequently infected with *P. aeruginosa*. Mice were treated i.m., s.c., i.p. or left untreated. In the untreated group only 6% of mice survived, whereas 28% and 22% of animals survived after i.m. or s.c. phage treatment, respectively. Intraperitoneal application yielded the highest effectiveness, resulting in 88% survival [24]. Oral application was shown to be effective in treating gastrointestinal infections [25,26], whereas topical application was successfully used to treat wound infections [27]. Nebulization of phages for inhaled application to treat lung infections has also been studied [28,29]. Huff et al. [30] reported that chickens, infected with *E. coli* into the thoracic air sac after pre-treatment with aerosolized phages showed significantly reduced mortality compared to untreated birds. However, Carmody et al. [31] demonstrated that intranasal inhalation of phages was less effective when compared to systemic application in a mouse model of lung infection caused by *Burkholderia cenocepacia*. Conversely, Semler et al. [32] observed that inhaled phage therapy was superior to i.p. injection in eliminating *Burkholderia cenocepacia* in murine lung infection.

Phages used for any medical application must be carefully selected and fully characterized [6]. Phages showing poor adsorption, replication and distribution should be excluded and exclusively obligate lytic phages should be applied [33]. Temperate phages may lead to the transfer of genes to the bacterial host, increasing its virulence by lysogenic conversion or transduction mechanisms or transferring virulence factors or antibiotic resistance genes from prophage genomes to the host bacteria [34,35]. The causative bacterial pathogen must be identified prior to phage selection, requiring fast and reliable pathogen detection and susceptibility screening [35]. Alternatively, bacteriophage cocktails including phages against the most common and typical pathogens in specific organs (e.g., "respiratory bacteria") could be employed [6]. In any case, phage therapy specific infrastructure, such as local, rapidly accessible phage libraries need to be established [6,36].

Whereas studies on effective phage therapy have been reported and extensively reviewed, there are hardly any reports on phage therapy failures in recent years. Reports of failures mainly date back to the early use of phage therapy [37]. In 2001, Sulakvelidze et al. [38] published a detailed overview of phage therapy in Eastern European countries starting in the 1920s. The authors stated that failures occur mostly due to poor phage preparations, limited knowledge regarding phage mode of action and inconsistencies between phages and host strains. Additionally, most studies were lacking placebo controls leading to controversial results [38]. Miedzybrodzki et al. [39] published a summary of 153 patients treated with phage therapy to different infections between 2008 and 2010 at the Hirszfeld Institute of Immunology and Experimental Therapy in Wrocław. 39.9% of all patients showed a good response to phage therapy and in 18.3% pathogen eradication and/or recovery was reported [39]. Moreover, there is one clinical trial from Bangladesh using phage cocktails targeting *Escherichia coli* (*E. coli*) in children with bacterial diarrhoea reporting no advantage of phage therapy [40]. Two different phage cocktails were tested and orally applied [40]. Phage therapy did not cause any side effects but also did not improve the clinical outcome compared to the control group receiving standard oral rehydration. The authors reported several limitations of the trial, including probable insensitivity of pathogenic *E. coli* strains to the applied phages and the possibility of low stomach pH affecting phage transport, as no antacid was given to the patients [40].

In this review we summarize the advantages and disadvantages of phage usage in terms of medical application and discuss challenges to the establishment of phage therapy as an approved treatment for MDR bacteria.

2. Advantages and Disadvantages of Phage Therapy

2.1. Phage Therapy Provides Several Advantages Over Conventional Antimicrobial Drugs Regarding Medical Application, Some of Which Are Addressed in the Following

2.1.1. Host Specificity and Potential to Spare Microbial Flora

Lytic bacteriophages are viruses that target and infect their specific host bacteria, replicate inside and destroy them [9]. Therefore, unlike indiscriminate antibiotics, bacteriophages are expected to spare the physiologically resident flora, thus avoiding the development of bacterial niches that typically result from antibiotic therapy and enable the settlement of antibiotic-resistant bacteria [41]. Moreover, an intact microbiome contributes to innate immunity vigilance [10,42,43]. Consequently, local gut immune response to bacterial challenge is dampened by preceding antimicrobial therapy, increasing susceptibility to intestinal colonization and infection with pathogenic microorganisms, including MDR bacteria [41]. Furthermore, disruption of the gut microbiome results in an impaired systemic immune response upon bacterial stimulation and ultimately insufficient bacterial elimination [44,45]. Thus, antibiotics [44] but not phage therapy may possibly compromise immunity through microbiome disruption, paving the way for subsequent MDR and non-MDR infections. However, this conjecture needs to be addressed in upcoming studies. Similarly interaction of resident phages (phageome) with commensals as well as function and dynamics of the phageome have not yet been completely unravelled [46].

2.1.2. Bacterial Phage Resistance

Phage infection and lysis occur independently of mechanisms used by antibiotics to kill bacteria. Therefore, antibiotic resistance does not imply phage resistance [47]. Resistance to phages, however, occurs naturally and to varying degrees in all bacterial cultures and communities. Different mechanisms of resistance development have been described, including phage adsorption to bacterial cell receptors, phage particle assembly in the bacterial cell or cell lysis processes [48,49]. Notably, resources in the environment to isolate new phages are abundant [50]. Thus, isolation of new phages for almost all bacterial species can be achieved quickly for example, by sampling the environment and phage screening overnight (so called phagogram) [6] for subsequent GMP production. Consequently, development of pan-phage resistant bacteria is very unlikely [51]. Interestingly, it has been reported that some bacteria evolving phage resistance might over time regain sensitivity to antibiotics [52,53] or lose their virulence [54,55].

2.1.3. Self-Replication, Self-Limitation and Anti-Biofilm Properties

Since bacteriophages can only target and infect their specific host bacteria, the lysis process is self-limiting [47]. Phages replicate as long as their host is accessible. Consequently, to initialize or continue the lytic cycle, phages need host bacterial cell contact, which is significantly impaired in the case of bacteria forming biofilms, remaining intracellularly or being less abundant. Phages encoding for depolymerases are able to degrade matrix exopolysaccharides of biofilms [56,57]. Consequently, phages and other antimicrobialsmight reach, infect and lyse bacteria inside the biofilm more easily [56,57], which is of particular interest in case of implants (e.g., vascular or joint devices) and airway infections [58–60]. Bedi et al. [61], showed a beneficial effect on the eradication of a biofilm formed by *Klebsiella pneumoniae* when antibiotic and phage were combined. Another study demonstrated that phage OMKO1 was able to reduce bacterial densities in a *P. aeruginosa in vitro* biofilm assay [62]. Moreover, this phage alone or in combination with an antibiotic was more effective in reducing the biofilm than antibiotics alone [62]. Nevertheless, the effectiveness depends on the phage and the host bacteria. Darch et al. [63] examined the ability of two phages to inhibit bacterial dissemination in a model of aggregate formation by *P. aeruginosa*. The two phages were able to kill *P. aeruginosa* and inhibit aggregate formation when applied simultaneously with the bacteria [63]. However, when applied after aggregate formation was already established, the authors did not observe complete elimination of aggregates, most likely due to exopolysaccharide production. Still phage application could prevent formation of new aggregates by proliferating bacteria [63]. Full elimination of a complex mature biofilm with one single phage seems unlikely but phage cocktails and combined therapy with antibiotics could be a potential strategy [56,57,64].

2.2. Some of the Disadvantages of Phage Therapy Are Addressed by the Following Aspects

2.2.1. Activity against Intracellular Pathogens

Phages are unlikely to be able to actively enter eukaryotic cells. Therefore, phages are less effective against intracellular bacteria for example, *Mycobacterium tuberculosis*, as well as against intracellularly-surviving and persistent clones of extracellular bacteria, for example, *A. baumannii* [38].

2.2.2. Liberation of Endotoxins

Although it seems unlikely that therapy with purified phages leads to relevant toxic side effects, major concerns encompass the potentially massive liberation of bacterial endotoxins after bacterial lysis. Similar observations have been made with the use of certain antibiotics [65], as well as immune reactions to bacterial components including endotoxin present in crude phage lysates [66].

Confrontation with large amounts of bacterial endotoxins could lead to clinical deterioration of septic patients [67,68]. However, Dufour et al [69] reported for two different *E. coli* phages fewer released endotoxins in vitro compared to β-lactams, while the phage-evoked endotoxin level was comparable to that evoked by amikacin [69].

2.2.3. Potential Risk of Anaphylaxis

Phages are members of microbial communities and are present in the environment as well as on and in the human body [70,71]. Despite this, therapy with phages requires a higher titre compared to their naturally occurring numbers. Moreover, the use of high phage titres in patients bears the theoretical risk of inducing extreme immune responses like anaphylaxis [38]. Although theoretically possible, anaphylaxis due to phage therapy has never been reported and does not seem to be a major concern in phage therapy [11,47].

2.2.4. Immune Response to Phages

Being composed of proteins and nucleic acids, phages in general are considered as innately non-toxic [47,72]. However, there is evidence for non-specific immunomodulatory characteristics of phages [73], as well as activation of phagocytosis and anti-inflammatory properties [74]. Roach et al. demonstrated that presence of neutrophils is necessary for phage therapy success against *P. aeruginosa* [75]. Moreover, a recent in vivo study revealed that an increased number of phages in the gut (applied via drinking water to mice) can aggravate colitis in a TLR9 and IFN-gamma dependent manner and that phages inside the gut could stimulate non-specific and phage-specific immunity [76].

It is also possible that the human immune system may recognize phages as foreign antigens and produce phage-neutralizing antibodies depending on the application route [77]. In order to minimize the risk of side-effects due to impurities, it is necessary in at least parenteral application routes to use highly purified phage preparations [78]. For a widespread use of human phage therapy according to Western European medical standards, more scientific evidence is needed. In particular further investigation is warranted in the areas of immunological reactions following single or repeated phage application, pharmacokinetics and -dynamics and interaction with bacterial biofilms and commensal flora.

3. Challenges in Clinical Use of Phage Therapy

3.1. Current State of Phage Therapy

Experience with human phage therapy dates back more than 100 years in Georgia, Russia and Poland. However, Jault et al [79] stated that these countries have not developed "evidence-based medical standards" so far and "if there are any, they are only available in Russian" [79]. Especially at the ELIAVA Institute of Bacteriophage in Tbilisi, significant effort has been put into the characterization and development of phage products. Phage cocktails are routinely used for treatment, including prescribed medicine and self-medication (over the counter products) [80,81]. In Poland, phage research is predominantly carried out at the Hirszfeld Institute of Immunology and Experimental Therapy in Wrocław [82]. The institute focuses on preparation of specific phage lysates for individual patients subsequent to identification of causative pathogens from patient's samples. After Poland's accession to the EU, the Phage Therapy Unit, an outpatient clinic working according to EU regulations, was established. In this unit phages are applied under terms of experimental treatment in accordance with the Declaration of Helsinki and Polish regulations [83,84]. Despite years of practical experience, numerous case reports [39,85–87] and data from clinical studies including investigations of immune response [20,88–93], the lack of peer-reviewed controlled clinical trials still renders an evidence-based evaluation of phage therapy by Western standards difficult [84]. Marketing authorization of phage therapy in Western Europe depends on several conditions, including production of phage preparations according to good manufacturing practice (GMP) conditions, the issue of patentability and official approval by European Authorities [35]. In the following section, we will discuss regulatory, production and clinical trial challenges to the establishment of phage therapy as a regulatory approved therapy.

3.2. Regulatory Challenges

The European Medicines Agency (EMA) held a workshop in 2015 together with relevant stakeholders including academia, industry, policy makers and patient organizations to identify possibilities for the development of bacteriophage-based therapies against bacterial infections [94]. Approximately 60 experts discussed practical and regulatory issues related to phage licensing pathways as opposed to conventional medicine, for example, whether or not the EU Directive 2001/83/EC (relating to medicinal products for human use) might be applicable for phages [84,94,95]. As EMA has not licensed any phage products so far, it is not clear which pathway to approval is most promising. Moreover, modification or updating of existing phage cocktails with new phages, necessities with regard to developing phage resistances or changing pathogens, are not yet covered by existing regulations [84]. Hence, currently new time and cost intensive re-production and re-approval under GMP conditions would be required [84]. As phage biology (and bacterial co-evolution) implies the necessity for fast turnover of specific phages in clinical use, the process of development and approval needs to be shortened, which could be achieved by approval of production processes rather than specific phage products. Additionally, it must be clarified whether each phage of a cocktail or the complete cocktail is considered a medicinal product and needs regulatory approval.

3.3. Production Challenges

For a broad medical application, phages have to be produced in large scale by pharmaceutically licensed facilities. Consequently, there will be a commercial interest in the optimization of processes and the reduction of costs, which is indeed not trivial as non-linear dynamics of phages and host bacteria have to be considered [96]. In order to scale up phage manufacturing Krysiak-Baltyn et al. [96] proposed a computational model appropriate for modelling phage production, including varying infection parameters. This model might be suitable to cut costs or to improve productivity [96]. With the increasing interest in phage therapy, the issue of intellectual property (IP) protection comes to the fore [97]. To date the options for IP protection of naturally occurring phages are limited because there are abundant resources to isolate phages from the environment [36]. Therefore, other possibilities should be considered. For example, there are several options to implement IP protection in the manufacturing process, for example, every new phage, new preparation (consisting of approved single phages) or the production method, in order to increase attractiveness of the field to economic stakeholders [39,98,99].

3.4. Clincial Trial Challenges and Ongoing Projects

Recently, results of the first European randomized, controlled phase 1/2 trial aimed at evaluating the efficacy and tolerability of a topical applied phage cocktail (PP1131) against *P. aeruginosa* in burn wounds (PhagoBurn, www.phagoburn.eu) have been published [79]. The authors reported that patients treated with the phage cocktail showed slower decrease of bacterial burden in the burn wounds compared to patients receiving standard therapy (1% sulfadiazine silver emulsion cream). This finding must be interpreted with caution, as the study had several limitations: The patient cohort was relatively small (standard therapy n = 13, PP1131 n = 12) and inhomogeneous, as patients treated with phages were older, were burned to a lesser extent and showed higher bacterial burden at therapy start compared to those treated with standard therapy [79]. Importantly, there have been some stability issues of the phage cocktail used, as the authors reported a decrease in plaque forming units (pfu) during the study, which was associated with the application of a lower than intended dose of active phages. Future studies should address stability and shelf-life of each phage product including phage cocktails. The relatively large cocktail of 12 phages resulted in double the expected production time and thus reduced time to recruit patients [79]. Although the study terminated prematurely due to insufficient efficacy, it was the first trial using a cocktail of phages purified according to GMP standards and approved by national health regulators [79].

The German Phage4Cure (http://phage4cure.de/) consortium aims to address the safety, tolerability and efficacy of a purified inhaled bacteriophage cocktail against chronic airway infection with *P. aeruginosa* [100]. The goal of the four project partners (Fraunhofer Institute for Toxicology and Experimental Medicine, ITEM; the Leibniz Institute DSMZ-German Collection of Microorganisms and Cell Cultures GmbH; Charité–Universitätsmedizin Berlin; and Charité Research Organisation GmbH, CRO) is to pave the way for clinical applications of bacteriophages in Germany and Western Europe by applying GMP standards in the entire production chain of the phage product and by getting approval for phage therapy from regulatory authorities [100]. The project gained governmental financial support (funded by the German Federal Ministry of Education and Research) and regulatory authorities, namely BfArM (Federal Institute for Drugs and Medical Devices), are closely involved [100]. The Phage4Cure team is focusing on bacteriophages targeting *P. aeruginosa*, which is characterized by high abundance, rapid growth, distinct ability to form biofilms and a highly flexible genome, aspects that contribute its wide distribution and difficulties in combatting this bacterium [101]. *P. aeruginosa* is intrinsically resistant to several classes of antibiotics and there is increasing evidence of strains resistant to antibiotics of last resort [101]. Immunocompromised patients and patients with pre-injured lungs, particularly those with cystic fibrosis (CF) are frequently colonized by bacteria with 80% of CF patients older than 18 years harbouring *P. aeruginosa* [102]. CF is a genetic disorder caused by a mutation in the cystic fibrosis transmembrane conductance regulator (CFTR) gene that affects the lungs, as well as the pancreas, liver, kidneys and intestine [103]. Non-cystic fibrosis (non-CF) bronchiectasis represents a chronic and heterogeneous airway disease with diverse aetiology. Like patients with CF, non-CF bronchiectasis patients are highly susceptible to pulmonary infections. Most patients with bronchiectasis are colonized with antibiotic resistant bacteria (e.g., *Acinetobacter, Pseudomonas, Burkholderia*), with *P. aeruginosa* infections being associated with poor prognosis [104–108]. As patients with bacterial colonization but not infection are clinically relatively stable, effectiveness and tolerability of inhaled phage therapy may well be tested. Consequently, *P. aeruginosa* colonization in CF and non-CF bronchiectasis patients was chosen by the consortium Phage4Cure as therapeutic target to pave the way to a clinical trial with phages produced under GMP conditions and, ultimately, to regulatory approval. On the basis of their expertise, each partner will work on different aspects of the project, ranging from phage selection and characterization (DSMZ; as discussed by Korf et al. in this issue of the journal) to drug production and stability testing (ITEM). Following phage production and thorough preclinical evaluation (Charité and ITEM), clinical phase 1 and 2 trials are planned to be performed at Charité. If successful, the project may lead to first-time establishment of phages as approved inhaled therapy for CF and non-CF bronchiectasis patients in Germany and may possibly provide a GMP-compliant phage purification platform process as a blueprint for phage therapy development with respect to other indications [6,100]. Further projects related to the implementation of phage therapy in Western Europe include PhagoMed and PhagoFlow. The biotech company PhagoMed Biopharma GmbH (https://www.phagomed.com/), based in Vienna, focuses on the development of phage-based therapies for bacterial infections [109]. Supported by grants and private investments, PhagoMed evaluates, inter alia, the treatment of infected prostheses with phages. PhagoFlow aims at testing a magistral prescription of phages in patients with wounds infected by MDR bacteria and is being carried out at the military hospital Berlin, together with DSMZ and Fraunhofer ITEM [110]. Magistral preparation in the EU is defined as "any medicinal product prepared in a pharmacy in accordance with a medical prescription for an individual patient" (Article 3 of Directive 2001/83 and Article 6 quarter, § 3 of the Law of 25 March 1964) [111] and is therefore a practical way to produce treatments adjusted to the special needs of an individual without being dependent on commercial manufacturing [111]. In Belgium, the magistral preparation is already used for phages [111] and could provide a solution for individual patients but cannot cover the requirements of a larger patient cohort. GMP and GCP guidelines also apply for magistral applications. To promote a European solution to conquer the regulatory challenges of personalized and phage based medicinal products, the idea of a "biological master file", a concept already existing for chemical drugs but not for biologically active substances, was suggested by Fauconnier [112]. In summary, the main prerequisites for the establishment of a bacteriophage-based therapy are specific regulations for phage-based pharmaceuticals, increased clinical trial evidence and an infrastructure for efficient and rapid phage provision [6,36,113].

After decades of sleeping like Rip van Winkle, phage therapy is currently awakening in Western Europe due to the noise made by antimicrobial resistance, causing relevant research activity in the field. New valuable data addressing current concerns regarding clinical use of phages can be expected. However, whether phages will be approved by regulatory authorities and get market access is currently unpredictable.

Author Contributions: S.M.W., J.L. and M.W. wrote the manuscript. All authors have approved the submitted version.

Funding: This work was supported by the German Research Foundation (SFB-TR84 projects C6, C9 to MW) and by the German Federal Ministry of Education and Research (e:Med CAPSyS-FKZ 01ZX1604B and Phage4Cure-FKZ 16GW0141 to MW).

Acknowledgments: The authors thank all partners involved in the Phage4Cure consortium for fruitful discussion and Alexander Taylor for proofreading the manuscript. We acknowledge the support of the German Research Foundation (DFG) and the Open Access Publication Fund of Charité–Universitätsmedizin Berlin.

Conflicts of Interest: The authors declare no conflict of interest.

References

1. Alanis, A.J. Resistance to antibiotics: Are we in the post-antibiotic era? *Arch. Med. Res.* **2005**, *36*, 697–705. [CrossRef] [PubMed]
2. Cassini, A.; Högberg, L.D.; Plachouras, D.; Quattrocchi, A.; Hoxha, A.; Simonsen, G.S.; Colomb-Cotinat, M.; Kretzschmar, M.E.; Devleesschauwer, B.; Cecchini, M.; et al. Attributable deaths and disability-adjusted life-years caused by infections with antibiotic-resistant bacteria in the EU and the European Economic Area in 2015: A population-level modelling analysis. *Lancet Infect. Dis.* **2019**, *19*, 56–66. [CrossRef]
3. Antibiotic/Antimicrobial Resistance (AR/AMR). Available online: https://www.cdc.gov/drugresistance/ (accessed on 14 February 2019).
4. Souli, M.; Galani, I.; Giamarellou, H. Emergence of extensively drug-resistant and pandrug-resistant Gram-negative bacilli in Europe. *Euro Surveill* **2008**, *13*, 19045. [PubMed]
5. Reardon, S. WHO Warns Against 'Post-Antibiotic' Era. 2014. Available online: https://www.nature.com/news/who-warns-against-post-antibiotic-era-1.15135 (accessed on 10 March 2019).
6. Rohde, C.; Wittmann, J.; Kutter, E. Bacteriophages: A Therapy Concept against Multi-Drug-Resistant Bacteria. *Surg. Infect. (Larchmt)* **2018**, *19*, 737–744. [CrossRef] [PubMed]
7. Norrby, S.R.; Nord, C.E.; Finch, R. Lack of development of new antimicrobial drugs: A potential serious threat to public health. *Lancet Infect. Dis.* **2005**, *5*, 115–119. [CrossRef]
8. Leaders' Declaration G7 Summit, 7–8 June 2015. Available online: https://www.bundesregierung.de/breg-de/service/datenschutzhinweis/g7-abschlusserklaerung-und-weitere-dokumente-387344 (accessed on 14 February 2019).
9. D'Herelle, F.; Smith. *G.H. The Bacteriophage and Its Behavior*; by F. d'Herelle Translated by George H. Smith; The Williams & Wilkins Company: Baltimore, MD, USA, 1926.
10. Duckworth, D.H.; Gulig, P.A. Bacteriophages: Potential treatment for bacterial infections. *BioDrugs* **2002**, *16*, 57–62. [CrossRef]
11. Kutateladze, M.; Adamia, R. Bacteriophages as potential new therapeutics to replace or supplement antibiotics. *Trends Biotechnol.* **2010**, *28*, 591–595. [CrossRef] [PubMed]
12. Kortright, K.E.; Chan, B.K.; Koff, J.L.; Turner, P.E. Phage Therapy: A Renewed Approach to Combat Antibiotic-Resistant Bacteria. *Cell Host Microbe* **2019**, *25*, 219–232. [CrossRef]
13. Takemura-Uchiyama, I.; Uchiyama, J.; Osanai, M.; Morimoto, N.; Asagiri, T.; Ujihara, T.; Daibata, M.; Sugiura, T.; Matsuzaki, S. Experimental phage therapy against lethal lung-derived septicemia caused by Staphylococcus aureus in mice. *Microbes Infect.* **2014**, *16*, 512–517. [CrossRef] [PubMed]

14. Witzenrath, M.; Schmeck, B.; Doehn, J.M.; Tschernig, T.; Zahlten, J.; Loeffler, J.M.; Zemlin, M.; Müller, H.; Gutbier, B.; Schütte, H.; et al. Systemic use of the endolysin Cpl-1 rescues mice with fatal pneumococcal pneumonia. *Crit. Care Med.* **2009**, *37*, 642–649. [CrossRef]
15. Doehn, J.M.; Fischer, K.; Reppe, K.; Gutbier, B.; Tschernig, T.; Hocke, A.C.; Fischetti, V.A.; Löffler, J.; Suttorp, N.; Hippenstiel, S.; et al. Delivery of the endolysin Cpl-1 by inhalation rescues mice with fatal pneumococcal pneumonia. *J. Antimicrob. Chemother.* **2013**, *68*, 2111–2117. [CrossRef]
16. Fischetti, V.A. Development of Phage Lysins as Novel Therapeutics: A Historical Perspective. *Viruses* **2018**, *10*, 310. [CrossRef] [PubMed]
17. Gerstmans, H.; Rodríguez-Rubio, L.; Lavigne, R.; Briers, Y. From endolysins to Artilysin®s: Novel enzyme-based approaches to kill drug-resistant bacteria. *Biochem. Soc. Trans.* **2016**, *44*, 123–128. [CrossRef] [PubMed]
18. Debarbieux, L.; Leduc, D.; Maura, D.; Morello, E.; Criscuolo, A.; Grossi, O.; Balloy, V.; Touqui, L. Bacteriophages can treat and prevent Pseudomonas aeruginosa lung infections. *J. Infect. Dis.* **2010**, *201*, 1096–1104. [CrossRef] [PubMed]
19. Shivshetty, N.; Hosamani, R.; Ahmed, L.; Oli, A.K.; Sannauallah, S.; Sharanbassappa, S.; Patil, S.A.; Kelmani, C.R. Experimental protection of diabetic mice against Lethal P. aeruginosa infection by bacteriophage. *Biomed. Res. Int.* **2014**, *2014*, 793242. [CrossRef] [PubMed]
20. Wright, A.; Hawkins, C.H.; Anggård, E.E.; Harper, D.R. A controlled clinical trial of a therapeutic bacteriophage preparation in chronic otitis due to antibiotic-resistant Pseudomonas aeruginosa; a preliminary report of efficacy. *Clin. Otolaryngol.* **2009**, *34*, 349–357. [CrossRef]
21. Parracho, H.M.; Burrowes, B.H.; Enright, M.C.; McConville, M.L.; Harper, D.R. The role of regulated clinical trials in the development of bacteriophage therapeutics. *J. Mol. Genet. Med.* **2012**, *6*, 279–286. [CrossRef]
22. Ryan, E.M.; Gorman, S.P.; Donnelly, R.F.; Gilmore, B.F. Recent advances in bacteriophage therapy: How delivery routes, formulation, concentration and timing influence the success of phage therapy. *J. Pharm. Pharmacol.* **2011**, *63*, 1253–1264. [CrossRef]
23. Qadir, M.I.; Mobeen, T.; Masood, A. Phage therapy: Progress in pharmacokinetics. *Braz. J. Pharm. Sci.* **2018**, *54*, 66. [CrossRef]
24. McVay, C.S.; Velásquez, M.; Fralick, J.A. Phage therapy of Pseudomonas aeruginosa infection in a mouse burn wound model. *Antimicrob. Agents Chemother.* **2007**, *51*, 1934–1938. [CrossRef]
25. Zhao, J.; Liu, Y.; Xiao, C.; He, S.; Yao, H.; Bao, G. Efficacy of Phage Therapy in Controlling Rabbit Colibacillosis and Changes in Cecal Microbiota. *Front. Microbiol.* **2017**, *8*, 957. [CrossRef] [PubMed]
26. Wagenaar, J.A.; van Bergen, M.A.P.; Mueller, M.A.; Wassenaar, T.M.; Carlton, R.M. Phage therapy reduces Campylobacter jejuni colonization in broilers. *Vet. Microbiol.* **2005**, *109*, 275–283. [CrossRef] [PubMed]
27. Kumari, S.; Harjai, K.; Chhibber, S. Topical treatment of Klebsiella pneumoniae B5055 induced burn wound infection in mice using natural products. *J. Infect. Dev. Ctries.* **2010**, *4*, 367–377.
28. Golshahi, L.; Seed, K.D.; Dennis, J.J.; Finlay, W.H. Toward modern inhalational bacteriophage therapy: Nebulization of bacteriophages of Burkholderia cepacia complex. *J. Aerosol Med. Pulm. Drug Deliv.* **2008**, *21*, 351–360. [CrossRef]
29. Turgeon, N.; Toulouse, M.-J.; Martel, B.; Moineau, S.; Duchaine, C. Comparison of five bacteriophages as models for viral aerosol studies. *Appl. Environ. Microbiol.* **2014**, *80*, 4242–4250. [CrossRef] [PubMed]
30. Huff, W.E.; Huff, G.R.; Rath, N.C.; Balog, J.M.; Donoghue, A.M. Prevention of Escherichia coli infection in broiler chickens with a bacteriophage aerosol spray. *Poult. Sci.* **2002**, *81*, 1486–1491. [CrossRef] [PubMed]
31. Carmody, L.A.; Gill, J.J.; Summer, E.J.; Sajjan, U.S.; Gonzalez, C.F.; Young, R.F.; LiPuma, J.J. Efficacy of bacteriophage therapy in a model of Burkholderia cenocepacia pulmonary infection. *J. Infect. Dis.* **2010**, *201*, 264–271. [CrossRef]
32. Semler, D.D.; Goudie, A.D.; Finlay, W.H.; Dennis, J.J. Aerosol phage therapy efficacy in Burkholderia cepacia complex respiratory infections. *Antimicrob. Agents Chemother.* **2014**, *58*, 4005–4013. [CrossRef] [PubMed]
33. Abedon, S.T.; Thomas-Abedon, C. Phage therapy pharmacology. *Curr. Pharm. Biotechnol.* **2010**, *11*, 28–47. [CrossRef]
34. Nobrega, F.L.; Costa, A.R.; Kluskens, L.D.; Azeredo, J. Revisiting phage therapy: New applications for old resources. *Trends Microbiol.* **2015**, *23*, 185–191. [CrossRef]
35. Pelfrene, E.; Willebrand, E.; Cavaleiro Sanches, A.; Sebris, Z.; Cavaleri, M. Bacteriophage therapy: A regulatory perspective. *J. Antimicrob. Chemother.* **2016**, *71*, 2071–2074. [CrossRef] [PubMed]

36. Sybesma, W.; Rohde, C.; Bardy, P.; Pirnay, J.-P.; Cooper, I.; Caplin, J.; Chanishvili, N.; Coffey, A.; de Vos, D.; Scholz, A.H.; et al. Silk Route to the Acceptance and Re-Implementation of Bacteriophage Therapy-Part II. *Antibiotics* **2018**, *7*, 35. [CrossRef]
37. Smith, J. The bacteriophage in the treatment of typhoid fever. *BMJ* **1924**, *2*, 47–49. [CrossRef] [PubMed]
38. Sulakvelidze, A.; Alavidze, Z.; Morris, J.G. Bacteriophage therapy. *Antimicrob. Agents Chemother.* **2001**, *45*, 649–659. [CrossRef] [PubMed]
39. Międzybrodzki, R.; Borysowski, J.; Weber-Dąbrowska, B.; Fortuna, W.; Letkiewicz, S.; Szufnarowski, K.; Pawełczyk, Z.; Rogóż, P.; Kłak, M.; Wojtasik, E.; et al. Clinical aspects of phage therapy. *Adv. Virus Res.* **2012**, *83*, 73–121. [PubMed]
40. Sarker, S.A.; Sultana, S.; Reuteler, G.; Moine, D.; Descombes, P.; Charton, F.; Bourdin, G.; McCallin, S.; Ngom-Bru, C.; Neville, T.; et al. Oral Phage Therapy of Acute Bacterial Diarrhea With Two Coliphage Preparations: A Randomized Trial in Children from Bangladesh. *EBioMedicine* **2016**, *4*, 124–137. [CrossRef]
41. Buffie, C.G.; Pamer, E.G. Microbiota-mediated colonization resistance against intestinal pathogens. *Nat. Rev. Immunol.* **2013**, *13*, 790–801. [CrossRef]
42. Maynard, C.L.; Elson, C.O.; Hatton, R.D.; Weaver, C.T. Reciprocal interactions of the intestinal microbiota and immune system. *Nature* **2012**, *489*, 231–241. [CrossRef] [PubMed]
43. Marsland, B.J.; Gollwitzer, E.S. Host-microorganism interactions in lung diseases. *Nat. Rev. Immunol.* **2014**, *14*, 827–835. [CrossRef]
44. Robak, O.H.; Heimesaat, M.M.; Kruglov, A.A.; Prepens, S.; Ninnemann, J.; Gutbier, B.; Reppe, K.; Hochrein, H.; Suter, M.; Kirschning, C.J.; et al. Antibiotic treatment-induced secondary IgA deficiency enhances susceptibility to Pseudomonas aeruginosa pneumonia. *J. Clin. Investig.* **2018**, *128*, 3535–3545. [CrossRef] [PubMed]
45. Belkaid, Y.; Hand, T.W. Role of the microbiota in immunity and inflammation. *Cell* **2014**, *157*, 121–141. [CrossRef]
46. Shkoporov, A.N.; Hill, C. Bacteriophages of the Human Gut: The "Known Unknown" of the Microbiome. *Cell Host Microbe* **2019**, *25*, 195–209. [CrossRef] [PubMed]
47. Loc-Carrillo, C.; Abedon, S.T. Pros and cons of phage therapy. *Bacteriophage* **2011**, *1*, 111–114. [CrossRef]
48. Labrie, S.J.; Samson, J.E.; Moineau, S. Bacteriophage resistance mechanisms. *Nat. Rev. Microbiol.* **2010**, *8*, 317–327. [CrossRef] [PubMed]
49. Rostøl, J.T.; Marraffini, L. (Ph)ighting Phages: How Bacteria Resist Their Parasites. *Cell Host Microbe* **2019**, *25*, 184–194. [CrossRef]
50. Haq, I.U.; Chaudhry, W.N.; Akhtar, M.N.; Andleeb, S.; Qadri, I. Bacteriophages and their implications on future biotechnology: A review. *Virol. J.* **2012**, *9*, 9. [CrossRef] [PubMed]
51. Ormälä, A.-M.; Jalasvuori, M. Phage therapy: Should bacterial resistance to phages be a concern, even in the long run? *Bacteriophage* **2013**, *3*, e24219. [CrossRef] [PubMed]
52. Rohde, C.; Resch, G.; Pirnay, J.-P.; Blasdel, B.G.; Debarbieux, L.; Gelman, D.; Górski, A.; Hazan, R.; Huys, I.; Kakabadze, E.; et al. Expert Opinion on Three Phage Therapy Related Topics: Bacterial Phage Resistance, Phage Training and Prophages in Bacterial Production Strains. *Viruses* **2018**, *10*, 178. [CrossRef] [PubMed]
53. Chan, B.K.; Sistrom, M.; Wertz, J.E.; Kortright, K.E.; Narayan, D.; Turner, P.E. Phage selection restores antibiotic sensitivity in MDR Pseudomonas aeruginosa. *Sci. Rep.* **2016**, *6*, 26717. [CrossRef]
54. Smith, H.W.; Huggins, M.B.; Shaw, K.M. The control of experimental Escherichia coli diarrhoea in calves by means of bacteriophages. *J. Gen. Microbiol.* **1987**, *133*, 1111–1126. [CrossRef]
55. Holst Sørensen, M.C.; van Alphen, L.B.; Fodor, C.; Crowley, S.M.; Christensen, B.B.; Szymanski, C.M.; Brøndsted, L. Phase variable expression of capsular polysaccharide modifications allows Campylobacter jejuni to avoid bacteriophage infection in chickens. *Front. Cell. Infect. Microbiol.* **2012**, *2*, 11. [PubMed]
56. Abedon, S.T. Ecology of Anti-Biofilm Agents I: Antibiotics versus Bacteriophages. *Pharmaceuticals* **2015**, *8*, 525–558. [CrossRef]
57. Abedon, S.T. Bacteriophage exploitation of bacterial biofilms: Phage preference for less mature targets? *FEMS Microbiol. Lett.* **2016**, *363*, fnv246. [CrossRef]
58. Curtin, J.J.; Donlan, R.M. Using bacteriophages to reduce formation of catheter-associated biofilms by Staphylococcus epidermidis. *Antimicrob. Agents Chemother.* **2006**, *50*, 1268–1275. [CrossRef] [PubMed]

59. Kwiatek, M.; Parasion, S.; Rutyna, P.; Mizak, L.; Gryko, R.; Niemcewicz, M.; Olender, A.; Łobocka, M. Isolation of bacteriophages and their application to control Pseudomonas aeruginosa in planktonic and biofilm models. *Res. Microbiol.* **2017**, *168*, 194–207. [CrossRef]
60. Fong, S.A.; Drilling, A.; Morales, S.; Cornet, M.E.; Woodworth, B.A.; Fokkens, W.J.; Psaltis, A.J.; Vreugde, S.; Wormald, P.-J. Activity of Bacteriophages in Removing Biofilms of Pseudomonas aeruginosa Isolates from Chronic Rhinosinusitis Patients. *Front. Cell. Infect. Microbiol.* **2017**, *7*, 418. [CrossRef] [PubMed]
61. Bedi, M.S.; Verma, V.; Chhibber, S. Amoxicillin and specific bacteriophage can be used together for eradication of biofilm of Klebsiella pneumoniae B5055. *World J. Microbiol. Biotechnol.* **2009**, *25*, 1145–1151. [CrossRef]
62. Chan, B.K.; Turner, P.E.; Kim, S.; Mojibian, H.R.; Elefteriades, J.A.; Narayan, D. Phage treatment of an aortic graft infected with Pseudomonas aeruginosa. *Evol. Med. Public Health* **2018**, *2018*, 60–66. [CrossRef] [PubMed]
63. Darch, S.E.; Kragh, K.N.; Abbott, E.A.; Bjarnsholt, T.; Bull, J.J.; Whiteley, M. Phage Inhibit Pathogen Dissemination by Targeting Bacterial Migrants in a Chronic Infection Model. *mBio* **2017**, *8*, e00240-17. [CrossRef]
64. Chaudhry, W.N.; Concepción-Acevedo, J.; Park, T.; Andleeb, S.; Bull, J.J.; Levin, B.R. Synergy and Order Effects of Antibiotics and Phages in Killing Pseudomonas aeruginosa Biofilms. *PLoS ONE* **2017**, *12*, e0168615. [CrossRef]
65. Holzheimer, R.G. Antibiotic induced endotoxin release and clinical sepsis: A review. *J. Chemother.* **2001**, *13* (Suppl. 4), 159–172. [CrossRef] [PubMed]
66. Skurnik, M.; Pajunen, M.; Kiljunen, S. Biotechnological challenges of phage therapy. *Biotechnol. Lett.* **2007**, *29*, 995–1003. [CrossRef] [PubMed]
67. Mignon, F.; Piagnerelli, M.; van Nuffelen, M.; Vincent, J.L. Effect of empiric antibiotic treatment on plasma endotoxin activity in septic patients. *Infection* **2014**, *42*, 521–528. [CrossRef]
68. Peng, Z.-Y.; Wang, H.-Z.; Srisawat, N.; Wen, X.; Rimmelé, T.; Bishop, J.; Singbartl, K.; Murugan, R.; Kellum, J.A. Bactericidal antibiotics temporarily increase inflammation and worsen acute kidney injury in experimental sepsis. *Crit. Care Med.* **2012**, *40*, 538–543. [CrossRef] [PubMed]
69. Dufour, N.; Delattre, R.; Ricard, J.-D.; Debarbieux, L. The Lysis of Pathogenic Escherichia coli by Bacteriophages Releases Less Endotoxin Than by β-Lactams. *Clin. Infect. Dis.* **2017**, *64*, 1582–1588. [CrossRef] [PubMed]
70. Breitbart, M.; Haynes, M.; Kelley, S.; Angly, F.; Edwards, R.A.; Felts, B.; Mahaffy, J.M.; Mueller, J.; Nulton, J.; Rayhawk, S.; et al. Viral diversity and dynamics in an infant gut. *Res. Microbiol.* **2008**, *159*, 367–373. [CrossRef]
71. Manrique, P.; Bolduc, B.; Walk, S.T.; van der Oost, J.; de Vos, W.M.; Young, M.J. Healthy human gut phageome. *Proc. Natl. Acad. Sci. USA* **2016**, *113*, 10400–10405. [CrossRef] [PubMed]
72. Kutter, E.; de Vos, D.; Gvasalia, G.; Alavidze, Z.; Gogokhia, L.; Kuhl, S.; Abedon, S.T. Phage therapy in clinical practice: Treatment of human infections. *Curr. Pharm. Biotechnol.* **2010**, *11*, 69–86. [CrossRef]
73. Górski, A.; Międzybrodzki, R.; Borysowski, J.; Dąbrowska, K.; Wierzbicki, P.; Ohams, M.; Korczak-Kowalska, G.; Olszowska-Zaremba, N.; Łusiak-Szelachowska, M.; Kłak, M.; et al. Phage as a modulator of immune responses: Practical implications for phage therapy. *Adv. Virus Res.* **2012**, *83*, 41–71. [PubMed]
74. Górski, A.; Jończyk-Matysiak, E.; Łusiak-Szelachowska, M.; Międzybrodzki, R.; Weber-Dąbrowska, B.; Borysowski, J. The Potential of Phage Therapy in Sepsis. *Front. Immunol.* **2017**, *8*, 1783. [CrossRef]
75. Roach, D.R.; Leung, C.Y.; Henry, M.; Morello, E.; Singh, D.; Di Santo, J.P.; Weitz, J.S.; Debarbieux, L. Synergy between the Host Immune System and Bacteriophage Is Essential for Successful Phage Therapy against an Acute Respiratory Pathogen. *Cell Host Microbe* **2017**, *22*, 38–47.e4. [CrossRef] [PubMed]
76. Gogokhia, L.; Buhrke, K.; Bell, R.; Hoffman, B.; Brown, D.G.; Hanke-Gogokhia, C.; Ajami, N.J.; Wong, M.C.; Ghazaryan, A.; Valentine, J.F.; et al. Expansion of Bacteriophages Is Linked to Aggravated Intestinal Inflammation and Colitis. *Cell Host Microbe* **2019**, *25*, 285–299.e8. [CrossRef] [PubMed]
77. Łusiak-Szelachowska, M.; Zaczek, M.; Weber-Dąbrowska, B.; Międzybrodzki, R.; Kłak, M.; Fortuna, W.; Letkiewicz, S.; Rogóż, P.; Szufnarowski, K.; Jończyk-Matysiak, E.; et al. Phage neutralization by sera of patients receiving phage therapy. *Viral Immunol.* **2014**, *27*, 295–304. [CrossRef]
78. Chan, B.K.; Abedon, S.T.; Loc-Carrillo, C. Phage cocktails and the future of phage therapy. *Future Microbiol.* **2013**, *8*, 769–783. [CrossRef]

79. Jault, P.; Leclerc, T.; Jennes, S.; Pirnay, J.P.; Que, Y.-A.; Resch, G.; Rousseau, A.F.; Ravat, F.; Carsin, H.; Le Floch, R.; et al. Efficacy and tolerability of a cocktail of bacteriophages to treat burn wounds infected by Pseudomonas aeruginosa (PhagoBurn): A randomised, controlled, double-blind phase 1/2 trial. *Lancet Infect. Dis.* **2019**, *19*, 35–45. [CrossRef]
80. Kutateladze, M.; Adamia, R. Phage therapy experience at the Eliava Institute. *Med. Mal. Infect.* **2008**, *38*, 426–430. [CrossRef] [PubMed]
81. Kutateladze, M. Experience of the Eliava Institute in bacteriophage therapy. *Virol. Sin.* **2015**, *30*, 80–81. [CrossRef]
82. Weber-Dąbrowska, B.; Jończyk-Matysiak, E.; Żaczek, M.; Łobocka, M.; Łusiak-Szelachowska, M.; Górski, A. Bacteriophage Procurement for Therapeutic Purposes. *Front. Microbiol.* **2016**, *7*, 1177. [CrossRef] [PubMed]
83. Phage Therapy Unit of the Medical Centre of the Institute of Immunology and Experimental Therapy PAS. Available online: https://www.iitd.pan.wroc.pl/en/OTF/ (accessed on 14 February 2019).
84. Hill, C.; Mills, S.; Ross, R.P. Phages & antibiotic resistance: Are the most abundant entities on earth ready for a comeback? *Future Microbiol.* **2018**, *13*, 711–726. [PubMed]
85. Chanishvili, N. *A Literature Review of the Practical Application of Bacteriophage Research*; Nova Biomedical Books: New York, NY, USA, 2012.
86. Hoyle, N.; Zhvaniya, P.; Balarjishvili, N.; Bolkvadze, D.; Nadareishvili, L.; Nizharadze, D.; Wittmann, J.; Rohde, C.; Kutateladze, M. Phage therapy against Achromobacter xylosoxidans lung infection in a patient with cystic fibrosis: A case report. *Res. Microbiol.* **2018**, *169*, 540–542. [CrossRef] [PubMed]
87. Zhvania, P.; Hoyle, N.S.; Nadareishvili, L.; Nizharadze, D.; Kutateladze, M. Phage Therapy in a 16-Year-Old Boy with Netherton Syndrome. *Front. Med. (Lausanne)* **2017**, *4*, 94. [CrossRef] [PubMed]
88. Rhoads, D.D.; Wolcott, R.D.; Kuskowski, M.A.; Wolcott, B.M.; Ward, L.S.; Sulakvelidze, A. Bacteriophage therapy of venous leg ulcers in humans: Results of a phase I safety trial. *J. Wound Care* **2009**, *18*, 237–243. [CrossRef] [PubMed]
89. Weber-Dabrowska, B.; Mulczyk, M.; Górski, A. Bacteriophages as an efficient therapy for antibiotic-resistant septicemia in man. *Transplant. Proc.* **2003**, *35*, 1385–1386. [CrossRef]
90. Markoishvili, K.; Tsitlanadze, G.; Katsarava, R.; Morris, J.G.; Sulakvelidze, A. A novel sustained-release matrix based on biodegradable poly(ester amide)s and impregnated with bacteriophages and an antibiotic shows promise in management of infected venous stasis ulcers and other poorly healing wounds. *Int. J. Dermatol.* **2002**, *41*, 453–458. [CrossRef]
91. Łusiak-Szelachowska, M.; Żaczek, M.; Weber-Dąbrowska, B.; Międzybrodzki, R.; Letkiewicz, S.; Fortuna, W.; Rogóż, P.; Szufnarowski, K.; Jończyk-Matysiak, E.; Olchawa, E.; et al. Antiphage activity of sera during phage therapy in relation to its outcome. *Future Microbiol.* **2017**, *12*, 109–117. [CrossRef]
92. Borysowski, J.; Międzybrodzki, R.; Wierzbicki, P.; Kłosowska, D.; Korczak-Kowalska, G.; Weber-Dąbrowska, B.; Górski, A. A3R Phage and Staphylococcus aureus Lysate Do Not Induce Neutrophil Degranulation. *Viruses* **2017**, *9*, 36. [CrossRef] [PubMed]
93. Międzybrodzki, R.; Borysowski, J.; Kłak, M.; Jończyk-Matysiak, E.; Obmińska-Mrukowicz, B.; Suszko-Pawłowska, A.; Bubak, B.; Weber-Dąbrowska, B.; Górski, A. In Vivo Studies on the Influence of Bacteriophage Preparations on the Autoimmune Inflammatory Process. *Biomed. Res. Int.* **2017**, *2017*, 3612015.
94. European Medicines Agency (EMA). Workshop on the Therapeutic Use of Bacteriophages. 2015. Available online: https://www.ema.europa.eu/documents/other/workshop-therapeutic-use-bacteriophages-summary_en.pdf (accessed on 14 February 2019).
95. European Commission. Directive 2001/83/EC of the European Parliament and of the Council of 6 November 2001 on the Community Code Relating to Medicinal Products for Human Use (consolidated version: 16/11/2012). In EudraLex—The Rules Governing Medicinal Products in the European Union (Volume 1), Pharmaceutical Legislation: Medicinal Products For Human Use. Available online: https://ec.europa.eu/health/sites/health/files/files/eudralex/vol-1/dir_2001_83_consol_2012/dir_2001_83_cons_2012_en.pdf (accessed on 14 February 2019).
96. Krysiak-Baltyn, K.; Martin, G.J.O.; Gras, S.L. Computational Modelling of Large Scale Phage Production Using a Two-Stage Batch Process. *Pharmaceuticals* **2018**, *11*, 31. [CrossRef]
97. Henein, A. What are the limitations on the wider therapeutic use of phage? *Bacteriophage* **2013**, *3*, e24872. [CrossRef] [PubMed]

98. Verbeken, G.; Pirnay, J.-P.; de Vos, D.; Jennes, S.; Zizi, M.; Lavigne, R.; Casteels, M.; Huys, I. Optimizing the European regulatory framework for sustainable bacteriophage therapy in human medicine. *Arch. Immunol. Ther. Exp. (Warsz)* **2012**, *60*, 161–172. [CrossRef] [PubMed]
99. Pirnay, J.-P.; de Vos, D.; Verbeken, G.; Merabishvili, M.; Chanishvili, N.; Vaneechoutte, M.; Zizi, M.; Laire, G.; Lavigne, R.; Huys, I.; et al. The phage therapy paradigm: Prêt-à-porter or sur-mesure? *Pharm. Res.* **2011**, *28*, 934–937. [CrossRef]
100. Bacteriophages Join the Fight Again Infection, Research Alliance Launched, Aimed at Establishing Bacteriophages as an Approved Drug. Press Release 09.11.2017. Available online: https://www.charite.de/service/pressemitteilung/artikel/detail/bakteriophagen_als_arzneimittel_im_kampf_gegen_infektionen/ (accessed on 14 February 2019).
101. Poole, K. Pseudomonas aeruginosa: Resistance to the max. *Front. Microbiol.* **2011**, *2*, 65. [CrossRef] [PubMed]
102. Pressler, T.; Bohmova, C.; Conway, S.; Dumcius, S.; Hjelte, L.; Høiby, N.; Kollberg, H.; Tümmler, B.; Vavrova, V. Chronic Pseudomonas aeruginosa infection definition: EuroCareCF Working Group report. *J. Cyst. Fibros.* **2011**, *10* (Suppl. 2), S75–S78. [CrossRef]
103. O'Sullivan, B.P.; Freedman, S.D. Cystic fibrosis. *Lancet* **2009**, *373*, 1891–1904. [CrossRef]
104. Tunney, M.M.; Einarsson, G.G.; Wei, L.; Drain, M.; Klem, E.R.; Cardwell, C.; Ennis, M.; Boucher, R.C.; Wolfgang, M.C.; Elborn, J.S. Lung microbiota and bacterial abundance in patients with bronchiectasis when clinically stable and during exacerbation. *Am. J. Respir. Crit. Care Med.* **2013**, *187*, 1118–1126. [CrossRef] [PubMed]
105. Rogers, G.B.; van der Gast, C.J.; Cuthbertson, L.; Thomson, S.K.; Bruce, K.D.; Martin, M.L.; Serisier, D.J. Clinical measures of disease in adult non-CF bronchiectasis correlate with airway microbiota composition. *Thorax* **2013**, *68*, 731–737. [CrossRef] [PubMed]
106. King, P.T.; Holdsworth, S.R.; Freezer, N.J.; Villanueva, E.; Holmes, P.W. Microbiologic follow-up study in adult bronchiectasis. *Respir. Med.* **2007**, *101*, 1633–1638. [CrossRef] [PubMed]
107. Angrill, J.; Agustí, C.; de Celis, R.; Rañó, A.; Gonzalez, J.; Solé, T.; Xaubet, A.; Rodriguez-Roisin, R.; Torres, A. Bacterial colonisation in patients with bronchiectasis: Microbiological pattern and risk factors. *Thorax* **2002**, *57*, 15–19. [CrossRef]
108. McDonnell, M.J.; Jary, H.R.; Perry, A.; MacFarlane, J.G.; Hester, K.L.M.; Small, T.; Molyneux, C.; Perry, J.D.; Walton, K.E.; de Soyza, A. Non cystic fibrosis bronchiectasis: A longitudinal retrospective observational cohort study of Pseudomonas persistence and resistance. *Respir. Med.* **2015**, *109*, 716–726. [CrossRef] [PubMed]
109. PhagoMed. Available online: http://www.phagomed.com/2018/08/28/viruses-against-bacteria/ (accessed on 14 February 2019).
110. Geförderte Projekte des Innovationsausschusses zur Förderbekanntmachung Versorgungsforschung vom 20.Oktober 2017, PhagoFlow (page 25). Available online: https://innovationsfonds.g-ba.de/downloads/media/112/Liste-gefoerderter-Projekte-VSF-FBK_20-10-2017.pdf (accessed on 14 February 2019).
111. Pirnay, J.-P.; Verbeken, G.; Ceyssens, P.-J.; Huys, I.; de Vos, D.; Ameloot, C.; Fauconnier, A. The Magistral Phage. *Viruses* **2018**, *10*, 64. [CrossRef]
112. Fauconnier, A. Regulating phage therapy: The biological master file concept could help to overcome regulatory challenge of personalized medicines. *EMBO Rep.* **2017**, *18*, 198–200. [CrossRef] [PubMed]
113. Debarbieux, L.; Pirnay, J.-P.; Verbeken, G.; de Vos, D.; Merabishvili, M.; Huys, I.; Patey, O.; Schoonjans, D.; Vaneechoutte, M.; Zizi, M.; et al. A bacteriophage journey at the European Medicines Agency. *FEMS Microbiol. Lett.* **2016**, *363*, fnv225. [CrossRef] [PubMed]

Review

Interactions between Bacteriophage, Bacteria, and the Mammalian Immune System

Jonas D. Van Belleghem [1,2,*], Krystyna Dąbrowska [3], Mario Vaneechoutte [1], Jeremy J. Barr [4] and Paul L. Bollyky [2]

1 Laboratory Bacteriology Research, Department of Clinical Chemistry, Microbiology and Immunology, Ghent University, 9000 Ghent, Belgium; Mario.vaneechoutte@ugent.be
2 Division of Infectious Diseases and Geographic Medicine, Department of Medicine, Stanford University School of Medicine, Stanford, CA 94305, USA; pbollyky@stanford.edu
3 Bacteriophage Laboratory, Institute of Immunology and Experimental Therapy, Polish Academy of Sciences, 53-114 Wrocław, Poland; dabrok@iitd.pan.wroc.pl
4 School of Biological Sciences, Monash University, Melbourne, VIC 3800, Australia; jeremybarr85@gmail.com
* Correspondence: Van.belleghem.jonas@gmail.com; Tel.: +(650)-723-1831

Received: 4 December 2018; Accepted: 21 December 2018; Published: 25 December 2018

Abstract: The human body is host to large numbers of bacteriophages (phages)–a diverse group of bacterial viruses that infect bacteria. Phage were previously regarded as bystanders that only impacted immunity indirectly via effects on the mammalian microbiome. However, it has become clear that phages also impact immunity directly, in ways that are typically anti-inflammatory. Phages can modulate innate immunity via phagocytosis and cytokine responses, but also impact adaptive immunity via effects on antibody production and effector polarization. Phages may thereby have profound effects on the outcome of bacterial infections by modulating the immune response. In this review we highlight the diverse ways in which phages interact with human cells. We present a computational model for predicting these complex and dynamic interactions. These models predict that the phageome may play important roles in shaping mammalian-bacterial interactions.

Keywords: bacteriophage; immunology; innate immunity; adaptive immunity; human host; phage-human host interaction

---

1. Introduction

Commensal microorganisms colonize and live in symbiosis with the human body and encompass diverse phyla from the three domains of life: Eukarya, Archaea, and Bacteria. Body surfaces that are in direct contact with the environment, including the intestine, skin, urogenital tract, and upper respiratory tract harbor most of these microorganisms. The bacterial component of the human microbiota and its associated genes have been a primary focus of research efforts over the past two decades [1–3]. These efforts have yielded a wealth of insight about the composition of human-associated bacterial communities, how these resident bacteria interact with the immune system and how bacterial-immune system interactions are altered in disease [1,4,5].

The microbiota of healthy humans also includes a large number of bacterial viruses, or bacteriophages (phages) [6]. Phages were previously regarded as bystanders that only impacted immunity indirectly via effects on the mammalian microbiome. However, it is becoming clearer that phages also impact immunity directly.

In this review we highlight the diverse ways in which phages interact with human cells: [1] effect of phages in the mammalian interface, [2] innate immune response, and [3] the adaptive immune response against the phages. We then present a computational model for predicting these complex and dynamic interactions. This model predicts that our phageome may play important roles in shaping mammalian-bacterial interactions, underlying the important effect of phage induced anti-inflammatory properties. Finally, the gaps in our knowledge and potential future lines of investigation are highlighted.

2. The Human Phageome

Phages colonize all body niches, including the skin [7,8], oral cavity [9–11], lungs [12–14], gut [15,16], and urinary tract [17]. However, phages are frequently overlooked in microbiome and metagenomic studies and their role is often unclear. Most phages present in these viromes are temperate phages that can integrate their DNA into the bacterial genomes (i.e., prophage) or be present as episomes, and as such can alter the phenotype of the host bacteria by lysogenic conversion [16,18]. Although human blood is considered to be sterile, metagenomic analysis has shown the presence of a viral community, most of which belonged to the *Myoviridae, Podoviridae, Siphoviridae, Microviridae,* and *Inoviridae* families [19–22]. Once present in the blood, these phages may interact with immune cells and induce innate and adaptive immune responses [23–26].

Of all the microbial communities within the body, the intestinal community is by far the most complex and dense. The human gut microbiome, as shown by metagenomic studies, includes many viral genes (the virome) [15,16,27,28]. Approximately 90% of the gut virome consists of phages [29], estimated at 10^9 viruses per gram of feces [30,31]. As new members of the bacterial community are introduced, the phage populations in the intestine diversify, suggesting that phage diversity and bacterial diversity are linked [32]. Furthermore, this relationship is very dynamic in infants and stabilizes in adults [33]. Although there is less variation of intestinal phage populations within individuals over time, there is substantial variation between individuals, even when those individuals have similar bacterial community structures [15,16].

Phages can supply bacteria with genes that are involved in toxin, polysaccharide, and carbohydrate metabolism, and, in rare cases, they represent a source of antibiotic resistance [34,35]. Some phages can modulate bacterial antigenicity through the production of enzymes capable of modifying the O-antigen component of LPS in microorganisms such as *Escherichia* coli, Salmonella spp., *Shigella* spp., and *Vibrio cholerae* [36–39].

It is thus important to consider whether phage interactions with commensal bacteria could alter community compositions in ways that impact the function of the immune system and influence the spread of pathogenic viruses, or even bacteria [1,40–42]. Among the mechanisms responsible for the recognition of microbial and viral structures are the Toll-like receptors (TLR) [43]. These TLR are able to recognize Pathogen Associated Molecular Patterns (PAMPs) (e.g., LPS, flagellin, or unmethylated CpG-DNA). Viral nucleic acids can be recognized by multiple TLR, notably TLR9 recognizes DNA, whereas TLR7 and 8 recognize ssRNA and TLR3 recognizes dsRNA [44–46]. These nucleic acid-sensing TLRs have the potential to promote, amongst others, the production of Type I IFN.

The virome continuously stimulates low-level immune responses without causing any overt symptoms [47,48]. Duerkop and Hooper hypothesized that commensal bacteriophages could activate one or more innate immune pathways, thereby stimulating antiviral immune responses and continuously inducing low cytokine production. These cytokines also exert their action on non-immune cells and may continuously induce inflammatory processes, thereby conferring constant protection against pathogenic viral infections [1,49].

It is clear that phages are omnipresent and form a major constituent of many microbiomes, nevertheless the interactions of phages with their human host warrants further research.

3. Phages Effects on the Bacterial - Mammalian Host Interface

3.1. Phages and Mucosal Tissues

Phages interact with host immunity at the mucosal surface. The mucosal surface (e.g., the human gut and respiratory tract) represents a critical immunological and physiological barrier within all animals that both protects against invading bacterial pathogens while also supporting large communities of commensal microorganisms [50,51]. The mucosal surface is predominantly composed of mucin glycoproteins that are secreted by the underlying epithelium. By offering both structure and nutrients, mucus layers influence the composition of the microbiota and select for commensal symbionts [52–54]. It has been shown that mucosal surfaces of the gut commonly support more abundant and stable bacterial populations than the surrounding environments (e.g., the luminal content of the gut) [55,56]. This is, in part, due to the degradation of mucins by gut microbes, but also in part due to host epithelial secretions that selectively shape the commensal microbiota [53,54,57]. These host secretions are diverse and can include antimicrobials, such as alpha-defensin and RegIIIγ [58,59]. Conversely, when mucosal surfaces are invaded by pathogenic bacterial species, the epithelium may respond by increasing the production of antimicrobial agents, hypersecretion of mucin, or alteration of mucin glycosylation patterns in an attempt to subvert microbial attachment and to increase physical removal of the invading bacterial species [60–62].

These mucosal layers also harbor large and diverse communities of phages (Figure 1A). Mucus-associated phage communities are significantly enriched compared to the surrounding non-mucosal environment [63]. Investigations across diverse mucosal surfaces ranging from those present in corals, fish, mice, and humans revealed an average 4.4-fold increase in phage numbers in mucus relative to bacterial cells [63–65]. This increase in phage abundance happens through an adherence mechanism whereby phages weakly bind mucin glycoproteins via immunoglobulin-like (Ig-like) protein domains displayed on their capsids. The Ig-like fold is one of the most common and widely dispersed in nature, present in antibodies and T-cell receptors where it mediates important binding interactions of the human adaptive immune system [66,67]. These Ig-like domains are found within approximately one quarter of sequenced *Caudovirales* genomes, and are typically displayed on the virion surface [68,69]. Most of these structurally displayed Ig-like domains are dispensable for phage growth in the laboratory, which led to the hypothesis that they aid the phage in the adsorption to their bacterial host under environmental conditions [68,70]. Phages that utilized Ig-like domains, which effectively bind to the mucus layer, would be under positive selection within the mucosa, leading to the proposal of a bacteriophage adherence to mucus (BAM) model as a non-host-derived layer of immunity, mediated by phages [63,71].

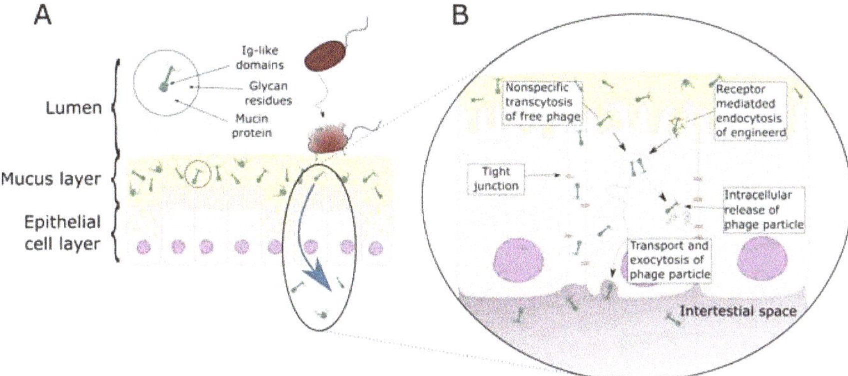

Figure 1. Schematic representation of direct interaction of phages with mammalian cells. (**A**) Bacteriophage adhering to mucus (BAM). Mucus is produced by the underlying epithelium. Phages of different morphologies (i.e., Myo-, Sipho-, and Podoviridae) can bind variable glycan residues displayed on mucin glycoproteins through variable capsid proteins, such as Ig-like domains. The adherence of phages to this mucus layer creates an antimicrobial layer that reduces bacterial attachment to and colonization of the mucus. This leads, in turn, to a reduction in epithelial cell death. Furthermore, these phages can migrate through theses epithelial cell layers subsequently ending up in the bloodstream. (**B**) Phage transcytosis. Binding interactions between phages and the membrane through transmembrane mucins, specific receptors, or through non-specific recognition, may allow signal transduction in the epithelial cell. Subsequently the phage particle is taken up by the epithelial cell. The internalized phage particles may be degraded leading to intracellular release of phage particles and DNA. Furthermore, it has been hypothesized that phage particles might cross the eukaryotic cell enabling phages to disseminate to the body. Phages may also gain access to the body via a "leaky gut", where they bypass the epithelial cell barrier at sites of cellular damage or punctured vasculature. Figure adapted from Barr et al. [63,72].

On top of their direct effects on bacterial populations, phages can also have an indirect effect on the colonization of their bacterial host to mammalian cells. In case of *Neisseria meningitidis* it has been shown that its filamentous phage (MDA]φ) increases its host-cell colonization [73]. The authors showed that the presence of this filamentous phage leads to a higher binding of the bacteria to the host epithelial cells. Furthermore, the phage also seemed to form a linker between the bacteria, further heightening its colonization. These effects were not observed for endothelial cells, indicating a specificity of the phage towards epithelial cells. In this case it is the phage itself that forms an additional virulence factor to the bacteria, promoting bacterial aggregation.

It can be further hypothesized whether there is a mutual benefit to phage and bacteria, whereby the phage interacts with the mucosal surface and binds the bacteria. Instead of infecting and lysing the bacteria, the phage would provide the bacteria with additional binding sites, thus, elevating the colonization frequency.

3.2. Phage Transcytosis

Below the mucosal surfaces, the cellular epithelium forms another physical barrier that separates the heavily colonized mucosa from the normally sterile regions of the body. Due to their ubiquity within the epithelial mucus layer, phages are in constant contact with the epithelial layers. The passage of commensal bacteria colonizing the intestine across the mucosal epithelium to local lymph nodes and internal organs is termed bacterial translocation and is a critical step in the pathology of various disorders [74,75]. While bacterial translocation is a well-described phenomenon, little is known about the translocation of bacterial viruses.

Low internalization of bacteriophages by enterocytes and other endothelial cells was demonstrated for M13 phages (empty vectors used as a control in phage display) in vivo [76] and in vitro [77]. In vitro uptake of phage M13 could be blocked by chloroquine, an inhibitor known to block clathrin-dependent endocytosis, suggesting this was the proposed pathway for internalization [77]. Since this type of endocytosis is strictly receptor-mediated (i.e., external objects must be bound to a membrane receptor to be dragged into the pits), there is reason to think that phage uptake can be a consequence of specific phage-to-epithelium interactions.

In vivo studies of oral administration of non-engineered phages demonstrated both effective [78–82] and ineffective [83–88] systemic dissemination. This demonstrates that natural phage translocation from gut to circulation is possible but suggests a range of other factors may regulate this process, such as physiological status of a host [24,89] and characteristics of the phage. To some extent, physical parameters of phage particles, like their size and shape, may influence the phage's ability to penetrate mammalian bodies. However, the most important factor seems to be the dose, which correlates strongly with the probability that an orally applied phage can be found in circulation or in tissues. This is in line with the fact that phages may differ in their ability to propagate on gut bacteria and this ability may further limit their systemic dissemination after application *per os* [86,90].

An important consideration regarding the translocation of orally administered phages is whether phages can cross the mucosal barrier in sufficient numbers to subsequently interact with and bypass the cellular epithelium. Recently, it has been demonstrated that phages can enter and cross epithelial cell layers by a non-specific transcytosis mechanism [91]. Phage-epithelial transcytosis seems to preferentially occur in an apical-to-basal direction and was shown to occur across different types of epithelial cell layers (e.g., gut, lung, liver, kidney, and brain cells) and for diverse phage types and morphologies (e.g., *Myoviridae*, *Siphoviridae*, and *Podoviridae*; Figure 1B). Microscopy revealed that roughly 10% of epithelial cells endocytosed phage particles, which appeared to be localized within membrane-bound vesicles. Interestingly, those few cells that did endocytose phage particles appeared to contain large numbers of such vesicles. Chemical inhibitor assays suggest that, once endocytosed, phage particles traffic via the Golgi apparatus before being functionally exocytosed at the basal cell layer. The transcytosis of phages across epithelial cell layers provides a mechanistic explanation for the systemic occurrence of phages within the human body in the absence of disease [91]. Contrary to these observations, others have observed the accumulation of phagocytosed phages near the cell nucleus of MAC-T cells [92]. The presence of phages close to or in the nucleus reassess the question as to whether phages might be able to have their genome replicated or translated. Furthermore, these data raise the question of whether the production of phage derived RNA induces cellular responses or whether the presence of the phage close to or in the nucleus have an effect on the cellular function of the phage "infected" mammalian cell.

4. Cell Perfusion and Access, Interaction with Intracellular Immune Response

The penetration of phages in higher organisms leads to direct contact of phages with eukaryotic cells. Therefore, it is important to know whether these phages can interact with or infect eukaryotic cells. Infection seems unlikely, because elements of the phage tail structure only bind to specific receptors on the surfaces of their target bacteria. Furthermore, it is generally recognized that phages cannot infect eukaryotic cells, because of major differences between eukaryotes and prokaryotes in regard to key intracellular machinery that are essential for translation and replication [93]. This was illustrated by Di Giovine et al. [94], who re-engineered the filamentous phage M13 to infect mammalian cells. Although subsequent binding and internalization of the engineered phage was observed, no multiplication of the phage was detected [94]. Further engineering of filamentous phages has shown the potential of these phages to produced RNAs in eukaryotic cells after their uptake [95,96]. Although most of these systems made use of eukaryotic gene promoters to drive transcription, these data demonstrate the potential for phage derived nucleic acids to be recognized by eukaryotic cellular pathways, including TLR and other induced (viral) immune responses.

Infection aside, it is feasible that phages can directly interact with eukaryotic cells, either extra- or intra-cellularly. Nguyen et al. [91] performed cellular fractionation of epithelial cells that had been incubated with phages and showed complete perfusion of the eukaryotic cell, with phage particles seen within all endomembrane compartments. From here, phage particles are likely degraded, shuffled, and transported throughout the cell, providing ample opportunities to interact with eukaryotic cellular components. The specific mechanisms involved remain largely uninvestigated, but could conceivably include recognition or binding with phage structural proteins or recognition, binding, transcription, or translation of phage nucleic acids [97].

It has recently been demonstrated that *E. coli* phage PK1A2 can actively bind and penetrate eukaryotic neuroblastoma cells in vitro. The interaction of the phage is attributed through the binding of cell surface polysialic acid by the phage, which shares structural similarity with the bacterial phage receptor [98]. The authors were able to show that these phage particles were able to be present in these cells for up to 24 h without affecting cell viability. Uptake of these phage particles may also lead to the activation of intracellular immunity, potentially priming the eukaryotic cell into an antimicrobial state or enhancing barrier function [99]. Further research is needed within this area to elucidate intracellular phage-eukaryote interactions.

5. Phage Innate Immune Response

5.1. Phage Phagocytosis

It is well established that phages can be phagocytosed by mammalian cells [100–102]. As such, the immune system plays a key role in phage clearance from animal and human bodies. Elements of the mononuclear phagocyte system (MPS) in the spleen and liver filter foreign objects, including phages, from the circulation. The spleen and liver have been identified as the major sites of phage accumulation, as phage titers are usually the highest there [103]. The MPS has been credited for the rapid removal of administered wild-type phage λ from the circulatory system in humans [104]. Moreover Merril et al. [105]) were able to identify certain phage λ mutants that were capable of circumventing the MPS immune response, whereby these mutants prevailed more than 24 h longer in the blood stream of mice than the wild-type phage. These phage λ mutants contained a single Glu-Lys substitution in the phage capsid protein E, leading to a charged change [105].

Both organs contain a large fraction of professional phagocytes. Phagocytosis by immune cells within the liver and spleen seem to be the major process of bacteriophage neutralization within the human body [26,78,80,104,106–108]. One should note that phagocytosis allows the removal of phage particles, even when no specific response to bacteriophages has been developed. Consequently, phagocytes are probably the major fraction of animal or human cells that interact with bacteriophages in vivo.

Clear evidence concerning the cooperation of phages with the innate immune system was first provided by Tiwari et al. [109], who showed the necessity of a neutrophil-phage cooperation in the resolution of *P. aeruginosa* infections [109]. The authors demonstrated that the presence of neutrophils is necessary to remove phage resistant bacteria, which emerge during the phage therapeutic treatment when only a single phage is used. This was later repeated by Roach et al. [110] and Pincus et al. [111] and converted into an in-silico model by Leung & Weitz [112].

Studies, in vitro [23,113,114] as well as in vivo [25,115], regarding the cellular immune response induced by phages have been conducted in recent years and revealed the potential of phages to interact with the mammalian immune system (Figure 2). However, it should be noted that many experiments [113,115] concerning the immune response induced by phages have been carried out using phage lysates containing remnants of lysed bacteria (e.g., LPS, cytosolic proteins, or membrane particles) or fragments of the host bacterial cell wall adhered to phage tails. This makes it extremely difficult to determine which components were truly responsible for the modulation of the immune response.

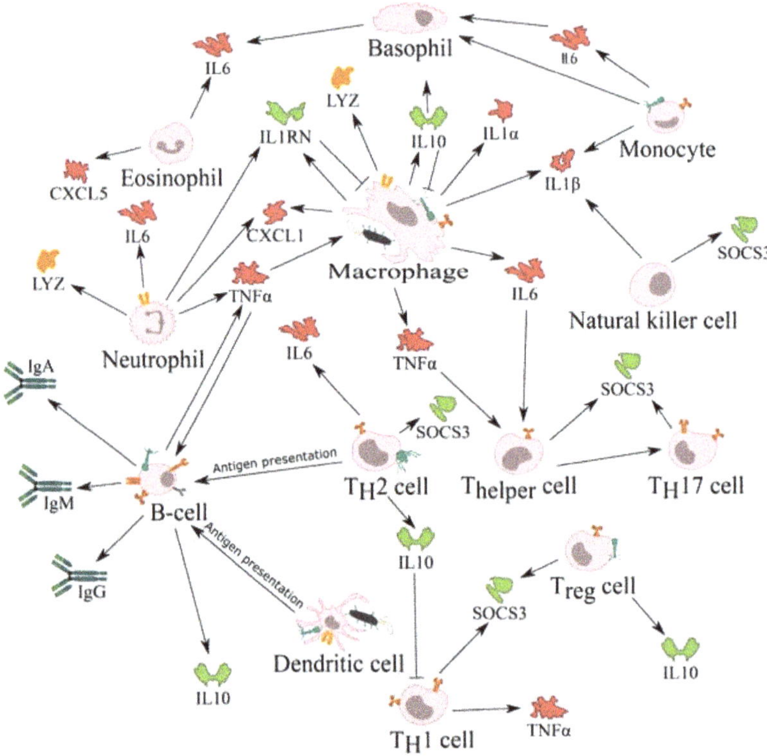

Figure 2. Interaction of bacteriophages with mammalian immune cells. Independent of the route of administration, phages can enter the bloodstream and tissues and encounter immune cells in the blood. Phages could encounter these immune cells whilst they are bound to their bacterial host and taken up together by either macrophages or dendritic cells. Alternatively, these phages can directly interact with any of these immune cells by either interacting with cell surface molecules or receptors, or taken up using a similar mechanism as observed with phage transcytosis. Once in contact with these immune cells, different pro- (red) or anti-inflammatory (green) cytokines are induced, giving the phage the opportunity to influence the immune response. For example, the induction of IL1RN by the phage blocks the pro-inflammatory signals induced by IL1α and IL1β. Although it is known that phages can induce cytokine response, the precise cells responsible are currently not known. Furthermore, the uptake of phages by antigen presenting cells (APC; e.g., dendritic cells) leads to the activation of B-cells and the production of specific antibodies against the phage.

5.2. Phage Induced Phagocytosis of Bacteria

Phages can also increase phagocytosis of bacteria by macrophages, since phages administered together with the host bacteria were able to stimulate bacterial phagocytosis [116] (Figure 2). This was attributed to opsonization of bacterial cells by phages, where the phage coats the bacteria and makes it more recognizable for the immune system. This opsonization is in addition to the direct lytic activity of phages, which may contribute to the effective elimination of pathogenic bacteria in vivo. As phages continue the process of infection when adsorbed onto their bacterial host, some authors have suggested that during phagocytosis, phages continue lysing the phagocytosed bacteria, helping the activity of phagocytic cells [117,118].

One of the possible responses of phagocytes to foreign objects is the production of reactive oxygen species (ROS). ROS mediate antibacterial activity of phagocytic cells, but excessive ROS production may cause oxidative stress and tissue damage. A preliminary study performed by Przerwa et al. [119] suggested that phage T4 influenced the phagocyte system and inhibited the ROS production in response to pathogenic bacteria (i.e., *Escherichia coli*). This phenomenon appeared to depend on specific phage-bacterium interactions, but the precise mechanism is currently not known. Furthermore, the host-specific effect could indicate that the ROS reduction is caused by a reduction of bacteria due to infection and lysis by the phage and not due to direct effects by the phage, per se.

A more comprehensive follow-up study was conducted, whereby polymorphonuclear leukocytes (PMN) were stimulated with one of three different R-type *E. coli* strains (i.e., *E. coli* B or *E. coli* J5, both susceptible for T4, or *E. coli* R4, resistant to T4) or with LPS derived from these three strains [120]. Through this setup, the authors could observe a reduction in ROS production when PMNs were stimulated, with either the live bacteria or their LPS in the presence of phage T4. The results provided by these authors indicate the potential of phages to directly modify functions of mammalian cells and to exert anti-inflammatory properties [120]. A possible explanation for a mechanism underlying phage ability to reduce bacteria-induced ROS production in phagocytes was proposed by Miernikiewicz et al. [120], who investigated T4 phage tail adhesin gp12, which specifically binds bacterial LPS and decreased the potency of LPS to induce an inflammatory response in vivo [121].

5.3. Cytokine Response against Phages

Several studies have been conducted to determine the potential of phages to induce a cytokine response. Often these studies make use of phage preparations that where not fully purified from bacterial endotoxins or proteins. For example, Park et al. [115] studied the cytokine production in mice induced by phage T7, after they were fed with a single dose of phage T7 every 24 hours for 10 days (an exact dose was not provided by the authors). The authors were able to demonstrate that phage T7 induced a very minor increase of inflammatory cytokine production in mice, although no histological changes were observed in the tissues or organs.

On the other hand, analysis of the cytokine production of mice treated intraperitoneally for 5.5 h with highly purified preparations of either whole phage T4 particles, or four phage T4 capsid proteins (i.e., gp23*, gp24*, Hoc, and Soc) showed no inflammatory mediating cytokines in mice [25].

The effect of phages on the production of TNF-α and IL-6 in human serum has also been studied, as well as the in vitro ability of blood cells to produce these cytokines in response to phage. Weber-Dąbrowska et al. [113] used blood derived from 51 patients with long-term suppurative infections of various tissues and organs caused by drug-resistant strains of bacteria. These patients were treated with phages and blood samples were collected and tested for the presence of TNF-α and IL-6. The authors were able to observe a reduction in the production of these cytokines after long-term treatment (i.e., 21 days). However, the observed normalization was likely influenced by the decreased number of pathogenic bacteria in the body following therapeutic application of the phage.

In vitro studies have indicated that phages could have anti-inflammatory properties. Using five highly purified phages targeting two different pathogens, *P. aeruginosa* and *S. aureus,* it was shown that these five phages induced comparable immune responses in PBMCs derived from healthy human donors. Anti-inflammatory markers such as suppressor of cytokine signaling 3 (SOSC3), IL-1 receptor antagonist (IL1RN), and IL-6 were similarly upregulated following treatment with the different phages [23]. The anti-inflammatory action of phages is also in line with some previous observations suggesting an immunosuppressive effect of phages in murine in vivo models of xenografts [122,123]. The anti-inflammatory characteristic of phages was further strengthened by the recent observation that another *S. aureus* phage, vB_SauM_JS25, is able to suppress LPS-induced inflammation [114]. Furthermore, the authors observed that this phage suppressed the phosphorylation of NF-κB p65. Whether this effect is due to a direct interaction of the phage with NF-κB is currently not clear.

Nevertheless, these studies clearly show the potential of phages to induce anti-inflammatory properties unrelated to their antibacterial activities.

It should, however, be emphasized that the potential anti-inflammatory or immunosuppressive action of bacteriophages should not be considered as comparable to physiological effects exerted by well-known anti-inflammatory or immunosuppressive drugs. The precise mechanism as to how phages are able to induce (anti-) inflammatory responses is currently not known, although the antimicrobial effect appears to be one of the factors.

5.4. *Phage Adaptive Immune Response*

Anti-Phage Antibody Production

Since phages consist of tightly packed DNA or RNA and a protein coat, formed by relatively large number of proteins or repeating protein units, it appears obvious that neutralizing antibodies should be produced in individuals subjected to phage therapy or exposed to naturally occurring phages [117,124–126] (Figure 2). Phage immunogenicity has been employed in medicine to test for immune competence of immunodeficient patients (e.g., HIV patients) [127]. In fact, immunization (intravenous administration) with bacteriophage ϕX174 is easy and has been used extensively to diagnose and monitor primary and secondary immunodeficiencies since the 1970s, without reported adverse events, even in patients in whom prolonged circulation of the phage in the bloodstream was observed. This suggests an intrinsically low toxicity of phage ϕX174, even in patients with a compromised immune system [128–130].

Naturally occurring bacteriophages also induce humoral immunity. Phage-neutralizing antibodies against naturally occurring phages (i.e., not therapeutically administered) were detected in the sera of different species (e.g., mice, horse, or human) [126,131–133]. Evaluating the anti-phage antibody production against phage T4 in 50 healthy volunteers who had never been subjected to phage therapy nor involved in phage work showed the presence of naturally occurring phage-antibodies [126]. Of the investigated sera, 81% significantly decreased phage activity, suggesting the presence of anti-phage antibodies. In these positive sera, natural IgG antibodies specific to the phage proteins gp23*, gp24*, Hoc, and Soc were identified (Figure 3). These results demonstrate that anti-T4 phage antibodies are frequent in the human population.

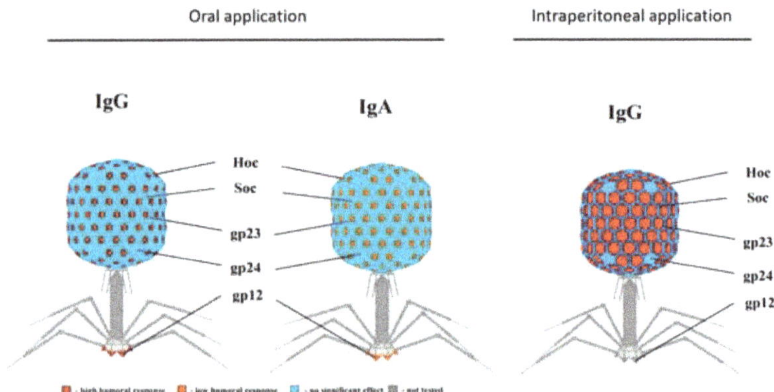

Figure 3. Antibody induction by phage T4 structural proteins. Individual contribution of T4 head proteins (Hoc, Soc, gp23, gp24, and gp12) to phage immunogenicity. Depending on the administration rote (i.e., oral or intraperitoneal), a difference in antibody response can be observed. When phages are administered orally, strong IgG or low IgA response towards Hoc can be observed, whereas intraperitoneal applications lead to high IgG responses towards Hoc and gp23. Modified Majewska et al. [24]. Permission was obtained for the reproduction of this figure.

Most studies suggest that it is very easy to generate phage antisera by immunization of humans or animals with phages [124,126,128,129]. Contrary to this, a safety study by Bruttin and Brüssow in 2005 administering T4 phages orally at very low doses to human patients revealed no antibody induction in phage-treated patients, potentially due to the very low doses of bacteriophages administered due to safety concerns or the lack of adjuvant. Recently, a study concerning the production of IgG, IgA, and IgM in human patients undergoing phage therapy was carried out by Żaczek et al. [134], who treated 20 patients, for an undisclosed time, with the MS-1 phage cocktail (containing three lytic *S. aureus* phages), either orally or locally [134]. For most patients, no antibodies could be detected. For the few patients that produced elevated levels of IgG or IgM, the presence of anti-phage antibodies did not translate into an unsatisfactory clinical result of the phage therapy. The low antibody production against the phage cocktail could be due to the small time-scale during which the patients were treated. On the other hand, the elevated antibody production in a few patients could be due to a previous encounter of one of the phages used in the cocktail and the presence of an immunological memory.

These reports demonstrate that the humoral response does not follow a simple scheme of induction [24,89,117,126,135]. This was further studied by Majewska et al. [24], who quantified the antibody production against a single phage (i.e., *E. coli* phage T4) in mice over a time period of 240 days [24]. Phage T4 was given orally to mice for 100 days, followed by 112 days without phage treatment. The treatment was then repeated with the same phage up to day 240. It was demonstrated that the long-term oral treatment of mice with phage T4 led to a humoral response. The authors observed that this response emerged from the secretion of IgA in the gut lumen and an IgG production in the blood. The intensity of this response and the time necessary for its induction depended on the exposure to phage antigens, which is related to the phage dose. The factor limiting phage activity in the gut was the production of specific IgA. If the secretory levels of IgA were low, phages remained present in the feces. When the IgA level increased (around day 80), there were no active phages present in the feces. On the other hand, when secretory IgA decreased with time (on day 213 it dropped to its initial levels), phages could be detected again, until phage-specific IgA levels increased again.

According to the same authors, the induction of serum IgG suggests that phages can translocate from the gut lumen to the circulation. This observation is further strengthened by recent data of transcytosing phages [91]. Furthermore, it was possible to isolate phages from murine blood after oral application of high phage doses (4×10^9 pfu/ml of drinking water), and this fact correlated with phage ability to induce a long-lasting secondary immune response. Lowering the phage dose ten-fold did not induce a significant increase of the adaptive immune response, nor did it allow for detection of active phages in the circulation. Besides considering the complete phage particle as a whole, it is of interest to evaluate the immune responses induced against individual phage proteins. It was demonstrated that phage T4 Hoc protein and gp12 strongly stimulated the IgG and IgA antibody production in the blood and gut respectively, while gp23*, gp24*, and Soc induced low responses [24].

5.5. In-Silico Modeling of the Immune Response Towards Phages

In-silico models predicting phage therapeutic interventions have been developed to better understand the immune response against phages and its impact on the outcome of phage therapeutic interventions [26,112,136–139]. These models are complicated by the fact that phages are protein-based biological agents that interact with the body's immune system, actively replicate, and even evolve during manufacture or use [140]. As such, phage applications have a vastly different pharmacology compared to conventional drugs [137–139,141,142]. In these mathematical models, the rate at which a bacterial population declines due to phage infection, the rate at which the phage population increases, and the levels at which they are maintained depends primarily on five parameters: the infectivity of the phage, the latency period, the burst size, the rate at which the phages are degraded or removed from the site of infection, and the bacterial growth rate. Besides these five parameters, two other variables need to be taken into account: the density of susceptible bacteria and the density of the phage [136]. In summary, these models describe phage pharmacokinetics as being analogous to the population dynamics of the phage-bacterial interaction [143], not taking into account potential interaction between bacteria and phages with the innate or adaptive immunity.

These mathematical models can be further extended to include the mammalian host response towards the phage [26]. Based on experimental data, a general scheme can be developed for the tripartite interactions between phage, bacteria, and mammalian immunity. This scheme summarizes the main reciprocal dependencies, specifically the limiting or inducing effects (Figure 4). There are three initial key assumptions on which this scheme is based. First, the innate immunity is activated by the bacteria and acts against the bacteria, but at the same time it also acts against the phage. The second assumption is that phages are not able to boost an innate immune response [25,115]. The third assumption is that the adaptive immunity specific to phages and the adaptive immunity specific to bacteria have no interfering cross-talk. This led to the development of a model with a set of immunity-representing variables; innate immunity (I), adaptive immunity specific to phages (A), and adaptive immunity specific to bacteria (B). A similar in-silico model described the outcome of a phage therapeutic intervention, taking into account the occurrence of phage resistant bacteria and a phage decay rate, which represents both the innate and adaptive immunity towards the phage [110,112]. This model indicated that neutrophils are necessary to completely clear a bacterial infection when phage resistance occurs, although it could be argued that phage resistance could partially be prevented when using a phage cocktail [109,110].

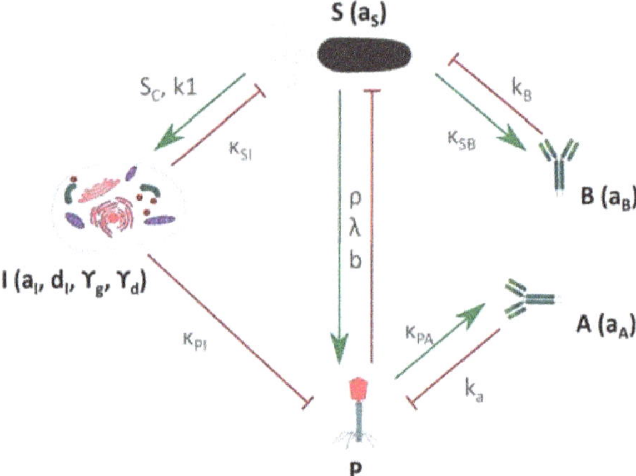

Figure 4. Schematic representation of the immune response against phages and bacteria. P–Phage, **S**–bacteria, **I**–innate immunity, **A**–adaptive immune response to phage, **B**–adaptive immune response to bacteria. Green arrows represent a stimulatory effect, red arrows represent an inhibitory effect. Variables and parameters used in these models are described in Tables S1 and S2. Adapted from Hodyra-Stefaniak et al. [26].

When no interaction occurs between the innate immune response (I) and the phage (P), the original Hodyra-Stefaniak model predicts a successful intervention of phages in the removal of a bacterial infection (Figure 5A) [26,112]. The inclusion of the variable for the innate immunity (I) demonstrates that the expected outcome of phage therapy could be abrogated by the innate immunity boosted by the bacteria (S) (Figure 5B; Hodyra-Stefaniak et al. [26]). Moreover, within the model, the removal of the phage (P) by the innate immune system (I) would lead to a secondary increase in bacterial (S) count, indicating an inefficacy of phage therapy. This is in contrast to the available phage therapy related data [144–148]. Alternatively, this failure could be counteracted by adjusting the phage dose or changing the timing, as long as the interaction with the innate immunity is considered (Figure 5D). Nevertheless, this indicates a shortcoming of the current model described by Hodyra-Stefaniak, indicating further adapting of the model is needed to more closely reflect current knowledge of phage therapeutic outcomes.

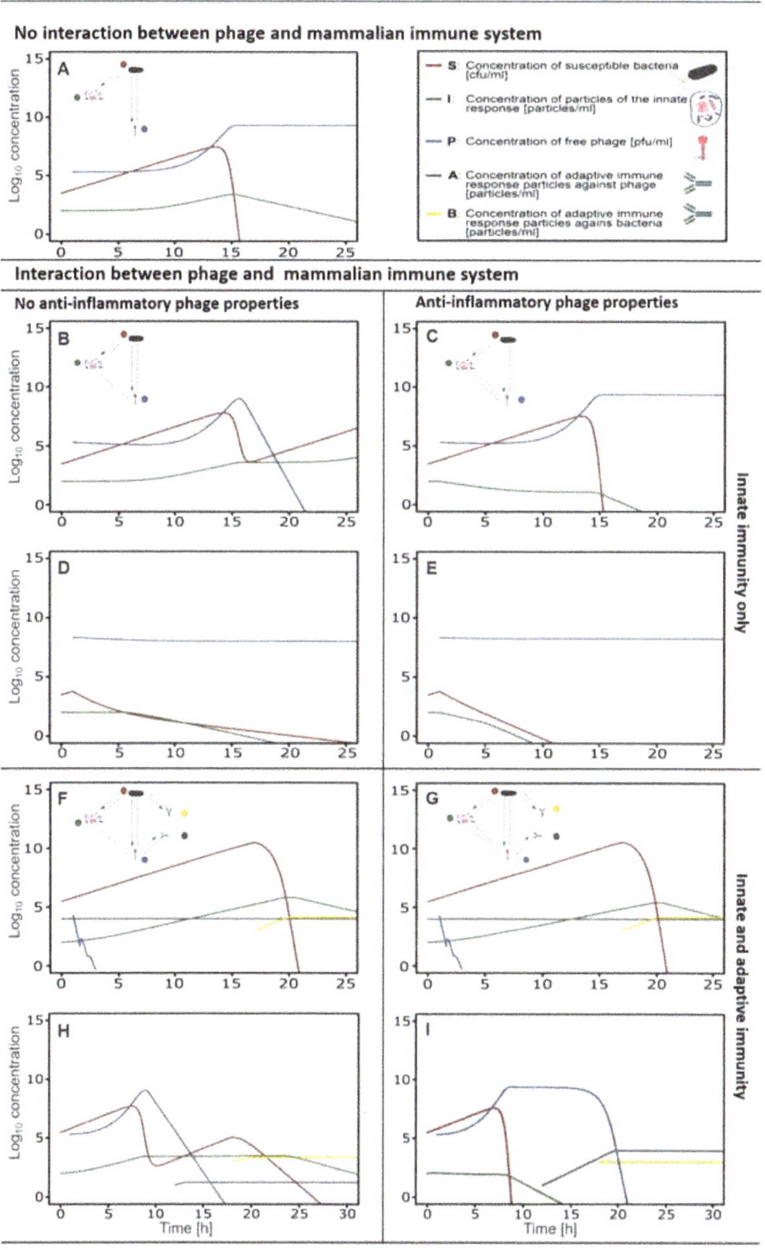

Figure 5. Effects of innate and adaptive immunity on the success or failure of phage antibacterial treatment, numerical simulations. Innate immune response. (**A**) No relation between innate immunity and phage viability. The survival of the phage is independent of the presence of an innate immune response.

(**B**) Phage susceptibility to the innate immune response. The innate immunity has a negative effect on the phage survival and leads to its removal. Subsequently the bacteria are no longer infected by the phage, and a rise in bacteria is observed. (**C**) Phage susceptibility to the innate immune response, considering the anti-inflammatory property of the phage. The anti-inflammatory characteristic of the phage leads to a decline in innate immune particles. This has as effect that the bacterial count diminishes, and the phage survives, similar to A. (**D**) Phage susceptibility to innate immune response accommodated and counteracted by an increased phage dose. The higher phage dose leads to the removal of the pathogen and the survival of the phage. (**E**) Phage susceptibility to innate immune response accommodated and counteracted by an increased phage dose, considering the anti-inflammatory property of the phage. The effect is the same as in D, but the innate immune response is diminished. Innate and adaptive immune response. (**F**) Phage susceptibility to the innate immune response and presence of pre-immunization towards the phage. Presence of pre-existing anti-phage antibodies lead to a rapid drop in phage concentration, hence the phage has no effect on the survival of the bacteria. Once an adaptive immune response towards the bacteria is present, bacterial count decreases. (**G**) Phage susceptibility to the innate immune response and no pre-immunization to the phage exists, considering the anti-inflammatory property of the phage. The anti-inflammatory response of the phage has no direct influence on the phage survival in the presence of an adaptive immune response towards the phage. Overall the response is similar to F. (**H**) Phage susceptibility to the innate immune response and no pre-immunization to the phage exists. The absence of a specific adaptive immune response towards the phage leads to a decrease in the bacterial population. The combined effect of innate and adaptive immunity towards the phage leads to a drop-in phage particle concentration. (**I**) Phage susceptibility to the innate immune response and no pre-immunization to the phage exists, considering the anti-inflammatory property of the phage. Once the phage reaches a critical concentration (Pc, the concentration of phages needed to induce an anti-inflammatory response), the innate immune response decreases, and the phage concentration grows until all bacteria are removed. Once an adaptive immune response is present against the phage, the phage concentration diminishes until completely removed. Variables and parameters used in these models are described in Tables S1 and S2.

6. Anti-Inflammatory Phage Properties Affect the Outcome of Phage Therapy

Most in-silico models miss one key feature—the interaction of the phage with the innate immune response. In theory, this interaction can be anti-inflammatory, leading to a suppression of the immune response, or pro-inflammatory, resulting in an increase of the immune response. Current literature states that phages are not able to induce pro-inflammatory responses [25]. Based on the recently described anti-inflammatory properties of phages [23,114], existing in-silico models can be further extended to include the interaction of phages with the innate immune response, as seen in the supplementary data Tables S1 and S2.

By including the anti-inflammatory property of phages in the model, the prediction of the phage therapeutic outcome becomes successful again (Figure 5C). The phage (P) can, partly, subdue the innate immune response (I) and hence clears the bacterial infection (S). When a bacterial infection is combated with an initial high phage dose, the effects of the innate immune response are negligible (Figure 5D,E). Yet, if the phage has anti-inflammatory properties, the bacterial clearance occurs much faster according to the model. Nevertheless, when anti-phage antibodies (A) are present prior to the phage therapeutic intervention, the intervention fails as the phages are rapidly removed (Figure 5F,G).

When no pre-immunization to the phage is present, and no anti-inflammatory phage properties are considered, the removal of the bacterial infection is attributed mainly to the adaptive immune response against the bacteria (B) (Figure 5H). Initially, the phages (P) lead to a reduction of the bacterial count (S) but are themselves removed by a combination of the innate and adaptive immune response against the phage. This leads to a second rise in the bacterial concentration (S). In a later stage, the bacterial infection is removed by the adaptive immune response against the bacteria, hence the clearance of the bacterial infection is not due to the presence of the phage but due to the adaptive immune response against the bacteria (for the modeling purposes, the time of induction of specific antibodies was shorter than in physiological conditions). According to this model, when the anti-inflammatory properties of the phage (P) are considered, the bacterial infection (S) is cleared much faster and this is attributed to the presence of the phage (Figure 5I).

7. Relevance of Phage-Mammalian Host Immune Responses

The diverse ways in which phages interact with the human host are clear, and recently more work is being focused on this. Phage adherence to mucosal surfaces provides a previously unrecognized antimicrobial defense that actively protects the mucosal surface from bacterial infection [63,71]. This extension of the human immune system to include the action of symbiotic phages within the mucosal surface provides the eukaryotic host with a number of potential benefits. The phages offer a selective antimicrobial defense that operates at a much finer spectrum than some other broad-spectrum host secretions, such as the antibacterial lectin RegIII-γ [59]. Additionally, the interaction of phages with the mucosal layers can also lead to a higher rate of bacterial colonization in case of non-lytic filamentous phages [73,149]. The ability of phages to bind to mucus layers would provide them with a higher probability to contact and transcytose across epithelial cells [91]. This not only raises the question of whether they can interact with intracellular immune pathways but also whether phages could interact with mitochondria, which originated from a bacterial origin, once they are taken up by the cell. Although the presence of phages in mammalian cells has been observed [91,94], replication of these viruses in theses cell types has not yet been observed.

Phages can induce intra-cellular interactions with Toll-like receptors (TLR). TLR are responsible for the recognition of microbial and viral structures [43]. Viral nucleic acids act as pathogen associated molecular patterns (PAMPs) and are recognized by TLRs. It could thus be postulated that phage DNA might be recognized by TLR9, which is responsible for the recognition of viral DNA [150], after phagocytosis or transcytosis of the phage.

The observation that phages can directly interact with human immune cells and induce certain cytokine productions [23,114] has important implications for their use. Our in-silico model shows the positive effect of phage anti-inflammatory properties on the outcome of a bacterial infection, but these phage immune responses could have a much broader effect. Based on the anti-inflammatory responses observed by certain phages, it could even be postulated that phages could have an impact on allergic disorders such as asthma, rhinitis, and atopic dermatitis. The anti-inflammatory properties observed in certain phages could heighten their bacterial host's fitness in an immunological context, creating potential microenvironments where the immune response is lowered [149], and the bacteria have a higher infection or survival rate. It is important to note that although the phage might have anti-inflammatory properties, this does not necessarily mean that the phage is able to effectively suppress the innate immune response. These anti-inflammatory properties do not seem comparable to typical immunosuppressive drugs or agents.

The most direct impact of phages might be during sepsis, where the lytic activity of the phage can reduce the bacterial burden and the immunomodulating properties of the phage could lead to a partial dampening of the inflammatory response induced by the bacteria or the bacterial lysis. Phage or phage-derived proteins that specifically interact with certain bacterial components (PAMPs) could even be used to moderate undesirable immune response (e.g., the use of phage T4 tail adhesin gp12 to capture and bind LPS in case of septic shock) [121]. The use of phages or phage-derived proteins as anti-inflammatory agents can lead to a possible new type of anti-inflammatory drugs with a new mode of action in comparison to the classic non-steroid anti-inflammatory drugs (NSAIDs). Possibly, these phages or phage-derived proteins might possess less side effects compared to NSAIDs. Phages can be engineered as nanocarriers for targeted drug delivery, or for the display of selected antigens and the subsequent stimulation of an immune response [24,151,152].

8. Conclusions and Areas of Future Investigation

The data reviewed here indicate that phages can interact with the mammalian immune system in a variety of ways that are both direct and indirect. However, the magnitude and nature of the influence that these viruses have on mammalian immunity are only beginning to come into focus. At present, the available data suggest that these interactions tend to be anti-inflammatory. If the observations by Van Belleghem et al. [23] and Zhang et al. [114] concerning the anti-inflammatory properties of phages can be further validated, it is conceivable that phages could influence both our interactions with our commensal flora as well as the outcome of phage therapeutic interventions.

However, the data on these interactions remains patchy, incomplete, and limited to small numbers of phages, cell types, and disease models. Further, definitive data indicating that phages impact human health or immunity, as opposed to cells or animal models, remains absent. Moreover, many of the specific mechanisms underlying the mammalian host immune response to phages remain unknown. Important areas of uncertainty include the following questions: How are phages taken up by cells? Is this an active or passive process? Is this uptake required to influence mammalian immunity or are cell surface interactions sufficient? Are these interactions specific to certain phages or phage families? Which parts of phages elicits the immune response? Do lytic and lysogenic phages influence host immunity in similar ways? Are these interactions primarily relevant to settings of immune interactions with commensal flora, microbial pathogens, or both? Knowing the answers to these and other questions could open many new fields of study and may facilitate the development of novel, phage-based therapies. We have much to learn but it is clear that phage and mammalian host interactions is an exciting and promising field of exploration.

Supplementary Materials: The following are available online at http://www.mdpi.com/1999-4915/11/1/10/s1, Table S1: Generalized model describing the phage-bacteria-immune response interaction, Table S2: State variable and parameters of the models described in Table 1.

Acknowledgments: Krystyna Dąbrowska is a recipient of grants from the National Science Centre in Poland UMO-2012/05/E/NZ6/03314 and UMO-2015/18/M/NZ6/00412. Jeremy J. Barr is the recipient of an Australia Research Council Discovery Early Career Researcher Award (project number DE170100525) and funded by the Australian Government. This work was supported by grants R21AI133370, R21AI133240, R01AI12492093, and grants from Stanford SPARK, the Falk Medical Research Trust and the Cystic Fibrosis Foundation (CFF) to P.L.B.

Conflicts of Interest: The authors declare no conflict of interest.

References

1. Duerkop, B.A.; Hooper, L.V. Resident viruses and their interactions with the immune system. *Nat. Immunol.* **2013**, *14*, 654–659. [CrossRef]
2. Proctor, L.; Sechi, S.; Di Giacomo, N.; Fettweis, J.; Jefferson, K.; Strauss, J., III; Rubens, C.; Brooks, J.; Girerd, P.; Huang, B.; et al. The Integrative Human Microbiome Project: Dynamic Analysis of Microbiome-Host Omics Profiles during Periods of Human Health and Disease. *Cell Host Microbe* **2014**, *16*, 276–289. [CrossRef]
3. Lloyd-Price, J.; Mahurkar, A.; Rahnavard, G.; Crabtree, J.; Orvis, J.; Hall, A.B.; Brady, A.; Creasy, H.H.; McCracken, C.; Giglio, M.G.; et al. Strains, functions and dynamics in the expanded Human Microbiome Project. *Nature* **2017**, *550*, 61–66. [CrossRef] [PubMed]
4. Hooper, L.V.; Littman, D.R.; Macpherson, A.J. Interactions between the microbiota and the immune system. *Science* **2012**, *336*, 1268–1273. [CrossRef]
5. Lozupone, C.A.; Stombaugh, J.I.; Gordon, J.I.; Jansson, J.K.; Knight, R. Diversity, stability and resilience of the human gut microbiota. *Nature* **2012**, *489*, 220–230. [CrossRef] [PubMed]
6. Reyes, A.; Semenkovich, N.P.; Whiteson, K.; Rohwer, F.; Gordon, J.I. Going viral: Next-generation sequencing applied to phage populations in the human gut. *Nat. Rev. Microbiol.* **2012**, *10*, 607–617. [CrossRef]
7. Oh, J.; Byrd, A.L.; Park, M.; Kong, H.H.; Segre, J.A. Temporal Stability of the Human Skin Microbiome. *Cell* **2016**, *165*, 854–866. [CrossRef] [PubMed]
8. Foulongne, V.; Sauvage, V.; Hebert, C.; Dereure, O.; Cheval, J.; Gouilh, M.A.; Pariente, K.; Segondy, M.; Burguière, A.; Manuguerra, J.C.; et al. Human skin Microbiota: High diversity of DNA viruses identified on the human skin by high throughput sequencing. *PLoS ONE* **2012**, *7*, e38499. [CrossRef]
9. Pride, D.T.; Salzman, J.; Haynes, M.; Rohwer, F.; Davis-Long, C.; White, R.A.; Loomer, P.; Armitage, G.C.; Relman, D.A. Evidence of a robust resident bacteriophage population revealed through analysis of the human salivary virome. *ISME J.* **2012**, *6*, 915–926. [CrossRef]

10. Wang, J.; Gao, Y.; Zhao, F. Phage-bacteria interaction network in human oral microbiome. *Environ. Microbiol.* **2016**, *18*, 2143–2158. [CrossRef]
11. Edlund, A.; Santiago-Rodriguez, T.M.; Boehm, T.K.; Pride, D.T. Bacteriophage and their potential roles in the human oral cavity. *J. Oral Microbiol.* **2015**, *7*, 1–12. [CrossRef]
12. Dickson, R.P.; Huffnagle, G.B. The Lung Microbiome: New Principles for Respiratory Bacteriology in Health and Disease. *PLoS Pathog.* **2015**, *11*, e1004923. [CrossRef]
13. Lim, Y.W.; Schmieder, R.; Haynes, M.; Willner, D.; Furlan, M.; Youle, M.; Abbott, K.; Edwards, R.; Evangelista, J.; Conrad, D.; et al. Metagenomics and metatranscriptomics: Windows on CF-associated viral and microbial communities. *J. Cyst. Fibros.* **2013**, *12*, 154–164. [CrossRef] [PubMed]
14. Willner, D.; Haynes, M.R.; Furlan, M.; Hanson, N.; Kirby, B.; Lim, Y.W.; Rainey, P.B.; Schmieder, R.; Youle, M.; Conrad, D.; et al. Case studies of the spatial heterogeneity of DNA viruses in the cystic fibrosis lung. *Am. J. Respir. Cell Mol. Biol.* **2012**, *46*, 127–131. [CrossRef]
15. Reyes, A.; Haynes, M.; Hanson, N.; Angly, F.E.; Heath, A.C.; Rohwer, F.; Gordon, J.I. Viruses in the faecal microbiota of monozygotic twins and their mothers. *Nature* **2010**, *466*, 334–338. [CrossRef]
16. Minot, S.; Sinha, R.; Chen, J.; Li, H.; Keilbaugh, S.A.; Wu, G.D.; Lewis, J.D.; Bushman, F.D. The human gut virome: Inter-individual variation and dynamic response to diet. *Genome Res.* **2011**, *21*, 1616–1625. [CrossRef] [PubMed]
17. Santiago-Rodriguez, T.M.; Ly, M.; Bonilla, N.; Pride, D.T. The human urine virome in association with urinary tract infections. *Front. Microbiol.* **2015**, *6*, 14. [CrossRef]
18. Brüssow, H.; Canchaya, C.; Hardt, W. Phages and the evolution of bacterial pathogens: From genomic rearrangements to lysogenic conversion. *Microbiol. Mol.* **2004**, *68*. [CrossRef]
19. Breitbart, M.; Rohwer, F. Method for discovering novel DNA viruses in blood using viral particle selection and shotgun sequencing. *Biotechniques* **2005**, *39*, 729–736. [CrossRef]
20. Dinakaran, V.; Rathinavel, A.; Pushpanathan, M.; Sivakumar, R.; Gunasekaran, P.; Rajendhran, J. Elevated levels of circulating DNA in cardiovascular disease patients: Metagenomic profiling of microbiome in the circulation. *PLoS ONE* **2014**, *9*, e105221. [CrossRef] [PubMed]
21. Li, S.K.; Leung, R.K.-K.; Guo, H.X.; Wei, J.F.; Wang, J.H.; Kwong, K.T.; Lee, S.S.; Zhang, C.; Tsui, S.K.-W. Detection and identification of plasma bacterial and viral elements in HIV/AIDS patients in comparison to healthy adults. *Clin. Microbiol. Infect.* **2012**, *18*, 1126–1133. [CrossRef] [PubMed]
22. Moustafa, A.; Xie, C.; Kirkness, E.; Biggs, W.; Wong, E.; Turpaz, Y.; Bloom, K.; Delwart, E.; Nelson, K.E.; Venter, J.C.; et al. The blood DNA virome in 8000 humans. *PLoS Pathog.* **2017**, *13*, e1006292. [CrossRef] [PubMed]
23. Van Belleghem, J.D.; Clement, F.; Merabishvili, M.; Lavigne, R.; Vaneechoutte, M. Pro- and anti-inflammatory responses of peripheral blood mononuclear cells induced by Staphylococcus aureus and Pseudomonas aeruginosa phages. *Sci. Rep.* **2017**, *7*, 8004. [CrossRef] [PubMed]
24. Majewska, J.; Beta, W.; Lecion, D.; Hodyra-Stefaniak, K.; Kłopot, A.; Kaźmierczak, Z.; Miernikiewicz, P.; Piotrowicz, A.; Ciekot, J.; Owczarek, B.; et al. Oral application of T4 phage induces weak antibody production in the gut and in the blood. *Viruses* **2015**, *7*, 4783–4799. [CrossRef] [PubMed]
25. Miernikiewicz, P.; Dąbrowska, K.; Piotrowicz, A.; Owczarek, B.; Wojas-Turek, J.; Kicielińska, J.; Rossowska, J.; Pajtasz-Piasecka, E.; Hodyra, K.; Macegoniuk, K.; et al. T4 phage and its head surface proteins do not stimulate inflammatory mediator production. *PLoS ONE* **2013**, *8*, e71036. [CrossRef] [PubMed]
26. Hodyra-Stefaniak, K.; Miernikiewicz, P.; Drapała, J.; Drab, M.; Jonczyk-Matysiak, E.; Lecion, D.; Kazmierczak, Z.; Beta, W.; Majewska, J.; Harhala, M.; et al. Mammalian Host-Versus-Phage immune response determines phage fate in vivo. *Sci. Rep.* **2015**, *5*, 3–8. [CrossRef]
27. Handley, S.A.; Thackray, L.B.; Zhao, G.; Presti, R.; Miller, A.D.; Droit, L.; Abbink, P.; Maxfield, L.F.; Kambal, A.; Duan, E.; et al. Pathogenic simian immunodeficiency virus infection is associated with expansion of the enteric virome. *Cell* **2012**, *151*, 253–266. [CrossRef] [PubMed]
28. McDaniel, L.; Breitbart, M.; Mobberley, J.; Long, A.; Haynes, M.; Rohwer, F.; Paul, J.H. Metagenomic analysis of lysogeny in Tampa Bay: Implications for prophage gene expression. *PLoS ONE* **2008**, *3*, e3263. [CrossRef]
29. Scarpellini, E.; Ianiro, G.; Attili, F.; Bassanelli, C.; De Santis, A.; Gasbarrini, A. The human gut microbiota and virome: Potential therapeutic implications. *Dig. Liver Dis.* **2015**, *47*, 1007–1012. [CrossRef]

30. Kim, M.-S.; Park, E.-J.; Roh, S.W.; Bae, J.-W. Diversity and abundance of single-stranded DNA viruses in human feces. *Appl. Environ. Microbiol.* **2011**, *77*, 8062–8070. [CrossRef]
31. Reyes, A.; Wu, M.; McNulty, N.P.; Rohwer, F.L.; Gordon, J.I. Gnotobiotic mouse model of phage–bacterial host dynamics in the human gut. *Proc. Natl. Acad. Sci. USA* **2013**, *110*, 20236–20241. [CrossRef] [PubMed]
32. Breitbart, M.; Haynes, M.; Kelley, S.; Angly, F.; Edwards, R.A.; Felts, B.; Mahaffy, J.M.; Mueller, J.; Nulton, J.; Rayhawk, S.; et al. Viral diversity and dynamics in an infant gut. *Res. Microbiol.* **2008**, *159*, 367–373. [CrossRef] [PubMed]
33. Lim, E.S.; Zhou, Y.; Zhao, G.; Bauer, I.K.; Droit, L.; Ndao, I.M.; Warner, B.B.; Tarr, P.I.; Wang, D.; Holtz, L.R. Early life dynamics of the human gut virome and bacterial microbiome in infants. *Nat. Med.* **2015**, *21*, 1228–1234. [CrossRef] [PubMed]
34. Rodriguez-Valera, F.; Martin-Cuadrado, A.B.; Rodriguez-Brito, B.; Pašić, L.; Thingstad, T.F.; Rohwer, F.; Mira, A. Explaining microbial population genomics through phage predation. *Nat. Rev. Microbiol.* **2009**, *7*, 828–836. [CrossRef] [PubMed]
35. Enault, F.; Briet, A.; Bouteille, L.; Roux, S.; Sullivan, M.B.; Petit, M.-A. Phages rarely encode antibiotic resistance genes: A cautionary tale for virome analyses. *ISME J.* **2016**, 053025. [CrossRef] [PubMed]
36. Verma, N.K.; Brandt, J.M.; Verma, D.J.; Lindberg, A.A. Molecular characterization of the O-acetyl transferase gene of converting bacteriophage SF6 that adds group antigen 6 to Shigella flexneri. *Mol. Microbiol.* **1991**, *5*, 71–75. [CrossRef]
37. Pieraerts, C.; Martin, V.; Jichlinski, P.; Nardelli-Haefliger, D.; Derré, L. Detection of functional antigen-specific T cells from urine of non-muscle invasive bladder cancer patients. *Oncoimmunology* **2012**, *1*, 694–698. [CrossRef]
38. Brüssow, H. Bacteriophage-host interaction: From splendid isolation into a messy reality. *Curr. Opin. Microbiol.* **2013**, *16*, 500–506. [CrossRef]
39. Davies, M.R.; Broadbent, S.E.; Harris, S.R.; Thomson, N.R.; van der Woude, M.W. Horizontally Acquired Glycosyltransferase Operons Drive Salmonellae Lipopolysaccharide Diversity. *PLoS Genet.* **2013**, *9*, e1003568. [CrossRef]
40. Ivanov, I.I.; Frutos R de, L.; Manel, N.; Yoshinaga, K.; Rifkin, D.B.; Sartor, R.B.; Finlay, B.B.; Littman, D.R. Specific microbiota direct the differentiation of IL-17-producing T-helper cells in the mucosa of the small intestine. *Cell Host Microbe* **2008**, *4*, 337–349. [CrossRef]
41. Mazmanian, S.K.; Cui, H.L.; Tzianabos, A.O.; Kasper, D.L. An immunomodulatory molecule of symbiotic bacteria directs maturation of the host immune system. *Cell* **2005**, *122*, 107–118. [CrossRef] [PubMed]
42. Przybylski, M.; Borysowski, J.; Jakubowska-Zahorska, R.; Weber-Dąbrowska, B.; Górski, A. T4 bacteriophage-mediated inhibition of adsorption and replication of human adenovirus in vitro. *Future Microbiol.* **2015**, *10*, 453–460. [CrossRef] [PubMed]
43. Kawai, T.; Akira, S. Toll-like receptors and their crosstalk with other innate receptors in infection and immunity. *Immunity* **2011**, *34*, 637–650. [CrossRef] [PubMed]
44. Alexopoulos, L.G.; Saez-Rodriguez, J.; Cosgrove, B.D.; Lauffenburger, D.A.; Sorger, P.K. Networks Inferred from Biochemical Data Reveal Profound Differences in Toll-like Receptor and Inflammatory Signaling between Normal and Transformed Hepatocytes. *Mol. Cell Proteom.* **2010**, *9*, 1849–1865. [CrossRef] [PubMed]
45. Blasius, A.L.; Beutler, B. Intracellular Toll-like receptors. *Immunity* **2010**, *32*, 305–315. [CrossRef] [PubMed]
46. Shi, Z.; Cai, Z.; Sanchez, A.; Zhang, T.; Wen, S.; Wang, J.; Yang, J.; Fu, S.; Zhang, D. A novel toll-like receptor that recognizes vesicular stomatitis virus. *J. Biol. Chem.* **2011**, *286*, 4517–4524. [CrossRef] [PubMed]
47. Focà, A.; Liberto, M.C.; Quirino, A.; Marascio, N.; Zicca, E.; Pavia, G. Gut inflammation and immunity: What is the role of the human gut virome? *Med. Inflamm.* **2015**, *2015*, 1–7. [CrossRef] [PubMed]
48. Foxman, E.F.; Iwasaki, A. Genome-virome interactions: Examining the role of common viral infections in complex disease. *Nat. Rev. Microbiol.* **2011**, *9*, 254–264. [CrossRef] [PubMed]
49. Farrar, M.A.; Schreiber, R.D. The Molecular Cell Biology of Interferon-gamma and its Receptor. *Annu. Rev. Immunol.* **1993**, *11*, 571–611. [CrossRef] [PubMed]
50. Linden, S.K.; Sutton, P.; Karlsson, N.G.; Korolik, V.; McGuckin, M.A. Mucins in the mucosal barrier to infection. *Mucosal. Immunol.* **2008**, *1*, 183–197. [CrossRef] [PubMed]

51. Johansson, M.E.V.; Phillipson, M.; Petersson, J.; Velcich, A.; Holm, L.; Hansson, G.C.; Petersson, J.; Velcich, A.; Holm, L.; Hansson, G.C.; et al. The inner of the two Muc2 mucin-dependent mucus layers in colon is devoid of bacteria. *PNAS* **2008**, *105*, 15064–15069. [CrossRef] [PubMed]
52. Sonnenburg, J.L. Glycan foraging in vivo by an intestine-adapted bacterial symbiont. *Science* **2005**, *307*, 1955–1959. [CrossRef] [PubMed]
53. Schluter, J.; Foster, K.R. The evolution of mutualism in gut microbiota via host epithelial selection. *PLoS Biol.* **2012**, *10*, e1001424. [CrossRef] [PubMed]
54. Hooper, L.V.; Xu, J.; Falk, P.G.; Midtvedt, T.; Gordon, J.I. A molecular sensor that allows a gut commensal to control its nutrient foundation in a competitive ecosystem. *Proc. Natl. Acad. Sci. USA* **1999**, *96*, 9833–9838. [CrossRef] [PubMed]
55. Martens, E.C.; Chiang, H.C.; Gordon, J.I. Mucosal glycan foraging enhances fitness and transmission of a saccharolytic human gut bacterial symbiont. *Cell Host Microbe* **2008**, *4*, 447–457. [CrossRef] [PubMed]
56. Poulsen, L.K.; Lan, F.; Kristensen, C.S.; Hobolth, P.; Molin, S.; Krogfelt, K.A. Spatial distribution of Escherichia coli in the mouse large intestine inferred from rRNA in situ hybridization. *Infect. Immun.* **1994**, *62*, 5191–5194. [PubMed]
57. Koropatkin, N.M.; Cameron, E.A.; Martens, E.C. How glycan metabolism shapes the human gut microbiota. *Nat. Rev. Microbiol.* **2012**, *10*, 323–335. [CrossRef]
58. Salzman, N.H.; Hung, K.; Haribhai, D.; Chu, H.; Karlsson-Sjöberg, J.; Amir, E.; Teggatz, P.; Barman, M.; Hayward, M.; Eastwood, D.; et al. Enteric defensins are essential regulators of intestinal microbial ecology. *Nat. Immunol.* **2010**, *11*, 76–83. [CrossRef]
59. Vaishnava, S.; Yamamoto, M.; Severson, K.M.; Ruhn, K.A.; Yu, X.; Koren, O.; Ley, R.; Wakeland, E.K.; Hooper, L.V. The Antibacterial Lectin RegIII Promotes the Spatial Segregation of Microbiota and Host in the Intestine. *Science* **2011**, *334*, 255–258. [CrossRef]
60. Gill, D.J.; Tham, K.M.; Chia, J.; Wang, S.C.; Steentoft, C.; Clausen, H.; Bard-Chapeau, E.A.; Bard, F.A. Initiation of GalNAc-type O-glycosylation in the endoplasmic reticulum promotes cancer cell invasiveness. *Proc. Natl. Acad. Sci. USA* **2013**, *110*, E3152–E3161. [CrossRef]
61. Jentoft, N. Why are proteins O-glycosylated? *Trends Biochem. Sci.* **1990**, *15*, 291–294. [CrossRef]
62. Schulz, B.L.; Sloane, A.J.; Robinson, L.J.; Prasad, S.S.; Lindner, R.A.; Robinson, M.; Bye, P.T.; Nielson, D.W.; Harry, J.L.; Packer, N.H.; et al. Glycosylation of sputum mucins is altered in cystic fibrosis patients. *Glycobiology* **2007**, *17*, 698–712. [CrossRef] [PubMed]
63. Barr, J.J.; Auro, R.; Furlan, M.; Whiteson, K.L.; Erb, M.L.; Pogliano, J.; Stotland, A.; Wolkowicz, R.; Cutting, A.S.; Doran, K.S.; et al. Bacteriophage adhering to mucus provide a non-host-derived immunity. *Proc. Natl. Acad. Sci. USA* **2013**, *110*, 10771–10776. [CrossRef]
64. Nguyen-Kim, H.; Bettarel, Y.; Bouvier, T.; Bouvier, C.; Doan-Nhu, H.; Nguyen-Ngoc, L.; Nguyen-Thanh, T.; Tran-Quang, H.; Brune, J. Coral mucus is a hot spot for viral infections. *Appl. Environ. Microbiol.* **2015**, *81*, 5773–5783. [CrossRef] [PubMed]
65. Nguyen-Kim, H.; Bouvier, T.; Bouvier, C.; Doan-Nhu, H.; Nguyen-Ngoc, L.; Rochelle-Newall, E.; Baudoux, A.C.; Desnues, C.; Reynaud, S.; Ferrier-Pages, C.; et al. High occurrence of viruses in the mucus layer of scleractinian corals. *Environ. Microbiol. Rep.* **2014**, *6*, 675–682. [CrossRef]
66. Bork, P.; Holm, L.; Sander, C. The Immunoglobulin Fold. Structural classification, sequence patterns and common core. *J. Mol. Biol.* **1994**, *242*, 309–320. [CrossRef]
67. Halaby, D.M.; Mornon, J.P.E. The immunoglobulin superfamily: An insight on its tissular, species, and functional diversity. *J. Mol. Evol.* **1998**, *46*, 389–400. [CrossRef]
68. Fraser, J.S.; Maxwell, K.L.; Davidson, A.R. Immunoglobulin-like domains on bacteriophage: Weapons of modest damage? *Curr. Opin. Microbiol.* **2007**, *10*, 382–387. [CrossRef]
69. Fraser, J.S.; Yu, Z.; Maxwell, K.L.; Davidson, A.R. Ig-like domains on bacteriophages: A tale of promiscuity and deceit. *J. Mol. Biol.* **2006**, *359*, 496–507. [CrossRef]
70. McMahon, S.A.; Miller, J.L.; Lawton, J.A.; Kerkow, D.E.; Hodes, A.; Marti-Renom, M.A.; Doulatov, S.; Narayanan, E.; Sali, A.; Miller, J.F.; et al. The C-type lectin fold as an evolutionary solution for massive sequence variation. *Nat. Struct. Mol. Biol.* **2005**, *12*, 886–892. [CrossRef]

71. Barr, J.J.; Auro, R.; Sam-Soon, N.; Kassegne, S.; Peters, G.; Bonilla, N.; Hatay, M.; Mourtada, S.; Bailey, B.; Youle, M.; et al. Subdiffusive motion of bacteriophage in mucosal surfaces increases the frequency of bacterial encounters. *Proc. Natl. Acad. Sci. USA* **2015**, *112*, 13675–13680. [CrossRef] [PubMed]
72. Barr, J.J. A bacteriophages journey through the human body. *Immunol. Rev.* **2017**, *279*, 106–122. [CrossRef] [PubMed]
73. Bille, E.; Meyer, J.; Jamet, A.; Euphrasie, D.; Barnier, J.P.; Brissac, T.; Larsen, A.; Pelissier, P.; Nassif, X. A virulence-associated filamentous bacteriophage of Neisseria meningitidis increases host-cell colonisation. *PLoS Pathog.* **2017**, *13*, 1–23. [CrossRef] [PubMed]
74. Guarner, F.; Malagelada, J.R. Gut flora in health and disease. *Lancet* **2003**, *361*, 512–519. [CrossRef]
75. Wiest, R.; Garcia-Tsao, G. Bacterial translocation (BT) in cirrhosis. *Hepatology* **2005**, *41*, 422–433. [CrossRef]
76. Costantini, T.W.; Putnam, J.G.; Sawada, R.; Baird, A.; Loomis, W.H.; Eliceiri, B.P.; Bansal, V.; Coimbra, R. Targeting the gut barrier: Identification of a homing peptide sequence for delivery into the injured intestinal epithelial cell. *Surgery* **2009**, *146*, 206–212. [CrossRef]
77. Ivanenkov, V.V.; Felici, F.; Menon, A.G. Uptake and intracellular fate of phage display vectors in mammalian cells. *Biochim. Biophys. Acta Mol. Cell Res.* **1999**, *1448*, 450–462. [CrossRef]
78. Keller, R.; Engley, F.B. Fate of Bacteriophage Particles Introduced into Mice by Various Routes. *Exp. Biol. Med.* **1958**, *98*, 577–580. [CrossRef]
79. Wolochow, H.; Hildebrand, G.J.; Lamanna, C. Translocation of microorganisms across the intestinal wall of the rat: Effect of microbial size and concentration. *J. Infect. Dis.* **1966**, *116*, 523–528. [CrossRef]
80. Reynaud, A.; Cloastre, L.; Bernard, J.; Laveran, H.; Ackermann, H.W.; Licois, D.; Joly, B. Characteristics and diffusion in the rabbit of a phage for Escherichia coli 0103. Attempts to use this phage for therapy. *Vet. Microbiol.* **1992**, *30*, 203–212. [CrossRef]
81. Jaiswal, A.; Koley, H.; Mitra, S.; Saha, D.R.; Sarkar, B. Comparative analysis of different oral approaches to treat Vibrio cholerae infection in adult mice. *Int. J. Med. Microbiol.* **2014**, *304*, 422–430. [CrossRef] [PubMed]
82. Jun, J.W.; Shin, T.H.; Kim, J.H.; Shin, S.P.; Han, J.E.; Heo, G.J.; De Zoysa, M.; Shin, G.W.; Chai, J.Y.; Park, S.C. Bacteriophage therapy of a Vibrio parahaemolyticus infection caused by a multiple-antibiotic-resistant O3:K6 pandemic clinical strain. *J. Infect. Dis.* **2014**, *210*, 72–78. [CrossRef] [PubMed]
83. Duerr, D.M.; White, S.J.; Schluesener, H.J. Identification of peptide sequences that induce the transport of phage across the gastrointestinal mucosal barrier. *J. Virol. Methods* **2004**, *116*, 177–180. [CrossRef] [PubMed]
84. Bruttin, A.; Brüssow, H. Human volunteers receiving Escherichia coli phage T4 orally: A safety test of phage therapy. *Antimicrob. Agents Chemother.* **2005**, *49*, 2874–2878. [CrossRef] [PubMed]
85. Denou, E.; Bruttin, A.; Barretto, C.; Ngom-Bru, C.; Brüssow, H.; Zuber, S. T4 phages against Escherichia coli diarrhea: Potential and problems. *Virology* **2009**, *388*, 21–30. [CrossRef] [PubMed]
86. Oliveira, A.; Sereno, R.; Nicolau, A.; Azeredo, J. The influence of the mode of administration in the dissemination of three coliphages in chickens. *Poult. Sci.* **2009**, *88*, 728–733. [CrossRef] [PubMed]
87. Letarova, M.; Strelkova, D.; Nevolina, S.; Letarov, A. A test for the "physiological phagemia" hypothesis—Natural intestinal coliphages do not penetrate to the blood in horses. *Folia Microbiol. (Praha)* **2012**, *57*, 81–83. [CrossRef]
88. McCallin, S.; Alam Sarker, S.; Barretto, C.; Sultana, S.; Berger, B.; Huq, S.; Krause, L.; Bibiloni, R.; Schmitt, B.; Reuteler, G.; et al. Safety analysis of a Russian phage cocktail: From MetaGenomic analysis to oral application in healthy human subjects. *Virology* **2013**, *443*, 187–196. [CrossRef]
89. Górski, A.; Ważna, E.; Weber-Dąbrowska, B.; Dąbrowska, K.; Switała-Jelen, K.; Miedzybrodzki, R.; Wazna, E.; Weber-Dabrowska, B.; Dabrowska, K.; Świtała-Jeleń, K.; et al. Bacteriophage translocation. *FEMS Immunol. Med. Microbiol.* **2006**, *46*, 313–319. [CrossRef]
90. Weiss, M.; Denou, E.; Bruttin, A.; Serra-Moreno, R.; Dillmann, M.L.; Brüssow, H. In vivo replication of T4 and T7 bacteriophages in germ-free mice colonized with Escherichia coli. *Virology* **2009**, *393*, 16–23. [CrossRef]
91. Nguyen, S.; Baker, K.; Padman, B.S.; Patwa, R.; Dunstan, R.A.; Weston, T.A.; Schlosser, K.; Bailey, B.; Lithgow, T.; Lazarou, M.; et al. Bacteriophage Transcytosis Provides a Mechanism To Cross Epithelial Cell Layers. *MBio* **2017**, *8*, e01874-17. [CrossRef] [PubMed]
92. Zhang, L.; Sun, L.; Wei, R.; Gao, Q.; He, T.; Xu, C.; Liu, X.; Wang, R. Intracellular Staphylococcus aureus control by virulent bacteriophages within MAC-T bovine mammary epithelial cells. *Antimicrob. Agents Chemother.* **2017**, *61*, AAC.01990-16. [CrossRef] [PubMed]

93. Sharp, R. Bacteriophages: Biology and history. *J. Chem. Technol. Biotechnol.* **2011**, 667–672. [CrossRef]
94. Di Giovine, M.; Salone, B.; Martina, Y.; Amati, V.; Zambruno, G.; Cundari, E.; Failla, C.M.; Saggio, I. Binding properties, cell delivery, and gene transfer of adenoviral penton base displaying bacteriophage. *Virology* **2001**, *282*, 102–112. [CrossRef] [PubMed]
95. Boulifard, D.A.; Labouvie, E. Gene Transfer into Mammalian Cells Using Targeted Filamentous Bacteriophage. *Int. J. Psychol.* **2009**, *19*, 1–20. [CrossRef]
96. Poul, M.A.; Marks, J.D. Targeted gene delivery to mammalian cells by filamentous bacteriophage. *J. Mol. Biol.* **1999**, *288*, 203–211. [CrossRef]
97. Lengeling, A.; Mahajan, A.; Gally, D. Bacteriophages as pathogens and immune modulators? *MBio* **2013**. [CrossRef]
98. Lehti, T.A.; Pajunen, M.I.; Skog, M.S.; Finne, J. Internalization of a polysialic acid-binding Escherichia coli bacteriophage into eukaryotic neuroblastoma cells. *Nat. Commun.* **2017**, *8*, 1915. [CrossRef]
99. Tam, J.C.H.; Jacques, D.A. Intracellular immunity: Finding the enemy within-how cells recognize and respond to intracellular pathogens. *J. Leukoc. Biol.* **2014**, *96*, 233–244. [CrossRef]
100. Barfoot, R.; Denham, S.; Gyure, L.A.; Hall, J.G.; Hobbs, S.M.; Jackson, L.E.; Robertson, D. Some properties of dendritic macrophages from peripheral lymph. *Immunology* **1989**, *68*, 233–239.
101. Wenger, S.L.; Turner, J.H.; Petricciani, J.C. The cytogenetic, proliferative and viability effects of four bacteriophages on human lymphocytes. *In Vitro* **1978**, *14*, 543–549. [CrossRef] [PubMed]
102. Aronow, R.; Danon, D. Electron microscopy of in vitro endocytosis of T2 phage by cells from rabbit peritoneal exudate. *J. Exp. Med.* **1964**, *120*, 943–954. [CrossRef] [PubMed]
103. Nungester, W.J.; Watrous, R.M. Accumulation of Bacteriophage in Spleen and Liver Following Its Intravenous Inoculation. *Exp. Biol. Med.* **1934**, *31*, 901–905. [CrossRef]
104. Geier, M.R.; Trigg, M.E.; Merril, C.R. Fate of bacteriophage lamba in non-immune germ-free mice. *Nature* **1973**, *246*, 221–223. [CrossRef] [PubMed]
105. Merril, C.R.; Biswas, B.; Carlton, R.; Jensen, N.C.; Creed, G.J.; Zullo, S.; Adhya, S. Long-circulating bacteriophage as antibacterial agents. *Proc. Natl. Acad. Sci. USA* **1996**, *93*, 3188–3192. [CrossRef] [PubMed]
106. Mukerjee, S.; Ghosh, S.N. Localization of cholera bacteriophage after intravenous injection. *Ann. Biochem. Exp. Med.* **1962**, *22*, 73–76. [PubMed]
107. Inchley, C.J. The activity of mouse Kupffer cells following intravenous injection of T4 bacteriophage. *Clin Exp. Immunol* **1969**, *5*, 173–187.
108. Cerveny, K.E.; DePaola, A.; Duckworth, D.H.; Gulig, P.A. Phage therapy of local and systemic disease caused by Vibrio vulnificus in iron-dextran-treated mice. *Infect. Immun.* **2002**, *70*, 6251–6262. [CrossRef]
109. Tiwari, B.R.; Kim, S.; Rahman, M.; Kim, J. Antibacterial efficacy of lytic Pseudomonas bacteriophage in normal and neutropenic mice models. *J. Microbiol.* **2011**, *49*, 994–999. [CrossRef]
110. Roach, D.R.; Leung, C.Y.; Henry, M.; Morello, E.; Singh, D.; Santo, P.D.; Weitz, J.S.; Debarbieux, L.; Unit, I.I. Immunophage synergy is essential for eradicating pathogens that provoke acute respiratory infections. *Cell Host Microbe* **2017**, *22*, 38–47. [CrossRef]
111. Pincus, N.B.; Reckhow, J.D.; Saleem, D.; Jammeh, M.L.; Datta, S.K.; Myles, I.A. Strain specific phage treatment for Staphylococcus aureus infection is influenced by host immunity and site of infection. *PLoS ONE* **2015**, *10*, e0124280. [CrossRef] [PubMed]
112. Leung, C.Y.; Weitz, J.S. Modeling the synergistic elimination of bacteria by phage and the innate immune system. *J. Theor. Biol.* **2017**, *429*, 241–252. [CrossRef] [PubMed]
113. Weber-Dąbrowska, B.; Zimecki, M.; Mulczyk, M. Effective phage therapy is associated with normalization of cytokine production by blood cell cultures. *Arch. Immunol. Ther. Exp.* **2000**, *48*, 31–37.
114. Zhang, L.; Hou, X.; Sun, L.; He, T.; Wei, R.; Pang, M.; Wang, R. Staphylococcus aureus Bacteriophage Suppresses LPS-Induced Inflammation in MAC-T Bovine Mammary Epithelial Cells. *Front. Microbiol.* **2018**, *9*, 1–8. [CrossRef]
115. Park, K.; Cha, K.E.; Myung, H. Observation of inflammatory responses in mice orally fed with bacteriophage T7. *J. Appl. Microbiol.* **2014**, *117*, 627–633. [CrossRef] [PubMed]
116. Kaur, S.; Harjai, K.; Chhibber, S. Bacteriophage-aided intracellular killing of engulfed methicillin-resistant Staphylococcus aureus (MRSA) by murine macrophages. *Appl. Microbiol. Biotechnol.* **2014**, *98*, 4653–4661. [CrossRef] [PubMed]

117. Górski, A.; Międzybrodzki, R.; Borysowski, J.; Dąbrowska, K.; Wierzbicki, P.; Ohams, M.; Korczak-Kowalska, G.; Olszowska-Zaremba, N.; Łusiak-Szelachowska, M.; Kłak, M.; et al. Phage as a modulator of immune responses: Practical implications for phage therapy. *Adv. Virus Res.* **2012**, *83*, 41–71. [CrossRef]
118. Jończyk-Matysiak, E.; Łusiak-Szelachowska, M.; Kłak, M.; Bubak, B.; Międzybrodzki, R.; Weber-Dąbrowska, B.; Zaczek, M.; Fortuna, W.; Rogóz, P.; Letkiewicz, S.; et al. The effect of bacteriophage preparations on intracellular killing of bacteria by phagocytes. *J. Immunol. Res.* **2015**, *2015*. [CrossRef]
119. Przerwa, A.; Zimecki, M.; Świtała-Jeleń, K.; Dąbrowska, K.; Krawczyk, E.; Łuczak, M.; Weber-Dąbrowska, B.; Syper, D.; Międzybrodzki, R.; Górski, A. Effects of bacteriophages on free radical production and phagocytic functions. *Med. Microbiol. Immunol.* **2006**, *195*, 143–150. [CrossRef]
120. Miedzybrodzki, R.; Switala-Jelen, K.; Fortuna, W.; Weber-Dąbrowska, B.; Przerwa, A.; Lusiak-Szelachowska, M.; Dąbrowska, K.; Kurzepa, A.; Boratynski, J.; Syper, D.; et al. Bacteriophage preparation inhibition of reactive oxygen species generation by endotoxin-stimulated polymorphonuclear leukocytes. *Virus Res.* **2008**, *131*, 233–242. [CrossRef]
121. Miernikiewicz, P.; Klopot, A.; Soluch, R.; Szkuta, P.; Keska, W.; Hodyra-Stefaniak, K.; Konopka, A.; Nowak, M.; Lecion, D.; Kazmierczak, Z.; et al. T4 phage tail Adhesin Gp12 counteracts LPS-induced inflammation In Vivo. *Front. Microbiol.* **2016**, *7*, 1–8. [CrossRef] [PubMed]
122. Górski, A.; Nowaczyk, M.; Weber-Dabrowska, B.; Kniotek, M.; Boratynski, J.; Ahmed, A.; Dabrowska, K.; Wierzbicki, P.; Switala-Jelen, K.; Opolski, A. New insights into the possible role of bacteriophages in transplantation. *Transplant. Proc.* **2003**, *35*, 2372–2373. [CrossRef]
123. Górski, A.; Miedzybrodzki, R.; Weber-Dabrowska, B.; Fortuna, W.; Letkiewicz, S.; Rogóz, P.; Jończyk-Matysiak, E.; Dabrowska, K.; Majewska, J.; Borysowski, J. Phage therapy: Combating infections with potential for evolving from merely a treatment for complications to targeting diseases. *Front. Microbiol.* **2016**, *7*, 1515. [CrossRef] [PubMed]
124. Kamme, C. Antibodies against staphylococcal bacteriophages in human sera. *Acta Pathol. Microbiol. Scand. B Microbiol. Immunol.* **1973**, *81*, 741–748. [CrossRef] [PubMed]
125. Smith, H.W.; Huggins, M.B.; Shaw, K.M. Factors influencing the survival and multiplication of bacteriophages in calves and in their environment. *J. Gen. Microbiol.* **1987**, *133*, 1127–1135. [CrossRef] [PubMed]
126. Dąbrowska, K.; Miernikiewicz, P.; Piotrowicz, A.; Hodyra, K.; Owczarek, B.; Lecion, D.; Ka mierczak, Z.; Letarov, A.; Górski, A. Immunogenicity studies of Proteins Forming the T4 Phage Head Surface. *J. Virol.* **2014**, *88*, 12551–12557. [CrossRef] [PubMed]
127. Fogelman, I.; Davey, V.; Ochs, H.D.; Elashoff, M.; Feinberg, M.B.; Mican, J.; Siegel, J.P.; Sneller, M.; Lane, H.C. Evaluation of CD4+ T cell function In vivo in HIV-infected patients as measured by bacteriophage phiX174 immunization. *J. Infect. Dis.* **2000**, *182*, 435–441. [CrossRef] [PubMed]
128. Ochs, H.D.; Davis, S.D.; Wedgwood, R.J. Immunologic responses to bacteriophage phi-X 174 in immunodeficiency diseases. *J. Clin. Investig.* **1971**, *50*, 2559–2568. [CrossRef]
129. Rubinstein, A.; Mizrachi, Y.; Bernstein, L.; Shliozberg, J.; Golodner, M.; Liu, G.Q.; Ochs, H.D. Progressive specific immune attrition after primary, secondary and tertiary immunizations with bacteriophage phi X174 in asymptomatic HIV-1 infected patients. *AIDS* **2000**, *14*, F55–F62. [CrossRef]
130. Shearer, W.T.; Lugg, D.J.; Rosenblatt, H.M.; Nickolls, P.M.; Sharp, R.M.; Reuben, J.M.; Ochs, H.D. Antibody responses to bacteriophage φX-174 in human subjects exposed to the Antarctic winter-over model of spaceflight. *J. Allergy Clin. Immunol.* **2001**, *107*, 160–164. [CrossRef]
131. Dąbrowska, K.; Switala-Jelen, K.; Opolski, A.; Weber-Dąbrowska, B.; Górski, A. Bacteriophage penetration in vertebrates. *J. Appl. Microbiol.* **2005**, *98*, 7–13. [CrossRef] [PubMed]
132. Jerne, N.K. The presence in normal serum of specific antibody against bacteriophage T4 and its increase during the earliest stages of immunization. *J. Immunol.* **1956**, *76*, 209–216. [PubMed]
133. Jerne, N.K. Bacteriophage inactivation by antiphage serum diluted in distilled water. *Nature* **1952**, *169*, 117–118. [CrossRef] [PubMed]
134. Żaczek, M.; Łusiak-Szelachowska, M.; Jończyk-Matysiak, E.; Weber-Dąbrowska, B.; Międzybrodzki, R.; Owczarek, B.; Kopciuch, A.; Fortuna, W.; Rogóż, P.; Górski, A.; et al. Antibody production in response to Staphylococcal MS-1 phage cocktail in patients undergoing phage therapy. *Front. Microbiol.* **2016**, *7*, 1–14. [CrossRef]

135. Łusiak-Szelachowska, M.; Żaczek, M.; Weber-Dąbrowska, B.; Międzybrodzki, R.; Kłak, M.; Fortuna, W.; Letkiewicz, S.; Rogóż, P.; Szufnarowski, K.; Jończyk-Matysiak, E.; et al. Phage Neutralization by Sera of Patients Receiving Phage Therapy. *Viral Immunol.* **2014**, *27*, 295–304. [CrossRef] [PubMed]
136. Levin, B.R.; Bull, J.J. Population and evolutionary dynamics of phage therapy. *Nat. Rev. Microbiol.* **2004**, *2*, 166–173. [CrossRef] [PubMed]
137. Payne, R.J.H.; Jansen, V.A.A. Pharmacokinetic principles of bacteriophage therapy. *Clin. Pharmacokinet.* **2003**, *42*, 315–325. [CrossRef] [PubMed]
138. Payne, R.J.; Jansen, V.A. Understanding bacteriophage therapy as a density-dependent kinetic process. *J. Theor. Biol.* **2001**, *208*, 37–48. [CrossRef]
139. Payne, R.J.H.; Jansen, V.A.A. Phage therapy: The peculiar kinetics of self-replicating pharmaceuticals. *Clin. Pharmacol. Ther.* **2000**, *68*, 225–230. [CrossRef]
140. Loc-Carrillo, C.; Abedon, S. Pros and cons of phage therapy. *Bacteriophage* **2011**, *1*, 111–114. [CrossRef]
141. Cairns, B.J.; Timms, A.R.; Jansen, V.A.A.; Connerton, I.F.; Payne, R.J.H. Quantitative Models of In Vitro Bacteriophage–Host Dynamics and Their Application to Phage Therapy. *PLoS Pathog.* **2009**, *5*, e1000253. [CrossRef] [PubMed]
142. Kasman, L.M.; Kasman, A.; Westwater, C.; Dolan, J.; Schmidt, M.G.; Norris, J.S. Overcoming the phage replication threshold: A mathematical model with implications for phage therapy. *J. Virol.* **2002**, *76*, 5557–5564. [CrossRef] [PubMed]
143. Abedon, S.; Thomas-Abedon, C. Phage therapy pharmacology. *Curr. Pharm. Biotechnol.* **2010**, *11*, 28–47. [CrossRef] [PubMed]
144. Ho, K. Bacteriophage Therapy for Bacterial Infections: Rekindling a Memory from the Pre-Antibiotics Era. *Perspect. Biol. Med.* **2001**, *44*, 1–16. [CrossRef] [PubMed]
145. Aleshkin, A.V.; Ershova, O.N.; Volozhantsev, N.V.; Svetoch, E.A.; Popova, A.V.; Rubalskii, E.O.; Borzilov, A.I.; Aleshkin, V.A.; Afanas'ev, S.S.; Karaulov, A.V.; et al. Phagebiotics in treatment and prophylaxis of healthcare-associated infections. *Bacteriophage* **2016**, *6*, e1251379. [CrossRef] [PubMed]
146. Kingwell, K. Bacteriophage therapies re-enter clinical trials. *Nat. Rev. Drug Discov.* **2015**, *14*, 515–516. [CrossRef] [PubMed]
147. Mattey, M.; Spencer, J. Bacteriophage therapy–cooked goose or phoenix rising? *Curr. Opin. Biotechnol.* **2008**, *19*, 608–612. [CrossRef] [PubMed]
148. Debarbieux, L.; Leduc, D.; Maura, D.; Morello, E.; Criscuolo, A.; Grossi, O.; Balloy, V.; Touqui, L. Bacteriophages can treat and prevent Pseudomonas aeruginosa lung infections. *J. Infect. Dis.* **2010**, *201*, 1096–1104. [CrossRef] [PubMed]
149. Secor, P.R.; Michaels, L.A.; Smigiel, K.S.; Rohani, M.G.; Jennings, L.K.; Hisert, K.B.; Arrigoni, A.; Braun, K.R.; Birkland, T.P.; Lai, Y.; et al. Filamentous bacteriophage produced by Pseudomonas aeruginosa alters the inflammatory response and promotes noninvasive infection in vivo. *Infect. Immun.* **2017**, *85*, e00648-16. [CrossRef]
150. Janeway, C.A.; Medzhitov, R. Innate Immune Recognition. *Annu. Rev. Immunol.* **2002**, *20*, 197–216. [CrossRef]
151. Eriksson, F.; Tsagozis, P.; Lundberg, K.; Parsa, R.; Mangsbo, S.M.; Persson, M.A.A.; Harris, R.A.; Pisa, P.; Culp, W.D.; Massey, R.; et al. Tumor-specific bacteriophages induce tumor destruction through activation of tumor-associated macrophages. *J. Immunol.* **2009**, *182*, 3105–3111. [CrossRef] [PubMed]
152. Eriksson, F.; Culp, W.D.; Massey, R.; Egevad, L.; Garland, D.; Persson, M.A.A.A.; Pisa, P. Tumor specific phage particles promote tumor regression in a mouse melanoma model. *Cancer Immunol. Immunother.* **2007**, *56*, 677–687. [CrossRef] [PubMed]

Review

Resistance Development to Bacteriophages Occurring during Bacteriophage Therapy

Frank Oechslin

Department of Fundamental Microbiology (DMF), University of Lausanne, CH-1015 Lausanne, Switzerland; frank.oechslin@gmail.com

Received: 10 June 2018; Accepted: 28 June 2018; Published: 30 June 2018

Abstract: Bacteriophage (phage) therapy, i.e., the use of viruses that infect bacteria as antimicrobial agents, is a promising alternative to conventional antibiotics. Indeed, resistance to antibiotics has become a major public health problem after decades of extensive usage. However, one of the main questions regarding phage therapy is the possible rapid emergence of phage-resistant bacterial variants, which could impede favourable treatment outcomes. Experimental data has shown that phage-resistant variants occurred in up to 80% of studies targeting the intestinal milieu and 50% of studies using sepsis models. Phage-resistant variants have also been observed in human studies, as described in three out of four clinical trials that recorded the emergence of phage resistance. On the other hand, recent animal studies suggest that bacterial mutations that confer phage-resistance may result in fitness costs in the resistant bacterium, which, in turn, could benefit the host. Thus, phage resistance should not be underestimated and efforts should be made to develop methodologies for monitoring and preventing it. Moreover, understanding and taking advantage of the resistance-induced fitness costs in bacterial pathogens is a potentially promising avenue.

Keywords: bacteriophage; phage; phage therapy; phage-resistance

1. Introduction

Antimicrobial resistance is a major public health problem that could possibly cause an estimated 10 million mortalities per year by 2050 [1]. For this reason, novel therapeutic strategies, beside traditional antibiotics, must be rapidly developed. One of these strategies is the use of bacteriophages (phages). Phages are nature's most abundant bacterial predators. They can be used alone or in combination with antibiotics against difficult-to-treat infections. Phage therapy has been used since the 1920s in the Soviet Union, and is still currently used in ex-Soviet countries like Poland, Russia, and Georgia [2]. Phage therapy is currently being revisited as a potential alternative to antibiotics in Western countries. However, challenging issues still exist, such as selecting the most adequate phage(s) against a given infection, the risk of phage resistance development, the immune response to phages by the host, as well as novel regulatory requirements [3,4].

Bacteria can resist phage attack through different mechanisms, including spontaneous mutations, restriction modification systems, and adaptive immunity via the CRISPR-Cas system [5]. Spontaneous mutations are the main mechanisms driving both phage resistance and phage–bacterial coevolution [6]. Spontaneous mutations may confer phage resistance by modifying the structure of bacterial surface components that act as phage receptors and that also determine phage specificity. These include lipopolysaccharides (LPS), outer membrane proteins, cell wall teichoic acids, capsules, and other bacterial appendices, such as flagella, many of which may all be part of virulence factors (e.g., LPS) [7].

However, phage resistance may also induce trade-off costs. Phage-resistant bacteria may become less virulent in case of mutations in surface virulence factors, such as LPS [8]. Likewise, the maintenance of anti-viral defence systems, such as for DNA restriction-modification enzymes and CRISPR-Cas adaptive immunity, also has its own cost [9,10].

This review discusses the implications of the development of phage resistance in the perspective of implementing phage therapy. Phage resistance is first considered in the context of population and phage–bacterial evolutionary dynamics. It is then considered in the frame of experimental therapy (summarized in Table 1), in order to determine its role in treatment failure or salvage therapy strategies.

Table 1. Principal in vivo studies investigating the relation between phage therapy and phage resistance.

	Bacterium	Model	Phage Type	Treatment Outcome	Resistant Found in after Treatment?	Impact of Resistance on Virulence	Receptor	Ref.
Intestinal colonization	*Campylobacter jejuni*	Chicken intestinal colonization	CP8 and CP34	Bacterial decrease between 0.5 and 5 log10 CFU/g of caecal contents compared to untreated controls over a 5-day period post-administration.	Yes, at a freq. of 4%	Less infective at low dose. Rapid phenotypic reversion when reintroduced in chicken.	ND	[11,12]
	Campylobacter jejuni	Chicken intestinal colonization	phiCcoIBB35, phiCcoIBB37, and phiCcoIBB12	Phage cocktail decreases the titre of *C. jejuni* in faeces by approximately 2 log10 CFU/g when administered orally.	Yes, at a freq. of 13%	Not less infective. No phenotypic reversion when reintroduced in chicken.	ND	[13]
	Escherichia coli	Calf, piglet, lamb Ta diarrhoea	B44/1, B44/2, B44/3, P433/1, and P433/2	Protected calves against a potentially lethal infection, cured diarrhoea in piglets, improved the course of disease in lambs.	Most calves that did not respond to phage treatment had a high number of phage-resistant variants. No phage-resistant mutants were isolated from lambs.	Decreased virulence	Capsular polysaccharides	[14]
	Escherichia coli	Calf diarrhoea	B41/1	Rapid reduction of bacterial titres to numbers that are harmless.	Yes	Reduced virulence	Capsular polysaccharides	[15]
	Escherichia coli	Sheep, mouse, steer intestinal colonization	KH1 and SH1	Oral phage treatment did not decrease intestinal *E. coli* in sheep. Decreased the number of *E. coli* CFU in cattle. Phage therapy cleared the bacteria in a mouse model of intestinal *E. coli* O157 carriage.	No	-	ND	[16]
	Escherichia coli	Mouse intestinal colonization	T4 phage, oral	ND	Phage resistant bacterial strains dominated gut after 92 days.	ND	ND	[17]
	Escherichia coli	Mouse intestinal colonization	cocktail made of phages CLB_P1, CLB_P2, and CLB_P3	No bacterial level change in the faeces after treatment.	No	-	ND	[18]
	Enterococcus faecalis	Gnotobiotic mouse intestinal colonization	φ VPE25	Threefold drop in *E. faecalis* total intestinal load after 24 h of VPE25 treatment.	Phage resistant variant replaced WT during treatment.	Resistant variants can colonize intestine.	Integral membrane protein PIPEF	[19]
	Salmonella enterica	Chicken intestinal colonization	cocktail of phages, EP2, MUT3, M4, and YP	Significant difference between phage-treated and untreated groups.	Yes	ND	ND	[20]
	Salmonella enterica	Chicken intestinal colonization	φ10, φ25, and φ151	Phages reduced caecal colonization.	Phage-resistance occurred at a frequency commensurate with the titre of phage being administered.	Colonization levels of resistant variants in the ceca did not differ from the controls. Reversion observed after infection.	ND	[21]
	Vibrio cholerae	Infant mouse and rabbit cholera model	ICP1, ICP2, and ICP3	Oral administration of phages up to 24 h before *V. cholerae* challenge reduced colonization of the intestinal tract and prevented cholera-like diarrhoea.	Yes	Variants can colonize intestine.	O-Antigen	[22]

Table 1. *Cont.*

	Bacterium	Model	Phage Type	Treatment Outcome	Resistant Found in after Treatment?	Impact of Resistance on Virulence	Receptor	Ref.
Meningitis	*Escherichia coli*	Mouse meningitis	phage R	One dose of phage was at least equivalent to multiple doses of antibiotics, whether administered intramuscularly or intrathecally.	Yes	Supposably reduced virulence as described in [14].	Capsular polysaccharides	[23]
Endocarditis	*Pseudomonas aeruginosa*	Rat infective endocarditis	cocktail made of phages 12 bacteriophages	3 log reduction or valve sterilisation when combined with antibiotics.	No	Reduced virulence	LPS and pilus	[24]
Sepsis	*Escherichia coli*	Rat neonatal sepsis	phage $EC200^{PP}$	Phage administered 7 h postinfection rescued 100% of the animals and 50% after 24 h.	Phage resistant variant were found when phage treatment was delayed for 24 h.	Avirulence	ND	[25]
	Klebsiella pneumoniae	Mouse liver abscess and bacteraemia	Phage φNK5	Intraperitoneal and intragastric administration of phage 30 min after infection protected mice from death in a dose-dependent manner. Decreased bacterial burden and liver damage.	No	Reduced virulence	ND	[26]
	Staphylococcus aureus	Experimental cow mastitis	Bacteriophage K	Decreased bacterial load after treatment.	Yes	ND	ND	[27]
	Pseudomonas plecoglossicida	Fish haemorrhagic ascites	PPpW-3 and PPpW-4. Oral	Protective effects of phage treatment with lower and delayed mortality 1 or 24 h after bacterial challenge.	No	Reduced virulence	ND	[28]
	Klebsiella pneumoniae	Mice acute bacteraemia	GH-K1, GH-K2, and GH-K3	Phage cocktail significantly enhanced the protection of bacteremic mice against lethal infection.	ND	Reduced virulence	ND	[29]
	Salmonella enterica Parathyphi B	Mouse sepsis	phage φ1	Phage given concurrently with a lethal dose of bacteria rescued 100% of the animals.	ND	Avirulence	O-Antigen	[30]
pneumonia	*Pseudomonas aeruginosa*	Mouse acute pneumonia	PAK_P1	Treatment failed to prevent fatality due to subsequent bacterial outgrowth after 24 h in immunocompromised mice. 100% of bacteria recovered from phage-treated at 24 h were resistant.	Yes, in immunocompromised mice	ND	ND	[31]

Table 1. *Cont.*

	Bacterium	Model	Phage Type	Treatment Outcome	Resistant Found in after Treatment?	Impact of Resistance on Virulence	Receptor	Ref.
Phage resistant variants for vaccine production and studying virulence factors	*Vibrio cholerae*	Infant mouse cholera model	ICP1	ND	ND	Attenuated in vivo.	O-Antigen	[32]
	Vibrio cholerae	Infant mouse cholera model	K139	ND	ND	Significantly reduced in its ability to colonize the mouse small intestine.	Core oligosaccharide	[33]
	Vibrio cholerae	Infant mouse cholera model	phage JA1	ND	ND	Impaired colonization	Capsule/O-antigen	[34]
	Staphylococcus aureus	Mouse vaccination	M^{Sa} phage	ND	ND	Avirulence	Teichoic acids	[35]
	Yersinia pestis	Mouse vaccination	L-413C, P2 vir1, φ JA1a, φ A1122, T7, $T7Yp^{e}$, Pokrovskaya, Y, PST, R^{h}	ND	ND	Atenuated or avirulent.	LPS	[36]
	Flavobacterium columnare	Zebrafish	FCL-1 and FCL-2	ND	ND	Avirulence	ND	[37]
	Bacillus thuringiensis	Cecropia moth	φ42, φ51,and φ64	ND	ND	Decreased virulence	ND	[38]
	Serratia marcescens	Cecropia moth, Drosophila	Phages φJ and φK	ND	ND	Decreased virulence	ND	[39]
	Serratia marcescens	Cecropia moth	Phages φJ	ND	ND	Decreased virulence	ND	[40]

2. The Evolution Dynamics of Resistance

Phage–bacteria coevolution can be defined as a process of reciprocal adaptation and counter-adaptation between the phage and its bacterial host. It is an important driving force for the ecology and evolution of microbial communities [6]. Phage–bacteria interactions are mediated first by phages using their tail fibres to adsorb onto the bacterial surface, through a lock-key mechanism. Since the complete phage replication life cycle, namely the lytic cycle, relies on the killing and lysis of the host bacteria, a strong reciprocal selection pressure evolves toward increased infectivity on the side of the phage, and phage-resistance on the side of the bacterium [41]. Bacteria can evolve phage-resistance by de novo chromosomal mutations, as well as through an arsenal of antiviral mechanisms targeting virtually all steps of the phage life cycle (reviewed in [5]). For example, bacteria can prevent phage adsorption by modifying the structure of their surface phage receptors, or by hindering the access of the phage to the receptor through the production of an excess of the extracellular matrix, or even by producing competitive inhibitors [42–44].

The development of phage-resistant bacteria was already described almost a century ago in a seminal paper by Luria and Delbrück, who observed that the initial phage-induced lysis of a bacterial population was followed by bacterial regrowth, due to the selection of a phage-resistant sub-populations [45]. Phage-resistant bacterial variants that were already present in the initial bacterial culture (at a rate of ca. 10^{-8}) were selected and led to the replacement of the entire culture with the resistant variant. This extreme situation leads to an evolutionary dead-end, for instance if the phage receptor is lost, and phage do not have the opportunity to develop a counter-resistance. A large number of studies concluded similarly on the absence of phage–bacterial coevolution following the emergence of phage-resistant bacteria (reviewed in [46]). This was the case in chemostat experiments using *Escherichia coli* infected with series of T or lambda phages. Phages could not interact anymore with the resistant *E. coli* variants [47]. However, phage-susceptible parent bacteria could still be recovered from sanctuary niches such as biofilms present on the chemostat's walls. These survivors allowed maintaining low levels of phage persistence.

Other studies observed more persistent cycles of coevolution between phages and the host bacteria. This was the case for *E. coli* O157:H7 and phage PP01 in a continuous chemostat culture [48]. In this experiment, phages could coexist with phage-resistant variants and evolve different host ranges for the phage-escape bacterial mutants. Phage-resistance was associated with a dual bacterial population carrying either LPS alterations or OmpC surface protein silencing. Moreover, a third type of mucoid colony mutant emerged and could coexist with phages until the end of the experiment. Eventually, phage mutants with different host ranges also appeared.

One possible explanation was that none of the three phage-resistant bacterial mutants were completely immune to phage infection and, thus, entered a phage coevolution cycle permitting parallel phage and bacterial expansion and selection for phage variants with broader host range.

In other experiments, using lambda phages, an arms race was observed when using minimal media and maltose as the only carbon source [49]. Since the lambda phage uses the maltose outer membrane porin, LamB, as a bacterial receptor, decreasing LamB synthesis decreased phage susceptibility. However, this also decreased bacterial fitness in the presence of a lactose-only carbon source. In parallel, phages selected variants with increased LamB affinity or new variants able to infect via alternative receptors. Phages shifted from using the LamB to OmpF receptor through amino acid substitutions in their tail fibre J protein [50]. This mutual counter-selection process between the phage and the bacterium that enabled each of them to survive, without eliminating the other.

Further confirmation of the arms race came from observations with *Pseudomonas fluroescens* and the Podovirus ϕ2 [51,52]. In this case, reciprocal evolution of infectivity and resistance was followed during >100 bacterial generations, with phages becoming more broadly infectious and bacteria more broadly resistant over time [41,52]. However, the arms race became progressively weaker, with increasing fitness cost due to generalist adaptive mutations on both sides [53].

A common paradigm in evolutionary biology is that evolution tends to maximize the adaptation ability (in this case phage-resistance) by allocating resources preferentially when they are limited. If resources are dedicated to phage-resistance, a fitness cost may be associated with a mutation conferring phage resistance that arises during coevolution, as demonstrated for the altered integrity of LPS and OmpC resulting in heterogeneous populations in the example of *E. coli* O157:H7 and phage PP01 [48], and altered maltose uptake with porin LamB mutation in the example of lambda phage and maltose restriction [49]. Numerous other examples exist regarding various phage–bacteria coevolution systems (see [54–57]). Environmental conditions can also greatly impact the coevolution of virulence and resistance by imposing limits to the arm race, as exemplified by the LamB example where maltose becomes a limiting factor [49]. This may result in different coevolution routes in natural settings from what is observed in the test tube. Such exogenous factors can include UV light-induced mutagenesis, which may induce phage-resistance mutations, but also cause additional mutations affecting in parallel [57]. Other examples include the influence of still or shaking culture conditions on population structures, which may influence particle collisions between phages and bacterial preys, or niche resource availability [58–60].

A recent work with *P. fluorescens* SBW25 showed that phage–bacterial interaction increased the bacterial mutation rates as well as its chance to adapt and survive both predation and altered environmental conditions. This indicates that phage-driven evolution may be ultimately beneficial for the bacteria [61].

Interestingly, while this phage-stimulated mutation rate of *P. fluorescens* SBW25 was observed in laboratory conditions [61], it was not observed for *P. fluorescens* communities living in the soil [62], most probably because selection for phage-resistance is higher in the soil than in vitro [63]. It is then assumed that, unlike in vitro coevolution that is characterized by the so-called arms race of increased resistance versus infectivity over time, coevolution in natural environments is largely driven by fluctuating selection, where cycles of phage-susceptible and phage-resistant bacterial populations intertwine with parent phages and evolved phages (mutant phages) that regain infectivity against the resistant bacteria [41].

3. Emergences of Phage Resistance in Animal Models

3.1. *E. coli Diarrhoea in Cattle*

Controlled studies on phage therapy and the emergence of phage resistance started in Western countries with a series of farm animal trials against experimental diarrhoea with enteropathogenic *E. coli* strains. In a first trial, oral phage therapy prevented *E. coli*-induced diarrhoea in colostrum-fed calves even when given 8 h after bacterial inoculation [14]. No phage-resistant variants were isolated from the calves. In another trial, phage therapy was administered at the onset of diarrhoeal symptoms, but resolved the intestinal symptoms in only half of the animals (14 out of 21 calves died). Phage-resistant variants were recovered from the small intestine of all calves failing clinical improvement. In parallel, the faecal bacterial content of 11 calves that responded to phage therapy was examined over a period of 14 days. *E. coli* titres progressively decreased over time and phage numbers increased during the first 48 h, therefore indicating phage replication. Phage-resistant bacteria emerged after 19 h, peaked at 2 days and decreased to undetectable levels after 10 days. Of note, these resistant mutants did not proliferate in the small intestine and did not cause diarrhoea when reinoculated to healthy colostrum-fed calves. The reduced virulence was associated with loss of the K antigen, which is known to be a virulence factor for enteropathogenic strains [64].

Prevention of diarrhoea by oral phages was also observed in piglets and lambs infected with *E. coli* [14]. Resistant variants were also isolated in the piglet faeces, but at relatively low rates. In contrast, no phage-resistant variants were observed in lambs. Similar results regarding the potential of phage therapy to prevent or treat *E. coli* diarrhoea were described in a second study done by Smith and Huggin [15]. This time, K-positive phage-resistant *E. coli* variants were isolated in addition to the already-observed K-negative variants. However, unlike the K-negative resistant variants, K-positive resistant variants were as virulent as the parent strain. These resistant variants were isolated only in vivo from the calves and not in vitro from a broth culture.

In another study in Holstein steers, phage therapy could reduce the average number of *E. coli* O157:H7 CFU in the faeces compared to controls, although it did not eliminate the bacteria from the majority of animals [16]. No phage-resistant *E. coli* O157:H7 mutants were observed.

Thus, phage-resistant variants are readily selected during phage therapy in vivo, but their pathological significance is unclear.

3.2. E. coli and Enterococcus faecalis Intestinal Colonization in Mice

Three studies using mouse models of *E. coli* intestinal colonization have also investigated the effect of phage therapy on the emergence of phage-resistance. In the first study, *E. coli* O157:H7 intestinal carriage could be eliminated within 48 h using three repeated phage oral doses. No resistant colonies were recovered and untreated control mice remained culture-positive for 10 days [16].

In the second study, an oral cocktail composed of three different phages was administered to mice colonized with enteroaggregative *E. coli* O104:H4 [18]. The bacterial titres did not decrease as compared to controls as expected and remained stable for 21 days, although phages were observed to replicate continuously during the time of the experiment. In addition, bacteria recovered on day 21 were still susceptible to each of the bacteriophages present in the cocktail when tested separately.

The third murine gut colonization study was a long-term 240 day protocol using oral T4 phages. Phage-resistant *E. coli* emerged only on day 92 and resistant variants constituted 100% of bacterial colonies isolated from phage-treated mice [17]. In comparison, only 20% of untreated mice carried phage-resistant variants. Moreover, when phage therapy was stopped at day 92, the presence of phage-resistant *E. coli* persisted over the 240 experimental days. The mechanism of phage-resistance was not described.

In an additional work, Duerkop et al. inoculated germ-free mice with *E. faecalis* V583 before starting phage therapy, which was first administered by oral gavage 6 h after colonization followed by administration in drinking water [19]. Phage therapy slightly decreased faecal bacterial loads by three-fold after 24 h. However, the level of colonization was not different from control animals after 48 h, even if phages were added to the drinking water. While 100% of the *E. faecalis* isolates remained phage-susceptible after 6 h of treatment, only 15% were still susceptible at 24 h and 100% were resistant at two days. Sequencing of 20 resistant bacterial variants revealed that they all had various mutations in the integral membrane protein PIPef, which was observed to promote phage infection.

Finally, using a model of gnotobiotic mouse intestinal colonization, Reyes et al. analysed the impact of viral predation on a consortium of 14 human bacterial symbionts [65]. When the community was subjected to virus-like particles purified from the faecal microbiota of human healthy donors, changes in the relative bacterial abundance was observed, but not with heat-killed viral-particle preparations. Especially, *Bacillus caccae* bacterial communities were observed to first decrease after phage attack, although they recovered later on. Evidence that phage resistance occurred due to genetic changes, like acquisition of CRISPR elements, could not be observed. The authors hypothesized that resistance was more the result of the expansion of an unexposed fraction of the population that could be protected in intestine microhabitat.

3.3. Control of Poultry Pathogens

A large part of the phage therapy application to control pathogenic bacteria in animals has been done in poultry, in order to prevent *Salmonella* spp. and *Campylobacter* spp. gut colonization and infection. *Salmonella enterica* is one of the major causes of foodborne infection in humans due to its symptomless carriage by chickens [66]. Up to now, the use of phage therapy to control *S. enterica* in poultry could reduce, but not eliminate, the bacteria. Sklar and Joerger reported that phage therapy did not significantly decrease *S. enterica* intestinal carriage in three animal trials [20].

Although phage resistance was observed after treatment, the authors speculate that other factors, like the salmonella intracellular lifestyle, could have contributed to the therapeutic failure. Atterbury et al. observed that low phage concentrations were not efficient in significantly decreasing bacterial loads [21]. They proposed that the low salmonella concentration in the chicken gut, associated with the complexity of the intestinal milieu—including physicochemical conditions, such as viscosity—was not suitable for phage amplification. Increasing the phage titre of phage preparations increase efficacy, but did not result in bacterial eradication. Interestingly, phage-resistance rates were higher following higher phage titres, indicating phage–bacteria interactions took place. Moreover, phage-resistant bacteria were not hampered in their ability to colonize the gut, and they often reverted to the susceptible parent phenotype. These observations were further confirmed by Carvalho et al., including the fact that phage-resistant mutants were able to colonize the gut, possibly by quick reversion to the parent phenotype [13].

Carrillo et al. reported somewhat analogous results with *Campylobacter jejuni* [11]. Phage treatment resulted in a decrease in the bacterial load ranging from 0.5 to 5 log CFU/g depending on the amount of phage and time of administration. As for *S. enterica*, certain phage-resistant isolates were observed to have decreased colonization abilities, but quickly recovered by reverting to the phage sensitive phenotype. Interestingly, phage-resistance in *C. jejuni* entailed a large (90 kb) genomic inversion at Mu-like prophage DNA sequences. The resulting cells demonstrated resistance to virulent phages, inefficient gut colonization and production of infectious bacteriophage CampMu particles [12]. Recovering gut colonization capability was associated with re-inversion of the DNA fragment. The observation revealed unprecedented phase-variation resistance mechanisms that could also occur in other bacteria. Later observations by Sørensen et al. suggested that phage-resistance phase variation was associated with excess capsular polysaccharide production in the case of chicken co-infection with *C. jejuni* and phage F336 [67]. However, resistant variants had kept their gut colonization capability.

Taken together, these different studies raise the question of the selection of phage-resistance in the gut environment and its implication for phage therapy. Indeed, the complexity of the gut environment, kinetics of resistance development with phage concentration dependency [21], resistance phenotype reversion [11,13] and selection for phase variable receptor structure leading to continuing co-evolution [12] must be taken into account when developing future phage strategies. One optimistic view is that, in contrast to antibiotic resistance, phage-resistance will be naturally kept under control via coevolution. In a longitudinal study on bacteria and phage interactions in a broiler house, Connerton et al. observed that although phage-resistant bacteria did emerge, they never outgrew and dominated the susceptible ones [68].

3.4. *Vibrio Cholerae*

Phages play a critical role in the evolution of pathogenic bacteria and especially that of *V. cholerae*, the causative agent of cholera epidemic diarrhoea. This is, for example, the case for the transmission of the cholera toxin into a nontoxigenic strain via integration of the lysogenic filamentous phage CTX_3 [69].

Phage-resistance appears to plays a central role in the evolution and regulation of this species in its natural environment. In a landmark three-year study in Bangladesh, Faruque et al. showed that the presence of phages infecting a given serogroup of *V. cholerae* was inversely correlated with the presence of viable *V. cholerae* of the same serogroup in the aquatic environment [70]. In addition, if a strain of a specific serogroup was observed in water samples with a phage infecting strains of the same serogroup, the strain was resistant to the coincidentally isolated phage in 73% of the case. During that period, the number of cholera patients correlated seasonally with the presence of *V. cholerae* in water samples devoid of cholera phages, and inter-epidemic periods correlated with water samples containing only cholera phages. These observations strongly confirmed the concept of fluctuating waves of different environmental *V. cholerae* serogroups existing in the aquatic environment, with successive rounds of phage amplification and selection of new resistant serogroups. The timing of phage peaks in the environment was also correlated with phage peaks in the stools of the patients, with increasing amount of phage particles in patients as the epidemic progressed [71]. In addition, phages excreted in cholera stools were the same as those found in the environment during the late stage of the epidemic and were expected to mediate *V. cholerae* elimination and antagonize its transmissibility. Indeed, *V. cholerae* populations recovered from phage-positive patient stools were significantly less infective than phage-negative stools in an animal model [72].

Interestingly, when phage-positive stools were cultured in rich nutrient medium, but not environmental water, a rapid emergence of phage-resistant variants that had lost the O1 antigen was observed. Since the O1 antigen is important for protection from the environmental stress and to escape host immune defences, it was suggested that the dominance of phage-resistant variants should not be able to sustain an ongoing epidemic. Indeed, phage-resistant *V. cholerae* variants having altered O1 antigens were significantly less able to colonize the small intestine of mice [33]. O1 antigen alteration was also observed to be phase variable due to single nucleotide deletions in two genes critical for O1 antigenic variation [32]. Indeed, modulation of O1 antigen in *V. cholerae* is important to escape O1 antigen specific phages in nature. Although O1 phage variants were attenuated in a mouse model of intestinal colonization, positive selection of revertants was shown in the intestinal tract. As a consequence, the intestinal environment favours O1 revertant that are infectious, but simultaneously susceptible to phages [32,72].

The phage content of patients' stools was analysed during a 10-year survey in Dhaka, Bangladesh [73]. One phage, ICP1, was present in all stools from cholera patient and used the O1 antigen of lipopolysaccharide as receptor. This suggests that ICP1 is extremely well adapted to its host with a high selective pressure to maintain its genomic structure. Two other phages (ICP2 and ICP3) were only transiently observed. ICP2 and ICP3 are not O1-specific, which explains why they were less frequent, since *V. cholerae* O1 is the predominant serotype.

The surface receptor of phage ICP2 is the OmpU outer membrane protein. *V. cholerae* resistant variants with decreased OmpU expression were described, and had attenuated virulence in an infant mouse colonization model in vivo [74].

Since a cocktail composed of phage targeting different bacterial receptor would reduce the chance of phage bacterial multi-resistance, Yen et al. reasoned that a cocktail composed of the three different ICP phages could be used to prevent cholera infection [22]. The tree-ICP cocktail could kill *V. cholerae* in vitro and prevent intestine colonization or cholera-like diarrhoea of infant mice and rabbit models. All isolates from mice having received the phage cocktail 6 or 12 h after bacterial challenge were sensitive to all three phages. Resistance was, however, observed for mice having received the cocktail 24 h before bacterial infection, raising the question of phage partial washout prior to bacterial inoculation, and thus incomplete efficacy. These resistant variants had a mutation in the O antigen gene for phage ICP 1 and ICP 3 resistant variants, and OmpU for phage ICP 2 resistant variants. None of the isolates were resistant to all three phages.

3.5. Experimental Meningitis and Endocarditis

The intrinsic bactericidal properties of anti-infective compounds such as phages can be reliably studied in models where host defences are poorly involved. Such models of therapeutic sanctuaries include experimental meningitis and experimental endocarditis (EE). Experimental meningitis implicates a special anatomical setting where drug distribution depends on the blood–brain barrier. In contrast, EE mirrors the general situation encountered in many deep-seated infections where pathogens on the cardiac valves surround themselves with amorphous aggregates of platelet-fibrin clots, which cellular host defences cannot penetrate (for review see [14]). Thus, the capability of antimicrobials to cross the blood-brain barrier for meningitis or to penetrate into cardiac valve lesions (also called vegetations) are critical issues in these models.

Early studies by Smith and Huggins used a mouse model of meningitis. Mice were infected with *E. coli* 018:K1:H7ColV+ and treated 16 h later with one intramuscular dose of anti K phage or 12 doses of tetracycline, ampicillin, chloramphenicol, or a mixture of trimethoprim and sulfamethoxazole. The mortality was significantly lower in phage-treated mice than in the different antibiotics groups. Isolates recovered from mouse brains were tested for phage and antibiotic resistance. No antibiotic resistance was detected. In contrast, 6 out of 360 independent colonies (observed in 5/36 of the mice) recovered from phage-treated mice were phage resistant. All six phage-resistant isolates were K1 antigen negative, predicting decreased infectivity.

Oechslin et al. examined the efficacy of an antipseudomonal cocktail of 12 phages, used alone or in combination with antibiotics, in a dual in vitro and in vivo model of *P. aeruginosa* experimental endocarditis [24]. First, ex vivo fibrin-platelet clots were inoculated with 10^8 log CFU of *P. aeruginosa*. Phage treatment rapidly decreased bacterial counts by 6 log CFU in 6 h. However, bacterial regrowth was observed after 24 h due to the selection of resistant variants. The rate of phage-resistance mutation in the original inoculum was of ca. 10^{-7}, and resistant mutants expectedly took over after initial phage-induced killing, as described by Luria and Delbrück [45]. Bacterial regrowth after 6 h was prevented by the addition of sub-inhibitory concentrations of antibiotics, namely ciprofloxacin or meropenem.

In rats with experimental aortic endocarditis, phage therapy decreased vegetation bacterial counts by 2.3 to 3 log CFU, depending on the mode of administration. Phage therapy alone was comparable to ciprofloxacin. However, combining both treatments resulted in a highly synergistic effect with 7/11 (64%) of rats having culture-negative vegetations after only 6 h, an unprecedented efficacy in this very experimental setting.

Most importantly, phage-resistant *pseudomonas* variants were not observed in in vivo endocarditis therapy, either before or after treatment, therefore suggesting that phage-resistance could result in altered virulence or altered fitness of bacteria in animals. The hypothesis was investigated by characterizing two phage-resistant variants recovered from the ex vivo fibrin clot experiments, which displayed either transient resistance to all phages present in the cocktail or total resistance against 10 of the 12 phages. Total genomic sequencing and comparison disclosed that one of the variants had a 15 bp deletion in the *pilt* ATPase gene involved in pilus retraction, thus resulting in altered twitching motility. The other phage-resistant variant had lost the O-antigen and LPS core due to a large 350 kb deletion encompassing the *galU* gene, which is involved in LPS synthesis. Both resistant variants were less able to infect sterile vegetations than the parent strain. Since both pilus and LPS are virulence factors, it was concluded that mutations conferring phage-resistance come at a high physiological cost in fitness and virulence.

3.6. Sepsis and Acute Infections

Pouillot et al. evaluated phage therapy in a murine model of fatal neonatal sepsis [25]. Rat pups received intraperitoneal injections with a virulent strain of *E. coli* O25b:H4-ST131 and were treated 7 h or 24 h post infection with subcutaneous injections of monophage $EC200^{PP}$. Phage therapy administered 7 h post infection rescued all the rats, whereas delaying therapy until 24 h rescued only 50% of the animals. Phage-resistant colonies with rough morphologies were recovered from the treatment failures. However, these variants were more susceptible to serum-induced killing and their virulence was dramatically attenuated in a sepsis model. Smith and Huggins made similar observations when injecting mice in one gastrocnemius muscle with *E. coli* 018:K1:H7 and injecting phages in the contralateral muscle [23]. Phages were efficient in decreasing bacterial muscle densities, and only very few phage-resistant variants were recovered at the inoculation site. These isolates were K1 antigen negative, which was previously shown to decrease virulence in mice [75].

Hung et al. used an experimental mouse model of *Klebsiella pneumoniae* liver abscess. Both intraperitoneal and intragastric administration of a single phage NK5 protected mice from death in a dose-dependent manner [26]. *K. pneumoniae*-induced liver injury and inflammatory cytokine production were significantly decreased by phage therapy. As in the experimental endocarditis study [24], phage-resistant variants emerged after 6 h or 12 h during phage time-kill curves in vitro, but no phage-resistant variants were observed in vivo. The resistant variants selected in vitro had lost the hypermucoviscosity characteristic of *K. pneumoniae* NK-5. Five individual resistant variants were tested for virulence in an intragastric model of infection and were significantly less virulent than the parent strain. The phage-resistant variants were more susceptible to phagocyte-induced killing. Gu et al. also observed the emergence of phage-resistant *K. pneumoniae* variants in vitro that were less virulent in vivo [29]. These variants exhibited colony morphology variations with a rough phenotype, as compared to the large and smooth wild-type colonies. This morphological feature of variant strains remained stable even after repeated subculture and storage at –80 °C. Variants also displayed much weaker virulence when intraperitoneally injected into mice.

Using a different type of model, Park et al. observed the effect of oral administration of phage-impregnated food (mixture of two different phages PPpW-3 and PPpW-4) to ayu fish infected by *Pseudomonas plecoglossicida* [28]. *P. plecoglossicida* were always detected in the kidneys of non-treated control fishes, while they were rapidly eradicated in fishes receiving phage therapy. Bacteria recovered from dying non-treated controls were susceptible to both phages. In contrast, phage-resistant variants were observed in liquid cultures after exposure to phages PPpW-3 and PPpW-4. Four individual variant isolates (three resistant to both phages and one resistant to phage PPpW-4 only) were tested in vivo by intramuscular injection. While the parent strain was highly virulent, all four resistant variants were avirulent, even at high inocula. In addition, one peculiar strain of *P. plecoglossicida*, which was highly virulent following intramuscular injection in ayu fish, was also tested and became poorly virulent after selection for phage-resistance. Moreover, bacterial growth in freshwater was observed to be lower in the presence of phages, and the number of phage PFUs increased rapidly, indicating phage predation and replication. These results are reminiscent of the *V. cholerae* phage ecology, and suggest that it might be possible to use phages to control *P. plecoglossicida*-induced disease in fish.

Finally, Lerodelle and Poutrel evaluated the potential of phage therapy to cure sub-clinical mastitis due to *Staphylococcus aureus* in lactating cows [27]. Udders were inoculated with *S. aureus* 106-6 and 107-59 via the mammary ducts and bacteriophage K lysates were administered by the same route once sub-clinical mastitis was confirmed. Phage therapy decreased *S. aureus* bacterial loads in 60% to 100% of the animals within 48 h of treatment, but could not sterilize all the udders. Treatment failures were attributed to the deep-seated and intracellular localization of *S. aureus* in mastitis, which hide bacteria from extracellular phages. Phage resistance was not responsible for treatment failure, as virtually no resistant variants were recovered.

4. Phage Resistant Variants for Vaccine Production and Studying Virulence Factors

The potential of phages to select for resistant variants with decreased in vivo virulence was used to generate vaccines against *S. enterica* or *S. aureus*. Capparelli et al. isolated a phage-resistant variant of *S. enterica* serovar Paratyphi B selected with phage φ1. The resistant variants formed smaller colonies and had lost their O-antigen; this phenotype was stable over many subculture passages. In addition, phage-resistance was also associated with impaired transcription of six virulence factors, resulting in an avirulent phenotype when inoculated intravenously into mice. Remarkably, immunization of mice with the resistant variant protected the animals against infection with the lethal parent strain, with 100% efficacy. As a control, vaccination with the heat-killed parent strain did not elicit protection.

The authors also observed that immunization of mice with phage-resistant *S. aureus* mutants conferred broad-spectrum immunity against this pathogen [35]. Acquisition of phage-resistance against phage M^{SA} resulted in several altered properties from the *S. aureus* parent strain, including teichoic acid alteration—which was responsible for resistance—reduced growth rate, decreased expression of several virulence factors, and increased production of capsular polysaccharides. All these features were stable during prolonged subculturing. Intramuscular administration of the phage-resistant variant protected mice from lethal doses of the wild-type parent strain in 90% of the animals.

Regarding virulence factor studies, Filippov et al. used site-directed mutagenesis of different LPS genes involved in the inner and outer core synthesis, followed by trans-complementation, to determine six *Yersinia pestis* phage receptors [36]. Phage-resistant mutants had attenuated virulence with increased LD50 and time-to-death in mice, including five mutants that became totally avirulent. Likewise, Lannto et al. reported loss of virulence driven by phage-resistant *Flavobacterium columnare* in a zebrafish model of infection [37]. Phage-resistant variants produced rough colony morphotypes and exhibited impaired gliding motility, a phenotype that was maintained over ten serial passages in liquid culture. Virulence of the parental morphotype was compared to the phage-resistant R type in a zebrafish infection model. The R type mutant became completely avirulent.

Heierson et al. reported that phage-resistant mutants of *Bacillus thuringiensis* had a decreased virulence phenotype in pupae of the Cecropia moths, which correlated with flagella loss and an increased susceptibility to methicillin [38]. In two studies performed by Flyg et al., phage-resistant variants of the insect pathogen *Serratia marcescens* were also observed to have decreased resistance to insect immunity and decreased virulence in a drosophila infection model, although the exact reason for this virulence decrease was not described [39,40]. Finally, Regeimbal et al. also observed that *Acinetobacter baumannii* phage-resistant variants that had lost their capsule became avirulent in a *Galleria mellonella* model [76].

While alterations of virulence features related to phage resistance might be useful for vaccination, they also help understand bacterial pathogenesis. In this regard, the above-mentioned studies support the fact that phage resistance may be accompanied by fitness costs for the bacteria that may benefit the host. As a result, the emergence of phage resistance during phage therapy is not always synonymous with treatment failure.

5. The Biological Cost of Antibiotic Resistance and the Combined Action of Phage and Antibiotics

As for phage–bacteria coevolution, antimicrobial resistance is an ancient process that results from the complex interaction between many microorganisms in their natural environment. Indeed, most antibiotics are naturally-produced toxic molecules against which bacteria had to evolve protective mechanisms in order to survive [77]. Antibiotic-resistant bacteria are an increasing problem in human and veterinary medicine, as well as in the farming industry, due to the overuse of antibiotics over the last half century. Antibiotic resistance may be intrinsic (i.e., bacteria may be naturally resistant to certain antibiotics) or may result from spontaneous mutations or from the acquisition of horizontally-transferred resistance genes [78]. Foreign gene acquisition may involve DNA transformation, cell-cell conjugation, and phage-mediated transduction.

Resistance mechanisms include structural alteration or decreased expression of the antibiotic target, decreased drug accumulation (via decreased permeability or increased drug efflux), or changes in global metabolic pathways (for review on the topic see [79]).

Since antibiotics target important physiological functions, such as protein synthesis, cell wall synthesis, or DNA replication, antibiotic resistance often implies a certain fitness cost (for a review on the topic see [80]). However, the fitness cost associated with resistance may sometimes become counterbalanced by compensatory mutations. This was the case in *E. coli*, were streptomycin resistance conferred by mutations in the ribosomal protein RpsL first decreased the speed of protein synthesis, but were compensated after several passages (evolved cultures) by neighbouring mutations that restored the speed of protein synthesis [81]. Likewise, acquisition of the tetracycline and chloramphenicol resistance plasmid pACYC184 by *E. coli* decreased its growth rate. The growth speed was recovered, and even surpassed in evolved cultures, thanks to adaptive mutations present on the bacterial chromosome (not on the plasmid), which took advantage of the tetracycline-resistance efflux pump [82]. Thus, it was the bacterium that took advantage of the presence of the plasmid, not the plasmid that took advantage of the bacterium. Numerous other examples of adaptive mutations exist both in vitro and in vivo [83,84]. In addition, the acquisition of mobile genetic elements can also lead to co-selection to more than one antibiotic resistance if different resistance genes are genetically linked [85], and such multi-resistance is not incompatible with the restoration of fitness, as exemplified in the pACYC184 experiments [82]. Therefore, as with phages, bacteria undergo dynamic evolutionary processes when challenged with antibiotics.

On the other hand, combining both phages and antibiotics could act in synergism to prevent resistance or increase therapeutic efficacy (for review on the topic see [86]). Verma et al. showed that combining ciprofloxacin and phages prevented the emergence of phage-resistant variants during treatment of *K. pneumoniae* biofilms, although no direct bactericidal synergism between phages and antibiotics was observed [87]. The emergence of phage-resistance was also prevented by treating *S. aureus* (in continuous culture) with a combination of phages and gentamicin [88]. Torres-Barcelo et al. confirmed the potential benefit of combining phages and antibiotics (in this case, streptomycin) against *P. aeruginosa* [89]. The phenomenon of phage-antibiotic synergism was also observed in animal experiments of *P. aeruginosa* endocarditis [24], as well as with the multi-resistant bacterium *Burkholderia cepacia* [90].

In addition, Chan et al. showed that selection of phage-resistance could also restore antibiotic susceptibility [91]. When using a lytic phage specifically targeting bacterial receptors that are part of the multidrug efflux systems, MexAB and MexXY—for instance, the outer membrane porin OprM—phage-resistance restored antibiotic susceptibility because the efflux pump, which confers resistance to several antibiotic classes, was no longer functional.

However, phage-antibiotic synergism may be dependent on experimental systems, and especially antibiotic dosages. Cairns et al. showed that using sub-inhibitory concentrations of streptomycin, as might be found in natural environments or sewage, could increase the rate of phage-resistance mutations in *Pseudomonas fluorescens*, and, conversely, phage exposure increased the rate of mutation to streptomycin resistance [92], which is compatible with phage-induced bacterial mutations described by Pal et al. [61]. Nevertheless, looking at the association between antibiotic and phage-resistance in a large collection of laboratory or clinical *E. coli* isolates, Allen et al. did not find a positive or systematic correlation between drug-resistance and phage-resistance, suggesting that antibiotics used in medicine or agriculture are unlikely to induce changes in phage resistance or phage-antibiotic cross-resistance in the environment [93].

Taken together, while different kinds of positive or negative phage–bacteria interactions can be observed in the laboratory or under natural conditions, potentially useful synergistic interactions do exist and could be valuable to use in specific clinical situations.

6. Phage-Resistance in the Setting of Phage Therapy

While the use of phages as therapeutic agents is conceptually simple, complex questions still exist regarding host range, route of administration, pharmacokinetic/pharmacodynamic parameters, and managing the risk of resistance. One of the main differences between phages and antibiotics is the ability of phages to self-replicate at the infection site. Therefore, the pharmacokinetics of phage therapy is closer to the population dynamics of predator-prey models described in co-evolutionary studies (see Section 1) than classical peak-distribution-elimination phases classically measured for antibiotics. Levin and Bull proposed theoretical predictions for modelling interactions between phages and bacteria during phage therapy of acute infections [94]. Assuming that there is a bacterial density threshold beyond which the patient dies, and a limit in host defences, below which bacterial growth cannot be controlled, the absence of therapy may lead to a situation where host defences cannot keep bacteria in check in order to prevent death. By combining host defence and phage therapy, the bacterial growth rate becomes negative before it reaches the lethal density threshold. In addition, the remaining host defences are likely to more easily hinder the delayed growth of phage-resistant bacteria before they reach the lethal threshold. In a recent study done by Roach et al., the effect of host immunity and phage-mediated bacterial clearance was investigated in a mouse model of acute *P. aeruginosa* pneumonia [31]. Phage therapy using healthy mice and mice with various immune defects revealed that neutrophil-phage synergism was essential for the resolution of disease. Indeed, phage therapy failed to prevent fatal outcomes in mice with neutrophil signalling defects due to the outgrowth of phage-resistant variants. In silico analysis also predicted that neutrophils were important to prevent the emergence of phage-resistant variants and to efficiently clear infection. Thus, without immune activation, phage-resistant mutants overwhelm the basal immune defences and lead to a resurgence of a phage-resistant population that ultimately causes mortality.

Two general models of phage therapy implementation were proposed in order to manage the risk of phage-resistance. First using phage cocktails and second adapting single phages to each patient condition, referred to as personalized phage therapy.

The main reason for combining multiple phages in cocktails is to broaden the phage host range and improve effectiveness by increasing the number of potential target pathogens. This results in a greater potential for empirical treatment [95,96]. Regarding the emergence of resistance, the different phages present in the cocktail are expected to synergize by targeting different receptors on the bacterial surface, resulting in a lower statistical chance of bacterial co-resistance, as with combined phage-antibiotic therapy. This was supported by Gu et al. who observed significantly lower frequencies of phage-resistant *K. pneumoniae* mutants using a cocktail of three phages, compared to monotherapy [29]. Similar observations were made with *E. coli*, where phage cocktails decreased the frequency of phage-resistance or delayed the emergence of phage-resistant variants [97,98]. This broad spectrum antimicrobial strategy is reminiscent of the model developed by pharmaceutical companies for antibiotics, with the risk of treatment failure in case of a lack of susceptible bacteria, as well as the risk of selecting resistance in fortuitous innocuous bacterial bystanders [96]. This approach is used in countries such as Georgia, where phage cocktails are administered as an empiric treatment, although the phage content may change over time in order to adapt to the most prevalent pathogens [99]. Alternatively, existing phages can also be adapted to existing phage-resistant strains [100].

The personalized phage strategy uses single phages or targeted phage cocktails directly formulated from a phage bank according to the pathogen isolated from the patient [96,100]. Although this strategy entails a higher cost associated with personalized treatment, it offers much more flexibility regarding the spectrum of the phage and can counter the emergence of bacterial resistance more efficiently. In one of the few well-documented phage therapy clinical trials that took the emergence of phage-resistance into account, Międzybrodzki et al. achieved ca. 40% of a positive clinical outcome with 20% pathogen eradication using phage monotherapy [101,102]. Following phage therapy, phage typing patterns of the pathogens were modified in 70% of the patients treated for *S. aureus* infection (53 patients in total), 91% for *P. aeruginosa* (11 patients in total), and 100% for *E. faecalis* (14 patients in total), and *E. coli* (14 patients in total). Resistance of the target pathogen to the therapeutic phage was also observed in up to 17% of *S. aureus* cases, 36% of *P. aeruginosa*, 43% of *E. faecalis*, and 86% of *E. coli*. The high frequency in the *E. coli* infection group was a cause of frequent change of the phage during the treatment. Complete resistance to any of the phages present in the phage collection of the Ludwik Hirszfeld Institute, Poland, was observed in 7% of the *S. aureus* cases, 27% of the *P. aeruginosa* cases, 21% *E. faecalis* cases, and 27% of the *E. coli* cases.

Emergence of resistance during phage therapy was also documented by Zhvania et al. in a recent case study of chronic *S. aureus* skin infection at the Eliava Phage Therapy Center, Georgia [103]. Treatment with two anti-staphylococcal products greatly improved the patients' symptoms starting from seven days posttreatment. However, phage-resistance to the phage cocktail (Pyobacteriophage) was observed after three months of treatment and an alternate phage cocktail had to be substituted.

The use of personalized phage therapy was also exemplified in a case report by Schooley et al., where personalized-based therapeutic phages were administered parenterally to successfully treat one patient with a disseminated multidrug resistant *A. baumannii* infection. Different phage cocktails were assembled based on time-kill assays using a library of 96 phages. In vitro tests by serial passages revealed a stepwise selection of resistance to two of the cocktails. A third phage cocktail was prepared using the resistant isolates, which was then administered to the patient until the successful outcome of the infection. Of note, the phage-resistant phenotype that arose over time was associated with increased antibiotic susceptibility when phage and antibiotics were simultaneously administered. In addition, differences in colony morphology were observed during the therapy, with the eventual loss of the capsule. The authors speculate that the capsule loss may have contributed to the phage-antibiotic synergy, which included decreased virulence that had also been observed by the same author in previous studies [76]. In addition, another case of successful personalized phage therapy was reported in a lung transplant patient suffering from multi-drug resistant *P. aeruginosa* pneumonia [104]. Different phage cocktails were administered to the patient intravenously or by inhalation. Susceptibility of the bacteria to the phage was monitored during the treatment. New phage cocktails were administered as bacteriophage resistance emerged. As above, a shift in the antibiotic susceptibility pattern was also observed during phage treatment. Thus, while phage resistance does emerge, it is not prohibitive to phage therapy as long as it is carefully monitored in order to adapt the phage composition, and additional synergistic interactions with host defences or antibiotics may occur.

Finally, Khawaldeh et al. reported a successful case of adjunctive bacteriophage therapy for a refractory *P. aeruginosa* urinary tract infection [105]. The phage cocktail used for the study was composed of six lytic bacteriophages coming from existing bacteriophage libraries at the Eliava Institute in Tbilisi and were selected based on several isolates of the infecting *P. aeruginosa*. Bacteriophage counts was observed to remain high until after the disappearance of the target organism and then diminished sharply. Urine samples remained sterile for six months after the completion of antibacterial treatment and no bacteriophage-resistant bacteria arose during the time of the treatment.

7. Conclusions and Perspectives

Early studies suggested that phage–bacterial coevolution was limited to a few rounds of infection cycles. Resistance emerges following the selection of bacterial subpopulations carrying preexisting mutations and results in alterations in envelope determinants used by phages to adsorb on the bacterial surface. Hence, phage-resistant variants were totally immune to further infection and coevolution was rapidly stopped. These initial observations raised doubts regarding the use of phages as therapeutic agents because such rapid emergence of phage-resistance could hamper treatment effectiveness. However, phage-resistance is often balanced with resulting fitness costs for the bacteria. Indeed, abiotic/biotic factors, including environmental conditions, multiple bacterial exploiters, and resource availability, can greatly impact the successful emergence or stability of phage-resistance in natural environments.

The altered fitness of phage-resistant bacteria is believed to be important in phage therapy, where resistance mechanisms have been shown to alter virulence factors. In this literature review, the cost of phage-resistance was associated with virulence reduction in 17/22 (78%) of the articles (summarized in Table 1). Phage-resistant variants emerged in up to 82% of cases during phage-induced gut decolonization (out of 11 studies). Resistant variants were also reported after treatment of acute infection such as meningitis or sepsis, in up to 50% of the studies (out of six studies). Regarding the studied organisms, only 3/28 studies assessed the emergence of phage-resistance in Gram-positive bacteria, including two in *S. aureus* and one in *E. faecalis*. This focus on Gram-negative bacteria raises the question as to whether Gram-negative bacteria are more problematic regarding resistance selection.

Several lessons can be extrapolated from the reviewed studies. First, phage-resistant variants are often recovered after experimental therapy. Second, the intestinal milieu seems to be more prone to the evolution of phage-resistance, possibly due to its complexity, including mechanical viscosity and limited host defences in the lumen. Third, although phage-resistance often has a cost for the bacteria, it is not always associated with decreased infectivity, at least in the intestinal milieu.

From the five studies that clearly linked the emergence of resistance during phage therapy with alteration of a known virulence factor, like the O-antigen or LPS [14,15,22–24], four still reported resistant variants after therapy. The question then arises as to whether these variants were mere innocuous bystanders on the way of being eliminated by host defences, or whether they could still produce infection.

In any case, the ideal experimental setting should be to apply the Koch postulate and inoculate the variants to the animals in order to re-evaluate their infectivity. Indeed, recovering phage-resistant variants from in vivo samples may not be automatically synonymous with therapeutic failure, a counter-intuitive concept that appears to apply to phage therapy.

Regarding phage therapy clinical trials in human, the emergence of resistance seems to be a serious case of therapy failure if not monitored correctly. In three out of four clinical studies that monitored resistance, phage resistance led to adaptation of the composition of administered phages. Moreover, additional factors other than spontaneous mutations could also impact clinical resistance, including the immune status of the patient, the presence of biofilm, bacterial persistors, a chronic type of disease, and the possibility that the pathogenic strain possesses acquired types of resistance, like CRISPR [5,31,94,106].

It remains unclear whether the widespread use of phages to treat infections might lead to a problematic increase in phage-resistant bacterial pathogens in an analogous way that resistance developed to antibiotics. Although fitness cost may be associated with phage-resistance, this may depend on the environment, e.g., less virulence reduction associated in the intestinal milieu. Fitness was mainly assessed in the context of virulence, not in the bacterial survival in the environment. The initial fitness cost associated with antibiotic resistance could be compensated by adaptive mutations that stabilized resistant bacteria in the environment. In addition, as for antibiotics, horizontal transfer of phage resistance by plasmid acquisition was observed, which could be a problem in the long-term [107].

The real question is whether or not phages will be as widely used as antibiotics in the future. Antibiotics are used in medicine not only to prevent and combat infection, but also for other industrial applications in agriculture, which used up to 63,000 tons of antibiotics for livestock production alone in 2010 [108]. For now, it is more likely that phage therapy will be utilized as a more personalized medicine. In this case, the emergence of resistance will be manageable thanks to careful monitoring during therapy.

It is interesting to note that in the hypothetical emergence of a phage-resistant superbug, coevolution studies suggest that new phages will always be available in nature. Indeed, from environmental perspectives, bacteria were observed to be more resistant to their contemporary phages than to past or future phages and that hard-to-infect bacteria were infected by generalist phages and not specialists [63,109].

Finally, in addition of the use of phage particles themselves, or phage-antibiotic combinations, it is also possible to use purified phage lysins as a potential therapy. Recombinant phage lysins demonstrate high antibacterial activity, although they are mainly restricted to Gram-positive pathogens (reviewed in [110]). Regarding resistance, the lysin PlyG was evaluated for the possible resistance emergence after repeated treatment of *Bacillus anthracis* [111]. Spontaneous resistant mutants could not be detected, even when a compound like ethyl methanesulfonate was used to increase the bacterial mutation rate. This suggests that phage lysins target essential cell wall components that are unlikely to be modified by the host bacteria. Similar observations were made for *Streptococcus pneumoniae* and the phage lysin Pal, where repeated exposure to low concentrations of enzyme did not lead to resistant mutants [112]. More recently, Totté et al. successfully treated three cases of chronic dermatoses due to *S. aureus* with topical applications of the Staphefekt SA.100 endolysin product [113]. For all cases, resistance induction was not observed during long-term treatment, which is usually observed with antibiotic therapy.

The reviewed studies highlight both the potential power and the limits of phage therapy. Phages and bacteria are longstanding partners that have learned how to respect each other and coevolve together. The use of phages for therapy might be highly efficacious to eradicate pathogens in well-defined and circumscribed infected niches, particularly if used in combination with antibiotics. Their great advantage over antibiotics alone is their extremely rapid killing kinetics, which surpasses any known antimicrobial molecules, and the fact that they can self-replicate at the infection site. Other advantages are that they may increase antibiotic susceptibility in specific cases, and that the emergence of phage-resistant escape mutants may be prevented by antibiotics, or may carry alterations in virulence factors. These developments are promising, but should follow a thorough step-by-step developmental process, in order to avoid creating a resistance dead-end like that of antibiotics.

On the other hand, large scale or open field utilization of phage therapy, such as gut decolonization for the agricultural industry, is less certain. The coevolution dynamics of phage and bacteria is extremely sophisticated in such complex environments. Moreover, although the fascinating example of cholera control is inspiring, it is clear that phages never eradicated *V. cholerae*, whereas *V. cholerae* never got rid of the phage. As enlightening as this example may be, it primarily underlines coevolution, but not eradication, and not even efficacious population control, as epidemics still proceed.

Funding: This research was partially funded by an unrestricted grand from the Foundation for Advances in Medical Microbiology and Infectious Diseases.

Acknowledgments: The author would like to thanks Philippe Moreillon and Shawna E. McCallin for their helpful comments on the manuscript.

Conflicts of Interest: The author declares no conflict of interest.

References

1. Sugden, R.; Kelly, R.; Davies, S. Combatting antimicrobial resistance globally. *Nat. Microbiol.* **2016**, *1*, 16187. [CrossRef] [PubMed]
2. Salmond, G.P.; Fineran, P.C. A century of the phage: Past, present and future. *Nat. Rev. Microbiol.* **2015**, *13*, 777–786. [CrossRef] [PubMed]
3. Roach, D.R.; Debarbieux, L. Phage therapy: Awakening a sleeping giant. *Emerg. Top. Life Sci.* **2017**, *1*, 93. [CrossRef]
4. Harper, D.R. Criteria for selecting suitable infectious diseases for phage therapy. *Viruses* **2018**, *10*, 177. [CrossRef] [PubMed]
5. Labrie, S.J.; Samson, J.E.; Moineau, S. Bacteriophage resistance mechanisms. *Nat. Rev. Microbiol.* **2010**, *8*, 317–327. [CrossRef] [PubMed]
6. Koskella, B.; Brockhurst, M.A. Bacteria–phage coevolution as a driver of ecological and evolutionary processes in microbial communities. *FEMS Microbiol. Rev.* **2014**, *38*, 916–931. [CrossRef] [PubMed]
7. Bertozzi Silva, J.; Storms, Z.; Sauvageau, D. Host receptors for bacteriophage adsorption. *FEMS Microbiol. Lett.* **2016**, *363*. [CrossRef] [PubMed]
8. León, M.; Bastías, R. Virulence reduction in bacteriophage resistant bacteria. *Front. Microbiol.* **2015**, *6*, 343. [CrossRef] [PubMed]
9. Vasu, K.; Nagaraja, V. Diverse functions of restriction-modification systems in addition to cellular defense. *Microbiol. Mol. Biol. Rev.* **2013**, *77*, 53–72. [CrossRef] [PubMed]
10. Vale, P.F.; Lafforgue, G.; Gatchitch, F.; Gardan, R.; Moineau, S.; Gandon, S. Costs of CRISPR-Cas-mediated resistance in *Streptococcus thermophilus*. *Proc. R. Soc. B* **2015**, *282*, 20151270. [CrossRef] [PubMed]
11. Loc Carrillo, C.; Atterbury, R.J.; el-Shibiny, A.; Connerton, P.L.; Dillon, E.; Scott, A.; Connerton, I.F. Bacteriophage therapy to reduce *Campylobacter jejuni* colonization of broiler chickens. *Appl. Environ. Microbiol.* **2005**, *71*, 6554–6563. [CrossRef] [PubMed]
12. Scott, A.E.; Timms, A.R.; Connerton, P.L.; Loc Carrillo, C.; Adzfa Radzum, K.; Connerton, I.F. Genome dynamics of *Campylobacter jejuni* in response to bacteriophage predation. *PLoS Pathog.* **2007**, *3*, e119. [CrossRef] [PubMed]
13. Carvalho, C.M.; Gannon, B.W.; Halfhide, D.E.; Santos, S.B.; Hayes, C.M.; Roe, J.M.; Azeredo, J. The in vivo efficacy of two administration routes of a phage cocktail to reduce numbers of *Campylobacter coli* and *Campylobacter jejuni* in chickens. *BMC Microbiol.* **2010**, *10*, 232. [CrossRef] [PubMed]
14. Smith, H.W.; Huggins, M.B. Effectiveness of phages in treating experimental *Escherichia coli* diarrhoea in calves, piglets and lambs. *J. Gen. Microbiol.* **1983**, *129*, 2659–2675. [CrossRef] [PubMed]
15. Smith, H.W.; Huggins, M.B.; Shaw, K.M. The control of experimental *Escherichia coli* diarrhoea in calves by means of bacteriophages. *Microbiology* **1987**, *133*, 1111–1126. [CrossRef] [PubMed]
16. Sheng, H.; Knecht, H.J.; Kudva, I.T.; Hovde, C.J. Application of bacteriophages to control intestinal *Escherichia coli* o157:H7 levels in ruminants. *Appl. Environ. Microbiol.* **2006**, *72*, 5359–5366. [CrossRef] [PubMed]
17. Majewska, J.; Beta, W.; Lecion, D.; Hodyra-Stefaniak, K.; Klopot, A.; Kazmierczak, Z.; Miernikiewicz, P.; Piotrowicz, A.; Ciekot, J.; Owczarek, B.; et al. Oral application of T4 phage induces weak antibody production in the gut and in the blood. *Viruses* **2015**, *7*, 4783–4799. [CrossRef] [PubMed]

18. Maura, D.; Morello, E.; du Merle, L.; Bomme, P.; Le Bouguenec, C.; Debarbieux, L. Intestinal colonization by enteroaggregative *Escherichia coli* supports long-term bacteriophage replication in mice. *Environ. Microbiol.* **2012**, *14*, 1844–1854. [CrossRef] [PubMed]
19. Duerkop, B.A.; Huo, W.; Bhardwaj, P.; Palmer, K.L.; Hooper, L.V. Molecular basis for lytic bacteriophage resistance in enterococci. *mBio* **2016**, *7*. [CrossRef] [PubMed]
20. Sklar, I.B.; Joerger, R.D. Attempts to utilize bacteriophage to combat salmonella enterica serovar entemtidis infection in chickens. *J. Food Saf.* **2001**, *21*, 15–29. [CrossRef]
21. Atterbury, R.J.; van Bergen, M.A.; Ortiz, F.; Lovell, M.A.; Harris, J.A.; de Boer, A.; Wagenaar, J.A.; Allen, V.M.; Barrow, P.A. Bacteriophage therapy to reduce salmonella colonization of broiler chickens. *Appl. Environ. Microbiol.* **2007**, *73*, 4543–4549. [CrossRef] [PubMed]
22. Yen, M.; Cairns, L.S.; Camilli, A. A cocktail of three virulent bacteriophages prevents vibrio cholerae infection in animal models. *Nat. Commun.* **2017**, *8*, 14187. [CrossRef] [PubMed]
23. Smith, H.W.; Huggins, M.B. Successful treatment of experimental escherichia coli infections in mice using phage: Its general superiority over antibiotics. *J. Gen. Microbiol.* **1982**, *128*, 307–318. [CrossRef] [PubMed]
24. Oechslin, F.; Piccardi, P.; Mancini, S.; Gabard, J.; Moreillon, P.; Entenza, J.M.; Resch, G.; Que, Y.A. Synergistic interaction between phage therapy and antibiotics clears *Pseudomonas aeruginosa* infection in endocarditis and reduces virulence. *J. Infect. Dis.* **2017**, *215*, 703–712. [CrossRef] [PubMed]
25. Pouillot, F.; Chomton, M.; Blois, H.; Courroux, C.; Noelig, J.; Bidet, P.; Bingen, E.; Bonacorsi, S. Efficacy of bacteriophage therapy in experimental sepsis and meningitis caused by a clone o25b:H4-st131 *Escherichia coli* strain producing CTX-M-15. *Antimicrob. Agents Chemother.* **2012**, *56*, 3568–3575. [CrossRef] [PubMed]
26. Hung, C.H.; Kuo, C.F.; Wang, C.H.; Wu, C.M.; Tsao, N. Experimental phage therapy in treating klebsiella pneumoniae-mediated liver abscesses and bacteremia in mice. *Antimicrob. Agents Chemother.* **2011**, *55*, 1358–1365. [CrossRef] [PubMed]
27. Lerondelle, C.; Poutrel, B. Bacteriophage treatment trials on staphylococcal udder infection in lactating cows. *Ann. Rech. Vet.* **1980**, *11*, 421–426. [PubMed]
28. Park, S.C.; Shimamura, I.; Fukunaga, M.; Mori, K.I.; Nakai, T. Isolation of bacteriophages specific to a fish pathogen, pseudomonas plecoglossicida, as a candidate for disease control. *Appl. Environ. Microbiol.* **2000**, *66*, 1416–1422. [CrossRef] [PubMed]
29. Gu, J.; Liu, X.; Li, Y.; Han, W.; Lei, L.; Yang, Y.; Zhao, H.; Gao, Y.; Song, J.; Lu, R.; et al. A method for generation phage cocktail with great therapeutic potential. *PLoS ONE* **2012**, *7*, e31698. [CrossRef] [PubMed]
30. Capparelli, R.; Nocerino, N.; Iannaccone, M.; Ercolini, D.; Parlato, M.; Chiara, M.; Iannelli, D. Bacteriophage therapy of salmonella enterica: A fresh appraisal of bacteriophage therapy. *J. Infect. Dis.* **2010**, *201*, 52–61. [CrossRef] [PubMed]
31. Roach, D.R.; Leung, C.Y.; Henry, M.; Morello, E.; Singh, D.; Di Santo, J.P.; Weitz, J.S.; Debarbieux, L. Synergy between the host immune system and bacteriophage is essential for successful phage therapy against an acute respiratory pathogen. *Cell Host Microbe* **2017**, *22*, 38–47. [CrossRef] [PubMed]
32. Seed, K.D.; Faruque, S.M.; Mekalanos, J.J.; Calderwood, S.B.; Qadri, F.; Camilli, A. Phase variable o antigen biosynthetic genes control expression of the major protective antigen and bacteriophage receptor in vibrio cholerae o1. *PLoS Pathog.* **2012**, *8*, e1002917. [CrossRef] [PubMed]
33. Nesper, J.; Kapfhammer, D.; Klose, K.E.; Merkert, H.; Reidl, J. Characterization of vibrio cholerae O1 antigen as the bacteriophage K139 receptor and identification of IS1004 insertions aborting O1 antigen biosynthesis. *J. Bacteriol.* **2000**, *182*, 5097–5104. [CrossRef] [PubMed]
34. Attridge, S.R.; Fazeli, A.; Manning, P.A.; Stroeher, U.H. Isolation and characterization of bacteriophage-resistant mutants of *Vibrio cholerae* O139. *Microb. Pathog.* **2001**, *30*, 237–246. [CrossRef] [PubMed]
35. Capparelli, R.; Nocerino, N.; Lanzetta, R.; Silipo, A.; Amoresano, A.; Giangrande, C.; Becker, K.; Blaiotta, G.; Evidente, A.; Cimmino, A.; et al. Bacteriophage-resistant *Staphylococcus aureus* mutant confers broad immunity against staphylococcal infection in mice. *PLoS ONE* **2010**, *5*, e11720. [CrossRef] [PubMed]
36. Filippov, A.A.; Sergueev, K.V.; He, Y.; Huang, X.Z.; Gnade, B.T.; Mueller, A.J.; Fernandez-Prada, C.M.; Nikolich, M.P. Bacteriophage-resistant mutants in yersinia pestis: Identification of phage receptors and attenuation for mice. *PLoS ONE* **2011**, *6*, e25486. [CrossRef] [PubMed]
37. Laanto, E.; Bamford, J.K.H.; Laakso, J.; Sundberg, L.R. Phage-driven loss of virulence in a fish pathogenic bacterium. *PLoS ONE* **2012**, *7*, e53157. [CrossRef] [PubMed]

38. Heierson, A.; Sidén, I.; Kivaisi, A.; Boman, H.G. Bacteriophage-resistant mutants of *Bacillus thuringiensis* with decreased virulence in pupae of *Hyalophora cecropia*. *J. Bacteriol.* **1986**, *167*, 18–24. [CrossRef] [PubMed]
39. Flyg, C.; Kenne, K.; Boman, H.G. Insect pathogenic properties of serratia marcescens: Phage-resistant mutants with a decreased resistance to cecropia immunity and a decreased virulence to drosophila. *J. Gen. Microbiol.* **1980**, *120*, 173–181. [CrossRef] [PubMed]
40. Flyg, C.; Xanthopoulos, K.G. Insect pathogenic properties of serratia marcescens. Passive and active resistance to insect immunity studied with protease-deficient and phage-resistant mutants. *J. Gen. Microbiol.* **1983**, *129*, 453–464. [CrossRef]
41. Buckling, A.; Rainey, P.B. Antagonistic coevolution between a bacterium and a bacteriophage. *Proc. R. Soc. B* **2002**, *269*, 931–936. [CrossRef] [PubMed]
42. Riede, I.; Eschbach, M.L. Evidence that trat interacts with ompa of *Escherichia coli*. *FEBS Lett.* **1986**, *205*, 241–245. [CrossRef]
43. Hanlon, G.W.; Denyer, S.P.; Olliff, C.J.; Ibrahim, L.J. Reduction in exopolysaccharide viscosity as an aid to bacteriophage penetration through *Pseudomonas aeruginosa* biofilms. *Appl. Environ. Microbiol.* **2001**, *67*, 2746–2753. [CrossRef] [PubMed]
44. Destoumieux-Garzón, D.; Duquesne, S.; Peduzzi, J.; Goulard, C.; Desmadril, M.; Letellier, L.; Rebuffat, S.; Boulanger, P. The iron–siderophore transporter fhua is the receptor for the antimicrobial peptide microcin J25: Role of the microcin val(11)–pro(16) β-hairpin region in the recognition mechanism. *Biochem. J.* **2005**, *389*, 869–876. [CrossRef] [PubMed]
45. Luria, S.E.; Delbruck, M. Mutations of bacteria from virus sensitivity to virus resistance. *Genetics* **1943**, *28*, 491–511. [PubMed]
46. Dennehy, J.J. What can phages tell us about host-pathogen coevolution? *Int. J. Evol. Biol.* **2012**, *2012*, 12. [CrossRef] [PubMed]
47. Lenski, R.E.; Levin, B.R. Constraints on the coevolution of bacteria and virulent phage: A model, some experiments, and predictions for natural communities. *Am. Nat.* **1985**, *125*, 585–602. [CrossRef]
48. Mizoguchi, K.; Morita, M.; Fischer, C.R.; Yoichi, M.; Tanji, Y.; Unno, H. Coevolution of bacteriophage pp01 and *Escherichia coli* o157:H7 in continuous culture. *Appl. Environ. Microbiol.* **2003**, *69*, 170–176. [CrossRef] [PubMed]
49. Spanakis, E.; Horne, M.T. Co-adaptation of *Escherichia coli* and coliphage lambda vir in continuous culture. *J. Gen. Microbiol.* **1987**, *133*, 353–360. [PubMed]
50. Meyer, J.R.; Dobias, D.T.; Weitz, J.S.; Barrick, J.E.; Quick, R.T.; Lenski, R.E. Repeatability and contingency in the evolution of a key innovation in phage lambda. *Science* **2012**, *335*, 428–432. [CrossRef] [PubMed]
51. Hall, A.R.; Scanlan, P.D.; Buckling, A. Bacteria-phage coevolution and the emergence of generalist pathogens. *Am. Nat.* **2011**, *177*, 44–53. [CrossRef] [PubMed]
52. Buckling, A.; Rainey, P.B. The role of parasites in sympatric and allopatric host diversification. *Nature* **2002**, *420*, 496–499. [CrossRef] [PubMed]
53. Hall, A.R.; Scanlan, P.D.; Morgan, A.D.; Buckling, A. Host-parasite coevolutionary arms races give way to fluctuating selection. *Ecol. Lett.* **2011**, *14*, 635–642. [CrossRef] [PubMed]
54. Lennon, J.T.; Khatana, S.A.; Marston, M.F.; Martiny, J.B. Is there a cost of virus resistance in marine cyanobacteria? *ISME J.* **2007**, *1*, 300–312. [CrossRef] [PubMed]
55. Quance, M.A.; Travisano, M. Effects of temperature on the fitness cost of resistance to bacteriophage t4 in *Escherichia coli*. *Evolution* **2009**, *63*, 1406–1416. [CrossRef] [PubMed]
56. Marston, M.F.; Pierciey, F.J., Jr.; Shepard, A.; Gearin, G.; Qi, J.; Yandava, C.; Schuster, S.C.; Henn, M.R.; Martiny, J.B. Rapid diversification of coevolving marine synechococcus and a virus. *Proc. Natl. Acad. Sci. USA* **2012**, *109*, 4544–4549. [CrossRef] [PubMed]
57. Buckling, A.; Wei, Y.; Massey, R.C.; Brockhurst, M.A.; Hochberg, M.E. Antagonistic coevolution with parasites increases the cost of host deleterious mutations. *Proc. R. Soc. B* **2006**, *273*, 45–49. [CrossRef] [PubMed]
58. Gómez, P.; Ashby, B.; Buckling, A. Population mixing promotes arms race host–parasite coevolution. *Proc. R. Soc. B* **2015**, *282*, 20142297. [CrossRef] [PubMed]
59. Friman, V.P.; Buckling, A. Effects of predation on real-time host-parasite coevolutionary dynamics. *Ecol. Lett.* **2013**, *16*, 39–46. [CrossRef] [PubMed]
60. Friman, V.P.; Buckling, A. Phages can constrain protist predation-driven attenuation of *Pseudomonas aeruginosa* virulence in multienemy communities. *ISME J.* **2014**, *8*, 1820–1830. [CrossRef] [PubMed]

61. Pal, C.; Macia, M.D.; Oliver, A.; Schachar, I.; Buckling, A. Coevolution with viruses drives the evolution of bacterial mutation rates. *Nature* **2007**, *450*, 1079–1081. [CrossRef] [PubMed]
62. Gomez, P.; Buckling, A. Coevolution with phages does not influence the evolution of bacterial mutation rates in soil. *ISME J.* **2013**, *7*, 2242–2244. [CrossRef] [PubMed]
63. Gomez, P.; Buckling, A. Bacteria-phage antagonistic coevolution in soil. *Science* **2011**, *332*, 106–109. [CrossRef] [PubMed]
64. Taylor, C.M.; Roberts, I.S. Capsular polysaccharides and their role in virulence. *Contrib. Microbiol.* **2005**, *12*, 55–66. [PubMed]
65. Reyes, A.; Wu, M.; McNulty, N.P.; Rohwer, F.L.; Gordon, J.I. Gnotobiotic mouse model of phage–bacterial host dynamics in the human gut. *Proc. Natl. Acad. Sci. USA* **2013**, *110*, 20236. [CrossRef] [PubMed]
66. Andino, A.; Hanning, I. Salmonella enterica: Survival, colonization, and virulence differences among serovars. *Sci. World J.* **2015**, *2015*, 520179. [CrossRef] [PubMed]
67. Sørensen, M.C.H.; van Alphen, L.B.; Fodor, C.; Crowley, S.M.; Christensen, B.B.; Szymanski, C.M.; Brøndsted, L. Phase variable expression of capsular polysaccharide modifications allows *Campylobacter jejuni* to avoid bacteriophage infection in chickens. *Front. Cell. Infect. Microbiol.* **2012**, *2*, 11. [CrossRef] [PubMed]
68. Connerton, P.L.; Loc Carrillo, C.M.; Swift, C.; Dillon, E.; Scott, A.; Rees, C.E.; Dodd, C.E.; Frost, J.; Connerton, I.F. Longitudinal study of *Campylobacter jejuni* bacteriophages and their hosts from broiler chickens. *Appl. Environ. Microbiol.* **2004**, *70*, 3877–3883. [CrossRef] [PubMed]
69. Waldor, M.K.; Mekalanos, J.J. Lysogenic conversion by a filamentous phage encoding cholera toxin. *Science* **1996**, *272*, 1910–1914. [CrossRef] [PubMed]
70. Faruque, S.M.; Naser, I.B.; Islam, M.J.; Faruque, A.S.G.; Ghosh, A.N.; Nair, G.B.; Sack, D.A.; Mekalanos, J.J. Seasonal epidemics of cholera inversely correlate with the prevalence of environmental cholera phages. *Proc. Natl. Acad. Sci. USA* **2005**, *102*, 1702–1707. [CrossRef] [PubMed]
71. Faruque, S.M.; Islam, M.J.; Ahmad, Q.S.; Faruque, A.S.G.; Sack, D.A.; Nair, G.B.; Mekalanos, J.J. Self-limiting nature of seasonal cholera epidemics: Role of host-mediated amplification of phage. *Proc. Natl. Acad. Sci. USA* **2005**, *102*, 6119–6124. [CrossRef] [PubMed]
72. Zahid, M.S.H.; Udden, S.M.N.; Faruque, A.S.G.; Calderwood, S.B.; Mekalanos, J.J.; Faruque, S.M. Effect of phage on the infectivity of vibrio cholerae and emergence of genetic variants. *Infect. Immun.* **2008**, *76*, 5266–5273. [CrossRef] [PubMed]
73. Seed, K.D.; Bodi, K.L.; Kropinski, A.M.; Ackermann, H.-W.; Calderwood, S.B.; Qadri, F.; Camilli, A. Evidence of a dominant lineage of vibrio cholerae-specific lytic bacteriophages shed by cholera patients over a 10-year period in Dhaka, Bangladesh. *mBio* **2011**, *2*, e00334-10. [CrossRef] [PubMed]
74. Seed, K.D.; Yen, M.; Shapiro, B.J.; Hilaire, I.J.; Charles, R.C.; Teng, J.E.; Ivers, L.C.; Boncy, J.; Harris, J.B.; Camilli, A. Evolutionary consequences of intra-patient phage predation on microbial populations. *eLife* **2014**, *3*, e03497. [CrossRef] [PubMed]
75. Smith, H.W.; Huggins, M.B. The association of the O18, K1 and H7 antigens and the CO1V plasmid of a strain of *Escherichia coli* with its virulence and immunogenicity. *J. Gen. Microbiol.* **1980**, *121*, 387–400. [PubMed]
76. Regeimbal, J.M.; Jacobs, A.C.; Corey, B.W.; Henry, M.S.; Thompson, M.G.; Pavlicek, R.L.; Quinones, J.; Hannah, R.M.; Ghebremedhin, M.; Crane, N.J.; et al. Personalized therapeutic cocktail of wild environmental phages rescues mice from *Acinetobacter baumannii* wound infections. *Antimicrob. Agents Chemother.* **2016**, *60*, 5806–5816. [CrossRef] [PubMed]
77. Clardy, J.; Walsh, C. Lessons from natural molecules. *Nature* **2004**, *432*, 829–837. [CrossRef] [PubMed]
78. Van Hoek, A.H.A.M.; Mevius, D.; Guerra, B.; Mullany, P.; Roberts, A.P.; Aarts, H.J.M. Acquired antibiotic resistance genes: An overview. *Front. Microbiol.* **2011**, *2*, 203. [CrossRef] [PubMed]
79. Munita, J.M.; Arias, C.A. Mechanisms of antibiotic resistance. *Microbiol. Spectr.* **2016**, *4*. [CrossRef]
80. Beceiro, A.; Tomás, M.; Bou, G. Antimicrobial resistance and virulence: A successful or deleterious association in the bacterial world? *Clin. Microbiol. Rev.* **2013**, *26*, 185–230. [CrossRef] [PubMed]
81. Schrag, S.J.; Perrot, V.; Levin, B.R. Adaptation to the fitness costs of antibiotic resistance in *Escherichia coli*. *Proc. R. Soc. B* **1997**, *264*, 1287–1291. [CrossRef] [PubMed]
82. Bouma, J.E.; Lenski, R.E. Evolution of a bacteria/plasmid association. *Nature* **1988**, *335*, 351–352. [CrossRef] [PubMed]
83. Marcusson, L.L.; Frimodt-Møller, N.; Hughes, D. Interplay in the selection of fluoroquinolone resistance and bacterial fitness. *PLoS Pathog.* **2009**, *5*, e1000541. [CrossRef] [PubMed]

84. Partridge, S.R. Analysis of antibiotic resistance regions in gram-negative bacteria. *FEMS Microbiol. Rev.* **2011**, *35*, 820–855. [CrossRef] [PubMed]
85. Enne, V.I.; Bennett, P.M.; Livermore, D.M.; Hall, L.M. Enhancement of host fitness by the sul2-coding plasmid p9123 in the absence of selective pressure. *J. Antimicrob. Chemother.* **2004**, *53*, 958–963. [CrossRef] [PubMed]
86. Torres-Barcelo, C.; Hochberg, M.E. Evolutionary rationale for phages as complements of antibiotics. *Trends Microbiol.* **2016**, *24*, 249–256. [CrossRef] [PubMed]
87. Verma, V.; Harjai, K.; Chhibber, S. Restricting ciprofloxacin-induced resistant variant formation in biofilm of *Klebsiella pneumoniae* b5055 by complementary bacteriophage treatment. *J. Antimicrob. Chemother.* **2009**, *64*, 1212–1218. [CrossRef] [PubMed]
88. Kirby, A.E. Synergistic action of gentamicin and bacteriophage in a continuous culture population of *Staphylococcus aureus*. *PLoS ONE* **2012**, *7*, e51017. [CrossRef] [PubMed]
89. Torres-Barcelo, C.; Arias-Sanchez, F.I.; Vasse, M.; Ramsayer, J.; Kaltz, O.; Hochberg, M.E. A window of opportunity to control the bacterial pathogen *Pseudomonas aeruginosa* combining antibiotics and phages. *PLoS ONE* **2014**, *9*, e106628. [CrossRef] [PubMed]
90. Kamal, F.; Dennis, J.J. Burkholderia cepacia complex phage-antibiotic synergy (PAS): Antibiotics stimulate lytic phage activity. *Appl. Environ. Microbiol.* **2015**, *81*, 1132–1138. [CrossRef] [PubMed]
91. Chan, B.K.; Sistrom, M.; Wertz, J.E.; Kortright, K.E.; Narayan, D.; Turner, P.E. Phage selection restores antibiotic sensitivity in MDR *Pseudomonas aeruginosa*. *Sci. Rep.* **2016**, *6*, 26717. [CrossRef] [PubMed]
92. Cairns, J.; Becks, L.; Jalasvuori, M.; Hiltunen, T. Sublethal streptomycin concentrations and lytic bacteriophage together promote resistance evolution. *Philos. Trans. R. Soc. B* **2017**, *372*, 20160040. [CrossRef] [PubMed]
93. Allen, R.C.; Pfrunder-Cardozo, K.R.; Meinel, D.; Egli, A.; Hall, A.R. Associations among antibiotic and phage resistance phenotypes in natural and clinical *Escherichia coli* isolates. *mBio* **2017**, *8*. [CrossRef] [PubMed]
94. Levin, B.R.; Bull, J.J. Population and evolutionary dynamics of phage therapy. *Nat. Rev. Microbiol.* **2004**, *2*, 166–173. [CrossRef] [PubMed]
95. Chan, B.K.; Abedon, S.T. Phage therapy pharmacology phage cocktails. *Adv. Appl. Microbiol.* **2012**, *78*, 1–23. [PubMed]
96. Chan, B.K.; Abedon, S.T.; Loc-Carrillo, C. Phage cocktails and the future of phage therapy. *Future Microbiol.* **2013**, *8*, 769–783. [CrossRef] [PubMed]
97. O'Flynn, G.; Ross, R.P.; Fitzgerald, G.F.; Coffey, A. Evaluation of a cocktail of three bacteriophages for biocontrol of *Escherichia coli* o157:H7. *Appl. Environ. Microbiol.* **2004**, *70*, 3417–3424. [CrossRef] [PubMed]
98. Tanji, Y.; Shimada, T.; Yoichi, M.; Miyanaga, K.; Hori, K.; Unno, H. Toward rational control of *Escherichia coli* O157:H7 by a phage cocktail. *Appl. Microbiol. Biotechnol.* **2004**, *64*, 270–274. [CrossRef] [PubMed]
99. Abedon, S.T.; Kuhl, S.J.; Blasdel, B.G.; Kutter, E.M. Phage treatment of human infections. *Bacteriophage* **2011**, *1*, 66–85. [CrossRef] [PubMed]
100. Pirnay, J.P.; De Vos, D.; Verbeken, G.; Merabishvili, M.; Chanishvili, N.; Vaneechoutte, M.; Zizi, M.; Laire, G.; Lavigne, R.; Huys, I.; et al. The phage therapy paradigm: Pret-a-porter or sur-mesure? *Pharm. Res.* **2011**, *28*, 934–937. [CrossRef] [PubMed]
101. Międzybrodzki, R.; Borysowski, J.; Weber-Dąbrowska, B.; Fortuna, W.; Letkiewicz, S.; Szufnarowski, K.; Pawełczyk, Z.; Rogóż, P.; Kłak, M.; Wojtasik, E.; et al. Chapter 3—clinical aspects of phage therapy. In *Advances in Virus Research*; Łobocka, M., Szybalski, W., Eds.; Academic Press: Cambridge, MA, USA, 2012; Volume 83, pp. 73–121.
102. Górski, A.; Międzybrodzki, R.; Weber-Dąbrowska, B.; Fortuna, W.; Letkiewicz, S.; Rogóż, P.; Jończyk-Matysiak, E.; Dąbrowska, K.; Majewska, J.; Borysowski, J. Phage therapy: Combating infections with potential for evolving from merely a treatment for complications to targeting diseases. *Front. Microbiol.* **2016**, *7*, 1515. [CrossRef] [PubMed]
103. Zhvania, P.; Hoyle, N.S.; Nadareishvili, L.; Nizharadze, D.; Kutateladze, M. Phage therapy in a 16-year-old boy with netherton syndrome. *Front. Med.* **2017**, *4*, 94. [CrossRef] [PubMed]
104. Aslam, S.; Yung, G.; Dan, J.; Reed, S.; LeFebvre, M.; Logan, C.; Taplitz, R.; Law, N.; Golts, E.; Afshar, K.; et al. (373)—Bacteriophage treatment in a lung transplant recipient. *J. Heart Lung Transplant.* **2018**, *37*, S155–S156. [CrossRef]

105. Khawaldeh, A.; Morales, S.; Dillon, B.; Alavidze, Z.; Ginn, A.N.; Thomas, L.; Chapman, S.J.; Dublanchet, A.; Smithyman, A.; Iredell, J.R. Bacteriophage therapy for refractory *Pseudomonas aeruginosa* urinary tract infection. *J. Med. Microbiol.* **2011**, *60*, 1697–1700. [CrossRef] [PubMed]
106. Harper, D.R.; Parracho, H.M.R.T.; Walker, J.; Sharp, R.; Hughes, G.; Werthén, M.; Lehman, S.; Morales, S. Bacteriophages and biofilms. *Antibiotics* **2014**, *3*, 270–284. [CrossRef]
107. Forde, A.; Daly, C.; Fitzgerald, G.F. Identification of four phage resistance plasmids from *Lactococcus lactis* subsp. Cremoris HO2. *Appl. Environ. Microbiol.* **1999**, *65*, 1540–1547. [PubMed]
108. Van Boeckel, T.P.; Brower, C.; Gilbert, M.; Grenfell, B.T.; Levin, S.A.; Robinson, T.P.; Teillant, A.; Laxminarayan, R. Global trends in antimicrobial use in food animals. *Proc. Natl. Acad. Sci. USA* **2015**, *112*, 5649. [CrossRef] [PubMed]
109. Flores, C.O.; Meyer, J.R.; Valverde, S.; Farr, L.; Weitz, J.S. Statistical structure of host–phage interactions. *Proc. Natl. Acad. Sci. USA* **2011**, *108*, E288. [CrossRef] [PubMed]
110. Fischetti, V.A. Bacteriophage lysins as effective antibacterials. *Curr. Opin. Microbiol.* **2008**, *11*, 393–400. [CrossRef] [PubMed]
111. Schuch, R.; Nelson, D.; Fischetti, V.A. A bacteriolytic agent that detects and kills *Bacillus anthracis*. *Nature* **2002**, *418*, 884. [CrossRef] [PubMed]
112. Loeffler, J.M.; Nelson, D.; Fischetti, V.A. Rapid killing of *Streptococcus pneumoniae* with a bacteriophage cell wall hydrolase. *Science* **2001**, *294*, 2170–2172. [CrossRef] [PubMed]
113. Totte, J.E.E.; van Doorn, M.B.; Pasmans, S. Successful treatment of chronic staphylococcus aureus-related dermatoses with the topical endolysin staphefekt sa.100: A report of 3 cases. *Case Rep. Dermatol.* **2017**, *9*, 19–25. [CrossRef] [PubMed]

Review

Delivering Phage Products to Combat Antibiotic Resistance in Developing Countries: Lessons Learned from the HIV/AIDS Epidemic in Africa

Tobi E. Nagel

Phages for Global Health, 383 62nd Street, Oakland, CA 94618, USA; tobi@phagesforglobalhealth.org; Tel.: +1 650-888-5522

Received: 30 May 2018; Accepted: 26 June 2018; Published: 27 June 2018

Abstract: The antimicrobial resistance (AMR) crisis and HIV/AIDS epidemic exhibit many parallels. In both, infectious diseases have caused millions of deaths worldwide, with AMR expected to kill even more people each year than HIV/AIDS did at its peak. In addition, both have required or will require new classes of drugs for control. For HIV/AIDS, development of vital antiretroviral drugs (ARVs) was accomplished in several stages: expanding public awareness about the disease, gathering commitment from the international community to tackle the problem, and eventually establishing policies and global funds to deliver new therapeutics. For AMR, the pursuit of new antimicrobials appears to be following a similar trajectory. This paper examines how lessons and processes leading to ARVs might be applied to developing AMR drugs, in particular bacteriophages (phages). These possess many essential characteristics: inexpensive manufacture, rapid drug development, and a ready means to prevent phage-resistant microbes from emerging. However, the broad application of phage-based products has yet to be fully demonstrated, and will require both international coordination and modified regulatory policies.

Keywords: bacteriophage; phage therapy; antimicrobial resistance; antibiotic; global health; developing countries; infectious disease

---

1. Introduction

HIV/AIDS spurred the deadliest epidemic in modern history. At its peak, the disease caused 1.9 million deaths per year worldwide [1]. The global antimicrobial resistance (AMR) crisis will be even bigger: by 2050, antibiotic-resistant infections are predicted to kill roughly 10 million people annually [2]. As with HIV/AIDS, the developing world will be hardest hit by AMR, with nearly 90% of expected deaths occurring in those countries. For example, in Africa alone, the annual loss of life from AMR is forecasted to be 4.15 million, surpassing the 1.54 million deaths caused by HIV/AIDS in Africa during 2005, the worst year of that crisis (see Figure 1).

Notably, when the HIV/AIDS epidemic began, there were no effective drugs available to treat the emerging infectious disease. Similarly, there are now no conventional treatment options in the face of antimicrobial resistance. Just as new drug classes—namely antiretrovirals (ARVs)—were needed to mitigate HIV/AIDS, new antimicrobial drugs will be required to tackle AMR.

Phages—viruses that can kill both antibiotic-sensitive and antibiotic-resistant bacteria—are a key drug class that could save many lives threatened by the AMR crisis. Phages have been used as antibacterial agents for nearly 100 years in the former Soviet Union, and they are now undergoing a renaissance in other countries due to the growing AMR problem [3–7]. In addition to being able to provide therapeutic options when no others exist, phages are inherently inexpensive to isolate, have relatively short product development time frames, and no major reported side effects, despite their decades of use. Unlike traditional antibiotics, phage products can readily be designed to thwart development of resistance. Nonetheless, phage therapy has yet to be fully proven and implemented in most regions of the world.

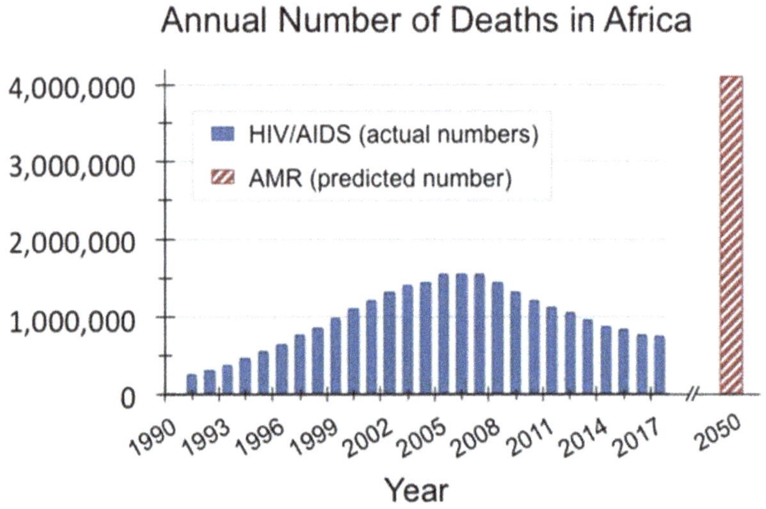

Figure 1. Africa was hardest hit by the HIV/AIDS crisis, and is expected to suffer the highest mortality per capita from AMR by 2050 [1,2].

Substantial funding and appropriate regulatory structures would be needed to develop and deploy phage products to the vast number of people who will be impacted, as was necessary for ARVs. This publication examines what lessons from the HIV/AIDS epidemic might be applied to phages and the AMR crisis; specifically, what funding programs and regulatory modifications enabled rapid research and development of a new class of drugs, followed by large-scale manufacturing and distribution of those drugs to developing countries.

2. Support for ARV Development and Delivery

The worldwide response to the HIV/AIDS crisis was unprecedented in terms of the speed with which ARVs were developed and applied clinically, especially in low and middle income countries (LMIC). These drugs ultimately transformed HIV/AIDS from a disease whose sufferers had a life expectancy of approximately one year to our current situation, in which HIV-positive individuals can experience a nearly normal life span [8]. An estimated 11 million lives have now been saved, and almost 2 million babies have been born HIV-free [9]. However, because of limitations in infrastructure and delivery capacity in most LMIC, many people globally have yet to fully benefit from ARVs.

The first AIDS case was diagnosed in 1981, and the first ARV to treat HIV/AIDS, azidothymidine (AZT), was approved in 1987, with much of the initial funding coming from the US government [10]. Once the huge scale of the HIV/AIDS epidemic became apparent, market incentives then motivated private investment into additional ARV drug development, particularly in the US. Over time, combinations of three different ARVs proved much more effective than AZT alone, with little or no resistance developing against the drugs, and these so-called highly active antiretroviral therapy (HAART) combinations became the standard of care. As will be discussed below, combinations of phages will undoubtedly also be required to stave off future bacterial resistance to phage products.

ARVs were eventually deployed to the developing world. The global community went through several stages of grappling with the problem before achieving some success: first comprehending the massive scale of the disease and disseminating that information; then mobilizing global resolve to address the issue; and eventually, establishing substantial international funding sources, particularly to deliver ARVs to LMIC [11]. Millions of deaths potentially could be prevented if a similar distribution of effective antibacterial agents were expedited. Let us first review how drug development and distribution was accomplished for the HIV/AIDS crisis, then consider how this could inform the global response to AMR.

2.1. Understanding the Scope of the Crisis

As early as 1983, the World Health Organization (WHO) convened its first meeting on HIV/AIDS and began formal international surveillance of the disease [11]. Two years later, the WHO and the US Department of Health and Social Services hosted the first International AIDS Conference, and in 1987, the WHO initiated the Global Program on AIDS, with a primary goal of raising awareness about the growing epidemic. That same year, the US Center for Disease Control and Prevention begin a large-scale public service campaign. This was followed by the first World AIDS Day in 1988, and the launching of the Red Ribbon Project in 1991, from which came the popular international symbol for AIDS. These and other awareness-raising efforts were eventually followed by formal resolutions from national and international organizations determined to combat the problem.

2.2. Generating Worldwide Commitment

New governmental organizations and processes were set up to coordinate policy development and global activities to tackle HIV/AIDS. Most notable of these was the Joint United Nations program on HIV/AIDS (UNAIDS), which began operations in 1996 [12]. In 2001 the UN General Assembly held a Special Session on AIDS—the first time that global body had ever focused a full session on a single disease [10]. A key outcome of the meeting was the formal Declaration of Commitment on HIV/AIDS, which included a call to establish an international global fund. Less than two years later, the WHO introduced the "3 by 5" initiative, a program so-named because it aimed to treat 3 million people in developing countries with ARVs by 2005 [13]. Importantly, international leaders also began to recognize the HIV/AIDS epidemic as both a health crisis and a global security issue, with formal statements issued by heads of state as well as by the UN Security Council [11]. With global will more fully engaged, steps were taken to find the monetary resources to achieve the specified goals.

2.3. Establishing Funding Sources for Delivering Medicines

Financing for the HIV/AIDS crisis came in stages of increasingly larger investments. The World Bank Multicountry AIDS Program (MAP), established in 2000, was the first international program, with disbursements totaling $500 million in the initial funding round [14]. After two or more years of international deliberations, the Global Fund to Fight AIDS, Tuberculosis, and Malaria (GFATM) was formalized in 2002 [15]. Instituted as a public-private partnership, the GFATM initially called for $1 billion, but received $1.9 billion in pledges by the time it became operational. Funding came from private individuals (typically in the order of hundreds of thousands of dollars), private corporations (the first being a $1 million donation), and private foundations and governments (with contributions ranging up to $200 million each). While this was a substantial leap forward in funding, it was still notably less than the $7–10 billion yearly investment that Kofi Annan, then UN Secretary-General, targeted as necessary to address the epidemic, especially in LMIC [16]. In 2003, the most significant funding materialized with the launch of the US President's Emergency Plan For AIDS Relief (PEPFAR), which allocated $15 billion for the first 5 years, then expanded to $48 billion in 2008 [17]. To date PEPFAR remains the largest public health investment program from a single country.

2.4. Inventing Mechanisms to Decrease Drug Prices in Developing Countries

In addition to gathering the financial resources needed, several regulatory modifications were enacted to facilitate faster and therefore more cost-effective delivery of ARVs to the developing world. In 1997, the US Congress passed the FDA Modernization Act in order to accelerate the drug approval process and loosen restrictions on communications regarding off-label use for potential HIV drugs [18]. And in 2004, the FDA issued new guidance policies to expedite approval of co-packaged and combination therapies aimed at developing countries [11]. These regulatory adaptations would ultimately prove to be significant in facilitating delivery of the drugs to the populations that needed them. Reforms in regulatory structures likely will be necessary for new AMR drugs as well, particularly for phage-based products.

New financial and legal resolutions were also put in place to help reduce the costs of ARVs. In 2000, the UNAIDS and WHO negotiated the Accelerating Access Initiative, an agreement with five major pharmaceutical companies to provide HIV/AIDS drugs to developing countries at decreased prices [10]. In addition, in 2001 the World Trade Organization established the Doha Declaration, which allowed developing countries access to generic drugs for public health crises, even without formal patent approval in each country [19]. Some companies also joined the Medicines Patent Pool, which facilitates licensing for manufacturing of generics for LMIC, speeding up the access to drugs in those countries [20]. Similar programs will undoubtedly facilitate the delivery of new AMR therapeutics to LMIC.

While these various support mechanisms for LMIC were initially funded by industrialized countries, developing countries themselves eventually began to underwrite the local delivery of ARVs. By 2013, roughly half of all costs for HIV treatment in sub-Saharan Africa were covered by in-country sources [21]. And in a few countries, particularly Angola, Botswana, and South Africa, more than 80% of the financing came from domestic funds. Some countries, such as Cape Verde and Cote d'Ivoire, are also utilizing creative funding schemes such as taxes on tobacco and alcohol to raise money for HIV/AIDS treatment. However, there is still less domestic funding across Western and Central Africa (15–29%).

3. Current Support for AMR Drug Development

As the world has been coming to grips with AMR, we have begun to progress through similar stages as with the HIV/AIDS epidemic. However, as discussed below, we are still at a relatively early point in this overall process. In order to effectively fund new drug development for AMR, including for phages, the global community will need to first more fully comprehend the scale of the crisis, which will then motivate international commitment to combat the problem, finally leading to the establishment of the financial resources and regulatory modifications needed.

3.1. Understanding the Scope of the Crisis

Resistant strains to the first small molecule antibiotic, penicillin, were identified even before penicillin was introduced in 1943 [22]. As new antibiotics have become available, resistance was recorded as early as one year after the first clinical use. This reality must be addressed as new antibacterials are developed for AMR.

Over the past decades, the scientific and public health communities have been documenting the growing rates of antibiotic resistance in specific bacterial strains and geographic regions. The WHO's first report on AMR, published in 2014, summarized the global resistance patterns for seven bacteria of major concern: *Escherichia coli*, *Klebsiella pneumoniae*, *Staphylococcus aureus*, *Streptococcus pneumoniae*, *Nontyphoidal Salmonella*, *Shigella*, and *Neisseria gonorrhoea* [23]. Strikingly, resistance to carbapenems, considered the last-resort antibiotics, was documented in the majority of the reporting countries, including some resistance rates up to 54%. The report highlighted that a post-antibiotic era is "far from being an apocalyptic fantasy, [but] is instead a very real possibility for the 21st century".

By July 2014, this data and others had prompted the UK Prime Minister to commission a study analyzing the key components of the crisis and proposing tangible steps that the international community could take to surmount it. Funded in partnership with the Wellcome Trust, the predictions from the resulting *Review on Antimicrobial Resistance* were astounding: by 2050, AMR is expected to cause over 10 million deaths each year and cost the global economy a total of $100 trillion [2,24]. In addition, a 2016 report from the World Bank Group estimates that AMR could push 28.3 million people into extreme poverty [25].

These reports served as wake-up calls for many, and the authors of the *Review on Antimicrobial Resistance* emphasized that there is still a primary need for AMR awareness-raising campaigns globally, despite the fact that numerous organizations have been spreading knowledge about AMR for decades. One of the earliest was the Alliance for the Prudent Use of Antibiotics, founded in 1981 by Dr. Stuart Levy, a leading researcher on molecular antibiotic efflux mechanisms [26]. In 2001 the European Union Council issued a formal recommendation on the judicious use of antimicrobial agents in human medicine [27]. This eventually led to the inauguration of European Antibiotic Awareness Day in 2008, now held yearly [28]. More recently, the WHO conducted a multi-country public awareness survey on AMR during the fall of 2015 to understand how best to deliver information and what topics to focus on [29]. Two months later, the first World Antibiotic Awareness Week was launched [30].

3.2. Generating Worldwide Commitment

Calls to action have come from various authoritative sources. In 2004 and 2008, the Infectious Diseases Society of America highlighted key causes and dangers of AMR, and proposed specific ways to stimulate investment in antibiotic research and development [31,32]. And in 2013, a panel of global experts published the Lancet Infectious Diseases Commission, a summary document on antibiotic resistance which called for coordinated international efforts to contain AMR, emphasizing that individual countries cannot effectively address the issue on their own [4]. A month after the WHO 2014 report on AMR, the Director of the Wellcome Trust, Dr. Jeremy Farrar, and Professor Mark Woolhouse of the University of Edinburgh published a Comment in *Nature* calling for the establishment of an intergovernmental panel on antimicrobial resistance [33].

To date, a global panel for AMR has not been formed. An oversight body of this type could more effectively coordinate international efforts, analogous to the role of UNAIDS for HIV/AIDS. In the meantime, worldwide forums have issued formal declarations on AMR, and both international and national bodies have created action plans. In May 2015, the World Health Assembly approved the Global Action Plan for Antimicrobial Resistance, which enumerates key strategies, including the need to increase AMR awareness globally and to develop policies for attracting more investment into new medical interventions [34]. It also called upon all Member States to establish national action plans for AMR by 2017. Thus far, 57 countries have formalized such plans [35]. Another major step forward was the 2016 meeting of the UN General Assembly focused on AMR, with plenary panel discussions on the need for multisectoral solutions to the problem [36]. Additionally, in July 2017, the G20 called for the creation of a Global R&D Collaboration Hub on AMR that would coordinate international funding efforts—the first G20 Declaration that has included R&D for public health [37]. The search for the appropriate individual to lead that hub began in early 2018.

3.3. Establishing Funding Sources for Delivering Medicines

In order to stall HIV/AIDS to the current level, PEPFAR, the largest funder for this disease, has spent over $70 billion since 2004 [38]. By comparison, the UK-commissioned *Review on Antimicrobial Resistance* estimated that $40 billion will be needed over 10 years to adequately address the global AMR problem. This represents only about 0.05% of the total amount that G20 countries currently spend on healthcare, and it is quite small compared to the projected $100 trillion that will be needed if AMR is not addressed. Of the total $40 billion proposed for AMR, likely $16 billion would be necessary to boost a new antibiotic pipeline, assuming a traditional drug development process and associated costs. However, phage manufacturing could be less expensive, and thus, do more with less.

As an incremental step toward gathering the financial support, the *Review on Antimicrobial Resistance* called for establishment of a Global Innovation Fund for AMR, with seed funding of $2 billion for 5 years. The UK and China responded by initiating such a fund in 2015, with each country pledging £50 million (equivalent to roughly $66 million); the Bill & Melinda Gates Foundation also agreed to contribute. While this falls far short of the recommended amount, it is a start. Recall that the analogous HIV funding began with $500 million before eventually growing to over $70 billion.

In early 2017, the Global Union for Antibiotics Research and Development (GUARD) published its own set of recommendations on specific steps for addressing AMR [39]. Its proposal included three separate funding mechanisms to stimulate a pipeline of antibiotics:

(1) Global Research Fund to build up infrastructure and increase the number of scientists working in the AMR field ($200 million/year for 10 years)
(2) Global Development Fund to provide forgivable loans primarily to small and medium-sized enterprises with the goal of pushing ten "high-need" products to market over a decade ($200 million/year for 10 years)
(3) Global Launch Reward of $1 billion for successfully delivering a commercial product that meets pre-specified AMR therapeutic goals

Both the *Review on Antimicrobial Resistance* and the GUARD report recommended a balance of so-called "pull" vs. "push" financial incentives. Push incentives, such as grants and forgivable loans, would help move the initial stages of R&D forward. Pull incentives, typified by the Global Launch Reward, would motivate companies to progress through the final, more costly stages of drug development by providing a monetary award that essentially lowers the financial risk. Economic experts generally agree that numerous push incentives are currently available, but pull incentives are sparse.

Other large AMR financing mechanisms have been established in Europe and the US. NewDrugs4BadBugs (ND4BB) was created in 2012 by the European public–private partnership Innovative Medicines Initiative (IMI). The goal of ND4BB is to bring together academic, industry and biotech groups to find new ways to overcome the practical challenges of developing new antimicrobial drugs [40]. To date, this program includes eight projects totaling EUR 650 million (approximately $758 million), with roughly half of the initial funding coming from IMI and the remainder from large pharmaceutical companies.

In the US, much of the funding for AMR comes through the Biomedical Advanced Research and Development Authority (BARDA), which was established in 2010 to address a number of public health emergencies. BARDA works through public-private partnerships with pharmaceutical and biotech companies, providing non-dilutive funding to help companies develop new antimicrobials. In 2016, BARDA partnered with the Wellcome Trust and the AMR Centre in the UK, as well as with the US NIAID, to launch a global business accelerator program called the Combating Antibiotic-Resistant Bacteria Biopharmaceutical Accelerator (CARB-X) [41]. To date, CARB-X has raised $500 million and funded 33 companies from 7 countries. Since 2004, the Wellcome Trust has also invested approximately $400 million in drug-resistant infection activities, and the US NIAID/NIH has contributed roughly $340 million per year since 2013 [42,43].

Another financial instrument, InnovFin Infectious Diseases, was founded in 2015 under the European Investment Bank, with estimates suggesting that it may provide up to $350 million [39]. And in 2016, the Global Antibiotic Research & Development Partnership (GARDP) was established by the WHO and the non-profit organization Drugs for Neglected Diseases initiative (DNDi) [44]. This program builds on DNDi's experience in developing drugs for neglected diseases that particularly impact LMIC, and it will include both push and pull financial incentives. The GARDP 2017–2023 plan includes raising the equivalent of $315 million and delivering four new therapeutics. While still shy of the projected $40 billion that will be needed, these and other smaller funds are important steps toward financing new therapeutics to overcome AMR.

3.4. *Inventing Mechanisms to Decrease Drug Prices in Developing Countries*

Several other strategies have been launched or proposed to foster new drug development. In 2012, the Generating Antibiotic Incentives Now (GAIN) legal provision was enacted in the US to incentivize pharmaceutical companies [45]. This law extended the exclusivity period by five years for specified antibiotic categories (those that target particularly concerning pathogens), thus prolonging the time that such drugs can be sold without competition from generics. In addition, the authors of the *Review on Antimicrobial Resistance* have proposed a tax on pharmaceutical companies, with the collected levies used to fund pull incentive awards for successful commercialization of new antimicrobials [24]. The rationale is that since antibiotics are such an integral component of modern medical systems, pharmaceutical companies which sell drugs other than antibiotics—such as those for oncological or surgical applications—indirectly benefit from having effective antibiotics available. The proposed tax would also directly incentivize drug companies to keep the antibiotic pipeline robust: companies that did invest in AMR drug development could receive tax credits and would be eligible for the market entry awards if their product delivery were successful. Another potential pull incentive is advance market commitments (AMCs), wherein drug developers would be guaranteed a set price and volume of sales for production of drugs that target specific pathogens of concern. In recent years, AMCs have been utilized to motivate development of a pneumonia vaccine [46].

Given that the first significant finances for AMR have only been committed in the past few years, the global community is still in the early stages of fundraising. Realistically, most of the money for LMIC will need to come from industrialized countries, as was the case for HIV/AIDS; developing countries simply do not have the resources. Of course, AMR truly is a global problem, since bacteria readily move across borders.

Nonetheless, the fundraising goals appear achievable, as evidenced by the worldwide response to HIV/AIDS. Now is the key time to strategize about how to use funds most efficiently. While conventional development of new antibiotics will certainly be important, that approach is typically an expensive and lengthy one. In addition, bacteria will undoubtedly continue to evolve resistance mechanisms to new classes of drugs. An ideal solution would be to invest in low-cost drugs that reduce development of resistance.

4. Attributes of Phage-Based Products

Phage-based drugs already provide such an option in the Eastern European countries of Georgia, Poland, and Russia [3,5,7]. Let us now discuss some of the currently available phages, and the general characteristics of those types of products that could be particularly beneficial for addressing the immense AMR crisis in the developing world, especially in the context of the funding environment described above.

4.1. *Inexpensive Drugs for Infectious Diseases*

Unlike ARVs for HIV/AIDS, which were very expensive drugs to develop, phages are inherently inexpensive to isolate. Indeed, therapeutic phages were first developed for patients in the 1920s using the relatively simple laboratory equipment that was available at that time [5,7]. In the former Soviet Union, phages have been utilized clinically for about 100 years. These include products targeting bacteria that underlie diarrhea, wound infections, and urinary tract infections, amongst others, including both antibiotic-sensitive and antibiotic-resistant strains. Today, phages can be bought over-the-counter in Georgia and Russia for as little as the equivalent of $1–2 per dose, though a full course of personalized phage treatment may cost as much as $1000–3000 per patient depending on the dose level and treatment duration [47–50].

Phages in Georgia, Poland, and Russia are manufactured under conditions that do not meet formal Western cGMP (Current Good Manufacturing Practice) requirements, but that do adhere to a different set of strict quality control standards—a key reason why phages are available in those countries at relatively low prices. These products have been used for decades with no reported major adverse events. Another non-cGMP regulatory system called *magistral preparations* was approved for phages in Belgium in early 2018 [51]. Under this system, phages are considered active pharmaceutical ingredients if they are produced as per a formally approved monograph, then quality tested in batches at a Belgian Approved Laboratory, accredited by the national regulatory authority. Both the Eastern European and Belgian manufacturing systems provide safe phage products at much lower costs than if they were prepared under conventional cGMP systems. Similar modifications to existing international manufacturing requirements would facilitate production of phages for AMR, just as the FDA adjusted policies to expedite regulatory approval for HIV/AIDS drugs. Given the huge quantities of antibacterial agents that will be needed to overcome antibiotic-resistant infections globally, this could be essential for providing quality products at reasonable costs.

4.2. Short Product Development Time Frames

Another key attribute of phages is that these products can be developed in very short time frames, which can both help to keep costs down and enable rapid responses to infection outbreaks. As an example, in 2016, a patient in San Diego, California, received emergency phage treatment after contracting a multi-drug resistant strain of *Acinetobacter baumannii* [52]. The search for appropriate phages for the patient began on February 21, and the first phage treatment was administered on March 15—just 23 days later [53]. During those few short weeks, previously isolated and characterized phages from several groups in the US and Europe were tested against bacteria isolated from the patient, then the selected phages went through two rounds of purification to ensure low endotoxin levels. In that same time period, the team managing the case submitted an Emergency Investigational New Drug application to the FDA and were given official approval to deliver the phages to the patient. After multiple phage doses were administered, the patient's infection completely cleared. This dramatic case highlights how rapidly an international community of scientific and clinical experts can address an acute infectious disease problem with phages. Also, while the quality of these phages was rigorously monitored by the FDA, they were not manufactured according to cGMP.

4.3. Decreased Probability of Resistance Development

Several tactics have been utilized to minimize the possibility that bacteria will eventually develop resistance to phages. The most common are mixing phages that target different bacterial epitopes and rotating the phage combinations at regular intervals [54]. Recall that combinations of ARVs were also required to address HIV/AIDS, and the FDA enacted guidelines for expediting the approval of such combinations, specifically with developing countries in mind [11]. Thus, there is precedence for revising standard regulatory processes to facilitate delivery of essential drug combinations.

Phage treatments are most effectively made of routinely shifting, carefully selected mixtures (termed "cocktails") that target multiple bacterial epitopes proliferating in a given locality [53]. Thus, what makes phages good at targeting antibiotic resistance is also what makes it impractical to manufacture them under a cGMP system—which works best for fixed chemical or biological matter, against which AMR can develop rapidly.

It is critically important to select appropriate phages for each cocktail. This includes avoiding phages that can transduce virulence genes through their local environment (e.g., lysogenic phages), as well as excluding phages that contain toxin genes which potentially could be transferred to the bacterial hosts. Regulatory systems must address these possible risks by rigorously controlling for them.

An ideal situation might be to have banks of pre-approved phages in each country or region. Centralized laboratories could regularly test those phages against bacterial populations currently circulating in each country, modifying the selected phage combinations as needed, and routinely adding new phages to the banks through certified approval processes. Local institutions and universities could potentially contribute phages to the banks, thereby providing incentives and opportunities for in-country development. Such a system would both address the need for altering phage mixtures to avoid resistance development, and also enable rapid responses to sudden disease outbreaks.

This is essentially the arrangement that has been in place the former Soviet Union for decades, and it has proven to be effective. It is notable that Georgia—which is officially categorized by the World Bank as an LMIC—is already showing how this system of delivering reasonably-priced, routinely-updated phage products can be accomplished in a developing country.

5. Final Thoughts

AMR knows no national boundaries, so overcoming it will take coordinated efforts from countries worldwide working together, likely overseen by a centralized international panel. Innovations in science, financing mechanisms, and regulatory processes will all be required to expedite the path forward. In this context, phage-based products could be an important part of the solution, since they can potentially provide effective and affordable options for killing antibiotic-resistant bacteria, with reduced probability of resistance development.

However, if the global response to AMR follows a similar pattern as that for HIV/AIDS, the worldwide community must focus more on raising awareness and motivating national and international commitment to the problem before sufficient funds will become available. Realistically, many other factors must also be in place to overcome AMR, such as improvements in water sanitation, appropriate use of existing antibiotics, suitable low-cost diagnostics, adequate surveillance, and effective local health systems, amongst others. It is an overwhelming, complex crisis, but the achievements that have been realized with the HIV/AIDS epidemic demonstrate that the global community is capable of bringing together the resources and creative solutions to surmount a problem as big as AMR.

Funding: This research received no external funding.

Acknowledgments: Discussions with numerous experts helped inform the writing of this manuscript, including Stefano Bertozzi, Sheri Hostetler, Bruno Marchon, Mark Offerhaus, Jean-Paul Pirnay, Daniel De Vos, Maia Merabishvili, Mzia Kutateladze, Heather Shane, Steve Karp, and Steffanie Strathdee.

Conflicts of Interest: Tobi Nagel has no conflicts of interest to report.

References

1. Number of Deaths due to HIV/AIDS; Estimates by WHO Region. Available online: http://apps.who.int/gho/data/view.main.22600REG?lang=en (accessed on 14 May 2018).
2. Review on Antimicrobial Resistance. *Antimicrobial Resistance: Tackling a Crisis for the Health and Wealth of Nations*; Review on Antimicrobial Resistance: London, UK, 2014.
3. Reardon, S. Phage therapy gets revitalized. *Nature* **2014**, *510*, 15–16. [CrossRef] [PubMed]
4. Laxminarayan, R.; Duse, A.; Wattal, C.; Zaidi, A.K.; Wertheim, H.F.; Sumpradit, N.; Vlieghe, E.; Hara, G.L.; Gould, I.M.; Goossens, H.; et al. Antibiotic resistance-the need for global solutions. *Lancet Infect. Dis.* **2013**, *13*, 1057–1098. [CrossRef]
5. Kutter, E.; De Vos, D.; Gvasalia, G.; Alavidze, Z.; Gogokhia, L.; Kuhl, S.; Abedon, S.T. Phage therapy in clinical practice: Treatment of human infections. *Curr. Pharm. Biotechnol.* **2010**, *11*, 69–86. [CrossRef] [PubMed]
6. Phage Therapy: Past, Present and Future (Ebook). Available online: https://www.frontiersin.org/research-topics/4111/phage-therapy-past-present-and-future (accessed on 14 May 2018).
7. Abedon, S.T.; Kuhl, S.J.; Blasdel, B.G.; Kutter, E.M. Phage treatment of human infections. *Bacteriophage* **2011**, *1*, 66–85. [CrossRef] [PubMed]
8. Antiretroviral Drug Discovery and Development. Available online: https://www.niaid.nih.gov/diseases-conditions/antiretroviral-drug-development (accessed on 14 May 2018).
9. PEPFAR Fact Sheet. Available online: https://www.pepfar.gov/documents/organization/264882.pdf (accessed on 14 May 2018).

10. Lange, J.M.; Ananworanich, J. The discovery and development of antiretroviral agents. *Antivir. Ther.* **2014**, *19* (Suppl. 3), 5–14. [CrossRef] [PubMed]
11. A Timeline of HIV and AIDS. Available online: https://www.hiv.gov/hiv-basics/overview/history/hiv-and-aids-timeline (accessed on 14 May 2018).
12. UNAIDS: The First 10 Years. Available online: http://data.unaids.org/pub/report/2008/jc1579_first_10_years_en.pdf (accessed on 14 May 2018).
13. Treating 3 Million by 2005: Making It Happen: The WHO Strategy: The WHO and UNAIDS Global Initiative to Provide Antiretroviral Therapy to 3 Million People with HIV/AIDS in Developing Countries by the End of 2005. Available online: http://www.who.int/3by5/publications/documents/en/Treating3millionby2005.pdf (accessed on 14 May 2018).
14. Multi-Country HIV/AIDS Program (MAP). Available online: http://web.worldbank.org/WBSITE/EXTERNAL/COUNTRIES/AFRICAEXT/EXTAFRHEANUTPOP/EXTAFRREGTOPHIVAIDS/0,,contentMDK:20415735~menuPK:1001234~pagePK:34004173~piPK:34003707~theSitePK:717148,00.html (accessed on 14 May 2018).
15. The Global Fund. Available online: https://www.theglobalfund.org/en/ (accessed on 14 May 2018).
16. The Global Fund to Fight AIDS, Tuberculosis, and Malaria: Background and Current Issues. Available online: https://fas.org/sgp/crs/misc/RL31712.pdf (accessed on 14 May 2018).
17. The United States President's Emergency Plan for AIDS Relief. Available online: https://www.pepfar.gov/about/270968.htm (accessed on 14 May 2018).
18. FDA Backgrounder on FDA Modernization Act of 1997. Available online: https://www.fda.gov/RegulatoryInformation/LawsEnforcedbyFDA/SignificantAmendmentstotheFDCAct/FDAMA/ucm089179.htm (accessed on 14 May 2018).
19. The Doha Declaration on the TRIPS Agreement and Public Health. Available online: http://www.who.int/medicines/areas/policy/doha_declaration/en/ (accessed on 14 May 2018).
20. Medicines Patent Poool. Available online: https://medicinespatentpool.org/ (accessed on 14 May 2018).
21. Acces to Antiretroviral Therapy in Africa: Status Report on Progress Toward the 2015 Goals. Available online: http://www.unaids.org/sites/default/files/media_asset/20131219_AccessARTAfricaStatusReportProgresstowards2015Targets_en_0.pdf (accessed on 14 May 2018).
22. Ventola, C.L. The antibiotic resistance crisis: Part 1: Causes and threats. *Pharm. Ther.* **2015**, *40*, 277–283.
23. Antimicrobial Resistance: Global Report on Surveillance. Available online: http://apps.who.int/iris/bitstream/handle/10665/112642/9789241564748_eng.pdf?sequence=1 (accessed on 14 May 2018).
24. Tackling Drug-Resistant Infections Globally: Final Report and Recommendations. Available online: https://amr-review.org/sites/default/files/160525_Final%20paper_with%20cover.pdf (accessed on 14 May 2018).
25. Drug-Resistant Infections: A Threat to Our Economic Future. Available online: http://pubdocs.worldbank.org/en/527731474225046104/AMR-Discussion-Draft-Sept18updated.pdf (accessed on 14 May 2018).
26. Alliance for the Prudent Use of Antibiotics. Available online: https://apua.org/ (accessed on 14 May 2018).
27. The Council of the European Union. Recommendation of 15 November 2001 on the Prudent Use of Antimicrobial Agents in Human Medicine (2002/77EC). Available online: https://eur-lex.europa.eu/legal-content/EN/TXT/?uri=celex%3A32002H0077 (accessed on 14 May 2018).
28. European Antibiotic Awareness Day: A European Health Initiative. Available online: https://antibiotic.ecdc.europa.eu/en (accessed on 14 May 2018).
29. Antibiotic Resistance: Multi-Country Public Awareness Survey. Available online: http://apps.who.int/iris/bitstream/handle/10665/194460/9789241509817_eng.pdf?sequence=1 (accessed on 14 May 2018).
30. World Antibiotic Awareness Week—All Years Archive. Available online: http://www.who.int/campaigns/world-antibiotic-awareness-week/archives/en/ (accessed on 14 May 2018).
31. Bad Bugs, No Drugs: As Antibiotic Discovery Stagnates—A Public Health Crisis Brews. Available online: https://www.idsociety.org/uploadedFiles/IDSA/Policy_and_Advocacy/Current_Topics_and_Issues/Advancing_Product_Research_and_Development/Bad_Bugs_No_Drugs/Statements/As%20Antibiotic%20Discovery%20Stagnates%20A%20Public%20Health%20Crisis%20Brews.pdf (accessed on 14 May 2018).

32. Spellberg, B.; Guidos, R.; Gilbert, D.; Bradley, J.; Boucher, H.W.; Scheld, W.M.; Bartlett, J.G.; Edwards, J., Jr.; Infectious Diseases Society of America. The epidemic of antibiotic-resistant infections: A call to action for the medical community from the Infectious Diseases Society of America. *Clin. Infect. Dis.* **2008**, *46*, 155–164. [CrossRef] [PubMed]
33. Woolhouse, M.; Farrar, J. Policy: An intergovernmental panel on antimicrobial resistance. *Nature* **2014**, *509*, 555–557. [CrossRef] [PubMed]
34. Global Action Plan on Antimicrobial Resistance. Available online: http://apps.who.int/iris/bitstream/handle/10665/193736/9789241509763_eng.pdf?sequence=1 (accessed on 14 May 2018).
35. Library of National Action Plans. Available online: http://www.who.int/antimicrobial-resistance/national-action-plans/library/en/ (accessed on 14 May 2018).
36. United Nations High-Level Meeting on Antimicrobial Resistance. Available online: http://www.who.int/antimicrobial-resistance/events/UNGA-meeting-amr-sept2016/en/ (accessed on 14 May 2018).
37. Wellcome and Gates Foundation to Support New Global Body to Tackle Superbugs. Available online: https://wellcome.ac.uk/press-release/wellcome-and-gates-foundation-support-new-global-body-tackle-superbugs (accessed on 14 May 2018).
38. PEPFAR Funding. Available online: https://www.pepfar.gov/documents/organization/252516.pdf (accessed on 14 May 2018).
39. Follow-Up Report for the German GUARD Initiative: Breaking through the Wall—A Call for Concerted Action on Antibiotics Research and Development. Full Report. February 2017. Available online: https://www.bundesgesundheitsministerium.de/fileadmin/Dateien/5_Publikationen/Gesundheit/Berichte/GUARD_Follow_Up_Report_Full_Report_final.pdf (accessed on 14 May 2018).
40. ND4BB: New Drugs for Bad Bugs. Available online: https://www.imi.europa.eu/projects-results/project-factsheets/nd4bb (accessed on 14 May 2018).
41. CARB-X: Combating Antibiotic Resistant Bacteria. Available online: https://carb-x.org/ (accessed on 14 May 2018).
42. New US-UK Partnership to Tackle Antibiotic Resistance. Available online: https://wellcome.ac.uk/news/new-us-uk-partnership-tackle-antibiotic-resistance (accessed on 14 May 2018).
43. Simpkin, V.L.; Renwick, M.J.; Kelly, R.; Mossialos, E. Incentivising innovation in antibiotic drug discovery and development: Progress, challenges and next steps. *J. Antibiot.* **2017**, *70*, 1087–1096. [CrossRef] [PubMed]
44. Global Antibiotic Research & Development Partnership (GARDP). Available online: https://www.dndi.org/diseases-projects/gardp/ (accessed on 14 May 2018).
45. Generating Antibiotic Incentives Now. Available online: https://www.fda.gov/downloads/AboutFDA/CentersOffices/OfficeofMedicalProductsandTobacco/CDER/UCM595188.pdf (accessed on 14 May 2018).
46. Cernuschi, T.; Furrer, E.; Schwalbe, N.; Jones, A.; Berndt, E.R.; McAdams, S. Advance market commitment for pneumococcal vaccines: Putting theory into practice. *Bull. World Health Organ.* **2011**, *89*, 913–918. [CrossRef] [PubMed]
47. Eliava BioPreparations. Available online: http://phage.ge/ (accessed on 14 May 2018).
48. Biochimpharm. Available online: http://biochimpharm.ge/en/ (accessed on 14 May 2018).
49. Microgen. Available online: http://www.microgen.ru/en/ (accessed on 14 May 2018).
50. Merabishvili, M.; Queen Astrid Military Hospital, Brussels, Belgium. Personal communication, 2018.
51. Pirnay, J.P.; Verbeken, G.; Ceyssens, P.J.; Huys, I.; De Vos, D.; Ameloot, C.; Fauconnier, A. The Magistral Phage. *Viruses* **2018**, *10*, 64. [CrossRef] [PubMed]
52. Schooley, R.T.; Biswas, B.; Gill, J.J.; Hernandez-Morales, A.; Lancaster, J.; Lessor, L.; Barr, J.J.; Reed, S.L.; Rohwer, F.; Benler, S.; et al. Development and Use of Personalized Bacteriophage-Based Therapeutic Cocktails to Treat a Patient with a Disseminated Resistant Acinetobacter baumannii Infection. *Antimicrob. Agents Chemother.* **2017**, *61*, e00954-17. [CrossRef] [PubMed]
53. Strathdee, S.A.; University of California, San Diego, CA, USA. Personal communication, 2018.
54. Chan, B.K.; Abedon, S.T.; Loc-Carrillo, C. Phage cocktails and the future of phage therapy. *Future Microbiol.* **2013**, *8*, 769–783. [CrossRef] [PubMed]

Review

The Diversity of Bacterial Lifestyles Hampers Bacteriophage Tenacity

Marta Lourenço [1,2], Luisa De Sordi [1] and Laurent Debarbieux [1,*]

1 Department of Microbiology, Institut Pasteur, F-75015 Paris, France; marta.mansos-lourenco@pasteur.fr (M.L.); luisa.de-sordi@pasteur.fr (L.D.S.)
2 Collège Doctoral, Sorbonne Université, F-75005 Paris, France
* Correspondence: laurent.debarbieux@pasteur.fr

Received: 18 May 2018; Accepted: 11 June 2018; Published: 15 June 2018

Abstract: Phage therapy is based on a simple concept: the use of a virus (bacteriophage) that is capable of killing specific pathogenic bacteria to treat bacterial infections. Since the pioneering work of Félix d'Herelle, bacteriophages (phages) isolated in vitro have been shown to be of therapeutic value. Over decades of study, a large number of rather complex mechanisms that are used by phages to hijack bacterial resources and to produce their progeny have been deciphered. While these mechanisms have been identified and have been studied under optimal conditions in vitro, much less is known about the requirements for successful viral infections in relevant natural conditions. This is particularly true in the context of phage therapy. Here, we highlight the parameters affecting phage replication in both in vitro and in vivo environments, focusing, in particular, on the mammalian digestive tract. We propose avenues for increasing the knowledge-guided implementation of phages as therapeutic tools.

Keywords: virus–host interactions; bacteriophage efficacy; gastrointestinal tract; phage therapy

1. Introduction

With the alarming worldwide increase in the prevalence of multidrug-resistant bacteria, phage therapy—the use of phages to target pathogenic bacteria [1]—has recently returned to the spotlight in the USA and Europe, although it had never fallen out of favour in countries such as Georgia [2]. The three main characteristics of phages that make phage therapy an appealing strategy are (i) the self-replication of phages, leading to a local increase in their concentration; (ii) the lack of broad off-target effects due to the narrow host specificity of phages and (iii) genomic flexibility making it possible to rapidly develop optimised variants. The recent publication of a successful compassionate clinical case treatment with phages has highlighted the potential value of phage therapy in the context of human health [3,4]. However, in modern phase II clinical trials, the efficacy of phage therapy was highly variable in a small number of patients with chronic otitis, and phage therapy was ineffective in a larger trial with children with diarrhoea [5,6]. This lack of success may partly reflect the paucity of data relating to the translation from in vitro to clinical settings [7]. We must, therefore, address the challenge of identifying the parameters characterising effective phage treatments. For example, in studies of several experimental models investigating the use of phages to target bacteria residing in the digestive tract of animals, treatment efficacy has been reported to range from complete inefficacy to highly successful [8–12]. These findings contrast strongly with in vitro observations in which most, if not all, phages are highly efficient at infecting their host. These discrepancies may be explained by the influence of the bacterial lifestyle on phage infection, as discussed below.

2. Bacteria Provide Essential Support for the Parasitic Lifestyle of Phages

Bacteria are among the most ubiquitous organisms on the planet and their high levels of diversity are regularly confirmed in metagenomics studies [13–15]. Bacteria colonise a multitude of environments, from oceans to deserts, demonstrating their great ability to thrive in different environments and to regulate major global processes, such as the biogeochemical cycles of essential elements (carbon, nitrogen, oxygen) [16].

From an anthropocentric point of view, most bacteria are harmless while a few are beneficial or pathogenic. Bacteria isolated from many body sites have been shown to survive in various conditions, such as the acidic medium of the stomach or the highly oxygenated respiratory tract. Even within a single species, bacteria may display considerable phenotypic flexibility. This is illustrated by the well-known model bacterium *Escherichia coli*, a facultative anaerobe able to survive in environmental conditions that are very different from its natural habitat, the digestive tract of warm-blooded animals [17].

Bacterial physiological responses play a crucial role in shaping the interactions of bacteria with their environment. The recent development of several techniques (membrane, chip, RNASeq), which facilitate the capture of mRNAs, has made a fundamental contribution to the description of global physiological responses in bacteria. These techniques have made it possible for researchers to describe the transcriptomic profile of bacteria growing in several different types of conditions [18–23]. For example, Denou et al. compared *Lactobacillus johnsonii* gene expression between in vitro (in flasks) and in vivo (mouse gastrointestinal tract) conditions and in different sections of the gastrointestinal tract (stomach, caecum and colon) [18]. Their observations confirmed that the animal host, either directly or indirectly via other microbes, influences gene expression in the bacterial populations colonizing different body sites.

Phages are obligate parasites and, as such, their distribution matches that of the bacteria they infect. Bacteria may be susceptible to phages or resistant via many mechanisms developed by bacteria during the course of their coevolution with phages. Bacteria can prevent phage adsorption by deleting phage receptors, modifying their conformation, or releasing factors that occupy the binding site or even mask it. Other mechanisms of protection involve the prevention of phage DNA injection, the digestion of phage DNA by restriction-modification enzymes or by the CRISPR-Cas machinery. For a more comprehensive and detailed description of these phage resistance mechanisms, we refer the reader to the review by Labrie, S.J., et al. [24]. In 2015, a novel system called BREX (bacteriophage exclusion) was described and reported to specifically prevent phage DNA replication [25]. Doron et al. (2018) recently used comparative genomics to predict an impressive list of 26 new putative antiphage systems, nine of which were experimentally validated [26]. In addition, environmental fluctuations driving bacterial modifications can directly or indirectly influence phage infection, as discussed in the chapters below focused on virulent phages and schematically illustrated in Figure 1.

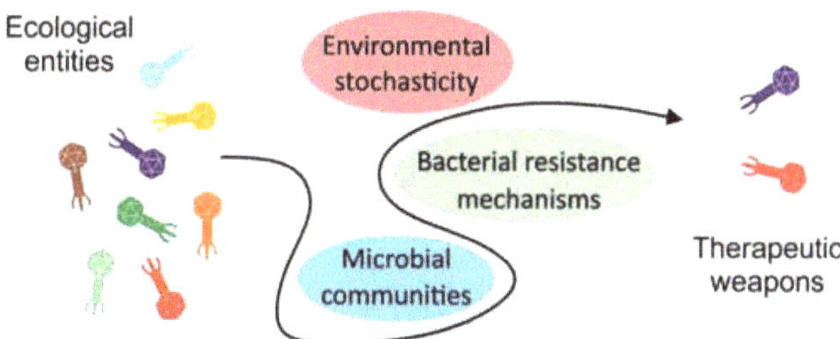

Figure 1. Schematic illustration summarising the obstacles that bacteriophages must overcome to be considered as antibacterial weapons.

3. Bacterial Physiology Affects the Outcome of Phage Infection

In optimal in vitro conditions, bacterial growth is characterised by four different phases: (i) the lag phase (initial phase) during which the bacteria are still adapting and adjusting to the growth conditions; (ii) the exponential growth or log phase during which the bacteria replicate rapidly; (iii) the stationary phase during which nutrients are depleted from the medium, limiting replication rates (during this phase, growth rate and death rate are usually matched); and (iv) death, which occurs when the nutrients are exhausted. The physiological state of a bacterium is linked to its growth conditions, which are, in turn, highly dependent on abiotic factors, such as nutrient variety and density, in particular [19]. Changes in growth conditions can affect the antibacterial activity of phages by preventing infection, replication or lysis. In vitro studies of phage–host interactions are typically performed in exponential phase cultures in liquid broth and little is known about these interactions in other conditions resembling those found in natural environments. The initial isolation of phages itself introduces a selection bias in that it often occurs in growth conditions that are optimal for the host (rich medium with shaking), i.e., those in which the bacteria are constantly in a planktonic state.

Many in vitro studies on the model system consisting of the phage T4 and its host, *E. coli*, have characterised the effects of host physiology on the infection efficiency of the phage. At high growth rates, phage T4 is absorbed and released more rapidly, its burst size increases and its eclipse and latent periods decrease [27–30]. These observations led to the suggestion that phage synthesis and assembly rates depend on the protein synthesis machinery of the host, whereas lysis time is correlated with cellular dimensions [29]. Other studies have shown that phages T4 and ms2 can enter a dormant state during the infection of stationary-phase cells. This state has been referred to as "hibernation" and is reversible. Some phage proteins are synthesised during hibernation but particle assembly is placed on hold until additional nutrients become available in the environment, which allows the phage infection processes to resume [27,31,32].

Bacteria may display various physiological states due to environmental stochasticity, which can convert a phage-susceptible bacterial host into a phage-resistant host. Indeed, stochastic differential gene expression can generate a heterogeneous population of cells within which a subpopulation may express lower levels of phage receptors, with consequences for the rate of phage adsorption. Such stochastic expression renders cells effectively resistant to phages without the need to acquire resistance through mutation. Although this phenomenon, known as phenotypic resistance, remains underappreciated and understudied, it may potentially account for the difference in infection efficiency between in vitro and in vivo conditions [33–35].

Another example of differences in phage infection efficiency due to shifts of environmental conditions is provided by phage T5. The infection efficiency of this phage has been shown to be dependent on temperature, which alters the host cell's membrane rigidity [36]. By contrast, *E. coli* phage infection efficiency seems to be independent of oxygen concentration, at least in vitro, as shown by studies in both aerobic and anaerobic conditions [11,12]. Nevertheless, it was shown that different aeration conditions imposed on *Bacillus thuringiensis* could affect the duration of the infectious cycle of phage BAM35 [37]. In 2004, Sillankorva et al. performed an extensive study with the phage US1 and its host, *Pseudomonas fluorescens* [38]. These authors showed that temperatures lower (4 °C) or higher (37 °C) than the optimal temperature (26 °C) had a major effect on phage infection efficiency, leading to an absence of phage amplification (37 °C) or rare (4 °C) phage infections. Furthermore, this phage cannot infect its host in a glucose medium despite its high infection efficiency in nutrient-rich conditions. Studies of the outer membrane protein profiles of cells grown in these two environments identified two proteins—17.5 and 99.0 kDa—with differential abundance under these growth conditions. These proteins were not detected in bacteria growing at 37 °C or in a glucose medium and the smaller protein was not detected at 4 °C, suggesting a possible role for these proteins as phage receptors. Environmental shifts can also, in some cases, trigger the production of capsules, which may mask phage receptors or allow other phages to use these same receptors [39–41]. In other cases, these environmental fluctuations can promote the induction (resumption of lytic cycle) of prophages present in the genome of bacteria, causing the destruction of their host [42]. Interestingly, prophage induction is frequent in the digestive tract of mammals as suggested by metagenomics data, however, their precise role waits to be defined [43,44].

4. Bacterial Community Lifestyle Influences Phage Infection

In any environment, including body sites, bacterial populations do not generally adopt the planktonic state of growth that is frequently observed in laboratory experiments. Instead, they tend to live in multilayer aggregates of cells that adhere to each other and frequently to surfaces via the production of a matrix of extracellular polymeric substances (EPSs) [45]. These EPSs include exopolysaccharides and proteins but also lipids and DNA. The resulting biofilms limit the efficacy of antibiotics, principally by decreasing their diffusion. As a result, the bacteria are not completely eradicated by such treatments, favouring the development of chronic bacterial infections [46]. In such situations, phages may constitute a potential solution given their impact on microbial communities [47]. However, the efficacy of phages against biofilms in vitro is variable and certain biofilm components may act as barriers against phage infection. For example, the presence of an amyloid fibre network of CsgA (curli polymer) can physically prevent phages from penetrating biofilms [48]. Phages can also attach to these amyloid fibres, preventing the viral binding to receptors [48]. On the other hand, some phages are equipped with enzymes that can degrade the polysaccharides produced by bacteria, thereby facilitating the diffusion of viral particles in biofilms [49,50]. The efficacy with which phages infect bacteria in biofilms is also strongly influenced by nutrient availability and nutrient concentrations that are highly heterogeneous within the biofilm structure [51].

An additional layer of complexity in interactions between phages and biofilms has been reported in studies of biofilms formed by the gut pathogen *Campylobacter jejuni*. Following phage infection, some of the cells in *C. jejuni* biofilms enter a carrier state. This involves phenotypic modifications to the bacterial cells, conferring advantages that enable them to survive in extraintestinal environments but preventing them from colonising the gut of chickens. Nevertheless, such carrier bacteria can import the phage into chickens that are already colonized by *C. jejuni*, providing the phage with opportunities to infect new cells following its release from the carrier [52,53].

Biofilms can also provide bacteria with a spatial refuge, reducing the probability of contact between a phage and its host, driving coexistence dynamics between the two populations without extinction of either the bacteria or the phage. This has been studied in vitro and modelled in silico. Spatially explicit individual-based stochastic models have shown that these structured refuges may maintain coexistence between the two populations within their boundaries, without the emergence of resistant clones [54]. In vitro experiments on populations of *P. aeruginosa* and bacteriophage PP7 in a heterogeneous artificial environment (static bacterial growth) showed a decrease in viral transmission and the emergence of refuges for the bacterial cells, stabilising interactions between the two antagonistic entities [55]. Similar observations were made when biofilms were grown on the wall of chemostats [56]. Finally, Eriksen et al. showed in a much more structured environment (solid agar in a Petri dish) that populations of phages and bacteria can co-exist in the long term but that this phenomenon is dependent on bacterial density, requiring the presence of at least 50,000 cells [57]. This threshold for phage replication is close to the threshold of 10,000 cells previously determined for well-mixed populations in several systems (*Bacillus subtilis*, *Escherichia coli* and *Staphylococcus aureus*), a phenomenon known as the "threshold for phage replication" or "proliferation threshold" [58,59].

5. Human Health and the Gut Phageome

Many aspects of phage biology, from initial adsorption to final lysis, can be affected by host behaviour, making it harder to reliably predict the overall efficacy of a phage in a given situation. This challenge is even greater when the complexity of viral species inhabiting the human gut is taken into account, as the cellular hosts of most of these viruses have yet to be identified [60,61].

The human gastrointestinal tract is a highly diverse and heterogeneous environment [62] that is inhabited by many different microorganisms [63]. It is also characterised by changes in conditions between sections, exposing its inhabitants to fluctuations in pH, nutrient levels, water and oxygen concentrations and even structure (ranging from liquid to semi-solid) [64–68].

It is now acknowledged that there are at least as many phages as bacterial cells in the mammalian gastrointestinal tract [69]. In healthy humans, only a small proportion of the phageome (phage community) is common to large numbers of individuals, with most of the phages present being subject specific [44]. Moreover, patients with inflammatory bowel disease (ulcerative colitis and Crohn's disease) or AIDS have been shown to have gut viral populations that are very different in size and diversity from those of healthy individuals [70,71]. Furthermore, changes in viral diversity have been shown to precede the appearance of type I diabetes in children [72]. Phageome variations are of course connected with bacteriome deviations, demonstrating the intimate but still poorly characterised link between these two antagonistic populations. These conditions of viral and cellular dysbiosis raise questions about whether certain diseases are caused by changes in the microbiome rather than a single pathogen, defining the new concept of a "pathobiome" [73]. This concept underlies a paradigm shift with a move away from targeting single pathogens to targeting whole communities. Within this framework, phages are potentially useful as modulators of the microbiome as a whole. A striking example of this approach is provided by the similar efficacies of treatments for recurrent *Clostridium difficile* infections based on faecal microbiota transfer or sterile faecal transfer with filtering to exclude bacteria (but not phages), highlighting the role of non-bacterial components of the microbiota in the clinical effect of treatment [74,75]. Interestingly, the virome composition of patients treated by sterile transfer was found to be similar to that in the donor [75].

Interesting features of these phages can be linked to their adaptation to this environment; for example, some phages carry specific motifs in their capsids that allow them to bind to the intestinal mucus, potentially creating an additional layer of protection against bacteria [76]. Moreover, a direct role of the microbiome in phage evolution has also been suggested by the results of a study reporting the evolution of an ability to infect new hosts through the use of a second strain as a stepping stone [9]. No such evolution was observed in vitro or in dixenic mice and it was, therefore, suggested that the gut microbiota can promote phage and bacterial population diversification [9,77].

In summary, each partner in this tripartite interaction (the phage, the bacterium and the mammalian host) plays an important role in phage–bacterium dynamics. It is therefore vital to consider these partners as an ecosystem rather than as two separate paired entities (phage/bacterium or bacterium/host) [78,79]. There are currently gaps in our knowledge that we need to overcome if we are to implement effective strategies based on phage treatments for intestinal pathogens or for the development of microbiota engineering strategies.

6. Overcoming the Limitations of Phage Infection Efficacy In Vivo

To optimise the output of applications based on phages, the gap between in vitro studies and in vivo conditions may be bridged in several ways. First, phages can be isolated and characterised in more realistic and ecologically relevant conditions than under the conditions for optimal bacterial growth that are typically used. For example, we can decide to start from in vitro biofilms consisting of single bacterial species or multi-species communities, and then proceed to ex-vivo conditions using organs [11,80] and, ultimately, in vivo environments [60]. Second, the precise identification of phage receptors and their expression profiles in ecologically relevant conditions will not only provide us with information about phage biology but will also guide the optimisation of conditions for in vivo efficacy. Adaptation of the phage to the targeted pathogen has also been shown to increase phage efficacy in some cases [81]. Moreover, the use of different doses and the localised release of microencapsulated phages may overcome some of the difficulties related to bacterial refuges and bacterial density thresholds [82].

Third, the use of phages together with other treatments (e.g., antibiotics) may improve overall treatment efficacy, an idea that has gained ground since the publication of the Phage Antibiotic Synergy system in 2007 [83]. Several studies have since confirmed the advantages of combining these two antibacterial weapons, although some of the mechanisms involved have yet to be identified (not all phage and antibiotic combinations display such synergy [84,85]). Such combinations may also be effective against biofilms, overcoming the limitations of each of these agents used separately [86–88]. The selection of resistant cells is a key concern in the use of both antibiotics and phages. However, there is no overall association between antibiotic resistance and phage resistance profiles supporting further their use in combination [89]. Nevertheless, double resistance or persister cells could provide a means for bacteria to protect themselves from these threats, however, this requires further studies. Interestingly, it was observed that the growth of phage-resistant bacteria during phage therapy in experimental models can be controlled with two independent allies: antibiotics, as demonstrated in an endocarditis model, and the innate immune response, as shown in a model of pulmonary infection [84,90].

About a century after their first use as an antibacterial agent for treating infections, phages have not yet revealed all their secrets. Phage biology is presenting scientists with new challenges every day. Many of the mechanisms involved in phage infection of bacteria remain unknown, hindering the effective use of phages as an ecological and sustainable alternative or complement to overcome the antibiotic resistance crisis and to tackle diseases caused by microbiome dysbiosis.

Acknowledgments: We thank Jorge Moura de Sousa for critically reading the manuscript and Dwayne Roach for valuable discussion. ML is part of the Pasteur-Paris University (PPU) International PhD Program. This project has received funding from the Institut Carnot Pasteur Maladie Infectieuse (ANR 11-CARN 017-01). LDS is founded by a Roux-Cantarini fellowship from the Institut Pasteur (Paris, France).

Conflicts of Interest: The authors declare no conflict of interest.

References

1. D'Herelle, F. Sur un microbe invisible antagoniste des bacilles dysentériques. *Comptes Rendus Acad. Sci. Paris* **1917**, *165*, 373–375.
2. Kutateladze, M. Experience of the Eliava Institute in bacteriophage therapy. *Virol. Sin.* **2015**, *30*, 80–81. [CrossRef] [PubMed]
3. Jennes, S.; Merabishvili, M.; Soentjens, P.; Pang, K.W.; Rose, T.; Keersebilck, E.; Soete, O.; Francois, P.M.; Teodorescu, S.; Verween, G.; et al. Use of bacteriophages in the treatment of colistin-only-sensitive *Pseudomonas aeruginosa* septicaemia in a patient with acute kidney injury-a case report. *Crit. Care* **2017**, *21*, 129. [CrossRef] [PubMed]
4. Schooley, R.T.; Biswas, B.; Gill, J.J.; Hernandez-Morales, A.; Lancaster, J.; Lessor, L.; Barr, J.J.; Reed, S.L.; Rohwer, F.; Benler, S.; et al. Development and use of personalized bacteriophage-based therapeutic cocktails to treat a patient with a disseminated resistant *Acinetobacter baumannii* infection. *Antimicrob. Agents Chemother.* **2017**, *61*. [CrossRef] [PubMed]
5. Wright, A.; Hawkins, C.H.; Anggard, E.E.; Harper, D.R. A controlled clinical trial of a therapeutic bacteriophage preparation in chronic otitis due to antibiotic-resistant *Pseudomonas aeruginosa*; a preliminary report of efficacy. *Clin. Otolaryngol.* **2009**, *34*, 349–357. [CrossRef] [PubMed]
6. Sarker, S.A.; Sultana, S.; Reuteler, G.; Moine, D.; Descombes, P.; Charton, F.; Bourdin, G.; McCallin, S.; Ngom-Bru, C.; Neville, T.; et al. Oral Phage Therapy of Acute Bacterial Diarrhea with Two Coliphage Preparations: A Randomized Trial in Children from Bangladesh. *EBioMedicine* **2016**, *4*, 124–137. [CrossRef] [PubMed]
7. Sarker, S.A.; Brussow, H. From bench to bed and back again: Phage therapy of childhood *Escherichia coli* diarrhea. *Ann. N. Y. Acad. Sci.* **2016**, *1372*, 42–52. [CrossRef] [PubMed]
8. Chibani-Chennoufi, S.; Sidoti, J.; Bruttin, A.; Dillmann, M.L.; Kutter, E.; Qadri, F.; Sarker, S.A.; Brussow, H. Isolation of *Escherichia coli* bacteriophages from the stool of pediatric diarrhea patients in Bangladesh. *J. Bacteriol.* **2004**, *186*, 8287–8294. [CrossRef] [PubMed]

9. De Sordi, L.; Khanna, V.; Debarbieux, L. The Gut Microbiota Facilitates Drifts in The Genetic Diversity and Infectivity of Bacterial Viruses. *Cell Host Microbe* **2017**, *22*, 801–808 e803. [CrossRef] [PubMed]
10. Galtier, M.; De Sordi, L.; Maura, D.; Arachchi, H.; Volant, S.; Dillies, M.A.; Debarbieux, L. Bacteriophages to reduce gut carriage of antibiotic resistant uropathogens with low impact on microbiota composition. *Environ. Microbiol.* **2016**, *18*, 2237–2245. [CrossRef] [PubMed]
11. Maura, D.; Galtier, M.; Le Bouguenec, C.; Debarbieux, L. Virulent bacteriophages can target O104:H4 enteroaggregative *Escherichia coli* in the mouse intestine. *Antimicrob. Agents Chemother.* **2012**, *56*, 6235–6242. [CrossRef] [PubMed]
12. Weiss, M.; Denou, E.; Bruttin, A.; Serra-Moreno, R.; Dillmann, M.L.; Brussow, H. In vivo replication of T4 and T7 bacteriophages in germ-free mice colonized with *Escherichia coli*. *Virology* **2009**, *393*, 16–23. [CrossRef] [PubMed]
13. Oh, J.; Byrd, A.L.; Deming, C.; Conlan, S.; Program, N.C.S.; Kong, H.H.; Segre, J.A. Biogeography and individuality shape function in the human skin metagenome. *Nature* **2014**, *514*, 59–64. [CrossRef] [PubMed]
14. Venter, J.C.; Remington, K.; Heidelberg, J.F.; Halpern, A.L.; Rusch, D.; Eisen, J.A.; Wu, D.; Paulsen, I.; Nelson, K.E.; Nelson, W.; et al. Environmental genome shotgun sequencing of the Sargasso Sea. *Science* **2004**, *304*, 66–74. [CrossRef] [PubMed]
15. Signori, C.N.; Thomas, F.; Enrich-Prast, A.; Pollery, R.C.; Sievert, S.M. Microbial diversity and community structure across environmental gradients in Bransfield Strait, western Antarctic Peninsula. *Front. Microbiol.* **2014**, *5*, 647. [CrossRef] [PubMed]
16. Falkowski, P.G.; Fenchel, T.; Delong, E.F. The microbial engines that drive Earth's biogeochemical cycles. *Science* **2008**, *320*, 1034–1039. [CrossRef] [PubMed]
17. Van Elsas, J.D.; Semenov, A.V.; Costa, R.; Trevors, J.T. Survival of *Escherichia coli* in the environment: Fundamental and public health aspects. *ISME J.* **2011**, *5*, 173–183. [CrossRef] [PubMed]
18. Denou, E.; Berger, B.; Barretto, C.; Panoff, J.M.; Arigoni, F.; Brussow, H. Gene expression of commensal *Lactobacillus johnsonii* strain NCC533 during in vitro growth and in the murine gut. *J. Bacteriol.* **2007**, *189*, 8109–8119. [CrossRef] [PubMed]
19. Feugeas, J.P.; Tourret, J.; Launay, A.; Bouvet, O.; Hoede, C.; Denamur, E.; Tenaillon, O. Links between transcription, environmental adaptation and gene variability in *Escherichia coli*: Correlations between gene expression and gene variability reflect growth efficiencies. *Mol. Biol. Evol.* **2016**, *33*, 2515–2529. [CrossRef] [PubMed]
20. Janoir, C.; Deneve, C.; Bouttier, S.; Barbut, F.; Hoys, S.; Caleechum, L.; Chapeton-Montes, D.; Pereira, F.C.; Henriques, A.O.; Collignon, A.; et al. Adaptive strategies and pathogenesis of *Clostridium difficile* from in vivo transcriptomics. *Infect. Immun.* **2013**, *81*, 3757–3769. [CrossRef] [PubMed]
21. Partridge, J.D.; Scott, C.; Tang, Y.; Poole, R.K.; Green, J. *Escherichia coli* transcriptome dynamics during the transition from anaerobic to aerobic conditions. *J. Biol. Chem.* **2006**, *281*, 27806–27815. [CrossRef] [PubMed]
22. Snyder, J.A.; Haugen, B.J.; Buckles, E.L.; Lockatell, C.V.; Johnson, D.E.; Donnenberg, M.S.; Welch, R.A.; Mobley, H.L. Transcriptome of uropathogenic *Escherichia coli* during urinary tract infection. *Infect. Immun.* **2004**, *72*, 6373–6381. [CrossRef] [PubMed]
23. Stintzi, A.; Marlow, D.; Palyada, K.; Naikare, H.; Panciera, R.; Whitworth, L.; Clarke, C. Use of genome-wide expression profiling and mutagenesis to study the intestinal lifestyle of *Campylobacter jejuni*. *Infect. Immun.* **2005**, *73*, 1797–1810. [CrossRef] [PubMed]
24. Labrie, S.J.; Samson, J.E.; Moineau, S. Bacteriophage resistance mechanisms. *Nat. Rev. Microbiol.* **2010**, *8*, 317–327. [CrossRef] [PubMed]
25. Goldfarb, T.; Sberro, H.; Weinstock, E.; Cohen, O.; Doron, S.; Charpak-Amikam, Y.; Afik, S.; Ofir, G.; Sorek, R. BREX is a novel phage resistance system widespread in microbial genomes. *EMBO J.* **2015**, *34*, 169–183. [CrossRef] [PubMed]
26. Doron, S.; Melamed, S.; Ofir, G.; Leavitt, A.; Lopatina, A.; Keren, M.; Amitai, G.; Sorek, R. Systematic discovery of antiphage defense systems in the microbial pangenome. *Science* **2018**, *359*. [CrossRef] [PubMed]
27. Bryan, D.; El-Shibiny, A.; Hobbs, Z.; Porter, J.; Kutter, E.M. Bacteriophage T4 infection of stationary phase *E. coli*: Life after log from a phage perspective. *Front. Microbiol.* **2016**, *7*, 1391. [CrossRef] [PubMed]
28. Golec, P.; Karczewska-Golec, J.; Los, M.; Wegrzyn, G. Bacteriophage T4 can produce progeny virions in extremely slowly growing *Escherichia coli* host: Comparison of a mathematical model with the experimental data. *FEMS Microbiol. Lett.* **2014**, *351*, 156–161. [CrossRef] [PubMed]

29. Hadas, H.; Einav, M.; Fishov, I.; Zaritsky, A. Bacteriophage T4 development depends on the physiology of its host *Escherichia coli*. *Microbiology* **1997**, *143*, 179–185. [CrossRef] [PubMed]
30. Nabergoj, D.; Modic, P.; Podgornik, A. Effect of bacterial growth rate on bacteriophage population growth rate. *Microbiol.Open* **2017**, *7*, e00558. [CrossRef] [PubMed]
31. Ricciuti, C.P. Host-virus interactions in *Escherichia coli*: Effect of stationary phase on viral release from MS2-infected bacteria. *J. Virol.* **1972**, *10*, 162–165. [PubMed]
32. Propst-Ricciuti, B. The effect of host-cell starvation on virus-induced lysis by MS2 bacteriophage. *J. Gen. Virol.* **1976**, *31*, 323–330. [CrossRef] [PubMed]
33. Bull, J.J.; Vegge, C.S.; Schmerer, M.; Chaudhry, W.N.; Levin, B.R. Phenotypic resistance and the dynamics of bacterial escape from phage control. *PLoS ONE* **2014**, *9*, e94690. [CrossRef] [PubMed]
34. Chapman-McQuiston, E.; Wu, X.L. Stochastic receptor expression allows sensitive bacteria to evade phage attack. Part I: Experiments. *Biophys. J.* **2008**, *94*, 4525–4536. [CrossRef] [PubMed]
35. Levin, B.R.; Moineau, S.; Bushman, M.; Barrangou, R. The population and evolutionary dynamics of phage and bacteria with CRISPR-mediated immunity. *PLoS Genet.* **2013**, *9*, e1003312. [CrossRef] [PubMed]
36. Labedan, B. Requirement for a fluid host cell membrane in injection of coliphage T5 DNA. *J. Virol.* **1984**, *49*, 273–275. [PubMed]
37. Daugelavicius, R.; Gaidelyte, A.; Cvirkaite-Krupovic, V.; Bamford, D.H. On-line monitoring of changes in host cell physiology during the one-step growth cycle of *Bacillus phage* Bam35. *J. Microbiol. Methods* **2007**, *69*, 174–179. [CrossRef] [PubMed]
38. Sillankorva, S.; Oliveira, R.; Vieira, M.J.; Sutherland, I.; Azeredo, J. Pseudomonas fluorescens infection by bacteriophage PhiS1: The influence of temperature, host growth phase and media. *FEMS Microbiol. Lett.* **2004**, *241*, 13–20. [CrossRef] [PubMed]
39. Dufour, N.; Clermont, O.; La Combe, B.; Messika, J.; Dion, S.; Khanna, V.; Denamur, E.; Ricard, J.D.; Debarbieux, L.; ColoColi, G. Bacteriophage LM33_P1, a fast-acting weapon against the pandemic ST131-O25B:H4 *Escherichia coli* clonal complex. *J. Antimicrob. Chemother.* **2016**, *71*, 3072–3080. [CrossRef] [PubMed]
40. Ohshima, Y.; Schumacher-Perdreau, F.; Peters, G.; Pulverer, G. The role of capsule as a barrier to bacteriophage adsorption in an encapsulated *Staphylococcus simulans* strain. *Med. Microbiol. Immunol.* **1988**, *177*, 229–233. [CrossRef] [PubMed]
41. Roach, D.R.; Sjaarda, D.R.; Castle, A.J.; Svircev, A.M. Host exopolysaccharide quantity and composition impact erwinia amylovora bacteriophage pathogenesis. *Appl. Environ. Microbiol.* **2013**, *79*, 3249–3256. [CrossRef] [PubMed]
42. Nanda, A.M.; Thormann, K.; Frunzke, J. Impact of spontaneous prophage induction on the fitness of bacterial populations and host-microbe interactions. *J. Bacteriol.* **2015**, *197*, 410–419. [CrossRef] [PubMed]
43. De Paepe, M.; Leclerc, M.; Tinsley, C.R.; Petit, M.A. Bacteriophages: An underestimated role in human and animal health? *Front. Cell. Infect. Microbiol.* **2014**, *4*, 39. [CrossRef] [PubMed]
44. Manrique, P.; Bolduc, B.; Walk, S.T.; van der Oost, J.; de Vos, W.M.; Young, M.J. Healthy human gut phageome. *Proc. Natl. Acad. Sci. USA* **2016**, *113*, 10400–10405. [CrossRef] [PubMed]
45. Flemming, H.C.; Wingender, J.; Szewzyk, U.; Steinberg, P.; Rice, S.A.; Kjelleberg, S. Biofilms: An emergent form of bacterial life. *Nat. Rev. Microbiol.* **2016**, *14*, 563–575. [CrossRef] [PubMed]
46. Costerton, J.W.; Stewart, P.S.; Greenberg, E.P. Bacterial biofilms: A common cause of persistent infections. *Science* **1999**, *284*, 1318–1322. [CrossRef] [PubMed]
47. Fernandez, L.; Rodriguez, A.; Garcia, P. Phage or foe: An insight into the impact of viral predation on microbial communities. *ISME J.* **2018**, *12*, 1171–1179. [CrossRef] [PubMed]
48. Vidakovic, L.; Singh, P.K.; Hartmann, R.; Nadell, C.D.; Drescher, K. Dynamic biofilm architecture confers individual and collective mechanisms of viral protection. *Nat. Microbiol.* **2018**, *3*, 26–31. [CrossRef] [PubMed]
49. Majkowska-Skrobek, G.; Latka, A.; Berisio, R.; Maciejewska, B.; Squeglia, F.; Romano, M.; Lavigne, R.; Struve, C.; Drulis-Kawa, Z. Capsule-Targeting Depolymerase, Derived from Klebsiella KP36 Phage, as a Tool for the Development of Anti-Virulent Strategy. *Viruses* **2016**, *8*, 324. [CrossRef] [PubMed]
50. Pires, D.P.; Oliveira, H.; Melo, L.D.; Sillankorva, S.; Azeredo, J. Bacteriophage-encoded depolymerases: Their diversity and biotechnological applications. *Appl. Microbiol. Biotechnol.* **2016**, *100*, 2141–2151. [CrossRef] [PubMed]

51. Simmons, M.; Drescher, K.; Nadell, C.D.; Bucci, V. Phage mobility is a core determinant of phage-bacteria coexistence in biofilms. *ISME J.* **2018**, *12*, 531–543. [CrossRef] [PubMed]
52. Brathwaite, K.J.; Siringan, P.; Connerton, P.L.; Connerton, I.F. Host adaption to the bacteriophage carrier state of *Campylobacter jejuni*. *Res. Microbiol.* **2015**, *166*, 504–515. [CrossRef] [PubMed]
53. Siringan, P.; Connerton, P.L.; Cummings, N.J.; Connerton, I.F. Alternative bacteriophage life cycles: The carrier state of *Campylobacter jejuni*. *Open Biol.* **2014**, *4*, 130200. [CrossRef] [PubMed]
54. Heilmann, S.; Sneppen, K.; Krishna, S. Coexistence of phage and bacteria on the boundary of self-organized refuges. *Proc. Natl. Acad. Sci. USA* **2012**, *109*, 12828–12833. [CrossRef] [PubMed]
55. Brockhurst, M.A.; Buckling, A.; Rainey, P.B. Spatial heterogeneity and the stability of host-parasite coexistence. *J. Evol. Biol.* **2006**, *19*, 374–379. [CrossRef] [PubMed]
56. Schrag, S.J.; Mittler, J.E. Host-parasite coexistence: The role of spatial refuges in stabilizing bacteria-phage interactions. *Am. Nat.* **1996**, *148*, 348–377. [CrossRef]
57. Eriksen, R.S.; Svenningsen, S.L.; Sneppen, K.; Mitarai, N. A growing microcolony can survive and support persistent propagation of virulent phages. *Proc. Natl. Acad. Sci. USA* **2018**, *115*, 337–342. [CrossRef] [PubMed]
58. Payne, R.J.; Phil, D.; Jansen, V.A. Phage therapy: The peculiar kinetics of self-replicating pharmaceuticals. *Clin. Pharmacol. Ther.* **2000**, *68*, 225–230. [CrossRef] [PubMed]
59. Wiggins, B.A.; Alexander, M. Minimum bacterial density for bacteriophage replication: Implications for significance of bacteriophages in natural ecosystems. *Appl. Environ. Microbiol.* **1985**, *49*, 19–23. [PubMed]
60. Reyes, A.; Wu, M.; McNulty, N.P.; Rohwer, F.L.; Gordon, J.I. Gnotobiotic mouse model of phage-bacterial host dynamics in the human gut. *Proc. Natl. Acad. Sci. USA* **2013**, *110*, 20236–20241. [CrossRef] [PubMed]
61. Yutin, N.; Makarova, K.S.; Gussow, A.B.; Krupovic, M.; Segall, A.; Edwards, R.A.; Koonin, E.V. Discovery of an expansive bacteriophage family that includes the most abundant viruses from the human gut. *Nat. Microbiol.* **2018**, *3*, 38–46. [CrossRef] [PubMed]
62. Donaldson, G.P.; Lee, S.M.; Mazmanian, S.K. Gut biogeography of the bacterial microbiota. *Nat. Rev. Microbiol.* **2016**, *14*, 20–32. [CrossRef] [PubMed]
63. Human Microbiome Project Consortium. Structure, function and diversity of the healthy human microbiome. *Nature* **2012**, *486*, 207–214.
64. He, G.; Shankar, R.A.; Chzhan, M.; Samouilov, A.; Kuppusamy, P.; Zweier, J.L. Noninvasive measurement of anatomic structure and intraluminal oxygenation in the gastrointestinal tract of living mice with spatial and spectral EPR imaging. *Proc. Natl. Acad. Sci. USA* **1999**, *96*, 4586–4591. [CrossRef] [PubMed]
65. Koziolek, M.; Grimm, M.; Becker, D.; Iordanov, V.; Zou, H.; Shimizu, J.; Wanke, C.; Garbacz, G.; Weitschies, W. Investigation of pH and temperature profiles in the GI tract of fasted human subjects using the intellicap((®)) system. *J. Pharm. Sci.* **2015**, *104*, 2855–2863. [CrossRef] [PubMed]
66. Maier, T.V.; Lucio, M.; Lee, L.H.; VerBerkmoes, N.C.; Brislawn, C.J.; Bernhardt, J.; Lamendella, R.; McDermott, J.E.; Bergeron, N.; Heinzmann, S.S.; et al. Impact of dietary resistant starch on the human gut microbiome, metaproteome, and metabolome. *mBio* **2017**, *8*, e01343-17. [CrossRef] [PubMed]
67. Marteyn, B.; West, N.P.; Browning, D.F.; Cole, J.A.; Shaw, J.G.; Palm, F.; Mounier, J.; Prevost, M.C.; Sansonetti, P.; Tang, C.M. Modulation of shigella virulence in response to available oxygen in vivo. *Nature* **2010**, *465*, 355–358. [CrossRef] [PubMed]
68. Wang, Y.; Holmes, E.; Comelli, E.M.; Fotopoulos, G.; Dorta, G.; Tang, H.; Rantalainen, M.J.; Lindon, J.C.; Corthesy-Theulaz, I.E.; Fay, L.B.; et al. Topographical variation in metabolic signatures of human gastrointestinal biopsies revealed by high-resolution magic-angle spinning 1H NMR spectroscopy. *J. Proteome Res.* **2007**, *6*, 3944–3951. [CrossRef] [PubMed]
69. Hoyles, L.; McCartney, A.L.; Neve, H.; Gibson, G.R.; Sanderson, J.D.; Heller, K.J.; van Sinderen, D. Characterization of virus-like particles associated with the human faecal and caecal microbiota. *Res. Microbiol.* **2014**, *165*, 803–812. [CrossRef] [PubMed]
70. Monaco, C.L.; Gootenberg, D.B.; Zhao, G.; Handley, S.A.; Ghebremichael, M.S.; Lim, E.S.; Lankowski, A.; Baldridge, M.T.; Wilen, C.B.; Flagg, M.; et al. Altered virome and bacterial microbiome in human immunodeficiency virus-associated acquired immunodeficiency syndrome. *Cell Host Microbe* **2016**, *19*, 311–322. [CrossRef] [PubMed]

71. Norman, J.M.; Handley, S.A.; Baldridge, M.T.; Droit, L.; Liu, C.Y.; Keller, B.C.; Kambal, A.; Monaco, C.L.; Zhao, G.; Fleshner, P.; et al. Disease-specific alterations in the enteric virome in inflammatory bowel disease. *Cell* **2015**, *160*, 447–460. [CrossRef] [PubMed]
72. Zhao, G.; Vatanen, T.; Droit, L.; Park, A.; Kostic, A.D.; Poon, T.W.; Vlamakis, H.; Siljander, H.; Harkonen, T.; Hamalainen, A.M.; et al. Intestinal virome changes precede autoimmunity in type I diabetes-susceptible children. *Proc. Natl. Acad. Sci. USA* **2017**, *114*, E6166–E6175. [CrossRef] [PubMed]
73. Vayssier-Taussat, M.; Albina, E.; Citti, C.; Cosson, J.F.; Jacques, M.A.; Lebrun, M.H.; Le Loir, Y.; Ogliastro, M.; Petit, M.A.; Roumagnac, P.; et al. Shifting the paradigm from pathogens to pathobiome: New concepts in the light of meta-omics. *Front. Cell. Infect. Microbiol.* **2014**, *4*, 29. [CrossRef] [PubMed]
74. Ott, S.J.; Waetzig, G.H.; Rehman, A.; Moltzau-Anderson, J.; Bharti, R.; Grasis, J.A.; Cassidy, L.; Tholey, A.; Fickenscher, H.; Seegert, D.; et al. Efficacy of sterile fecal filtrate transfer for treating patients with *Clostridium difficile* infection. *Gastroenterology* **2017**, *152*, 799–811 e797. [CrossRef] [PubMed]
75. Zuo, T.; Wong, S.H.; Lam, K.; Lui, R.; Cheung, K.; Tang, W.; Ching, J.Y.L.; Chan, P.K.S.; Chan, M.C.W.; Wu, J.C.Y.; et al. Bacteriophage transfer during faecal microbiota transplantation in *Clostridium difficile* infection is associated with treatment outcome. *Gut* **2018**, *67*, 634–643. [CrossRef] [PubMed]
76. Barr, J.J.; Auro, R.; Furlan, M.; Whiteson, K.L.; Erb, M.L.; Pogliano, J.; Stotland, A.; Wolkowicz, R.; Cutting, A.S.; Doran, K.S.; et al. Bacteriophage adhering to mucus provide a non-host-derived immunity. *Proc. Natl. Acad. Sci. USA* **2013**, *110*, 10771–10776. [CrossRef] [PubMed]
77. De Sordi, L.; Lourenço, M.; Debarbieux, L. I will survive: A tale of bacteriophage-bacteria coevolution in the gut. *Gut Microbes* **2018**. [CrossRef]
78. Debarbieux, L. Bacterial sensing of bacteriophages in communities: The search for the *Rosetta stone*. *Curr. Opin. Microbiol.* **2014**, *20*, 125–130. [CrossRef] [PubMed]
79. Mirzaei, M.K.; Maurice, C.F. Menage a trois in the human gut: Interactions between host, bacteria and phages. *Nat. Rev. Microbiol.* **2017**, *15*, 397–408. [CrossRef] [PubMed]
80. Galtier, M.; De Sordi, L.; Sivignon, A.; de Vallee, A.; Maura, D.; Neut, C.; Rahmouni, O.; Wannerberger, K.; Darfeuille-Michaud, A.; Desreumaux, P.; et al. Bacteriophages targeting adherent invasive *Escherichia coli* strains as a promising new treatment for crohn's disease. *J. Crohns Colitis* **2017**, *11*, 840–847. [CrossRef] [PubMed]
81. Morello, E.; Saussereau, E.; Maura, D.; Huerre, M.; Touqui, L.; Debarbieux, L. Pulmonary bacteriophage therapy on *Pseudomonas aeruginosa* cystic fibrosis strains: First steps towards treatment and prevention. *PLoS ONE* **2011**, *6*, e16963. [CrossRef] [PubMed]
82. Vinner, G.K.; Vladisavljevic, G.T.; Clokie, M.R.J.; Malik, D.J. Microencapsulation of *Clostridium difficile* specific bacteriophages using microfluidic glass capillary devices for colon delivery using pH triggered release. *PLoS ONE* **2017**, *12*, e0186239. [CrossRef] [PubMed]
83. Comeau, A.M.; Tetart, F.; Trojet, S.N.; Prere, M.F.; Krisch, H.M. Phage-Antibiotic Synergy (PAS): Beta-lactam and quinolone antibiotics stimulate virulent phage growth. *PLoS ONE* **2007**, *2*, e799. [CrossRef] [PubMed]
84. Oechslin, F.; Piccardi, P.; Mancini, S.; Gabard, J.; Moreillon, P.; Entenza, J.M.; Resch, G.; Que, Y.A. Synergistic interaction between phage therapy and antibiotics clears *Pseudomonas aeruginosa* infection in endocarditis and reduces virulence. *J. Infect. Dis.* **2017**, *215*, 703–712. [CrossRef] [PubMed]
85. Torres-Barcelo, C.; Arias-Sanchez, F.I.; Vasse, M.; Ramsayer, J.; Kaltz, O.; Hochberg, M.E. A window of opportunity to control the bacterial pathogen *Pseudomonas aeruginosa* combining antibiotics and phages. *PLoS ONE* **2014**, *9*, e106628. [CrossRef] [PubMed]
86. Chaudhry, W.N.; Concepcion-Acevedo, J.; Park, T.; Andleeb, S.; Bull, J.J.; Levin, B.R. Synergy and order effects of antibiotics and phages in killing *Pseudomonas aeruginosa* biofilms. *PLoS ONE* **2017**, *12*, e0168615. [CrossRef] [PubMed]
87. Ryan, E.M.; Alkawareek, M.Y.; Donnelly, R.F.; Gilmore, B.F. Synergistic phage-antibiotic combinations for the control of *Escherichia coli* biofilms in vitro. *FEMS Immunol. Med. Microbiol.* **2012**, *65*, 395–398. [CrossRef] [PubMed]
88. Verma, V.; Harjai, K.; Chhibber, S. Restricting ciprofloxacin-induced resistant variant formation in biofilm of *Klebsiella pneumoniae* B5055 by complementary bacteriophage treatment. *J. Antimicrob. Chemother.* **2009**, *64*, 1212–1218. [CrossRef] [PubMed]

89. Allen, R.C.; Pfrunder-Cardozo, K.R.; Meinel, D.; Egli, A.; Hall, A.R. Associations among antibiotic and phage resistance phenotypes in natural and clinical *Escherichia coli* isolates. *mBio* **2017**, *8*, e01341-17. [CrossRef] [PubMed]
90. Roach, D.R.; Leung, C.Y.; Henry, M.; Morello, E.; Singh, D.; Di Santo, J.P.; Weitz, J.S.; Debarbieux, L. Synergy between the Host Immune System and Bacteriophage is Essential for Successful Phage Therapy Against an Acute Respiratory Pathogen. *Cell Host Microbe* **2017**, *22*, 38–47 e34. [CrossRef] [PubMed]

Review

Phage Therapy: What Have We Learned?

Andrzej Górski [1,2,3,*], Ryszard Międzybrodzki [1,2,3], Małgorzata Łobocka [4,5], Aleksandra Głowacka-Rutkowska [4], Agnieszka Bednarek [4], Jan Borysowski [3], Ewa Jończyk-Matysiak [1], Marzanna Łusiak-Szelachowska [1], Beata Weber-Dąbrowska [1,2], Natalia Bagińska [1], Sławomir Letkiewicz [2,6], Krystyna Dąbrowska [1,7] and Jacques Scheres [8,†]

1 Bacteriophage Laboratory, Ludwik Hirszfeld Institute of Immunology and Experimental Therapy, Polish Academy of Sciences, Rudolfa Weigla Street 12, 53-114 Wroclaw, Poland; mbrodzki@iitd.pan.wroc.pl (R.M.); ewa.jonczyk@iitd.pan.wroc.pl (E.J.-M.); marzena@iitd.pan.wroc.pl (M.Ł.-S.); weber@iitd.pan.wroc.pl (B.W.-D.); natalia.baginska@iitd.pan.wroc.pl (N.B.); dabrok@iitd.pan.wroc.pl (K.D.)

2 Phage Therapy Unit, Ludwik Hirszfeld Institute of Immunology and Experimental Therapy, Polish Academy of Sciences, Rudolfa Weigla Street 12, 53-114 Wroclaw, Poland; letkiewicz1@o2.pl

3 Department of Clinical Immunology, Transplantation Institute, Medical University of Warsaw, Nowogrodzka Street 59, 02-006 Warsaw, Poland; jborysowski@interia.pl

4 Institute of Biochemistry and Biophysics, Polish Academy of Sciences, Pawińskiego Street 5 A, 02-106 Warsaw, Poland; lobocka@ibb.waw.pl (M.Ł.); glowacka@ibb.waw.pl (A.G.-R.); a.kielan@ibb.waw.pl (A.B.)

5 Autonomous Department of Microbial Biology, Faculty of Agriculture and Biology, Warsaw University of Life Sciences, Nowoursynowska Street 159, 02-776 Warsaw, Poland

6 Medical Sciences Institute, Katowice School of Economics, Harcerzy Września Street 3, 40-659 Katowice, Poland

7 Research and Development Center, Regional Specialized Hospital, Kamieńskiego 73a, 51-124 Wrocław, Poland

8 National Institute of Public Health NIZP, Chocimska Street 24, 00-971 Warsaw, Poland; jscheres@icloud.com

* Correspondence: agorski@ikp.pl; Tel.: +48-71-370-99-05

† Current Address: Department of Medical Microbiology, University Medical Centre Groningen, Hanzeplein 1, 9713 GZ Groningen, The Netherlands

Received: 18 April 2018; Accepted: 22 May 2018; Published: 28 May 2018

Abstract: In this article we explain how current events in the field of phage therapy may positively influence its future development. We discuss the shift in position of the authorities, academia, media, non-governmental organizations, regulatory agencies, patients, and doctors which could enable further advances in the research and application of the therapy. In addition, we discuss methods to obtain optimal phage preparations and suggest the potential of novel applications of phage therapy extending beyond its anti-bacterial action.

Keywords: phage therapy; experimental therapy; phage cocktails; anti-phage antibodies; prophage; immunomodulation

---

The intention of this article is to highlight the current events and issues related to phage therapy (PT) which seem to be most relevant for its further progress. These issues correspond to two main topics addressed in our article: the regulatory/ethical/awareness raising topic, which will subsequently yield to the topic of lysogeny/immunity/optimal use of phage preparations. These issues appear to be especially timely and relevant from the perspective of our team with leading expertise in PT among the EU countries.

1. More Room for Phage Therapy on the Horizon?

After decades of being kept out of the mainstream infectious disease armamentarium of the Western world, there now appears to be a silver lining on the horizon for phage therapy. PT is shedding its dubious associations with alternative and fringe medicine. Triggered by the growing threat of antibiotic resistance, there is a slow but substantial change in the appreciation of PT and a more permissive attitude of the main stakeholders in the infectious disease arena. Reviews on PT covered by PubMed appear almost every month. According to Web of Science, their average citation number per annum in recent years has been around 1100, and increased to approximately 1400 in 2017. There is also a growing understanding of the ethical, legal, and administrative rules relevant to experimental therapy which currently allow such treatment to be provided to patients for whom all other available therapies have failed. Below are some observations and reflections on the present attitudes of doctors, patients, academia, policymakers, media, and industry towards PT.

2. Doctors, Pharmacists, and Academia

The professionals in the fight against serious infections are doctors, general practitioners, infectiologists or medical microbiologists, and pharmacists. In the case of a serious infection, they have to choose the most appropriate remedy. From the plenitude of available antibiotics, they select those for which the pathogen in question tests sensitive, and standard application protocols are followed. However, almost every day doctors and pharmacists are confronted with pathogens that are increasingly resistant to certain or even a long list of antibiotics. More and more they feel the urgent need for new antibiotics or other instruments to help them improve or even save the lives of their critically ill patients. Without effective antibiotics (and thus effectively standing helpless), doctors eagerly look for alternatives. Phage therapy might represent such an alternative, at least in certain cases. In the last decade, many publications on bacteriophages and their possible applicability have appeared regularly in the clinical, applied, and fundamental scientific microbiological literature [1–3]. PT is often a specific subject on the programme of clinical and fundamental microbiological conferences, and is sometimes even the sole focus of dedicated PT symposia. As a result, a growing group of physicians and pharmacists in Western countries are acquainted with the potency and the pros and cons of PT as a possible alternative or an auxiliary therapy in cases of untreatable antibiotic-resistant infections, which is applied in neighbouring non-EU countries on the continent. Publications of successful and sometimes spectacular phage therapy cases trigger this interest, and in their aftermath often lead to a flow of requests by doctors to phage laboratories for help in analogous cases. Such requests keep coming, even from places where phages are not officially registered medical products, and PT still is generally not available in most of the West, although it is sometimes available experimentally. Therefore, in some cases doctors refer or mediate their patients to recognized PT centres elsewhere, such as in Poland (Wroclaw) and Georgia (Tbilisi).

In general, Western medical professionals show signs of increased openness towards PT as a possibly valuable additional tool in the fight against resistant and seriously threatening or disabling infections. At the same time, quite a long road of broad basic research, robust clinical trials, adjustment of the regulatory systems, education, and training still lies ahead before PT becomes practical, optimally effective, and compatible with the rules. Nevertheless, it may be advisable for doctors and medical students, pharmacists and pharmaceutical students to inform themselves in anticipation of the possible role of bacteriophages in infectious disease treatment. Is it not amazing that most medical professionals do not know about bacteriophages, these evolutionarily important creatures which are at least ten times more frequent in the microbiome than all bacteria and also greatly outnumber them within our body?

In this field, certain non-governmental organizations (NGOs) of professionals sometimes arise and intend to fulfil a role in closing the existing knowledge gap and building the bridge to formal recognition of PT. For instance, P.H.A.G.E. (Phages for Human Applications Europe Group) is a multidisciplinary group of doctors with practical experience or strong interest in PT, basic and applied phage researchers, and policymakers [4]. The exchange of phages, knowledge, and technology, participation in projects, organizing conferences and presentations all over Europe, publication, and education are among its main activities.

In 2015, a number of attendants of a bacteriophage conference in Tbilisi (Georgia) composed a multidisciplinary and intercontinental expert panel to establish an academic and medical initiative for the re-implementation of PT. The papers on the "Silk Route to the acceptance and re-implementation of bacteriophage therapy" which have recently been produced by this expert round-table are a significant contribution to the development of international guidelines and frameworks which are needed for a legal and effective application of bacteriophage therapy by physicians and the receiving patients [5,6].

Phages for Global Health is another very interesting multidisciplinary organization. Its mission is "to bring phage expertise to the developing world". Developing countries are disproportionally impacted by infectious diseases (e.g., *Campylobacter* infection has a fatality rate of about 0.1% in wealthy countries, but 8.8% in Kenya, mostly children) [7]. Phages for Global Health provides laboratory training workshops, teaching phage biology to scientists on location in developing countries where the need for alternatives to antibiotics (e.g., PT) is felt especially [8]. In addition, product development projects are performed in which international multidisciplinary teams are built that co-develop phage products for specific applications in developing countries [9]. In June and July 2018, the Second East African Phage Workshop will be held at Pwani University in Kilifi, Kenya. The participants will learn how to isolate and characterize phages as antibiotic alternatives for use against antibiotic-resistant bacteria.

3. CRISPR-Cas: From Phages to Eukaryotes

An additional important referral should be made to the recent development of simplified methods for high-efficiency gene-editing. This spectacular innovative technology is based on the CRISPR-Cas mechanisms which bacteria developed during their evolution in order to protect themselves against infections by phages. This has once again made clear how interesting and important the study of the very old relationship between phages and bacteria can be, and that it can lead to unexpected benefits and great leaps forward for science and its practical applications, including great promises for the prevention or treatment of genetic and complex diseases [10–12].

4. Patients, the Media, and PT

The patient, not the doctor, is the primary stakeholder in health and health care. Stimulating patient empowerment, health literacy, shared decision-making, and personal responsibility are core elements of health policy in almost all countries. Especially when the doctors can neither heal nor help with the existing medical means (e.g., in cases of incurable cancer), it is often the patient who opens the question of alternative therapies and asks for a referral to any other centre that might be able to help them, wherever on Earth, with whatever therapy, and at whatever costs. Sometimes the patient or their relatives are, via the internet, well-informed about possible alternatives. Asking for a second opinion has become the generally accepted standard. This pattern also applies to phage therapy. Though still quite exceptional, there are patients with chronic untreatable threatening resistant infections who indeed know about the option of bacteriophages, and ask their doctor to try phage therapy or to refer them for it. A growing number of patients find their way to bacteriophage centres abroad, staying there several weeks for phage selection and initial therapy, and are willing to bear the total costs of treatment, travel, and accommodation themselves. This medical tourism for phage therapy has grown especially since the media have taken their own responsibility in the national campaigns against the inappropriate use of antibiotics and have also informed the general public about PT as an alternative. They often mention PT as being applied in Central and Eastern European countries, and have reported spectacular cases of wound healing and the prevention of diabetic limb amputations with phages. Phage stories with basic information and successful cases of PT including the places where and how you can access it appear on TV [13,14] and in a broad range of societal magazines, ranging from knowledge magazines such as Der Spiegel Wissen in Germany [15] or Elsevier Weekblad [16] to the popular women's magazine Libelle [17] in the Netherlands. So, thanks to some pioneering patients and with the help of the media, PT has gained a place on the stage for the general public in the Western world—almost a century later than in the East.

5. Industry and SMEs

To make its way from the experimental level towards registration for safe application in human medicine, PT needs the engagement of a dedicated industry which is willing to produce phages following the safety and quality requirements [18], requiring high investments. So far, very few firms, usually SMEs, have chosen to engage in the production of phages ready for use in clinical trials and human application, usually in the context of developmental projects performed in cooperation with research institutes, academia, and or state laboratories. This contrasts somewhat with the food, disinfection, cosmetic, and veterinary sector, where phages and phage products (lysins) have already reached consumers. The US Food and Drug Administration (FDA) has approved a small number of products for these markets, and several applications are in the pipeline for approval. Very recently, the phage-producing SME, Phage Technology Center GmbH [19], was present at the international Anuga FoodTec International Food Technology Fair (Cologne, Germany, March 2018), presenting its phages against *Salmonella* and *E. coli* for various food applications. According to its Senior Manager Research & Development, the market for phages is going to boom in this sector, which is certainly not yet the case in human medicine.

6. Authorities and PT

Globally, national authorities consider antibiotic resistance to be a profound threat to health. Their national strategies, action plans, and preventive campaigns focus on a more appropriate use of antibiotics and the search for alternatives. The development of vaccines, innovative diagnostic tests, and novel interventions are usually mentioned as alternatives. Only very exceptionally are the words bacteriophage, PT, or phage products (lysins or endolysins) found in the action plans. The main reason is the current lack of positive clinical trials with PT. The reputation and successes of PT in countries with longstanding application of PT are distrusted and considered to be poorly documented, not convincing and not proven, and serious adverse effects of PT are feared or at least not to be excluded. The dictum *primum non nocere* (first do not harm) and quality assurance, both based on solid clinical trials according to the standard rules, are indeed strong pillars of drug policy. For similar reasons, there is no mention of PT in the five-year action plan of the European Commission against antibiotic resistance launched in 2012 and updated in 2017 [20]. The words "phage therapy" and "phages" are also lacking in the global action plan to tackle antimicrobial resistance which was endorsed in May 2015 by the World Health Assembly in Geneva [21]. This action of the Assembly was truly a unique one, showing the United Nations' serious concern that antibiotic resistance "threatens the very core of modern medicine and the sustainability of an effective, global public health response to the enduring threat from infectious diseases". Is this emergency situation still not serious enough to allow a little more place for phage therapy, a method which a century ago was effectively applied and appreciated in curative care and public health, in the East and in the West before we used antibiotics?

Fortunately, the position of the authorities appears to be shifting, albeit slowly, towards more latitude for phage therapy. This may be illustrated by the following selection of interesting formal actions and documents of authorities in the US and/or the EU:

In mid-2017, the U.S. Food and Drug Agency (FDA), the National Institutes of Health, the National Institute of Allergy and Infectious Diseases, and the Center for Biologics Evaluation and Research co-organized a two-day workshop to facilitate the development of a rigorous clinical assessment of bacteriophage therapy [22].

In late 2017, the FDA also gave the status of Emergency Investigational New Drug to phages specifically active against a multidrug-resistant *Acinetobacter baumannii*, which were applied in a patient with septic shock who improved within days and survived, being the first case of intravenous use for systemic infection. The phages were obtained from the US Navy and Texas A&M University, in combination with the San Diego biotech firm AmpliPhi [23].

Later, the FDA gave its seal of approval to a new phase I/II clinical trial in humans at Mount Sinai Hospital in New York City to test a new bacteriophage treatment for Crohn's disease [24].

In 2013, the European Commission funded the PHAGOBURN project co-ordinated by the French Ministry of Defence, with partners from France, Belgium, and Switzerland. The main objective of the project is "to assess the safety, effectiveness and pharmacodynamics of two therapeutic phage cocktails to treat either *E. coli* or *P. aeruginosa* burn wound infections" [25].

In 2015, the White House National Action Plan for combating Antibiotic-Resistant Bacteria launched by the White House in 2015 listed "the use of phage and phage derived lysins to kill specific bacteria while preserving the microbiota" among the non-traditional therapeutics which should be further developed [26].

The Transatlantic Taskforce on Antimicrobial Resistance (TATFAR) was created in 2009 to enhance synergy and communication between government agencies on both sides of the Atlantic Ocean. The first partners were the US and the EU, and Canada and Norway joined later . Action no. 3.6 of TATFAR's updated action plan is: "Exchange information on possible regulatory approaches to development of alternative approaches for managing bacterial infections, such as bacteriophage therapy and vaccines for health care associated infections (joint action by FDA, EMA, HC, and NMA)"[27]. In a message from the recent TATFAR meeting (Atlanta CDC 7–9 of March 2018), according to Marco Cavaleri from EMA (personal communication, March 12, 2018) it was reiterated that in the discussion on the alternatives to antibiotics, phages should be on the radar as an option that deserves to be discussed across the Atlantic. The biggest problem is that not many companies are interested in discussing the topic or in considering how to approach clinical development.

In its German Antimicrobial Resistance Strategy entitled "DART 2020: Fighting antibiotic resistance for the good of both humans and animals", the Federal Ministry of Food and Agriculture announced plans to assess the "Possible positive effects of bacteriophages and other substances to reduce or eliminate bacteria on carcasses as a supplement to process hygiene". Though this action is clearly meant to improve food hygiene and not as phage therapy for human patients, it is nevertheless noted here because it is one of the very few governmental documents mentioning bacteriophages as a means to fight antibiotic resistance, for the good of humanity and animals [28].

The same action point was proposed by the Federal Government of Germany in the report "Combating Antimicrobial Resistance. Examples of Best-Practices of the G7 countries" of the G7 GERMANY 2015 meeting in Berlin [29].

Very recently, a major, hope-giving and possibly historical step for the applicability of PT was taken by the Belgian Federal Government (January 2018) [30]. In cooperation with academia (including ethicists), researchers and experts from the care sector the Federal Agency for Medicines and Health Products succeeded in developing a regulation for phage production and the clinical application of PT. The procedure, which obtained its legal approval at the end of January 2018, is based on the legal possibilities in Belgium for a pharmacist to prepare a medical product (including phages) for an individual patient. The active ingredients used in this so-called magistral preparation (in the US, "compound prescription drug preparation") must meet the requirements of the European, Belgian, or another official Pharmacopoeia. If this magistral route would be copied *mutatis mutandis* by other countries, it would truly represent a breakthrough for the application of PT, especially in individual life-threatening situations (based on the Declaration of Helsinki, WMA, 1964) [31]. In fact, a similar approach has long been in use in Poland at the Phage Therapy Center of the Institute of Immunology and Experimental Therapy [32,33].

7. National Regulations Enabling Experimental Therapy (Including PT)

In view of these developments, it might be useful to summarize the current status of experimental therapy in Europe and elsewhere.

Generally, every medicinal product must be approved by a relevant regulatory agency before it can be used in clinical practice. However, in response to the needs of patients who cannot be treated satisfactorily with authorized drugs, many countries have introduced regulations which enable doctors to use experimental treatments.

In the European Union (EU), the legal framework for treatment with unauthorized medicinal products (termed compassionate use—CU) was introduced by Article 83 [34] of Regulation (EC) No. 726/2004 of the European Parliament and of the Council. This article permits the use of unauthorized medicinal products in groups of patients, provided that two main requirements are met: (1) the patient has a chronically or seriously debilitating disease, or a life-threatening disease which cannot be treated satisfactorily with an authorized medicinal product; and (2) the medicinal product must be either the subject of an application for a centralized marketing authorization or be undergoing clinical trials. Specific CU programs are to be implemented and governed by individual Member States (MSs) [34]. As of 2016, 18 out of 28 MSs had specific CU regulations and 20 had implemented CU programmes [35]. Moreover, Article 5 of Directive 2001/83/EC of the European Parliament and of the Council allows the use of unauthorized medicinal products in individual patients under the direct responsibility of a healthcare professional (i.e., named-patient basis treatment) [36].

In the US, according to the terminology adopted by the FDA, the use of unauthorized drugs outside of clinical trials is called expanded access (EA). General requirements for EA include the following: (1) a serious or immediately life-threatening disease where no comparable or satisfactory alternative therapy is available; (2) the potential benefits justify the potential risks and the potential risks are not unreasonable in the context of the disease; (3) there is no threat to the initiation, conduct, or completion of clinical trials; (4) informed consent of the patient; (5) Institutional Review Board (IRB) review [37,38]. Independently of the existing FDA regulations, 38 states have recently introduced so-called right-to-try laws which are to facilitate access of terminally ill patients to investigational drugs that have completed phase I of a clinical trial. However, these laws have been heavily criticized by experts for offering "false hope" to patients without providing any actual improvements in access to investigational drugs [39]. Nevertheless, at the time of this writing, US Congress has passed a relevant bill which has been a priority of President Trump [40]. If approved by the US Senate, the law would allow patients to sidestep FDA approval once they have received permission from a company [41].

In Canada, the use of unauthorized drugs is legally permissible in Special Access Programmes (SAPs). Basic information about these programmes is available in the Guidance Document for Industry and Practitioners—Special Access Programme for Drugs developed by the Canadian regulatory agency Health Canada [42]. Under SAP rules, an unauthorized drug can be used in patients with serious or life-threatening diseases, especially in emergency cases when conventional therapies have failed, are unsuitable, or are unavailable. The use of an unauthorized drug must be supported by some credible evidence of its safety and efficacy, and a doctor should obtain informed consent from the patient.

In Australia, there are two schemes that enable doctors to use unauthorized drugs: the Authorized Prescriber Scheme (APS) and the Special Access Scheme (SAS) [43]. In APS, an application for the use of an unauthorized drug needs to be approved by a bioethics committee or endorsed by a specialist in a discipline relevant to the proposed treatment. Important issues that are evaluated include the qualifications and experiences of the doctor, access to facilities necessary to perform the treatment, evidence to support the proposed treatment, clinical justification including whether other therapeutic alternatives have been tried, and an explanation of why the unauthorized drug is proposed. In addition, informed consent of the patient is required. Under this scheme, the doctor can be granted permission to prescribe a specified unauthorized drug to specific patients (or groups of patients) with a particular disease. In the other major Australian scheme (the SAS), unauthorized drugs can be used in single patients on a case-by-case basis. It is expected that before use of an unauthorized drug, all authorized treatment options will be considered. The doctor must also obtain informed consent from the patient. Moreover, in cases when the treated disease is not life-threatening and the unauthorized drug does not have an established history of use, clinical justification for the use of an unauthorized drug must also be provided.

8. Important Issues Which Need Addressing to Enable Further Progress and Optimization of PT and Relevant Clinical Trials

In this part of our article, we wish to briefly discuss the issues pertinent to PT that have not been dealt with adequately so far, and where the advancement of our knowledge may lead to a faster introduction of phages to the health market.

The long-lasting effects of PT confirm its safety. Even though the therapeutic value of PT still awaits confirmation by clinical trials—in line with the requirements of evidence-based medicine—as pointed out by a former FDA commissioner: "Although randomized trials perform an essential role in the development of therapies, we should not neglect the crucial and complementary role than can be played by high-quality observational studies" [44]. In this regard, our results of suggested PT efficacy appear to be quite encouraging (>50% success rate using purified phage preparations), while the safety of the therapy is remarkable [32]. This has been confirmed by our recent preliminary analysis of remote observations in a group of 33 patients who completed PT up to 7 years ago. When questioned, two-thirds of those patients were satisfied with therapy results and, importantly, none of them reported any complications that could be related to PT [45].

9. PT and Antibody Responses against Phages

Our studies in animals and patients have provided interesting and potentially useful information on anti-phage antibody responses during PT. Among healthy donors, 29–82% may be positive for serum anti-phage antibodies depending on phage type (anti-T4 coliphage antibodies being most common) [46]. Antibody responses during PT have been described by us in detail. In patients awaiting PT, very low levels of anti-phage antibodies were detectable (mean K index in 60 patients was 0.17), while the index could reach values as high as 200 during PT. Furthermore, purified phage preparations seem to induce higher antibody responses than do the lysates. In addition, identical phages can elicit different levels of antibody responses in patients, which may depend on the immune reactivity of those patients. The most important finding has been that a good clinical outcome of PT may be observed in patients with high antibody responses [47]. Our recent analysis suggests that there is an association between the duration of therapy and antibody responses (for *Staphylococcus* phages, the Spearman correlation was 0.856, $p < 0.0001$). Similar data were obtained in mice [46]. While high antibody responses do not appear to affect the outcome of PT, we prefer to terminate the therapy if the antibody levels are high to avoid possible complications in the future (e.g., the unknown effect of phage–antibody complexes).

10. Monotherapy vs. Phage Cocktails

The issue of phage cocktails vs. monovalent phage preparations remains undecided: our preliminary data might suggest that there is no significant difference in the therapeutic efficacy between these preparations, while the frequency of high antibody responses was higher in patients treated with cocktails compared to those on monotherapy [48].

11. Optimal Clinical Models for PT and Prognosis of Therapy

One of the key questions asked by the Guest Editor of this volume, Prof. H. Brüssow, was: is it possible to formulate a set of rules with respect to infection type, which predict successful interventions? [49] Our experience so far suggests that intrarectal PT of chronic bacterial prostatitis offers the highest success rate [50]. Several factors could be responsible for those results, among them possible good penetration of phages from the rectum to the prostatic tissue (phage ability to penetrate cell layers has recently been demonstrated) [51,52], eradication of rectal carriage of a pathogen, as well as low anti-phage antibody responses elicited by this mode of phage administration [47]. Our data on patients' immunomonitoring suggest that an increase in phagocytosis may be a good prognostic sign of PT success [53].

12. Mouse Model of Acute Urinary Tract Infection Confirms Neutrophil–Phage Synergy

The value of this parameter has been confirmed by an experimental study in mice. The experiments were performed on a mouse model of acute urinary tract infection [54] caused by transurethral bacterial inoculation with uropathogenic strain isolated from patients: *E. faecalis* 15/P or *P. aeruginosa* 119×. Spleen mononuclear cells were isolated according to the method described by Kruisbeck (2000) [55] using a density gradient (Histopaque-1083, Sigma-Aldrich, St. Louis, MO, USA). Intracellular killing of bacteria by splenic macrophages was tested according to the method described by Buisman et al. (1991) and Leijh et al. (1982) [56,57]. The obtained value corresponded to the percentage of killed phagocytosed bacteria, and it was examined both 3 and 6 days after the infection. In the infected group of DBA1/LAC J mice (n = 6) (without phage treatment), significantly lower (Mann–Whitney U-test, p = 0.004) intracellular killing of a pathogenic bacterial strain (the same as the cause of infection) by splenic mononuclear cells (63.2% ± 7.1 for mice infected with (*P. aeruginosa*) was observed when compared to the bactericidal capacity of healthy animals (82.8% ± 8.0). Reduced intracellular killing was observed in infected mice on days 3 and 6 after the infection, regardless of the uropathogenic strain used. Importantly, the intraperitoneal administration of the phage lysate (at a concentration of 5×10^{10} pfu/mL) exerted a stimulatory effect on the spleen phagocytes in the group of mice with experimentally-induced infection by *E. faecalis* 6 days after sequential application of three doses (1 h, 24 h, and 48 h after bacterial inoculation) of specific enterococcal phage lysate Ent 15/P (86.7% ± 3.8) when compared to non-treated mice (74.1% ± 9.2) (Mann–Whitney U-test, p = 0.014). An improvement in bactericidal activity of splenic mononuclear cells was also obtained for a group of mice treated with three doses of the phage lysate (86.7% ± 3.8) after 6 days of infection when compared to the same group tested 3 days after bacterial inoculation (72.2% ± 6.7, Mann–Whitney U-test, p = 0.004). The improvement of splenic macrophage anti-bacterial function was paralleled by a significant fall of bacteria counts in liver, kidneys, and urinary bladder of phage-treated mice [58]. Recent data fully confirm this assumption by showing that neutrophil–phage synergy is needed for successful PT of experimental pneumonia in mice [59].

13. Prophages in Bacterial Strains Used for Therapeutic Phage Propagation: Their Significance, Detection, and Elimination

Bacterial Strains for the Propagation of Therapeutic Phages

Sources of phages for therapeutic use are lysates of cells that serve for the propagation of those phages. In addition to the desired phage, they contain bacterial cell components and may contain contaminating phages that are produced as a result of prophage induction if the phage propagation strain is a lysogen [60,61]. Genome analysis of bacterial strains used for phage propagation reveals not only genes that encode toxins or other virulence determinants, but also mobile genetic elements, including plasmids, transposons, and prophages. The presence of toxins in lysates increases the cost of lysate purification. The presence of mobile genetic elements poses a risk of uncontrolled spread of bacterial virulence or antibiotic-resistance genes. The most problematic lysate contaminants are temperate phages. Due to the physico-chemical similarity of contaminating temperate phages and lytic phages, the former are practically inseparable from the main phage population in a lysate. Despite the possibilities of their detection in lysates and even the estimation of what fraction of the total phage population is represented by them [61], the only way to eliminate them is the construction of phage propagation strains that are depleted of prophages [60,62].

A key argument for the removal of active prophages from the genomes of bacteria that serve as therapeutic phage propagation strains is the prophage genetic load. Temperate phages are major driving forces of horizontal gene transfer and bacterial evolution [63–65]. They typically carry genes that encode functions which are adaptive for their bacterial hosts, and in that way decrease the probability of overgrowth of the bacterial population by cells that have lost them. In the case of prophages and plasmids of bacterial pathogens, the adaptive functions encoded by these elements are nearly always associated with better adaptation of the bacteria to pathogenicity [63,65–72]. In addition to virulence factors, certain prophages encode homologs of error-prone DNA polymerase *V subunits* [73,74], and were proposed to play roles in the diversification of bacterial strains (e.g., by facilitating the acquisition of resistance to toxins or antimicrobials by mutations) [75]. Temperate phage virions that contaminate therapeutic phage preparations act not only as vectors of their own DNA, but can also act as vectors of bacterial, plasmid, or pathogenicity island DNA [76–83]. For instance, a spontaneous intraspecies transfer of the blaNDM-1 carbapenemase gene from a carbapenem-resistant strain containing two active prophages to a carbapenem-sensitive *Acinetobacter baumannii* strain was attributed to the transduction mediated by a prophage-derived temperate phage [84]. Undoubtedly, the release to the environment of temperate phages containing their own DNA or sometimes even the DNA of plasmids or bacteria derived from contaminated therapeutic phage preparations can contribute to the spread of virulence or antibiotic-resistance genes. In the worst-case scenario, the contaminating phages could be acquired by the infecting bacteria during phage therapy and make these bacteria more pathogenic, negatively influencing the treatment outcome. Although incidents of adverse effects of phage therapy have been surprisingly rare, the possibility of such a scenario should be taken into consideration and avoided when possible, especially in view of the emergence of strains resistant to certain therapeutic phages in the course of phage therapy [32].

Although only about a half of the sequenced bacteria are lysogens, prophages are more frequent in pathogens [85–87]. Their abundance varies among different species of pathogenic bacteria. However, bacteria that are known for especially good adaptation to pathogenicity and for their fast acquisition of antibiotic resistance, including ESKAPE pathogens (*Escherichia coli*, *Staphylococcus aureus*, *Klebsiella pneumoniae*, *Acinetobacter baumannii*, *Pseudomonas aeruginosa,* and *Enterococcus* spp.), are often or even in most cases polylysogens [86,88–103]. Active and defective prophages in the genomes of certain pathogenic bacterial strains (e.g., *E. coli* O157:H7 strain Sakai, or highly virulent *S. pyogenes* strain MGAS315) can occupy as much as about 15% of total genomic DNA [104,105].

The ubiquity of lysogeny among bacterial pathogens makes the selection of non-lysogenic bacteria for phage propagation from environmental samples either difficult or impossible. Hence, the identification of active prophages in the genomes of efficient phage propagation strains and their subsequent removal is a strategy of choice in ensuring the monoclonality and safety of therapeutic phage preparations, as well as in decreasing the cost of their production and the evaluation of their purity [60].

Prophage-free strains may be acquired from among natural isolates of a given bacterial species or selected from laboratory cultures of prophage-carrying phage propagation strains upon the induction of prophage lytic development and the selection of surviving cells, as reviewed by [60]. Which of these strategies may be optimal depends on several factors. The task may not be simple, as a propagation strain should have all the features of the target bacteria that allow a phage released from this strain to infect the target pathogenic bacterial strain efficiently.

The stability of lysogeny is associated with numerous factors. In general, the rate of prophage loss by induction increases under conditions of decreased host viability, such as upon exposure to UV, reactive oxygen species, or other mutagenic factors that trigger the SOS response (for review see [106–113]), under high temperatures [114], as well as in the response to certain bacteriocins [115], certain antibiotics that block the action of essential enzymes [93,116] or interfere with intracellular regulatory processes [117,118] or to quorum-sensing signalling molecules [119–121]. Typically, induction also occurs spontaneously in a variable fraction of a population of cells [122–129], being responsible for the presence of relevant free temperate phages in the cultures of lysogens [62,130,131]. Thus, derivatives of lysogens that are depleted of certain prophages are expected to occur in nature and in laboratory cultures, although their number may be low, as together with the prophage they lose the prophage-mediated immunity to the infection by the relevant phage.

14. Prophage Detection Methods

Several bioinformatic methods have been developed to identify prophages in bacterial genomes. Programs that implement them can be downloaded from internet resources or are accessible online (e.g., PHAST, PHASTER and PHASTEST [132,133]; Prophinder [134]; Phage_Finder [105]; Prophage Finder [135]; PhiSpy [136]; VirSorter [137]). Their performance is in the range 64–85% for sensitivity and 74–93% for precision when tested with known prophage sequences in complete bacterial genomes [137]. Prophages in the phage propagation strain of a known sequence can also be identified by comparing the sequence of this strain with sequences of other species representatives and by the identification of genome regions that are interrupted by insertions of prophage-size elements [60]. Prophages in the genomes of *S. aureus* or *Salmonella enterica* serovar Typhimurium can be detected by the analysis of PCR reaction products with total genomic DNA of these bacteria and pairs of primers complementary to the conserved DNA regions of their species-specific prophages [94,138–140]. The main disadvantage of the aforementioned methods is the distinction of active and defective prophages, which is not always accurate. While defective prophages may be a source of toxins or virulence factors, they are unable to contaminate therapeutic phage preparations in a phage form unless their DNA is not packed into capsids of other phages. To detect active prophages, one should design pairs of primers complementary to the prophage sequences identified in a given strain and use them to amplify the relevant temperate phage DNA with the total virion DNA of a lysate as a template. In our hands, this method works sufficiently well to quickly distinguish active prophages from prophages that cannot produce viable progeny [62]. A necessary condition is to degrade host DNA in a lysate prior to the amplification experiments.

The sensitivity of contaminating phage detection may be increased by inducing prophage lytic development, with the most commonly used inducing factors such as mitomycin C or UV light. Upon treatment with these factors, bacteria can be grown in a liquid medium until signs of lysis (if any) are observable. Lysate that has been treated with DNase can be used as a source of phages to prepare phage DNA for PCR amplification with prophage-specific primer pairs. The inducible factor-treated cells can also be streaked on a soft agar medium with suspended phage-sensitive cells (in a Petri dish). If the prophage was induced, the lysis zone in the underlying sensitive cell layer should surround each growing colony of lysogen. However, a limitation of the latter method is often the lack of a prophage-free strain able to serve as an indicator.

15. Elimination of Prophages from Phage Propagation Strains

Traditional phage curing methods have been based on the selection of bacteria that have lost the prophage spontaneously or in response to inducing factors. If the prophage excision system is functional, prophage induction can be used to cure bacteria from that prophage [60]. Following prophage induction, cells are plated on a solid medium and tested for lysogeny. Prophage insertion in a chromosome may be associated with a specific phenotype, if it interrupts a gene of easily recognizable function. Curing from such prophages is associated with recovery of the wild-type strain phenotype, which may help to recognize prophage-free cells [62,94,141]. However, of the approximately 60% of phages that use intragenic regions as their attachment sites, over half have the attachment sites in tRNA encoding genes [105]. Additionally, other genes interrupted by prophages rarely have an easily recognizable phenotype. An additional difficulty may be "prophage jumping"—certain prophages excised from the primary attachment site can temporarily integrate into a secondary attachment site in the same cell, and thus the loss of phage conversion phenotype is not always associated with phage loss [94]. In such cases, the loss of prophage can be verified by testing cells' sensitivity to a parental strain phage or by PCR with a prophage-specific primer pair. If factors that induce the excision and lytic development of a given prophage cannot be identified, one can search for colonies of spontaneously cured cells in a population of lysogens by plating lysogen culture cells onto a solid medium, growing them, and testing by colony blot for the presence of prophage [142]. An amplicon of any prophage-specific gene can serve as a probe in blotting tests.

The overexpression of a cloned prophage excisionase gene in a respective lysogen can increase the frequency of prophage cured cell formation, as was shown in the case of lambda or KplE1 phage lysogens [143,144]. In certain cases, one prophage supports the excision of another prophage in the same cell by providing a helper function [145]. The removal of all active prophages from such cells using traditional methods is impossible. Thus, more reliable methods of prophage-free bacteria construction rely on recombineering techniques. For example, the *S. aureus* strain Newman was cured of four prophages by recombinational replacements of prophage-containing regions with the prophage-free regions of attachment sites for these phages cloned in temperature-sensitive replicon-based suicidal plasmids [146]. A curable plasmid expressing phage λ Red recombination system genes was used to replace four prophages in the *E. coli* chromosome with a PCR-amplified antibiotic resistance cassette, which was then eliminated with the help of another curable plasmid [128].

16. Future Possibilities to Produce Industrial Phage Propagation Strains

The construction of new phage propagation hosts using traditional approaches might be a never-ending story possibly requiring hundreds of strains to be cured of plasmids, active prophages, and possibly other mobile genetic elements. However, taking into account recent achievements in synthetic biology as well as the progress in recombineering and genome editing methods, this need not be the case.

Whether a given phage infects a given bacterial strain from a susceptible species depends on the features of the bacterium and the phage. Metabolic compatibility of a bacterium with a phage to support the phage propagation in already-established infection appears to be species-specific, but sometimes it is extended to more than one bacterial species of the same or different genera [147,148]. Differential phage susceptibility determinants that are encoded by various strains of the same species include genes encoding phage receptors or pathways of their synthesis and phage-compatible restriction-modification systems [149–155]. Additionally, bacteria encode phage defence mechanisms, but these mechanisms protect the bacterium by itself either from infection with certain phages or from phage propagation, or induce apoptosis to protect the population from spread of the infection [156–163]. The differential phage susceptibility determinants are exchangeable between strains of a given species. Bacteria can gain or lose sensitivity to a given phage or the ability to support this phage development by mutation-, recombination-, or horizontal gene transfer-driven changes in their phage susceptibility or phage defence determinants [151,164–175]. Several genes associated with phage resistance or susceptibility are carried by mobile genetic elements [120,158,175–187].

Phage features important for the successful infection of a metabolically-compatible host include the compatibility of phage receptor binding proteins with receptors at the surface of a bacterial cell, the compatibility of phage genome modifications with the restriction-modification system of a bacterium, or the ability to prevent the action of bacterial restriction-modification systems either by avoiding sites that are recognized by the bacterial restriction-modification systems or by encoding efficient anti-restriction mechanisms [149,188]. Additionally, to productively infect bacteria, phages encode proteins that allow them to overcome bacterial phage resistance mechanisms, such as anti-CRISPR proteins and proteins that prevent the action of bacterial Abi or toxin–antitoxin (TA) systems [189,190].

The structure of each phage and its infectivity for particular hosts are determined by the genome of this phage. The only host-determined features of a phage seem to be certain epigenetic modifications, namely host-specific DNA methylation patterns [191,192]. They strongly influence the efficiency of infection of new hosts by a phage, being responsible for the limitations of horizontal gene transfer by bacteriophages [86,191,193,194]. Thus, in addition to species-specific basic metabolic pathways supporting the efficient propagation of a given phage, a phage propagation strain should be equipped with surface receptors for this phage attachment, cell envelope structures susceptible to the action of given phage lytic proteins, and a restriction-modification system that will allow the phage released from this strain to infect a desired set of clinical strains. The removal from such a strain of genetic determinants of other phage defence mechanisms (e.g., CRISPR/Cas, Abi, or TA loci), if any are encoded by its genome, could extend the number of phages able to propagate in its cells to phages infecting strains of the same species and using the same host receptors, but unable to overcome the respective phage-defence mechanisms. The acquisition of sensitivity to certain phages upon the abolishment of various bacterial phage defence systems has been demonstrated in several cases [120,195–198].

An optimal future strategy to acquire therapeutic phage propagation strains of desired properties may be the construction of a bacterial chassis of selected clinically relevant pathogenic species. In synthetic biology, a chassis refers to the organism serving as a foundation to physically house genetic components and support them by providing the resources for basic functions, such as replication, transcription, and translation machinery [199]. The bacterial chassis strains to serve as basic platforms for the construction of industrial phage propagation strains should have genomes reduced in their complexity and the content of undesired genes by the depletion of most of the mobile genetic elements as well as virulence and phage resistance determinants—a procedure that is known as a top-down strategy of the genome reduction process [200]. Additionally, they should be ready for the introduction or exchange of genomic modules (e.g., an appropriate restriction-modification system or phage receptors determining gene cassettes), enabling these strains to serve as microbial cell factories for the propagation of selected therapeutic phages. Methodologies enabling the abolishment of mobile genetic elements and other genome fragments using genome shuffling, recombineering, oligo-mediated allelic replacement, or genome editing using CRISPR/Cas-assisted selection of desired clones have been developed for model bacteria, even on a genome-wide scale [201–209]. The repertoire of genetic engineering tools that extend the ability of genomic manipulations to bacteria other than *E. coli* using the newest strategies has been constantly increasing, providing means to edit genomes belonging to genera represented by the most problematic bacterial pathogens, including potential phage propagation strains [210–218].

The results of studies on bacteria that were cured of some or most of the recombinogenic or mobile genetic elements (including prophages) indicate that they have several advantages. For instance, *Escherichia coli* K-12 with a genome reduced by 15% by the removal of mobile DNA and cryptic virulence genes preserved good growth profiles and protein production as well as the accurate propagation of recombinant genes and plasmids that could not be stably propagated in other strains [219]. The growth properties and endurance of environmental stresses of a *Pseudomonas putida* KT2440 derivative which was cured of prophages, some transposons, and some restriction-modification cassettes was found to be superior to its wild-type parent [220,221]. Curing a *Corynebacterium glutamicum* industrial strain of prophages caused an increase of strain fitness, stress tolerance, transformability, and protein production yield [222]. Thus, in our opinion, the construction for the propagation of therapeutic phages, of chassis strains equipped with certain phage susceptibility determinants and depleted of phage resistance determinants as well as certain mobile genetic elements or virulence determinants will not only ensure the safety of therapeutic phage preparations, but will also reduce the cost of phage production substantially. This reduction will be a result of: (i) minimizing the number of strains required for the production of different phages; (ii) eliminating the need of evaluating phage preparations for the content of undesired elements, including temperate phages and toxins; and (iii) increasing the fitness and stability of such strains in the industrial production of therapeutic phages. Additionally, one foundation strain constructed for a bacterial species can serve as a platform for the enrichment of its genome with various gene cassettes required for the propagation of various phages. We have already constructed basic prophage- or plasmid-free strains to start the development of a chassis of *S. aureus* and *E. faecalis* strains. They serve for the production of monoclonal preparations of certain *S.*

aureus and *E. faecalis* phages [62,223]. Further work to remove additional undesired genomic elements from the genomes of these strains is in progress.

17. Surrogate Hosts for the Propagation of Therapeutic Phages

The use of non-pathogenic relatives of pathogenic strains enabling therapeutic phage propagation was proposed to eliminate the problem of phage preparations' contaminants derived from virulent phage propagation hosts [224,225]. Unfortunately, suitable "surrogate" hosts can be found only in a limited number of cases, and not all of them enable the efficient propagation of therapeutic phages [226–231]. Additionally, long-term effects of the enrichment of a pathogenic strain population with prophages released from strains believed to be non-pathogenic are impossible to predict, especially in view of documented cases of infections caused by certain strains belonging to the surrogate host species [232–239] and cross-species transfer of mobile genetic elements between representatives of surrogate host species and their pathogenic relatives [240–246]. Moreover, genomic analysis of pathogenic strains of certain species and their relatives representing non-pathogenic species indicates that the latter may function as reservoirs of accessory genes for the former [103]. Thus, even when using surrogate non-pathogenic hosts for the propagation of therapeutic phages, the removal of prophages from such hosts may be a wise strategy to avoid unpredicted problems in the future.

18. Economic Aspects of the Industrial Construction of Phage Propagation Strains

In nature, prophages are temporary components of bacterial genomes which can enter, exit, or change their location in the genome. Their loss is a natural process that occurs with various frequencies, as long as the mobility of a prophage is not abolished by deletions or other rearrangements that make the prophage remnants a permanent part of the genome. Thus, in most cases, the major cost of acquiring cells that are depleted of active prophages is the cost of screening (labour, media, and blotting or PCR reactions), and sometimes the cost of recombineering and genome editing techniques, provided the availability of tools. Economic aspects argue for going further and constructing species-specific bacterial chassis for the production of therapeutic phages by the removal of plasmids, if any, and chromosomal elements that cause genome mutability, phage resistance, or encode virulence factors. The construction of such strains could be done based on recombineering and genome editing methods analogous to those that have been used in the process of modification of bacterial producers of various compounds for industry [89,222,247–255]. Subsequently, such a chassis strain could be used as a platform for the exchange of particular phage-sensitivity determinants in its genome with selected strains sensitive to certain phages. The economic benefits of such an approach would be associated not only with the increased safety of phage preparations produced with the use of these strains, but also with a switch from many different strains of various properties to fewer strains of the same core genome and only a few gene cassettes to be exchanged. Results of studies on certain model or industrially-applicable bacteria that were depleted of prophages and certain other mobile elements as well as certain determinants of mutability indicate that such strains have a better genomic stability and are more efficient producers of certain compounds than their wild-type parents [89,199,219,252,254,255]. Engineering of their genomes does not need to be associated with the permanent presence of heterologous DNA, as markerless gene knock-out or gene replacement systems have been developed for a number of pathogenic bacterial species and are in constant further development [254–276].

19. PT: Beyond the Antibacterial Action

In recent years, data have been accumulating indicating that phages may also interact with mammalian cells, thus "crossing the border to eukaryotic cells"—binding to their surface receptors and penetrating into them. Phages can therefore pass across confluent epithelial cell layers and migrate to blood, lymph, and other tissues [51]. These findings essentially confirm our hypothesis of "phage translocation" from the intestines [277] extended by Barr, who used the term "journey" to suggest that phages travel through the human body [278]. Phages have been shown to mediate anti-inflammatory and immunomodulating properties [279]; therefore, such phenomena may be relevant for the maintenance of immunological homeostasis. Consequently, we recently hypothesized that phage therapy may be considered for treating disorders such as inflammatory bowel disease, autoimmune hepatitis, allergy, as well as some viral infections [280–283]. Evidently, this requires further work and confirmation by relevant clinical trials. While the most trustworthy advances come through the performance of well-designed trials, sometimes experimental treatments based on theoretical considerations alone may lead to major breakthroughs [284]. As stated, "the potential for broader application of phage therapy is evident and it is certainly worthy of further studies" [285].

20. Conclusions

Almost a century after its consolidation in Eastern countries, a silver lining is appearing on the horizon for phage therapy in the Western world. The increased threat of antibiotic resistance makes all stakeholders in the sector of infectious disease feel a high pressure to find new antibiotics and search for safe alternatives. In this situation, phage therapy is increasingly considered as a potential alternative or auxiliary tool. More and more patients, doctors, pharmacists, media, authorities, and industry show their active interest and signs of a more open mind to assess the possible benefits of phage therapy. This is especially triggered by an increasing number of publications of patient cases where spectacular results were achieved with bacteriophages. It is now essential that the efficacy and safety of phage application be demonstrated in rigorous clinical trials. National and international authorities are opening their doors to such trials, and are prone to regulate phage therapy if it is found to be effective and safe. Furthermore, progress in research on phage biology suggests that other applications of phages unrelated to their anti-bacterial action may be on the horizon.

Author Contributions: A.G., M.Ł., J.B., J.S., and R.M. wrote the manuscript; A.G, A.G.-R., A.B., E.J.-M., M.Ł.-S., B.W.-D., N.B., S.L., and K.D. contributed to the design of the work, acquisition and interpretation of data. All authors have approved the submitted version.

Funding: This work was supported by statutory funds from the Ludwik Hirszfeld Institute of Immunology and Experimental Therapy of the Polish Academy of Sciences, Warsaw Medical University, and statutory funds from the Institute of Biochemistry and Biophysics of the Polish Academy of Sciences.

Acknowledgments: We thank A. Ajdukiewicz-Tarkowska, Head of Scientific Information, Main Library of the Medical University of Warsaw for her help in accessing information from the Web of Science.

Conflicts of Interest: A.G., R.M., M.Ł., A.G.-R , J.B., B.W.-D. and K.D. are co-inventors of patents owned by the Institute and covering phage preparations. Other authors declare that they have no conflict of interest.

References

1. Borysowski, J.; Międzybrodzki, R.; Górski, A. *Phage Therapy: Current Research and Application*; Caister Academic Press: Norfolk, UK, 2014.
2. Azeredo, J.; Sillankorva, J. (Eds.) *Bacteriophage Therapy: From Lab to Clinical Practice*; Springer Nature, Humana Press: New York, NY, USA, 2018; ISBN 978-1-4939-7395-8.
3. Alvarez, D.R.; Abedon, S.T. An online phage therapy bibliography: Separating under-indexed wheat from overly indexed chaff. *AIMS Microbiol.* **2017**, *3*, 525–528.
4. P.H.A.G.E. Phages for Human Applications Europe Group. Available online: www.p-h-a-g-e.org (accessed on 5 April 2018).
5. Sybesma, W.; Rohde, C.; Bardy, P.; Pirnay, J.P.; Cooper, I.; Caplin, J.; Chanishvili, N.; Coffey, A.; De Vos, D.; Scholz, A.H.; et al. Silk route to the acceptance and re-implementation of bacteriophage therapy. *Biotechnol. J.* **2016**, *11*, 595–600. [CrossRef]
6. Sybesma, W.; Rohde, C.; Bardy, P.; Pirnay, J.P.; Cooper, I.; Caplin, J.; Chanishvili, N.; Coffey, A.; De Vos, D.; Scholz, A.H.; et al. Silk Route to the Acceptance and Re-Implementation of Bacteriophage Therapy-Part II. *Antibiotics* **2018**, *7*, 35. [CrossRef]
7. O'Reilly, C.E.; Jaron, P.; Ochieng, B.; Nyaguara, A.; Tate, J.E.; Parsons, M.B.; Bopp, C.A.; Williams, K.A.; Vinje, J.; Blanton, E.; et al. Risk factors for death among children less than 5 years old hospitalized with

diarrhea in rural western Kenya, 2005–2007: A cohort study. *PLoS Med.* **2012**, *9*, e1001256. [CrossRef] [PubMed]

8. Nagel, T.E.; Chan, B.K.; De Vos, D.; El-Shibiny, A.; Kang'ethe, E.; Makumi, A.; Pirnay, J.P. The developing world urgently needs phages to combat pathogenic bacteria. *Front. Microbiol.* **2016**, *7*, 882. [CrossRef] [PubMed]
9. Available online: https://www.phagesforglobalhealth.org/ (accessed on 24 May 2018).
10. Deltcheva, E.; Chylinski, K.; Sharma, C.M.; Gonzales, K.; Chao, Y.; Pirzada, Z.A.; Eckert, M.R.; Vogel, J.; Charpentier, E. CRISPR RNA maturation by trans-encoded small RNA and host factor RNase III. *Nature* **2011**, *471*, 602–607. [CrossRef] [PubMed]
11. Makarova, K.S.; Haft, D.H.; Barrangou, R.; Brouns, S.J.; Charpentier, E.; Horvath, P.; Moineau, S.; Mojica, F.J.; Wolf, Y.I.; Yakunin, A.F.; et al. Evolution and classification of the CRISPR-Cas systems. *Nat. Rev. Microbiol.* **2011**, *9*, 467–477. [CrossRef] [PubMed]
12. Mali, P.; Yang, L.; Esvelt, K.M.; Aach, J.; Guell, M.; DiCarlo, J.E.; Norville, J.E.; Church, G.M. RNA-guided human genome engineering via Cas9. *Science* **2013**, *339*, 823–826. [CrossRef] [PubMed]
13. Hertsenberg, A.; AVROTROS TV. Dokters van Morgen over bacteriën (Update of emission of 21-03-2017 Bacteriofagen: een alternatief voor antibiotica?). Emission 24-10-2017. Available online: https://zorgnu.avrotros.nl/uitzending/24-10-2017/ (accessed on 24 May 2018).
14. Das Erste. Phagen—hilfreiche Viren gegen bakterielle Infektionen. Available online: www.daserste.de/information/wissen-kultur/w-wie-wissen/videos/phagen-hilfreiche-viren-gegen-bakterielle-infektionen-100.html (accessed on 24 May 2018).
15. Pranz, S.; Weiss, F. PhagenWagen. *Der Spiegel Wissen* **2017**, *6*, 64–69.
16. Van Zonneveld, B. Getreuzel met de faag. *Elsevier Weekblad* **2017**, *45*, 70–71.
17. Melchior, M. Update antibiotica. Kunnen we in de toekomstzonder? *Libelle* **2018**, *6*, 58–61.
18. Pirnay, J.P.; Blasdel, B.G.; Bretaudeau, L.; Buckling, A.; Chanishvili, N.; Clark, J.R.; Corte-Real, S.; Debarbieux, L.; Dublanchet, A.; De Vos, D.; et al. Quality and safety requirements for sustainable phage therapy products. *Pharm. Res.* **2015**, *32*, 2173–2179. [CrossRef] [PubMed]
19. BioIndustry-ein Service-Verbund mit internationalem Anspruch. Available online: https://www.bioindustry.de/nc/mitglieder/details.html?mtg=38 (accessed on 24 May 2018).
20. European Commission. Action plan against the rising threats from antimicrobial resistance. COM (2011)748 and Road Map (Updated 16/11/2016). Available online: https://ec.europa.eu/health/amr/ (accessed on 24 May 2018).
21. World Health Organization (WHO). Global action plan on antimicrobial resistance (document WHA68/2015/REC/1, Annex 3). Available online: http://apps.who.int/gb/ebwha/pdf_files/WHA68-REC1/A68_R1_REC1-en.pdf (accessed on 24 May 2018).
22. The Food and Drug Administration (FDA). Bacteriophage Therapy: Scientific and Regulatory Issues Public Workshop. Available online: https://www.fda.gov/BiologicsBloodVaccines/NewsEvents/WorkshopsMeetingsConferences/ucm544294.htm (accessed on 24 May 2018).
23. Schooley, R.T.; Biswas, B.; Gill, J.J.; Hernandez-Morales, A.; Lancaster, J.; Lessor, L.; Bar, J.J.; Reed, S.L.; Rohwer, F.; Benler, S.; et al. Development and use of personalized bacteriophage-based therapeutic cocktails to treat a patient with a disseminated resistant Acinetobacter baumanniiinfection. *Antimicrob. Agents Chemother.* **2017**, *61*, e00954-17. [CrossRef] [PubMed]
24. Intralytix Safety by Nature. Intralytix Receives FDA Clearance to Initiate Phase I/IIa Clinical Trials. Available online: http://www.intralytics.com (accessed on 30 March 2018).
25. PhagoBurn Project Funded by the European Union under the 7th Framework Programme for Research and Development. Available online: www.phagoburn.eu (accessed on 4 May 2018).
26. The White House. National Action Plan for Combating Antibiotic Resistant Bacteria. *The White House Washington*. March 2015, p. 44. Available online: https://obamawhitehouse.archives.gov/sites/default/files/docs/national_action_plan_for_combating_antibotic-resistant_bacteria.pdf (accessed on 24 May 2018).
27. Transatlantic Taskforce Antimicrobial Resistance (TATFAR). Actions and Recommendations, Action 3.6. Available online: https://www.cdc.gov/drugresistance/tatfar/tatfar-recomendations.html (accessed on 24 May 2018).

28. Federal Government. *Deutsche Antimicrobiale Resistance Strategie DART 2020*; Fighting Antibiotic Resistance for the Good of Both Humans and Animal; Decision by the Federal Cabinet of 13th of May 2015; Federal Government: Berlin, Germany, 2015; p. 24.
29. Federal Government Germany. *CombatingAntimicrobial Resistance; Examples of Best-Practices of the G7 Countries*; Report of the G7 Meeting; Federal Government Germany: Berlin, Germany, 2015; p. 99.
30. Pirnay, J.P.; Verbeken, G.; Ceyssens, P.J.; Huys, I.; De Vos, D.; Ameloot, C.; Fauconnier, A. The Magistral Phage. *Viruses* **2018**, *10*, 64. [CrossRef] [PubMed]
31. World Medical Association (WMA). *Declaration of Helsinki—Ethical Principles for Medicalresearch Involving Human Subjects*; Article 37. Unproven interventions in clinical practice. Adopted by the 18th WMA General Assembly, Helsinki, Finland, June 1964 and Current Version as Amended 64th WMA General Assembly; WMA: Fortaleza, Brazil, 2013.
32. Międzybrodzki, R.; Borysowski, J.; Weber-Dąbrowska, B.; Fortuna, W.; Letkiewicz, S.; Szufnarowski, K.; Pawełczyk, Z.; Rogóż, P.; Kłak, M.; Wojtasik, E.; et al. Clinical aspects of phage therapy. *Adv. Virus Res.* **2012**, *83*, 73–121. [CrossRef] [PubMed]
33. Górski, A.; Międzybrodzki, R.; Weber-Dąbrowska, B.; Fortuna, W.; Letkiewicz, S.; Rogóż, P.; Jończyk-Matysiak, E.; Dąbrowska, K.; Majewska, J.; Borysowski, J. Phage therapy: Present and future. *Front. Microbiol.* **2016**, *7*, 1515. [CrossRef] [PubMed]
34. European Commission Directorate-General for Health and Food Safety. Health Systems, Medical Products and Innovation. Medicines: Policy, Authorisation and Monitoring. STAMP Commission Expert Group. STAMP 4/22, Subject: Compassionate Use Programmes Agenda Item 6. 13 March 2016. Available online: http://www.ema.europa.eu/docs/en_GB/document_library/Regulatory_and_procedural_guideline/2009/10/WC500004075.pdf (accessed on 4 May 2018).
35. Balasubramanian, G.; Morampudi, S.; Chhabra, P.; Gowda, A.; Zomorodi, B. An overview of Compassionate Use Programs in the European Union member states. *Intractable Rare Dis. Res.* **2016**, *5*, 244–254. [CrossRef] [PubMed]
36. Directive 2001/83/EC of the European Parliament and of the Council. Available online: https://ec.europa.eu/health/sites/health/files/files/eudralex/vol-1/dir_2001_83_consol_2012/dir_2001_83_cons_2012_en.pdf (accessed on 24 May 2018).
37. US Food and Drug Administration FDA. Expanded Access (Compassionate Use). Available online: https://www.fda.gov/NewsEvents/PublicHealthFocus/ExpandedAccessCompassionateUse/default.htm (accessed on 27 April 2018).
38. US Food and Drug Administration FDA. Expanded Access to Investigational Drugs for Treatment Use. Questions and Answers. Guidance for Industry. June 2016; updated October 2017. Available online: https://www.fda.gov/downloads/Drugs/GuidanceComplianceRegulatoryInformation/Guidances/UCM351261.pdf (accessed on 4 May 2018).
39. Joffe, S.; Lynch, H.F. Federal right-to-try legislation—Threatening the FDA's public health mission. *N. Engl. J. Med.* **2018**, *378*, 695–697. [CrossRef] [PubMed]
40. House of Representatives, 115th CONGRESS 2d Session. H.R. 5247. To authorize the use of eligible investigational drugs by eligible patients who have been diagnosed with a stage of a disease or condition in which there is reasonable likelihood that death will occur within a matter of months, or with another eligible illness, and for other purposes. Available online: https://www.congress.gov/bill/115th-congress/house-bill/5247/text/eh (accessed on 24 May 2018).
41. Thomas, K. Why can't dying patients get the drugs they want? *NYT*, 28 March 2018.
42. Available online: https://www.canada.ca/content/dam/hc-sc/migration/hc-sc/dhp-mps/alt_formats/hpfb-dgpsa/pdf/acces/sapg3_pasg3-eng.pdf (accessed on 27 March 2018).
43. Australian Government, Department of Health. Therapeutic Goods Administration. Accessing Unapproved Products. Available online: https://www.tga.gov.au/accessing-unapproved-products (accessed on 4 May 2018).
44. Califf, R.M.; Ostroff, S. FDA as a catalyst for translation. *Sci. Transl. Med.* **2015**, *7*, 296ed9. [CrossRef] [PubMed]
45. Międzybrodzki, R.; Hoyle, N.; Zhvaniya, F.; Gogokhia, L. Current updates from the long-standing phage research centers in Georgia, Poland, and Russia. In *Bacteriophages*; Harper, D., Abedon, S., Burrowes, B., McConville, M., Eds.; Springer: Cham, Switzerland, 2018; ISBN 978-3-319-40598-8.

46. Majewska, J.; Beta, W.; Lecion, D.; Hodyra-Stefaniak, K.; Kłopot, A.; Kaźmierczak, Z.; Miernikiewicz, P.; Piotrowicz, A.; Ciekot, J.; Owczarek, B.; et al. Oral Application of T4 Phage Induces Weak Antibody Production in the Gut and in the Blood. *Viruses* **2015**, *7*, 4783–4799. [CrossRef] [PubMed]
47. Łusiak-Szelachowska, M.; Żaczek, M.; Weber-Dąbrowska, M.; Międzybrodzki, R.; Letkiewicz, S.; Fortuna, W.; Rogóż, P.; Szufnarowski, K.; Jończyk-Matysiak, E.; Olchawa, E.; et al. Antiphage Activity of Sera in Patients During Phage Therapy in Relation to its Outcome. *Future Microbiol.* **2017**, *12*, 109–117. [CrossRef] [PubMed]
48. Persn com Międzybrodzki, R.; Rogóż, P.; Fortuna, W.; Wójcik, E.; Letkiewicz, A.; Weber-Dąbrowska, B.; Górski, A. Wrocław, Poland. The first retrospective analysis of long term results of the application of phage preparations in patients with chronic bacterial infections. *Viruses Microbes* **2018**, in press.
49. Brüssow, H. Special Issue Information. Available online: www.mdpi.com/journal/viruses/special_issues/Phagetherapy (accessed on 10 May 2018).
50. Letkiewicz, S.; Miedzybrodzki, R.; Fortuna, W.; Weber-Dabrowska, B.; Górski, A. Eradication of *Enterococcus faecalis* by phage therapy in chronic bacterial prostatitis—Case report. *Folia Microbiol.* **2009**, *54*, 457–461. [CrossRef] [PubMed]
51. Nguyen, S.; Baker, K.; Padman, B.S.; Patwa, R.; Dunstan, R.A.; Weston, T.A.; Schlosser, K.; Bailey, B.; Lithgow, T.; Lazarou, M.; et al. Bacteriophage transcytosis provides a mechanism to cross epithelial cell layers. *mBio* **2017**, *8*, e01874-17. [CrossRef] [PubMed]
52. Lehti, T.A.; Pajunen, M.I.; Skog, M.S.; Finne, J. Internalization of a polysialic acid-binding *Escherichia coli* bacteriophage into eukaryotic neuroblastoma cells. *Nat. Commun.* **2017**, *8*, 1915. [CrossRef] [PubMed]
53. Górski, A.; Międzybrodzki, R.; Borysowski, J.; Dąbrowska, K.; Wierzbicki, P.; Ohams, M.; Korczak-Kowalska, G.; Olszowska-Zaremba, N.; Łusiak-Szelachowska, M.; Kłak, M.; et al. Phage as a modulator of immune responses: Practical implications for phage therapy. *Adv. Virus Res.* **2012**, *83*, 41–71. [CrossRef] [PubMed]
54. Hopkins, W.J. Mouse model of ascending urinary tract infection. In *Handbook of Animal Models of Infection*; Zak, O., Sande, M.A., Eds.; Academic Press: London, UK, 1999; pp. 435–439. ISBN 0-12-776390-7.
55. Kruisbeck, A.M. Isolation and fractionation of mononuclear cell populations. In *Current Protocols in Immunology*; Coligan, J.E., Kruisbeck, A.M., Margulies, D.H., Shevach, E.M., Strober, W., Eds.; Wiley: New York, NY, USA, 2000; pp. 3.1.1–3.1.5.
56. Buisman, H.P.; Buys, L.F.M.; Langermans, J.A.M.; van den Broek, P.J.; van Furth, R. Effect of probenecid on phagocytosis and intracellular killing of *Staphylococcus aureus* and *Escherichia coli* by human monocytes and granulocytes. *Immunology* **1991**, *74*, 338–341. [PubMed]
57. Leijh, P.C.J.; Van Zwet, T.L.; Van Furth, R. Effect of concanavalin A on intracellular killing of *Staphylococcus aureus* by human phagocytes. *Clin. Exp. Immunol.* **1984**, *58*, 557–565. [PubMed]
58. Jończyk-Matysiak, E. The Effect of Bacteriophage Preparations on Intracellular Killing of Bacteria by Phagocytes. Ph.D. Thesis, Ludwik Hirszfeld Institute of Immunology and Experimental Therapy Polish Academy of Sciences, Wrocław, Poland, 2015; pp. 61–118.
59. Roach, D.R.; Leung, C.Y.; Henry, M.; Morello, E.; Singh, D.; Di Santo, J.P.; Weitz, J.S.; Debarbieux, L. Synergy between the Host Immune System and Bacteriophage Is Essential for Successful Phage Therapy against an Acute Respiratory Pathogen. *Cell Host Microbe* **2017**, *22*, 38.e4–47.e4. [CrossRef] [PubMed]
60. Łobocka, M.; Hejnowicz, M.S.; Gągała, U.; Weber-Dąbrowska, B.; Węgrzyn, G.; Dadlez, M. The first step to bacteriophage therapy—How to choose the correct phage. In *Phage Therapy: Current Research and Applications*; Borysowski, J., Międzybrodzki, R., Górski, A., Eds.; Caister Academic Press: Norfolk, UK, 2014; pp. 23–69.
61. Rohde, C.; Resch, G.; Pirnay, J.P.; Blasdel, B.G.; Debarbieux, L.; Gelman, D.; Górski, A.; Hazan, R.; Huys, I.; Kakabadze, E.; et al. Expert Opinion on Three Phage Therapy Related Topics: Bacterial Phage Resistance, Phage Training and Prophages in Bacterial Production Strains. *Viruses* **2018**, *10*, E178. [CrossRef] [PubMed]
62. Łobocka, M.; Hejnowicz, M.S.; Dąbrowski, K.; Izak, D.; Gozdek, A.; Głowacka, A.; Gawor, J.; Kosakowski, J.; Gromadka, R.; Weber-Dąbrowska, B.; et al. Staphylococcus aureus Strains for the Production of Monoclonal Bacteriophage Preparations Deprived of Contamination with Plasmid DNA. U.S. Patent WO 2016/030871 A1, 16 March 2016.
63. Brussow, H.; Canchaya, C.; Hardt, W.D. Phages and the evolution of bacterial pathogens: From genomic rearrangements to lysogenic conversion. *Microbiol. Mol. Biol. Rev.* **2004**, *68*, 560–602. [CrossRef] [PubMed]
64. Bossi, L.; Fuentes, J.A.; Mora, G.; Figueroa-Bossi, N. Prophage contribution to bacterial population dynamics. *J. Bacteriol.* **2003**, *185*, 6467–6471. [CrossRef] [PubMed]

65. Fortier, L.C.; Sekulovic, O. Importance of prophages to evolution and virulence of bacterial pathogens. *Virulence* **2013**, *5*, 354–365. [CrossRef] [PubMed]
66. Abedon, S.T.; Lejeune, J.T. Why bacteriophage encode exotoxins and other virulence factors. *Evol. Bioinform. Online* **2005**, *1*, 97–110. [CrossRef]
67. Varani, A.M.; Monteiro-Vitorello, C.B.; Nakaya, H.I.; Van Sluys, M.A. The role of prophage in plant-pathogenic bacteria. *Annu. Rev. Phytopathol.* **2013**, *51*, 429–451. [CrossRef] [PubMed]
68. Davies, E.V.; Winstanley, C.; Fothergill, J.L.; James, C.E. The role of temperate bacteriophages in bacterial infection. *FEMS Microbiol. Lett.* **2016**, *363*. [CrossRef] [PubMed]
69. Trost, E.; Blom, J.; Soares Sde, C.; Huang, I.H.; Al-Dilaimi, A.; Schröder, J.; Jaenicke, S.; Dorella, F.A.; Rocha, F.S.; Miyoshi, A.; et al. Pangenomic study of *Corynebacterium diphtheriae* that provides insights into the genomic diversity of pathogenic isolates from cases of classical diphtheria, endocarditis, and pneumonia. *J. Bacteriol.* **2012**, *194*, 3199–3215. [CrossRef] [PubMed]
70. The, H.C.; Thanh, D.P.; Holt, K.E.; Thomson, N.R.; Baker, S. The genomic signatures of *Shigella* evolution, adaptation and geographical spread. *Nat. Rev. Microbiol.* **2016**, *4*, 235–250. [CrossRef] [PubMed]
71. Gyles, C.; Boerlin, P. Horizontally transferred genetic elements and their role in pathogenesis of bacterial disease. *Vet. Pathol.* **2014**, *51*, 328–340. [CrossRef] [PubMed]
72. Mai-Prochnow, A.; Hui, J.G.; Kjelleberg, S.; Rakonjac, J.; McDougald, D.; Rice, S.A. Big things in small packages: The genetics of filamentous phage and effects on fitness of their host. *FEMS Microbiol. Rev.* **2015**, *39*, 465–487. [CrossRef] [PubMed]
73. Hare, J.M.; Ferrell, J.C.; Witkowski, T.A.; Grice, A.N. Prophage induction and differential RecA and UmuDAb transcriptome regulation in the DNA damage responses of *Acinetobacter baumannii* and *Acinetobacter baylyi*. *PLoS ONE* **2014**, *9*, e93861. [CrossRef] [PubMed]
74. Repizo, G.D.; Viale, A.M.; Borges, V.; Cameranesi, M.M.; Taib, N.; Espariz, M.; Brochier-Armanet, C.; Gomes, J.P.; Salcedo, S.P. The Environmental Acinetobacter baumannii Isolate DSM30011 Reveals Clues into the Preantibiotic Era Genome Diversity, Virulence Potential, and Niche Range of a Predominant Nosocomial Pathogen. *Genome Biol. Evol.* **2017**, *9*, 2292–2307. [CrossRef] [PubMed]
75. Touchon, M.; Cury, J.; Yoon, E.J.; Krizova, L.; Cerqueira, G.C.; Murphy, C.; Feldgarden, M.; Wortman, J.; Clermont, D.; Lambert, T.; et al. The genomic diversification of the whole *Acinetobacter* genus: Origins, mechanisms, and consequences. *Genome Biol. Evol.* **2014**, *10*, 2866–2882. [CrossRef] [PubMed]
76. Bearson, B.L.; Brunelle, B.W. Fluoroquinolone induction of phage-mediated gene transfer in multidrug-resistant *Salmonella*. *Int. J. Antimicrob. Agents* **2015**, *46*, 201–204. [CrossRef] [PubMed]
77. Mašlaňová, I.; Stříbná, S.; Doškař, J.; Pantůček, R. Efficient plasmid transduction to Staphylococcus aureus strains insensitive to the lytic action of transducing phage. *FEMS Microbiol. Lett.* **2016**, *363*, fnw211. [CrossRef] [PubMed]
78. Chen, J.; Novick, R.P. Phage-mediated intergeneric transfer of toxin genes. *Science* **2009**, *5910*, 139–141. [CrossRef] [PubMed]
79. Zeman, M.; Mašlaňová, I.; Indráková, A.; Šiborová, M.; Mikulášek, K.; Bendíčková, K.; Plevka, P.; Vrbovská, V.; Zdráhal, Z.; Doškař, J.; et al. Staphylococcus sciuri bacteriophages double-convert for staphylokinase and phospholipase, mediate interspecies plasmid transduction, and package mecA gene. *Sci. Rep.* **2017**, *13*, 46319. [CrossRef] [PubMed]
80. Moon, B.Y.; Park, J.Y.; Robinson, D.A.; Thomas, J.C.; Park, Y.H.; Thornton, J.A.; Seo, K.S. Mobilization of Genomic Islands of *Staphylococcus aureus* by Temperate Bacteriophage. *PLoS ONE* **2016**, *11*, e0151409. [CrossRef] [PubMed]
81. Matilla, M.A.; Salmond, G.P. Bacteriophage φMAM1, a viunalikevirus, is a broad-host-range, high-efficiency generalized transducer that infects environmental and clinical isolates of the enterobacterial genera Serratia and Kluyvera. *Appl. Environ. Microbiol.* **2014**, *80*, 6446–6457. [CrossRef] [PubMed]
82. Varga, M.; Kuntová, L.; Pantůček, R.; Mašlaňová, I.; Růžičková, V.; Doškař, J. Efficient transfer of antibiotic resistance plasmids by transduction within methicillin-resistant *Staphylococcus aureus* USA300 clone. *FEMS Microbiol. Lett.* **2012**, *332*, 146–152. [CrossRef] [PubMed]
83. Valero-Rello, A.; López-Sanz, M.; Quevedo-Olmos, A.; Sorokin, A.; Ayora, S. Molecular Mechanisms That Contribute to Horizontal Transfer of Plasmids by the Bacteriophage SPP1. *Front. Microbiol.* **2017**, *22*, 1816. [CrossRef] [PubMed]

84. Krahn, T.; Wibberg, D.; Maus, I.; Winkler, A.; Bontron, S.; Sczyrba, A.; Nordmann, P.; Pühler, A.; Poirel, L.; Schlüter, A. Intraspecies Transfer of the Chromosomal *Acinetobacter baumannii* blaNDM-1 Carbapenemase Gene. *Antimicrob. Agents Chemother.* **2016**, *60*, 3032–3040. [CrossRef] [PubMed]
85. Touchon, M.; Bernheim, A.; Rocha, E.P. Genetic and life-history traits associated with the distribution of prophages in bacteria. *ISME J.* **2016**, *11*, 2744–2754. [CrossRef] [PubMed]
86. McCarthy, A.J.; Witney, A.A.; Lindsay, J.A. *Staphylococcus aureus* temperate bacteriophage: Carriage and horizontal gene transfer is lineage associated. *Front. Cell. Infect. Microbiol.* **2012**, *2*, 6. [CrossRef] [PubMed]
87. Fogg, P.C.; Saunders, J.R.; McCarthy, A.J.; Allison, H.E. Cumulative effect of prophage burden on Shiga toxin production in *Escherichia coli*. *Microbiology* **2012**, *158*, 488–497. [CrossRef] [PubMed]
88. Boyd, E.F.; Brüssow, H. Common themes among bacteriophage-encoded virulence factors and diversity among the bacteriophages involved. *Trends Microbiol.* **2002**, *11*, 521–529. [CrossRef]
89. Li, Y.; Zhu, X.; Zhang, X.; Fu, J.; Wang, Z.; Chen, T.; Zhao, X. Characterization of genome-reduced *Bacillus subtilis* strains and their application for the production of guanosine and thymidine. *Microb. Cell Fact.* **2016**, *15*, 94. [CrossRef] [PubMed]
90. Solheim, M.; Brekke, M.C.; Snipen, L.G.; Willems, R.J.; Nes, I.F.; Brede, DA. Comparative genomic analysis reveals significant enrichment of mobile genetic elements and genes encoding surface structure-proteins in hospital-associated clonal complex 2 *Enterococcus faecalis*. *BMC Microbiol.* **2011**, *11*, 3. [CrossRef] [PubMed]
91. Paulsen, I.T.; Banerjei, L.; Myers, G.S.; Nelson, K.E.; Seshadri, R.; Read, T.D.; Fouts, D.E.; Eisen, J.A.; Gill, S.R.; Heidelberg, J.F.; et al. Role of mobile DNA in the evolution of vancomycin-resistant *Enterococcus faecalis*. *Science* **2003**, *5615*, 2071–2074. [CrossRef] [PubMed]
92. McBride, S.M.; Fischetti, V.A.; Leblanc, D.J.; Moellering, R.C., Jr.; Gilmore, M.S. Genetic diversity among *Enterococcus faecalis*. *PLoS ONE* **2007**, *2*, e582. [CrossRef] [PubMed]
93. Matos, R.C.; Lapaque, N.; Rigottier-Gois, L.; Debarbieux, L.; Meylheuc, T.; Gonzalez-Zorn, B.; Repoila, F.; Lopes Mde, F.; Serror, P. *Enterococcus faecalis* prophage dynamics and contributions to pathogenic traits. *PLoS Genet.* **2013**, *9*, e1003539. [CrossRef] [PubMed]
94. Goerke, C.; Pantucek, R.; Holtfreter, S.; Schulte, B.; Zink, M.; Grumann, D.; Bröker, B.M.; Doskar, J.; Wolz, C. Diversity of prophages in dominant *Staphylococcus aureus* clonal lineages. *J. Bacteriol.* **2009**, *191*, 3462–3468. [CrossRef] [PubMed]
95. Rahimi, F.; Bouzari, M.; Katouli, M.; Pourshafie, M.R. Prophage and antibiotic resistance profiles of methicillin-resistant *Staphylococcus aureus* strains in Iran. *Arch. Virol.* **2012**, *157*, 1807–1811. [CrossRef] [PubMed]
96. Satta, G.; Pruzzo, C.; Debbia, E.; Fontana, R. Close association between shape alteration and loss of immunity to superinfection in a wild-type *Klebsiella pneumoniae* stable lysogen which can be both immune and nonimmune to superinfection. *J. Virol.* **1978**, *28*, 772–785. [PubMed]
97. Kwon, T.; Jung, Y.H.; Lee, S.; Yun, M.R.; Kim, W.; Kim, D.W. Comparative genomic analysis of *Klebsiella pneumoniae subsp. pneumoniae* KP617 and PittNDM01, NUHL24835, and ATCC BAA-2146 reveals unique evolutionary history of this strain. *Gut Pathog.* **2016**, *8*, 34. [CrossRef] [PubMed]
98. Huang, W.; Wang, G.; Sebra, R.; Zhuge, J.; Yin, C.; Aguero-Rosenfeld, M.E.; Schuetz, A.N.; Dimitrova, N.; Fallon, J.T. Emergence and Evolution of Multidrug-Resistant *Klebsiella pneumoniae* with both bla(KPC) and bla(CTX-M) Integrated in the Chromosome. *Antimicrob. Agents Chemother.* **2017**, *61*, e00076-17. [CrossRef] [PubMed]
99. Wang, X.; Xie, Y.; Li, G.; Liu, J.; Li, X.; Tian, L.; Sun, J.; Ou, H.Y.; Qu, H. Whole-Genome-Sequencing characterization of bloodstream infection-causing hypervirulent *Klebsiella pneumoniae* of capsular serotype K2 and ST374. *Virulence* **2018**, *1*, 510–521. [CrossRef] [PubMed]
100. Chen, L.; Chavda, K.D.; DeLeo, F.R.; Bryant, K.A.; Jacobs, M.R.; Bonomo, R.A.; Kreiswirth, B.N. Genome Sequence of a *Klebsiella pneumoniae* Sequence Type 258 Isolate with Prophage-Encoded *K. pneumoniae* Carbapenemase. *Genome Announc.* **2015**, *3*, e00659-15. [CrossRef] [PubMed]
101. Bi, D.; Jiang, X.; Sheng, Z.K.; Ngmenterebo, D.; Tai, C.; Wang, M.; Deng, Z.; Rajakumar, K.; Ou, H.Y. Mapping the resistance-associated mobilome of a carbapenem-resistant *Klebsiella pneumoniae* strain reveals insights into factors shaping these regions and facilitates generation of a 'resistance-disarmed' model organism. *J. Antimicrob. Chemother.* **2015**, *10*, 2770–2774. [CrossRef] [PubMed]

102. Zautner, A.E.; Bunk, B.; Pfeifer, Y.; Spröer, C.; Reichard, U.; Eiffert, H.; Scheithauer, S.; Groß, U.; Overmann, J.; Bohne, W. Monitoring microevolution of OXA-48-producing *Klebsiella pneumoniae* ST147 in a hospital setting by SMRT sequencing. *J. Antimicrob. Chemother.* **2017**, *72*, 2737–2744. [CrossRef] [PubMed]
103. Di Nocera, P.P.; Rocco, F.; Giannouli, M.; Triassi, M.; Zarrilli, R. Genome organization of epidemic *Acinetobacter baumannii* strains. *BMC Microbiol.* **2011**, *11*, 224. [CrossRef] [PubMed]
104. Ohnishi, M.; Kurokawa, K.; Hayashi, T. Diversification of *Escherichia coli* genomes: Are bacteriophages the major contributors? *Trends Microbiol.* **2001**, *10*, 481–485. [CrossRef]
105. Fouts, D.E. Phage_Finder: Automated identification and classification of prophage regions in complete bacterial genome sequences. *Nucleic Acids Res.* **2006**, *34*, 5839–5851. [CrossRef] [PubMed]
106. Ptashne, M. *Genetic Switch: Phage Lambda and Higher Organisms*, 2nd ed.; Blackwell: Cambridge, MA, USA, 1992.
107. Cavalcanti, S.M.; Siqueira, J.P., Jr. Cure of prophage *in Staphylococcus aureus* by furocoumarin photoadditions. *Microbios* **1995**, *327*, 85–91.
108. Duval-Iflah, Y. Lysogenic conversion of the lipase gene in *Staphylococcus pyogenes* group III strains. *Can. J. Microbiol.* **1972**, *18*, 1491–1497. [CrossRef] [PubMed]
109. Gasson, M.J.; Davies, F.L. Prophage-cured derivatives of *Streptococcus lactis* and *Streptococcus cremoris*. *Appl. Environ. Microbiol.* **1980**, *40*, 964–966. [PubMed]
110. Waldor, M.K.; Friedma, D.I. Phage regulatory circuits and virulence gene expression. *Curr. Opin. Microbiol.* **2005**, *8*, 459–465. [CrossRef] [PubMed]
111. Raya, R.R.; H'bert, E.M. Isolation of phage via induction of lysogens. *Methods Mol. Biol.* **2009**, *501*, 23–32. [PubMed]
112. Selva, L.; Viana, D.; Regev-Yochay, G.; Trzcinski, K.; Corpa, J.M.; Lasa, I.; Novick, R.P.; Penadés, J.R. Killing niche competitors by remote-control bacteriophage induction. *Proc. Natl. Acad. Sci. USA* **2009**, *106*, 1234–1238. [CrossRef] [PubMed]
113. Banks, D.J.; Lei, B.; Musser, J.M. Prophage induction and expression of prophage-encoded virulence factors in group A *Streptococcus* serotype M3 strain MGAS315. *Infect. Immun.* **2003**, *71*, 7079–7086. [CrossRef] [PubMed]
114. Bertani, G. Studies on lysogenesis. III. Superinfection of lysogenic Shigella dysenteriae with temperate mutants of the carried phage. *J. Bacteriol.* **1954**, *67*, 696–707. [PubMed]
115. Madera, C.; García, P.; Rodríguez, A.; Suárez, J.E.; Martínez, B. Prophage induction in *Lactococcus lactis* by the bacteriocin Lactococcin 972. *Int. J. Food Microbiol.* **2009**, *129*, 99–102. [CrossRef] [PubMed]
116. Affolter, M.; Parent-Vaugeois, C.; Anderson, A. Curing and induction of the Fels 1 and Fels 2 prophages in the Ames mutagen tester strains of *Salmonella typhimurium*. *Mutat. Res.* **1983**, *110*, 243–262. [CrossRef]
117. Menouni, R.; Champ, S.; Espinosa, L.; Boudvillain, M.; Ansaldi, M. Transcription termination controls prophage maintenance in *Escherichia coli* genomes. *Proc. Natl. Acad. Sci. USA* **2013**, *110*, 14414–14419. [CrossRef] [PubMed]
118. Allen, H.K.; Looft, T.; Bayles, D.O.; Humphrey, S.; Levine, U.Y.; Alt, D.; Stanton, T.B. Antibiotics in feed induce prophages in swine fecal microbiomes. *mBio* **2011**, *6*, e00260-11. [CrossRef] [PubMed]
119. Ghosh, D.; Roy, K.; Williamson, K.E.; Srinivasiah, S.; Wommack, K.E.; Radosevich, M. Acyl-homoserine lactones can induce virus production in lysogenic bacteria: An alternative paradigm for prophage induction. *Appl. Environ. Microbiol.* **2009**, *75*, 7142–7152. [CrossRef] [PubMed]
120. Miller, S.T.; Xavier, K.B.; Campagna, S.R.; Taga, M.E.; Semmelhack, M.F.; Bassler, B.L.; Hughson, F.M. *Salmonella typhimurium* recognizes a chemically distinct form of the bacterial quorum-sensing signal AI-2. *Mol. Cell* **2004**, *15*, 677–687. [CrossRef] [PubMed]
121. Rossmann, F.S.; Racek, T.; Wobser, D.; Puchalka, J.; Rabener, E.M.; Reiger, M.; Hendrickx, A.P.; Diederich, A.K.; Jung, K.; Klein, C.; et al. Phage-mediated dispersal of biofilm and distribution of bacterial virulence genes is induced by quorum sensing. *PLoS Pathog.* **2015**, *11*, e1004653. [CrossRef] [PubMed]
122. Lwoff, A. Lysogeny. *Bacteriol. Rev.* **1953**, *17*, 269–337. [PubMed]
123. Birdsell, D.C.; Hathaway, G.M.; Rutberg, L. Characterization of Temperate *Bacillus* Bacteriophage phi105. *J. Virol.* **1969**, *4*, 264–270. [PubMed]
124. Garro, A.J.; Law, M.F. Relationship between lysogeny, spontaneous induction, and transformation efficiencies in *Bacillus subtilis*. *J. Bacteriol.* **1974**, *120*, 1256–1259. [PubMed]

125. Livny, J.; Friedman, D.I. Characterizing spontaneous induction of Stx encoding phages using a selectable reporter system. *Mol. Microbiol.* **2004**, *51*, 1691–1704. [CrossRef] [PubMed]
126. Bullwinkle, T.J.; Koudelka, G.B. The lysis-lysogeny decision of bacteriophage 933W: A 933W repressor-mediated long-distance loop has no role in regulating 933W P(RM) activity. *J. Bacteriol.* **2011**, *193*, 3313–3323. [CrossRef] [PubMed]
127. Carrolo, M.; Frias, M.J.; Pinto, F.R.; Melo-Cristino, J.; Ramirez, M. Prophage spontaneous activation promotes DNA release enhancing biofilm formation in *Streptococcus pneumoniae*. *PLoS ONE* **2010**, *5*, e15678. [CrossRef] [PubMed]
128. Wang, X.; Kim, Y.; Ma, Q.; Hong, S.H.; Pokusaeva, K.; Sturino, J.M.; Wood, T.K. Cryptic prophages help bacteria cope with adverse environments. *Nat. Commun.* **2010**, *1*, 147. [CrossRef] [PubMed]
129. Nanda, A.M.; Heyer, A.; Krämer, C.; Grünberger, A.; Kohlheyer, D.; Frunzke, J. Analysis of SOS-induced spontaneous prophage induction in *Corynebacterium glutamicum* at the single-cell level. *J. Bacteriol.* **2014**, *196*, 180–188. [CrossRef] [PubMed]
130. Rippon, J.E. The classification of bacteriophages lysing staphylococci. *J. Hyg.* **1956**, *54*, 213–226. [CrossRef] [PubMed]
131. Allué-Guardia, A.; García-Aljaro, C.; Muniesa, M. Bacteriophage-encoding cytolethal distending toxin type V gene induced from nonclinical *Escherichia coli* isolates. *Infect. Immun.* **2011**, *79*, 3262–3272. [CrossRef]
132. Zhou, Y.; Liang, Y.; Lynch, K.H.; Dennis, J.J.; Wishart, D.S. PHAST: A fast phage search tool. *Nucleic Acids Res.* **2011**, *39*, W347–W352. [CrossRef] [PubMed]
133. Arndt, D.; Marcu, A.; Liang, Y.; Wishart, D.S. PHAST, PHASTER and PHASTEST: Tools for finding prophage in bacterial genomes. *Brief Bioinform.* **2017**. [CrossRef] [PubMed]
134. Lima-Mendez, G.; Van Helden, J.; Toussaint, A.; Leplae, R. Prophinder: A computational tool for prophage prediction in prokaryotic genomes. *Bioinformatics* **2008**, *24*, 863–865. [CrossRef] [PubMed]
135. Bose, M.; Barber, R.D. Prophage Finder: A prophage loci prediction tool for prokaryotic genome sequences. *In Silico Biol.* **2006**, *6*, 223–227. [PubMed]
136. Akhter, S.; Aziz, R.K.; Edwards, R.A. PhiSpy: A novel algorithm for finding prophages in bacterial genomes that combines similarity- and composition-based strategies. *Nucleic Acids Res.* **2012**, *40*, 1–13. [CrossRef] [PubMed]
137. Roux, S.; Enault, F.; Hurwitz, B.L.; Sullivan, M.B. VirSorter: Mining viral signal from microbial genomic data. *PeerJ* **2015**, *3*, e985. [CrossRef] [PubMed]
138. Pantůcek, R.; Doskar, J.; Růzicková, V.; Kaspárek, P.; Orácová, E.; Kvardová, V.; Rosypal, S. Identification of bacteriophage types and their carriage in *Staphylococcus aureus*. *Arch. Virol.* **2004**, *149*, 1689–1703. [CrossRef] [PubMed]
139. Kahánková, J.; Pantůček, R.; Goerke, C.; Růžičková, V.; Holochová, P.; Doškař, J. Multilocus PCR typing strategy for differentiation of *Staphylococcus aureus* siphoviruses reflecting their modular genome structure. *Environ. Microbiol.* **2010**, *12*, 2527–2538. [CrossRef] [PubMed]
140. Ross, I.L.; Heuzenroeder, M.W. Discrimination within phenotypically closely related definitive types of Salmonella enterica serovar typhimurium by the multiple amplification of phage locus typing technique. *J. Clin. Microbiol.* **2005**, *43*, 1604–1611. [CrossRef] [PubMed]
141. Lee, C.Y.; Iandolo, J.J. Mechanism of bacteriophage conversion of lipase activity in *Staphylococcus aureus*. *J. Bacteriol.* **1985**, *164*, 288–293. [PubMed]
142. Loeffler, J.M.; Fischetti, V.A. Lysogeny of *Streptococcus pneumoniae* with MM1 phage: Improved adherence and other phenotypic changes. *Infect. Immun.* **2006**, *74*, 4486–4495. [CrossRef] [PubMed]
143. Leffers, G.G., Jr.; Gottesman, S. Lambda Xis degradation in vivo by Lon and FtsH. *J. Bacteriol.* **1998**, *180*, 1573–1577. [PubMed]
144. Panis, G.; Méjean, V.; Ansaldi, M. Control and regulation of KplE1 prophage site-specific recombination: A new recombination module analyzed. *J. Biol. Chem.* **2007**, *282*, 21798–21809. [CrossRef] [PubMed]
145. Figueroa-Bossi, N.; Coissac, E.; Netter, P.; Bossi, L. Unsuspected prophage-like elements in *Salmonella typhimurium*. *Mol. Microbiol.* **1997**, *25*, 161–173. [CrossRef] [PubMed]
146. Bae, T.; Baba, T.; Hiramatsu, K.; Schneewind, O. Prophages of *Staphylococcus aureus* Newman and their contribution to virulence. *Mol. Microbiol.* **2006**, *62*, 1035–1047. [CrossRef] [PubMed]
147. Koskella, B.; Meaden, S. Understanding bacteriophage specificity in natural microbial communities. *Viruses* **2013**, *5*, 806–823. [CrossRef] [PubMed]

148. Hyman, P.; Abedon, S.T. Bacteriophage host range and bacterial resistance. *Adv. Appl. Microbiol.* **2010**, *70*, 217–248. [CrossRef] [PubMed]
149. Rakhuba, D.V.; Kolomiets, E.I.; Dey, E.S.; Novik, G.I. Bacteriophage receptors, mechanisms of phage adsorption and penetration into host cell. *Pol. J. Microbiol.* **2010**, *59*, 145–155. [PubMed]
150. Xia, G.; Maier, L.; Sanchez-Carballo, P.; Li, M.; Otto, M.; Holst, O.; Peschel, A. Glycosylation of wall teichoic acid in *Staphylococcus aureus* by TarM. *J. Biol. Chem.* **2010**, *285*, 13405–13415. [CrossRef] [PubMed]
151. Winstel, V.; Xia, G.; Peschel, A. Pathways and roles of wall teichoic acid glycosylation in *Staphylococcus aureus*. *Int. J. Med. Microbiol.* **2014**, *304*, 215–221. [CrossRef] [PubMed]
152. Kim, J.W.; Dutta, V.; Elhanafi, D.; Lee, S.; Osborne, J.A.; Kathariou, S. A novel restriction-modification system is responsible for temperature-dependent phage resistance in *Listeria monocytogenes* ECII. *Appl. Environ. Microbiol.* **2012**, *78*, 1995–2004. [CrossRef] [PubMed]
153. Dowah, A.S.A.; Clokie, M.R.J. Review of the nature, diversity and structure of bacteriophage receptor binding proteins that target Gram-positive bacteria. *Biophys. Rev.* **2018**. [CrossRef] [PubMed]
154. Bertozzi, S.J.; Storms, Z.; Sauvageau, D. Host receptors for bacteriophage adsorption. *FEMS Microbiol. Lett.* **2016**, *363*, fnw002. [CrossRef] [PubMed]
155. Ainsworth, S.; Sadovskaya, I.; Vinogradov, E.; Courtin, P.; Guerardel, Y.; Mahony, J.; Grard, T.; Cambillau, C.; Chapot-Chartier, M.P.; van Sinderen, D. Differences in lactococcal cell wall polysaccharide structure are major determining factors in bacteriophage sensitivity. *mBio* **2014**, *6*, e00880-14. [CrossRef] [PubMed]
156. Ardissone, S.; Fumeaux, C.; Bergé, M.; Beaussart, A.; Théraulaz, L.; Radhakrishnan, S.K.; Dufrêne, Y.F.; Viollier, P.H. Cell cycle constraints on capsulation and bacteriophage susceptibility. *Elife* **2014**, *3*. [CrossRef] [PubMed]
157. Dy, R.L.; Richter, C.; Salmond, G.P.; Fineran, P.C. Remarkable Mechanisms in Microbes to Resist Phage Infections. *Annu. Rev. Virol.* **2014**, *1*, 307–331. [CrossRef] [PubMed]
158. Seed, K.D. Battling Phages: How Bacteria Defend against Viral Attack. *PloS Pathog.* **2015**, *11*, e1004847. [CrossRef] [PubMed]
159. Zschach, H.; Larsen, M.V.; Hasman, H.; Westh, H.; Nielsen, M.; Międzybrodzki, R.; Jończyk-Matysiak, E.; Weber-Dąbrowska, B.; Górski, A. Use of a Regression Model to Study Host-Genomic Determinants of Phage Susceptibility in MRSA. *Antibiotics* **2018**, *7*, E9. [CrossRef] [PubMed]
160. Shabbir, M.A.; Hao, H.; Shabbir, M.Z.; Wu, Q.; Sattar, A.; Yuan, Z. Bacteria vs. Bacteriophages: Parallel Evolution of Immune Arsenals. *Front. Microbiol.* **2016**, *7*, 1292. [CrossRef] [PubMed]
161. Goldfarb, T.; Sberro, H.; Weinstock, E.; Cohen, O.; Doron, S.; Charpak-Amikam, Y.; Afik, S.; Ofir, G.; Sorek, R. BREX is a novel phage resistance system widespread in microbial genomes. *EMBO J.* **2015**, *34*, 169–183. [CrossRef] [PubMed]
162. Chopin, M.C.; Chopin, A.; Bidnenko, E. Phage abortive infection in lactococci: Variations on a theme. *Curr. Opin. Microbiol.* **2005**, *8*, 473–479. [CrossRef] [PubMed]
163. Labrie, S.J.; Samson, J.E.; Moineau, S. Bacteriophage resistance mechanisms. *Nat. Rev. Microbiol.* **2010**, *8*, 317–327. [CrossRef] [PubMed]
164. Li, X.; Gerlach, D.; Du, X.; Larsen, J.; Stegger, M.; Kühner, P.; Peschel, A.; Xia, G.; Winstel, V. An accessory wall teichoic acid glycosyltransferase protects *Staphylococcus aureus* from the lytic activity of *Podoviridae*. *Sci. Rep.* **2015**, *5*, 17219. [CrossRef] [PubMed]
165. Tzipilevich, E.; Habusha, M.; Ben-Yehuda, S. Acquisition of Phage Sensitivity by Bacteria through Exchange of Phage Receptors. *Cell* **2017**, *168*, 186.e2–199.e2. [CrossRef] [PubMed]
166. Deng, K.; Fang, W.; Zheng, B.; Miao, S.; Huo, G. Phenotypic, fermentation characterization, and resistance mechanism analysis of bacteriophage-resistant mutants of *Lactobacillus delbrueckiissp. bulgaricus* isolated from traditional Chinese dairy products. *J. Dairy Sci.* **2018**, *101*, 1901–1914. [CrossRef] [PubMed]
167. Zago, M.; Orrù, L.; Rossetti, L.; Lamontanara, A.; Fornasari, M.E.; Bonvini, B.; Meucci, A.; Carminati, D.; Cattivelli, L.; Giraffa, G. Survey on the phage resistance mechanisms displayed by a dairy *Lactobacillus helveticus* strain. *Food Microbiol.* **2017**, *66*, 110–116. [CrossRef] [PubMed]
168. Suárez, V.B.; Maciel, N.; Guglielmotti, D.; Zago, M.; Giraffa, G.; Reinheimer, J. Phage-resistance linked to cell heterogeneity in the commercial strain *Lactobacillus delbrueckii* subsp. *lactis* Ab1. *Int. J. Food Microbiol.* **2008**, *128*, 401–405. [CrossRef]

169. Koczula, A.; Willenborg, J.; Bertram, R.; Takamatsu, D.; Valentin-Weigand, P.; Goethe, R. Establishment of a Cre recombinase based mutagenesis protocol for markerless gene deletion in *Streptococcus suis*. *J. Microbiol. Methods* **2014**, *107*, 80–83. [CrossRef] [PubMed]
170. Zaleski, P.; Wojciechowski, M.; Piekarowicz, A. The role of Dam methylation in phase variation of *Haemophilus influenzae* genes involved in defence against phage infection. *Microbiology* **2005**, *151*, 3361–3369. [CrossRef] [PubMed]
171. Styriak, I.; Pristas, P.; Javorský, P. Lack of surface receptors not restriction-modification system determines F4 phage resistance in *Streptococcus bovis* II/1. *Folia Microbiol.* **1998**, *43*, 35–38. [CrossRef]
172. Sanders, M.E. Phage resistance in lactic acid bacteria. *Biochimie* **1988**, *70*, 411–422. [CrossRef]
173. Faruque, S.M.; Bin Naser, I.; Fujihara, K.; Diraphat, P.; Chowdhury, N.; Kamruzzaman, M.; Qadri, F.; Yamasaki, S.; Ghosh, A.N.; Mekalanos, J.J. Genomic sequence and receptor for the *Vibrio cholerae* phage KSF-1phi: Evolutionary divergence among filamentous vibriophages mediating lateral gene transfer. *J. Bacteriol.* **2005**, *187*, 4095–4103. [CrossRef] [PubMed]
174. Smarda, J.; Doroszkiewicz, W.; Lachowicz, T.M. Sensitivity of *Shigella flexneri* and *Escherichia coli* bacteria to bacteriophages and to colicins, lost or established by the acquisition of R plasmids. *Acta Microbiol. Pol.* **1990**, *39*, 23–35. [PubMed]
175. Ram, G.; Chen, J.; Ross, H.F.; Novick, R.P. Precisely modulated pathogenicity island interference with late phage gene transcription. *Proc. Natl. Acad. Sci. USA* **2014**, *111*, 14536–14541. [CrossRef] [PubMed]
176. Hofer, B.; Ruge, M.; Dreiseikelmann, B. The superinfection exclusion gene (sieA) of bacteriophage P22: Identification and overexpression of the gene and localization of the gene product. *J. Bacteriol.* **1995**, *177*, 3080–3086. [CrossRef] [PubMed]
177. Dempsey, R.M.; Carroll, D.; Kong, H.; Higgins, L.; Keane, C.T.; Coleman, D.C. Sau42I, a BcgI-like restriction-modification system encoded by the *Staphylococcus aureus* quadruple-converting phage Phi42. *Microbiology* **2005**, *151*, 1301–1311. [CrossRef] [PubMed]
178. Coleman, D.C.; Sullivan, D.J.; Russell, R.J.; Arbuthnott, J.P.; Carey, B.F.; Pomeroy, H.M. *Staphylococcus aureus* bacteriophages mediating the simultaneous lysogenic conversion of beta-lysin, staphylokinase and enterotoxin A: Molecular mechanism of triple conversion. *J. Gen. Microbiol.* **1989**, *135*, 1679–1697. [PubMed]
179. Makarova, K.S.; Wolf, Y.I.; Snir, S.; Koonin, E.V. Defense islands in bacterial and archaeal genomes and prediction of novel defense systems. *J. Bacteriol.* **2011**, *193*, 6039–6056. [CrossRef] [PubMed]
180. O'Driscoll, J.; Glynn, F.; Cahalane, O.; O'Connell-Motherway, M.; Fitzgerald, G.F.; Van Sinderen, D. Lactococcal Plasmid pNP40 Encodes a Novel, Temperature-Sensitive Restriction-Modification System. *Appl. Environ. Microbiol.* **2004**, *70*, 5546–5556. [CrossRef] [PubMed]
181. Trotter, M.; Ross, R.P.; Fitzgerald, G.F.; Coffey, A. Lactococcus lactis DPC5598, a plasmid-free derivative of a commercial starter, provides a valuable alternative host for culture improvement studies. *J. Appl. Microbiol.* **2002**, *93*, 134–143. [CrossRef] [PubMed]
182. Boucher, I.; Emond, E.; Parrot, M.; Moineau, S. DNA sequence analysis of three *Lactococcus lactis* plasmids encoding phage resistance mechanisms. *J. Dairy Sci.* **2001**, *84*, 1610–1620. [CrossRef]
183. Burrus, V.; Bontemps, C.; Decaris, B.; Guédon, G. Characterization of a novel type II restriction-modification system, Sth368I, encoded by the integrative element ICESt1 of *Streptococcus thermophilus* CNRZ368. *Appl. Environ. Microbiol.* **2001**, *67*, 1522–1528. [CrossRef] [PubMed]
184. Forde, A.; Daly, C.; Fitzgerald, G.F. Identification of four phage resistance plasmids from *Lactococcus lactis* subsp. *cremoris* HO_2. *Appl. Environ. Microbiol.* **1999**, *65*, 1540–1547. [PubMed]
185. Mohammed, M.; Cormican, M. Whole genome sequencing provides possible explanations for the difference in phage susceptibility among two Salmonella Typhimurium phage types (DT8 and DT30) associated with a single foodborne outbreak. *BMC* **2015**, *8*, 728. [CrossRef] [PubMed]
186. Deng, Y.M.; Harvey, M.L.; Liu, C.Q.; Dunn, N.W. A novel plasmid-encoded phage abortive infection system from *Lactococcus lactis* biovar. diacetylactis. *FEMS Microbiol Lett.* **1997**, *146*, 149–154. [CrossRef] [PubMed]
187. Millen, A.M.; Horvath, P.; Boyaval, P.; Romero, D.A. Mobile CRISPR/Cas-mediated bacteriophage resistance in *Lactococcus lactis*. *PLoS ONE* **2012**, *12*, e51663. [CrossRef] [PubMed]
188. Tock, M.R.; Dryden, D.T. The biology of restriction and anti-restriction. *Curr. Opin. Microbiol.* **2005**, *8*, 466–472. [CrossRef] [PubMed]
189. Ofir, G.; Sorek, R. Contemporary Phage Biology: From Classic Models to New Insights. *Cell* **2018**, *172*, 1260–1270. [CrossRef] [PubMed]

190. Samson, J.E.; Magadán, A.H.; Sabri, M.; Moineau, S. Revenge of the phages: Defeating bacterial defences. *Nat. Rev. Microbiol.* **2013**, *11*, 675–687. [CrossRef] [PubMed]
191. Suárez, V.; Zago, M.; Giraffa, G.; Reinheimer, J.; Quiberoni, A. Evidence for the presence of restriction/modification systems in *Lactobacillus delbrueckii*. *J. Dairy Res.* **2009**, *76*, 433–440. [CrossRef] [PubMed]
192. Akatov, A.K.; Zueva, V.S.; Dmitrenko, O.A. A new approach to establishing the set of phages for typing methicillin-resistant *Staphylococcus aureus*. *J. Chemother.* **1991**, *5*, 275–278. [CrossRef]
193. Waldron, D.E.; Lindsay, J.A. Sau1: A novel lineage-specific type I restriction-modification system that blocks horizontal gene transfer into *Staphylococcus aureus* and between *S. aureus* isolates of different lineages. *aureus isolates of different lineages. J. Bacteriol.* **2006**, *188*, 5578–5585. [CrossRef] [PubMed]
194. Corvaglia, A.R.; François, P.; Hernandez, D.; Perron, K.; Linder, P.; Schrenzel, J. A type III-like restriction endonuclease functions as a major barrier to horizontal gene transfer in clinical *Staphylococcus aureus* strains. *Proc. Natl. Acad. Sci. USA* **2010**, *107*, 11954–11958. [CrossRef] [PubMed]
195. Cady, K.C.; Bondy-Denomy, J.; Heussler, G.E.; Davidson, A.R.; O'Toole, G.A. The CRISPR/Cas adaptive immune system of *Pseudomonas aeruginosa* mediates resistance to naturally occurring and engineered phages. *J. Bacteriol.* **2012**, *194*, 5728–5738. [CrossRef] [PubMed]
196. Fineran, P.C.; Blower, T.R.; Foulds, I.J.; Humphreys, D.P.; Lilley, K.S.; Salmond, G.P. The phage abortive infection system, ToxIN, functions as a protein-RNA toxin-antitoxin pair. *Proc. Natl. Acad. Sci. USA* **2009**, *106*, 894–899. [CrossRef] [PubMed]
197. Haaber, J.; Moineau, S.; Fortier, L.C.; Hammer, K. AbiV, a novel antiphage abortive infection mechanism on the chromosome of *Lactococcus lactis* subsp. *cremoris* MG1363. *Appl. Environ. Microbiol.* **2008**, *74*, 6528–6537. [CrossRef] [PubMed]
198. Durmaz, E.; Klaenhammer, T.R. Genetic analysis of chromosomal regions of *Lactococcus lactis* acquired by recombinant lytic phages. *Appl. Environ. Microbiol.* **2000**, *66*, 895–903. [CrossRef] [PubMed]
199. Adams, B.L. The Next Generation of Synthetic Biology Chassis: Moving Synthetic Biology from the Laboratory to the Field. *ACS Synth. Biol.* **2016**, *12*, 1328–1330. [CrossRef] [PubMed]
200. Szathmary, E. Life—In search of the simplest cell. *Nature* **2005**, *433*, 469–470. [CrossRef] [PubMed]
201. Umenhoffer, K.; Draskovits, G.; Nyerges, Á.; Karcagi, I.; Bogos, B.; Tímár, E.; Csörgő, B.; Herczeg, R.; Nagy, I.; Fehér, T.; et al. Genome-Wide Abolishment of Mobile Genetic Elements Using Genome Shuffling and CRISPR/Cas-Assisted MAGE Allows the Efficient Stabilization of a Bacterial Chassis. *ACS Synth. Biol.* **2017**, *8*, 1471–1483. [CrossRef] [PubMed]
202. Lauritsen, I.; Porse, A.; Sommer, M.O.A.; Nørholm, M.H.H. A versatile one-step CRISPR-Cas9 based approach to plasmid-curing. *Microb. Cell Fact.* **2017**, *16*, 135. [CrossRef] [PubMed]
203. Ellis, H.M.; Yu, D.; DiTizio, T.; Court, D.L. High efficiency mutagenesis, repair, and engineering of chromosomal DNA using single-stranded oligonucleotides. *Proc. Natl. Acad. Sci. USA* **2001**, *98*, 6742–6746. [CrossRef] [PubMed]
204. Martínez-García, E.; de Lorenzo, V. Molecular tools and emerging strategies for deep genetic/genomic refactoring of Pseudomonas. *Curr. Opin. Biotechnol.* **2017**, *47*, 120–132. [CrossRef] [PubMed]
205. Wang, H.H.; Isaacs, F.J.; Carr, P.A.; Sun, Z.Z.; Xu, G.; Forest, C.R.; Church, G.M. Programming cells by multiplex genome engineering and accelerated evolution. *Nature* **2009**, *7257*, 894–898. [CrossRef] [PubMed]
206. Zheng, X.; Xing, X.H.; Zhang, C. Targeted mutagenesis: A sniper-like diversity generator in microbial engineering. *Synth. Syst. Biotechnol.* **2017**, *2*, 75–86. [CrossRef] [PubMed]
207. Yu, D.; Ellis, H.M.; Lee, E.C.; Jenkins, N.A.; Copeland, N.G.; Court, D.L. An efficient recombination system for chromosome engineering in *Escherichia coli*. *Proc. Natl. Acad. Sci. USA* **2000**, *97*, 5978–5983. [CrossRef] [PubMed]
208. Zerbini, F.; Zanella, I.; Fraccascia, D.; König, E.; Irene, C.; Frattini, L.F.; Tomasi, M.; Fantappiè, L.; Ganfini, L.; Caproni, E.; et al. Large scale validation of an efficient CRISPR/Cas-based multi gene editing protocol in *Escherichia coli*. *Microb. Cell Fact.* **2017**, *16*, 68. [CrossRef] [PubMed]
209. Liu, Q.; Jiang, Y.; Shao, L.; Yang, P.; Sun, B.; Yang, S.; Chen, D. CRISPR/Cas9-based efficient genome editing in *Staphylococcus aureus*. *Acta Biochim. Biophys. Sin.* **2017**, *49*, 764–770. [CrossRef] [PubMed]
210. Ricaurte, D.E.; Martínez-García, E.; Nyerges, Á.; Pál, C.; de Lorenzo, V.; Aparicio, T. A standardized workflow for surveying recombinases expands bacterial genome-editing capabilities. *Microb. Biotechnol.* **2018**, *11*, 176–188. [CrossRef] [PubMed]

211. Aparicio, T.; Jensen, S.I.; Nielsen, A.T.; de Lorenzo, V.; Martinez-Garcia, E. The Ssr protein (T1E_1405) from *Pseudomonas putida* DOT-T1E enables oligonucleotide-based recombineering in platform strain *P. putida* EM42. *Biotechnol. J.* **2016**, *11*, 1309–1319. [CrossRef] [PubMed]
212. Chen, J.; Ram, G.; Yoong, P.; Penadés, J.R.; Shopsin, B.; Novick, R.P. An rpsL-based allelic exchange vector for *Staphylococcus aureus*. *Plasmid* **2015**, *79*, 8–14. [CrossRef] [PubMed]
213. Prax, M.; Lee, C.Y.; Bertram, R. An update on the molecular genetics toolbox for staphylococci. *Microbiology* **2013**, *159*, 421–435. [CrossRef] [PubMed]
214. Li, Y.; Tian, P. Contemplating 3-Hydroxypropionic Acid Biosynthesis in *Klebsiella pneumoniae*. *Indian J. Microbiol.* **2015**, *55*, 131–139. [CrossRef] [PubMed]
215. Penewit, K.; Holmes, E.A.; McLean, K.; Ren, M.; Waalkes, A.; Salipante, S.J. Efficient and Scalable Precision Genome Editing in Staphylococcus aureus through Conditional Recombineering and CRISPR/Cas9-Mediated Counterselection. *mBio* **2010**, *9*, e00067-18. [CrossRef] [PubMed]
216. Tucker, A.T.; Nowicki, E.M.; Boll, J.M.; Knauf, G.A.; Burdis, N.C.; Trent, M.S.; Davies, B.W. Defining gene-phenotype relationships in *Acinetobacter baumannii* through one-step chromosomal gene inactivation. *mBio* **2014**, *5*, e01313-14. [CrossRef] [PubMed]
217. Dalia, T.N.; Yoon, S.H.; Galli, E.; Barre, F.X.; Waters, C.M.; Dalia, A.B. Enhancing multiplex genome editing by natural transformation (MuGENT) via inactivation of ssDNA exonucleases. *Nucleic Acids Res.* **2017**, *45*, 7527–7537. [CrossRef] [PubMed]
218. Prathapam, R.; Uehara, T. A temperature-sensitive replicon enables efficient gene inactivation in Pseudomonas aeruginosa. *J. Microbiol. Methods* **2018**, *144*, 47–52. [CrossRef] [PubMed]
219. Pósfai, G.; Plunkett, G., 3rd; Fehér, T.; Frisch, D.; Keil, G.M.; Umenhoffer, K.; Kolisnychenko, V.; Stahl, B.; Sharma, S.S.; de Arruda, M.; et al. Emergent properties of reduced-genome *Escherichia coli*. *Science* **2006**, *5776*, 1044–1046. [CrossRef] [PubMed]
220. Martinez-Garcia, E.; Nikel, P.I.; Aparicio, T.; de Lorenzo, V. *Pseudomonas* 2.0: Genetic upgrading of *P. putida* KT2440 as an enhanced host for heterologous gene expression. *Microb. Cell Fact.* **2014**, *13*, 159. [CrossRef] [PubMed]
221. Martínez-García, E.; Jatsenko, T.; Kivisaar, M.; de Lorenzo, V. Freeing Pseudomonas putida KT2440 of its proviral load strengthens endurance to environmental stresses. *Environ. Microbiol.* **2015**, *17*, 76–90. [CrossRef] [PubMed]
222. Baumgart, M.; Unthan, S.; Rückert, C.; Sivalingam, J.; Grünberger, A.; Kalinowski, J.; Bott, M.; Noack, S.; Frunzke, J. Construction of a prophage-free variant of *Corynebacterium glutamicum* ATCC 13032 for use as a platform strain for basic research and industrial biotechnology. *Appl. Environ. Microbiol.* **2013**, *79*, 6006–6015. [CrossRef] [PubMed]
223. Łobocka, M.; Gozdek, A.; Gozdek, A.; Izak, D.; Zalewska, A.; Gawor, J.; Dąbrowski, K.; Gromadka, R.; Weber-Dąbrowska, B.; Górski, A. Enterococcus faecalis strains for the production of Bacteriophage Preparations. PCT Patent Application WO 2016/030872 A1, 16 March 2016.
224. Balogh, B.; Jones, J.B.; Iriarte, F.B.; Momol, M.T. Phage therapy for plant disease control. *Curr. Pharm. Biotechnol.* **2010**, *11*, 48–57. [CrossRef] [PubMed]
225. Gill, J.J.; Hyman, P. Phage choice, isolation, and preparation for phage therapy. *Curr. Pharm. Biotechnol.* **2010**, *11*, 2–14. [CrossRef] [PubMed]
226. Hatfull, G.F. The secret lives of mycobacteriophages. *Adv. Virus Res.* **2012**, *82*, 179–288. [PubMed]
227. Carlton, R.M.; Noordman, W.H.; Biswas, B.; de Meester, E.D.; Loessner, M.J. Bacteriophage P100 for control of Listeria monocytogenes in foods: Genome sequence, bioinformatic analyses, oral toxicity study, and application. *Regul. Toxicol. Pharm.* **2005**, *43*, 301–312. [CrossRef] [PubMed]
228. Lehman, S.M.; Kropinski, A.M.; Castle, A.J.; Svircev, A.M. Complete genome of the broad-host-range Erwinia amylovora phage phiEa21-4 and its relationship to Salmonella phage Félix O1. *Appl. Environ. Microbiol.* **2009**, *75*, 2139–2147. [CrossRef] [PubMed]
229. Santos, S.B.; Fernandes, E.; Carvalho, C.M.; Sillankorva, S.; Krylov, V.N.; Pleteneva, E.A.; Shaburova, O.V.; Nicolau, A.; Ferreira, E.C.; Azeredo, J. Selection and characterization of a multivalent Salmonella phage and its production in a nonpathogenic *Escherichia coli* strain. *Appl. Environ. Microbiol.* **2010**, *76*, 7338–7342. [CrossRef] [PubMed]

230. El Haddad, L.; Ben Abdallah, N.; Plante, P.L.; Dumaresq, J.; Katsarava, R.; Labrie, S.; Corbeil, J.; St-Gelais, D.; Moineau, S. Improving the safety of *Staphylococcus aureus* polyvalent phages by their production on a *Staphylococcus xylosus* strain. *PLoS ONE* **2014**, *9*, e102600. [CrossRef] [PubMed]
231. González-Menéndez, E.; Arroyo-López, F.N.; Martínez, B.; García, P.; Garrido-Fernández, A.; Rodríguez, A. Optimizing Propagation of Staphylococcus aureus Infecting Bacteriophage vB_SauM-philPLA-RODI on Staphylococcus xylosus Using Response Surface Methodology. *Viruses* **2018**, *10*, E153. [CrossRef] [PubMed]
232. Karli, A.; Sensoy, G.; Unal, N.; Yanik, K.; Cigdem, H.; Belet, N.; Sofuoglu, A. Ventriculoperitoneal shunt infection with *Listeria innocua*. *Pediatr. Int.* **2014**, *56*, 621–623. [CrossRef] [PubMed]
233. Moreno, L.Z.; Paixão, R.; Gobbi, D.D.; Raimundo, D.C.; Ferreira, T.P.; Hofer, E.; Matte, M.H.; Moreno, A.M. Characterization of atypical *Listeria innocua* isolated from swine slaughterhouses and meat markets. *Res. Microbiol.* **2012**, *163*, 268–271. [CrossRef] [PubMed]
234. Reyrat, J.M.; Kahn, D. *Mycobacterium smegmatis*: An absurd model for tuberculosis? *Trends Microbiol.* **2001**, *10*, 472–474. [CrossRef]
235. Büyükcam, A.; Tuncer, Ö.; Gür, D.; Sancak, B.; Ceyhan, M.; Cengiz, A.B.; Kara, A. Clinical and microbiological characteristics of *Pantoea agglomerans* infection in children. *J. Infect. Public Health* **2018**, *11*, 304–309. [CrossRef] [PubMed]
236. Dutkiewicz, J.; Mackiewicz, B.; Kinga Lemieszek, M.; Golec, M.; Milanowski, J. *Pantoea agglomerans*: A mysterious bacterium of evil and good. Part III. Deleterious effects: Infections of humans, animals and plants. *Ann. Agric. Environ. Med.* **2016**, *23*, 197–205. [CrossRef] [PubMed]
237. Kaur, G.; Arora, A.; Sathyabama, S.; Mubin, N.; Verma, S.; Mayilraj, S.; Agrewala, J.N. Genome sequencing, assembly, annotation and analysis of *Staphylococcus xylosus* strain DMB3-Bh1 reveals genes responsible for pathogenicity. *Gut Pathog.* **2016**, *8*, 55. [CrossRef] [PubMed]
238. Thornton, V.B.; Davis, J.A.; St Clair, M.B.; Cole, M.N. Inoculation of *Staphylococcus xylosus* in SJL/J mice to determine pathogenicity. *Contemp. Top. Lab. Anim. Sci.* **2003**, *42*, 49–52. [PubMed]
239. Almeida, R.A.; Oliver, S.P. Interaction of coagulase-negative *Staphylococcus* species with bovine mammary epithelial cells. *Microb. Pathog.* **2001**, *31*, 205–212. [CrossRef] [PubMed]
240. Clayton, E.M.; Daly, K.M.; Guinane, C.M.; Hill, C.; Cotter, P.D.; Ross, P.R. Atypical Listeria innocua strains possess an intact LIPI-3. *BMC Microbiol.* **2014**, *14*, 58. [CrossRef] [PubMed]
241. Moreno, L.Z.; Paixão, R.; de Gobbi, D.D.; Raimundo, D.C.; Porfida Ferreira, T.S.; Micke Moreno, A.; Hofer, E.; dos Reis, C.M.F.; Matté, G.R.; Matté, M.H. Phenotypic and genotypic characterization of atypical *Listeria monocytogenes* and *Listeria innocua* isolated from swine slaughterhouses and meat markets. *Biomed. Res. Int.* **2014**, *2014*, 742032. [CrossRef] [PubMed]
242. Coros, A.; DeConno, E.; Derbyshire, K.M. IS6110, a *Mycobacterium tuberculosis* complex-specific insertion sequence, is also present in the genome of *Mycobacterium smegmatis*, suggestive of lateral gene transfer among mycobacterial species. *J. Bacteriol.* **2008**, *190*, 3408–3410. [CrossRef] [PubMed]
243. Derbyshire, K.M.; Gray, T.A. Distributive Conjugal Transfer: New Insights into Horizontal Gene Transfer and Genetic Exchange in Mycobacteria. *Microbiol. Spectr.* **2014**, *2*. [CrossRef] [PubMed]
244. Naum, M.; Brown, E.W.; Mason-Gamer, R.J. Phylogenetic evidence for extensive horizontal gene transfer of type III secretion system genes among enterobacterial plant pathogens. *Microbiology* **2009**, *155*, 3187–3199. [CrossRef] [PubMed]
245. Kirzinger, M.W.; Butz, C.J.; Stavrinides, J. Inheritance of Pantoea type III secretion systems through both vertical and horizontal transfer. *Mol. Genet. Genom.* **2015**, *290*, 2075–2088. [CrossRef] [PubMed]
246. Tormo, M.A.; Knecht, E.; Götz, F.; Lasa, I.; Penadés, J.R. Bap-dependent biofilm formation by pathogenic species of *Staphylococcus*: Evidence of horizontal gene transfer? *Microbiology* **2005**, *151*, 2465–2475. [CrossRef] [PubMed]
247. Näsvall, J.; Knöppel, A.; Andersson, D.I. Duplication-Insertion Recombineering: A fast and scar-free method for efficient transfer of multiple mutations in bacteria. *Nucleic Acids Res.* **2017**, *45*, e33. [CrossRef] [PubMed]
248. Becker, J.; Schäfer, R.; Kohlstedt, M.; Harder, B.J.; Borchert, N.S.; Stöveken, N.; Bremer, E.; Wittmann, C. Systems metabolic engineering of *Corynebacterium glutamicum* for production of the chemical chaperone ectoine. *Microb. Cell Fact.* **2013**, *12*, 110. [CrossRef] [PubMed]
249. Aubert, D.F.; Hamad, M.A.; Valvano, M.A. A markerless deletion method for genetic manipulation of *Burkholderia cenocepacia* and other multidrug-resistant gram-negative bacteria. *Methods Mol. Biol.* **2014**, *1197*, 311–327. [CrossRef] [PubMed]

250. Unthan, S.; Baumgart, M.; Radek, A.; Herbst, M.; Siebert, D.; Brühl, N.; Bartsch, A.; Bott, M.; Wiechert, W.; Marin, K.; et al. Chassis organism from *Corynebacterium glutamicum*—A top-down approach to identify and delete irrelevant gene clusters. *Biotechnol. J.* **2015**, *10*, 290–301. [CrossRef] [PubMed]
251. Zhou, L.; Tian, K.M.; Niu, D.D.; Shen, W.; Shi, G.Y.; Singh, S.; Wang, Z.X. Improvement of D-lactate productivity in recombinant *Escherichia coli* by coupling production with growth. *Biotechnol. Lett.* **2012**, *34*, 1123–1130. [CrossRef] [PubMed]
252. Liu, L.; Liu, Y.; Shin, H.D.; Chen, R.R.; Wang, N.S.; Li, J.; Du, G.; Chen, J. Developing *Bacillus spp.* as a cell factory for production of microbial enzymes and industrially important biochemicals in the context of systems and synthetic biology. *Appl. Microbiol. Biotechnol.* **2013**, *97*, 6113–6127. [CrossRef] [PubMed]
253. Zhang, W.; Gao, W.; Feng, J.; Zhang, C.; He, Y.; Cao, M.; Li, Q.; Sun, Y.; Yang, C.; Song, C.; et al. A markerless gene replacement method for *B. amyloliquefaciens* LL3 and its use in genome reduction and improvement of poly-γ-glutamic acid production. *Appl. Microbiol. Biotechnol.* **2014**, *98*, 8963–8973. [CrossRef] [PubMed]
254. Leprince, A.; van Passel, M.W.J.; Dos Santos, V.A.P.M. Streamlining genomes: Toward the generation of simplified and stabilized microbial systems. *Curr. Opin. Biotechnol.* **2012**, *23*, 651–658. [CrossRef] [PubMed]
255. Lieder, S.; Nikel, P.I.; de Lorenzo, V.; Takors, R. Genome reduction boosts heterologous gene expression in *Pseudomonas putida*. *Microb. Cell Fact.* **2015**, *14*, 23. [CrossRef] [PubMed]
256. Sabri, S.; Steen, J.A.; Bongers, M.; Nielsen, L.K.; Vickers, C.E. Knock-in/Knock-out (KIKO) vectors for rapid integration of large DNA sequences, including whole metabolic pathways, onto the *Escherichia coli* chromosome at well-characterised loci. *Microb. Cell Fact.* **2013**, *12*, 60. [CrossRef] [PubMed]
257. Choi, K.H.; Schweizer, H.P. An improved method for rapid generation of unmarked *Pseudomonas aeruginosa* deletion mutants. *BMC Microbiol.* **2005**, *5*, 30. [CrossRef] [PubMed]
258. Zhang, S.; Zou, Z.; Kreth, J.; Merritt, J. Recombineering in *Streptococcus* mutans Using Direct Repeat-Mediated Cloning-Independent Markerless Mutagenesis (DR-CIMM). *Front. Cell. Infect. Microbiol.* **2017**, *7*, 202. [CrossRef] [PubMed]
259. Bauer, R.; Mauerer, S.; Grempels, A.; Spellerberg, B. The competence system of *Streptococcus anginosus* and its use for genetic engineering. *Mol. Oral Microbiol.* **2018**, *33*, 194–202. [CrossRef] [PubMed]
260. Yan, M.Y.; Yan, H.Q.; Ren, G.X.; Zhao, J.P.; Guo, X.P.; Sun, Y.C. CRISPR-Cas12a-Assisted Recombineering in Bacteria. *Appl. Environ. Microbiol.* **2017**, *83*, e00947-17. [CrossRef] [PubMed]
261. Kato, F.; Sugai, M. A simple method of markerless gene deletion in *Staphylococcus aureus*. *J. Microbiol. Methods* **2011**, *87*, 76–81. [CrossRef] [PubMed]
262. Oh, M.H.; Lee, J.C.; Kim, J.; Choi, C.H.; Han, K. Simple Method for Markerless Gene Deletion in Multidrug-Resistant *Acinetobacter baumannii*. *Appl. Environ. Microbiol.* **2015**, *81*, 3357–3368. [CrossRef] [PubMed]
263. Junges, R.; Khan, R.; Tovpeko, Y.; Åmdal, H.A.; Petersen, F.C.; Morrison, D.A. Markerless Genome Editing in Competent *Streptococci*. *Methods Mol. Biol.* **2017**, *1537*, 233–247. [PubMed]
264. Van Dam, V.; Bos, M.P. Generating knock-out and complementation strains of *Neisseria meningitidis*. *Methods Mol. Biol.* **2012**, *799*, 55–72. [CrossRef] [PubMed]
265. Trebosc, V.; Gartenmann, S.; Royet, K.; Manfredi, P.; Tötzl, M.; Schellhorn, B.; Pieren, M.; Tigges, M.; Lociuro, S.; Sennhenn, P.C.; et al. A Novel Genome-Editing Platform for Drug-Resistant *Acinetobacter baumannii* Reveals an AdeR-Unrelated Tigecycline Resistance Mechanism. *Antimicrob. Agents Chemother.* **2016**, *60*, 7263–7271. [PubMed]
266. White, A.P.; Allen-Vercoe, E.; Jones, B.W.; DeVinney, R.; Kay, W.W.; Surette, M.G. An efficient system for markerless gene replacement applicable in a wide variety of enterobacterial species. *Can. J. Microbiol.* **2007**, *53*, 56–62. [CrossRef] [PubMed]
267. Geng, S.; Tian, Q.; An, S.; Pan, Z.; Chen, X.; Jiao, X. High-Efficiency, Two-Step Scarless-Markerless Genome Genetic Modification in *Salmonella enterica*. *Curr. Microbiol.* **2016**, *72*, 700–706. [CrossRef] [PubMed]
268. Pyne, M.E.; Bruder, M.R.; Moo-Young, M.; Chung, D.A.; Chou, C.P. Harnessing heterologous and endogenous CRISPR-Cas machineries for efficient markerless genome editing in *Clostridium*. *Sci. Rep.* **2016**, *6*, 25666. [CrossRef] [PubMed]
269. Plaut, R.D.; Stibitz, S. Improvements to a Markerless Allelic Exchange System for *Bacillus anthracis*. *PLoS ONE* **2015**, *10*, e0142758. [CrossRef] [PubMed]
270. Hossain, M.J.; Thurlow, C.M.; Sun, D.; Nasrin, S.; Liles, M.R. Genome modifications and cloning using a conjugally transferable recombineering system. *Biotechnol. Rep.* **2015**, *8*, 24–35. [CrossRef] [PubMed]

271. Zhang, L.; Li, Y.; Dai, K.; Wen, X.; Wu, R.; Huang, X.; Jin, J.; Xu, K.; Yan, Q.; Huang, Y.; et al. Establishment of a Successive Markerless Mutation System in *Haemophilus parasuis* through Natural Transformation. *PLoS ONE* **2015**, *10*, e0127393. [CrossRef] [PubMed]
272. Gómez, E.; Álvarez, B.; Duchaud, E.; Guijarro, J.A. Development of a markerless deletion system for the fish-pathogenic bacterium *Flavobacterium psychrophilum*. *PLoS ONE* **2015**, *10*, e0117969. [CrossRef] [PubMed]
273. Sun, X.; Yang, D.; Wang, Y.; Geng, H.; He, X.; Liu, H. Development of a markerless gene deletion system for *Streptococcus zooepidemicus*: Functional characterization of hyaluronan synthase gene. *Appl. Microbiol. Biotechnol.* **2013**, *97*, 8629–8636. [CrossRef] [PubMed]
274. Horzempa, J.; Shanks, R.M.; Brown, M.J.; Russo, B.C.; O'Dee, D.M.; Nau, G.J. Utilization of an unstable plasmid and the I-SceI endonuclease to generate routine markerless deletion mutants in Francisella tularensis. *J. Microbiol. Methods* **2010**, *80*, 106–108. [CrossRef] [PubMed]
275. Sun, W.; Wang, S.; Curtiss, R. Highly efficient method for introducing successive multiple scarless gene deletions and markerless gene insertions into the Yersinia pestis chromosome. *Appl. Environ. Microbiol.* **2008**, *74*, 4241–4245. [CrossRef] [PubMed]
276. Kristich, C.J.; Chandler, J.R.; Dunny, G.M. Development of a host-genotype-independent counterselectable marker and a high-frequency conjugative delivery system and their use in genetic analysis of *Enterococcus faecalis*. *Plasmid* **2007**, *57*, 131–144. [CrossRef] [PubMed]
277. Górski, A.; Ważna, E.; Weber-Dąbrowska, B.; Dąbrowska, K.; Świtała-Jeleń, K.; Międzybrodzki, R. Bacteriophage translocation. *FEMS Immunol. Med. Microbiol.* **2006**, *46*, 313–319. [CrossRef] [PubMed]
278. Barr, J.J. A bacteriophage journey through the human body. *Immunol. Rev.* **2017**, *279*, 106–122. [CrossRef] [PubMed]
279. Górski, A.; Dąbrowska, K.; Międzybrodzki, R.; Weber-Dąbrowska, B.; Łusiak-Szelachowska, M.; Jończyk-Matysiak, E.; Borysowski, J. Phages and immunomodulation. *Future Microbiol.* **2017**. [CrossRef]
280. Górski, A.; Jończyk-Matysiak, E.; Łusiak-Szelachowska, M.; Międzybrodzki, R.; Weber-Dąbrowska, B.; Borysowski, J. Bacteriophages targeting intestinal epithelial cells: A potential novel form of immunotherapy. *Cell. Mol. Life Sci.* **2017**. [CrossRef] [PubMed]
281. Górski, A.; Jończyk-Matysiak, E.; Łusiak-Szelachoeska, M.; Weber-Dąbrowska, B.; Międzybrodzki, R.; Borysowski, J. Therapeutic potential of phages in autoimmune liver diseases. *Clin. Exp. Immunol.* **2017**. [CrossRef] [PubMed]
282. Górski, A.; Jończyk-Matysiak, E.; Łusiak-Szelachowska, M.; Międzybrodzki, R.; Weber-Dąbrowska, B.; Borysowski, J. Phage therapy in allergic disorders? *Exp. Biol. Med.* **2018**, *1*. [CrossRef] [PubMed]
283. Górski, A.; Jończyk-Matysiak, E.; Łusiak-Szelachowska, M.; Międzybrodzki, R.; Weber-Dąbrowska, B.; Borysowski, J. The potential of phage therapy in sepsis. *Front. Immunol.* **2017**. [CrossRef] [PubMed]
284. Truog, R.D. The UK sets limits on experimental treatments. *JAMA* **2017**, *318*, 1001–1002. [CrossRef] [PubMed]
285. Górski, A.; Jończyk-Matysiak, E.; Międzybrodzki, R.; Weber-Dąbrowska, B.; Łusiak-Szelachowska, M.; Bagińska, N.; Borysowski, J.; Lobocka, M.B.; Węgrzyn, A.; Wegrzyn, G. Phage therapy: Beyond the antibacterial action. *Front. Med.* **2018**. [CrossRef]

Review

Framing the Future with Bacteriophages in Agriculture

Antonet Svircev [1,*], Dwayne Roach [2] and Alan Castle [3]

1 Agriculture and Agri-Food Canada, Vineland Station, ON L0R 2E0, Canada
2 Department of Microbiology, Pasteur Institute, 75015 Paris, France; dwayne.roach@pasteur.fr
3 Department of Biological Sciences, Brock University, St. Catharines, ON L2S 3A1, Canada; acastle@brocku.ca
* Correspondence: Antonet.Svircev@agr.gc.ca; Tel.: +905-562-2018

Received: 27 March 2018; Accepted: 22 April 2018; Published: 25 April 2018

Abstract: The ability of agriculture to continually provide food to a growing world population is of crucial importance. Bacterial diseases of plants and animals have continually reduced production since the advent of crop cultivation and animal husbandry practices. Antibiotics have been used extensively to mitigate these losses. The rise of antimicrobial resistant (AMR) bacteria, however, together with consumers' calls for antibiotic-free products, presents problems that threaten sustainable agriculture. Bacteriophages (phages) are proposed as bacterial population control alternatives to antibiotics. Their unique properties make them highly promising but challenging antimicrobials. The use of phages in agriculture also presents a number of unique challenges. This mini-review summarizes recent development and perspectives of phages used as antimicrobial agents in plant and animal agriculture at the farm level. The main pathogens and their adjoining phage therapies are discussed.

Keywords: bacteriophage; phage therapy; sustainable agriculture; zoonosis; antibiotic resistance

---

1. Introduction

The goal of sustainable agriculture is to implement practices that will attain healthy disease-free plants and animals, provide safe food for a growing global population, and minimize the impact of agricultural practices on the environment [1–3]. Conversely, agricultural practices are impacted by economic and disease pressures, consumer preferences, geographic location, weather conditions, and government regulations. Following the Second World War, antibiotics have been incorporated into animal husbandry [4,5] and for the control of plant pathogens [6–10]. Important strides in phage therapy were overshadowed by the widespread usage of antibiotics to treat diseases in humans, animal husbandry, and the control of bacterial plant pathogens. The overuse in medicine and animal husbandry has contributed to the rise of worldwide antimicrobial resistant (AMR) bacteria. Using *Erwinia amylovora* as an example, antimicrobial resistance is present in a number of geographic locations where antibiotics have been overused in apple and pear orchards [6–8,10]. The debate and scientific discussion on the impact and consequences of the presence of streptomycin resistant *E. amylovora* in orchards still continues in scientific literature [11].

Most antibiotics are non-specific, acting not only against the target pathogen, but also against other bacteria naturally present in the environment or plant and animal microflora. Drug-resistant infections result in millions of people being affected from drug-resistant bacteria each year, with an estimated 700,000 deaths worldwide each year, a number that could increase to 10 million by 2050 if the drug resistance trend continues [12]. Imprudent use of antimicrobials in agriculture may result in reduced efficacy of antibiotics due to facilitated emergence of antibiotic resistant human pathogens, increased human morbidity and mortality, increased healthcare costs, and increased potential for carriage and dissemination of pathogens. Together with consumers' calls for antibiotic-free products, popularity of organic products and the removal of antibiotics for agricultural use in certain jurisdictions have led to the search for alternatives. Use of phages, which infect and destroy bacteria, could significantly reduce the environmental impact of antibiotic use in agriculture, while potentially increasing profitability by lowering crop loss or animal mortality in early stages of the breeding process.

2. Phages in Agriculture

"What is the impact of phages on agricultural environments where sustainable agriculture is being practiced?" becomes an important and intriguing question. Phages are inherently highly specific towards bacterial hosts. This characteristic has both negative and positive aspects in that it is beneficial in terms of avoiding negative effects on the host microbiota and a hindrance when it comes to detection and elimination of the target pathogen. This mini-review will focus primarily on the progress of phage-based biocontrol in food production systems covering the past 10 years. Phage-based laboratory studies that include phage isolation, host range determination, molecular characterization, genomic and proteomics analyses are well described in the recent review articles on plant and animal-associated phage therapy [13–17]. The development of phages as antimicrobial agents in animal and plant production systems follows a similar path in the initial discovery stage however the processes become divergent in the implementation processes. In the following sections we discuss the progress made in the use of phages in plant and animal farming, focusing on the challenges and success stories reported in scientific literature.

3. Bacteriophages in Food Animal Production

By volume, the vast majority of antibiotics consumed worldwide are for veterinary purposes, predominantly in intensive and large-scale animal production systems, such as dairy, livestock, poultry, and aquaculture [18,19]. Animal husbandry practices widely use antibiotics therapeutically to treat infectious diseases, as well as non-therapeutically to prevent the spread of disease (prophylaxis) and to promote growth. Controversy, however, surrounds the widespread use of antibiotics for animal production, as their overuse and possible misuse is driving antibiotic microbial resistance. For instance, the practice of prolonged exposure to sub-therapeutic antibiotic doses, the context in which prophylactic and growth-promoting antibiotics are administered, exerts an inestimable amount of selective pressure toward the emergence of AMR [20,21]. Furthermore, AMR bacteria and AMR genes of animal origin can then be transmitted to humans through environmental contamination, food distribution, or direct contact with farm animals [22–24]. Intensive animal production systems necessitate antibiotics to keep animals healthy and maintain productivity, and with rising incomes in transitioning countries expected to boost antibiotic consumption by 67% by 2030 [25], this presents a major health risk to humans and animals.

The World Health Organization, the European Commission, the Centers for Disease Control and Prevention, and Health Canada, to name a few, all support immediate antimicrobial stewardship in animal food production, aimed primarily at reducing or eliminating the nontherapeutic use of medically important antibiotics. Eliminating prophylactic antimicrobials outright may not be feasible in intensive animal production systems due to increasing worldwide demand for protein, the potential compromise in animal welfare and health, and in human health and food safety. Phages instead of antibiotics are a promising option in food animal production to maintain animal health and limit the transfer of AMR and zoonotic pathogens that may be harmful to consumers. This section will focus only on application of phages as alternatives to antibiotic growth promoters, prophylaxis, and zoonotic pathogen animal decolonization at the farm level.

3.1. Phages as Growth Promoters

Antibiotics in subtherapeutic doses have played important roles in the promotion of growth, enhancement of feed efficiency and improvement of the quality of animal products [20]. To combat the increased rate of mortality and morbidity due to reduction of in-feed antibiotics, phages have been proposed as the replacement, particularly in the early stages when vaccination is not possible and the maintenance of the bacterial ecosystem is crucial [26]. The studies reviewed in this subsection highlight the addition of phages in feed rather than for clinical treatment. The distinction between growth promotion and prevention or treatment of diseases is subtle and further work is needed to see if phages do offer growth promotion effects other than simply reducing disease incidence.

Clostridium perfringens is a major problem for the poultry industry, resulting in both clinical and subclinical infections. A cocktail of five phages could effectively control necrotic enteritis in chicken broilers and thus improve feed conversion ratios and weight gain [27]. This efficacy was independent of whether the phages were administered in feed or in drinking water. Dietary supplementation with phages has also been shown to improve on growth performance in pigs [28]. Feed supplemented with a commercial phage product, which contained a mixture of phages targeting several pathogens, including *Salmonella* spp., *Escherichia coli*, *Staphylococcus aureus*, and *C. perfringens*, improved different aspects of grower pig's performance, such as average daily feed intake [28]. It was determined that barrow gut health improved with a higher abundance of commensal bacterial and lower pathogen load in pig faeces. However, the success of phages against pathogenic bacteria could be related to the method of its addition. Administering phages in drinking water may be disease and/or pathogen dependent. Huff et al. [29] found that phage administered in drinking water could not cure experimental *E. coli* respiratory infections in broilers. Phages were only effective in reducing respiratory bacterial load when they were administered via direct intratracheal administration [30].

For dairy herds, mastitis is the most important disease worldwide [31]. *S. aureus*, one of the etiological agents for mastitis, which has a propensity to recur chronically, causes a potentially fatal inflammatory response in gland tissues. In an experimental model, lactating mice intramammarily infected with a clinical bovine *S. aureus* strain showed significant improvement in mammary gland pathology and a 4-log reduction in bacterial load after phage treatment [32]. However, compared to the antibiotic cefalonium, the phage treatment was far less effective. Gill et al. [33] also found that multiday high-titre intramammary infusions of phage K did not lead to a reduction in *S. aureus* load in the utter of lactating cows with pre-existing subclinical mastitis [34]. In this latter study, the adsorption of milk whey proteins to the *S. aureus* cell surface inhibited phage infection in vitro, suggesting this was the cause for treatment failure [33]. It should be noted that antibiotic treatment success is also highly variable with mastitis cure rates as low as 4% [35].

Phages have the potential to be a viable and eco-friendly alternative to antibiotics in aquaculture. Aquaculture is the fastest growing food production sector, providing over fifty percent of the world's supply of fish and seafood. Antibiotics in feed are commonly used as prophylactics to decrease the corresponding heavy economic losses due to bacterial diseases worldwide. Vibriosis is one of the most prevalent diseases of marine and estuarine fish in both natural and commercial production [36–38]. *Vibrio anguillarum* is the etiologic agent of vibriosis, a fatal haemorrhagic septicaemia that affects more than 50 fresh- and salt-water fish species including several important food species, such as the Atlantic salmon, rainbow trout, turbot, sea bass, and sea bream [36]. A single phage treatment protected 100% of Atlantic salmon against experimentally induced *V. anguillarum* infection [39]. Vibriosis also causes high mortality rates in fish larvae. Phages administered in culture water of zebrafish larvae experimentally infected with *V. anguillarum* significantly lowered larvae mortality [38]. Likewise, phages added to culture water of shrimp larvae improved survival after experimental infection with *Vibrio harveyi* [40]. Thus, directly supplying phages to the culture water could be an effective and economical approach toward reducing the negative impact of vibriosis in aquaculture, in particular when vaccines are not an option to protect larvae.

Other experimental aquaculture models have also shown promising phage efficacy. For instance, phages in-feed has been shown to protect against water-borne *Pseudomonas plecoglossicida* infection, the etiological agent of bacterial haemorrhagic ascites in freshwater fish, including ayu, pejerrey, rainbow trout, and large yellow croaker [41]. In a field trial, phage-impregnated feed was added to the fishpond where *P. plecoglossicida* was naturally present and daily ayu mortality of fish decreased by 30%

after multiple weeks of prophylaxis. Moreover, neither phage-resistant bacteria nor phage-neutralizing antibodies were detected in infected or cured fish [42].

3.2. Phages that Combat Zoonotic Pathogens

Phages offer a non-antibiotic method to improve food safety as a preharvest intervention to reduce zoonotic pathogens from the food supply. For instance, contaminated poultry, pork, beef, and fish have led to food poisoning and food-related disease. Often, food-borne pathogen contamination of meat products occurs during processing when carcasses are exposed to infected animal faeces. Campylobacteriosis caused by *Campylobacter jejuni*, is the most frequent food-borne human enteritis in developed countries, the major source being tainted poultry meat. Loc Carrillo et al. [43] showed that an antacid solution containing phages given orally could effectively decolonize the gut of birds experimentally colonized with *C. jejuni*. Under commercial conditions, however, phage decontamination success was highly variable. When a phage cocktail was added to the drinking water at three commercial farms with broilers confirmed to be colonized with *Campylobacter* spp., only one farm experienced a reduction in bacterial load (<50 CFU/g) in faecal samples [44]. For the other two farms, no significant reduction occurred for undetermined reasons.

Salmonellosis is another common cause of gastroenteritis in humans. Pigs can become colonized with *Salmonella* spp. from contaminated trailers and holding pens, resulting in increased pathogen shedding just prior to processing. Wall et al. [19] showed that administration of a phage cocktail at the time of experimental inoculation with *Salmonella enterica* serovar Typhimurium reduced bacterial load to almost undetectable limits in the tonsils, ileum, and cecum of infected small pigs. A phage cocktail significantly reduced cecal and ileal *Salmonella* concentrations by up to 95% after being in a highly contaminated holding pen. *S. enteritidis* is also a prevalent foodborne pathogen, its main reservoir being the eggshell. Use of a mixture of three different *Salmonella*-specific phages to reduce *S. enteritidis* colonization in the ceca of laying hens resulted in a significant decrease in bacterial prevalence of incidence of up to 80% [45].

E. coli is typically a commensal member of human and animal microbiota. However, certain strains can cause a variety of human diseases, including urinary tract infections, haemorrhagic colitis, appendicitis and septicaemia. The most notorious zoonotic strains are those referred to as Vero-Toxigenic *E. coli* (VTEC). The most common member of this group is strain O157:H7 and the natural reservoir is the cattle gut. A cocktail of phages isolated from cattle faeces was able to reduce O157:H7 populations in the gut of experimentally inoculated sheep, with a 1:1 ratio of phage to bacteria found to be more effective than higher phage ratios [46]. Upon necropsy, *E. coli* populations were found to be reduced in both the cecum and colon, while ruminal load was not significantly changed, likely due to a relatively low starting population [46].

4. Bacteriophages in Crop Production

The discovery research on phages and plant pathogens took place nine years following the highly disputed discovery of phages by Frederic Twort in 1915 and Felix D'Herelle in 1917 [47]. The first experimental evidence that phages may be associated with plant pathogenic bacteria occurred when it was demonstrated that a filtrate obtained from decomposing cabbage was able to inhibit cabbage-rot caused by *Xanthomonas campestris* pv. *campestris* [48]. The following year, Kotila and Coons [49] demonstrated that the exposure of *Pectobacterium atrosepticum* to phages prevented the development of soft rot in potatoes. The first recorded field trial occurred in 1935, when Stewart's wilt disease of corn, caused by *Pantoea stewartii*, was reduced by pre-treatment of seeds by phages [50].

In a 2012 survey, bacterial pathologists that read the Journal Molecular Plant Pathology were asked to list three important plant pathogens [51]. The top 10 plant pathogens listed in descending order were *Pseudomonas syringae*, *Ralstonia solanacearum*, *Agrobacterium tumefaciens*, *Xanthomonas* spp., *Erwinia amylovora*, *Xylella fastidiosa*, *Dickeya* spp., and *Pectobacterium* spp. Lack of chemical control options and development of antibiotic resistance in many plant pathogens combined with consumers' preference for organic and antibiotic free products has led to a phage therapy renaissance in agriculture.

4.1. Soft Rot, Bacterial Wilt, and Blight

Dickeya solani and *Pectobacterium* spp. are pathogens associated with potato tuber soft rot in storage and blackleg disease in the field [52,53]. Adriaenssens [52] used two *Dickeya* sp. *Myoviridae* phages as biological control agents for the control of soft rot/blackleg in potato. Potato tubers were vacuum infiltrated with the pathogen and the phages were sprayed (nebulised) at MOI of 10 and/or 100 over the infested tubers. Treated tubers were planted in the field and the disease progression was monitored through the growing season. There was no significant difference between the treated control, untreated controls, and the phage/bacteria treatments. To study the ability of the phage mixtures to control multiple pathogen species associated with bacterial soft rot, two broad host range phages [53] and 9 phage mixtures [14] were tested on potato slices but not in the field. Czajkowski [14] provides in a recent review a detailed summary on the advances in research on the phages of the soft rot bacteria.

Pre-treatment with a *Podoviridae* phage PE204 under growth chamber conditions did not achieve control of *R. solanacearum*, the cause of bacterial wilt of tomato [54]. Single phage and/or cocktails composed of commercial phage mixtures were applied as a soil drench with an attenuated *Xanthomonas perforans* isolate. Phage populations were followed in the treatments in the root zone and the inside of plants. Partial translocation of phages occurred into the lower portions of the tomato plant and greenhouse and field trials demonstrated that in the presence of *X. perforans* mutant phage populations increased on leaf surfaces and in the soil [55].

Pseudomonas syringae pathovars are responsible for a large number of plant diseases in agriculture [51,56]. The recent serious global outbreak of *P. syringae* pv. *actinidae* in kiwifruit production and the lack of control options has re-focused research onto phages [57,58]. To date, this research has focused on phage characterization by host range and genomic studies. Two parallel field trials in three locations were conducted for the control of *P. syringae* pv. *porri*, bacterial blight of leek [59]. The treatments involved a 6-phage cocktail and plants that were either pre-treated with phage and then infected by the pathogen or treated with pathogen followed by the phage cocktail at 10^9 pfu/ml. Statistically significant difference between treatment and control were not obtained and the results were highly variable between the locations.

4.2. Citrus Bacterial Canker and Spot

Balogh (2008) studied phage-mediated control of *Xanthomonas axonopodis* pv. *citri*, Asiatic citrus canker (ACC) and *X. axonopodis* pv. *citrumelo*, citrus bacterial spot (CBS). Treatments without skim milk, used to stabilise the phage, additive significantly reduced ACC disease severity. In nursery trials, the ability of phage mixtures, copper-mancozeb, and the combination of phage-copper-mancozeb to control CBS and ACC were tested. Phages reduced ACC disease significantly but were not as effective as the copper-macozeb treatment alone. The phage-copper-mancozeb combined treatment failed. Similar results were seen for CBS, where phage control was significantly different from the control in Valencia oranges but not in grapefruit under low disease pressures [60]. Ibrahim et al. [61] obtained successful control of Asiatic citrus canker in greenhouse and field trials, by combing a compound which induced the plants systemic acquired resistance and phage mixtures formulated in skim milk and sugar.

4.3. Pierce's Disease of Grape

Xylella fastidiosa is a pathogen of a number of plants but it has the greatest economic impact in grapes. Disease control options are limited and challenging since the pathogen is limited to the xylem of the grape [62]. Recently, two lytic phages of *X. fastidiosa* subsp. *fastidiosa* have been isolated and fully characterised [63]. Phage cocktails in grape using therapeutic and prophylactic treatments were able to significantly control the pathogen and symptom development in greenhouse trials.

4.4. Fire Blight in Apples and Pears

The causal organism of fire blight, *Ewinia amylovora*, is a major pathogen in commercially grown apples and pears. The pathogen can exist in asymptomatic tissue or as an epiphyte in the orchard ecosystem [9]. All commercially desirable apple and pear cultivars are moderately to highly susceptible to this pathogen and resistant germplasm is not available. In Canada and the US, streptomycin and kasugamycin are applied during open bloom to obtain control of the fire blight pathogen [6,11]. In growing regions where streptomycin resistance is present and/or organic fruit is grown, the use of antibiotics is prohibited and alternative control strategies for integrated pest management practices are urgently needed.

Phages combined with *Pantoea agglomerans*, non-pathogenic host, belonging to the *Myoviridae* and *Podoviridae* have been tested under greenhouse [64] and field conditions [16,65] for their ability to control the pathogen during open bloom. The highly variable seasonal variation in biological control is not uncommon and it serves as one of the biggest challenges to the commercial development of phages in agriculture.

4.5. Impact of Host Exopolysaccharides and Phage Family on Efficacy

E. amylovora pathogenicity is largely determined by the presence of amylovoran, a capsular exopolysaccharide (EPS), while virulence is associated with levan, a secondary component of the bacterial capsule [66]. Roach et al. [67] showed that the structure of the host cell surface plays a very important role in phage pathogenesis. Isolates of *E. amylovora* characterised as producing relatively large amounts of EPS, were called high EPS producers (HEPs) and low producers were labelled low EPS producers (LEPs). Phages in the *Myoviridae* grew better on LEPs than HEPs. In contrast, most but not all *Podoviridae* phages exhibited improved replication on HEPs hosts as measured by efficiency of plating. Deletion of genes required for the production of amylovoran and levan provided further insight into the function of the cell surface in phage growth. Deletion of the *rcsB* gene, which prevented synthesis of the EPS component amylovoran, resulted in almost complete resistance to most podoviruses tested. The effect of this deletion on myoviruses was variable with one phage showing a reduction in the efficiency of plating (EOP) and two others showing an increase, suggesting that amylovoran is likely not a significant contributor to phage pathogenesis for these phages. In contrast, deletion of the levansucrase gene, *lsc*, had little impact on the pathogenesis of podoviruses but resulted in a reduction of EOP by one to two orders of magnitude for myoviruses. These observations have three important implications on the impact of the use of these phages in a program to control *E. amylovora*. First, phages in the *Podoviridae* will likely have little to no impact on other bacterial epiphytes in the orchard because amylovoran synthesis is limited to *E. amylovora*. This prediction has been validated by tests on *P. agglomerans*. This epiphytic species does not produce amylovoran and the majority of the isolates do not support the growth of *E. amylovora* podoviruses). Second, one of the likely mechanisms by which *E. amylovora* could become resistant would be through a mutation that prevents amylovoran production. This would result in an avirulent bacterium that would greatly reduce the chance of survival. Thus, podoviruses should be included in any biocontrol formulation with the goal of reducing fire blight. Third, myoviruses should also be included in the formulation to increase the probability of inhibiting growth of all *E. amylovora* strains, including the LEPs.

5. Potential Problems with Phages as Biocontrol Agents

The development of phage resistance in the bacterial host is a major concern in phage therapy. Just as bacteria may become resistant to antibiotics they may also become resistant to phages by a variety of mechanisms. These include modification of the phage surface receptors on the bacterial cell such as conversion to mucoidy [68], integration of the phage genome into the bacterial chromosome [57,69], restriction/modification systems [70], CRISPR/Cas systems [71,72], BREX [73], DISARM [74], and up to 9 new defense systems [75]. To prevent the development of bacterial resistance to phages the standard adopted practice has been to use a mixtures or cocktails that may contain combinations of phages with host ranges that are narrow, wide and/or composed of host range mutants [27,53,59,60,76]. One intriguing possible outcome of the use of a phage mixture is bacteria that are resistant to a particular phage can still be lysed by that phage through the acquisition of phage receptors from lysed sensitive cells. This effect has been observed during infection of *Bacillus subtilis* with phage SPP1 [77]. It will be important to investigate if the transfer of receptors is a phenomenon that extends well beyond this one example.

Another potential hurdle with the use of phages as biological agents is the production of lysogens or pseudolysogens. Persistence of the phage genome in the host cell would provide superinfection immunity that would negate the efficacy of the biological and possibly impart novel characteristics to the target bacterium. For example, ΦRSS1, a phage that exists in a persistent infective state in *R. solanacearum* increases virulence of the bacterial host on tomato [78]. Although this risk clearly exists, the scope of the problem remains poorly understood. Roach et al. [69] examined the prevalence of lysogens of myoviruses and podoviruses in 161 isolates of *E. amylovora* and 82 of *P. agglomerans*. None was detected. Use of phages to recover bacteriophage insensitive mutants (BIMs), however, showed that lysogeny was possible with the recovery of one stable lysogen. In addition, PCR analysis indicated that phage DNAs could be detected in subcultures of numerous BIMs for up to a year after selection although the association of phage and host was unstable. The authors concluded that though lysogeny could occur, it was likely to be selected against in the resource rich environment of the apple or pear blossom. As such, the risks associated with lysogeny were low. Nonetheless, this possibility should be considered for any application of phages for biocontrol.

A third potential hurdle is that phages could serve as vectors for mobile genetic elements, including antibiotic resistance genes [79,80]. Colavecchio et al. [81] recently reviewed the literature on the role of phages in the spread of AMR genes amongst members of the *Enterobacteriaceae*. These genes could certainly be transferred horizontally by transducing phage particles. The contribution of transduction to AMR spread, however, may be low as compared to conjugation or transformation. This issue is currently unresolved and deserves further attention.

Many bacterial pathogens form biofilms, which in turn impact phage therapy. In *E. amylovora*, amylovoran and levan contribute to the formation of a biofilm [82], yet phage efficacy bioassays continue to be carried out in liquid cultures. Today, models of phage–host interactions should take into consideration that biofilms form a spatial environment where resources are concentrated and bacterial materials and debris build up as cell numbers increase [83]. All these factors will influence the ability of the phage to adsorb and kill the bacterial host. Laboratory studies that use liquid cultures to study phage-host interactions are poor indicators of phage efficacy under greenhouse and/or field conditions. In a recent publication, Abedon [84] provides an excellent treatise on how phage therapy can be improved by incorporating important standards such as the Poisson distribution curve when reporting on the infection of host cells by phage and the avoidance of the commonly used MOIs to report phage dosages.

6. Discussion

Present-day research indicates that phages have the potential as an alternative control mechanism for eliminating pathogens posing a threat to animals and plants (Table 1), particularly with the increased risk of AMR and regulatory restrictions on the use of antibiotics in agriculture. Phages developed for the control of plant and animal pathogenic, zoonotic, and problematic bacteria exploit the multiple and complex host–microbe interactions to significantly reduce disease, reduce economic losses, and minimize the effect on the environment and on non-target microorganisms. In animal production, the focus of using phages as antimicrobial agents has been on controlling human and zoonotic pathogens.

Table 1. Experimental studies using bacteriophages to control bacterial pathogens.

Target Species	Disease/Issue	Animal/Plant	Study
Clostridium perfringens	necrotic enteritis	poultry	[27]
C. perfringens, E. coli, S. aureus	weight gain	swine	[28]
Escherichia coli	respiratory infection	poultry	[29,30]
Staphylococcus aureus	mastitis	bovine	[32,33]
Vibrio anguillarum	vibriosis	fish	[38,39]
Vibrio harveyi	vibriosis	shrimp	[40]
Pseudomonas plecoglossicida	haemorrhagic ascites	fish	[41,42]
Campylobacter jejuni	zoonotic	poultry	[43,44]
Salmonella enterica serovar Typhimurium	zoonotic	swine	[19]
Salmonella enteriditis	zoonotic	poultry	[45]
Escherichia coli	colitis	sheep	[46]
Xanthomonas campestris pv. *campestris*	cabbage rot	cabbage	[48]
Pectobacterium atrosepticum	soft rot	potato	[49,52]
Pantoea stewartii	Stewart's wilt	corn	[50]
Dickeya solani, Pectobacterium spp.	soft rot/blackleg	potato	[52,53]
Ralstonia solanacearum	bacterial wilt	tomato	[54]
Pseudomonas syringae pv. *actinidae*	canker	kiwifruit	[57,58]
Pseudomonas syringae pv. *porri*	bacterial blight	leak	[59]
Xanthomonas axonopodis pv. *citrumelo*	bacterial spot	citrus	[60]
Xanthomonas axonopodis pv. *citri*	canker	citrus	[61]
Xylella fastidiosa	Pierce's disease	grape	[63]
Erwinia amylovora	fire blight	apple/pear	[16,64,65]

In plant agriculture, control with phages has been difficult to implement due to a number of challenges. These include development of formulations to effectively treat hectares of plants grown in monoculture and/or in greenhouse conditions, assessing susceptible hosts including both bacterial pathogen and plant interactions as well as phage–bacterium matches, persistent pathogen presence, transmission of the pathogen by wind, rain, and insects, modern day farming practices that rely on chemical pesticides that may be deleterious to the phage, and unpredictable weather patterns within and between growing seasons. Timing of the biocontrol delivery is crucially important. Therapeutic treatments may involve phage application to reduce a pre-existing pathogen population or an application timed to the expected arrival of the pathogen [52,55,59,64,85]. For prophylactic treatment, phages are introduced prior to the anticipated appearance of the pathogen [59,85]. Efficacy of both options should be evaluated as part of a biocontrol development program. Aerial phage applications require formulations that will ensure the survival of the phage in the environment [60,61,76,86]. The alternative application methodology is to utilize a living bacterial cell delivery system that ensures survival and continued replication of the phages prior to the arrival of the pathogen [16,65]. For example, live cells of an attenuated bacterial strain of *Xanthomonas perforans* were used to improve the persistence of the phage populations in and on the soil [55].

In animal production, much of the focus of using phages as antimicrobial agents has been on controlling bacterial infection. The benefits of antibiotics in animal feed have added benefits in production. For instance, Thomke and Elwinger [87] hypothesize that cytokines released during the immune response may also stimulate the release of catabolic hormones, which reduce muscle mass. In addition, there is evidence that antibiotics suppress microbial fermentation in the gastrointestinal tract improving feed conversion by up to 6% (Jensen, 1998). Recent studies showed that a sub-therapeutic antibiotic correlates with the decreased activity of bile salt hydrolase, an intestinal bacteria-produced enzyme that exerts negative impact on host fat digestion and utilization [88]. Regardless of the mechanism of action, the use of animal growth promoters can improve daily growth rates between 1% and 10% resulting in meat of better quality with less fat and increased protein content. It will be important to explore whether phages provide similar growth enhancing effects beyond the benefits of controlling infectious diseases. Phages can also be used in post-slaughter or later processing systems as decontaminants, including the FDA approved commercial products ListShield™ (Intralytix, Baltimore, MD, USA) and PhageGuard L™ (formerly Listex™) (Micreos Food Safety B.V., Wageningen, Netherlands) as food additives for prevention of meat contamination with *Listeria monocytogenes* [89].

EcoShield™ (Intralytixs) for *E. coli* and SalmoFresh™ (Intralytix) for *Salmonella* spp. are also FDA approved to decontaminate ready-to-eat meat and poultry, fish and seafood, and dairy products.

Plant and animal phage development systems in food agriculture have their own distinct and specialised processes, protocols, and challenges. Regardless of the agricultural application, the process itself should be better defined, organized, and laid out. The science innovation chain for the development of biologicals or biopesticides was developed by Boyetchko [90]. This model defines and designates specific steps and processes that workers should address in the developed of phage biologicals (synonym in agriculture biopesticide). The project deliverables, arranged in continuous and ascending order, include acquisition of scientific knowledge, greenhouse/field/animal efficacy trials, fermentation/formulation, defining of markets, license agreements, large scale field test, manufacturing/process engineering, production of phage product, and product sales/client adoption. Concurrent with the deliverables and in the same ascending order, a series of stages and/or gates include discovery and selection of phages, proof of concept that the therapy works, technology development, market identification, technology transfer, commercial scale up, registration/regulatory processes, and technology adaptation by end users. This type of a model takes into consideration work beyond the laboratory and basic science and provides basic guidelines to the processes and decision points that need to be addressed during the development of a phage biological that can be successfully used in agriculture.

Author Contributions: Antonet Svircev, Dwayne Roach and Alan Castle contributed equally to the concept and writing of the mini-review. Darlene Nesbitt (Agriculture Agri-Food Canada) edited the manuscript.

Funding: Dwayne Roach was supported by a European Respiratory Society Fellowship (RESPIRE2-2015-8416). Alan Castle is funded by the RGPIN-2016-05590 Natural Sciences and Engineering Research Council of Canada. Antonet Svircev was supported by Agriculture and Agri-Food Canada Growing Forward II grant.

Conflicts of Interest: The authors declare no conflict of interest.

References

1. Ramankutty, N.; Mehrabi, Z.; Waha, K.; Jarvis, L.; Kremen, C.; Herreo, M.; Rieseberg, L.H. Trends in global agriculturl land use: Implications for environmental health and food safety. *Ann. Rev. Plant Biol.* **2018**, *69*. [CrossRef] [PubMed]
2. Muller, A.; Schader, C.; El-Hage Scialabba, N.; Bruggemann, J.; Isensee, A.; Erb, K.H.; Smith, P.; Klocke, P.; Leiber, F.; Stolze, M.; et al. Strategies for feeding the world more sustainably with organic agriculture. *Nat. Commun.* **2017**, *8*, 1290. [CrossRef] [PubMed]
3. Pingali, P.L. Green revolution: Impacts, limits, and the path ahead. *Proc. Natl. Acad. Sci. USA* **2012**, *109*, 12302–12308. [CrossRef] [PubMed]
4. Moore, P.; Evenson, A.; Luckey, T.; McCoy, E.; Elvehjem, C.; Hart, E. Studies with the chick streptomycin in nutritional streptothricin, and use of sulfasuxidine. *J. Biol. Chem.* **1946**, *165*, 437–441. [PubMed]
5. Cheng, G.; Hao, H.; Xie, S.; Wang, X.; Dai, M.; Huang, L.; Yuan, Z. Antibiotic alternatives: The substitution of antibiotics in animal husbandry? *Front. Microbiol.* **2014**, *5*, 217. [CrossRef] [PubMed]
6. McManus, P.S.; Stockwell, V.O.; Sundin, G.W.; Jones, A.L. Antibiotic use in plant agriculture. *Annu. Rev. Phytopathol.* **2002**, *40*, 443–465. [CrossRef] [PubMed]
7. Sholberg, P.L.; Bedford, K.E.; Haag, P.; Randall, P. Survey of *Erwinia amylovora* isolates from British Columbia for resistance to bactericides and virulence on apple. *Can. J. Plant Pathol.* **2001**, *23*, 60–67. [CrossRef]
8. Förster, H.; McGhee, G.C.; Sundin, G.W.; Adaskaveg, J.E. Characterization of streptomycin resistance in isolates of *Erwinia amylovora* in California. *Phytopathology* **2015**, *105*, 1302–1310. [CrossRef] [PubMed]
9. Tancos, K.A.; Borejsza-Wysocka, E.; Kuehne, S.; Breth, D.; Cox, K.D. Fire blight symptomatic shoots and the presence of *Erwinia amylovora* in asymptomatic apple budwood. *Plant Dis.* **2017**, *101*, 186–191. [CrossRef]
10. Tancos, K.A.; Villani, S.; Kuehne, S.; Borejsza-Wysocka, E.; Breth, D.; Carol, J.; Aldwinckle, H.S.; Cox, K.D. Prevalence of streptomycin-resistant *Erwinia amylovora* in New York apple orchards. *Plant Dis.* **2016**, *100*, 802–809. [CrossRef]
11. McManus, P.S. Does a drop in the bucket make a splash? Assessing the impact of antibiotic use on plants. *Curr. Opin. Microbiol.* **2014**, *19*, 76–82. [CrossRef] [PubMed]

12. O'Neil, J. *Tracking a Global Health Crisis: Initial Steps*, 2015th ed.; Review of Antimicrobial Resitance; Welcome Trist and UK Government: London, UK, 2015.
13. Buttimer, C.; McAuliffe, O.; Ross, R.P.; Hill, C.; O'Mahony, J.; Coffey, A. Bacteriophages and bacterial plant diseases. *Front. Microbiol.* **2017**, *8*, 34. [CrossRef] [PubMed]
14. Czajkowski, R. Bacteriophages of Soft Rot *Enterobacteriaceae*—A minireview. *FEMS Microbiol. Lett.* **2016**, *363*. [CrossRef] [PubMed]
15. Nagy, J.K.; Király, L.; Schwarczinger, I. Phage therapy for plant disease control with a focus on fire blight. *Cent. Eur. J. Biol.* **2012**, *7*, 1–12. [CrossRef]
16. Svircev, A.M.; Castle, A.J.; Lehman, S.M. Bacteriophages for control of phytopathogens in food production systems. In *Bacteriophages in the Control of Food- and Waterborne Pathogens*; Sabour, P.M., Griffiths, M.W., Eds.; ASM Press: Washington, DC, USA, 2010; pp. 79–102.
17. Wittebole, X.; de Roock, S.; Opal, S.M. A historical overview of bacteriophage therapy as an alternative to antibiotics for the treatment of bacterial pathogens. *Virulence* **2014**, *5*, 226–235. [CrossRef] [PubMed]
18. Aarestrup, F. Get pigs off antibiotics. *Nature* **2012**, *486*, 465–466. [CrossRef] [PubMed]
19. Wall, S.K.; Zhang, J.; Rostagno, M.H.; Ebner, P.D. Phage therapy to reduce preprocessing *Salmonella* infections in market-weight swine. *Appl. Environ. Microbiol.* **2010**, *76*, 48–53. [CrossRef] [PubMed]
20. Nosanchuk, J.D.; Lin, J.; Hunter, R.P.; Aminov, R.I. Low-dose antibiotics: current status and outlook for the future. *Front. Microbiol.* **2014**, *5*, 478. [CrossRef] [PubMed]
21. Goneau, L.W.; Hannan, T.J.; MacPhee, R.A.; Schwartz, D.J.; Macklaim, J.M.; Gloor, G.B.; Razvi, H.; Reid, G.; Hultgren, S.J.; Burton, J.P. Subinhibitory antibiotic therapy alters recurrent urinary tract infection pathogenesis through modulation of bacterial virulence and host immunity. *mBio* **2015**, *6*, e00356-15. [CrossRef] [PubMed]
22. Graham, J.P.; Evans, S.L.; Price, L.B.; Silbergeld, E.K. Fate of antimicrobial-resistant enterococci and staphylococci and resistance determinants in stored poultry litter. *Environ. Res.* **2009**, *109*, 682–689. [CrossRef] [PubMed]
23. Robinson, T.P.; Bu, D.P.; Carrique-Mas, J.; Fevre, E.M.; Gilbert, M.; Grace, D.; Hay, S.I.; Jiwakanon, J.; Kakkar, M.; Kariuki, S.; et al. Antibiotic resistance is the quintessential One Health issue. *Trans. R. Soc. Trop. Med. Hyg.* **2016**, *110*, 377–380. [CrossRef] [PubMed]
24. Li, D.; Wu, C.; Wang, Y.; Fan, R.; Schwarz, S.; Zhang, S. Identification of multiresistance gene *cfr* in methicillin-resistant *Staphylococcus aureus* from pigs: Plasmid location and integration into a staphylococcal cassette chromosome *mec* complex. *Antimicrob. Agents Chemother.* **2015**, *59*, 3641–3644. [CrossRef] [PubMed]
25. Gelband, H.; Miller-Petrie, M.; Pant, S.; Gandra, S.; Levinson, J.; Barter, D.; White, A.; Laxminarayan, R. The state of the world's antibiotics 2015. *Medpharm* **2015**, *8*, 30–34.
26. Seal, B.S.; Lillehoj, H.S.; Donovan, D.M.; Gay, C.G. Alternatives to antibiotics: A symposium on the challenges and solutions for animal production. *Anim. Health Res. Rev.* **2013**, *14*, 78–87. [CrossRef] [PubMed]
27. Miller, R.W.; Skinner, E.J.; Sulakvelidze, A.; Mathis, G.F.; Hofacre, C.L. Bacteriophage therapy for control of necrotic enteritis of broiler chickens experimentally infected with *Clostridium perfringens*. *Avian Dis.* **2010**, *54*, 33–40. [CrossRef] [PubMed]
28. Kim, K.H.; Ingale, S.L.; Kim, J.S.; Lee, S.H.; Lee, J.H.; Kwon, I.K.; Chae, B.J. Bacteriophage and probiotics both enhance the performance of growing pigs but bacteriophage are more effective. *Anim. Feed Sci. Technol.* **2014**, *196*, 88–95. [CrossRef]
29. Huff, W.E.; Huff, G.R.; Rath, N.C.; Balog, J.M.; Xie, H.; Moore, P.A.; Donoghue, A.M. Prevention of *Escherichia coli* respiratory infection in broiler chickens with bacteriophage (SPR02). *Poult. Sci.* **2002**, *81*, 437–441. [CrossRef] [PubMed]
30. Huff, W.E.; Huff, G.R.; Rath, N.C.; Donoghue, A.M. Method of administration affects the ability of bacteriophage to prevent colibacillosis in 1-day-old broiler chickens. *Poult. Sci.* **2013**, *92*, 930–934. [CrossRef] [PubMed]
31. Fessler, A.; Scott, C.; Kadlec, K.; Ehricht, R.; Monecke, S.; Schwarz, S. Characterization of methicillin-resistant *Staphylococcus aureus* ST398 from cases of bovine mastitis. *J. Antimicrob. Chemother.* **2010**, *65*, 619–625. [CrossRef] [PubMed]
32. Breyne, K.; Honaker, R.W.; Hobbs, Z.; Richter, M.; Zaczek, M.; Spangler, T.; Steenbrugge, J.; Lu, R.; Kinkhabwala, A.; Marchon, B.; et al. Efficacy and safety of a bovine-associated *Staphylococcus aureus* phage cocktail in a murine model of mastitis. *Front. Microbiol.* **2017**, *8*, 2348. [CrossRef] [PubMed]

33. Gill, J.J.; Sabour, P.M.; Leslie, K.E.; Griffiths, M.W. Bovine whey proteins inhibit the interaction of *Staphylococcus aureus* and bacteriophage K. *J. Appl. Microbiol.* **2006**, *101*, 377–386. [CrossRef] [PubMed]
34. Fernandez, L.; Escobedo, S.; Gutierrez, D.; Portilla, S.; Martinez, B.; Garcia, P.; Rodriguez, A. Bacteriophages in the dairy environment: From enemies to allies. *Antibiotics (Basel)* **2017**, *6*, 27. [CrossRef] [PubMed]
35. Barkema, H.; Schukken, Y.; Zadoks, R. Invited review: The role of cow, pathogen, and treatment regimen in the therapeutic success of bovine *Staphylococcus aureus* mastitis. *J. Dairy Sci.* **2006**, *89*, 1877–1895. [CrossRef]
36. Toranzo, A.E.; Magariños, B.; Romalde, J.L. A review of the main bacterial fish diseases in mariculture systems. *Aquacul* **2005**, *246*, 37–61. [CrossRef]
37. Rao, B.; Lalitha, K. Bacteriophages for aquaculture: Are they beneficial or inimical. *Aquacul* **2015**, *437*, 146–154.
38. Silva, Y.J.; Costa, L.; Pereira, C.; Mateus, C.; Cunha, A.; Calado, R.; Gomes, N.C.; Pardo, M.A.; Hernandez, I.; Almeida, A. Phage therapy as an approach to prevent *Vibrio anguillarum* infections in fish larvae production. *PLoS ONE* **2014**, *9*, e114197. [CrossRef] [PubMed]
39. Higuera, G.; Bastías, R.; Tsertsvadze, G.; Romero, J.; Espejo, R.T. Recently discovered *Vibrio anguillarum* phages can protect against experimentally induced vibriosis in Atlantic salmon, *Salmo salar*. *Aquaculture* **2013**, *392–395*, 128–133. [CrossRef]
40. Karunasagar, I.; Shivu, M.; Girisha, S.; Krohne, G.; Karunasagar, I. Biocontrol of pathogens in shrimp hatcheries using bacteriophages. *Aquacul* **2007**, *268*, 288–292. [CrossRef]
41. Mao, Z.; Li, M.; Chen, J. Draft genome sequence of pseudomonas plecoglossicida strain NB2011, the causative agent of white nodules in large yellow croaker (*Larimichthys crocea*). *Genome Announc.* **2013**, *1*, e00586-13. [CrossRef] [PubMed]
42. Park, S.; Nakai, T. Bacteriophage control of *Pseudomonas plecoglossicida* infection in ayu *Plecoglossus altivelis*. *Dis. Aquat. Org.* **2003**, *53*, 33–39. [CrossRef] [PubMed]
43. Loc Carrillo, C.; Atterbury, R.J.; El-Shibiny, A.; Connerton, P.L.; Dillon, E.; Scott, A.; Connerton, I.F. Bacteriophage therapy to reduce *Campylobacter jejuni* colonization of broiler chickens. *Appl. Environ. Microbiol.* **2005**, *71*, 6554–6563. [CrossRef] [PubMed]
44. Kittler, S.; Fischer, S.; Abdulmawjood, A.; Glunder, G.; Klein, G. Effect of bacteriophage application on *Campylobacter jejuni* loads in commercial broiler flocks. *Appl. Environ. Microbiol.* **2013**, *79*, 7525–7533. [CrossRef] [PubMed]
45. Borie, C.; Sanchez, M.L.; Navarro, C.; Ramirez, S.; Morales, M.A.; Retamales, J.; Robeson, J. Aerosol spray treatment with bacteriophages and competitive exclusion reduces *Salmonella enteritidis* infection in chickens. *Avian Dis.* **2009**, *53*, 250–254. [CrossRef] [PubMed]
46. Callaway, T.R.; Edrington, T.S.; Brabban, A.D.; Anderson, R.C.; Rossman, M.L.; Mike, J.; Engler, M.J.; Carr, M.A.; Genovese, K.J.; Keen, J.E.; Looper, M.L.; et al. Bacteriophage isolated from feedlot cattle can reduce *Escherichia coli* O157:H7 populations in ruminant gastrointestinal tracts. *Foodborne Pathog. Dis.* **2008**, *5*, 183–191. [CrossRef] [PubMed]
47. Duckworth, D. Who discovered bacteriophage? *Bacteriol. Rev.* **1976**, *40*, 793–802. [PubMed]
48. Mallmann, W.; Hemstreet, C. Isolation of an inhibitory substance from plants. *Agric. Res.* **1924**, *28*, 599–602.
49. Kotila, J.; Coons, G. *Investigations on the Black Leg Disease of Potato*; Michigan Agri. Exp. Station Technical Bulletin; Michigan Agricultural College: East Lansing, MI, USA, 1925; Volume 67, pp. 3–29.
50. Thomas, R. A bacteriophage in relationto Stewart's disease of corn. *Phytopathology* **1935**, *25*, 371–372.
51. Mansfield, J.; Genin, S.; Magori, S.; Citovsky, V.; Sriariyanum, M.; Ronald, P.; Dow, M.; Verdier, V.; Beer, S.V.; Machado, M.A.; et al. Top 10 plant pathogenic bacteria in molecular plant pathology. *Mol. Plant Pathol.* **2012**, *13*, 614–629. [CrossRef] [PubMed]
52. Adriaenssens, E.M.; van Vaerenbergh, J.; Vandenheuvel, D.; Dunon, V.; Ceyssens, P.J.; de Proft, M.; Kropinski, A.M.; Noben, J.P.; Maes, M.; Lavigne, R. T4-related bacteriophage LIMEstone isolates for the control of soft rot on potato caused by "Dickeya solani". *PLoS ONE* **2012**, *7*, e33227. [CrossRef] [PubMed]
53. Czajkowski, R.; Ozymko, Z.; de Jager, V.; Siwinska, J.; Smolarska, A.; Ossowicki, A.; Narajczyk, M.; Lojkowska, E. Genomic, proteomic and morphological characterization of two novel broad host lytic bacteriophages PhiPD10.3 and PhiPD23.1 infecting pectinolytic *Pectobacterium* spp. and *Dickeya* spp. *PLoS ONE* **2015**, *10*, e0119812. [CrossRef] [PubMed]

54. Fujiwara, A.; Fujisawa, M.; Hamasaki, R.; Kawasaki, T.; Fujie, M.; Yamada, T. Biocontrol of *Ralstonia solanacearum* by treatment with lytic bacteriophages. *Appl. Environ. Microbiol.* **2011**, *77*, 4155–4162. [CrossRef] [PubMed]
55. Iriarte, F.B.; Obradovic, A.; Wernsing, M.H.; Jackson, L.E.; Balogh, B.; Hong, J.A.; Momol, M.T.; Jones, J.B.; Vallad, G.E. Soil-based systemic delivery and phyllosphere in vivo propagation of bacteriophages: Two possible strategies for improving bacteriophage persistence for plant disease control. *Bacteriophage* **2012**, *2*, 215–224. [CrossRef] [PubMed]
56. Hirano, S.; Upper, C. Population biology and epidemiology of *Pseudomonas syringae*. *Annu. Rev. Phytopathol.* **1990**, *28*, 155–177. [CrossRef]
57. Frampton, R.A.; Taylor, C.; Holguín Moreno, A.V.; Visnovsky, S.B.; Petty, N.K.; Pitman, A.R.; Fineran, P.C. Identification of bacteriophages for biocontrol of the kiwifruit canker phytopathogen *Pseudomonas syringae* pv. *actinidiae*. *Appl. Environ. Microbiol.* **2014**, *80*, 2216–2228. [CrossRef] [PubMed]
58. Di Lallo, G.; Evangelisti, M.; Mancuso, F.; Ferrante, P.; Marcelletti, S.; Tinari, A.; Superti, F.; Migliore, L.; D'Addabbo, P.; Frezza, D.; et al. Isolation and partial characterization of bacteriophages infecting *Pseudomonas syringae* pv. *actinidiae, causal agent of kiwifruit bacterial canker. J. Basic Microbiol.* **2014**, *54*, 1210–1221. [CrossRef] [PubMed]
59. Rombouts, S.; Volckaert, A.; Venneman, S.; Declercq, B.; Vandenheuvel, D.; Allonsius, C.N.; van Malderghem, C.; Jang, H.B.; Briers, Y.; Noben, J.P.; et al. Characterization of novel bacteriophages for biocontrol of bacterial blight in leek caused by *Pseudomonas syringae* pv. *porri*. *Front. Microbiol.* **2016**, *7*, 279. [CrossRef] [PubMed]
60. Balogh, B.; Canteros, B.I.; Stall, R.E.; Jones, J.B. Control of citrus canker and citrus bacterial spot with bacteriophages. *Plant Dis.* **2008**, *92*, 1048–1052. [CrossRef]
61. Ibrahim, Y.E.; Saleh, A.A.; Al-Saleh, M.A. Management of asiatic citrus canker under field conditions in Saudi Arabia using bacteriophages and acibenzolar-*S*-methyl. *Plant Dis.* **2017**, *101*, 761–765. [CrossRef]
62. Chatterjee, S.; Almeida, R.P.; Lindow, S. Living in two worlds: The plant and insect lifestyles of *Xylella fastidiosa*. *Annu. Rev. Phytopathol.* **2008**, *46*, 243–271. [CrossRef] [PubMed]
63. Ahern, S.J.; Das, M.; Bhowmick, T.S.; Young, R.; Gonzalez, C.F. Characterization of novel virulent broad-host-range phages of *Xylella fastidiosa* and *Xanthomonas*. *J. Bacteriol.* **2014**, *196*, 459–471. [CrossRef] [PubMed]
64. Boulé, J.; Sholberg, P.L.; Lehman, S.M.; O'Gorman, D.T.; Svircev, A.M. Isolation and characterization of eight bacteriophages infecting *Erwinia amylovora* and their potential as biological control agents in British Columbia, Canada. *Can. J. Plant Pathol.* **2011**, *33*, 308–317. [CrossRef]
65. Lehman, S.M. Development of a Bacteriophage-Based Biopesticide for Fire Blight. Ph.D. Thesis, Brock University, St. Catharines, ON, Canada, 2007.
66. Piqué, N.; Miñana-Galbis, D.; Merino, S.; Tomás, J.M. Virulence factors of *Erwinia amylovora*: A review. *Int. J. Mol. Sci.* **2015**, *16*, 12836–12854. [CrossRef] [PubMed]
67. Roach, D.R.; Sjaarda, D.R.; Castle, A.J.; Svircev, A.M. Host exopolysaccharide quantity and composition impact *Erwinia amylovora* bacteriophage pathogenesis. *Appl. Environ. Microbiol.* **2013**, *79*, 3249–3256. [CrossRef] [PubMed]
68. Scanlan, P.D.; Hall, A.R.; Blackshields, G.; Friman, V.P.; Davis, M.R., Jr.; Goldberg, J.B.; Buckling, A. Coevolution with bacteriophages drives genome-wide host evolution and constrains the acquisition of abiotic-beneficial mutations. *Mol. Biol. Evol.* **2015**, *32*, 1425–1435. [CrossRef] [PubMed]
69. Roach, D.R.; Sjaarda, D.R.; Sjaarda, C.P.; Ayala, C.J.; Howcroft, B.; Castle, A.J.; Svircev, A.M. Absence of lysogeny in wild populations of *Erwinia amylovora* and *Pantoea agglomerans*. *Microb. Biotechnol.* **2015**, *8*, 510–518. [CrossRef] [PubMed]
70. Tock, M.R.; Dryden, D.T. The biology of restriction and anti-restriction. *Curr. Opin. Microbiol.* **2005**, *8*, 466–472. [CrossRef] [PubMed]
71. Mojica, F.J.; Diez-Villasenor, C.; Garcia-Martinez, J.; Soria, E. Intervening sequences of regularly spaced prokaryotic repeats derive from foreign genetic elements. *J. Mol. Evol.* **2005**, *60*, 174–182. [CrossRef] [PubMed]
72. Barrangou, R.; Fremaux, C.; Deveau, H.; Richards, M.; Boyaval, P.; Moineau, S.; Romero, D.A.; Horvath, P. CRISPR provides acquired resistance against viruses in prokaryotes. *Science* **2007**, *315*, 1709–1712. [CrossRef] [PubMed]

73. Goldfarb, T.; Sberro, H.; Weinstock, E.; Cohen, O.; Doron, S.; Charpak-Amikam, Y.; Afik, S.; Ofir, G.; Sorek, R. BREX is a novel phage resistance system widespread in microbial genomes. *EMBO J.* **2015**, *34*, 169–183. [CrossRef] [PubMed]
74. Ofir, G.; Melamed, S.; Sberro, H.; Mukamel, Z.; Silverman, S.; Yaakov, G.; Doron, S.; Sorek, R. DISARM is a widespread bacterial defence system with broad anti-phage activities. *Nat. Microbiol.* **2018**, *3*, 90–98. [CrossRef] [PubMed]
75. Doron, S.; Melamed, S.; Ofir, G.; Leavitt, A.; Lopatina, A.; Keren, M.; Amitai, G.; Sorek, R. Systematic discovery of antiphage defense systems in the microbial pangenome. *Science* **2018**, *359*. [CrossRef] [PubMed]
76. Jones, J.B.; Jackson, L.E.; Balogh, B.; Obradovic, A.; Iriarte, F.B.; Momol, M.T. Bacteriophages for plant disease control. *Annu. Rev. Phytopathol.* **2008**, *45*, 245–262. [CrossRef] [PubMed]
77. Tzipilevich, E.; Habusha, M.; Ben-Yehuda, S. Acquisition of phage sensitivity by bacteria through exchange of phage receptors. *Cell* **2017**, *168*, 186–199. [CrossRef] [PubMed]
78. Addy, H.S.; Askora, A.; Kawasaki, T.; Fujie, M.; Yamada, T. Loss of virulence of the phytopathogen *Ralstonia solanacearum* through infection by ΦRSM filamentous phages. *Phytopathology* **2012**, *102*, 469–477. [CrossRef] [PubMed]
79. Muniesa, M.; Colomer-Lluch, M.; Jofre, J. Could bacteriophages transfer antibiotic resistance genes from environmental bacteria to human-body associated bacterial populations? *Mob. Genet. Elem.* **2013**, *3*, e25847. [CrossRef] [PubMed]
80. Muniesa, M.; Colomer-Lluch, M.; Jofre, J. Potential impact of environmental bacteriophages in spreading antibiotic resistance genes. *Future Microbiol.* **2013**, *8*, 739–751. [CrossRef] [PubMed]
81. Colavecchio, A.; Cadieux, B.; Lo, A.; Goodridge, L.D. Bacteriophages contribute to the spread of antibiotic resistance genes among foodborne pathogens of the *Enterobacteriaceae* family—A Review. *Front. Microbiol.* **2017**, *8*, 1108. [CrossRef] [PubMed]
82. Koczan, J.M.; Lenneman, B.R.; McGrath, M.J.; Sundin, G.W. Cell surface attachment structures contribute to biofilm formation and xylem colonization by *Erwinia amylovora*. *Appl. Environ. Microbiol.* **2011**, *77*, 7031–7039. [CrossRef] [PubMed]
83. Bull, J.J.; Christensen, K.A.; Scott, C.; Jack, B.R.; Crandall, C.J.; Krone, S.M. Phage-bacterial dynamics with spatial structure: Self organization around phage sinks can promote increased cell densities. *Antibiot* **2018**, *7*, 8. [CrossRef] [PubMed]
84. Abedon, S.T. Phage therapy: Various perspectives on how to improve the art. *Method Mol. Biol.* **2018**, *1734*, 113–127.
85. Das, M.; Bhowmick, T.S.; Ahern, S.J.; Young, R.; Gonzalez, C.F. Control of Pierce's disease by phage. *PLoS ONE* **2015**, *10*, e0128902. [CrossRef] [PubMed]
86. Born, Y.; Bosshard, L.; Duffy, B.; Loessner, M.J.; Fieseler, L. Protection of *Erwinia amylovora* bacteriophage Y2 from UV-induced damage by natural compounds. *Bacteriophage* **2015**, *5*, e1074330. [CrossRef] [PubMed]
87. Thomke, S.; Elwinger, K. Growth promotants in feeding pigs and poultry. I. Growth and feed efficiency responses to antibiotic growth promotants. *Ann. Zootech.* **1998**, *47*, 85–97. [CrossRef]
88. Lin, J. Antibiotic growth promoters enhance animal production by targeting intestinal bile salt hydrolase and its producers. *Front. Microbiol.* **2014**, *5*, 33. [CrossRef] [PubMed]
89. Migueis, S.; Saraiva, C.; Esteves, A. Efficacy of LISTEX P100 at different concentrations for reduction of *Listeria monocytogenes* inoculated in Sashimi. *J. Food Prot.* **2017**, *80*, 2094–2098. [CrossRef] [PubMed]
90. Boyetchko, S.; Svircev, A.M. A novel approach for developing microbial biopesticides. In *Biological Control Programmes in Canada 2001–2012*; Mason, P., Gillespie, D., Eds.; CAB International: Wallingford, UK, 2013; pp. 37–43.

Review

Bacteriophage Applications for Food Production and Processing

Zachary D. Moye *, Joelle Woolston and Alexander Sulakvelidze

Intralytix, Inc., The Columbus Center, 701 E. Pratt Street, Baltimore, MD 21202, USA; jwoolston@intralytix.com (J.W.); asulakvelidze@intralytix.com (A.S.)

* Correspondence: zmoye@intralytix.com

Received: 19 March 2018; Accepted: 11 April 2018; Published: 19 April 2018

Abstract: Foodborne illnesses remain a major cause of hospitalization and death worldwide despite many advances in food sanitation techniques and pathogen surveillance. Traditional antimicrobial methods, such as pasteurization, high pressure processing, irradiation, and chemical disinfectants are capable of reducing microbial populations in foods to varying degrees, but they also have considerable drawbacks, such as a large initial investment, potential damage to processing equipment due to their corrosive nature, and a deleterious impact on organoleptic qualities (and possibly the nutritional value) of foods. Perhaps most importantly, these decontamination strategies kill indiscriminately, including many—often beneficial—bacteria that are naturally present in foods. One promising technique that addresses several of these shortcomings is bacteriophage biocontrol, a green and natural method that uses lytic bacteriophages isolated from the environment to specifically target pathogenic bacteria and eliminate them from (or significantly reduce their levels in) foods. Since the initial conception of using bacteriophages on foods, a substantial number of research reports have described the use of bacteriophage biocontrol to target a variety of bacterial pathogens in various foods, ranging from ready-to-eat deli meats to fresh fruits and vegetables, and the number of commercially available products containing bacteriophages approved for use in food safety applications has also been steadily increasing. Though some challenges remain, bacteriophage biocontrol is increasingly recognized as an attractive modality in our arsenal of tools for safely and naturally eliminating pathogenic bacteria from foods.

Keywords: bacteriophages; phages; food safety; foodborne illness

---

1. Introduction

From leaves of lettuce and cheddar cheese in a Cobb salad to frozen pre-cooked meals, the foods we eat remain under constant threat of contamination by microbial pathogens, which can subsequently be transmitted to the consumer. Recently, the Foodborne Disease Burden Epidemiology Reference Group (FERG) was established by the World Health Organization (WHO) to monitor foodborne illness across the world. FERG monitored the 31 foodborne pathogens that caused the highest morbidity and mortality in humans. In their most recent (2015) estimate of the global burden of foodborne illness, FERG approximated that 600 million foodborne infections occurred in 2010, resulting in over 400,000 deaths. Of the top five microorganisms causing foodborne illness, four were bacteria: *Escherichia coli* (~111 million), *Campylobacter* spp. (~96 million), non-typhoid *Salmonella enterica* (~78 million), and *Shigella* spp. (~51 million), with estimates for the number of foodborne-related deaths caused by these bacteria ranging from ~15,000 for *Shigella* spp. to ~63,000 for *E. coli* [1]. Strikingly, children under five years old were disproportionally impacted; they account for 40% of deaths while representing just 9% of the world population [1]. These foodborne illnesses are also a tremendous drain on the economy of nations; for example, in the United States the average incident is estimated to cost ~$1500/person, with the total annual estimated cost of these foodborne diseases reaching over $75 billion [2].

Several approaches are used to help improve the safety of our foods. Heat pasteurization is commonly used to reduce bacterial numbers in liquids and dairy items, most notably milk. However, pasteurization is not suitable for many fresh food items, as the process results in the items being cooked. Another method used to reduce pathogens in foods is High Pressure Processing (HPP) which exposes foods to high pressure to inactivate microbes. This technique has been successfully used on liquid products and pre-cooked meals, meant to be frozen; however, as with heat pasteurization, it is generally not used with fresh meats and produce, as it can affect the appearance (color) and/or nutritional content of these products [3,4]. Irradiation is also an effective means for reducing the burden of pathogenic organisms in foods. However, irradiation can deleteriously affect the organoleptic qualities of foods; in addition, customer acceptance of this method is low and is compounded by a labelling requirement for many food items treated with radiation [5,6]. Finally, chemical sanitizers, such as chlorine and peracetic acid (PAA), are commonly utilized to reduce microbial contaminants of many fresh fruits and vegetables as well as Ready-To-Eat (RTE) food products [7,8]. While they are, in general, effective, many of these chemicals are corrosive and can damage food processing equipment. Chemical sanitizers can also deleteriously affect the environment (i.e., not environmentally-friendly) and, with the current trends toward chemical-free, organic foods, consumer acceptance of chemical additives in foods (particularly in fresh produce) is declining rapidly. One common downside shared by all of these techniques is that they kill microbes indiscriminately; in other words, both the pathogenic as well as potentially advantageous normal flora bacteria are targeted equally. Additionally, even with the variety of methods available, foodborne outbreaks still occur relatively frequently. These factors combined illustrate the need for a targeted antimicrobial approach, one that can be used alone or in combination with the techniques described above, to establish additional barriers in a multi-hurdle approach to preventing foodborne bacterial pathogens from reaching consumers. One such technique is the use of lytic bacteriophages for targeting specific foodborne bacteria in our foods, without deleteriously impacting their normal—and often beneficial—microflora. This approach is termed "bacteriophage biocontrol" or "phage biocontrol".

Phage biocontrol is increasingly accepted as a natural and green technology, effective at specifically targeting bacterial pathogens in various foods, in order to safeguard the food chain (Table 1). Bacteriophages were first identified by Felix d'Herelle in 1917, and the usefulness of these "bacteria eaters" for combating bacterial diseases was quickly exploited [9]. In the context of food safety, bacteriophages address many of the concerns voiced by consumers. For example, because of the specificity of bacteriophages, phage biocontrol offers a unique opportunity to target pathogenic bacteria in foods without disturbing the normal microflora of foods. Of note, the United States Army recently initiated a project (W911QY-18-C-0010) to further elucidate the impact of phage application versus traditional chemical antimicrobials on the normal microbiota of fresh produce and how these interventions may impact the nutritional value of foods. Also, phage biocontrol is arguably the most environmentally-friendly antimicrobial intervention available today. Most, if not all, currently-available commercial phage biocontrol products contain natural phages, i.e., phages isolated from the environment, that are not genetically modified. Many of these preparations also do not contain any additives or preservatives; they are typically water-based solutions consisting of purified phages and low levels of salts. Several phage preparations on the market are also certified Kosher and Halal and are available for use in organic foods (OMRI-listed in USA; SKAL in EU) (Table 2). Although there is limited testing, work conducted by our group suggests that bacteriophages do not alter the organoleptic (i.e., sensory) properties of foods [10]. Finally, compared to other food safety interventions, the cost of applying bacteriophages is relatively low and is typically in the range of 1–4 cents per pound of food treated; whereas HPP treatment and irradiation typically cost 10–30 cents per pound [11]. It is important to note that these figures represent the cost of each intervention alone, and do not account for situations where a multi-hurdle approach may be required for food safety purposes (e.g., foods are feared to be contaminated by more than one foodborne pathogen) or for considerations apart from food safety (e.g., food spoilage which is typically caused by multiple different microorganisms).

The biological properties of lytic bacteriophages and other qualities of commercial phage biocontrol products as explained above make phage biocontrol a very attractive modality for further improving the safety of our foods, and an increasing number of companies worldwide are engaging in their development and commercialization [12] (Table 2). However, phage biocontrol does have its limitations and drawbacks. For example, phage preparations require refrigerated storage (typically 2–8 °C), and if used in conjunction with chemical sanitizers, may need to be applied separately, as harsh chemicals can also inactivate the phage particles and render phage biocontrol less effective. Also, because of their high natural specificity, phage preparations can effectively address targeted pathogens in foods, but if food items happen to be contaminated with two or more foodborne bacterial pathogens, a phage preparation targeted against a single pathogen will not be effective in removing non-targeted pathogenic bacteria from foods. As a final consideration, care must be taken to use lytic phages and exclude temperate phages from bacteriophage preparations. Temperate phages are typically less effective than lytic phages at killing their bacterial hosts. Moreover, temperate phages are capable of integrating their DNA into the bacterial chromosome, and therefore, they can potentially promote the transfer of virulence genes or other undesirable genes (e.g., antibiotic-resistance encoding genes) among bacterial strains, which could lead to the emergence of new pathogenic strains. The risk of such emergence is significantly lower when lytic phages are utilized.

This review is focused on applications of wild type bacteriophages for improving the safety of foods. We do not discuss other possible phage-related methods such as, for example, the use of phage endolysins for targeting foodborne pathogens, or using bacteriophages to manage food spoilage. Those topics have been discussed by other authors previously and respective reviews are available [13,14]. In the context of food safety applications, wild type lytic bacteriophages can be used both pre-harvest (e.g., in live animals, administered via animal feed or spray-applied prior to slaughter) and/or post-harvest (e.g., applied directly to food surfaces, either via direct spraying, via packaging materials, or by some other means) to reduce contamination by pathogenic bacteria [12,15]. Bacteriophage biocontrol could also be a means to disinfect surfaces used in the production and processing of foods [16,17]. In previous reviews [12,14,18,19], we and others have assembled a general overview of the industries and products where bacteriophages are used in food safety applications. Here, we provide an updated review (and an extended summary table) describing studies where bacteriophages have been applied to predominantly post-harvest foods, particularly meats, fresh produce and RTE foods (Table 1). In the next section, we review selected studies from the last five years where bacteriophage biocontrol was used to combat four major foodborne pathogens. Finally, we also discuss the regulation of bacteriophages for food safety applications and some of the challenges of phage biocontrol.

Table 1. A summary of studies of direct phage application onto a variety of foods.

Bacterium *	Phages	Notes	Ref.
Bacillus cereus	BCP1-1	*Bacillus cereus* counts decreased after treatment with a single phage in fermented soya bean paste without affecting *Bacillus subtilis*, a critical component of the fermentation process.	[20]
Campylobacter jejuni	Φ2	Counts of *Campylobacter* were reduced by ~1 log on the surface of chicken skin stored at 4 °C after the application of a single phage.	[21]
Campylobacter jejuni; *Salmonella* spp.	*C. jejuni* typing page 12673, P22, 29C; *Salmonella* typing phage 12	*C. jejuni* levels decreased ~2 logs on experimentally-contaminated chicken skin after application of phage at a MOI of 100:1 or 1000:1. *Salmonella* levels were reduced by ~2 logs on chicken skin treated with phage at an MOI of either 100:1 or 1000:1 and stored for 48 h; bacterial counts were reduced below the limit of detection when lower levels of bacteria were used to contaminate the chicken.	[22]
Campylobacter jejuni; *Salmonella* spp.	Cj6; P7	*Campylobacter* levels significantly decreased in beef after application of the phage Cj6, and decreases in bacterial levels were not significant at low levels of bacterial contamination (~100 CFU/cm^2). *Salmonella* counts were decreased ~2–3 logs at 5 °C and >5.9 logs at 24 °C in raw and cooked beef after P7 phage application. Surviving *Salmonella* colonies were still sensitive to P7. For both phages, the killing of bacteria was higher at an MOI of 10,000:1 and ~10,000 CFU/cm^2 of bacteria.	[23]
Cronobacter sakazakii	ESP 1-3, ESP 732-1	In infant formula, *Cronobacter sakazakii* (formerly *Enterobacter sakazakii*) levels were decreased after phage addition. The reduction was dependent on the phage concentration, and the phages were more effective at 24 °C than 37 °C or 12 °C.	[24]
Cronobacter sakazakii	Five phages	Growth of 36 of 40 test strains was inhibited by a phage cocktail tested in infant formula experimentally contaminated with *C. sakazakii*. Furthermore, both high and low concentrations (10^6 and 10^2 CFU/mL) of bacteria were eliminated from liquid culture medium treated with the individual phage (10^8 PFU/mL).	[25]
Escherichia coli O157:H7	e11/2, pp01, e4/1c	After incubation at 37 °C, a three-phage cocktail used to treat the surface of beef that was contaminated (10^3 CFU/g) with *E. coli* O157:H7 eliminated the bacterium from a majority of the treated specimens.	[26]
Escherichia coli O157:H7	EcoShield™ (formerly ECP-100)	*E. coli* 0157:H7 levels decreased by ~1–3 logs, or were reduced below the limits of detection, on tomatoes, broccoli or spinach after treatment with a phage cocktail, while *E. coli* O157:H7 levels were decreased by ~1 log when the phages were applied to ground beef.	[17]
Escherichia coli O157:H7	EcoShield™ (formerly ECP-100)	A phage cocktail applied to experimentally contaminated lettuce and cut cantaloupe significantly reduced *E. coli* O157:H7 levels by up to 1.9 and 2.5 logs, respectively.	[27]
Escherichia coli O157:H7	Cocktail BEC8	At various temperatures (4, 8, 23 and 37 °C), the phage cocktail significantly reduced the level of *E. coli* O157:H7 on leafy green vegetables by ~2–4 logs. The inclusion of an essential oil (*trans*-cinnamaldehyde) increased this effect.	[28]
Escherichia coli O157:H7	EcoShield™ (formerly ECP-100)	The levels of *E. coli* O157:H7 were reduced by ≥94% and ~87% on the surface of experimentally contaminated beef and lettuce, respectively, after addition of the phage cocktail; however, the single treatment did not protect foods after recontamination with the same bacteria (i.e., phage biocontrol had no continued technical effect on the foods).	[29]
Escherichia coli O157:H7	EcoShield™ (formerly ECP-100)	After a 30 min phage treatment at both 4 and 10 °C, levels of *E. coli* O157:H7 decreased by >2 logs on leafy greens under both ambient and modified atmosphere packaging storage.	[30]

Table 1. *Cont.*

Bacterium *	Phages	Notes	Ref.
Escherichia coli	FAHEc1	Contamination of raw and cooked beef decreased by 2–4 logs at 5, 24 and 37 °C in a concentration dependent manner after phage application. The *E. coli* displayed regrowth at higher temperatures.	[31]
Escherichia coli O157:H7	EcoShield™ (formerly ECP-100)	A phage cocktail was applied to lettuce by spraying and dipping. A larger initial reduction (~0.8–1.3 logs) in *E. coli* O157:H7 counts was observed after spraying. Dipping required submerging the lettuce for as long as 2 min, and the initial reductions were not significant. After 1 day of storage at 4 °C, dipping in the highest concentration of the phage cocktail reduced *E. coli* by ~0.7 log.	[32]
Escherichia coli	EC6, EC9, EC11	Two *E. coli* strains were eradicated from raw and UHT milk after treatment with a three-phage cocktail at 5–9 °C and 25 °C. For a third *E. coli* strain, phage treatment eliminated the bacteria from UHT milk; however, after an initial reduction, regrowth occurred in the raw milk after 144 or 9 h for 5–9 °C and 25 °C storage, respectively.	[33]
Escherichia coli, Salmonella, Shigella	EcoShield™ (formerly ECP-100), SalmoFresh™, ShigActive™	Phage cocktails were as effective or more effective than chlorine wash at reducing targeted pathogenic bacteria from broccoli, cantaloupe and strawberries in samples containing a large amount of organic content. Combination of the phage cocktail and a produce wash generated a synergistic effect, i.e., higher reductions of bacteria.	[34]
Listeria monocytogenes	ListShield™ (formerly LMP-102)	*Listeria* counts decreased by ~2 logs and ~0.4 log after application of a phage cocktail on melon and apple slices, respectively; a synergistic effect was observed when phage and nisin were used, decreasing levels of *Listeria* on the fruit ~5.7 logs and ~2.3 logs, respectively.	[35]
Listeria monocytogenes	ListShield™ (formerly LMP-102)	Application of a phage cocktail 1, 0.5 or 0 h before honeydew melon tissue were contaminated with the bacterium was most effective at reducing *Listeria* counts. This effect depended on the concentration of phage applied. *Listeria* counts decreased by ~5–7 logs after 7 days, when the phages were applied at the times described above.	[36]
Listeria monocytogenes	PhageGuard Listex™ (formerly Listex™; P100)	Levels of *Listeria* were reduced by at least 3.5 logs after a single phage was administered to the surface of ripened red-smear soft cheese. The surviving *Listeria* colonies isolated from the cheese after phage treatment were not resistant to the phage.	[37]
Listeria monocytogenes	A511, PhageGuard Listex™ (formerly Listex™; P100)	Levels of *Listeria* in experimentally contaminated chocolate milk and mozzarella cheese brine were eradicated after phage treatment at 6 °C. When the phage cocktail was applied to various solid foods, including sliced cabbage, iceberg lettuce leaves, smoked salmon, mixed seafood, hot dogs, and sliced turkey meat, a reduction of *Listeria* of up to 5 logs was observed.	[38]
Listeria monocytogenes	PhageGuard Listex™ (formerly Listex™; P100)	*Listeria* counts decreased by 1.8–3.5 logs after application of a single phage at ~10^8 PFU/g to the surface of raw salmon fillets that were stored at 4 °C or 22 °C.	[39]
Listeria monocytogenes	PhageGuard Listex™ (formerly Listex™; P100)	Levels of *Listeria* decreased by 1.4–2.0 logs CFU/g at 4 °C, 1.7–2.1 logs CFU/g at 10 °C, and 1.6–2.3 logs CFU/g at room temperature (22 °C) after application a single phage to the surface of raw catfish fillets. Regrowth was not observed after ten days of storage at either 4 °C or 10 °C.	[40]

Table 1. *Cont.*

Bacterium *	Phages	Notes	Ref.
Listeria monocytogenes	A511	The natural microbial community on soft cheese was maintained after addition of the phage. Levels of *Listeria* on experimentally contaminated cheese decreased by 2 logs and additional phage administrations did not improve the reduction of *Listeria*.	[41]
Listeria monocytogenes	FWLLm1	*Listeria* levels decreased by 1–2 logs on the surface of experimentally contaminated chicken stored in vacuum packages at 4 °C or 30 °C. Subsequent regrowth of *Listeria* was observed at 30 °C, but not at 4 °C.	[42]
Listeria monocytogenes	PhageGuard Listex™ (formerly Listex™; P100)	Counts of *Listeria* decreased by ~3 logs in experimentally contaminated queso fresco cheese after the addition of a single phage; however, subsequent growth was observed. Regrowth was prevented, and a similar log reduction was observed when PL + SD were included with the phage. Reduction of *Listeria* was lower, and regrowth occurred when LAE was included with phage.	[43]
Listeria monocytogenes	PhageGuard Listex™ (formerly Listex™; P100)	Compared to PL or PL + SD, a single phage was most effective at decreasing *Listeria* levels on RTE roast beef and turkey after storage at 4 °C or 10 °C, and subsequent bacterial growth was observed at both temperatures. Similar log reductions occurred when PL or PL + SD were used in conjunction with the phage, and regrowth was prevented or diminished at both 4 °C and 10 °C.	[44]
Listeria monocytogenes	PhageGuard Listex™ (formerly Listex™; P100)	Counts of *Listeria* decreased by ~1.5 logs on experimentally contaminated melon and pear slices, but not apple slices after two days at 10 °C. Additionally, treatment with phage did not impact *Listeria* levels in apple juice but decreased bacterial contamination by ~4 and ~2.5 logs in melon and pear juice, respectively.	[45]
Listeria monocytogenes	PhageGuard Listex™ (formerly Listex™; P100)	*Listeria* levels on soft cheese decreased by ~2–3 logs after 30 min and ~0.8–1 log after storage for 7 days at 10 °C.	[46]
Listeria monocytogenes	ListShield™ (formerly LMP-102)	Counts of *Listeria* decreased by 0.7 and 1.1 log on experimentally contaminated cheese and lettuce, respectively, after a 5 min treatment with phage and decreased the bacteria 1.1 log on the surface of apple slices after 24 h when combined with an antibrowning solution. The phage cocktail also virtually eliminated *Listeria* from experimentally contaminated frozen entrees that were frozen and thawed after treatment and was effective at eliminating environmental contamination by *Listeria* at a smoked salmon preparation plant.	[10]
Listeria monocytogenes	PhageGuard Listex™ (formerly Listex™; P100)	When applied to the surface of experimentally contaminated sliced pork ham, the phage reduced *Listeria* counts below the limit of detection after 72 h, and performed better than nisin, sodium lactate, or combinations of these antibacterial measures.	[47]
Mycobacterium smegmatis	Six phages	*M. smegmatis* counts were reduced below the limit of detection in milk treated with a six-phage cocktail or each component phage. Subsequent bacterial growth occurred when the component phages were used, but no growth was observed after 96 h at 37 °C, when the cocktail was applied.	[48]
Salmonella spp.	SJ2	*Salmonella* levels were reduced by 1–2 logs in raw and pasteurized cheeses created using milk that was treated with phage, while cheese made from milk without phage saw *Salmonella* counts rise ~1 log.	[49]

Table 1. *Cont.*

Bacterium *	Phages	Notes	Ref.
Salmonella spp.	SCPLX-1	Counts of *Salmonella* decreased by ~3.5 logs at 5 and 10 °C and ~2.5 logs at 20 °C on melon slices after application of a four-phage cocktail; treatment of apple slices with phage showed no reduction of bacteria.	[50]
Salmonella spp.	Felix-O1	*Salmonella* counts decreased by 1.8–2.1 logs after phage application to chicken frankfurters.	[51]
Salmonella spp.	PHL4	The levels of *Salmonella* recovered from experimentally contaminated broiler and naturally contaminated turkey carcasses were reduced by as high as 100% or 60%, respectively, after phage administration.	[52]
Salmonella spp.		Levels of *Salmonella* decreased by ~3 logs after application of a phage cocktail to sprouts; addition of an antagonistic bacteria to the phage cocktail increased this reduction to ~6 logs.	[53]
Salmonella spp.	FO1-E2	In chocolate milk and mixed seafood, *Salmonella* levels were reduced to undetectable levels after phage treatment and storage for 24 h at 8 °C and remained below the limit of detection. When foods were phage-treated and stored at 15 °C, *Salmonella* counts were reduced to undetectable levels within 24–48 h for hot dogs, sliced turkey breast, and chocolate milk, but regrowth occurred after 5 days. *Salmonella* levels were initially inhibited at ~0.5–2 logs and ~1–3 logs in egg yolk and mixed seafood, respectively, after phage addition; but bacterial recovery matched controls in egg yolks after two days, while the log reduction was maintained in seafood.	[54]
Salmonella spp.	UAB_Phi 20, UAB_Phi78, UAB_Phi87	*Salmonella* counts decreased by ~1 log on the shells of fresh eggs and by 2–4 logs on lettuce 60 min after application of the phage. After an initial reduction of 1–2 logs, when chicken breasts were dipped in a phage cocktail, no further decrease in the bacterial counts was observed over the next seven days at 4 °C. The levels of *Salmonella* were reduced by 2–4 logs on pig skin after phage application and storage for 6 h at 33 °C.	[55]
Salmonella spp.	wksl3	*Salmonella* counts decreased by ~3 logs on chicken skin after application of a single phage, and no further decrease in bacterial levels was observed over the next seven days at 8 °C. Further, mice that received a single dose of phage orally displayed no adverse effects.	[56]
Salmonella spp.	Five phages	The levels of *Salmonella* decreased by ~1 log on chicken skin after application of a five-phage cocktail comprised of closely related phages. The reduction of bacteria achieved by the phages was comparable to three different chemical antimicrobials.	[57]
Salmonella spp.	P22	After the administration of a single temperate phage and storage at 4 °C, levels of *Salmonella* decreased by ~0.5–2 logs on chicken, below the limits of detection in whole and skimmed milk, ~3 logs in apple juice, ~2 logs in liquid egg, and ~2 logs in an energy drink.	[58]
Salmonella spp.	SalmoFresh™	The stability of a *Salmonella*-specific phage preparation was determined in various chemical antimicrobials. Treatment of chicken breast fillets with a combination of phages and individual chemical antimicrobials did not produce a synergistic effect on the reduction of *Salmonella*; however, application of chlorine or PAA followed by spraying with phages significantly reduced *Salmonella* from chicken skin by up to 2.5 logs, compared to use of chlorine, low levels of PAA, or phage alone (0.5–1.5 logs).	[59]

Table 1. *Cont.*

Bacterium *	Phages	Notes	Ref.
Salmonella spp.	SalmoFresh™	Treatment of chicken breast fillets by dipping or surface application of a *Salmonella*-specific bacteriophage preparation and storage at 4 °C significantly reduced *Salmonella* contamination by up to 0.9 log; further, storing the meat in modified atmospheric packaging after surface application produced a higher reduction in bacterial counts (up to 1.2 logs).	[60]
Salmonella spp.	SalmoLyse®	A phage cocktail was sprayed onto experimentally contaminated raw pet food ingredients, including chicken, tuna, turkey, cantaloupe, and lettuce, and reduced the levels of the targeted bacteria by ~0.4–1.1 logs.	[61]
Salmonella spp.	SJ2	Application of the phage SJ2 significantly reduced *Salmonella* colonies recovered from experimentally contaminated ground pork and eggs with a larger reduction observed at room temperature, compared to 4 °C. After treatment, *Salmonella* colonies were screened for phage resistance, and significantly more phage-resistant *Salmonella* isolates were recovered from eggs, compared with ground pork.	[62]
Salmonella spp.	PhageGuard S™ (formerly Salmonelex™)	Boneless chicken thighs and legs were experimentally contaminated with *Salmonella* serovars isolated from ground chicken or other sources. A larger reduction of *Salmonella* was achieved when the bacteriophage preparation was diluted in tap water, compared to filtered water prior to application, and the phage cocktail was more effective against *Salmonella* isolated from other sources, compared to those from ground chicken.	[63]
Salmonella spp.	PhageGuard S™ (formerly Salmonelex™)	Treatment with a bacteriophage cocktail or irradiation significantly reduced (~1 log) the level of *Salmonella* on experimentally contaminated ground beef trim, and a combination of these methods decreased bacterial contamination by ~2 logs.	[64]
Shigella spp.	SD-11, SF-A2, SS-92	*Shigella* counts were reduced by ~1–4 logs on pieces of spiced chicken after application of a phage cocktail or each of the component phages and storage at 4 °C.	[65]
Shigella sonnei	ShigaShield™	Application of a five-phage, *Shigella*-specific cocktail to various RTE foods, including lettuce, melon, smoked salmon, corned beef and pre-cooked chicken, reduced the recovery of *Shigella* ~1.0–1.4 logs at the highest phage concentration applied compared to control.	[66]
Staphylococcus aureus	Φ88, Φ35	*S. aureus* levels decreased below the limit of detection in experimentally contaminated whole milk after treatment, with a two-phage cocktail and storage at 37 °C. After phage treatment, *S. aureus* was not recovered from the acid curd after storage for 4 h at 25 °C, and was eliminated from the renneted curd after 1 h at 30 °C.	[67]
Staphylococcus aureus	vB_SauS-phi-IPLA35, vB_SauS-phi-SauS-IPLA88	Counts of *S. aureus* were significantly decreased in cheese made using milk treated with phages compared to milk without the addition of phages. The microbiota of the cheese was not impacted by the addition of the phages.	[68]

* Listed in alphabetical order by bacteria and then chronologically. In cases where multiple bacteria were examined, the study is listed with the alphabetically first bacteria. Adapted and modified from Sulakvelidze 2013 and Woolston and Sulakvelidze 2015 [12,18]. h, hour; LAE, lauric arginate; log(s), logarithmic unit(s); min, minutes; PAA, peracetic acid; PL, potassium lactate; PL + SD, potassium lactate + sodium diacetate; RTE, Ready-To-Eat; UHT, ultra-high temperature.

Table 2. Phage products approved for food safety applications.

Company	Phage Product	Target Organism(s)	Regulatory	Certifications	References
FINK TEC GmbH (Hamm, Germany)	Secure Shield E1	*E. coli*	FDA, GRN 724 *pending as of 19 March 2018*		
Intralytix, Inc. (Baltimore, MD, USA)	Ecolicide® (EcolicidePX™)	*E. coli* O157:H7	USDA, FSIS Directive 7120.1		
	EcoShield™	*E. coli* O157:H7	FDA, FCN 1018; Israel Ministry of Health; Health Canada	Kosher; Halal	[17,27,29,30,32,34]
	ListShield™	*L. monocytogenes*	FDA, 21 CFR 172.785; FDA, GRN 528; EPA Reg. No. 74234-1; Israel Ministry of Health; Health Canada	Kosher; Halal; OMRI	[10,35,36]
	SalmoFresh™	*Salmonella* spp.	FDA, GRN 435; USDA, FSIS Directive 7120.1; Israel Ministry of Health; Health Canada	Kosher; Halal; OMRI	[59,60]
	ShigaShield™ (ShigActive™)	*Shigella* spp.	FDA, GRN 672		[66,69]
Micreos Food Safety (Wageningen, Netherlands)	PhageGuard Listex™	*L. monocytogenes*	FDA, GRN 198/218; FSANZ; EFSA; Swiss BAG; Israel Ministry of Health; Health Canada	Kosher; Halal; OMRI; SKAL	[37–40,43–47]
	PhageGuard S™	*Salmonella* spp.	FDA, GRN 468; FSANZ; Swiss BAG; Israel Ministry of Health; Health Canada	Kosher; Halal; OMRI; SKAL	[63,64]
		E. coli O157:H7	FDA, GRN 757 *pending as of 19 March 2018*		
Passport Food Safety Solutions (West Des Moines, IA, USA)	Finalyse®	*E. coli* O157:H7	USDA, FSIS Directive 7120.1		
Phagelux (Shanghai, China)	AgriPhage™	*Xanthomonas campestris* pv. vesicatoria, *Pseudomonas syringae* pv. tomato	EPA Reg. No. 67986-1		
	SalmoPro®	*Salmonella* spp.	FDA, GRN 603		
		Salmonella spp.	FDA, GRN 752 *pending as of March 19, 2018*		

Adapted and modified from Woolston and Sulakvelidze 2015 [18]. This is not meant to be an exhaustive list of phage products or approvals and listings. Some of the information included in this table was obtained from company webpages and promotional material and has not been independently verified. BAG, Bundesamt für Gesundheit; CFR, Code of Federal Regulations; FSIS, Food Safety and Inspection Service; GRN, GRAS Notice.

2. Phage Biocontrol for Targeting Common Foodborne Bacterial Pathogens

2.1. *Listeria monocytogenes*

Listeria monocytogenes is a rod-shaped, Gram-positive, facultative anaerobe. Consumption of foods contaminated with *L. monocytogenes* causes a range of symptoms in humans such as initial flu-like or gastrointestinal symptoms which, in some cases, progress to encephalitis or cervical symptoms, and possibly stillbirth in pregnant mothers. It was estimated that in 2010, global cases of foodborne infection with *L. monocytogenes* exceeded 14,000 and resulted in more than 3000 deaths [1]. *L. monocytogenes* is able to survive and grow at refrigerated temperatures (2–8 °C) commonly used during the distribution and storage of many foods; therefore, the detection and elimination of *L. monocytogenes* is critically important to ensuring the safety of the food chain, especially in RTE foods. In this context, the application of bacteriophages to assorted foods (including RTE foods) has been shown, by several investigators, to be effective at reducing contamination with *L. monocytogenes* (Table 1). For example, a commercial monophage preparation (i.e., phage preparation consisting of one single phage) targeting *Listeria* was reported to be effective in reducing the levels of *L. monocytogenes* in sliced ham, and to be superior to nisin and sodium lactate, when compared at the storage abuse temperature of 6–8 °C [47]. A similar study by Chibeu and colleagues (2013) demonstrated that the same monophage preparation was also able to reduce *L. monocytogenes* on the surface of other deli meats [44]. The meats (cooked sliced turkey and roast beef) were stored at 4 °C and the abuse temperature of 10 °C. The *Listeria*-specific phage was effective against *L. monocytogenes* when used alone, and it enhanced the effectiveness of other antimicrobials when used together with sodium diacetate or potassium lactate. All these studies utilized a single phage preparation. A phage cocktail prepared with multiple bacteriophages compared to a single phage preparation may be superior, both in terms of providing broader coverage of the target species and of reducing the risk of resistant bacteria emerging. One such commercially available six-phage cocktail targeting *L. monocytogenes* has been tested on a number of foods experimentally contaminated with *L. monocytogenes*, including lettuce, hard pasteurized cheese, smoked salmon, and Gala apple slices; application of this bacteriophage cocktail reduced *L. monocytogenes* levels in all these foods by ~0.7–1.1 logs [10]. The same study examined the application of the *L. monocytogenes*-specific cocktail on prepackaged, frozen meals. The meals were experimentally contaminated with *L. monocytogenes*, treated with the phage cocktail, and subjected to freezing and thawing cycles. The results showed a 2.2 log reduction of *L. monocytogenes*, which suggests that phage biocontrol can be an effective means to control *L. monocytogenes* in foods under "storage abuse" conditions when the frozen meals are intentionally or unintentionally thawed multiple times during their storage [10].

In many of the above-reviewed studies, despite the initial significant reduction in *L. monocytogenes* levels in the foods, the targeted bacterial populations were not completely eradicated, and viable *L. monocytogenes* cells could still be recovered, albeit in much lower numbers. However, the bacteriophage preparations were still effective against randomly selected colonies of the recovered bacteria, suggesting that phage-resistance was not the primary reason for the incomplete eradication of *L. monocytogenes* [23,37,44]. There could be several possible explanations for this observation. For example, the *L. monocytogenes* cells could be exhibiting temporal resistance to phage infection, as reported previously [70,71]. Another possible explanation is that the phages did not come into direct contact with some *L. monocytogenes* cells after the phages were sprayed onto the foods (e.g., due to using an insufficient volume of spray, particularly on foods with complex topography), which resulted in those bacterial cells not being lysed by phages. In this latter scenario, using larger spray volumes, fine (mist-like) sprays, rotating/tumbling foods during phage application, and otherwise ensuring thorough surface coverage with phages may help enhance the effectiveness of phage biocontrol.

2.2. *Salmonella spp.*

The non-typhoid serotypes of *Salmonella enterica* account for many incidents of gastroenteritis worldwide each year. The disease caused by these Gram-negative, rod-shaped bacteria is often self-limiting, with symptoms typically including stomach cramps, fever, nausea and diarrhea. However, life-threatening instances can occur in cases of dehydration and when the bacteria invade beyond the gastrointestinal tract. Estimates indicate that globally over 78 million cases of foodborne infection were caused by non-typhoid *Salmonella* in 2010, leading to almost 60,000 deaths [1]. During the processing and packaging of foods, *Salmonella*, as well as other pathogens, can adhere to the surfaces where food is prepared, leaving them contaminated. These factors place RTE foods, such as fresh fruits and vegetables that are not cooked before eating, at a particularly high risk for transmitting bacterial pathogens and causing food poisoning.

At least two FDA-cleared *Salmonella*-targeting phage preparations are currently on the market (Table 2). Several publications are available describing their applications (and that of other noncommercial phage preparations) in various foods. Brief summaries of those studies are given in Table 1. One study is of particular interest, as it demonstrates an example of how phage-resistance could be managed if and when it hinders the efficacy of a bacteriophage preparation. In that study, a GRAS-listed (Generally Recognized as Safe) six-phage cocktail targeting *Salmonella* was examined for its ability to reduce the levels of *Salmonella* on surfaces mimicking those commonly found in food processing establishments, e.g., stainless steel and glass [16]. Initial studies demonstrated that the *Salmonella*-specific bacteriophage cocktail significantly reduced the population of susceptible *Salmonella* strains on all surfaces examined by ~2–4 logs; at the same time, it was ineffective in reducing the levels of another strain of *Salmonella* (*Salmonella* Paratyphi B S661) that was resistant to the phage cocktail *in vitro* [16]. However, when the phage cocktail was adjusted to include phages specifically targeting this resistant strain, the updated preparation showed a significant reduction (~2 logs) of S. Paratyphi B S661 from the surfaces, while also maintaining effectiveness against the previously susceptible isolates [16]. This study provides compelling evidence that phage cocktails can easily be modified to target specific bacterial strains, e.g., if phage-resistant mutants emerge, or to specifically target the problem strains prevalent in particular food-manufacturing facilities.

In addition to their usefulness in decontaminating food preparation surfaces, bacteriophage cocktails have also been effective at eliminating *Salmonella* directly from the foods. For example, the same *Salmonella*-specific cocktail discussed above reduced the levels of *Salmonella* on experimentally contaminated chicken parts when applied alone, and this effect was enhanced when the phage was applied in combination with conventional chemical sanitizers [59]. In the case of chicken breast fillets, the bacteriophage cocktail significantly reduced the numbers of a mixture of *Salmonella* species when applied to the surface of the fillets or when the fillets were dipped into a vessel containing the phage solution [60]. Furthermore, this phage cocktail significantly reduced the number of *Salmonella* when the fillets were stored under aerobic or modified atmospheric conditions [60]. This latter finding may have direct practical implications as food manufacturers often use modified atmospheric conditions to discourage growth of bacteria and increase the shelf life of foods. Another study found that a single phage, SJ2, significantly reduced the amount of *Salmonella* in liquid egg and ground pork, and this reduction was more pronounced at higher temperatures [62]. The authors screened remaining *Salmonella* colonies for resistance; while there was no difference in the number of resistant clones from phage treated and untreated samples of ground pork, there was a significantly higher number of resistant clones found in the phage treated samples of egg liquid [62]. The authors suggested both the food matrix (solid versus liquid) and the differences in the microbiome of the two foods could have contributed to this difference in the number of resistant *Salmonella* isolates [62].

Foodborne illnesses caused by non-typhoid serotypes of *Salmonella* are also a health risk for companion animals (e.g., dogs and cats), and the close association of these animals with their owners raises the possibility of illness in humans. Indeed, human *Salmonella* outbreaks have been associated with contaminated dry cat and dog food, and approximately one third of commercial raw and natural pet foods sampled have been found to contain *Salmonella* [72,73]. In an effort to address this health risk, phage biocontrol has recently been examined as a technique to reduce or eliminate *Salmonella* in pet foods. The six-phage *Salmonella*-specific cocktail discussed above was found to reduce the levels of *Salmonella* in experimentally contaminated dry dog food by 1 log [74]; when cats and dogs were fed dry kibble treated with the same phage cocktail, it appeared to be safe and did not noticeably impact any of the major health metrics recorded for either of the animals [61].

An alternative to dry pet chow that is gaining increasing popularity is raw pet food. These pet meals consist of meats, such as chicken, duck or tuna, combined with vegetables, including lettuce, blueberries and broccoli, which are sold and served raw [61]. Raw pet foods are gaining increased popularity due to their superb nutritional values; at the same time, because they are uncooked, there is a heightened possibility of foodborne pathogens being present in them, which can be transferred to pets as well as to unsuspecting consumers during the feeding process. At least one report was recently published in which the authors examined the value of using phages to control *Salmonella* in raw pet food ingredients. Reductions of bacterial contamination ranged from 0.4 log to 1.1 logs, the efficacy was concentration-dependent, and the largest reduction was achieved when high doses of the bacteriophage preparation were used [61] (Table 1).

2.3. *Escherichia coli*

Many strains of the Gram-negative, rod-shaped bacteria *Escherichia coli* are naturally found in the human gut and are beneficial for our health and wellbeing; for example, they aid in the digestion of food and maintenance of a robust immune system. However, some *E. coli* strains can and do cause illnesses in humans. For example, the Shiga toxin producing *E. coli* serotype O157:H7, which is sometimes found in contaminated water or foods, especially beef, can invade the human gastrointestinal tract and trigger disease, with symptoms including abdominal cramping and hemorrhagic diarrhea. These infections are typically self-limiting in immunocompetent individuals but can potentially be life-threatening in very young or old patients. It has been estimated that globally more than one million cases of foodborne illness and over one hundred deaths could be attributed to Shiga toxin-producing *E. coli*, including the O157:H7 serotype [1].

Recent work has demonstrated that *E. coli*-specific phage preparations were effective when used to treat fresh vegetables [75] and both Ultra-High-Temperature (UHT) treated and raw milk contaminated with *E. coli* [33]. In the first study, the levels of *E. coli* O157:H7 on green pepper slices and spinach leaves were reduced by a single phage by ~1–4 logs, and the initial reduction was maintained at 4 °C while some regrowth was seen at 25 °C. In the second study, the levels of *E. coli* were reduced to undetectable levels in both UHT and raw milk when a cocktail of two or three phages was used. Of note, in all samples treated with the three-phage preparation, this reduction was maintained over storage at both 4 and 25 °C; in contrast, there was regrowth of the *E. coli* strain in the samples treated with the two-phage cocktail. While the underlying reasons are not fully understood, it is possible that the three-phage cocktail provided better management of resistance versus a two-phage cocktail, and the enhanced efficacy of multi-phage cocktails has been demonstrated for other phage preparations previously [76]. Although the underlying reasons for this phenomenon have not been rigorously determined, it is possible that having multiple phages in a phage cocktail reduces the risk of the emergence of phage-resistant mutants, because multiple mutations would be required to render a given bacterial cell resistant to not one, but multiple phages in the cocktail, assuming the phages target distinct cellular structures. This concept is essentially the same as the multi-hurdle approach, which proposes using a combination of antibacterial strategies to discourage the development of bacterial resistance [77]. These and some additional studies using *E. coli*-specific phages in food safety applications are briefly summarized in Table 1.

2.4. Shigella spp.

Species of the Gram-negative, rod-shaped bacterial genus *Shigella* cause a self-limiting gastrointestinal infection with symptoms including hemorrhagic diarrhea and stomach pain. Globally, the incidence of foodborne infection caused by *Shigella* species in 2010 was recently estimated to be over 50 million, resulting in over 15,000 deaths [1]. The vast majority of these infections occurred in developing countries, with the highest number of the infections and death occurring in children under the age of 5 [1,78].

Only one FDA-cleared food safety phage preparation is currently available to target *Shigella* spp. [66,69]. This five-phage cocktail was granted the GRAS status in 2017 (GRN 672) (Table 2), and it was shown to reduce the levels of *Shigella* by approximately 1 log in a variety of foods, including melons, lettuce, yogurt, deli corned beef, smoked salmon, and chicken breast meat [66]. In another study, the same *Shigella*-specific bacteriophage cocktail was used to compare the safety and efficacy of phage administration to antibiotic treatment in mice challenged with a *Shigella sonnei* strain [69]. This study demonstrated that, while the *Shigella* specific bacteriophage cocktail was as effective as a standard antibiotic treatment at reducing the bacterial load in mice, treatment with the antibiotic significantly altered the diversity of the mouse intestinal community, while the phage treatment did not i.e., phage administration had a much milder impact on the normal gut microbiota of mice, compared to the antibiotic treatment [69]. The authors did not observe any deleterious side effects in the mice after phage administration, that is, the phage did not alter the composition of the blood or urine of the mice, nor did it have any detrimental effect on the morbidity or mortality, weight or any other physiological parameters of the animals [69]. Although not directly relevant to food safety applications, the study did suggest that these bacteriophages, when administered orally (mimicking a scenario when they would be consumed when eating foods treated with them) did not deleteriously impact the normal gut microflora (in contrast to antibiotics) and triggered no side effects in any of the animals examined.

2.5. Campylobacter jejuni

Campylobacter spp., Gram-negative, curved rod-shaped bacteria, are major foodborne pathogens of humans, causing gastrointestinal symptoms that can include stomach pain, fever and diarrhea. It a recent (2015) report, FERG estimated that in 2010, the global cases of foodborne illness caused by *Campylobacter* spp. exceed 95 million and resulted in over 21,000 deaths [1]. The intestinal microflora of many fowl and other livestock animals include species of *Campylobacter*. Additionally, though the route of entry is not fully understood, *Campylobacter* can frequently be isolated from both the surface of and internally within chicken livers. Zoonotic infections commonly occur in humans when contaminated animal products, such as meats, are handled or consumed. Thus, humans are at an elevated risk for *Campylobacter* infection when minimally cooked preparations, e.g., pâté, are prepared.

Several *Campylobacter* bacteriophages have been isolated from chickens, including the fecal matter as well as the surface and internal tissues of chicken livers, and some of them have been examined for their ability to reduce contamination of various foods by *Campylobacter* [79–82]. For example, Hammerl and colleagues [80] used the phages as a pre-harvest treatment, and showed significant reduction (~3 logs) in *Campylobacter* fecal counts when 20-day-old chickens were treated with two phages in successive application (a Group III phage, then a Group II phage). Interestingly, dosing of the Group III phage alone or in conjunction with another Group III phage was not effective, suggesting that a combination of different phages (Group II and III) was required for optimal efficacy. The isolation of *Campylobacter*-specific phages has historically been done on a limited number of *Campylobacter* isolates, with many studies utilizing just one *C. jejuni* NCTC 12662 isolate as a host strain for phage isolation. Phages isolated using that one strain are almost exclusively Group III phages that target a particular receptor, the capsular polysaccharide [83]. In contrast, phages isolated on *C. jejuni* RM1221 are typically Group II phages that utilize the flagella as a route of entry [83]. As the above study suggests [80], phage cocktail consisting of phages that target different receptors could potentially lead to a broader target range and more effective cocktails.

3. Bacteriophage Preparations as Commercial Products

3.1. Regulation of Bacteriophage Preparations

In the last approximately 12 years, the number of regulatory approvals issued for bacteriophage preparations and their use for improving food safety has been steadily increasing (Table 2). In 2006, the first approval for a bacteriophage preparation to be used directly in the food supply was issued by the FDA for the *L. monocytogenes*-specific cocktail ListShield™ as a food additive (FDA does not "approve" any products, phage-based or otherwise; however, the term "approval" is commonly used to indicate obtaining FDA clearance to use products for their intended applications). Later that year, the FDA issued a no-objection letter for the *Listeria*-specific preparation Listex™ (currently PhageGuard Listex™) as a Generally Recognized as Safe (GRAS) substance. In recent years, a number of phage products (e.g., SalmoFresh™ and PhageGuard S™) have been granted GRAS designation by the FDA, and application for GRAS designation now appears to be the standard route of approval for phage products intended to treat post-harvest foods. As wild-type (i.e., not genetically modified) lytic bacteriophages are all natural and already inherently present in the food supply, the GRAS designation does seem to be an appropriate regulatory avenue for such preparations. In addition, the USDA has included several phage preparations in their issued guidelines for safe and suitable ingredients used in the production of meat, poultry, and egg products. For example, under FSIS Directive 7120.1, the application of phage to livestock animals prior to slaughter (e.g., *E. coli* O157:H7-targeted phages to the hides of cattle) and food (e.g., *Salmonella*-targeted phage onto poultry or meat) is permitted. These guideless were developed using specific phage preparations, but, in general, any phage product that meets the description in the directive may be considered to be compliant. Following the lead of regulators in the United States, several health agencies in countries around the world have issued approvals of phage products for use on foods; some examples include Israel, Canada, Switzerland, Australia, New Zealand, and the European Union (Table 2).

3.2. Challenges for Bacteriophage Biocontrol

As described in the previous sections, bacteriophage biocontrol is being increasingly used for targeting specific pathogenic bacteria in various foods, with a growing body of literature attesting to the utility of bacteriophages to reduce or eradicate their targeted pathogenic bacteria in foods. However, some challenges still remain before bacteriophage biocontrol is more widely accepted, including technical constraints and the general consumer acceptance of phage application on foods. Some of these challenges are briefly discussed below.

3.2.1. Technical Challenges

Arguably the biggest technical challenge with phage biocontrol is its efficacy. One common observation in studies using bacteriophages on foods is that levels of contaminating bacteria drop initially, and very little or no further reduction in bacteria occurs afterwards [54,56]. In other words, phages can effectively reduce the levels of their targeted bacteria in foods, but they do not always eliminate them fully. Bacteriophage must come into contact with susceptible bacterial cells to lyse them. Given the nature of phage replication cycle (which starts with one phage infecting one bacterial cell and resulting in 100–200 progeny phages bursting from that cell at the end of each replication cycle; i.e., an exponential effect), one could expect that the reduction in bacterial cells will exponentially increase with more replication cycles, as more progeny phages are generated as the result of ongoing phage-mediated lysis of the targeted bacteria. However, several reports have suggested that the concentration of phages does not substantially increase after application to foods [43–45], strongly suggesting that "autodosing" (exponential increases in the population of phages due to repetitive lytic replication cycles) does not occur, at least under the conditions tested to date. It is likely that the progeny phages are unable to reach and invade additional bacteria in foods, especially in drier food matrixes, where passive movement of phages across food surfaces is limited due to the lack of moisture. In this context, it has been suggested that fewer phage particles may be required to significantly reduce bacterial contamination on moist food surfaces and in liquids compared to drier food matrices, presumably because of the increased "mobility" of phages in the presence of moisture (e.g., natural juices of some foods) [84]. One potential answer to this challenge is to use a phage solution with higher concentrations of phage particles, to increase the likelihood of phages coming into contact with their targeted bacteria upon application [17,21,36,66]; however, a more concentrated solution will be more expensive, to the point that it may be prohibitively expensive for food processors to implement. Another option is the use of larger spray volumes applied via fine mist sprays, to more efficiently disperse the phage particles across the surface of the food, increasing their likelihood of encountering a target bacterium, which could be especially important under circumstances when foodborne pathogens are present in very low concentrations or when the infective dose of the pathogen is extremely low. Proper application of bacteriophages onto foods to ensure thorough surface coverage and optimal efficacy is one of the main technical challenges for phage biocontrol, and it encompasses a range of issues from phage dosing (i.e., effective concentration of phage delivered in an optimal volume and how these could be verified in food processing facilities) to proper equipment (both to provide accurate dosing, as just mentioned, and appropriate mixing or tumbling during phage application to ensure that the entire surface of food is thoroughly treated with the phage solution).

An additional efficacy-related issue is that phage biocontrol typically reduces the levels of targeted bacteria by 1–3 logs (with rare exceptions: in one study, a reduction of *Listeria* of up to 5 logs was reported as a result of phage treatment [36]), and this is considerably lower than the up to 5 logs reduction reported for some other, more harsh interventions, e.g., irradiation. Although this may be more a perception problem than a real technical issue (since very few, if any, foods are contaminated with 5 logs of foodborne pathogens per gram), the lower reduction may be considered by the food industry to be inferior. Even when the targeted bacterium is not totally eliminated from foods and is only reduced by 1 or 2 logs, it may still render the food safer for consumption. For example, and to put this into a broader perspective, in 2003 the FDA and USDA's FSIS jointly authored a risk-assessment study in which they modeled a series of "what if" scenarios, including one in which reductions in deli meat contamination would affect the mortality rate of elderly people. According to that analysis, a 10-fold reduction (1 log) and 100-fold reduction (2 logs) in pre-retail contamination with *L. monocytogenes* would reduce the mortality rate by ca. 50% and 74%, respectively in that segment of population [85]. Thus, implementation of phage biocontrol protocols—even if they do not eradicate (i.e., totally eliminate) the targeted foodborne pathogens from foods but reduce them by 1–3 logs—may yield significant improvements in food safety and public health.

Another technical challenge is related to the way phage biocontrol is implemented. Phage biocontrol provides an effective tool for improving food safety, but it does not eliminate the need for safe food handling practices. For example, regrowth of bacteria was observed after phage treatment if the foods are stored at abuse temperatures [33,48,54]. Also, some planning is required to maintain optimal efficacy of phage biocontrol when combining bacteriophages with some other food safety interventions, such as using phages in conjunction with chemical sanitizers [59]. For example, a number of chemical sanitizers are capable of inactivating phages, and therefore, they must be applied separately to ensure that phages retain viability in order to achieve the largest reductions of bacteria [59]. In this context, some investigators have reported that combinations of bacteriophage and preservatives are less effective than either treatment alone [86]. However, when proper synergistic combinations of phage preparations with other sanitizers are identified, the efficacy of each could be improved. For example, in the presence of high organic loads, the efficacy of a levulinic acid produce wash was enhanced (by up to 2 logs) when the fruits and vegetables were pretreated with a bacteriophage preparation [34].

Finally, another application-related (and efficacy-affecting) technical challenge is the possible emergence of phage-resistant bacterial isolates. Researchers do recover bacteria resistant to phage treatments [62], and there is the concern that widespread use of this treatment may eventually select for phage-resistant bacteria. Phages utilize a variety of bacterial structures to initiate the invasion of bacterial cells, including surface polysaccharides and proteins, as well as the flagella [87–89]. The use of phage cocktails containing multiple, diverse phages (e.g., phages that use different receptors on the surface of bacteria) versus a single monophage may provide a mechanism to reduce the risk/likelihood of bacterial resistance. Also, the intervention strategy itself can play a key role in managing the emergence of phage-resistant mutants. For instance, applying phages at the end of the food processing cycle (e.g., when phages are sprayed onto food immediately before packaging) reduces the overall "selective pressure" in the environment as bacterial exposure to the phages is limited. Consequently, there is less risk of phage-resistant mutants emerging when compared to, for example, spraying chicken houses or similar complex environments with phages in an effort to reduce the contamination of livestock animals. Finally, if and when resistance does arise, phage cocktails could be modified to include phages targeting previously resistant bacteria; one example of such an approach was published previously and discussed elsewhere in this review [16].

3.2.2. Customer Acceptance

In recent years, consumers have increasingly demonstrated a reluctance to purchase foods treated with chemical sanitizers and antibiotics or foods that are "genetically modified", while simultaneously the demand for organic foods and products produced locally, such as at local farmer's markets and Community Supported Agriculture (CSA), has been on the rise [90,91]. This trend bodes well for phage biocontrol, which offers a non-chemical, green, and targeted antimicrobial approach for improving the safety of foods. However, the public may be disinclined to purchase foods processed with unfamiliar techniques, and the idea of "spraying viruses onto their food" could cause discomfort. Furthermore, food producers are generally reluctant to modify their practices, especially if there is a chance the public will react negatively. Thus, for phage biocontrol to be more widely utilized, it will be critical to provide education to the public and food processors, to explain the safety, efficacy, and ubiquity of bacteriophages.

Phages are the most abundant organisms on the planet with approximately 10^{31} particles existing (ten times that of the total global bacterial population) [92], and approximately 10^{15} phage particles populating the human intestinal tract [93]. Phages are part of the normal microflora of all fresh foods [94], and they have been isolated from a variety of foods, from fruits and vegetables to meat and dairy products, often in very high numbers, e.g., up to 1×10^9 PFU/mL in yogurt [95,96]. Phage biocontrol is also likely to be one of the most environmentally-friendly interventions available. In a previous review [18], we estimated that if phages were applied at the *maximum* approved amount (10^9 PFU/g for one phage product, all other current approvals are for up to 10^7–10^8 PFU/g) to *all* the approved food consumed by the average American in one day, the phages consumed would represent <0.2% of the number of phages already present in the human intestinal tract. This calculation is a gross overestimate, especially considering several GRAS approvals permit an application of up to 10^8 PFU/g (reducing the daily intake of phage to ~0.02% of the phage in the human intestinal tract). Also, this estimate assumes that (1) all possible food is treated, (2) all the applied phages survive the stomach acid and make it into the small intestine (yet most of the phages are usually destroyed when exposed to the acidic pH of the stomach), (3) the maximum approved amount of phages is applied, and (4) bacteriophage biocontrol is universally used by all relevant food industries in the United States. In short, the number of phages added to the environment and introduced into the human intestinal tract as a result of phage biocontrol is negligible, especially when compared to naturally present phage populations. Moreover, the phages in all currently available commercial products (Table 2) are not genetically modified and originated from the environment, potentially even from foods, in the first place. However, the general public is often not aware of these facts. Thus, proper understanding of the safe nature and ubiquity of lytic phages and the *pros* and *cons* of phage biocontrol by consumers and food processors alike will be critical for further successful implementation of this promising approach. In at least one recent study, consumers appeared to be willing to pay more for bacteriophage-treated fresh produce after the science behind phage biocontrol and the advantages of this technique were explained to them [97].

4. Concluding Remarks

Though some challenges remain, bacteriophage biocontrol is increasingly accepted as a safe and effective method to eliminate, or significantly reduce the levels of, specific bacterial pathogens from foods. Commercial bacteriophage products are currently available and have been approved for use in a growing number of countries. These products can be used to address contamination by specific bacterial pathogens at a variety of timepoints during food production, including spraying on produce, applying to livestock animals before processing, rinsing of food contact surfaces in processing facilities, and treatment of post-harvest food products, including RTE foods. Despite the progress made in improving the safety of our foods, foodborne illnesses remain a constant threat, especially for individuals with weaker immune systems, e.g., children, the elderly, and pregnant women. Bacteriophage biocontrol can serve as an additional tool in a multi-hurdle approach to prevent foodborne pathogens from reaching consumers, and this method is especially promising under circumstances when food processors strive to preserve the natural, and often beneficial, microbial population of foods and to only remove the bacteria that may cause illness in humans.

Acknowledgments: This material is based upon work supported, in part, by the U.S. Army Contracting Command—APG, Natick Contracting Division, Natick, MA, USA, under Contract No. #W911QY-18-C-0010 (to Alexander Sulakvelidze). The funders had no role in the conception of this literature review, decision to publish, or preparation of the manuscript.

Conflicts of Interest: Joelle Woolston and Alexander Sulakvelidze hold an equity stake in Intralytix, Inc., a Maryland corporation developing bacteriophage preparations for various applications, including food safety.

References

1. Havelaar, A.H.; Kirk, M.D.; Torgerson, P.R.; Gibb, H.J.; Hald, T.; Lake, R.J.; Praet, N.; Bellinger, D.C.; de Silva, N.R.; Gargouri, N.; et al. World Health Organization global estimates and regional comparisons of the burden of foodborne disease in 2010. *PLoS Med.* **2015**, *12*, e1001923. [CrossRef] [PubMed]
2. Scharff, R.L. Economic burden from health losses due to foodborne illness in the United States. *J. Food Prot.* **2012**, *75*, 123–131. [CrossRef] [PubMed]
3. Wolbang, C.M.; Fitos, J.L.; Treeby, M.T. The effect of high pressure processing on nutritional value and quality attributes of *Cucumis melo* L. *Innov. Food Sci. Emerg.* **2008**, *9*, 196–200. [CrossRef]
4. Bajovic, B.; Bolumar, T.; Heinz, V. Quality considerations with high pressure processing of fresh and value added meat products. *Meat Sci.* **2012**, *92*, 280–289. [CrossRef] [PubMed]
5. Suklim, K.; Flick, G.J.; Vichitphan, K. Effects of gamma irradiation on the physical and sensory quality and inactivation of *Listeria monocytogenes* in blue swimming crab meat (*Portunas pelagicus*). *Radiat. Phys. Chem.* **2014**, *103*, 22–26. [CrossRef]
6. Wheeler, T.L.; Shackelford, S.D.; Koohmaraie, M. Trained sensory panel and consumer evaluation of the effects of gamma irradiation on palatability of vacuum-packaged frozen ground beef patties. *J. Anim. Sci.* **1999**, *77*, 3219–3224. [CrossRef] [PubMed]
7. Beuchat, L.R.; Ryu, J.H. Produce handling and processing practices. *Emerg. Infect. Dis.* **1997**, *3*, 459–465. [CrossRef] [PubMed]
8. Sohaib, M.; Anjum, F.M.; Arshad, M.S.; Rahman, U.U. Postharvest intervention technologies for safety enhancement of meat and meat based products; a critical review. *J. Food Sci. Technol.* **2016**, *53*, 19–30. [CrossRef] [PubMed]
9. Sulakvelidze, A.; Alavidze, Z.; Morris, J.G., Jr. Bacteriophage therapy. *Antimicrob. Agents Chemother.* **2001**, *45*, 649–659. [CrossRef] [PubMed]

10. Perera, M.N.; Abuladze, T.; Li, M.R.; Woolston, J.; Sulakvelidze, A. Bacteriophage cocktail significantly reduces or eliminates *Listeria monocytogenes* contamination on lettuce, apples, cheese, smoked salmon and frozen foods. *Food Microbiol.* **2015**, *52*, 42–48. [CrossRef] [PubMed]
11. Viator, C.L.; Muth, M.K.; Brophy, J.E. *Costs of Food Safety Investments*; Report; RTI International: Research Triangle Park, NC, USA, 2015. Available online: https://www.fsis.usda.gov/wps/wcm/connect/0cdc568e-f6b1-45dc-88f1-45f343ed0bcd/Food-Safety-Costs.pdf?MOD=AJPERES (accessed on 19 March 2018).
12. Sulakvelidze, A. Using lytic bacteriophages to eliminate or significantly reduce contamination of food by foodborne bacterial pathogens. *J. Sci. Food Agric.* **2013**, *93*, 3137–3146. [CrossRef] [PubMed]
13. Schmelcher, M.; Loessner, M.J. Bacteriophage endolysins: Applications for food safety. *Curr. Opin. Biotechnol.* **2016**, *37*, 76–87. [CrossRef] [PubMed]
14. Greer, G.G. Bacteriophage control of foodborne bacteria. *J. Food Prot.* **2005**, *68*, 1102–1111. [CrossRef] [PubMed]
15. Lone, A.; Anany, H.; Hakeem, M.; Aguis, L.; Avdjian, A.C.; Bouget, M.; Atashi, A.; Brovko, L.; Rochefort, D.; Griffiths, M.W. Development of prototypes of bioactive packaging materials based on immobilized bacteriophages for control of growth of bacterial pathogens in foods. *Int. J. Food Microbiol.* **2016**, *217*, 49–58. [CrossRef] [PubMed]
16. Woolston, J.; Parks, A.R.; Abuladze, T.; Anderson, B.; Li, M.; Carter, C.; Hanna, L.F.; Heyse, S.; Charbonneau, D.; Sulakvelidze, A. Bacteriophages lytic for *Salmonella* rapidly reduce *Salmonella* contamination on glass and stainless steel surfaces. *Bacteriophage* **2013**, *3*, e25697. [CrossRef] [PubMed]
17. Abuladze, T.; Li, M.; Menetrez, M.Y.; Dean, T.; Senecal, A.; Sulakvelidze, A. Bacteriophages reduce experimental contamination of hard surfaces, tomato, spinach, broccoli, and ground beef by *Escherichia coli* O157:H7. *Appl. Environ. Microbiol.* **2008**, *74*, 6230–6238. [CrossRef] [PubMed]
18. Woolston, J.; Sulakvelidze, A. Bacteriophages and food safety. In *eLS*; Chichester, Ed.; John Wiley & Sons Ltd.: Hoboken, NJ, USA, 2015.
19. Endersen, L.; O'Mahony, J.; Hill, C.; Ross, R.P.; McAuliffe, O.; Coffey, A. Phage therapy in the food industry. *Annu. Rev. Food Sci. Technol.* **2014**, *5*, 327–349. [CrossRef] [PubMed]
20. Bandara, N.; Jo, J.; Ryu, S.; Kim, K.P. Bacteriophages BCP1-1 and BCP8-2 require divalent cations for efficient control of *Bacillus cereus* in fermented foods. *Food Microbiol.* **2012**, *31*, 9–16. [CrossRef] [PubMed]
21. Atterbury, R.J.; Connerton, P.L.; Dodd, C.E.; Rees, C.E.; Connerton, I.F. Application of host-specific bacteriophages to the surface of chicken skin leads to a reduction in recovery of *Campylobacter jejuni*. *Appl. Environ. Microbiol.* **2003**, *69*, 6302–6306. [CrossRef] [PubMed]
22. Goode, D.; Allen, V.M.; Barrow, P.A. Reduction of experimental *Salmonella* and *Campylobacter* contamination of chicken skin by application of lytic bacteriophages. *Appl. Environ. Microbiol.* **2003**, *69*, 5032–5036. [CrossRef] [PubMed]
23. Bigwood, T.; Hudson, J.A.; Billington, C.; Carey-Smith, G.V.; Heinemann, J.A. Phage inactivation of foodborne pathogens on cooked and raw meat. *Food Microbiol.* **2008**, *25*, 400–406. [CrossRef] [PubMed]
24. Kim, K.P.; Klumpp, J.; Loessner, M.J. *Enterobacter sakazakii* bacteriophages can prevent bacterial growth in reconstituted infant formula. *Int. J. Food Microbiol.* **2007**, *115*, 195–203. [CrossRef] [PubMed]
25. Zuber, S.; Boissin-Delaporte, C.; Michot, L.; Iversen, C.; Diep, B.; Brussow, H.; Breeuwer, P. Decreasing *Enterobacter sakazakii* (*Cronobacter* spp.) food contamination level with bacteriophages: Prospects and problems. *Microb. Biotechnol.* **2008**, *1*, 532–543. [CrossRef] [PubMed]
26. O'Flynn, G.; Ross, R.P.; Fitzgerald, G.F.; Coffey, A. Evaluation of a cocktail of three bacteriophages for biocontrol of *Escherichia coli* O157:H7. *Appl. Environ. Microbiol.* **2004**, *70*, 3417–3424. [CrossRef] [PubMed]
27. Sharma, M.; Patel, J.R.; Conway, W.S.; Ferguson, S.; Sulakvelidze, A. Effectiveness of bacteriophages in reducing *Escherichia coli* O157:H7 on fresh-cut cantaloupes and lettuce. *J. Food Prot.* **2009**, *72*, 1481–1485. [CrossRef] [PubMed]
28. Viazis, S.; Akhtar, M.; Feirtag, J.; Diez-Gonzalez, F. Reduction of *Escherichia coli* O157:H7 viability on leafy green vegetables by treatment with a bacteriophage mixture and *trans*-cinnamaldehyde. *Food Microbiol.* **2011**, *28*, 149–157. [CrossRef] [PubMed]
29. Carter, C.D.; Parks, A.; Abuladze, T.; Li, M.; Woolston, J.; Magnone, J.; Senecal, A.; Kropinski, A.M.; Sulakvelidze, A. Bacteriophage cocktail significantly reduces *Escherichia coli* O157:H7 contamination of lettuce and beef, but does not protect against recontamination. *Bacteriophage* **2012**, *2*, 178–185. [CrossRef] [PubMed]

30. Boyacioglu, O.; Sharma, M.; Sulakvelidze, A.; Goktepe, I. Biocontrol of *Escherichia coli* O157:H7 on fresh-cut leafy greens. *Bacteriophage* **2013**, *3*, e24620. [CrossRef] [PubMed]
31. Hudson, J.A.; Billington, C.; Wilson, T.; On, S.L. Effect of phage and host concentration on the inactivation of *Escherichia coli* O157:H7 on cooked and raw beef. *Food Sci. Technol. Int.* **2013**, *21*, 104–109. [CrossRef] [PubMed]
32. Ferguson, S.; Roberts, C.; Handy, E.; Sharma, M. Lytic bacteriophages reduce *Escherichia coli* O157:H7 on fresh cut lettuce introduced through cross-contamination. *Bacteriophage* **2013**, *3*, e24323. [CrossRef] [PubMed]
33. McLean, S.K.; Dunn, L.A.; Palombo, E.A. Phage inhibition of *Escherichia coli* in ultrahigh-temperature-treated and raw milk. *Foodborne Pathog. Dis.* **2013**, *10*, 956–962. [CrossRef] [PubMed]
34. Magnone, J.P.; Marek, P.J.; Sulakvelidze, A.; Senecal, A.G. Additive approach for inactivation of *Escherichia coli* O157:H7, *Salmonella*, and *Shigella* spp. on contaminated fresh fruits and vegetables using bacteriophage cocktail and produce wash. *J. Food Prot.* **2013**, *76*, 1336–1341. [CrossRef] [PubMed]
35. Leverentz, B.; Conway, W.S.; Camp, M.J.; Janisiewicz, W.J.; Abuladze, T.; Yang, M.; Saftner, R.; Sulakvelidze, A. Biocontrol of *Listeria monocytogenes* on fresh-cut produce by treatment with lytic bacteriophages and a bacteriocin. *Appl. Environ. Microbiol.* **2003**, *69*, 4519–4526. [CrossRef] [PubMed]
36. Leverentz, B.; Conway, W.S.; Janisiewicz, W.; Camp, M.J. Optimizing concentration and timing of a phage spray application to reduce *Listeria monocytogenes* on honeydew melon tissue. *J. Food Prot.* **2004**, *67*, 1682–1686. [CrossRef] [PubMed]
37. Carlton, R.M.; Noordman, W.H.; Biswas, B.; de Meester, E.D.; Loessner, M.J. Bacteriophage P100 for control of *Listeria monocytogenes* in foods: Genome sequence, bioinformatic analyses, oral toxicity study, and application. *Regul. Toxicol. Pharmacol.* **2005**, *43*, 301–312. [CrossRef] [PubMed]
38. Guenther, S.; Huwyler, D.; Richard, S.; Loessner, M.J. Virulent bacteriophage for efficient biocontrol of *Listeria monocytogenes* in ready-to-eat foods. *Appl. Environ. Microbiol.* **2009**, *75*, 93–100. [CrossRef] [PubMed]
39. Soni, K.A.; Nannapaneni, R. Bacteriophage significantly reduces *Listeria monocytogenes* on raw salmon fillet tissue. *J. Food Prot.* **2010**, *73*, 32–38. [CrossRef] [PubMed]
40. Soni, K.A.; Nannapaneni, R.; Hagens, S. Reduction of *Listeria monocytogenes* on the surface of fresh channel catfish fillets by bacteriophage Listex P100. *Foodborne Pathog. Dis.* **2010**, *7*, 427–434. [CrossRef] [PubMed]
41. Guenther, S.; Loessner, M.J. Bacteriophage biocontrol of *Listeria monocytogenes* on soft ripened white mold and red-smear cheeses. *Bacteriophage* **2011**, *1*, 94–100. [CrossRef] [PubMed]
42. Bigot, B.; Lee, W.J.; McIntyre, L.; Wilson, T.; Hudson, J.A.; Billington, C.; Heinemann, J.A. Control of *Listeria monocytogenes* growth in a ready-to-eat poultry product using a bacteriophage. *Food Microbiol.* **2011**, *28*, 1448–1452. [CrossRef] [PubMed]
43. Soni, K.A.; Desai, M.; Oladunjoye, A.; Skrobot, F.; Nannapaneni, R. Reduction of *Listeria monocytogenes* in queso fresco cheese by a combination of listericidal and listeriostatic GRAS antimicrobials. *Int. J. Food Microbiol.* **2012**, *155*, 82–88. [CrossRef] [PubMed]
44. Chibeu, A.; Agius, L.; Gao, A.; Sabour, P.M.; Kropinski, A.M.; Balamurugan, S. Efficacy of bacteriophage LISTEX™ P100 combined with chemical antimicrobials in reducing *Listeria monocytogenes* in cooked turkey and roast beef. *Int. J. Food Microbiol.* **2013**, *167*, 208–214. [CrossRef] [PubMed]
45. Oliveira, M.; Viñas, I.; Colàs, P.; Anguera, M.; Usall, J.; Abadias, M. Effectiveness of a bacteriophage in reducing *Listeria monocytogenes* on fresh-cut fruits and fruit juices. *Food Microbiol.* **2014**, *38*, 137–142. [CrossRef] [PubMed]
46. Silva, E.N.; Figueiredo, A.C.; Miranda, F.A.; de Castro Almeida, R.C. Control of *Listeria monocytogenes* growth in soft cheeses by bacteriophage P100. *Braz. J. Microbiol.* **2014**, *45*, 11–16. [CrossRef] [PubMed]
47. Figueiredo, A.C.L.; Almeida, R.C.C. Antibacterial efficacy of nisin, bacteriophage P100 and sodium lactate against *Listeria monocytogenes* in ready-to-eat sliced pork ham. *Braz. J. Microbiol.* **2017**, *48*, 724–729. [CrossRef] [PubMed]
48. Endersen, L.; Coffey, A.; Neve, H.; McAuliffe, O.; Ross, R.P.; O'Mahony, J.M. Isolation and characterisation of six novel mycobacteriophages and investigation of their antimicrobial potential in milk. *Int. Dairy J.* **2013**, *28*, 8–14. [CrossRef]
49. Modi, R.; Hirvi, Y.; Hill, A.; Griffiths, M.W. Effect of phage on survival of *Salmonella* Enteritidis during manufacture and storage of cheddar cheese made from raw and pasteurized milk. *J. Food Prot.* **2001**, *64*, 927–933. [CrossRef] [PubMed]

50. Leverentz, B.; Conway, W.S.; Alavidze, Z.; Janisiewicz, W.J.; Fuchs, Y.; Camp, M.J.; Chighladze, E.; Sulakvelidze, A. Examination of bacteriophage as a biocontrol method for *Salmonella* on fresh-cut fruit: A model study. *J. Food Prot.* **2001**, *64*, 1116–1121. [CrossRef] [PubMed]
51. Whichard, J.M.; Sriranganathan, N.; Pierson, F.W. Suppression of *Salmonella* growth by wild-type and large-plaque variants of bacteriophage Felix O1 in liquid culture and on chicken frankfurters. *J. Food Prot.* **2003**, *66*, 220–225. [CrossRef] [PubMed]
52. Higgins, J.P.; Higgins, S.E.; Guenther, K.L.; Huff, W.; Donoghue, A.M.; Donoghue, D.J.; Hargis, B.M. Use of a specific bacteriophage treatment to reduce *Salmonella* in poultry products. *Poult. Sci.* **2005**, *84*, 1141–1145. [CrossRef] [PubMed]
53. Ye, J.; Kostrzynska, M.; Dunfield, K.; Warriner, K. Control of *Salmonella* on sprouting mung bean and alfalfa seeds by using a biocontrol preparation based on antagonistic bacteria and lytic bacteriophages. *J. Food Prot.* **2010**, *73*, 9–17. [CrossRef] [PubMed]
54. Guenther, S.; Herzig, O.; Fieseler, L.; Klumpp, J.; Loessner, M.J. Biocontrol of *Salmonella* Typhimurium in RTE foods with the virulent bacteriophage FO1-E2. *Int. J. Food Microbiol.* **2012**, *154*, 66–72. [CrossRef] [PubMed]
55. Spricigo, D.A.; Bardina, C.; Cortes, P.; Llagostera, M. Use of a bacteriophage cocktail to control *Salmonella* in food and the food industry. *Int. J. Food Microbiol.* **2013**, *165*, 169–174. [CrossRef] [PubMed]
56. Kang, H.W.; Kim, J.W.; Jung, T.S.; Woo, G.J. wksl3, a new biocontrol agent for *Salmonella enterica* serovars Enteritidis and Typhimurium in foods: Characterization, application, sequence analysis, and oral acute toxicity study. *Appl. Environ. Microbiol.* **2013**, *79*, 1956–1968. [CrossRef] [PubMed]
57. Hungaro, H.M.; Mendonça, R.C.S.; Gouvêa, D.M.; Vanetti, M.C.D.; Pinto, C.L.D. Use of bacteriophages to reduce *Salmonella* in chicken skin in comparison with chemical agents. *Food Res. Int.* **2013**, *52*, 75–81. [CrossRef]
58. Zinno, P.; Devirgiliis, C.; Ercolini, D.; Ongeng, D.; Mauriello, G. Bacteriophage P22 to challenge *Salmonella* in foods. *Int. J. Food Microbiol.* **2014**, *191*, 69–74. [CrossRef] [PubMed]
59. Sukumaran, A.T.; Nannapaneni, R.; Kiess, A.; Sharma, C.S. Reduction of *Salmonella* on chicken meat and chicken skin by combined or sequential application of lytic bacteriophage with chemical antimicrobials. *Int. J. Food Microbiol.* **2015**, *207*, 8–15. [CrossRef] [PubMed]
60. Sukumaran, A.T.; Nannapaneni, R.; Kiess, A.; Sharma, C.S. Reduction of *Salmonella* on chicken breast fillets stored under aerobic or modified atmosphere packaging by the application of lytic bacteriophage preparation SalmoFresh™. *Poult. Sci.* **2016**, *95*, 668–675. [CrossRef] [PubMed]
61. Soffer, N.; Abuladze, T.; Woolston, J.; Li, M.; Hanna, L.F.; Heyse, S.; Charbonneau, D.; Sulakvelidze, A. Bacteriophages safely reduce *Salmonella* contamination in pet food and raw pet food ingredients. *Bacteriophage* **2016**, *6*, e1220347. [CrossRef] [PubMed]
62. Hong, Y.; Schmidt, K.; Marks, D.; Hatter, S.; Marshall, A.; Albino, L.; Ebner, P. Treatment of *Salmonella*-contaminated eggs and pork with a broad-spectrum, single bacteriophage: Assessment of efficacy and resistance development. *Foodborne Pathog. Dis.* **2016**, *13*, 679–688. [CrossRef] [PubMed]
63. Grant, A.; Parveen, S.; Schwarz, J.; Hashem, F.; Vimini, B. Reduction of *Salmonella* in ground chicken using a bacteriophage. *Poult. Sci.* **2017**, *96*, 2845–2852. [CrossRef] [PubMed]
64. Yeh, Y.; de Moura, F.H.; Van Den Broek, K.; de Mello, A.S. Effect of ultraviolet light, organic acids, and bacteriophage on *Salmonella* populations in ground beef. *Meat Sci.* **2018**, *139*, 44–48. [CrossRef] [PubMed]
65. Zhang, H.; Wang, R.; Bao, H.D. Phage inactivation of foodborne *Shigella* on ready-to-eat spiced chicken. *Poult. Sci.* **2013**, *92*, 211–217. [CrossRef] [PubMed]
66. Soffer, N.; Woolston, J.; Li, M.; Das, C.; Sulakvelidze, A. Bacteriophage preparation lytic for *Shigella* significantly reduces *Shigella sonnei* contamination in various foods. *PLoS ONE* **2017**, *12*, e0175256. [CrossRef] [PubMed]
67. Garcia, P.; Madera, C.; Martinez, B.; Rodriguez, A. Biocontrol of *Staphylococcus aureus* in curd manufacturing processes using bacteriophages. *Int. Dairy J.* **2007**, *17*, 1232–1239. [CrossRef]
68. Bueno, E.; García, P.; Martínez, B.; Rodríguez, A. Phage inactivation of *Staphylococcus aureus* in fresh and hard-type cheeses. *Int. J. Food Microbiol.* **2012**, *158*, 23–27. [CrossRef] [PubMed]
69. Mai, V.; Ukhanova, M.; Reinhard, M.K.; Li, M.; Sulakvelidze, A. Bacteriophage administration significantly reduces *Shigella* colonization and shedding by *Shigella*-challenged mice without deleterious side effects and distortions in the gut microbiota. *Bacteriophage* **2015**, *5*, e1088124. [CrossRef] [PubMed]

70. Tokman, J.I.; Kent, D.J.; Wiedmann, M.; Denes, T. Temperature significantly affects the plaguing and adsorption efficiencies of *Listeria* phages. *Front Microbiol.* **2016**, *7*, 631. [CrossRef] [PubMed]
71. Hoskisson, P.A.; Smith, M.C. Hypervariation and phase variation in the bacteriophage 'resistome'. *Curr. Opin. Microbiol.* **2007**, *10*, 396–400. [CrossRef] [PubMed]
72. Freeman, L.M.; Chandler, M.L.; Hamper, B.A.; Weeth, L.P. Current knowledge about the risks and benefits of raw meat-based diets for dogs and cats. *J. Am. Vet. Med. Assoc.* **2013**, *243*, 1549–1558. [CrossRef] [PubMed]
73. Behravesh, C.B.; Ferraro, A.; Deasy, M., 3rd; Dato, V.; Moll, M.; Sandt, C.; Rea, N.K.; Rickert, R.; Marriott, C.; Warren, K.; et al. Human *Salmonella* infections linked to contaminated dry dog and cat food, 2006–2008. *Pediatrics* **2010**, *126*, 477–483. [CrossRef] [PubMed]
74. Heyse, S.; Hanna, L.F.; Woolston, J.; Sulakvelidze, A.; Charbonneau, D. Bacteriophage cocktail for biocontrol of *Salmonella* in dried pet food. *J. Food Prot.* **2015**, *78*, 97–103. [CrossRef] [PubMed]
75. Snyder, A.B.; Perry, J.J.; Yousef, A.E. Developing and optimizing bacteriophage treatment to control enterohemorrhagic *Escherichia coli* on fresh produce. *Int. J. Food Microbiol.* **2016**, *236*, 90–97. [CrossRef] [PubMed]
76. Tomat, D.; Quiberoni, A.; Mercanti, D.; Balagué, C. Hard surfaces decontamination of enteropathogenic and Shiga toxin-producing *Escherichia coli* using bacteriophages. *Food Res. Int.* **2014**, *57*, 123–129. [CrossRef]
77. Bower, C.K.; Daeschel, M.A. Resistance responses of microorganisms in food environments. *Int. J. Food Microbiol.* **1999**, *50*, 33–44. [CrossRef]
78. Kotloff, K.L.; Winickoff, J.P.; Ivanoff, B.; Clemens, J.D.; Swerdlow, D.L.; Sansonetti, P.J.; Adak, G.K.; Levine, M.M. Global burden of *Shigella* infections: Implications for vaccine development and implementation of control strategies. *Bull. World Health Organ.* **1999**, *77*, 651–666. [PubMed]
79. Firlieyanti, A.S.; Connerton, P.L.; Connerton, I.F. *Campylobacters* and their bacteriophages from chicken liver: The prospect for phage biocontrol. *Int. J. Food Microbiol.* **2016**, *237*, 121–127. [CrossRef] [PubMed]
80. Hammerl, J.A.; Jäckel, C.; Alter, T.; Janzcyk, P.; Stingl, K.; Knüver, M.T.; Hertwig, S. Reduction of *Campylobacter jejuni* in broiler chicken by successive application of group II and group III phages. *PLoS ONE* **2014**, *9*, e114785. [CrossRef] [PubMed]
81. Zampara, A.; Sørensen, M.C.H.; Elsser-Gravesen, A.; Brøndsted, L. Significance of phage-host interactions for biocontrol of *Campylobacter jejuni* in food. *Food Control.* **2017**, *73*, 1169–1175. [CrossRef]
82. Kittler, S.; Fischer, S.; Abdulmawjood, A.; Glunder, G.; Klein, G. Effect of bacteriophage application on *Campylobacter jejuni* loads in commercial broiler flocks. *Appl. Environ. Microbiol.* **2013**, *79*, 7525–7533. [CrossRef] [PubMed]
83. Sorensen, M.C.; Gencay, Y.E.; Birk, T.; Baldvinsson, S.B.; Jackel, C.; Hammerl, J.A.; Vegge, C.S.; Neve, H.; Brondsted, L. Primary isolation strain determines both phage type and receptors recognised by *Campylobacter jejuni* bacteriophages. *PLoS ONE* **2015**, *10*, e0116287. [CrossRef] [PubMed]
84. Hudson, J.A.; McIntyre, L.; Billington, C. Application of bacteriophages to control pathogenic and spoilage bacteria in food processing and distribution. In *Bacteriophages in the Control of Food- and Waterborne Pathogens*; Sabour, P.M., Griffiths, M.W., Eds.; ASM Press: Washington, DC, USA, 2010; pp. 119–135.
85. US Food and Drug Administration Center for Food Safety and Applied Nutrition (FDA). *Quantitative Assessment of Relative Risk to Public Health from Foodborne Listeria Monocytogenes Among Selected Categories of Ready-to-Eat Foods*; US Food and Drug Administration Center for Food Safety and Applied Nutrition: College Park, MD, USA, 2003. Available online: https://www.fda.gov/downloads/food/scienceresearch/researchareas/riskassessmentsafetyassessment/ucm197330.pdf (accessed on 19 March 2018).
86. Rodríguez, E.; Seguer, J.; Rocabayera, X.; Manresa, A. Cellular effects of monohydrochloride of l-arginine, Nα-lauroyl ethylester (LAE) on exposure to *Salmonella typhimurium* and *Staphylococcus aureus*. *J. Appl. Microbiol.* **2004**, *96*, 903–912. [CrossRef] [PubMed]
87. Zhang, H.; Li, L.; Zhao, Z.; Peng, D.; Zhou, X. Polar flagella rotation in *Vibrio parahaemolyticus* confers resistance to bacteriophage infection. *Sci. Rep.* **2016**, *6*, 26147. [CrossRef] [PubMed]
88. Marti, R.; Zurfluh, K.; Hagens, S.; Pianezzi, J.; Klumpp, J.; Loessner, M.J. Long tail fibres of the novel broad-host-range T-even bacteriophage S16 specifically recognize *Salmonella* OmpC. *Mol. Microbiol.* **2013**, *87*, 818–834. [CrossRef] [PubMed]
89. Lindberg, A.A.; Holme, T. Influence of O side chains on the attachment of the Felix O-1 bacteriophage to *Salmonella* bacteria. *J. Bacteriol.* **1969**, *99*, 513–519. [PubMed]

90. Reganold, J.P.; Wachter, J.M. Organic agriculture in the twenty-first century. *Nat. Plants* **2016**, *2*, 15221. [CrossRef] [PubMed]
91. Woods, T.; Ernst, M.; Tropp, D. *Community Supported Agriculture–New Models for Changing Markets*; U.S. Department of Agriculture, Agricultural Marketing Service: Washington, DC, USA, 2017.
92. Hatfull, G.F. Bacteriophage genomics. *Curr. Opin. Microbiol.* **2008**, *11*, 447–453. [CrossRef] [PubMed]
93. Dalmasso, M.; Hill, C.; Ross, R.P. Exploiting gut bacteriophages for human health. *Trends Microbiol.* **2014**, *22*, 399–405. [CrossRef] [PubMed]
94. Sulakvelidze, A.; Barrow, P. Phage therapy in animals and agribusiness. In *Bacteriophages: Biology and Applications*; Kutter, E., Sulakvelidze, A., Eds.; CRC Press: Boca Raton, FL, USA, 2005; pp. 335–380.
95. Suárez, V.B.; Quiberoni, A.; Binetti, A.G.; Reinheimer, J.A. Thermophilic lactic acid bacteria phages isolated from Argentinian dairy industries. *J. Food Prot.* **2002**, *65*, 1597–1604. [CrossRef] [PubMed]
96. Gautier, M.; Rouault, A.; Sommer, P.; Briandet, R. Occurrence of *Propionibacterium freudenreichii* bacteriophages in swiss cheese. *Appl. Environ. Microbiol.* **1995**, *61*, 2572–2576. [PubMed]
97. Naanwaab, C.; Yeboah, O.A.; Ofori Kyei, F.; Sulakvelidze, A.; Goktepe, I. Evaluation of consumers' perception and willingness to pay for bacteriophage treated fresh produce. *Bacteriophage* **2014**, *4*, e979662. [CrossRef] [PubMed]

Review

Criteria for Selecting Suitable Infectious Diseases for Phage Therapy

David R. Harper

Evolution Biotechnologies, Colworth Science Park, Sharnbrook, Bedfordshire MK44 1LZ, UK; drh@evolutionbiotech.com; Tel.: +44-1234-818312

Received: 16 March 2018; Accepted: 30 March 2018; Published: 5 April 2018

Abstract: One of the main issues with phage therapy from its earliest days has been the selection of appropriate disease targets. In early work, when the nature of bacteriophages was unknown, many inappropriate targets were selected, including some now known to have no bacterial involvement whatsoever. More recently, with greatly increased understanding of the highly specific nature of bacteriophages and of their mechanisms of action, it has been possible to select indications with an increased chance of a successful therapeutic outcome. The factors to be considered include the characteristics of the infection to be treated, the characteristics of the bacteria involved, and the characteristics of the bacteriophages themselves. At a later stage all of this information then informs trial design and regulatory considerations. Where the work is undertaken towards the development of a commercial product it is also necessary to consider the planned market, protection of intellectual property, and the sourcing of funding to support the work. It is clear that bacteriophages are not a "magic bullet". However, with careful and appropriate selection of a limited set of initial targets, it should be possible to obtain proof of concept for the many elements required for the success of phage therapy. In time, success with these initial targets could then support more widespread use.

Keywords: bacteriophage; therapy; phage therapy; bacterial disease; infection; target selection

---

1. Introduction

When bacteriophages were first used as therapeutic agents from 1919 onwards [1] this was done in the context of an extremely limited knowledge of bacteriophage biology. Given the toxicity of antibacterial agents at the time, which included both mercury and arsenic, there was a clear need for new approaches. Unfortunately, many early uses of phage therapy were driven more by commercial pressures than by science. As a result, bacteriophages were used for a wide variety of indications, many of which had no bacterial component (Figure 1).

Unsurprisingly, with such indications as urticaria and herpes, there was unlikely to be therapeutic benefit due to the effects of bacteriophages. In addition, early bacteriophage preparations contained a large amount of bacterial debris which had a range of immunomodulatory effects, which were considered likely to be responsible for many of the observed clinical effects [2,3]. When data from 100 papers published in the early years of phage therapy was analysed for the American Medical Association, it was concluded that only a very few indications showed good evidence of beneficial effects, with convincing data only for *Staphylococcal* skin disease and some instances of cystitis [2]. With a very limited understanding of the nature of the agent, controversy over its effects, and complications arising from the crude nature of the early therapeutic preparations, it is perhaps unsurprising that phage therapy fell into disuse once chemical antibiotics became widely available. However, as antibiotic resistance changed from an abstract concern to a full-blown crisis [4], interest in phage therapy has been revived [5,6]. This drew on the much greater understanding of bacteriophages resulting from their use during that intervening period in experimental studies, notably in the area of molecular biology.

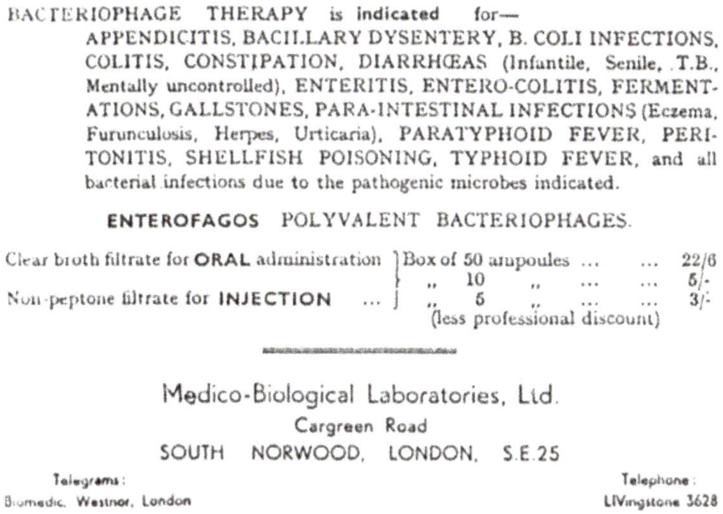

BACTERIOPHAGE THERAPY is indicated for—
APPENDICITIS, BACILLARY DYSENTERY, B. COLI INFECTIONS, COLITIS, CONSTIPATION, DIARRHŒAS (Infantile, Senile, T.B., Mentally uncontrolled), ENTERITIS, ENTERO-COLITIS, FERMENTATIONS, GALLSTONES, PARA-INTESTINAL INFECTIONS (Eczema, Furunculosis, Herpes, Urticaria), PARATYPHOID FEVER, PERITONITIS, SHELLFISH POISONING, TYPHOID FEVER, and all bacterial infections due to the pathogenic microbes indicated.

ENTEROFAGOS POLYVALENT BACTERIOPHAGES.

Clear broth filtrate for ORAL administration } Box of 50 ampoules 22/6
„ 10 „ 5/-
Non-peptone filtrate for INJECTION ... } „ 5 „ 3/-
(less professional discount)

Medico-Biological Laboratories, Ltd.
Cargreen Road
SOUTH NORWOOD, LONDON, S.E.25

Telegrams: Biomedic, Westnor, London
Telephone: LIVingstone 3628

Figure 1. An advertisement for therapeutic bacteriophages from the 1920s (reprinted with the kind permission of Dr. J. Soothill).

With this increased understanding of bacteriophage biology came the opportunity for far more precision in their use [5]. Key properties of bacteriophages that had been clarified by this point included many aspects of their biology, such as their nature as bacterial viruses, their highly specific host requirements, and the molecular processes of infection. Technical advances also allowed the purification of bacteriophages away from bacterial contaminants, including the removal of highly active host cell components such as endotoxins. While we are still a long way from a full understanding of the biology of these complex viruses, we can now build on what we do know to use these unique agents far more effectively.

2. Selection of Therapeutic Approaches

2.1. Clinical Need

When a serious disease cannot be controlled by current therapies there is a need for new approaches. Much of the current interest in phage therapy is driven by concerns over antimicrobial drug resistance, and a working group evaluating a broad range of alternative approaches has concluded that phage therapy is a promising way to counter this issue [7]. There is also some evidence of both natural and engineered bacteriophages driving bacteria towards the loss of existing antibiotic resistance [8–10], suggesting a possible role for combinations of bacteriophages with conventional antibiotics.

The initial motivation for the selection of a therapeutic target is typically that there is an unmet need for control of a bacterial infection—that is, a lack of effective antibacterial approaches for a disease with serious medical or economic consequences. If effective controls exist there is rarely a driver for the development of novel therapies, except perhaps to provide an alternative due to commercial drivers. A therapy can be targeted at infections of humans, animals, or plants; conditions addressed by the use of bacteriophages range from ear infections in humans [11] to rot in harvested potatoes [12]. However, the focus of this article is on therapeutic uses and, thus, on unmet clinical need in humans (and, to an extent, animals). Lack of an effective therapy can arise from multiple factors, including evolved antimicrobial resistance or inherent resistance determinants, such as growth in biofilms [13,14].

Thus, unmet need underpins the development of new therapies and generates funding (whether as grants or commercial investment) that is necessary to undertake such work.

No phage therapy has yet been approved for market by the European Medicines Agency (EMA) in the EU or the Food and Drugs Administration (FDA) in the USA. Until this is achieved, phage therapy remains an experimental approach in these jurisdictions, however accepted it is in some other areas such as Eastern Europe. Thus, it is of critical importance to maximise the risk of success in development efforts, since failures can delay, or even destroy, the prospects for developing such approaches. An example of this is gene therapy, which is only now recovering from some early high-profile setbacks [15].

2.2. Key Elements of the Disease Target

Sufficient unmet need: While individual patients might be in dire need of a particular therapy, there generally need to be enough potential users to justify the costs of development and commercialisation. As an example, many early gene therapy treatments were directed at indications where there were only very small numbers of affected individuals. These "ultra-orphan" targets can help with approval, but the first gene therapy to be approved in Europe, Glybera, targeted a disease which occurs in less than one in a million people and had a consequently high price. The drug was withdrawn five years after launch, having been used just once [16].

Antimicrobial resistance: The driver for much of the interest in phage therapy is antimicrobial resistance, where infections may be largely or even completely resistant to conventional antibiotics. Bacteriophages, with their entirely different modes of action, are unaffected by such resistance. Bacteriophages can and will evolve to counter the evolution of bacteriophage resistance by the target bacteria [17]. Another major cause of resistance to conventional antibiotics is bacterial growth in biofilms, which can decrease bacterial sensitivity to antibiotics by over a thousand-fold [13,14]. Bacteriophages have unique capabilities in attacking bacteria in such a setting by targeting both the biofilm matrix and specialized cells growing within it [14]. Efficacy against biofilms has been demonstrated both in vitro and in vivo [11,14] and has the potential to be a significant driver for the adoption of phage therapy.

Disease results from bacterial infection: It (nowadays) goes without saying that the target for phage therapy should be bacterial in nature. Bacteriophages are highly specific, with only a relatively small number even crossing species boundaries. They are almost completely inert towards other cells. The days when herpes or urticaria could be regarded as a viable target are, fortunately, long gone.

Infection is caused by one type of bacteria or a small number of types: Bacteriophages are often referred to as "exquisitely specific" since most are able to infect and replicate in only a subset of strains within a single bacterial species, meaning that a mixture of bacteriophages (usually referred to as a "cocktail") is usually needed to target even a single bacterial species, though there are exceptions [5,18]. Thus, a polymicrobial infection is poorly suited to phage therapy. This is a very important difference to the historical use of broad spectrum chemical antibiotics. Although some broad spectrum bacteriophage treatments have been used in Eastern Europe, these contain extremely high numbers of relatively undefined bacteriophages, which would make approval by EMA or FDA complex. An early polymicrobial infection may change during treatment with conventional antibiotics, with a much more limited range of bacteria in the later stages of infection [11]. Such infections offer both unmet need (having failed to resolve with conventional treatments) and a limited range of bacterial targets. They are, thus, well suited to phage therapy approaches.

Bacteria causing the infection are identified: Unlike chemical antibiotics, it is unlikely that there will be broad spectrum phage therapeutic cocktails without an extremely complex regulatory process and prohibitive production costs. This is likely to limit first-line use, unless such a treatment is paired with a rapid point-of-care diagnostic test. A simpler approach, at least initially, is to focus on cases where the bacterial target has been identified but is resistant to clearance by existing methods. This was the approach taken in the only phase 2 trial to date to report positive results [11] by targeting late stage ear infections where one bacterial species (*Pseudomonas aeruginosa*) predominates. Additionally, testing must be able to identify and quantify the target bacteria in the intended recipients, both prior to and as part of the trial process.

Bacteria targeted are responsible for the clinical pathology: Given the specificity of bacteriophages, simply removing one component of a polymicrobial infection may not result in a positive clinical outcome. While regulators may accept microbiological endpoints (reduction of target bacterial numbers) in initial trials, in later work, improvement in clinical symptoms is likely to be required for trial success. This has been an issue with some work where positive outcomes were not observed [19].

Potential for useful preclinical work: While in vitro data is useful, good preclinical data derived from in vivo work is extremely useful in making the case for progression into human trials. This can either be from model systems [20,21] or from analogous infections in animals [11,22,23]. Where no suitable system is available, data collection to permit trials is likely to be more complex.

Suitability for clinical trials: Patients must be both available and accessible, in line with current ethical practices. If the patient group requires simultaneous treatment with other antibacterial agents (as is often the case with seriously ill patients, for example, under "expanded access" single patient uses) this can greatly complicate the interpretation of results, reducing the value of the data generated. There are, of course, also many issues relating to the design and conduct of clinical trials, from liaison with regulators to selection of endpoints. However, these fall outside the scope of this article. Nevertheless, it should be noted that even the most promising approach will not succeed with a poorly designed trial.

Site of infection is accessible: Bacteriophages are large nucleoprotein structures in the megadalton to gigadalton range. Chemical antibiotics are far smaller. This can lead to misunderstandings over how best to administer bacteriophages among those more used to existing approaches. Given the simplicity of delivery, attention has been focused on topical administration where it is possible to deliver bacteriophages directly to the infected site [11,22–26]. Unlike conventional antibiotics, bacteriophages are unlikely to be able to cross many barriers within the body with useful levels of efficiency, at least when these barriers are intact. These can be normal elements of the body, such as the gut wall or the blood-brain barrier, or part of the pathology of the infection, such as lung tubercles in tuberculosis [27] or closed comedones in acne [28]. However, the ability of bacteriophages to amplify exponentially from very small initial doses [23] can allow even limited bacteriophage numbers to produce a strong localized therapeutic effect.

Bacterial numbers are sufficient to support amplification: Localised amplification where their target is present is a unique feature of bacteriophages and underlies their proposed use as therapeutics in most cases. However, in order to support such amplification, there needs to be a sufficient supply of susceptible bacteria [29]. While this is more complex in vivo, it is clear that in some applications, particularly those with cleaned or disinfected sites of infection, bacterial densities may be insufficient to rely on in situ phage amplification [26]. Any therapeutic effect would thus be reduced. In contrast, even very low levels of bacteriophage can produce rapid and dramatic effects when bacterial numbers are high [23]. Thus, phage therapy is better suited to high density bacterial infections, which is, again, counterintuitive for those used to conventional antibiotics.

Limitations of oral dosing: Oral dosing is considered highly desirable for conventional antibiotics but is inherently limited for bacteriophages. One issue is the degradation of bacteriophages in the acid environment of the stomach, leading to limited oral bioavailability unless the administered bacteriophages are encapsulated [30] or the stomach acid is neutralized [19]. There is also evidence that bacteria that have established infection within the gut may be poorly accessible to bacteriophages, probably because they are located within the coating of the intestinal walls [31], leaving administered bacteriophages to simply pass by them on their way through.

Issues with systemic delivery: Although there is considerable evidence of efficacy in model systems [20,21], bacteriophages delivered via the circulation are challenged by a number of issues [32].

Both innate and adaptive immunity are major concerns. Bacteriophages are prime candidates for those elements of the innate response intended to provide a first response to invading viruses, notably phagocytosis [33]. In addition, the adaptive immune response will, in time, respond to bacteriophages, particularly after repeated administration. However, the ablative effect of such responses on bacteriophage efficacy appears to be limited [34]. Another major issue is delivery to the site of infection. Although bacteriophages have the unique ability to amplify locally even from a very small initial dose [11,23], they nevertheless have to reach and infect their target bacteria in order to be able to do so. Barriers that can be crossed by conventional antibiotics may be impervious to bacteriophages, favouring topical uses or use in body activities. However, a recent in-depth review of the subject noted bacteriophages as one of the most promising approaches to combating antimicrobial resistance even in systemic applications [7].

3. Bacteriophages

Intellectual property issues: It has long been argued that the patenting of phage therapy approaches is challenging. It is undeniable that the basic approach of phage therapy is firmly in the public domain. However, this is also true for monoclonal antibodies, and these have led to drugs with current market values in the tens of billions of dollars. It is also true that novel bacteriophages can be identified for almost any bacterial target, potentially bypassing patent protection. Again, this is also true for monoclonal antibodies. From a commercial point of view a patent is useful, but not actually essential since the progress of a drug or phage mix through the regulatory process underpins much of its value. However, patent protection is highly desirable, particularly in the early stages of development. There is considerable confusion over the patenting of naturally occurring biological materials [35], but it is clear that bacteriophage mixtures exhibiting activities different from those seen in nature can form the basis for patent awards.

Bacteriophages are available: It should go without saying that the bacteriophages used must be able to target the infecting bacteria, whether as part of a broadly effective cocktail [11] or by selection for the bacterial strain(s) present [36]. While it is thought that bacteriophages exist for all bacteria, they are sometimes hard to find, as with certain members of the *Streptococci* [37]. Alternatively, properties of the host bacteria can make isolation of bacteriophages difficult, as with slow-growing members of the *Mycobacteria*. However, for many bacterial species, isolation can be both rapid and simple.

Bacteriophages are effective in killing the target bacteria: While some bacteriophages cause rapid killing of their host, others do not. Selection of those which produce rapid lysis and liberate large numbers of progeny bacteriophage (high burst size) is usually assessed in vitro. The normal method for this initially is to monitor the formation of plaques, selecting those bacteriophages that form large, clear plaques [38]. Growth in liquid culture, often using plate-based optical density systems, is also used for this purpose.

Bacteriophages must target the bacteria responsible for the infection: Most bacteriophages kill only a subset of strains within a single bacterial species. In order to kill sufficient members of a representative panel of strains of that species (a diversity panel), the activity of individual bacteriophages is tested against that panel and a mixture selected which provides broad coverage. This may be a mixture of broadly effective bacteriophages along with those selected to cover specific strains from a diversity panel, for example those with concerning levels of resistance to conventional antibiotics. While 100% coverage of a large diversity panel is not usually required (for antibiotics or for bacteriophages), there are minimum levels (analogous to those defined for antibiotic development) which need to be attained for a generally applicable therapy. It is also advantageous to ensure that as many strains as possible are targeted by multiple bacteriophages, minimising the potential for resistance.

Cross-resistance: Generation of resistance in vitro may be used to confirm that cross-resistance to candidate therapeutic bacteriophages in a cocktail (implying similar biology of infection) is minimized. This involves the selection of bacteria which do not show similar resistance profiles against bacteriophage-resistant mutants of the target bacteria, and (as with antibiotics) minimizes the potential for the development of resistance in vivo.

Coverage: Isolates from different pathologies and geographic locations must be covered by the bacteriophages selected if a standardized cocktail is to be used [11,19,22,24–26], and this can be difficult to attain with some highly variable bacterial species. An alternative approach, of having a panel of bacteriophages with defined activities, from which personalized cocktails are prepared for individual patients [36,39] faces a number of currently unresolved regulatory challenges, as well as significant resourcing issues.

Bacteriophages are obligately lytic, not temperate: Obligately lytic bacteriophages that infect a permissive host produce rapid killing. The bacterial life cycle is simply "kill or be killed". However, in many cases bacteriophages are temperate. That is, they are capable of entering a latent state within the host bacteria (lysogeny) in which they are usually integrated into the bacterial genome. In such a state, host bacteria are often rendered resistant to superinfection with the same, or even closely related, bacteriophages [40,41]. As and when a lysogenic bacteriophage reactivates, it may carry bacterial genes with it to a new host, a process known as specialized transduction. In addition, it is now becoming clear that lysogenic bacteriophages may modulate genetic activity and actually benefit their bacterial host [40]. Thus, while lysogenic bacteriophages may reduce bacterial numbers both in vitro and in vivo [42], they are generally considered unsuitable for therapeutic use. Traditional tests for lysogeny include the formation of turbid plaques, where the clearance zone is clouded by surviving bacteria. More recently, genetic analysis can identify markers of lysogeny such as the presence in the bacteriophage genome of a functional integrase gene or known repressors linked to the establishment and maintenance of a lysogenic infection state. This then allows exclusion from the development pathway. For some bacteria, such as *Clostridium difficle*, all bacteriophages identified to date appear to be temperate [43], limiting the options for phage therapy in these settings.

Transduction and toxins: As well as the specialized transduction seen in lysogenic infections, bacteriophages can pick up a wide range of segments of bacterial DNA essentially at random and transfer them to a new host. This is considered undesirable in candidate therapeutic bacteriophages since virulence genes could be transferred by this route. In addition, some bacteriophages actually carry such genes, including both antibiotic resistance genes [44] and clinically significant bacterial toxins [45], while others have adapted forms that transfer virulence-associated genes [46]. It is necessary to exclude such bacteriophages from therapeutic development, usually by genetic analysis.

Stability: The most promising therapeutic bacteriophage is of no value if it loses activity too fast before it can be used to treat patients. Assessment of stability in suitable forms for therapeutic use is, thus, a vital part of development [47].

Quality of product: In order to be used as therapeutics, bacteriophages must be able to grow to sufficiently high titres in a suitable bacterial host. Such hosts are often selected for their non-pathogenic nature, including absence of toxins and even of lysogenic bacteriophage genomes. It is however necessary to ensure that the bacteriophages produced by such compliant hosts retain their virulence against target bacteria. Unlike early therapeutic products, it is also necessary to purify bacteriophages away from host bacterial components, in particular the endotoxins produced by Gram-negative bacteria for which strict limits are applied [48,49]. The whole issue of producing bacteriophages to the required standards (as a drug substance) then combining them to make a therapeutic cocktail (i.e., a drug product) while complying with necessary quality control standards (cGMP, except for all but the earliest trials) is both demanding and expensive. However, if such standards cannot be attained, perhaps because of a persistent contaminant, then all of the preceding work has been in vain.

Bacteriophages as biological control agents: Bacteriophages are not chemical antibiotics. Although this statement seems obvious, it has not stopped some groups from assuming that the same processes used in the development of chemical antibiotics must of necessity be applied a priori to therapeutic bacteriophages. While bacteriophages do need to conform with existing rules, the regulators show a significant degree of flexibility, and indeed enthusiasm, in working with these new paradigms [50–53]. This is an area which is likely to see significant changes. However, these will rely on good science and, in particular, on data from fully regulated clinical trials, which are still sparse in this area. One important difference to conventional antibiotics is that bacteriophages are unlikely to produce sterilising effects in vivo since, when bacterial numbers fall below the replication threshold [29], they will lose much of their effect. However, in such situations, lowering bacterial numbers far enough can inhibit their replication [52] and reduce toxin production, thus allowing the immune system of an infected patient to aid in resolving the infection [11,22].

4. Stages of target selection

The following stages need to be considered (see Figure 2):

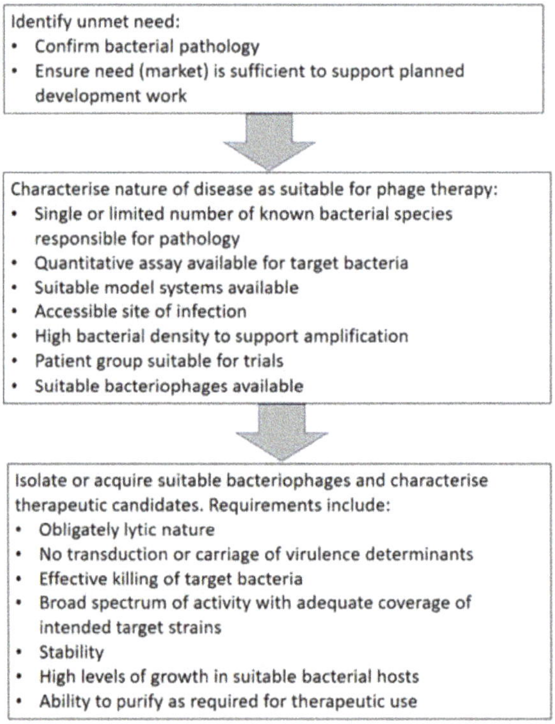

Identify unmet need:
- Confirm bacterial pathology
- Ensure need (market) is sufficient to support planned development work

Characterise nature of disease as suitable for phage therapy:
- Single or limited number of known bacterial species responsible for pathology
- Quantitative assay available for target bacteria
- Suitable model systems available
- Accessible site of infection
- High bacterial density to support amplification
- Patient group suitable for trials
- Suitable bacteriophages available

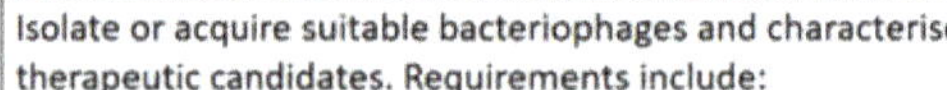

Isolate or acquire suitable bacteriophages and characterise therapeutic candidates. Requirements include:
- Obligately lytic nature
- No transduction or carriage of virulence determinants
- Effective killing of target bacteria
- Broad spectrum of activity with adequate coverage of intended target strains
- Stability
- High levels of growth in suitable bacterial hosts
- Ability to purify as required for therapeutic use

Figure 2. Steps in the selection of an initial target disease for phage therapy.

5. A Worked Example

As an example of the practical application of such considerations, *P. aeruginosa* appears to highly suitable as a phage therapy target, as outlined below and elsewhere [54]. As a result, it is proving a popular target for such intervention [11,22–26]. Indeed, the only successful phase 2 trial reported to date targeted late stage ear infections caused by *P. aeruginosa* [11].

In this study [11] the bacterial pathology had been confirmed both generally and in the specific cases by repeated microbiological assessment. The individual clinical need was severe (aiding patient recruitment and ethical review) and the market need was assessed as sufficient to support development, with the potential for expansion into other types of topical infection thereafter. While ear infections are initially polymicrobial, use of existing therapies that fail to clear the infection typically results in a "late-stage" infection dominated by *P. aeruginosa*, which was assessed prior to trial entry. A suitable trial centre was identified with strong clinical expertise, at which patients were accessible and willing to participate. *P. aeruginosa* is relatively simple to culture and to enumerate, and a previous veterinary field trial [22,55] had provided useful preclinical data. The ear is accessible, allowing direct application of bacteriophage onto the infected site, and had a high density of *P. aeruginosa* (2.3×10^6 to 4.5×10^{10} CFU/gram at the start of the trial, averaging well over 10^9 CFU/gram) [11]. Bacteriophages specific for *P. aeruginosa* are relatively widespread and simple to isolate from available sources [54]. Obligately lytic forms are common, showing rapid bacterial killing and releasing high numbers of progeny bacteriophages. Levels of transduction or toxin carriage are low, and broad activity of bacteriophages against *P. aeruginosa* is obtainable. The bacteriophages could be grown in culture to the required level and purified from such cultures. Though the presence of endotoxins in lysate of Gram-negative bacteria such as *P. aeruginosa* does require careful consideration of levels remaining in a candidate therapy, the low dosing levels used in the study (a single input dose of 2.4 ng) aided in this. With due consideration given to the above elements, the phase 1/2 clinical trial was conducted (albeit at a small scale) [11] and, uniquely to date, produced promising and positive results.

6. Summary

Phage therapy remains experimental in both human and veterinary medicine. With antibiotic resistance now acknowledged as a worldwide crisis [4], the need for new approaches to antibacterial therapy makes the development of this powerful approach a priority. Despite this, there is a real shortage of high quality clinical work on which to base progression of phage therapy through the regulatory approvals required to permit widespread use.

The selection of appropriate disease targets is an essential step in progressing this important technology. Trial failures are expensive in both time and resources and benefit nobody. Such failures carry the risk of causing serious damage to confidence in the field as a whole [15]. In order to maximise the chances of success, the selection of disease targets must be informed by sound knowledge of the disease to be treated, of the infecting bacteria, and of the nature and interactions of the bacteriophages to be used. Success against carefully selected initial targets is required to build the confidence to allow later work in more challenging applications

Based on the work to date, while it seems likely that bacteriophages will be used systemically and even orally in time, topical treatments where the bacteriophage can be placed onto an infection site with a high bacterial density offer the best chance of success. Such success is vital to moving phage therapy into clinical use, to save lives, and to improve lives.

Acknowledgments: The author would like to thank Ben Burrowes for his helpful comments on the manuscript.

Conflicts of Interest: David R. Harper is a director of and shareholder in Evolution Biotechnologies, a company with an active interest in phage therapy, and is a shareholder in AmpliPhi Biosciences Corporation, a company working in the area of phage therapy.

References

1. D'Herelle, F. Sur le role du microbe bacteriophage dans la typhose aviare. *Comptes Rendus de l'Académie des Sciences Paris* **1919**, *169*, 932–934.
2. Eaton, M.D.; Bayne-Jones, S. Bacteriophage therapy: Review of the principles and results of the use of bacteriophage in the treatment of infections. *JAMA* **1934**, *103*, 1769–1776, 1847–1853, 1934–1939. [CrossRef]
3. Krueger, A.P.; Scribner, E.J. The bacteriophage: Its nature and its therapeutic use. *JAMA* **1941**, *116*, 2160–2167. [CrossRef]
4. Chan, M. WHO Director-General briefs UN on Antimicrobial Resistance. 2016. Available online: http://www.who.int/dg/speeches/2016/antimicrobial-resistance-un/en/ (accessed on 9 March 2018).
5. Harper, D.R.; Burrowes, B.H.; Kutter, E.M. Bacteriophage: Therapeutic Uses. In *Encyclopedia of Life Sciences*; John Wiley and Sons: Chichester, UK, 2014.

6. Harper, D.; Abedon, S.; Burrowes, B.; McConville, M. (Eds.) *Bacteriophgages: Biology, Technology, Therapy*; Springer: Cham, Switzerland, 2018; ISBN 978-3-319-40598-8. Available online: https://link.springer.com/referencework/10.1007/978-3-319-40598-8#about (accessed on 9 March 2018).
7. Czaplewski, L.; Bax, R.; Clokie, M.; Dawson, M.; Fairhead, H.; Fischetti, V.A.; Foster, S.; Gilmore, B.F.; Hancock, R.E.; Harper, D.; et al. Alternatives to antibiotics—A pipeline portfolio review. *Lancet Infect. Dis.* **2016**, *16*, 239–251. [CrossRef]
8. Hagens, S.; Habel, A.; Bläsi, U. Augmentation of the Antimicrobial Efficacy of Antibiotics by Filamentous Phage. *Microb. Drug Resist.* **2006**, *12*, 164–168. [CrossRef] [PubMed]
9. Chan, B.K.; Sistrom, M.; Wertz, J.E.; Kortright, K.E.; Narayan, D.; Turner, P.E. Phage selection restores antibiotic sensitivity in MDR Pseudomonas aeruginosa. *Sci. Rep.* **2016**, *6*, 26717. [CrossRef] [PubMed]
10. Lu, T.K.; Collins, J.J. Engineered bacteriophage targeting gene networks as adjuvants for antibiotic therapy. *PNAS* **2009**, *106*, 4629–4634. [CrossRef] [PubMed]
11. Wright, A.; Hawkins, C.; Änggård, E.; Harper, D.A. Controlled clinical trial of a therapeutic bacteriophage in chronic otitis due to antibiotic-resistant *Pseudomonas aeruginosa*; A preliminary report of efficacy. *Clin. Otolaryngol.* **2009**, *34*, 349–357. [CrossRef] [PubMed]
12. Blackwell, A. The Potential for Bacteriophage to Control Soft Rot Development in Store. Available online: https://potatoes.ahdb.org.uk/sites/default/files/publication_upload/APS%20Biocontrol%20Ltd.pdfDevelopmentDevelopment (accessed on 9 March 2018).
13. Ceri, H.; Olson, M.E.; Stremick, C.; Read, R.R.; Morck, D.; Buret, A. The Calgary biofilm device: New technology for rapid determination of antibiotic susceptibilities of bacterial biofilms. *J. Clin. Microbiol.* **1999**, *37*, 1771–1776. [PubMed]
14. Harper, D.R.; Parracho, H.M.; Walker, J.; Sharp, R.; Hughes, G.; Werthén, M.; Lehman, S.; Morales, S. Bacteriophages and biofilms. *Antibiotics* **2014**, *3*, 270–284. [CrossRef]
15. Kumar, S.R.; Markusic, D.M.; Biswas, M.; High, K.A.; Herzog, R.W. Clinical development of gene therapy: Results and lessons from recent successes. *Mol. Ther. Methods Clin. Dev.* **2016**, *3*, 16034. [CrossRef] [PubMed]
16. FiercePharma. With Its Launch Fizzling Out, UniQure Gives Up on $1M+ Gene Therapy Glybera. Available online: https://www.fiercepharma.com/pharma/uniqure-gives-up-1m-gene-therapy-glybera (accessed on 9 March 2018).
17. Brockhurst, M.A.; Koskella, B.; Zhang, Q.G. Bacteria-Phage Antagonistic Coevolution and the Implications for Phage Therapy. In *Bacteriophages: Biology, Technology, Therapy*; Harper, D., Abedon, S., Burrowes, B., McConville, M., Eds.; Springer Nature: Cham, Switzerland, 2017.
18. Sillankorva, S.M.; Oliveira, H.; Azeredo, J. Bacteriophages and Their Role in Food Safety. *Int. J. Microbiol.*, **2012**, *2012*, 863945. [CrossRef] [PubMed]
19. Sarker, S.A.; Sultana, S.; Reuteler, G.; Moine, D.; Descombes, P.; Charton, F.; Bourdin, G.; McCallin, S.; Ngom-Bru, C.; Neville, T.; et al. Oral Phage Therapy of Acute Bacterial Diarrhea with Two Coliphage Preparations: A Randomized Trial in Children From Bangladesh. *EBioMedicine* **2016**, *4*, 124–137. [CrossRef] [PubMed]
20. Soothill, J.S. Treatment of experimental infections of mice with bacteriophages. *J. Med. Microbiol.* **1992**, *37*, 258–261. [CrossRef] [PubMed]
21. Watanabe, R.; Matsumoto, T.; Sano, G.; Ishii, Y.; Tateda, K.; Sumiyama, Y.; Uchiyama, J.; Sakurai, S.; Matsuzaki, S.; Imai, S. Efficacy of bacteriophage therapy against gut-derived sepsis caused by Pseudomonas aeruginosa in mice. *Antimicrob. Agents Chemother.* **2007**, *51*, 446–452. [CrossRef] [PubMed]
22. Hawkins, C.; Harper, D.; Burch, D.; Änggård, E.; Soothill, J. Topical treatment of *Pseudomonas aeruginosa* otitis of dogs with a bacteriophage mixture: A before/after clinical trial. *Vet. Microbiol.* **2010**, *146*, 309–313. [CrossRef] [PubMed]
23. Marza, J.A.; Soothill, J.S.; Boydell, P.; Collyns, T.A. Multiplication of therapeutically administered bacteriophages in *Pseudomonas aeruginosa* infected patients. *Burns* **2006**, *32*, 644–646. [CrossRef] [PubMed]
24. Phagoburn. Available online: http://www.phagoburn.eu/ (accessed on 9 March 2018).
25. Rhoads, D.D.; Wolcott, R.D.; Kuskowski, M.A.; Wolcott, B.M.; Ward, L.S.; Sulakvelidze, A. Bacteriophage therapy of venous leg ulcers in humans: Results of a phase I safety trial. *J. Wound Care* **2009**, *18*, 237–243. [CrossRef] [PubMed]

26. Rose, T.; Verbeken, G.; De Vos, D.; Merabishvili, M.; Vaneechoutte, M.; Lavigne, R.; Jennes, S.; Zizi, M.; Pirnay, J.P. Experimental phage therapy of burn wound infection: Difficult first steps. *Int. J. Burns Trauma* **2014**, *4*, 66–73. [PubMed]
27. Cole, S.T.; Eisenach, K.D.; McMurray, D.N.; Jacobs, W.R. (Eds.) *Tuberculosis and the Tubercle Bacillus*; ASM Press: Washington, DC, USA, 2005; ISBN 1-55581-295-3.
28. PubMed Health. Closed Comedones. Available online: https://www.ncbi.nlm.nih.gov/pubmedhealth/PMHT0025363/ (accessed on 9 March 2018).
29. Payne, R.J.H.; Vincent, A.A. Evidence for a Phage Proliferation Threshold? *J. Virol.* **2002**, *76*, 13123–13124. [CrossRef] [PubMed]
30. Vinner, G.K.; Vladisavljević, G.T.; Clokie, M.R.J.; Malik, D.J. Microencapsulation of Clostridium difficile specific bacteriophages using microfluidic glass capillary devices for colon delivery using pH triggered release. *PLoS ONE* **2017**, *12*, e0186239. [CrossRef] [PubMed]
31. Chibani-Chennoufi, S.; Sidoti, J.; Bruttin, A.; Kutter, E.; Sarker, S.; Brüssow, H. In vitro and in vivo bacteriolytic activities of *Escherichia coli* phages: Implications for phage therapy. *Antimicrob. Agents Chemother.* **2004**, *48*, 2558–2569. [CrossRef] [PubMed]
32. Ryan, E.M.; Gorman, S.P.; Donnelly, R.F.; Gilmore, B.F. Recent advances in bacteriophage therapy: How delivery routes, formulation, concentration and timing influence the success of phage therapy. *J. Pharm. Pharmacol.* **2011**, *63*, 1253–1264. [CrossRef] [PubMed]
33. Jończyk-Matysiak, E.; Weber-Dąbrowska, B.; Owczarek, B.; Międzybrodzki, R.; Łusiak-Szelachowska, M.; Łodej, N.; Górski, A. Phage-Phagocyte Interactions and Their Implications for Phage Application as Therapeutics. *Viruses* **2017**, *9*, 150. [CrossRef] [PubMed]
34. Łusiak-Szelachowska, M.; Żaczek, M.; Weber-Dąbrowska, B.; Międzybrodzki, R.M.; Letkiewicz, S.; Fortuna, W.; Rogóż, P.; Szufnarowski, K.; Jończyk-Matysiak, E.; Olchawa, E.; et al. Antiphage activity of sera during phage therapy in relation to its outcome. *Future Microbiol.* **2017**, *12*, 109–117.
35. Ledford, H. Myriad ruling causes confusion. *Nature* **2013**, *498*, 281–282. [CrossRef] [PubMed]
36. Schooley, R.T.; Biswas, B.; Gill, J.J.; Hernandez-Morales, A.; Lancaster, J.; Lessor, L.; Barr, J.J.; Reed, S.L.; Rohwer, F.; Benler, S.; et al. Development and Use of Personalized Bacteriophage-Based Therapeutic Cocktails to Treat a Patient with a Disseminated Resistant Acinetobacter baumannii Infection. *Antimicrob. Agents Chemother.* **2017**, *61*, e00954-17. [CrossRef] [PubMed]
37. Dalmasso, M.; de Haas, E.; Neve, H. Isolation of a Novel Phage with Activity against Streptococcus mutans Biofilms. *PLoS ONE* **2015**, *10*, e0138651. [CrossRef] [PubMed]
38. Abedon, S.T.; Thomas-Abedon, C. Phage therapy pharmacology. *Curr. Pharm. Biotechnol.* **2010**, *11*, 28–47. [CrossRef] [PubMed]
39. Pirnay, J.P.; De Vos, D.; Verbeken, G.; Merabishvili, M.; Chanishvili, N.; Vaneechoutte, M.; Zizi, M.; Laire, G.; Lavigne, R.; Huys, I.; et al. The phage therapy paradigm: Prêt-à-porter or sur-mesure? *Pharm. Res.* **2011**, *28*, 934–937. [CrossRef] [PubMed]
40. Feiner, R.; Argov, T.; Rabinovich, L.; Sigal, N.; Borovok, I.; Herskovits, A.A. A new perspective on lysogeny: Prophages as active regulatory switches of bacteria. *Nat. Rev. Microbiol.* **2015**, *13*, 641–650. [CrossRef] [PubMed]
41. Hyman, P.; Abedon, S.T. Bacteriophage host range and bacterial resistance. *Adv. Appl. Microbiol.* **2010**, *70*, 217–248. [PubMed]
42. Nale, J.Y.; Spencer, J.; Hargreaves, K.R.; Buckley, A.M.; Trzepiński, P.; Douce, G.R.; Clokie, M.R. Bacteriophage Combinations Significantly Reduce Clostridium difficile Growth in Vitro and Proliferation in Vivo. *Antimicrob. Agents Chemother.* **2015**, *60*, 968–981. [CrossRef] [PubMed]
43. Hargreaves, K.R.; Clokie, M.R. Clostridium difficile phages: Still difficult? *Front. Microbiol.* **2014**, *28*, 184. [CrossRef] [PubMed]
44. Colavecchio, A.; Cadieux, B.; Lo, A.; Goodridge, L.D. Bacteriophages Contribute to the Spread of Antibiotic Resistance Genes among Foodborne Pathogens of the *Enterobacteriaceae* Family—A Review. *Front. Microbiol.* **2017**, *20*, 1108. [CrossRef] [PubMed]
45. Strauch, E.; Lurz, R.; Beutin, L. Characterization of a Shiga toxin-encoding temperate bacteriophage of Shigella sonnei. *Infect. Immun.* **2001**, *69*, 7588–7595. [CrossRef] [PubMed]
46. Christie, G.E.; Dokland, T. Pirates of the Caudovirales. *Virology* **2012**, *434*, 210–221. [CrossRef] [PubMed]

47. Ackermann, H.-W.; Tremblay, D.; Moineau, S. Long-term bacteriophage preservation. *W.F.C.C. Newsl.* **2004**, *38*, 35–40.
48. European Pharmacopoeia Online. Available online: http://online.edqm.eu/EN/entry.htm (accessed on 9 March 2018).
49. USP-NF. Available online: http://www.uspnf.com/?_ga=2.80419031.1739394632.1520820978-1625466886.1520820978 (accessed on 9 March 2018).
50. Verbeken, G.; Pirnay, J.P.; De Vos, D.; Jennes, S.; Zizi, M.; Lavigne, R.; Casteels, M.; Huys, I. Optimizing the European regulatory framework for sustainable bacteriophage therapy in human medicine. *Arch. Immunol. Ther. Exp. (Warsz.)* **2012**, *60*, 161–172. [CrossRef] [PubMed]
51. Verbeken, G.; Pirnay, J.P.; Lavigne, R.; Jennes, S.; De Vos, D.; Casteels, M.; Huys, I. Call for a dedicated European legal framework for bacteriophage therapy. *Arch. Immunol. Ther. Exp. (Warsz.)* **2014**, *62*, 117–129. [CrossRef] [PubMed]
52. Huys, I.; Pirnay, J.P.; Lavigne, R.; Jennes, S.; De Vos, D.; Casteels, M.; Verbeken, G. Paving a regulatory pathway for phage therapy. Europe should muster the resources to financially, technically and legally support the introduction of phage therapy. *EMBO Rep.* **2013**, *14*, 951–954. [CrossRef] [PubMed]
53. Defoirdt, T. Quorum-Sensing Systems as Targets for Antivirulence Therapy. *Trends Microbiol.* **2017**, *26*, 313–328. [CrossRef] [PubMed]
54. Harper, D.R.; Enright, M.C. Bacteriophages for the treatment of Pseudomonas aeruginosa infections. *J. Appl. Microbiol.* **2010**, *111*, 1–7. [CrossRef] [PubMed]
55. Soothill, J.S.; Hawkins, C.; Anggard, E.A.; Harper, D.R. Therapeutic use of bacteriophages (letter). *Lancet Infect. Dis.* **2004**, *4*, 544–545. [CrossRef]

Review

In Vitro Characteristics of Phages to Guide 'Real Life' Phage Therapy Suitability

Eoghan Casey, Douwe van Sinderen * and Jennifer Mahony

School of Microbiology and APC Microbiome Ireland, University College Cork, T12 YT20 Cork, Ireland; eoghan.casey@ucc.ie (E.C.); j.mahony@ucc.ie (J.M.)

* Correspondence: d.vansinderen@ucc.ie; Tel.: +353-21-4901365

Received: 14 March 2018; Accepted: 29 March 2018; Published: 30 March 2018

Abstract: The increasing problem of antibiotic-resistant pathogens has put enormous pressure on healthcare providers to reduce the application of antibiotics and to identify alternative therapies. Phages represent such an alternative with significant application potential, either on their own or in combination with antibiotics to enhance the effectiveness of traditional therapies. However, while phage therapy may offer exciting therapeutic opportunities, its evaluation for safe and appropriate use in humans needs to be guided initially by reliable and appropriate assessment techniques at the laboratory level. Here, we review the process of phage isolation and the application of individual pathogens or reference collections for the development of specific or "off-the-shelf" preparations. Furthermore, we evaluate current characterization approaches to assess the in vitro therapeutic potential of a phage including its spectrum of activity, genome characteristics, storage and administration requirements and effectiveness against biofilms. Lytic characteristics and the ability to overcome anti-phage systems are also covered. These attributes direct phage selection for their ultimate application as antimicrobial agents. We also discuss current pitfalls in this research area and propose that priority should be given to unify current phage characterization approaches.

Keywords: pH stability; phage-host interactions; genomics; antibiotic-resistance; phage preparation; lysins; biofilms

---

1. Introduction

Frederick Twort and Felix d'Hérelle are accredited as the founding fathers of phage biology having recognized that bacteriophages or "bacteria eaters" are present wherever bacteria are found [1,2]. During World War I, the unsanitary conditions in the trenches on the battle-fronts caused infections such as dysentery affecting numerous soldiers. Felix d'Hérelle studied the bacterial cause of these infections and identified *Shigella* as the etiologic agent of this rampaging infectious disease. Frederick Twort demonstrated the existence of a propagatable bactericidal agent in 1915, while independently (or not) Felix d'Hérelle co-incubated fecal filtrates with the isolated bacteria in petri dishes resulting in the killing of the bacteria (d'Hérelle's prior knowledge of Twort's research remains uncertain, reviewed here [3]). The first description of the co-existence and relationship between bacteria and their infecting phages was a historic one that created the opportunity to develop treatments against bacterial infections for which therapies were not available at that time. During the 1920's and 1930's d'Hérelle blazed the trail of phage therapy by successfully developing phage-based treatments against a range of human infections including those caused by *Shigella dysenteriae*, *Salmonella typhi*, *Escherichia coli*, *Pasteurella multocida*, and *Vibrio cholerae*.

The discovery of mold-derived penicillin production (by *Penicillium notatum*) by Alexander Fleming in 1928 and the subsequent discovery of its broad-spectrum of antibacterial activity and relative ease of large scale production delivered a major blow to developments in phage therapy.

The need for rapid production of agents that could effectively treat wound infections during World War II accelerated the rise of antibiotic-based therapies, which were highly effective and were considered a wonder drug at that time. However, little was known about the modes of action of either phages or antibiotics, and phage therapy research was almost completely abandoned in favor of further development of the antibiotic industry. The lack of adequate controls in these early phage experiments combined with the lack of clarity regarding phage efficacy were among the reasons that antibiotic research and development out-paced that of phage therapy research. While antibiotics were successfully exploited for the ensuing decades, the rapid rise of antibiotic resistance and the emergence of so-called "superbugs", such as methicillin-resistant *Staphylococcus aureus* (MRSA), carbapenem-resistant *Acinetobacter baumannii* and *Enterobacteriaceae* and *Clostridium difficile*, among others, are crippling to modern healthcare facilities. Furthermore, the World Health Organization (WHO) has stated that alternatives to antibiotics are required to treat such superbugs, which are estimated to cause 25,000 deaths in Europe and 23,000 deaths in the U.S.A. each year (WHO statement, February 2017). Plasmid-encoded antibiotic resistance represents a particularly challenging threat to society due to possible transfer or mobilization of such extra-chromosomal elements between bacterial strains or even species. Therefore, phage therapy or phage-based treatments may provide a solution to this pressing health issue in modern society, just over a century since their initial identification.

Both antibiotics and phages have inherent associated risks, draw-backs, and benefits. Phage therapy presents the possibility of applying a more targeted, narrow spectrum treatment, thereby providing greater specificity than most antibiotics. Furthermore, phages may be delivered as intact phage particles/phage cocktails, while phage-derived proteins may also be produced to overcome many of the concerns associated with delivering intact phage particles into humans. Considerable work has been undertaken in recent years to isolate and characterize phages with therapeutic potential against a range of pathogenic bacteria. While many phage/phage cocktails have proven effective in lab-scale trials, this has not always been matched by successful in vivo trials, cautioning against too much optimism with regards to the success of phage therapy when applied in real life settings. Careful evaluation of the in vitro phenotypes of phages is therefore required before they may be considered suitable for clinical trials for the purpose of animal or human therapy [4,5]. In this review, we assess the methods by which phages are isolated and characterized for therapeutic consideration, the benefits and pitfalls of current approaches, while also considering alternative approaches for the future. Furthermore, we explore the advantages and disadvantages of phage cocktails and phage-derived products, and the most recent advances in combination therapies as a 21st century approach to deal with bacterial infections.

2. Phage Isolation

As bacteriophages represent the most abundant biological entity in the biosphere [6], isolation of phages infecting pathogenic strains of interest should not represent a significant hurdle. Isolation of phages against pathogenic species requires identification of areas where the pathogenic host is abundant. In particular, sewage samples represent a plentiful source of phages for use in phage therapy due to the presence of many human pathogens [7]; however, any environment where the pathogenic host is present represents a potential source of phages. After identification of these environments, methods can be employed to isolate phages of particular value in phage therapy.

In general, the utilization of a diverse range of strains as isolating hosts should lead to the isolation of a diverse range of infecting bacteriophages. The use of reference collections as isolation hosts for pathogens of interest should be the gold standard for phage isolation studies. Collections such as the ECOR collection of *E. coli* [8] and the *Salmonella* SARA and SARB collections [9,10], purported to represent the full intra-species diversity have been used in phage isolation studies and can be helpful in generating an arsenal of phages with a high likelihood of infecting strains of clinical relevance with respect to the targeted species (as such phages are presumed to recognize any of the diverse range of cell surface receptors that may be present among such strains) [11]. Even more preferable are studies utilizing reference collections in addition to strains of clinical relevance, leading to the generation of highly diverse phage collections [12]. A further development in this area is the SBS (step-by-step) method, which can be applied when attempting to isolate a phage against single strains, and which was first developed for *Klebsiella pneumoniae*. This involves isolation of phages against the strain of interest in addition to bacteriophage-insensitive mutants of the same strain. In short, a phage is isolated against a strain of interest, then a resistant variant of this strain is generated through exposure to the isolated phage. Following this, a new phage is isolated against this resistant variant and so on until the latest phage-resistant variant is sensitive to the original phage. The mechanisms through which host resistance is achieved were not investigated, although presumably it relies on efficient CRISPR (Clustered Regularly Interspaced Short Palindromic Repeats) spacer incorporation. In essence, this method will generate a cocktail of phages infecting both the original bacterium and many phage-resistant variants, reducing the possibility of variants emerging which are resistant to the cocktail during application [13]. Furthermore, it has been theorized that standard bacteriophage isolation and enrichment protocols involving single strains are biased against broad range bacteriophages, (unintentionally) selecting for those with most avid adsorption to host cells. The use of a mixed host enrichment procedure has previously led to the isolation of broad host range bacteriophages, which arguably are the most valuable candidates for phage therapy [14].

At the isolation step, care should be taken to ensure that only virulent phages are selected as candidates for phage therapy. It is advised that phages capable of lysogenic conversion be avoided, as these will easily convert hosts into (phage-resistant) lysogens, thus making them incapable of causing immediate lysis [15]. Avoidance of plaques with turbidity, typical of lysogenic phages, will assist in the selection of lytic phages, but as some temperate phages produce clear plaques, further methods are required for lifestyle validation [11]. With the rapid advancement of genome sequencing technologies, it is now feasible to sequence the genomes of all candidates, therefore allowing for more definitive exclusion of phages that encode integrases, site-specific recombinases, and repressors of the lytic cycle [7]. Indeed, a genomic approach to guide phage selection has in recent years gained in popularity and is covered in more detail below.

Long-term storage and stability of selected phages is desirable, and a good candidate for phage therapy should be one that maintains infective ability upon storage. Several methods have been proposed for maintenance of phage stocks including storage at 4 °C [16], storage at −80 °C with the addition of glycerol [17], and lyophilization of phage stocks, with the latter proposed to be a particularly effective method for long term storage [18]. A novel method involving the storage of phages as injected DNA in freshly infected frozen cells has shown promise, although some strains were shown to be poor storage hosts [17]. It has been noted that due to variation amongst bacteriophages a "gold-standard" phage storage condition remains to be identified, and long-term storage of phages is likely to be phage dependent [19]. Ackermann et al. have detailed the stability and instability of various phages, and for example described that *Bacillus* phage CP-54Ber suffers a 7-log reduction in infective ability following three months of storage at 4 °C (representing the most common phage storage temperature), exemplifying this diversity [16]. Therefore, in vitro assays to determine optimum storage conditions for each phage of interest should be undertaken [20].

3. Characterization of Phages for Phage Therapy Applications

3.1. Genomic and Morphological Characteristics

A lack of uniformity in the approaches to assess phage isolates in the laboratory for their suitability as a therapeutic agent is possibly one of the biggest stumbling blocks to phage therapy. Recent phage therapy studies that report genome characterization are typically partnered with morphological analysis, which is a routine analysis step. While the isolation of phages has been a continuous and ongoing effort of phage biologists globally [21–25], it should be noted that considerable collections of phages that have the potential for therapeutic trialing and application, exist. One such collection of nine *E. coli* phages has recently been assessed to identify the common microbiological and genomic characteristics of phage isolates exhibiting therapeutic potential in vivo [26]. This historic collection of *E. coli* O18:K1:H7 phages was characterized as being comprised of six *Podoviridae* phages and three *Siphoviridae* phages. Four of the podophages exhibited the most significant host-infecting efficacy among the collection, and genomic analysis revealed that these isolates were all members of the same species bearing at least 99% identity to each other. The presence of a DNA-dependent RNA polymerase was present only in the genomes of "fast replicating" phages. This protein is purported to cause a shutdown of host transcription (5 min after infection), resulting in a much shorter latent period than non-RNA polymerase-encoding phages. This confers a more efficient replication cycle on the phages, a property suggested to underpin the success of these isolates in in vivo trials. Indeed, the relatively simple structures associated with the podophage may lend itself to shorter latent times and larger burst sizes compared to phages that possess elaborate and decorative structures. In contrast, two recently isolated, *Clostridium difficile*-infecting *Myoviridae* phages were shown to exhibit a broad host range, infecting strains from approximately half of the strains that represent the 20 ribotypes of *C. difficile* [24]. The more extensive host range of these isolates relative to other *Myo*- and *Siphoviridae* phages from the same study highlights that large phages such as myophages may have therapeutic potential, while the infective success of various phage morphologies appears to be host-specific. Therefore, identification of genetic and morphological features associated with highly infective phages may represent useful markers to guide the selection of phages for subsequent therapeutic application. However, these markers may be genus- or species-specific, and therefore caution should be taken not to make generic recommendations for the identification process of phages with therapeutic potential.

Notwithstanding the above-mentioned limitations, genome sequencing permits the rapid identification of undesirable features that would quickly "rule out" unsuitable phages. For example, as mentioned in Section 2, lysogenic phages are generally not considered suitable for phage therapy, since their integration into the host genome may confer alterations on the host phenotype (lysogenic conversion). Additionally, in the integrated state, they may recombine with genetic elements of other (pro)phage or bacterial genomes and acquire undesirable features such as pathogenicity islands or antibiotic resistance markers, among others. Genome characterization permits the identification of lysogeny-associated functions including repressors and integrases, which are often readily identifiable. Furthermore, the presence of recombination-related functions in phage genomes may present issues in terms of genetic instability. Analysis of genome stability is an area that has not yet been deeply investigated, although recent advances in protein function prediction tools such as Pfam [27] and HHpred [28] have significantly improved the quality and accuracy of genome annotations, which may contribute to the development of this aspect.

Morphological assessment by electron microscopy is a standard characterization step of almost all newly reported isolated phages. Although this requires specialized equipment and expertise, it is an essential characterization step, and undoubtedly will continue to be the cornerstone of phage characterizations. Electron microscopy may be particularly useful in assessing the stability of stored phage isolates. Relying on microbiological assays such as the standard plaque assay for phage enumeration may not reveal the true extent of the stability of the produced phages, whereas electron microscopy may reveal the presence of additional "ghost" or degraded particles.

3.2. Host Receptor Identification

One of the concerns associated with human phage therapy is the emergence of phage-resistant variants of pathogenic bacteria with increased fitness. The host range of a phage reflects its ability to (lytically) infect strains within a given test panel where narrow host range phages infect a small number of strains and broad host range phages infect a wide range of strains. Phages may exhibit narrow or broad host range depending on (i) the presence of anti-phage mechanisms in the test strains and; (ii) the presence of generalized (highly conserved) or specialized (variable, non-conserved) host-encoded phage receptors. While broad host range phages are generally more acceptable due to the increased likelihood that clinical isolates that emerge will be infected, narrow host range phages may be useful in certain scenarios. In contrast to antibiotics, which are broad spectrum antimicrobial agents, the use of narrow host range phages presents a new opportunity. It allows the possibility of isolating phages against prevalent and specific strains of pathogenic bacteria and delivering a treatment without the problem of host dysbiosis. The identification of narrow spectrum phages combined with the SBS approach (mentioned in Section 2) could reduce the risk of the phages becoming defunct upon the emergence of resistant variants of the target strains and present a useful approach to developing next generation therapies. However, as with all treatments, narrow host range phages have their limitations. As pathogens evolve and populations diversify, it is necessary to have monitoring programmes to continually evaluate prevalent strains of problematic pathogens. Therefore, corresponding phage screening programmes should be initiated against pathogenic strain collections including the most recent clinical isolates on a continuous basis. This would require a concerted approach by various phage research groups and associated funding from government and other agencies to ensure the "future-proofing" of this approach. The peak-and-trough profile of phage research over the past century indicates that this is difficult, however, with the growing demand for alternative antimicrobial therapies, it is expected to gain support, at least in the foreseeable future.

While targeted narrow host range phage therapies may have potential for specific applications, it is likely that broad host range phages will continue to be the preferred option as they possess a more powerful destructive potential against a wider range of pathogenic isolates. Furthermore, even those phages that are classified as exhibiting a broad host range would still be considered to possess a narrow activity spectrum relative to antibiotics. Antibiotics may be effective against multiple genera of bacteria, while phages are rarely genus-specific, but mostly species- or strain-specific. During the past two decades, interactions between phages and their host bacteria have become the subject of intense research scrutiny [29–37]. The availability of bacterial and phage genome sequences has facilitated studies in which phage-resistant derivatives of bacterial host strains are sequenced to uncover the genetic basis of phage-resistance and by inference the receptors that phages recognize [38,39]. Conversely, the isolation of phage mutants that can circumvent such phage-resistant derivatives have been characterized to understand the molecular basis of phage infection and the genes that encode host recognition functions [40,41]. Through such analyses, the interactions of a wide range of phages and their bacterial hosts are now well defined and in theory the methods can be adapted to study any phage-host combination. To alleviate concerns regarding the emergence of phage-resistant variants of pathogenic bacterial strains, phage cocktails incorporating phages that employ different receptors is advisable. However, this requires knowledge of the receptor types employed by the incorporated phages. For this reason, phage-host interaction studies are vital to the design of robust phage cocktails.

Phage-host interactions require both a host-encoded receptor(s) and a phage-encoded receptor binding protein (RBP). The receptor presented on the cell surface may be a carbohydrate, protein or (lipo-)teichoic acid moiety, or a combination of these. The fundamental differences in the composition of Gram-negative and Gram-positive cell walls dictate the types and range of interactions that may occur between phages and their respective hosts. For example, the majority of Gram-positive bacteria-infecting phages are reported to recognize saccharidic moieties on the cell surface, while those of Gram negative-infecting phages such as those infecting *E. coli* and *Salmonella* are less biased, with several coliphages known to recognize proteinaceous and saccharidic receptors according to a recent review of phage receptors [42]. This extensive review of phage receptors has culminated in the generation of the Phage Receptor Database (PhReD, https://phred.herokuapp.com/), which is a highly informative resource for phage biologists.

For many phages, a single receptor is required for the recognition and attachment stages of the infection process while for others, two (or more) components are required. Coliphages are undoubtedly the best studied Gram-negative-infecting group of phages with model *Myoviridae* and *Siphoviridae* phages such as T4 and lambda, respectively, representing paradigms of infection of this bacterial species (For reviews, see [43,44]). Lambda, with its long non-contractile tail, is an example of a phage that requires a single receptor, i.e., the protein LamB [29], while the long tail fibers of the myophage T4 bind reversibly to the protein OmpC before the short tail fibers commit phage binding to the heptose moiety of the lipopolysaccharide (LPS) [45,46]. In the review of phage receptors mentioned above, 26 coliphage interactions were covered, of which eight require both cell envelope-associated proteins and sugar moieties, while ten and eight coliphages, respectively, require proteins or carbohydrate moieties alone for host adsorption [42]. A similar analysis of currently characterized *Salmonella* phages tells us that 11 phages employ proteinaceous receptors, seven recognize carbohydrate moieties on the cell surface, while one requires both moieties to adsorb to their host. In contrast, all *Pseudomonas* phage-host systems characterized to date attach to saccharidic receptors. Therefore, it is clear that there is a diverse array of interactions at play among these phage-host combinations.

In contrast, the interactions of the majority of characterized Gram positive-infecting phages involve saccharidic molecules [42,47]. Among the model phage-host systems of Gram-positive bacteria are those of *Bacillus subtilis* and its infecting phage SPP1 [31,35] and those of the dairy bacterium *Lactococcus lactis* and members of the 936 and P335 phage groups [34,48,49]. While these phage-host interactions bear no direct clinical relevance, studies of these two Gram-positive hosts and their phages have consolidated studies of the interactions of clinically relevant Gram-positive hosts such as *Listeria monocytogenes*, *Bacillus anthracis* and *Staphylococcus aureus*, among others. The level of structural detail that now exists for SPP1, and the 936 and P335 lactococcal phages have provided insights into the nature and interactions of phages of other Gram-positive bacteria and those with clinical relevance. Module shuffling to accommodate host interaction flexibility is not a new concept; however, the structural characterization of phages and phage components (particularly the distal tail components including the host recognition devices) have generated data pertaining to modules that may be present and conserved among phages of different bacterial genera. An example of this is the so-called "evolved" distal tail protein (Dit) of phages of *Lactobacillus* and *Lactococcus* [48,50], in which a carbohydrate-binding domain (CBD) is inserted to expand/enhance the binding capabilities of the carrying phage. Furthermore, another CBD was structurally characterized in the accessory baseplate (distal tail appendage) protein (BppA) of the lactococcal phage Tuc2009 [33] and based on HHpred analysis has been identified in a number of other lactococcal, *Lactobacillus* and streptococcal phage proteins, highlighting that this domain is found in phages infecting other bacterial genera [48,50]. Taken together, knowledge on the diversity of interaction types will guide the selection of phages for inclusion as therapeutic agents either alone or as part of a cocktail.

3.3. Stable Storage, Administration and Effectiveness in Trials

In order to be acceptable for therapeutic application, a phage must retain its infectivity under storage conditions over extended periods of time, and be able to withstand the administration process and route. To evaluate this, phages are often tested for robustness to pH treatments and simulated gastric juices to mimic the oral route of administration [51,52]. Additionally, other studies have evaluated the stability of phages upon freeze drying, spray drying and cold storage to define their appropriateness for the clinical setting [53,54]. Such studies have demonstrated that the stability is pH-, excipient- and phage-dependent. Therefore, it is essential that each candidate phage is assessed using a range of conditions to define the appropriate preservation/storage conditions. In addition to preserving the phage, the cultivation medium, scale up and harvesting of phage preparations are all aspects that require careful scrutiny in order to ensure that the phage is capable of transitioning from the laboratory to the clinic. A recent study examined such characteristics of T4-like coliphages [55]. Here, it was demonstrated that the employed growth medium had minimal impact on the phage titers achieved, while growth phase (i.e., early exponentially growing cells were more effective than lag or late exponential phase cells) and the choice of host strain were important factors in producing a high phage titer. The authors also evaluated different propagation approaches including 2 L Erlenmeyer flasks, 16 L stirred fermentation tanks and 10 L wave bags, as well as different methods of purification including ultrafiltration, chromatography and ultracentrifugation. Phages were shown to exhibit flexibility to scaling up and purification, permitting inexpensive and relatively fast production of lysates of coliphages for application. Since such therapies are also increasingly aimed at providing cost-effective and practical solutions to commonplace infections in developing countries, the propagation and purification techniques need to be considered as specialized laboratory equipment such as ultracentrifuges may not be readily available [55].

The phage (or phage-derived product) administration route also dictates the type of characteristics that are required. For example, if the phage product is to be administered orally to treat enteric infections, the phage should be able to withstand low pH, gastric conditions and/or be suitable for encapsulation for delivery to the desired site. Encapsulation trials have been undertaken with phages of various pathogens including *C. difficile* and enterohemorrhagic *E. coli* using a range of materials including Eudragit® S100 (a pH responsive polymer), and in the presence of alginate or pectin as base polymers, among others [56,57]. Such microencapsulation was shown to retain infection efficacy of the tested phages upon exposure to simulated gastric conditions and the associated acidity. For treatment of respiratory infections and chronic conditions, an aerosol-based application may be the preferred route of administration requiring that the phage should retain infectivity following freeze- or spray-drying and administration [58].

A phage's potential for application is commonly based on animal trials prior to consideration of human trials. Animal trials most often involve a murine model system although larger animals have been employed in the assessment of phages destined for the treatment of animals in the food chain [59]. Trials in sheep and cattle inoculated with *E. coli* O157:H7 highlighted the need for adequate controls as endemic phages contributed to the phage population that was shed, interfering with the results of the therapeutic analysis. Animal trial-based assessment of phage therapy applications is a contentious issue and must be limited where possible. Some recent studies have performed preliminary trials using wax moth (*Galleria mellonella*) larvae in order to allay such ethical concerns. The *Galleria* larval model is a useful system since these organisms possess complex innate immune systems, and, additionally, similarities have been identified between insect larval epithelial cells and intestinal mammalian cells [60]. This model system is inexpensive and requires little specialist training, and is therefore useful to define which phages display genuine therapeutic potential and are suitable for further testing.

Relatively few studies have investigated the interactions between bacteriophages and the immune system, however, persistence in the face of the host immune system represents an essential characteristic of candidates for phage therapy. It is known that phages are immunogenic, with studies of T4 coliphage showing stimulation of antibody production through both oral and subcutaneous injection with significantly lower immunogenicity observed via the oral administration route [61]. However, persistent injection at a dose higher than what would normally be administered was required to elicit a marked immune response [61]. Continual exposure of the body to the intra-body phageome presumably has a role to play in this, and it has been suggested that the apparent lack of immune response is due to chronic phage exposure during evolution [62]. Indeed, phage exposure causes an immunomodulatory response, displaying an inhibitory effect of T-cell proliferation and downregulation of antibody production contributing towards the homeostasis of the immune system [63,64].

While phages and phage cocktails present an opportunity to combat the current shortage of antimicrobial therapies, phage therapy remains an unattractive option to many regulatory bodies. However, advances in phage biology, genome sequencing and molecular biology may provide the knowledge to overcome such regulatory concerns through the exploitation of phage-derived proteins as we will discuss below.

4. Phage Particles Versus Phage-Derived Products

Endolysins are produced by phages at the end of their infection cycle in order to release progeny phages from the host cell. They are characterized by their ability to hydrolyze the peptidoglycan layer of the bacterial cell wall [65]. Endolysins represent the most promising lytic enzymes used in phage therapy, possessing several advantages over other candidates. They are effective immediately (in contrast to the lag time exhibited by bacteriophage particles), lysing target cells within seconds of first contact. For example, a streptococcal lysin specific for groups A, C and E was shown to completely sterilize a culture containing 10^7 colony forming units of the group A *Streptococcus* strain D471 in vitro. This in vitro activity was a good indicator of in vivo effectivity as mice which had been heavily colonized in the upper respiratory tract with a streptomycin-resistant group A *Streptococcus* (T14/46) showed complete eradication of said strain two hours after administration of purified lysin [66]. A further advantage of endolysin treatment is that, although purified enzymes are antigenic, they are capable of avoiding the humoral immune response in a study of endolysin application to methicillin resistant *S. aureus* [67]. Here, endolysin LysGh15 administration was shown to elicit specific-IgG antibody production, although incubation of LysGh15 with anti-LysGH15 serum showed no difference in bactericidal activity when compared to LysGH15 incubated with normal mouse serum. This indicates that purified lysins are suitable for application without inactivated by the host immune system. Endolysins also display a higher target specificity compared to antibiotics (thereby reducing the potential for resistance development). Purportedly, no resistance to endolysin activity has yet been observed due to the evolutionary link between phage endolysin and host cell autolysin [68]. A significant disadvantage associated with the use of purified endolysins is the lack of efficacy against most Gram-negative pathogens due to the presence of the outer membrane preventing access to the peptidoglycan layer [68]. Here, a combinatorial approach, i.e., treatment of Gram-negative cells with an antibiotic to rupture the membrane, thereby allowing access to the peptidoglycan layer for lysin degradation, may be advantageous. This synergy has been observed in the treatment of *C. difficile* infection with the purified lysin PlyCD, where use in conjunction with vancomycin (inhibiting cell wall synthesis) was shown to have a higher inhibitory effect than either substance alone [69] (combination therapy discussed in more detail below). In the treatment of Gram-positive infections, however, it remains a powerful tool and as most phages will encode an endolysin to release progeny, a plentiful source of lytic enzymes against bacterial strains is undoubtedly available.

Endolysin research has led to the identification of useful enzymes in targeting many "drug resistant" pathogenic bacteria, as the need for alternative emerging therapies increases. Lysins have shown significant promise in murine models in the treatment of multidrug resistant bacteria, such as rescuing mice from *Acinetobacter baumanii* bacteremia [70], protecting them against systemic MRSA infection [67,71], and ex vivo treatment of *C. difficile* in the colon [69].

This has led to significant commercial interest in these enzymes for use in a clinical setting. Particular efforts have been made to engineer specialized lysins with increased efficacy against pathogens or displaying multifunctional activity. Due to the modular nature of phage-encoded lysins, consisting of cell wall binding domains (CWBD) and enzymatically active domains (EAD) [72], swapping of modules from other sources, or fusing of modules to other proteins can be undertaken, generating a lysin with new functional characteristics. These engineered lysins can be divided into two broad classes, chimeric lysins and artilysins. Chimeric lysins (or chimeolysins) are those which consist of domains from differing sources which have been engineered to possess improved characteristics for use. For example, ClyF active against MRSA was engineered through fusion of a staphylococcal and streptococcal lysin was found to possess improved thermostability and pH tolerance [73], fusion of the EAD from a Plyy187 (*S. aureus*) to a CWBD domain of PLyV12 (*Enterococcus*) resulted in a lysin with a wider lytic spectrum capable of infecting staphylococci, enterococci and streptococci [74]. Furthermore, it is possible to increase the lytic activity of a lysin against its desired target, as in the case of Ply187AN, a fusion of the EAD domain from Ply187 and the CWBD of LysK (Staphylococcal phage K), which shows an increased lytic capacity when compared to the original Ply187 lysin [75].

Artilysins represent a more targeted approach to engineering lysins for the purpose of overcoming the outer-membrane barrier of Gram-negative strains. It involves the fusion of an LPS-degrading peptide with the N-terminus of a lysin, puncturing the LPS layer and providing access to the peptidoglycan layer for lysin activity [76]. An example of this represents the fusion of the sheep myeloid antimicrobial peptide (SMAP29) to the KZ144 endolysin, creating Art-175 which confers the ability to puncture the outer membrane and cleave the peptidoglycan layer of *P. aeruginosa* PAO1, thereby causing a one log reduction after two minutes of treatment and a four-log reduction after 30 min [77]. Art-175 appears to have a wide lytic spectrum, capable of lysing *Klebsiella pneumoniae* [77], while also showing significant promise in lysing and disrupting persistent multidrug resistant *A. baumannii* strains [78].

Biofilms, microbial communities adhered to surfaces, are involved in many chronic infections and are noted for their resistance to host immune systems and medical treatments. Phages encoding depolymerases are of particular interest in the treatment of biofilm forming cultures. Due to the limited success of antibiotic therapy in treating biofilms [79], the potential to use bacteriophages or derived enzymes to treat biofilms has been under investigation. This ability is usually due to the expression of depolymerases capable of dispersing the biofilm through enzymatic digestion of extracellular polymeric substances, the main obstacle to antibiotic treatment or phage therapy [80]. The ability of phages to disrupt this biofilm is a valuable phenotype of phage therapy candidates.

The depolymerases expressed by phages digest these polymeric substances so as to obtain access to cell surface receptors [81,82], however it has been noted that the depolymerase activity alone may not be sufficient to disrupt the biofilm and the ability of the phage to amplify in the biofilm is crucial for biofilm treatment [83]. Phage-associated depolymerase activity can easily be identified in phages of interest through analysis of plaque morphology where depolymerase-expressing phages usually form a plaque surrounded by a large halo indicative of its degrading activity [84]. This phenotype has been observed for phages infecting members of several genera including *Pseudomonas* [81], *Klebsiella* [85], *Staphylococcus* [86] and *Escherichia* [87], thus representing a useful in vitro marker for phages of interest.

5. Overcoming Host-Encoded Phage-Resistance Mechanisms

Across phage therapy studies, the almost inevitable development of phage resistance in the targeted cells, analogous to the development of antibiotic resistance, is a primary weakness of the phage therapy concept. Therefore, the ability to overcome host resistance mechanisms represents the most valuable in vitro phenotype for a phage therapy candidate. Host resistance mechanisms fall primarily into four categories, (i) DNA degradation as provided by restriction-modification (R-M) and CRISPR-Cas systems, (ii) prevention of phage adsorption, (iii) superinfection exclusion, and (iv) abortive infection [88], along with additional hurdles to phage infection, such as the inherent resistance of biofilm communities. A summary of the following methods utilized by bacteriophages to overcome and bypass host-encoded phage resistance mechanisms is provided in Table 1.

Table 1. Identification and application of methods to bypass phage resistance in target strains.

Resistance Mechanism	Method of Bypass	Application	Reference(s)
Restriction modification	Phage-encoded methyltransferases	Protein homology query for identification in candidates	[89]
	Enhancement of host methylation	Protein homology query for identification in candidates	[90,91]
	Base modification	Protein homology query for identification in candidates	[92,93]
CRISPR	Mutation of protospacers	High MOI to encourage mutation of protospacers	[94–96]
	Phage-encoded anti CRISPR systems	Protein homology query for identification in candidates	[97–100]
Prevention of adsorption	Mutation of receptor binding protein	High MOI to encourage mutation in RBP	[40,101]
	Selection of multiple RBP type phages	Target a diverse range of receptors on target surface	[102]
Biofilm	Antibiotic combination therapy	Dual-pronged inhibition of target decreasing likelihood of resistance emergence	[103–107]
Emergence of phage resistant variants	Informed cocktail development (SBS method, serial enrichment)	Selection of phages capable of infecting "future" resistant variants	[13,108]
	Selection of multiple phages infecting a single strain	Target a diverse range of receptors on target surface	[12,109,110]

5.1. DNA Degradation by R-M Systems

One of the most prevalent bacteriophage resistance mechanisms are R-M systems being present on approximately 90% of available bacterial sequences [111]. The general function of these systems is to degrade invading (unmethylated) exogenous phage DNA by an endonuclease, while providing protecting of its own DNA through methylation [111,112]. Some striking adaptations to this infection barrier are evident in the genomes of infecting phages, mostly concerning restriction inhibition. Some bacteriophages encode "orphan" methyltransferases, i.e., those lacking a restriction endonuclease partner [89]. These methylases allow self-methylation of the phage DNA, thus eliminating the ability of the host to degrade injected DNA as host encoded restriction enzymes no longer recognize the methylated restriction sites [89]. Interestingly, orphan methyltransferases in *Bacillus* phages have been observed to be multispecific, i.e., they methylate multiple recognition sites (through possession of multiple target recognition domains) thus conferring protection against a wider range of host-encoded restriction enzymes [113,114]. It should be noted that methyltransferases have been proposed to play other roles in bacteriophages. A methyltransferase in φLM21, a temperate phage of *Sinorhizobium*, has been shown to protect against restriction enzyme activity while also mimicking the activity of a host encoded regulatory methyltransferase thus suggesting a regulatory role in the phage life cycle [115]. Orphan methylases should be easily identifiable in putative therapeutic phage genomes through protein homology searches. Other less prevalent approaches involve increasing the methylase activity of hosts to encourage phage DNA methylation, typified by Ral (restriction alleviation) activity in lambda phage, which enhances the activity of host type I methyltransferases to more efficiently methylate phage DNA [90,91] or base modification of phage DNA, such as that seen in T4 which utilizes hydroxymethylcytosine (HMC) instead of cytosine which is then further modified by alpha and beta glucosylation thus rendering it impervious to many restriction enzymes through prevention of restriction site recognition [92,93].

5.2. DNA Degradation by CRISPR-Cas Systems

CRISPR-Cas loci encoded by many bacteria provide an adaptive response to invading bacteriophages through the incorporation of non-host DNA (protospacers), into the CRISPR array. This array acts as the memory for targeted defense against subsequent infection by protospacer-containing phages [116]. Bypass of this mechanism can be achieved in two ways, namely, mutation of protospacers and specific anti-CRISPR activity. Mutation of protospacers represents the more common strategy of CRISPR escape. High mutation rates in protospacer regions are evident during exposure to CRISPRs in the dairy bacterium *Streptococcus thermophilus* containing the corresponding spacer, here a single base change has been observed to be sufficient to return the bacterium to a phage susceptible state [94]. This phenomenon has subsequently been observed in phage infection *E. coli* [95]. Here, CRISPR spacer arrays were constructed targeting various regions of coliphages, with phage escape mutants freely isolated through standard plaque assay displaying various point mutations, deletions and insertions in the protospacer regions allowing infection of previously resistant strains [95]. The application of a high multiplicity of infection cocktail of phages should increase the likelihood of obtaining the necessary mutation in a protospacer necessary to bypass the CRISPR sequence, while application of a cocktail of phages has been seen to reduce the efficiency of a host CRISPR system to eliminate a single phage [96]. Therefore, the application of a high titer cocktail targeting a single strain should have an increased chance of success in lysing a pathogenic host.

Phage-encoded anti-CRISPR activity was first observed in *Pseudomonas aeruginosa* Mu-like phages. Here, five protein families have been identified which inhibit class 1 CRISPR type I-F systems and four protein families inhibiting the type I-E system [97,117]. These proteins are a product of distinct anti-CRISPR modules, which would be advantageous in the genomes of any phage therapy candidate [97]. The mechanisms of action of three of these proteins AcrF1, AcrF2 and AcrF3 have been elucidated, showing that they all interfere with the function of the Csy complex which facilitates the targeted recognition and cleavage of target DNA sequences in *Pseudomonas aeruginosa* [118]. Both AcrF1 and AcrF2 bind to the Csy complex of the CRISPR system, competing with crRNA for DNA binding activity [98]. AcrF3 interacts with the Cas-3 helicase nuclease protein interfering with its recruitment to the Csy complex. Homology studies have led to the identification of ten anti-CRISPR Type I-F and four anti-CRISPR type I-E genes across the Proteobacteria phylum [100]. Recently, anti-CRISPR proteins (AcrIIA) targeting Class 2 CRISPRs have been identified in *Listeria monocytogenes* prophages through identification of self-targeting spacers, i.e., a protospacer in the prophage which matches a spacer in the CRISPR array. Two phage-encoded proteins (AcrIIA1 and AcrIIA2) were found to inhibit Cas9 function allow stable co-existence of the self-targeting spacer-protospacer pair. Homologues of *acrIIA* are also present in genomes of phages infecting *Streptococcus* indicating that anti-class II CRISPR genes may be prevalent across the *Firmicutes* [99].

5.3. Prevention of Adsorption

Antagonistic co-evolution between phages and respective hosts is a well-documented phenomenon, defined as the reciprocal evolution of bacterial resistance and phage infectivity [119]. It has been shown for *P. fluorescens* SBW25 and its phage φ2 that coevolution leads to significantly increased divergence of phage genes predicted to encode the adhesion device, presumably in response to receptor changes on the host cell surface, thus preventing infection of the ancestral genotype [120]. This interaction between receptor and phage receptor binding protein is highly specific, and it has been observed in phage lambda that a combination of only four mutations allows the phage to utilize an alternative receptor [101], and a single amino acid change leads to an altered RBP specificity [40]. Lessons can be learned from this for the selection of phages for application in phage therapy. Here an attempt can be made to shift the balance in favor of bacteriophages through inclusion of multiple phages encoding a diverse range of RBPs, increasing the hurdles required for the host to acquire the necessary mutations to confer resistance against multiple receptor binding protein types. This can be achieved through large scale sequencing and phylogenetic analysis of the receptor binding protein encoding genes of candidate phages, or alternatively by individually testing phages against a panel of strains differing in cell surface carbohydrates/outer membrane molecules [102].

5.4. Cocktails—the Power of Many

In practice, phage therapy is typically applied in two forms. "Monophage therapy" consisting of a single phage, usually with a broad host range for application against a single species, or a phage cocktail consisting of a mixture of phages (multiphage). Phage cocktails can consist of a mixture of phages targeting a single species or a broad range of pathogenic hosts [121]. There are pros and cons associated with monophage and multiphage therapeutics. For example, a cocktail targeting a single species requires proof of an etiological agent before selection and application of the cocktail, but represents the most specific treatment available. In contrast to this, a cocktail against a range of pathogenic hosts can be applied presumptively, however, it may have a negative effect on non-target bacteria at the site of application [121]. A second advantage of cocktails is in overcoming of resistance mechanisms in the target strains, where the target would have to develop resistance to all phages in the cocktail in order to survive. For example, a study on the application of a five-different phage-containing cocktail targeting *Klebsiella pneumoniae* isolated from infected burn wounds, observed a full log reduction in target load when compared to any of the phages individually. Furthermore, the cocktail had the lowest incidence of emergence of phage resistant variants [109]. Cocktail development can be optimized to this effect. As previously mentioned, a range of different phages capable of infecting a strain and phage resistant variants can be isolated from environmental samples using the SBS method by using wild-type and phage resistant variants as hosts allowing for the formation of an effective phage cocktail [13]. In addition to this, another method first applied to *Staphylococcus aureus* strains involves the serial passaging of an isolated phage of interest against its host strain of interest followed by phage-resistant variants to enrich for broad host-range phage mutants [108]. These two approaches both lead to the development of optimized phage cocktails capable of infecting strains of interest and potential phage resistant variants. In vitro analyses of phage cocktails should be undertaken to ensure the desired phenotype. This analysis usually takes the form of time-course killing experiments comparing single phages to cocktails, while also monitoring for the emergence of resistance variants [12,110]. Another noted concept in phage cocktails is that of phage synergy where the action of one phage augments the properties of a second phage in the cocktail [110]. This has been observed under in vitro conditions, where two phages infecting *E. coli* exhibited a 10-fold greater ability to lyse their host when applied together compared to either phage alone. The increased capacity for lysis was theorized to be due to the stripping of colonic acid from the host cell surface by one phage (J8-85) allowing increased access to receptors on the cell surface for the other (T7-61) [110]. However, in contrast to this, the opposite effect has also been noted, the prospect of viral interference, where the possibility remains for phages to interfere with each other following co-infection [6], where co-infection with two phages may reduce observed burst sizes.

5.5. Combination Therapy

A further development to the power of multiple bacteriophages, is the efficacy of multiple bactericidal agents. It is theorized that resistance to bacteriophages (and antibiotics) will evolve less frequently with a combination of phage and antibiotic therapy because a strain which is resistant to a phage will be inhibited by the antibiotic and vice versa, necessitating multiple independent mutations to overcome both [103]. This phenomenon has been investigated in vitro studying *Pseudomonas fluorescens* SBW25 and its infecting phage SBW25φ2, where combined treatment with lethal concentrations of kanamycin prevented resistance development in treated samples [104]. However caution must be exercised as the opposite effect has been noted in utilizing sub-lethal concentrations of streptomycin in combination with the same phage host combination [105]. Here, an increase in resistance development was observed as well as extinction of the phage [105], clearly suggesting that combination therapy requires a high antibiotic concentration.

There is significant promise in utilizing combination therapy in the eradication of biofilms where (as mentioned above) phage-encoded, exopolyscaccharide (EPS)/capsule-degrading depolymerase activity provides access for the antibiotic to target cells. This method has shown encouraging promise in the treatment of biofilms of *E. coli* [106] and *P. aeruginosa* [107]. The previously mentioned prevention of resistant variants has also been observed in the biofilm environment where a study of combined therapy on a *K. pneumoniae* biofilm noted that combined treatment with ciproflaxin resulted in a reduced emergence of resistant variants [122]. This suggests that combination therapy has a dual benefit in biofilm treatment, increasing the capacity of antibiotics to eradicate the biofilm and preventing emergence of resistance to both components of the treatment.

Combination with antibiotic can also lead to phage-antibiotic synergy (PAS), the tendency of phages to appear more virulent in the presence of sub-lethal antibiotic concentrations first observed in uropathogenic *E. coli* [123]. The potential for PAS can be easily ascertained in vitro through simple one step growth curves and in vitro biofilm eradication trials [106], however due to the potential for increased resistance, caution should be observed with use of sub lethal antibiotic concentrations. In addition to this, in all antibiotic combination therapy the potential for development of antibiotic resistance in non-target cells is a considerable risk, though it should not be considered to be more probable than the risk associated with antibiotic treatment alone.

6. Conclusions

Phage therapy has the potential to alleviate the ever-growing problem of antibiotic-resistance and the development of so-called "superbugs", either as an alternative to antibiotics, or in combination with traditional antibiotic therapies to enhance their effectiveness. Despite the regulatory concerns associated with phages as therapeutic agents, phage biologists have continued in the search for phages with therapeutic potential, resulting in the isolation of countless phages that could represent an endless arsenal against a range of human and animal pathogens. Furthermore, the development of phage products such as chimeric lysins and artilysins, highlights the reservoir of antimicrobial agents that may be harnessed from phages without the need for direct application of intact phages. However, we must endeavor to overcome regulatory concerns regarding the application of intact phages, since phages are ubiquitous and are innate residents of humans. While numerous studies have been performed regarding the isolation of novel phages and their characterization, a more unified approach to the assessment of phages may be required to ensure the ultimate success of phage therapy into the future. Here, through a review of desirable phage phenotypes that phage biologists may seek out when isolating phages, we propose a workflow for selection of candidates for therapeutic purposes (Figure 1). Firstly, in isolation of phages attempts should be made to target isolation towards broad or narrow host range phages (as desired) through implementation of various isolation methods (Figure 1(1.1)). The next step in the workflow should be genome sequencing (Figure 1(1.2)) representing one of the most important steps in candidate selection for several of reasons. Firstly, it will enable identification of putative phage-encoded lytic proteins for lytic enzyme therapy (Figure 1(1.3)), and secondly it will allow assessment of the genomic characteristics (both favorable and unfavorable) of candidates for implementation in phage particle therapy. Here, we can ensure the lytic nature of candidates through identification of proteins (in particular integrase/resolvase and repressor proteins) likely to be involved in the lysogenic lifestyle preventing inefficient lysis due to lysogenic conversion. Additionally, identification of receptor binding proteins, leading to the prediction of cell surface receptors, should guide informed selection of phages for generation of a phage cocktail targeting different cell surface moieties. Genome sequencing has other added advantages in the identification of other desirable such as depolymerase activity or the ability to overcome phage resistance as well as undesirable traits such as antibiotic resistance genes, pathogenicity islands, and genome instability. Sequencing at an early stage in candidate selection is advisable, as identification of these traits would render a phage unsuitable for use in therapy, thus rendering all other characterizations redundant. Following confirmation of suitability for phage therapy (Figure 1(1.4)), phages should be assessed for suitable in vitro characteristics as discussed above before selection for use in a therapeutic setting (perhaps first by means of an animal model prior to a human clinical trial). As we face the current antibiotic crisis, this workflow should prove to be useful in the isolation and identification of phage therapy candidates, which we hope will become a viable and widespread alternative to antibiotic therapy.

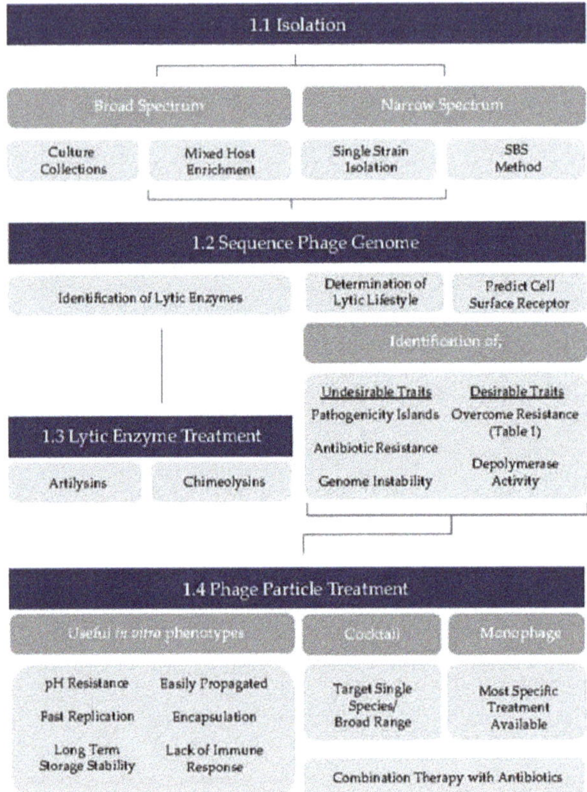

Figure 1. Suggested workflow for selection of phage therapy candidates, from isolation to implementation including desirable in vitro phenotypes.

Acknowledgments: This publication was supported in part by a research grant from Science Foundation Ireland (SFI) under Grant Number SFI/12/RC/2273. Eoghan Casey and Douwe van Sinderen are funded by the EU Joint Programming Initiative—A Healthy Diet for a Healthy Life (JPI HDHL, http://www.healthydietforhealthylife.eu/). Jennifer Mahony and Douwe van Sinderen are funded by the Bill and Melinda Gates Foundation under the Grand Challenges xplorations scheme (Ref. No. OPP1150567). Jennifer Mahony is the recipient of a Starting Investigator Research Grant funded by Science Foundation Ireland (SFI) (Ref. No. 15/SIRG/3430). Douwe van Sinderen is the recipient of an SFI Investigator award (Ref. No.13/IA/1953).

Author Contributions: Eoghan Casey, Douwe van Sinderen and Jennifer Mahony were involved in the design and layout of the review; Eoghan Casey and Jennifer Mahony prepared the manuscript; and Douwe van Sinderen was involved in reviewing and editing the manuscript.

Conflicts of Interest: The authors declare no conflicts of interest.

References

1. D'Hérelle, F. Sur un microbe invisible antagoniste des bacilles dysentérique. *Acad. Sci. Paris* **1917**, *165*, 373–375.
2. Twort, F. An investigation on the nature of ultra-microscopic viruses. *Lancet* **1915**, *186*, 1241–1243. [CrossRef]
3. Duckworth, D.H. Who discovered bacteriophage? *Bacteriol. Rev.* **1976**, *40*, 793. [PubMed]

4. Sabouri, S.; Sepehrizadeh, Z.; Amirpour-Rostami, S.; Skurnik, M. A minireview on the in vitro and in vivo experiments with anti-*Escherichia coli* O157:H7 phages as potential biocontrol and phage therapy agents. *Int. J. Food Microbiol.* **2017**, *243*, 52–57. [CrossRef] [PubMed]
5. Rose, T.; Verbeken, G.; De Vos, D.; Merabishvili, M.; Vaneechoutte, M.; Lavigne, R.; Jennes, S.; Zizi, M.; Pirnay, J.-P. Experimental phage therapy of burn wound infection: Difficult first steps. *Int. J. Burns Trauma* **2014**, *4*, 66. [PubMed]
6. Abedon, S.T.; Thomas-Abedon, C. Phage therapy pharmacology. *Curr. Pharm. Biotechnol.* **2010**, *11*, 28–47. [CrossRef] [PubMed]
7. Weber-Dąbrowska, B.; Jończyk-Matysiak, E.; Żaczek, M.; Łobocka, M.; Łusiak-Szelachowska, M.; Górski, A. Bacteriophage procurement for therapeutic purposes. *Front. Microbiol.* **2016**, *7*, 1177. [CrossRef] [PubMed]
8. Ochman, H.; Selander, R.K. Standard reference strains of *Escherichia coli* from natural populations. *J. Bacteriol.* **1984**, *157*, 690–693. [PubMed]
9. Beltran, P.; Plock, S.A.; Smith, N.H.; Whittam, T.S.; Old, D.C.; Selander, R.K. Reference collection of strains of the *Salmonella typhimurium* complex from natural populations. *Microbiology* **1991**, *137*, 601–606. [CrossRef] [PubMed]
10. Boyd, E.F.; Wang, F.-S.; Beltran, P.; Plock, S.A.; Nelson, K.; Selander, R.K. *Salmonella* reference collection B (SARB): Strains of 37 serovars of subspecies I. *Microbiology* **1993**, *139*, 1125–1132. [CrossRef] [PubMed]
11. Mirzaei, M.K.; Nilsson, A.S. Isolation of phages for phage therapy: A comparison of spot tests and efficiency of plating analyses for determination of host range and efficacy. *PLoS ONE* **2015**, *10*, e0118557. [CrossRef] [PubMed]
12. Nale, J.Y.; Spencer, J.; Hargreaves, K.R.; Buckley, A.M.; Trzepiński, P.; Douce, G.R.; Clokie, M.R. Bacteriophage combinations significantly reduce *Clostridium difficile* growth in vitro and proliferation in vivo. *Antimicrob. Agents Chemother.* **2016**, *60*, 968–981. [CrossRef] [PubMed]
13. Gu, J.; Liu, X.; Li, Y.; Han, W.; Lei, L.; Yang, Y.; Zhao, H.; Gao, Y.; Song, J.; Lu, R. A method for generation phage cocktail with great therapeutic potential. *PLoS ONE* **2012**, *7*, e31698. [CrossRef] [PubMed]
14. Jensen, E.C.; Schrader, H.S.; Rieland, B.; Thompson, T.L.; Lee, K.W.; Nickerson, K.W.; Kokjohn, T.A. Prevalence of Broad-Host-Range Lytic Bacteriophages of *Sphaerotilus natans*, *Escherichia coli*, and *Pseudomonas aeruginosa*. *Appl. Environ. Microbiol.* **1998**, *64*, 575–580. [PubMed]
15. Lin, D.M.; Koskella, B.; Lin, H.C. Phage therapy: An alternative to antibiotics in the age of multi-drug resistance. *World J. Gastrointest. Pharmacol. Ther.* **2017**, *8*, 162. [CrossRef] [PubMed]
16. Ackermann, H.-W.; Tremblay, D.; Moineau, S. Long-term bacteriophage preservation. *WFCC Newsl.* **2004**, *38*, 35–40.
17. Golec, P.; Dąbrowski, K.; Hejnowicz, M.S.; Gozdek, A.; Łoś, J.M.; Węgrzyn, G.; Łobocka, M.B.; Łoś, M. A reliable method for storage of tailed phages. *J. Microbiol. Methods* **2011**, *84*, 486–489. [CrossRef] [PubMed]
18. Fortier, L.-C.; Moineau, S. Phage production and maintenance of stocks, including expected stock lifetimes. *Bacteriophages* **2009**, 203–219.
19. Bonilla, N.; Rojas, M.I.; Cruz, G.N.F.; Hung, S.-H.; Rohwer, F.; Barr, J.J. Phage on tap—A quick and efficient protocol for the preparation of bacteriophage laboratory stocks. *PeerJ* **2016**, *4*, e2261. [CrossRef] [PubMed]
20. Pirnay, J.-P.; Blasdel, B.G.; Bretaudeau, L.; Buckling, A.; Chanishvili, N.; Clark, J.R.; Corte-Real, S.; Debarbieux, L.; Dublanchet, A.; De Vos, D. Quality and safety requirements for sustainable phage therapy products. *Pharm. Res.* **2015**, *32*, 2173–2179. [CrossRef] [PubMed]
21. Dalmasso, M.; Strain, R.; Neve, H.; Franz, C.M.; Cousin, F.J.; Ross, R.P.; Hill, C. Three new *Escherichia coli* phages from the human gut show promising potential for phage therapy. *PLoS ONE* **2016**, *11*, e0156773. [CrossRef] [PubMed]
22. Hoyles, L.; Murphy, J.; Neve, H.; Heller, K.J.; Turton, J.F.; Mahony, J.; Sanderson, J.D.; Hudspith, B.; Gibson, G.R.; McCartney, A.L.; et al. *Klebsiella pneumoniae* subsp. *pneumoniae*-bacteriophage combination from the caecal effluent of a healthy woman. *PeerJ* **2015**, *3*, e1061. [PubMed]
23. Pallavali, R.R.; Degati, V.L.; Lomada, D.; Reddy, M.C.; Durbaka, V.R.P. Isolation and in vitro evaluation of bacteriophages against MDR-bacterial isolates from septic wound infections. *PLoS ONE* **2017**, *12*, e0179245. [CrossRef] [PubMed]
24. Rashid, S.J.; Barylski, J.; Hargreaves, K.R.; Millard, A.A.; Vinner, G.K.; Clokie, M.R. Two novel Myoviruses from the north of Iraq reveal insights into *Clostridium difficile* phage diversity and biology. *Viruses* **2016**, *8*, 310. [CrossRef] [PubMed]

25. Zhang, G.; Zhao, Y.; Paramasivan, S.; Richter, K.; Morales, S.; Wormald, P.J.; Vreugde, S. Bacteriophage effectively kills multidrug resistant *Staphylococcus aureus* clinical isolates from chronic rhinosinusitis patients. *Int. Forum Allergy Rhinol.* **2018**, *8*, 406–414. [CrossRef] [PubMed]
26. Baig, A.; Colom, J.; Barrow, P.; Schouler, C.; Moodley, A.; Lavigne, R.; Atterbury, R. Biology and genomics of an historic therapeutic *Escherichia coli* bacteriophage collection. *Front. Microbiol.* **2017**, *8*, 1652. [CrossRef] [PubMed]
27. Finn, R.D.; Coggill, P.; Eberhardt, R.Y.; Eddy, S.R.; Mistry, J.; Mitchell, A.L.; Potter, S.C.; Punta, M.; Qureshi, M.; Sangrador-Vegas, A.; et al. The Pfam protein families database: Towards a more sustainable future. *Nucleic Acids Res.* **2016**, *44*, D279–D285. [CrossRef] [PubMed]
28. Soding, J.; Biegert, A.; Lupas, A.N. The HHpred interactive server for protein homology detection and structure prediction. *Nucleic Acids Res.* **2005**, *33*, W244–W248. [CrossRef] [PubMed]
29. Chatterjee, S.; Rothenberg, E. Interaction of bacteriophage l with its *E. coli* receptor, LamB. *Viruses* **2012**, *4*, 3162–3178. [CrossRef] [PubMed]
30. Hu, B.; Margolin, W.; Molineux, I.J.; Liu, J. The bacteriophage T7 virion undergoes extensive structural remodeling during infection. *Science* **2013**, *339*, 576–579. [CrossRef] [PubMed]
31. Baptista, C.; Santos, M.A.; Sao-Jose, C. Phage SPP1 reversible adsorption to *Bacillus subtilis* cell wall teichoic acids accelerates virus recognition of membrane receptor YueB. *J. Bacteriol.* **2008**, *190*, 4989–4996. [CrossRef] [PubMed]
32. Collins, B.; Bebeacua, C.; Mahony, J.; Blangy, S.; Douillard, F.P.; Veesler, D.; Cambillau, C.; van Sinderen, D. Structure and functional analysis of the host recognition device of lactococcal phage Tuc2009. *J. Virol.* **2013**, *87*, 8429–8440. [CrossRef] [PubMed]
33. Legrand, P.; Collins, B.; Blangy, S.; Murphy, J.; Spinelli, S.; Gutierrez, C.; Richet, N.; Kellenberger, C.; Desmyter, A.; Mahony, J.; et al. The Atomic Structure of the Phage Tuc2009 Baseplate Tripod Suggests that Host Recognition Involves Two Different Carbohydrate Binding Modules. *MBio* **2016**, *7*, e01781-15. [CrossRef] [PubMed]
34. Mahony, J.; Kot, W.; Murphy, J.; Ainsworth, S.; Neve, H.; Hansen, L.H.; Heller, K.J.; Sorensen, S.J.; Hammer, K.; Cambillau, C.; et al. Investigation of the relationship between lactococcal host cell wall polysaccharide genotype and 936 phage receptor binding protein phylogeny. *Appl. Environ. Microbiol.* **2013**, *79*, 4385–4392. [CrossRef] [PubMed]
35. Veesler, D.; Blangy, S.; Spinelli, S.; Tavares, P.; Campanacci, V.; Cambillau, C. Crystal structure of Bacillus subtilis SPP1 phage gp22 shares fold similarity with a domain of lactococcal phage p2 RBP. *Protein Sci.* **2010**, *19*, 1439–1443. [CrossRef] [PubMed]
36. Koc, C.; Xia, G.; Kuhner, P.; Spinelli, S.; Roussel, A.; Cambillau, C.; Stehle, T. Structure of the host-recognition device of *Staphylococcus aureus* phage varphi11. *Sci. Rep.* **2016**, *6*, 27581. [CrossRef] [PubMed]
37. Xia, G.; Maier, L.; Sanchez-Carballo, P.; Li, M.; Otto, M.; Holst, O.; Peschel, A. Glycosylation of wall teichoic acid in *Staphylococcus aureus* by TarM. *J. Biol. Chem.* **2010**, *285*, 13405–13415. [CrossRef] [PubMed]
38. Denes, T.; den Bakker, H.C.; Tokman, J.I.; Guldimann, C.; Wiedmann, M. Selection and characterization of phage-resistant mutant strains of *Listeria monocytogenes* reveal host genes linked to phage adsorption. *Appl. Environ. Microbiol.* **2015**, *81*, 4295–4305. [CrossRef] [PubMed]
39. Dupont, K.; Janzen, T.; Vogensen, F.K.; Josephsen, J.; Stuer-Lauridsen, B. Identification of *Lactococcus lactis* genes required for bacteriophage adsorption. *Appl. Environ. Microbiol.* **2004**, *70*, 5825–5832. [CrossRef] [PubMed]
40. Tremblay, D.M.; Tegoni, M.; Spinelli, S.; Campanacci, V.; Blangy, S.; Huyghe, C.; Desmyter, A.; Labrie, S.; Moineau, S.; Cambillau, C. Receptor-binding protein of *Lactococcus lactis* phages: Identification and characterization of the saccharide receptor-binding site. *J. Bacteriol.* **2006**, *188*, 2400–2410. [CrossRef] [PubMed]
41. Blower, T.R.; Chai, R.; Przybilski, R.; Chindhy, S.; Fang, X.; Kidman, S.E.; Tan, H.; Luisi, B.F.; Fineran, P.C.; Salmond, G.P. Evolution of *Pectobacterium* bacteriophage PhiM1 to escape two bifunctional Type III toxin-antitoxin and abortive infection systems through mutations in a single viral gene. *Appl. Environ. Microbiol.* **2017**, *83*, e03229-16. [CrossRef] [PubMed]
42. Bertozzi Silva, J.; Storms, Z.; Sauvageau, D. Host receptors for bacteriophage adsorption. *FEMS Microbiol. Lett.* **2016**, *363*, fnw002. [CrossRef] [PubMed]
43. Casjens, S.R.; Hendrix, R.W. Bacteriophage lambda: Early pioneer and still relevant. *Virology* **2015**, *479–480*, 310–330. [CrossRef] [PubMed]

44. Yap, M.L.; Rossmann, M.G. Structure and function of bacteriophage T4. *Future Microbiol.* **2014**, *9*, 1319–1327. [CrossRef] [PubMed]
45. Montag, D.; Riede, I.; Eschbach, M.L.; Degen, M.; Henning, U. Receptor-recognizing proteins of T-even type bacteriophages. Constant and hypervariable regions and an unusual case of evolution. *J. Mol. Biol.* **1987**, *196*, 165–174. [CrossRef]
46. Riede, I. Receptor specificity of the short tail fibres (gp12) of T-even type *Escherichia coli* phages. *Mol. Gen. Genet. MGG* **1987**, *206*, 110–115. [CrossRef] [PubMed]
47. Mahony, J.; van Sinderen, D. Gram-positive phage-host interactions. *Front. Microbiol.* **2015**, *6*, 61. [CrossRef] [PubMed]
48. Mahony, J.; Oliveira, J.; Collins, B.; Hanemaaijer, L.; Lugli, G.A.; Neve, H.; Ventura, M.; Kouwen, T.R.; Cambillau, C.; van Sinderen, D. Genetic and functional characterisation of the lactococcal P335 phage-host interactions. *BMC Genomics* **2017**, *18*, 146. [CrossRef] [PubMed]
49. Ainsworth, S.; Sadovskaya, I.; Vinogradov, E.; Courtin, P.; Guerardel, Y.; Mahony, J.; Grard, T.; Cambillau, C.; Chapot-Chartier, M.P.; van Sinderen, D. Differences in lactococcal cell wall polysaccharide structure are major determining factors in bacteriophage sensitivity. *MBio* **2014**, *5*, e00880-14. [CrossRef] [PubMed]
50. Dieterle, M.E.; Spinelli, S.; Sadovskaya, I.; Piuri, M.; Cambillau, C. Evolved distal tail carbohydrate binding modules of *Lactobacillus* phage J-1: A novel type of anti-receptor widespread among lactic acid bacteria phages. *Mol. Microbiol.* **2017**, *104*, 608–620. [CrossRef] [PubMed]
51. Koo, J.; DePaola, A.; Marshall, D.L. Effect of simulated gastric fluid and bile on survival of *Vibrio vulnificus* and *Vibrio vulnificus* phage. *J. Food Prot.* **2000**, *63*, 1665–1669. [CrossRef] [PubMed]
52. O'Flynn, G.; Coffey, A.; Fitzgerald, G.F.; Ross, R.P. The newly isolated lytic bacteriophages st104a and st104b are highly virulent against *Salmonella enterica*. *J. Appl. Microbiol.* **2006**, *101*, 251–259. [CrossRef] [PubMed]
53. Lang, R.; Winter, G.; Vogt, L.; Zurcher, A.; Dorigo, B.; Schimmele, B. Rational design of a stable, freeze-dried virus-like particle-based vaccine formulation. *Drug Dev. Ind. Pharm.* **2009**, *35*, 83–97. [CrossRef] [PubMed]
54. Vandenheuvel, D.; Singh, A.; Vandersteegen, K.; Klumpp, J.; Lavigne, R.; Van den Mooter, G. Feasibility of spray drying bacteriophages into respirable powders to combat pulmonary bacterial infections. *Eur. J. Pharm. Biopharm.* **2013**, *84*, 578–582. [CrossRef] [PubMed]
55. Bourdin, G.; Schmitt, B.; Marvin Guy, L.; Germond, J.E.; Zuber, S.; Michot, L.; Reuteler, G.; Brussow, H. Amplification and purification of T4-like *Escherichia coli* phages for phage therapy: From laboratory to pilot scale. *Appl. Environ. Microbiol.* **2014**, *80*, 1469–1476. [CrossRef] [PubMed]
56. Dini, C.; Islan, G.A.; de Urraza, P.J.; Castro, G.R. Novel biopolymer matrices for microencapsulation of phages: Enhanced protection against acidity and protease activity. *Macromol. Biosci.* **2012**, *12*, 1200–1208. [CrossRef] [PubMed]
57. Vinner, G.K.; Vladisavljevic, G.T.; Clokie, M.R.J.; Malik, D.J. Microencapsulation of *Clostridium difficile* specific bacteriophages using microfluidic glass capillary devices for colon delivery using pH triggered release. *PLoS ONE* **2017**, *12*, e0186239. [CrossRef] [PubMed]
58. Semler, D.D.; Lynch, K.H.; Dennis, J.J. The promise of bacteriophage therapy for *Burkholderia cepacia* complex respiratory infections. *Front. Cell. Infect. Microbiol.* **2011**, *1*, 27. [CrossRef] [PubMed]
59. Kropinski, A.M.; Lingohr, E.J.; Moyles, D.M.; Ojha, S.; Mazzocco, A.; She, Y.M.; Bach, S.J.; Rozema, E.A.; Stanford, K.; McAllister, T.A.; et al. Endemic bacteriophages: A cautionary tale for evaluation of bacteriophage therapy and other interventions for infection control in animals. *Virol. J.* **2012**, *9*, 207. [CrossRef] [PubMed]
60. Nale, J.Y.; Chutia, M.; Carr, P.; Hickenbotham, P.T.; Clokie, M.R. 'Get in Early'; Biofilm and Wax Moth (*Galleria mellonella*) Models Reveal New Insights into the Therapeutic Potential of *Clostridium difficile* Bacteriophages. *Front. Microbiol.* **2016**, *7*, 1383. [CrossRef] [PubMed]
61. Majewska, J.; Beta, W.; Lecion, D.; Hodyra-Stefaniak, K.; Kłopot, A.; Kaźmierczak, Z.; Miernikiewicz, P.; Piotrowicz, A.; Ciekot, J.; Owczarek, B. Oral application of T4 phage induces weak antibody production in the gut and in the blood. *Viruses* **2015**, *7*, 4783–4799. [CrossRef] [PubMed]
62. Abedon, S.T.; Kuhl, S.J.; Blasdel, B.G.; Kutter, E.M. Phage treatment of human infections. *Bacteriophage* **2011**, *1*, 66–85. [CrossRef] [PubMed]
63. Górski, A.; Kniotek, M.; Perkowska-Ptasinska, A.; Mróz, A.; Przerwa, A.; Gorczyca, W.; Dabrowska, K.; Weber-Dabrowska, B.; Nowaczyk, M. Bacteriophages and transplantation tolerance. *Transplant. Proc.* **2006**, *38*, 331–333. [CrossRef] [PubMed]

64. Górski, A.; Międzybrodzki, R.; Borysowski, J.; Dąbrowska, K.; Wierzbicki, P.; Ohams, M.; Korczak-Kowalska, G.; Olszowska-Zaremba, N.; Łusiak-Szelachowska, M.; Kłak, M. Phage as a modulator of immune responses: Practical implications for phage therapy. *Adv. Virus Res.* **2012**, *83*, 41–71. [PubMed]
65. Loessner, M.J. Bacteriophage endolysins—Current state of research and applications. *Curr. Opin. Microbiol.* **2005**, *8*, 480–487. [CrossRef] [PubMed]
66. Nelson, D.; Loomis, L.; Fischetti, V.A. Prevention and elimination of upper respiratory colonization of mice by group A streptococci by using a bacteriophage lytic enzyme. *Proc. Natl. Acad. Sci. USA* **2001**, *98*, 4107–4112. [CrossRef] [PubMed]
67. Zhang, L.; Li, D.; Li, X.; Hu, L.; Cheng, M.; Xia, F.; Gong, P.; Wang, B.; Ge, J.; Zhang, H. LysGH15 kills *Staphylococcus aureus* without being affected by the humoral immune response or inducing inflammation. *Sci. Rep.* **2016**, *6*, 29344. [CrossRef]
68. Viertel, T.M.; Ritter, K.; Horz, H.-P. Viruses versus bacteria—Novel approaches to phage therapy as a tool against multidrug-resistant pathogens. *J. Antimicrob. Chemother.* **2014**, *69*, 2326–2336. [CrossRef] [PubMed]
69. Wang, Q.; Euler, C.W.; Delaune, A.; Fischetti, V.A. Using a novel lysin to help control *Clostridium difficile* infections. *Antimicrob. Agents Chemother.* **2015**, *59*, 7447–7457. [CrossRef] [PubMed]
70. Oliveira, H.; Vilas Boas, D.; Mesnage, S.; Kluskens, L.D.; Lavigne, R.; Sillankorva, S.; Secundo, F.; Azeredo, J. Structural and enzymatic characterization of ABgp46, a novel phage endolysin with broad anti-Gram-negative bacterial activity. *Front. Microbiol.* **2016**, *7*, 208. [CrossRef] [PubMed]
71. Schmelcher, M.; Shen, Y.; Nelson, D.C.; Eugster, M.R.; Eichenseher, F.; Hanke, D.C.; Loessner, M.J.; Dong, S.; Pritchard, D.G.; Lee, J.C. Evolutionarily distinct bacteriophage endolysins featuring conserved peptidoglycan cleavage sites protect mice from MRSA infection. *J. Antimicrob. Chemother.* **2015**, *70*, 1453–1465. [CrossRef] [PubMed]
72. Schmelcher, M.; Donovan, D.M.; Loessner, M.J. Bacteriophage endolysins as novel antimicrobials. *Future Microbiol.* **2012**, *7*, 1147–1171. [CrossRef] [PubMed]
73. Yang, H.; Zhang, H.; Wang, J.; Yu, J.; Wei, H. A novel chimeric lysin with robust antibacterial activity against planktonic and biofilm methicillin-resistant *Staphylococcus aureus*. *Sci. Rep.* **2017**, *7*, 40182. [CrossRef] [PubMed]
74. Dong, Q.; Wang, J.; Yang, H.; Wei, C.; Yu, J.; Zhang, Y.; Huang, Y.; Zhang, X.E.; Wei, H. Construction of a chimeric lysin Ply187N-V12C with extended lytic activity against *staphylococci* and *streptococci*. *Microb. Biotechnol.* **2015**, *8*, 210–220. [CrossRef] [PubMed]
75. Mao, J.; Schmelcher, M.; Harty, W.J.; Foster-Frey, J.; Donovan, D.M. Chimeric Ply187 endolysin kills *Staphylococcus aureus* more effectively than the parental enzyme. *FEMS Microbiol. Lett.* **2013**, *342*, 30–36. [CrossRef] [PubMed]
76. Gerstmans, H.; Rodríguez-Rubio, L.; Lavigne, R.; Briers, Y. From endolysins to Artilysin®s: Novel enzyme-based approaches to kill drug-resistant bacteria. *Biochem. Soc. Trans.* **2016**, *44*, 123–128. [CrossRef] [PubMed]
77. Briers, Y.; Walmagh, M.; Grymonprez, B.; Biebl, M.; Pirnay, J.-P.; Defraine, V.; Michiels, J.; Cenens, W.; Aertsen, A.; Miller, S. Art-175 is a highly efficient antibacterial against multidrug-resistant strains and persisters of *Pseudomonas aeruginosa*. *Antimicrob. Agents Chemother.* **2014**, *58*, 3774–3784. [CrossRef] [PubMed]
78. Defraine, V.; Schuermans, J.; Grymonprez, B.; Govers, S.K.; Aertsen, A.; Fauvart, M.; Michiels, J.; Lavigne, R.; Briers, Y. Efficacy of artilysin Art-175 against resistant and persistent *Acinetobacter baumannii*. *Antimicrob. Agents Chemother.* **2016**, *60*, 3480–3488. [CrossRef] [PubMed]
79. Høiby, N.; Bjarnsholt, T.; Givskov, M.; Molin, S.; Ciofu, O. Antibiotic resistance of bacterial biofilms. *Int. J. Antimicrob. Agents* **2010**, *35*, 322–332. [CrossRef] [PubMed]
80. Flemming, H.-C.; Wingender, J. The biofilm matrix. *Nat. Rev. Microbiol.* **2010**, *8*, 623. [CrossRef] [PubMed]
81. Cornelissen, A.; Ceyssens, P.-J.; Krylov, V.N.; Noben, J.-P.; Volckaert, G.; Lavigne, R. Identification of EPS-degrading activity within the tail spikes of the novel Pseudomonas putida phage AF. *Virology* **2012**, *434*, 251–256. [CrossRef] [PubMed]
82. Majkowska-Skrobek, G.; Łątka, A.; Berisio, R.; Maciejewska, B.; Squeglia, F.; Romano, M.; Lavigne, R.; Struve, C.; Drulis-Kawa, Z. Capsule-targeting depolymerase, derived from *Klebsiella* KP36 phage, as a tool for the development of anti-virulent strategy. *Viruses* **2016**, *8*, 324. [CrossRef] [PubMed]

83. Pires, D.P.; Melo, L.D.; Boas, D.V.; Sillankorva, S.; Azeredo, J. Phage therapy as an alternative or complementary strategy to prevent and control biofilm-related infections. *Curr. Opin. Microbiol.* **2017**, *39*, 48–56. [CrossRef] [PubMed]
84. Kostakioti, M.; Hadjifrangiskou, M.; Hultgren, S.J. Bacterial biofilms: Development, dispersal, and therapeutic strategies in the dawn of the postantibiotic era. *Cold Spring Harb. Perspect. Med.* **2013**, *3*, a010306. [CrossRef] [PubMed]
85. Hsu, C.-R.; Lin, T.-L.; Pan, Y.-J.; Hsieh, P.-F.; Wang, J.-T. Isolation of a bacteriophage specific for a new capsular type of *Klebsiella pneumoniae* and characterization of its polysaccharide depolymerase. *PLoS ONE* **2013**, *8*, e70092. [CrossRef] [PubMed]
86. Gutiérrez, D.; Martínez, B.; Rodríguez, A.; García, P. Genomic characterization of two *Staphylococcus epidermidis* bacteriophages with anti-biofilm potential. *BMC Genomics* **2012**, *13*, 228. [CrossRef] [PubMed]
87. Guo, Z.; Huang, J.; Yan, G.; Lei, L.; Wang, S.; Yu, L.; Zhou, L.; Gao, A.; Feng, X.; Han, W. Identification and Characterization of Dpo42, a Novel Depolymerase Derived from the *Escherichia coli* Phage vB_EcoM_ECOO78. *Front. Microbiol.* **2017**, *8*, 1460. [CrossRef] [PubMed]
88. Labrie, S.J.; Samson, J.E.; Moineau, S. Bacteriophage resistance mechanisms. *Nat. Rev. Microbiol.* **2010**, *8*, 317. [CrossRef] [PubMed]
89. Murphy, J.; Mahony, J.; Ainsworth, S.; Nauta, A.; van Sinderen, D. Bacteriophage orphan DNA methyltransferases: Insights from their bacterial origin, function, and occurrence. *Appl. Environ. Microbiol.* **2013**, *79*, 7547–7555. [CrossRef] [PubMed]
90. Loenen, W.A.; Murray, N.E. Modification enhancement by the restriction alleviation protein (Real) of bacteriophage λ. *J. Mol. Biol.* **1986**, *190*, 11–22. [CrossRef]
91. King, G.; Murray, N.E. Restriction alleviation and modification enhancement by the Rac prophage of Escherichia coli K-12. *Mol. Microbiol.* **1995**, *16*, 769–777. [CrossRef] [PubMed]
92. Kaplan, D.A.; Nierlich, D. Cleavage of Nonglucosylated Bacteriophage T4 deoxyribonucleic acid by Restriction Endonuclease Eco RI. *J. Biol. Chem.* **1975**, *250*, 2395–2397. [PubMed]
93. Borgaro, J.G.; Zhu, Z. Characterization of the 5-hydroxymethylcytosine-specific DNA restriction endonucleases. *Nucleic Acids Res.* **2013**, *41*, 4198–4206. [CrossRef] [PubMed]
94. Levin, B.R.; Moineau, S.; Bushman, M.; Barrangou, R. The population and evolutionary dynamics of phage and bacteria with CRISPR–mediated immunity. *PLoS Genet.* **2013**, *9*, e1003312. [CrossRef] [PubMed]
95. Strotskaya, A.; Savitskaya, E.; Metlitskaya, A.; Morozova, N.; Datsenko, K.A.; Semenova, E.; Severinov, K. The action of *Escherichia coli* CRISPR–Cas system on lytic bacteriophages with different lifestyles and development strategies. *Nucleic Acids Res.* **2017**, *45*, 1946–1957. [PubMed]
96. Paez-Espino, D.; Sharon, I.; Morovic, W.; Stahl, B.; Thomas, B.C.; Barrangou, R.; Banfield, J.F. CRISPR immunity drives rapid phage genome evolution in *Streptococcus thermophilus*. *MBio* **2015**, *6*, e00262-15. [CrossRef] [PubMed]
97. Maxwell, K.L. Phages fight back: Inactivation of the CRISPR-Cas bacterial immune system by anti-CRISPR proteins. *PLoS Pathog.* **2016**, *12*, e1005282. [CrossRef] [PubMed]
98. Bondy-Denomy, J.; Garcia, B.; Strum, S.; Du, M.; Rollins, M.F.; Hidalgo-Reyes, Y.; Wiedenheft, B.; Maxwell, K.L.; Davidson, A.R. Multiple mechanisms for CRISPR–Cas inhibition by anti-CRISPR proteins. *Nature* **2015**, *526*, 136. [CrossRef] [PubMed]
99. Rauch, B.J.; Silvis, M.R.; Hultquist, J.F.; Waters, C.S.; McGregor, M.J.; Krogan, N.J.; Bondy-Denomy, J. Inhibition of CRISPR-Cas9 with bacteriophage proteins. *Cell* **2017**, *168*, 150–158. [CrossRef] [PubMed]
100. Pawluk, A.; Staals, R.H.; Taylor, C.; Watson, B.N.; Saha, S.; Fineran, P.C.; Maxwell, K.L.; Davidson, A.R. Inactivation of CRISPR-Cas systems by anti-CRISPR proteins in diverse bacterial species. *Nat. Microbiol.* **2016**, *1*, 16085. [CrossRef] [PubMed]
101. Meyer, J.R.; Dobias, D.T.; Weitz, J.S.; Barrick, J.E.; Quick, R.T.; Lenski, R.E. Repeatability and contingency in the evolution of a key innovation in phage lambda. *Science* **2012**, *335*, 428–432. [CrossRef] [PubMed]
102. Goodridge, L.D. Designing phage therapeutics. *Curr. Pharm. Biotechnol.* **2010**, *11*, 15–27. [CrossRef] [PubMed]
103. Burrowes, B.; Harper, D.R.; Anderson, J.; McConville, M.; Enright, M.C. Bacteriophage therapy: Potential uses in the control of antibiotic-resistant pathogens. *Expert Rev. Anti-Infect. Ther.* **2011**, *9*, 775–785. [CrossRef] [PubMed]
104. Zhang, Q.G.; Buckling, A. Phages limit the evolution of bacterial antibiotic resistance in experimental microcosms. *Evol. Appl.* **2012**, *5*, 575–582. [CrossRef] [PubMed]

105. Cairns, J.; Becks, L.; Jalasvuori, M.; Hiltunen, T. Sublethal streptomycin concentrations and lytic bacteriophage together promote resistance evolution. *Philos. Trans. R. Soc. B* **2017**, *372*, 20160040. [CrossRef] [PubMed]
106. Ryan, E.M.; Alkawareek, M.Y.; Donnelly, R.F.; Gilmore, B.F. Synergistic phage-antibiotic combinations for the control of *Escherichia coli* biofilms in vitro. *Pathog. Dis.* **2012**, *65*, 395–398.
107. Knezevic, P.; Curcin, S.; Aleksic, V.; Petrusic, M.; Vlaski, L. Phage-antibiotic synergism: A possible approach to combatting *Pseudomonas aeruginosa*. *Res. Microbiol.* **2013**, *164*, 55–60. [CrossRef] [PubMed]
108. Kelly, D.; McAuliffe, O.; O'Mahony, J.; Coffey, A. Development of a broad-host-range phage cocktail for biocontrol. *Bioeng. Bugs* **2011**, *2*, 31–37. [CrossRef] [PubMed]
109. Chadha, P.; Katare, O.P.; Chhibber, S. In vivo efficacy of single phage versus phage cocktail in resolving burn wound infection in BALB/c mice. *Microb. Pathog.* **2016**, *99*, 68–77. [CrossRef] [PubMed]
110. Schmerer, M.; Molineux, I.J.; Bull, J.J. Synergy as a rationale for phage therapy using phage cocktails. *PeerJ* **2014**, *2*, e590. [CrossRef] [PubMed]
111. Mohapatra, S.S.; Fioravanti, A.; Biondi, E.G. DNA methylation in Caulobacter and other Alphaproteobacteria during cell cycle progression. *Trends Microbiol.* **2014**, *22*, 528–535. [CrossRef] [PubMed]
112. Marinus, M.G.; Casadesus, J. Roles of DNA adenine methylation in host–pathogen interactions: Mismatch repair, transcriptional regulation, and more. *FEMS Microbiol. Rev.* **2009**, *33*, 488–503. [CrossRef] [PubMed]
113. Gunthert, U.; Reiners, L. *Bacillus subtilis* phage SPR codes for a DNA methyttransferase with triple sequence specificity. *Nucleic Acids Res.* **1987**, *15*, 3689–3702. [CrossRef] [PubMed]
114. Wilke, K.; Rauhut, E.; Noyer-Weidner, M.; Lauster, R.; Pawlek, B.; Behrens, B.; Trautner, T. Sequential order of target-recognizing domains in multispecific DNA-methyltransferases. *EMBO J.* **1988**, *7*, 2601–2609. [PubMed]
115. Decewicz, P.; Radlinska, M.; Dziewit, L. Characterization of *Sinorhizobium* sp. LM21 Prophages and Virus-Encoded DNA Methyltransferases in the Light of Comparative Genomic Analyses of the Sinorhizobial Virome. *Viruses* **2017**, *9*, 161. [CrossRef] [PubMed]
116. Makarova, K.S.; Wolf, Y.I.; Alkhnbashi, O.S.; Costa, F.; Shah, S.A.; Saunders, S.J.; Barrangou, R.; Brouns, S.J.; Charpentier, E.; Haft, D.H. An updated evolutionary classification of CRISPR–Cas systems. *Nat. Rev. Microbiol.* **2015**, *13*, 722. [CrossRef] [PubMed]
117. Bondy-Denomy, J.; Pawluk, A.; Maxwell, K.L.; Davidson, A.R. Bacteriophage genes that inactivate the CRISPR/Cas bacterial immune system. *Nature* **2013**, *493*, 429. [CrossRef] [PubMed]
118. Wiedenheft, B.; van Duijn, E.; Bultema, J.B.; Waghmare, S.P.; Zhou, K.; Barendregt, A.; Westphal, W.; Heck, A.J.; Boekema, E.J.; Dickman, M.J. RNA-guided complex from a bacterial immune system enhances target recognition through seed sequence interactions. *Proc. Natl. Acad. Sci. USA* **2011**, *108*, 10092–10097. [CrossRef] [PubMed]
119. Buckling, A.; Rainey, P.B. Antagonistic coevolution between a bacterium and a bacteriophage. *Proc. R. Soc. Lond. B Biol. Sci.* **2002**, *269*, 931–936. [CrossRef] [PubMed]
120. Paterson, S.; Vogwill, T.; Buckling, A.; Benmayor, R.; Spiers, A.J.; Thomson, N.R.; Quail, M.; Smith, F.; Walker, D.; Libberton, B. Antagonistic coevolution accelerates molecular evolution. *Nature* **2010**, *464*, 275. [CrossRef] [PubMed]
121. Chan, B.K.; Abedon, S.T.; Loc-Carrillo, C. Phage cocktails and the future of phage therapy. *Future Microbiol.* **2013**, *8*, 769–783. [CrossRef] [PubMed]
122. Verma, V.; Harjai, K.; Chhibber, S. Restricting ciprofloxacin-induced resistant variant formation in biofilm of Klebsiella pneumoniae B5055 by complementary bacteriophage treatment. *J. Antimicrob. Chemother.* **2009**, *64*, 1212–1218. [CrossRef] [PubMed]
123. Comeau, A.M.; Tétart, F.; Trojet, S.N.; Prere, M.-F.; Krisch, H. Phage-antibiotic synergy (PAS): β-lactam and quinolone antibiotics stimulate virulent phage growth. *PLoS ONE* **2007**, *2*, e799. [CrossRef] [PubMed]

Opinion

Phage Therapy Regulation: From Night to Dawn

Alan Fauconnier

Culture In Vivo ASBL, rue du Progrès, 4, boîte 7, 1400 Nivelles, Belgium; alan.fauconnier@invivo.be

Received: 28 February 2019; Accepted: 7 April 2019; Published: 17 April 2019

Abstract: After decades of disregard in the Western world, phage therapy is witnessing a return of interest. However, the pharmaceutical legislation that has since been implemented is basically designed for regulating industrially-made pharmaceuticals, devoid of any patient customization and intended for large-scale distribution. Accordingly, the resulting regulatory framework is hardly reconcilable with the concept of sustainable phage therapy, involving tailor-made medicinal products in the global perspective of both evolutionary and personalized medicine. The repeated appeal for a dedicated regulatory framework has not been heard by the European legislature, which, in this matter, features a strong resistance to change despite the precedent of the unhindered implementation of advanced therapy medicinal product (ATMPs) regulation. It is acknowledged that in many aspects, phage therapy medicinal products are quite unconventional pharmaceuticals and likely this lack of conformity to the canonical model hampered the development of a suitable regulatory pathway. However, the regulatory approaches of countries where phage therapy traditions and practice have never been abandoned are now being revisited by some Western countries, opening new avenues for phage therapy regulation. As a next step, supranational and international organizations are urged to take over the initiatives originally launched by national regulatory authorities.

Keywords: phage therapy; PTMP; ATMP; regulatory framework; pharmaceutical paradigm shift; clinical trial; magistral formula; personalized medicine

---

The idea of using bacteriophages to cure patients originally emerged in d'Hérelle's mind one hundred years ago [1]. However, in the middle of the last century, the introduction of antibiotics led to the banishment of phage therapy from mainstream medical practice, whereas it remained in use in Eastern Europe, for instance in several institutes in Russia, in the Eliava Institute of Bacteriophage, Microbiology and Virology in Tbilisi (Georgia) and in the Hirszfeld Institute in Wroclaw (Poland). After decades of neglect in the Western world, phage therapy has witnessed a remarkable return to interest, as evidenced by the profile of PubMed search results, which features an increase in the late nineties to the start of the millennium [2]. This renewed interest is essentially due to the growing incidence of antibiotic resistance. However, this re-emerging therapy now faces the regulation that has been implemented since the days of d'Hérelle, entailing serious difficulties. Modern pharmaceutical legislation has been pointed out as a hindrance to phage therapy implementation [3] and has been consistently blamed for obstructing its deployment [4–7]. The regulatory issue impacts not only the market placement but also the conduct of clinical trials.

For these reasons, the phage community has called for a switch in the mindset of regulatory agencies [7] or even in the pharmaceutical paradigm [6,8]. Clearly, phage therapy turns the conventional rules and established codes upside down. For instance, because bacteriophages are self-replicating, phage therapy has been reported as an active treatment, referring to a concept developed in agricultural biocontrol [9]. This means that the drug may amplify in the body, depending on the bacterial density, which, in turn, is evolving in response to the phage density. This makes phages fundamentally different from passive pharmaceuticals such as antibiotics, whose concentrations decline by a combination of metabolism or excretion processes, according to more canonical pharmacokinetic behaviors.

The pharmacology of phage therapy is thus quite unusual [9,10] and, according to models, might give rise to unexpected therapeutic outcomes such as a reduction or failure of efficacy when inoculation is given too early or because of the adjuvant use of antibiotics [11]. Phage therapy also differentiates in the evolutionary considerations it elicits. As observed with antibiotics, phage therapy will likely entail the selection of phage-resistant bacteria. However, given the narrow host range of phages as compared to the broader therapeutic spectrum of antibiotics, the selection pressure for resistance is exerted on only a limited number of bacterial types [12]. Moreover, it has been reported that phage resistance may reduce the virulence of bacteria [13,14]. Finally, phages are themselves evolving, giving rise to co-evolutionary dynamic patterns [15], which will never happen with antibiotics and poses a unique challenge for regulators [16]. The dynamics of the phage–bacterial interaction are becoming increasingly complex owing to the interference of a third intervener, which is the patient's body. Thus, in contrast to the classical mechanistic approach of medicine, phage therapy will only reveal its full dimensions in a Darwinian medicine perspective, which takes evolutionist and ecological prospects into account [2,17,18]. Another specific feature of phage therapy relates to its economic viability. The current business model of pharmaceuticals, requiring large and costly randomized, double-blind clinical trials is hardly applicable to phage therapy. This clinical development, normally attainable for medicinal products used to treat chronic medical conditions, becomes tricky for "ordinary" antibacterial compounds intended to be used for short durations [19], and even more arduous for medicinal products made of natural phages, which can only benefit from limited intellectual property protection and whose poor return on investment would likely not balance the resource expenditure [2,7,20]. In this regard, a Supreme Court jurisprudence analyzing patentable subject matter questions the eligibility of phage therapeutics for strong patent protection [21,22]. In spite of this, patents covering the use of phage therapy have been granted [23] and some clinical trials have been conducted or are still ongoing in Europe [24–26] and the United States [27–29], but so far they have not contributed to any licensing. Given that no phage product is currently marketed, authorities have little incentive to develop regulatory schemes and guidelines specific to bacteriophages [30] and, through a negative feedback loop, the absence thereof constitutes a hindrance to phage therapy development. Further clinical evidence would thus help to foster regulatory advance. In the European Union (EU), investigational medicinal products (IMP) defined as "a pharmaceutical form of an active substance or placebo being tested or used as a reference in a clinical trial" must be manufactured and checked in compliance with the principles and guidelines of good manufacturing practice (GMP) [31]. However, GMP compliance represents a real challenge and requires extensive financial resources [16,25,32], which may constitute an insurmountable obstacle for phage therapy sponsors when they are hospitals or non-for-profit phage therapy centers.

Besides these atypical features, there is one additional singularity making phage therapy medicinal products (PTMPs) unconventional, namely their qualitative and quantitative composition, which may be subject to variations. Indeed, PTMPs are either ready-prepared medicines intended for large scale distribution or patient-specific, tailor-made preparations issued from local small-scale productions. While the former, which have a fixed composition, match the current regulatory framework, the latter, with their moving target formulation, do not. In Europe, the pharmaceutical legislation was basically launched in the early sixties, following the thalidomide tragedy. It was designed to control industrially-made pharmaceuticals. The amended European Directive 2001/83/EC related to medicinal products leaves no doubt in this regard since it applies to "medicinal products for human use intended to be placed on the market in Member States and either prepared industrially or manufactured by a method involving an industrial process" [33], as opposed to the "medicinal products prepared in a pharmacy in accordance with a medical prescription for an individual patient (commonly known as the magistral formula)", which are beyond the scope of the directive. Still, in the early 2000s, autologous cell-based therapeutics first began to create problems in this regard. As they are patient-specific and given that some of them are prepared locally in the hospital pharmacy, they relate to the magistral formula. However, their manufacture may involve an industrial process, especially when part of their manufacturing process takes place in a biotech company. As such, they should fall within the scope of Directive 2001/83/EC. Moreover, assessing the quality and the benefit/risk balance of these innovative therapeutics is sensitive, thus amply justifying their tight regulatory control and, accordingly, their licensing as prescribed in the directive. To overcome this contradiction, the European legislature coined the concept of advanced therapy medicinal products (ATMPs) and designed a specific regulatory framework. Autologous somatic cell therapy medicinal products and tissue engineered products are both ATMPs. Strictly speaking, these medicinal products change from one patient to another since they stem from a patient's own cells, although they share a common manufacturing process. Therefore, the basis on which the marketing authorization is issued switched, and instead of focusing on the product itself, it became process-driven. Interestingly, the extent of quality, non-clinical and clinical data to be included in the marketing authorization application may be determined using a risk-based approach. Furthermore, ATMPs "which are prepared on a nonroutine basis according to specific quality standards, and used within the same Member State in a hospital under the exclusive professional responsibility of a medical practitioner, in order to comply with an individual medical prescription for a custom-made product for an individual patient" [33] may be discharged of the marketing authorization obligation under the umbrella of the so-called "hospital exemption" procedure. The similarity to phage therapy is undeniable. Custom-made PTMPs have more to do with cottage factories than with big pharma, as they are patient-specific, and, at the same time, they may share a common industrial process. This analogy did not escape the attention of regulators [16] and researchers engaged in the field, who advocate for a specific regulatory framework for phage therapy [34–36] that could, for instance, take advantage of the hospital exemption [37].

Autologous ATMPs are not the only pharmaceuticals that face a conflict between tailor-made production and industrial manufacturing. For instance, custom-made, anti-sense, oligonucleotide medicinal products also fall between the two. Linked to this, a call for new pharmaceutical legislation has been issued [38]. Similarly, as observed in cancer management, the concept of large disease groups being administered "one-size-fits-all" blockbuster drugs is gradually being replaced by the stratification of patients into small sub-groups, each treated with a different medication [39]. These medicinal products, as well as the PTMPs, share with autologous ATMPs the fact that they relate to personalized medicine. However, they are not considered ATMPs, and therefore they cannot benefit from the exceptions foreseen for ATMPs [16].

Encompassing phage therapy within personalized medicines is a direct consequence of the narrow therapeutic spectrum of bacteriophages. However, while high specificity has its advantages, it also carries drawbacks [16]. Prior to patient treatment, the phage susceptibility of the infecting bacteria must be determined by performing a phagogram. The expected time frame for performing such an analysis is expected to be similar to the turn-around-time for antibiogram results. However, the selected phage(s) must then be amplified, a process that may take an additional 18 h [40,41]. This can make a difference in the management of a bacterial infection.

In view of the underlying trend toward personalized medicine, the difference in the regulatory treatment of the custom-made medicines is difficult to understand. Indeed, whereas the European legislature implemented rather quickly the specific regulation for ATMPs, it has since showed a defensive position resistant to regulatory change. The design of the adaptive pathway is an instructive example in this respect [42]. To meet the need of critically ill patients, the European Medicines Agency (EMA) implemented this pathway, which relies on the procedures of scientific advice, compassionate use, conditional approval mechanism, and pharmacovigilance tools. Interestingly, it is clearly mentioned that this approach makes use of regulatory processes already in place within the existing EU legal framework. No new regulatory pathway has been implemented. The EMA therefore had to develop a creative regulatory avenue, while respecting the status quo attitude of the European lawmaker. However, in terms of regulatory affairs, the most intense creativity has its legal limitations. Thus, a workshop organized by the EMA in 2015, aimed at facilitating the development of bacteriophage therapy by reviewing regulatory aspects, eventually failed to deliver tangible openness towards an alternate regulatory scheme [16], primarily because of the EU decision-makers' conservatism. The European Commission made clear that the existing regulatory framework is adequate for bacteriophage therapy and that PTMPs can be regulated like any other medicinal product [43], whereas the stakeholders repeatedly expressed their disagreement with this stance [7]. It has thus proved necessary to turn to the Member States to find the beginnings of a solution.

Recently, the Belgian authorities have opened a gateway to phage therapy regulation by taking advantage of the national regulation of magistral preparation (compounding pharmacy in the US) [44]. The procedure relies on two cornerstones, namely (i) the issuing of a monograph serving as a written standard for assessing the quality of the phage active substance to be used as raw material for the preparation of the PTMP, and (ii) the availability of a Belgian approved laboratory that is able to test the phage stock and, where applicable, may issue a certificate of analysis stating that the tested phage complies with the monograph, in line with the current state of technical and scientific knowledge. The pharmacist can then use this certified material for preparing a customized medicinal product based on the prescription of a physician. This regulatory scheme is probably not optimal, since it places all the responsibility on the prescriber and the pharmacist, exempting the manufacturers and the regulatory authorities from the liability that they normally have for authorized medicinal products [36]. Therefore, it should be regarded as transitional [45]. However, even though it has some shortcomings, this process has at least the virtue of existing and of breaking down the regulatory barrier. As such, it was welcomed as a breakthrough that nurtures hope for the implementation of phage therapy in accepted therapeutic practices [46]. Changes are also taking place in France, where a specialized scientific temporary committee on phage therapy issued recommendations for using PTMPs under the umbrella of the so-called nominative Temporary Authorization for Use (ATUn, standing for Authorisation Temporaire d'Utilisation nominative) subject to certain conditions [47]. The ATUn of a medicinal product is issued for a single named patient who cannot participate in a clinical trial, at the request and under the responsibility of the prescribing physician. The ATUn is an exceptional authorization procedure, issued by way of derogation, which allows, in the absence of any appropriate alternative treatment, a medicinal product with no marketing authorization to be made available provided that its efficacy/safety balance is presumed to be favorable for these patients based on the available data. Medicinal products with ATUns can only be dispensed by hospital pharmacies.

Regulatory change can also be identified across the Atlantic. In the United States, some patients were treated with phages following the emergency investigational new drug (eIND) pathway of the Food and Drug Administration (FDA) [48,49]. Indeed, patients may have access to non-approved drugs or biological products under the expanded access program. Among the different categories of expanded access, the individual patient expanded access IND for emergency use appeared suitable for personalized phage products, which are regulated as biologics in the jurisdiction of the Office of Vaccines Research Review, in the FDA Center for Biologics Evaluation and Research (FDA/CBER/OVRR).

Interestingly, the Western world now implements regulatory principles that are reminiscent of the ones that apply in the countries where phage therapy traditions and practice have never been abandoned. In Georgia, regarded as a stronghold for bacteriophage therapy [50], phage products are considered to be pharmaceuticals. Bacteriophage ready-to-use medicines require a marketing authorization according to regular legislation. As for customized phage preparations, they may be prepared as a magistral preparation in an authorized pharmacy that has been granted a special license issued by the Georgian Ministry of Healthcare on the preparation of extempore medications. In Russia, which also has a longstanding practice of phage therapy, there is a precedent for the Belgian monograph [51], since the Russian pharmacopeia includes a monograph on bacteriophages for prophylactic and therapeutic use [52].

Phage therapy is not only moving forward in receiving regulatory approval. Progress may also be expected in clinical trial applications. With this in mind, it is worth mentioning that there is a new provision in the EU regulatory framework that may have gone unnoticed by the phage therapy sponsors, although it could facilitate PTMP clinical development. Whereas the former EU provisions relating to the conduct of clinical trials prescribe that the principles of GMP should be applied to IMP [15], some flexibility is foreseen in Regulation 536/2014, repealing Directive 2001/20/EC [53]. Indeed, according to Art 61(5) and 63 of this regulation, the preparation of IMPs "where this process is carried out in hospitals, health centres or clinics legally authorised in the Member State concerned to carry out such process and if the IMPs are intended to be used exclusively in hospitals, health centres or clinics taking part in the same clinical trial in the same Member State" may be exempted from GMP requirements. This provision markedly reshapes the EU landscape of clinical trial applications and may help meet the repeated demand for scientific evidence from human trials conducted to modern standards [16].

The regulation of PTMPs is evolving slowly but is moving in the right direction. The appeal for a paradigm change is beginning to be heard at least at the national level where recent initiatives are overcoming regulatory obstacles to a certain extent. However, despite this progress, there is still a way to go before a fully practicable regulation is implemented. The next step might come from international organizations. In the European Union, an initiative needs to be taken at the Community level to provide a genuine and harmonized regulation for PTMPs. In more general terms, considering the profound changes occurring in therapeutic practices, and especially the increasing personalization of medicine, it is the author's opinion that the EU lawmakers can no longer maintain their position resisting change without facing the risk of hampering innovation and, more critically, ignoring patients' needs. At a higher regional level, the Council of Europe's Directorate for the Quality of Medicines and Health Care could also be involved through the European Pharmacopoeia (Ph.Eur.). Indeed, elaborating a Ph.Eur. text on phage therapy would foster harmonization and strengthen the scientific base of what would then become an official public standard. Lastly, at a global level, the involvement of the World Health Organization appears essential for the development of phage therapy in general [54], and especially for its implementation in the low- and middle-income countries where it is urgently needed [55].

To conclude, it is worth emphasizing that the issue of PTMP regulation extends well beyond the area of phage therapy, since the debate is fundamentally related to the customization of medicinal products tailored to an individual patient. From this perspective, we like to think that while bacteriophages were a prominent model for uncovering the nature of genes, they remain so in the separate but promising context of personalized medicine.

Acknowledgments: The author would like to thank Charlotte Fauconnier for critical reading of the manuscript and valuable editorial advice.

Conflicts of Interest: Although the author works for regulatory bodies, the views expressed in this article are his personal opinion. As such, they may not be understood, interpreted, or quoted as being made on behalf of, or reflecting the position of, any authority, agency or organization.

References

1. D'Hérelle, F. Sur le rôle du microbe filtrant bactériophage dans la dysentérie bacillaire. *C. R. Acad. Sci. Paris* **1918**, *167*, 970–972.
2. Pirnay, J.-P.; Verbeken, G.; Rose, T.; Jennes, S.; Zizi, M.; Huys, I.; Lavigne, R.; Merabishvili, M.; Vaneechoutte, M.; Buckling, A.; et al. Introducing yesterday's phage therapy in today's medicine. *Future Virol.* **2012**, *7*, 379–390. [CrossRef]
3. Withington, R. Regulatory issues for phage-based clinical products. *J. Chem. Technol. Biotechnol.* **2001**, *76*, 673–676. [CrossRef]
4. Verbeken, G.; De Vos, D.; Vaneechoutte, M.; Merabishvili, M.; Zizi, M.; Pirnay, J.-P. European regulatory conundrum of phage therapy. *Future Microbiol.* **2007**, *2*, 485–491. [CrossRef] [PubMed]
5. Brüssow, H. What is needed for phage therapy. *Virology* **2012**, *434*, 138–142. [CrossRef]
6. Henein, A. What are the limitations on the wider therapeutic use of phage? *Bacteriophage* **2013**, *3*, 2. [CrossRef] [PubMed]
7. Cooper, C.J.; Mirzaei, M.K.; Nilsson, A.S. Adapting drugs approval pathway for bacteriophage-based therapeutics. *Front. Microbiol.* **2016**, *7*, 1209. [CrossRef] [PubMed]

8. Pirnay, J.-P.; De Vos, D.; Verbeken, G.; Merabishvili, M.; Chanishvili, N.; Vaneechoutte, M.; Zizi, M.; Laire, G.; Lavigne, R.; Huys, I.; et al. The phage therapy paradigm: *Prêt-à-porter* or *sur-mesure*? *Pharm. Res.* **2011**, *28*, 934–937. [CrossRef]
9. Payne, R.J.; Jansen, V.A. Phage therapy: The peculiar kinetics of self-replicating pharmaceuticals. *Clin. Pharmacol. Ther.* **2000**, *68*, 225–229. [CrossRef]
10. Abedon, S.T.; Thomas-Abedon, C. Phage therapy pharmacology. *Curr. Pharm. Biotechnol.* **2010**, *11*, 28–47. [CrossRef] [PubMed]
11. Payne, R.J.; Jansen, V.A. Pharmacokinetic principles of bacteriophage therapy. *Clin. Pharmacokinet.* **2003**, *42*, 315–325. [CrossRef]
12. Loc-Carrillo, C.; Abedon, S.T. Pros and cons of phage therapy. *Bacteriophage* **2011**, *1*, 111–114. [CrossRef]
13. Levin, B.R.; Bull, J.J. Population and evolutionary dynamics of phage therapy. *Nat. Rev. Microbiol.* **2004**, *2*, 166–173. [CrossRef] [PubMed]
14. Capparelli, R.; Nocerino, N.; Iannaccone, M.; Ercolini, D.; Parlato, M.; Chiara, M.; Iannelli, D. Bacteriophage therapy of *Salmonella enterica*: A fresh appraisal of bacteriophage therapy. *J. Infect. Dis.* **2010**, *201*, 52–61. [CrossRef]
15. Koskella, B.; Brockhurst, M.A. Bacteria–phage coevolution as a driver of ecological and evolutionary processes in microbial communities. *FEMS Microbiol. Rev.* **2014**, *38*, 916–931. [CrossRef] [PubMed]
16. Pelfrene, E.; Willebrand, E.; Cavaleiro Sanches, A.; Sebris, Z.; Cavaleri, M. Bacteriophage therapy: A regulatory perspective. *J. Antimicrob. Chemother.* **2016**, *71*, 2071–2074. [CrossRef] [PubMed]
17. De Vos, D.; Verbeken, G.; Dublanchet, A.; Jennes, S.; Pirnay, J.-P. La phagothérapie durable: Une question d'évolution. *Biofutur* **2016**, *35*, 44–47.
18. Torres-Barceló, C. Phage therapy faces evolutionary challenges. *Viruses* **2018**, *10*, 323. [CrossRef]
19. Bax, R.; Green, S. Antibiotics: The changing regulatory and pharmaceutical industry paradigm. *J. Antimicrob. Chemother.* **2015**, *70*, 1281–1284. [CrossRef]
20. Reardon, S. Phage therapy gets revitalized. *Nature* **2014**, *510*, 15–16. [CrossRef]
21. Minssen, T. The revival of phage therapy to fight antimicrobial resistance – Part II: What about patent protection and alternative incentives? Available online: http://blog.petrieflom.law.harvard.edu/2014/08/07/the-revival-of-phage-therapy-to-fight-antimicrobial-resistance-part-ii-what-about-patent-protection-and-alternative-incentives/ (accessed on 22 March 2019).
22. Todd, K. The promising viral threat to bacterial resistance: The uncertain patentability of phage therapeutics and the necessity of alternative incentives. *Duke Law J.* **2019**, *68*, 767–805. [PubMed]
23. Anonymous. AmpliPhi Biosciences to be Granted European Patent Covering the Use of Phage Therapy to Resensitize Bacterial Infections to Antibiotics. Available online: https://www.businesswire.com/news/home/20160602005456/en/AmpliPhi-Biosciences-Granted-European-Patent-Covering-Phage (accessed on 22 March 2019).
24. Wright, A.; Hawkins, C.H.; Anggård, E.E.; Harper, D.R. A controlled clinical trial of a therapeutic bacteriophage preparation in chronic otitis due to antibiotic-resistant *Pseudomonas aeruginosa*; a preliminary report of efficacy. *Clin. Otolaryngol.* **2009**, *34*, 349–357. [CrossRef] [PubMed]
25. Jault, P.; Leclerc, T.; Jennes, S.; Pirnay, J.-P.; Que, Y.A.; Resch, G.; Rousseau, A.F.; Ravat, F.; Carsin, H.; Le Floch, R.; et al. Efficacy and tolerability of a cocktail of bacteriophages to treat burn wounds infected by *Pseudomonas aeruginosa* (PhagoBurn): A randomised, controlled, double-blind phase 1/2 trial. *Lancet Infect. Dis.* **2019**, *19*, 35–45. [CrossRef]
26. Anonymous. Phagoburn. Available online: http://www.phagoburn.eu/ (accessed on 22 March 2019).
27. Rhoads, D.D.; Wolcott, R.D.; Kuskowski, M.A.; Wolcott, B.M.; Ward, L.S.; Sulakvelidze, A. Bacteriophage therapy of venous leg ulcers in humans: Results of a phase I safety trial. *J. Wound Care* **2009**, *18*, 237–238, 240–243. [CrossRef] [PubMed]
28. Anonymous. Intralytix Receives FDA clearance to initiate Phase I / IIa clinical trials. Available online: http://www.intralytix.com/index.php?page=news&id=87 (accessed on 20 March 2019).
29. Furfaro, L.L.; Payne, M.S.; Chang, B.J. Bacteriophage therapy: Clinical trials and regulatory hurdles. *Front. Cell. Infect. Microbiol.* **2018**, *8*, 376. [CrossRef] [PubMed]
30. Parracho, H.M.; Burrowes, B.H.; Enright, M.C.; McConville, M.L.; Harper, D.R. The role of regulated clinical trials in the development of bacteriophage therapeutics. *J. Mol. Genet. Med.* **2012**, *6*, 279–286. [CrossRef]

31. Directive 2001/20/EC of the European Parliament and of the Council on the on the approximation of the laws, regulations and administrative provisions of the Member States relating to the implementation of good clinical practice in the conduct of clinical trials on medicinal products for human use. Available online: https://ec.europa.eu/health/sites/health/files/files/eudralex/vol-1/dir_2001_20/dir_2001_20_en.pdf (accessed on 27 February 2019).
32. Jault, P.; Gabard, J.; Boisteau, O.; Meichenin, M.; Pirnay, J.-P.; Jennes, S.; Que, Y.A. Final report summary - PHAGOBURN (evaluation of phage therapy for the treatment of *Escherichia coli* and *Pseudomonas aeruginosa* burn wound infections (Phase I-II clinical trial)). Available online: https://cordis.europa.eu/project/rcn/108695/reporting/en (accessed on 22 March 2019).
33. Directive 2001/83/EC of the European Parliament and the Council on the Community code relating to medicinal products for human use. Available online: https://ec.europa.eu/health/sites/health/files/files/eudralex/vol-1/dir_2001_83_consol_2012/dir_2001_83_cons_2012_en.pdf (accessed on 27 February 2019).
34. Huys, I.; Pirnay, J.-P.; Lavigne, R.; Jennes, S.; De Vos, D.; Casteels, M.; Verbeken, G. Paving a regulatory pathway for phage therapy. *EMBO Rep.* **2013**, *14*, 951–954. [CrossRef]
35. Verbeken, G.; Pirnay, J.-P.; Lavigne, R.; Jennes, S.; De Vos, D.; Casteels, M.; Huys, I. Call for a Dedicated European Legal Framework for Bacteriophage Therapy. *Arch. Immunol. Ther. Exp.* **2014**, *62*, 117–129. [CrossRef]
36. Fauconnier, A. Regulating phage therapy: The biological master file concept could help to overcome regulatory challenge of personalized medicines. *EMBO Rep.* **2017**, *18*, 198–200. [CrossRef]
37. Verbeken, G.; Pirnay, J.-P.; De Vos, D.; Jennes, S.; Zizi, M.; Lavigne, R.; Huys, I. Optimizing the European Regulatory Framework for Sustainable Bacteriophage Therapy in Human Medicine. *Arch. Immunol. Ther. Exp.* **2012**, *60*, 161–172. [CrossRef]
38. Johnston, J.D.; Feldschreiber, P. Proposal for new European pharmaceutical legislation to permit access to custom-made anti-sense oligonucleotide medicinal products. *Br. J. Clin. Pharmacol.* **2014**, *77*, 939–946. [CrossRef]
39. Leyens, L.; Richer, E.; Melien, O.; Ballensiefen, W.; Brand, A. Available Tools to Facilitate Early Patient Access to Medicines in the EU and the USA: Analysis of conditional approvals and the implications for personalized medicine. *Public Health Genomics* **2015**, *18*, 249–259. [CrossRef]
40. Moelling, K.; Broecker, F.; Willy, C. A wak-up call: We need phage therapy now. *Viruses* **2018**, *10*, 688. [CrossRef] [PubMed]
41. Bourdin, G.; Schmitt, B.; Marvin Guy, L.; Germond, J.E.; Zuber, S.; Michot, L.; Reuteler, G.; Brüssow, H. Amplification and purification of T4-like *Escherichia coli* phages for phage therapy: From laboratory to pilot scale. *Appl. Environ. Microbiol.* **2014**, *80*, 1469–1476. [CrossRef] [PubMed]
42. European Medicines Agency. Adaptative Pathways. Available online: https://www.ema.europa.eu/en/human-regulatory/research-development/adaptive-pathways (accessed on 28 February 2019).
43. European Parliament. Parliamentary questions, answer given by Mr Dalli on behalf of the Commission. Available online: http://www.europarl.europa.eu/sides/getAllAnswers.do?reference=E-2011-001144&language=EN (accessed on 28 February 2019).
44. Pirnay, J.-P.; Verbeken, G.; Ceyssens, P.-J.; Huys, I.; De Vos, D.; Ameloot, C.; Fauconnier, A. The Magistral Phage. *Viruses* **2018**, *10*, 64. [CrossRef] [PubMed]
45. Fauconnier, A. Guidelines for bacteriophage product certification. *Methods Mol. Biol.* **2018**, *1693*, 253–268. [PubMed]
46. Górski, A.; Miedzybrodzki, R.; Lobocka, M.; Glowacka-Rutkowska, A.; Bednarek, A.; Borysowki, J.; Jonczyk-Matysiak, E.; Lusiak-Szelachowska, M.; Weber-Dabrowska, B.; Baginska, N.; et al. Phage therapy: What have we learned? *Viruses* **2018**, *10*, 288. [CrossRef]
47. Agence nationale de sécurité du médicament et des produits de santé. Comité scientifique spécialisé temporaire: Phagothérapie. Available online: http://ansm.sante.fr/content/download/91159/1144681/version/1/file/CR_CSST_Phagotherapie_CSST201611013_24-03-2016.pdf (accessed on 28 February 2019).
48. Schooley, R.T.; Biswas, B.; Gill, J.J.; Hernandez-Morales, A.; Lancaster, J.; Lessor, L.; Bar, J.J.; Reed, S.L.; Rohwer, F.; Benler, S.; et al. Development and use of personalized bacteriophage-based therapeutic cocktails to treat a patient with a disseminated resistant *Acinetobacter baumannii* infection. *Antimicrob. Agents Chemother.* **2017**, *61*. [CrossRef]

49. LaVergne, S.; Hamilton, T.; Biswas, B.; Kumaraswamy, M.; Schooley, R.T.; Wooten, D. Phage therapy for a multidrug-resistant *Acinetobacter baumannii* craniectomy site infection. *Open Forum Infect. Dis.* **2018**, *5*. [CrossRef]
50. Parfitt, T. Georgia: An unlikely stronghold for bacteriophage therapy. *Lancet* **2005**, *365*, 2166–2167. [CrossRef]
51. General Monograph. Phage active pharmaceutical ingredients. Available online: http://www.mdpi.com/1999-4915/10/2/64/s1 (accessed on 28 February 2019).
52. Russian Pharmacopoeia. OFS.1.7.1.0002.15 Bacteriophages are therapeutic and prophylactic. Available online: http://pharmacopoeia.ru/ofs-1-7-1-0002-15-bakteriofagi-lechebno-profilakticheskie/ (accessed on 28 February 2019).
53. Regulation 536/2014 of the European Parliament and of the Council of 16 April 2014 on Clinical Trials on Medicinal Products for Human Use, and Repealing Directive 2001/20/EC. Available online: https://ec.europa.eu/health/sites/health/files/files/eudralex/vol-1/reg_2014_536/reg_2014_536_en.pdf (accessed on 28 February 2019).
54. Fauconnier, C.; Fauconnier, A. The role WHO is called to play in implementing phage therapy. Submitted for publication.
55. Nagel, T.E.; Chan, B.K.; De Vos, D.; El-Shibiny, A.; Kang'ethe, E.K.; Makumi, A.; Pirnay, J.-P. The developing world urgently needs phages to combat pathogenic bacteria. *Front. Microbiol.* **2016**, *7*, 882. [CrossRef] [PubMed]

Opinion

Clinical Indications and Compassionate Use of Phage Therapy: Personal Experience and Literature Review with a Focus on Osteoarticular Infections

Olivier Patey [1], Shawna McCallin [2], Hubert Mazure [3], Max Liddle [4], Anthony Smithyman [5] and Alain Dublanchet [1,*]

1 Service of Infectious and Tropical Diseases, CHI Lucie et Raymond Aubrac, 94190 Villeneuve Saint Georges, France; opatey@aol.com
2 Department of Musculoskeletal Medicine DAL, Centre Hospitalier Universitaire Vaudois CHUV, Service of Plastic, Reconstructive & Hand Surgery, Regenerative Therapy Unit (UTR), CHUV-EPCR/Croisettes 22, 1066 Epalinges, Switzerland; shawna.mccallin@gmail.com
3 HGM Consultants, 63 Rebecca Parade, Winston Hills, NSW 2153, Australia; humazure@hotmail.com
4 School of Life Sciences, University of Technology, Ultimo, NSW 2007, Australia; max.liddle17@gmail.com
5 Cellabs Pty Ltd, and Founder Special Phage Services Pty Ltd, both of 7/27 Dale St, Brookvale, NSW 2100, Australia; tonysmithyman@gmail.com
* Correspondence: adublanchet@orange.fr

Received: 27 November 2018; Accepted: 21 December 2018; Published: 28 December 2018

Abstract: The history of phage therapy started with its first clinical application in 1919 and continues its development to this day. Phages continue to lack any market approval in Western medicine as a recognized drug, but are increasingly used as an experimental therapy for the compassionate treatment of patients experiencing antibiotic failure. The few formal experimental phage clinical trials that have been completed to date have produced inconclusive results on the efficacy of phage therapy, which contradicts the many successful treatment outcomes observed in historical accounts and recent individual case reports. It would therefore be wise to identify why such a discordance exists between trials and compassionate use in order to better develop future phage treatment and clinical applications. The multitude of observations reported over the years in the literature constitutes an invaluable experience, and we add to this by presenting a number of cases of patients treated compassionately with phages throughout the past decade with a focus on osteoarticular infections. Additionally, an abundance of scientific literature into phage-related areas is transforming our knowledge base, creating a greater understanding that should be applied for future clinical applications. Due to the increasing number of treatment failures anticipatedfrom the perspective of a possible post-antibiotic era, we believe that the introduction of bacteriophages into the therapeutic arsenal seems a scientifically sound and eminently practicable consideration today as a substitute or adjuvant to antibiotic therapy.

Keywords: bacterial infection; antibiotic resistance; bacteriophage; antibiotic therapy; phage therapy; cases report

---

1. Introduction

In 1917, Félix d'Hérelle observed a phenomenon in stool cultures from convalescent patients with bacillary dysentery [1], which took the form of perfectly round clear areas in the bacterial lawn. He made the assumption that these clear zones were caused by an "invisible microbe" capable of killing bacteria, to which he gave the name bacteriophage. Two years later (1919), he demonstrated that the oral administration of bacteriophages in humans is harmless and causes the healing of bacterial enteritis caused by *Shigella* sp. (bacillary dysentery). Based on a large number of published cases in the years that followed, the interest and use of this new treatment in various infections spread rapidly across the world, reaching nearly every continent [2,3]. This was the situation until the discovery of antibiotics; when faced with their easier use, phage therapy was gradually abandoned in Western countries until it finally disappeared completely in France with the closure in 1990 of the elast remaining sources of therapeutic bacteriophages from the two Pasteur Institutes (Paris and Lyon). However, phage therapy continued uninterrupted in the Soviet Union during this time and is still practiced in Russia, Poland, Georgia and some other former Soviet States today, in accordance with specific national regulations.

We have been witnessing the worldwide spread of multidrug-resistant (MDR) bacteria in recent years. As new and truly innovative antibiotics are rare, the increasing frequency of therapeutic failures are raising fears of a new pre-antibiotic era [4]. To respond to this worrying situation, the return of phage therapy seems to be an answer not only as an alternative [5], but also a complementary treatment, to faltering antibiotic therapy [6–8]. This renewed interest in phage therapy is manifested by the motivation to conduct several clinical trials since 2009 that have used phages for a variety of indications, including chronic otitis, burn wound or urinary tract infections (UTI), and *Escherichia coli* diarrhea [9–12]. Indeed, phage therapy must be proven to be therapeutically effective through experimental clinical trials in order to obtain marketing approval, which is required for use in Western medicine. While studies have repeatedly documented its safety, it is unfortunate that no marketing approval has been attributed to a phage product to date as a result of these resource-intensive studies; three trials were unable to statistically prove efficacy [9,11,13], even if clinical benefit was achieved for some patients, and the only trial that was successful has not been further pursued for commercialization [12].

Many researchers and medical doctors have voiced the need to revise the regulatory classification of phage therapy products in order to facilitate their clinical evaluation. Natural phages are currently classified as Medicinal Products (MP) under European Union (EU) legislation [14] and as a drug by the Food and Drug Administration (FDA) in the United States, which necessitate that phages be produced under Good Manufacturing Practice (GMP) guidelines and infrastructure. While such criteria do not completely inhibit the ability to conduct trials, they do render formal phage trials more difficult and more expensive to conduct. Substantial financial investment is required to conduct clinical trials, and an inconvenient amount of time is needed to procure results in order to address current clinical needs.

Phage therapy is now at a state where it is not officially recognized as a legitimate treatment, but has been increasingly granted emergency approvals for addressing antibiotic treatment failures. There are more than 10 published case reports [15] and a fast-growing number of undocumented compassionate cases that report successful treatment outcomes with phage therapy. Within only the last year, two experimental phage therapy centers have opened in addition to the long-established Phage Therapy Unit at the Ludwik Hirszfeld Institute in Poland (Box 1). From this perspective, and in conjunction with a century of publications on this subject for different bacterial infections treated by phage therapy, compassionate use and case reports constitute an invaluable source of knowledge that help to elucidate best practices for phage therapy. Many original and historical texts published in French or Russian have been unfortunately excluded from contributing to this large body of information and should enter into consideration. While case reports and historical accounts do not substitute for formal clinical trials, the findings and remarks they contain are useful to set up modern therapeutic protocols and hopefully to avoid conducting additional therapeutically-futile clinical trials. This is what we propose to report here, both by summarizing findings from the literature and by adding our own experience of cases, particularly for osteoarticular infections, treated under compassionate protocols, as a means to advise on the progression of phage therapy into modern medicine.

Box 1. Experimental phage therapy centers established in Western countries.

Ludwik Hirszfeld Institute of Immunology and Experimental Therapy, Polish Academy of Sciences, Phage Therapy Unit in Wroclaw, Poland (IIET PAS PTU): This is the oldest and most established experimental center in central Europe, which has been preparing phage formulations for hospital use in Poland since the 1970s, before it became a member state of the European Union (EU). Phage therapy was and is continued under the national regulatory framework as an experimental therapy for specific medical conditions and in specific centers under Article 37 of the Helsinki Declaration. They have an in-house bank of phages against 15 different bacterial pathogens (*Staphylococcus, Enterococcus, Pseudomonas, Escherichia, Klebsiella, Serratia, Proteus, Acinetobacter, Citrobacter, Enterobacter, Stenotrophomonas, Shigella, Salmonella, Burkholderia, Morganella*). Treatment is proceeded by phage susceptibility testing (phage typing procedure) and preparations are used for outpatient treatment. The PTU periodically publishes summaries of their experiences [16–20] that provide factual justification for using phage therapy and useful information for clinical applications.

Magistral preparations in Belgium (also known as a compounded prescription drug in the US) [14]. Phage therapy can be provided as a magistral preparation in Belgium since 2018, after several years of discussion involving public health and federal regulatory authorities, in order to facilitate physician-prescribed treatment for individual patients. Phages are considered active pharmaceutical ingredients (APIs) that must be produced according to an internal monograph (set of instructions) and that are subsequently certified by competent laboratories before they are mixed or put into formulation under the supervision of a pharmacist and delivered to a specific patient.

Center for Innovative Phage Applications and Therapeutics (IPATH) University of California San Diego, School of Medicine: This center announced its opening in June 2018 following several successful treatments with phage. Their first case used phage to treat an MDR systemic infection caused by *Acinetobacter baumannii*, which was initiated and coordinated by the wife of the patient, a global health professor, and his physician [21,22]. While this was the first American patient with a systemic MDR infection to be successfully treated intravenous (iv) by phage therapy, more than five patients have been treated since under the FDAs compassionate use program and IPATH is planning to conduct clinical trials in the near future.

2. General Prerequisites for the Medical Use of Bacteriophages

Due to the unfamiliarity with and particularities of phage therapy, it is worthwhile to touch upon several general aspects of clinical use: product availability, production, formulation and administration, dosage, and evaluation. The permission to use phage therapy for compassionate or experimental treatment, at the patient, physician, hospital, and health and regulatory authority levels, are beyond the scope of this publication, but are evidently necessary to proceed with treatment and requirements may vary country-to-country. Approvals are now often granted on an individual bases for emergency use or in the case of antibiotic treatment failures, mostly in France, Belgium, Poland, Australia, and the US.

2.1. Availability

The first condition for use of phage therapy is simply to have bacteriophages available for treatment, which is often complicated at this stage of phage development. This implies having access to phages that are both biologically active against the patient's bacterial isolate and satisfy regulatory requirements (purity, traceability, characterization). A single phage may be used (monophage preparation) or several phages may be combined against one or more bacterial species (phage cocktail).

As phage therapy is not currently a recognized medicine in the West and no registered products exist, phages are either being prepared specifically for a patient infection (personalized or custom approach) or treatment can be done with commercial phage preparations from Russian or Georgian suppliers, which have pre-defined phage compositions ("ready-to-use") [23]. Such commercial bacteriophage preparations that are available for purchase may, or may not, encounter importation difficulties into Western countries due to product traceability or a lack of certification or analytical information. Access can be accomplished by patients traveling to countries where phage therapy is an approved practice (medical tourism), and the Eliava Institute in Georgia treats a number of foreign patients onsite each year. This last option, however, is dependent upon patient mobility and financial ability to pay for treatment.

Alternatively, phages have been prepared for compassionate cases by small biotech and academic institutions for individual patients. Indeed, a large number of different bacteriophages are deposited in different collections that target clinically relevant bacteria, and it would be desirable that the collections held in these "phage banks" organize themselves into a network to facilitate exchanges. A networking initiative, known as the Phage Directory [24], is attempting to facilitate phage sharing for emergency or compassionate clinical needs. If an active phage is not present or available from such an organization, it is normally still possible to isolate one from the environment, although this is pathogen dependent, sufficient characterization is still required, and is difficult to achieve for acute life-threatening infections [25].

2.2. Production

Phage products must be produced with an acceptable level of purification for clinical use in order to remove remaining endotoxin and bacterial contaminants. If phage preparations are viewed as medicinal products, they will be subject to GMP compliance, which are standards intended to guarantee the quality of a medicine [26]. This requires that a procedure be defined for their manufacture and stipulates a combination of physicochemical and biological tests, as well as stringent production facilities. The quality (i.e., the stability and consistency) of a biological drug, such as phages, is harder to guarantee and control than that of a chemical, and GMP requirements have put a strain on the clinical development of phage therapy in Western medicine, as well as greatly increasing production costs. Indeed, GMP constraints both delayed patient enrollment for the Phagoburn clinical study and negatively impacted the phage titer of the final product [9]. An adaptation of the regulation is necessary [27–32] and, in particular, will have to take into account the use of individualized preparations [33] as a personalized medicine [28,34], and modification of phage components throughout treatment to counteract bacterial–phage resistance. Production considerations must take into account the sustainability of the phage treatment approach and patient safety.

Phage therapy is currently implemented for compassionate use and individual patients by by-passing GMP-requirements. Belgium has opted to facilitate phage therapy by presenting the phage as magistral preparations, which are individually prepared by prescription for individual patients by a qualified hospital pharmacist, and the quality of phage preparations are verified by accredited laboratories (Box 1) [14]. Even without such a systematic approval system, phage biotech companies (MicroGen, Eliava, Pherecydes Pharma, Advanced Phage Therapeutics, AmpliPhi Biosciences), as well as academic institutions and military research institutions, have helped in the production process and/or supply of phages for emergency use.

2.3. Formulation and Administration

The administration of a drug is dependent on the vectorization/formulation of the active phage component [35,36]. Local application is easiest to apply, and tolerance has been repeatedly documented for this route [9,37]. Phages may be applied topically either in cream/gel formulations or by contact with soaked bandages on the wound surface. Bacteriophages, being of a protein nature, raise the concern of an anaphylactic reaction following repeated administration. However, severe reactions have only been rarely reported, and today the risk is further reduced by advanced purification methods [38–40].

While oral administration is easy and without side effects, gastric acidity is a hostile barrier to ingested bacteriophages. To overcome this drawbackthere are two possible approaches: alkalinisation, by administration of an alkaline liquid (bicarbonate water, carbonated water), or gastro-resistant vectorization (i.e., in release capsules or pills). Alkaline neutralization may come with an increased risk of opportunistic infections for patients, and vectorization comes at an increased cost of production; more clinical data are required to determine the best strategy for oral phage application.

Inhalation seems to be effective in delivering lyophilized bacteriophages to the lungs in the form of powder propelled by inhalers [41]. The bronchopulmonary tree is indeed easily accessible by air, and thus, it is conceivable to spray bacteriophage suspensions (nebulization, misting) or dry forms (spray) [42]. However, few cases have been published to date [43–48], and using appropriate vectorization for inhalation remains to be evaluated.

Diffusion is rapid after systemic ie intravenous administration, though circulating phages are sequestered by the reticuloendothelial system in the spleen and liver. In the absence of bacterial target hosts, phages are quickly eliminated. On the contrary, if bacterial targets are present, phages multiply to a degree dependent on a multitude of bacterial (metabolic activity, sensitivity) and mammalian factors, making the estimation of pharmacokinetics [49] variable and difficult to estimate between patients.

2.4. Dosage

The required dosage, rhythm, and duration of treatment have been poorly studied. Theoretically, in situ multiplication requires only one application; while in practice, repetition is often the rule. Unlike conventional drug treatments, the pharmacological parameters are poorly defined and understood at present, which presents the main difficulty in being unable to predict the extent of in vivo multiplication.

2.5. Therapeutic Evaluation

Like any drug, a biomedicine must be studied experimentally to appreciate its positive and negative effects on a living organism. Although many publications (individual cases and clinical series) have shown positive results of phage therapy and presented few adverse effects, it is necessary to respond to modern requirements and to carry out randomized, double-blind controlled trials [50,51]. Nonetheless, simpler hospital observational studies, despite their drawbacks and inadequacies, would make it possible to provide highly valuable information for pressing questions while satisfying prerequisites (i.e., the number of patients likely to be included within a defined period of time) that have often been difficult for modern trials to achieve to date.

Many case studies today evaluate phage therapy by the most essential factor: the clinical improvement of the patient. However, information documenting phage activity within the patient, such as phage amplification or phage sensitivity, are often lacking, and therefore claims that phage therapy causes clinical amelioration are not data-supported. Much more information could and should be obtained from compassionate and emergency-use treatments to further our knowledge-base of phage therapy in humans.

3. Clinical Indications in the Literature

Inherently, phages are able to treat any clinical presentation of bacterial infections. Reports have been published using phage therapy for a large array of clinical indications, including gastro-intestinal [2,11,52–54], localized [3,37,55–59], burn wound [9,60], systemic [21,39,61–72], urogenital [10,73–82], respiratory [44,45,47,82–85], oto-rhino-laryngeal (ORL) [12,86–95], and osteoarticular infections (see section below). These clinical indications include infections that are acute or chronic, sensitive or resistant to antibiotics, and are caused by highly variable common or opportunistic pathogens.

Acute systemic infections, such for septicemia or meningitis, have been treated with phages with some success. It seems premature to consider such indications initially for phage therapy for at least two reasons: the urgency of treatment and the need for parenteral administration. Both aspects require readily-available, highly-purified phages, and rapid approval processes that, while not insurmountable, are not feasible for broad implementation at this time. Indeed, the few published cases of systemic treatments are the result of a few geographical competency centers and close collaboration between phage researchers and clinicians.

Chronic infections, however, are increasingly frequent and have gained attention as a target for phage therapy. Chronicity is supported by the formation of bacterial biofilms, intracellular bacterial persisters, or tolerant bacteria that are particularly problematic for UTIs, bacterial prostatitis, prosthetic joint infections (PJI), osteomyelitis, and respiratory conditions such as cystic fibrosis (CF). They require long-term antibiotic treatment that disrupts healthy microbiomes and selects for antimicrobial resistance. Such infections, if not constantly suppressed, risk development into bacterial sepsis. CF, although not an infectious disease itself, is the subject of special attention for phage therapy because of the chronic state of repetitive superinfections in these patients, which are usually caused by mucosal *Pseudomonas aeruginosa* strains resistant to many antimicrobials and capable of forming biofilms [96].

In addition to classical pathogens, opportunistic bacteria are often multidrug-resistant and cause infections that are difficult to control [96], for which the question of the interest of phage therapy is repeatedly raised. Infections with some bacteria, such as mycobacteria, present additional biological obstacles, such as preferential intracellular location of bacteria (macrophage or epithelial cells) and a slow growth rate, as in tuberculosis. Ready-to-use bacteriophage suspensions are generally not available for such situations, and a few teams have looked at some of them, although it is still too early to draw any conclusions. These include, more specifically, infections caused by *Helicobacter* [97,98]; *Borrelia* (Lyme disease [99]), as well as *Brucella*, *Yersinia pestis* and *Bacillus anthracis* [100] in the context of biological-weapon risks [101]. Other bacterial species (*Haemophilus influenzae*, *Streptococcus pneumoniae*) do not generate much interest today for phage therapy application. It should be noted that *Campylobacter* bacteriophages are mainly studied in poultry farms in a preventive context rather than therapeutically.

Osteoarticular infections are a particular form of deep-seated, localized infection that are a prime target for phage therapy given their frequency and poor response to antibiotic therapy. The diffusion of antibiotics into bone tissue is often mediocre and impaired by the presence of bacterial biofilms that form *in vivo* at the contact between bone and prosthetic material. The recurrence and transition to chronicity is more and more common for many reasons, including the presence of MDR bacteria. Today, the number of post-surgical bone infections on fracture or joint prosthesis continues to increase [102]. Conventional antibiotic treatments are long and costly, with frequent repeat surgery, and sometimes amputation is the only infection control option [103].

Phage therapy has been used very early and frequently for this type of infection, as evidenced by many publications from Northern America [104] and in Eastern European countries [105–107]. In France, the surgeon André Raiga [66,108] made several assessments of his long experience in this field. Clinical cases in Strasbourg, France were published in 1979, which document positive outcomes in bone infections with phages (Box 2) [109]. A review of Soviet literature has also indicated complete recovery from osteomyelitis using phages alone or in combination with antibiotics [110]. More recently, two cases of PJI (*Staphylococcus aureus*) and one case of *P. aeruginosa* osteomyelitis were treated with direct application of phages in France [111].

The potential of phage therapy to treat such post-accidental, surgical osteitis, or peri-prosthetic joint infections is likely rooted in phage activity against bacterial biofilms and potentially against intracellular bacteria. An experimental model [112] has demonstrated that a treatment combining bacteriophages and antibiotics helps to dissolve biofilms with a pronounced effect on biofilms of *Staphylococcus* sp. compared to those of *P. aeruginosa*. Indeed, there has been a very large number of experimental studies for several years on this subject not only in vitro, but also in vivo [113,114]. While bacterial infections begin by biofilm formation on prosthetic surfaces, they can become chronic by establishing an intracellular life-style within mammalian cells that shields them from antibiotic treatment and then causes recurrent active infections. A recent model documented the ability of phages to kill intracellular *S. aureus* [115].

All of the above provide substantial evidence that osteoarticular infections are a sound target for phage therapy. With this logic, a budget has been attributed for a future clinical trial in France, "Phagos," for PJIs caused by *S. aureus*, which will begin as soon as GMP-compliant phage suspensions are achieved [116]. Our experience with the compassionate treatment of oecsteoarticular, as well as other, infections is presented herein.

Box 2. Conclusion of Lang et al. 1979 [109].

Seven orthopedic surgery cases were treated with bacteriophages between 1975 and 1976. Of the treated patients, six were male and one was female. The age of patients ranged from 19–70 years of age. The cases presented by authors were chronic, having exhausted the usual therapeutic arsenal, and phage was added to other treatments in order to maximize patient benefit.

Five treatments resulted in good clinical outcomes, which was supported by radiological and bacteriological examination. A condition was considered improved if symptoms were ameliorated and radiological examination was positive, but problems persisted with scarring and positive bacterial cultures (one case). Treatment failure with added phages occurred for one patient and caused a change in treatment plan, comprising first local and general antibiotic therapy (ampicillin, cephalosporin, gentamicin), then hyperbaric oxygen therapy, and finally surgical intervention, which ultimately resulted in a favorable outcome. In conclusion, the use of suitable bacteriophages in the treatment of antibiotic-resistant chronic bone infections seemed to be an interesting therapeutic alternative for authors, and the results of these cases encouraged continuation in this therapeutic direction.

4. Compassionate Phage Use in France and at Villeneuve Saint Georges

Phage therapy was used to treat patients compassionately during the 1970s and 80s in France, at a time when it was possible obtain suspensions of therapeutic phage for the pathogenic bacterium of a patient from the Pasteur Institute. The clinical outcomes during this time with the treatment of frequent, high-risk infections with phages have been summarized previously in a short paper outlining conclusions and new indications for phage therapy (Box 3) [117]. At that time, phage therapy was routinely performed in some hospitals, such as in Lyon, Paris, and Strasbourg. A surgical service at the latter had published a small clinical study of seven cases and concluded that phage therapy was promising, particularly in bone infections (Box 2) [109]. Several patients with bone infections in the hospital of Villeneuve Saint Georges, for whom conventional treatment had failed, also benefited from such phage therapy treatment during this time (unpublished results). However, by 1990, phage therapy and its practice in France became impossible after phage production was ceased at the Pasteur Institute. There followed a period of about 15 years during which phage therapy was totally inaccessible in France.

Box 3. Conclusion of Vieu et al. 1979. [117].

This article, published in French, highlighted how and why phage therapy was used in France at this time. In particular, the growing importance of opportunistic bacteria resistant to antibiotics in infectious pathology oriented the therapeutic applications of bacteriophages to three new areas: (1) the curative treatment of postoperative surgical infections; (2) suppression of the infectious process during gram-negative pediatric epidemics, caused notably by *Salmonella*, *Klebsiella*, *E. coli*, and *Serratia*, via oral phage administration; and (3) curative treatment of chronic UTIs. The authors noted that a close collaboration between phage scientists and clinicians was absolutely necessary to treat patients with phages, from identifying phages to following clinical progression over time. The success of phage therapy was dependent upon verifying in vitro susceptibility prior to treatment. If treatment failure occurred, it was attributable to low titers of the phage, pH environment of the GI or urinary tracts, inactivation of the phage by simultaneously-prescribed local antiseptics, or the involvement of several pathogens not identified at diagnosis outside the bacterial host range of the phage preparation. In conclusion, the authors affirmed that phage therapy was a merited treatment option due to the frequent clinical successes it produced.

In 2004, we were able to buy over-the-counter commercial preparations of bacteriophages from pharmacies in Moscow for a few dozen Euros. After an evaluation (for sterility, activity, specificity) of these preparations [118], those capable of responding to the clinical problems at hand were retained and used. The first case we treated was a particularly worrying case of an evolving infection of the external auditory canal, where a bacteriophage suspension against *S. aureus* was used to treat chronic otitis externa (Box 4; Patient 1 in Table 1). With this experience, and in the face of the increasing therapeutic failures that we were confronted with, especially in orthopedic surgery, some of us decided to reintroduce phage therapy more routinely from 2008 in the hospital in which we practiced, and it is still occasionally used as needed at the hospital of Villeneuve Saint Georges. We will briefly outline the process of using phages for compassionate use and present several clinicals cases of our experiences in phage therapy.

Box 4. Treatment of an external otitis.

A young patient was examined for chronic otitis after episodes of repeated otitis treated with various antibiotics. The specialist noted an otorrhea and decided to treat it medically (cefpodoxime and ofloxacin) before surgery. Repair of the tympanic membrane was performed. The immediate treatment outcome was obvious: symptoms (pain, drainage) rapidly disappeared with no complications or side effects.

After three months, the otorrhea reappeared. The examination was particularly difficult because of very sharp local pain, as the eardrum was inflamed and wet. The resumption of local antibiotic therapy (bacitracin) helped reduce pain. During one year, the patient experienced several treated otorrheas (ofloxacin). During an outpatient consultation with acute pain and under general anaesthesia, a specimen was collected showing the presence in pure culture of *S. aureus* (penicillin-R, methicillin-R, erythromycin-R and ofloxacin-R). Despite antibiotic therapy being immediately prescribed (not specified), the purulent flow and pain persisted, and the *Staphylococcus* was still present.

It was then decided to carry out a more precise examination and to collect multiple specimens (tympanic membrane, cutaneous coating of the external duct) before the local application of a bacteriophage suspension, active in vitro against the patient isolate, in combination with pristinamycin. Within 48 hours, the patient noticed a clear improvement: the cessation of purulent flow and pain. Subsequent consultations confirmed a favorable course: the absence of otorrhea or pain and disappearance of *Staphylococcus*. After three months, the ear examination was still very satisfactory and the treatment was stopped.

5. Protocol for Compassionate Use of Phage Therapy

Before patient admission, the decision to use phage therapy is made by a multi-disciplinary hospital team (surgeon, infectious disease specialist, microbiologist), who conduct a complete examination of the patient and patient file. In addition to the biological assessment, one or more preliminary specimens is taken to isolate the bacterium and test its sensitivity to available phages. Patients are informed about phages and the possibility of treatment. Phages are administered by a treating physician who exercises their ethical right to use an experimental treatment in the best interest of the patient, without an elaborate regulatory or administrative framework.

During therapeutic care, if necessary, the infectious foci are excised (debridement) and cleaned in the operating room. One or more intraoperative specimens are collected to confirm the initial bacteriological diagnosis. At the end of surgery and before closure of the operative field, the preparation of bacteriophages is used to flood the operative field (5 to 10 ml according to the surface of the field). Access to the treatment site (opening or drain) allows a bacteriological control and the introduction of the same phage preparation in the days following the intervention.

Antibiotic therapy reflecting the pathogen's antibiotic resistance profile is used in combination with phage therapy, and the patient is kept under surveillance for several days (less than one week) to ensure that there was no evidence of infection (local, biological, or bacteriological). The postoperative course has presented no complications, and no side effects have been reported.

Regarding follow-up, ambulatory monitoring is performed in our facility for several months at a variable frequency, as deemed necessary. The evaluation of each case is performed clinically, as well as biologically and radiologically. Some patients provide us periodically with their health status, which so far has been excellent.

All cases reported here (Table 1) have benefited from compassionate phage therapy for the duly recorded treatment failure. The phage therapy treatments were carried out between 2006 and 2018 after a long evolution, generally several years, of a conventional treatment according to official medical guidelines. All patients had benefited from multiple attempts at treatment (surgical interventions and antibiotic therapy) and had been in therapeutic failure for months or even years. Some had previously tried treatment at the Eliava Institute in Tbilisi. All presented cases were treated in France at the Villeneuve Saint Georges Hospital, unless otherwise noted. The authors have also been involved in the treatment with phage therapy for a case for a refractory UTI in Australia, published previously [78].

Table 1. Summary of patients treated with compassionate use of bacteriophages from 2006–2018.

N	Age; Sex	Symptom Onset; PT Start	Clinical Symptoms	Bacteria	Phage Therapy	Outcome
1	20; F	2004; 2006	Suppurating chronic otitis; intense pain	*S. aureus*	Commercial anti-*S. aureus* suspension; ear drop instillations (15 days)	2006 Complete cure
2	44; M	2005; 2008	Accidental fall; multiple fractures ($n = 37$); amputation considered	*S. aureus*	Commercial anti-*S. aureus* and Pyophage suspensions; administered peroperatively over several weeks	2009 Wound closure and complete cure
3	25; M	2007; 2008	Road accident causing multiple trauma; uncontrolled pelvic bone infection	*S. aureus* *P. aeruginosa*	Anti-*S. aureus* and anti-*P. aeruginosa* phage suspension; administered peroperatively and via catheter in days following operation (Belgium).	2010 Complete cure
4	40; F	1995; 2009	Fall leading to complex fracture of the right foot; Planned amputation	*S. aureus*	Commercial anti-*S. aureus* suspension administered peroperatively and via catheter in the days following operation	2009 Wound closure and complete cure
5	60; M	2008; 2009	Fistulised abdominal plaque infection; continuous suppressive antibiotic administration	Methicillin resistant *S. aureus* (MRSA)	Commercial anti-*S. aureus* suspension administered via fistula	2010 No recurrence without any antibiotic over 4 years
6	80; F	2008; 2010	Knee prosthesis infection unsuitable for surgery	*P. aeruginosa*	Commercial broad spectrum multi-bacteriophage suspension; Knee joint injection	2012 *P. aeruginosa* clearance, but appearance of *Enterococcus* sp.
7	61; F	1995/2005; 2010	Operated tongue cancer; Dental extraction, jaw fracture, osteo-synthesis and fistulised infection	*S. aureus* (MRSA)	Commercial anti-*S. aureus* suspension administered peroperatively	2011 Complete cure
8	90; F	2009/2010; 2010	Femoral fracture under hip prosthesis; Drained hematoma and antibiotherapy-infection	*S. aureus* (MRSA)	Commercial anti-*S. aureus* suspension administered peroperatively by flooding the infection site and via catheter in the 10 days following the operation	2011 Complete cure, rapid recovery without recurrence after 1 year with retention of the hip prosthesis and osteosynthesis material *in situ*
9	20; M	2012; 2012	Chronic Ulcerative Colitis with liver complications. Severe weight loss (54 kg down from 80 kg). Poor digestion of food.	*E. coli*, *Proteus* spp. *S. aureus* (Urine) *S. aureus* (skin) *E. coli*, *Proteus vulgaris*, *Proteus mirabilis* (stool)	Treatment in Tbilisi (Georgia) with 2 commercially available phage suspensions plus special customised phage suspension. Probiotics, enzymes and Camelyn immune stimulant also given. Treatment lasted 1 month.	2012 Healing with sterilisation of urine, reduction of *E. coli* and *P. vulgaris* growth from high (10^8) to low ($<10^2$) in stool. Weight gain to 72 kg by end of treatment. Digestion improved but still poor
10	72; F	2009; 2013	Left knee prosthesis infection	*Staphylococcus* sp.	Commercial anti-*S. aureus* suspension administered peroperatively by flooding the infection site	2013 Initial partial disinfection with closure of several fistula followed by stabilisation

Table 1. *Cont.*

N	Age; Sex	Symptom Onset; PT Start	Clinical Symptoms	Bacteria	Phage Therapy	Outcome
11	84; M	1943/2012; 2013	Osteomyelitis of the left tibia; Fistula next to the wound	*S. aureus* (MRSA)	Initial phage therapy treatment in Tbilisi via fistula with temporary improvement, followed by surgical follow up intervention in France in 2013; Commercial anti-*S. aureus* suspension administered peroperatively by flooding the infection site	2013 Complete cure
12	58; F	2000; 2013	Acoustic neuroma with nosocomial infection of the ENT and ophthalmic regions	*S. aureus*	Treatment in Tbilisi with locally produced phage suspensions administered locally and orally	2013 Complete cure allowing an ophthalmic intervention of the retina that had been delayed for several years
13	68; F	1973; 2015	Operated left tibia fracture, followed by re-opened bone infection 2013: Travel to Phage Therapy Center (Tbilisi)	*S. aureus*	Surgery, phage therapy with commercial staphylococcal phage suspension, and antibiotherapy	2016 Disappearance of *S. aureus* replaced by *P. aeruginosa* & *Streptococcus constellatus*, followed by complete cure without recurrence
14	84; M	2006 & 2015; 2016	Prostate adenectomy with chronic urinary infection and bacteraemia	Extended-spectrum beta-lacatamase *E. coli* (ESBL)	Anti-*E. coli* phage suspension administered per os and rectally	2018 Complete cure
15	86; M	2016; 2018	Recurring prostatitis with bacteraemia	*P. aeruginosa*	Commercial multi-phage suspension administered orally and rectally	2018 Complete cure with disappearance of any urinary infection for the first time in 2 years

The infectious sites were predominantly osteoarticular (9/15), but also included two cases that involved the prostate and other four various infections (two ENT, one abdominal, and one GI tract). The predominantly targeted bacterial species was *S. aureus* (12/15). More rarely, *P. aeruginosa* (three instances) and two instances of *E. coli* were the causative pathogens or were present in polymicrobial infections. Most often, this was a mono-microbial infection (13/15). Suspensions of bacteriophages were mainly from commercial sources (Microgen in Russia and the Eliava Institute in Georgia). In the absence of commercially available preparations, two cases were treated with personalized bacteriophage suspensions.

This small series of cases calls for some remarks. We found that the local application of bacteriophages is completely safe, and no accidents or incidents have been reported. We have also observed highly satisfactory results, and often with rapid improvement. In fact, 12/15 cases resulted in a complete recovery (a secondary problematic pathogen emerged in one case, only a stable condition was achieved for one patient, and one case of a GI infection improved after phage treatment, but for which the condition was not fully resolved). The administration of bacteriophages had always been accompanied by antibiotic therapy with the aim of obtaining a possible synergy. Note that these were chronic cases which had exhausted the usual therapeutic resources, and whose clinical condition was worrying with a poorly functional prognosis. The focus was not to try to experiment or optimize phage therapy, but instead to treat patients with all available resources.

The pathologies that have been treated are varied. In our small case study, bone infections were the most frequent and generally evolved favorably within a few weeks. If there were fistulas, they disappeared, and bone consolidation was observed both clinically and functionally and was confirmed by imaging. Bacterial pathogens became quickly undetectable by microbiology after phage therapy began. After a follow-up for some patients of over 10 years, no relapse has been observed, and it is possible to conclude that patients were completely healed. In two cases where amputation of the lower limb was being considered, this option was avoided. Treatments that prevented the ablation of prosthetic material were also clinically satisfying.

To emphasize the treatment of two prostatitis cases, which constitute the most recent that we have taken care of, infections were caused by *E. coli* in one case and *P. aeruginosa* in the other. They affected elderly people who had been undergoing antibiotic therapy for several months. Concomitant oral and rectal administration over two consecutive days in one case and over three days for the other quickly resolved the recurrent infectious problem.

6. Recent Knowledge to be Taken into Consideration for Phage Therapy

The number of phage-related in vitro and in vivo studies, combined with newer areas of research such as the microbiome, has never been greater and provides a wealth of knowledge to keep in consideration when approaching clinical application. The activity of phages against biofilms, their ability to block bacterial receptors, and their synergy with conventional antibiotics has important implications for clinical treatment. Beyond bacterial lysis, phages have also been shown to interact in different ways with the immune system of the patient and their overall microbial community. The role that phages play naturally in the microbiome ecosystem is only starting to be discovered. Awareness and incorporation of these aspects provides a greater understanding of phage therapy and its clinical utility.

Regarding antibacterial aspects, the most pertinent aspect of new knowledge is the exploration of phage-antibiotic synergy (PAS). Several recent studies both in vitro [7,8,119,120] and in vivo on numerous experimental animal models [121] have confirmed the potential of combined use by showing the synergy of specific bacteriophage–antibiotic combinations at sometimes sub-inhibitory doses [122,123]. This could be a function of reducing the development of bacterial clones resistant to traditional antibiotics, by separate killing mechanisms, or other additive functions.

In almost all compassionate use cases, phages have been used in conjunction with antibiotics. It was even shown that phage administration changed the antibiotic resistance profile during the treatment of an *A. baumanii* infection, which led to the inclusion of the antibiotic in the treatment regimen [21]. Additionally, it has been shown [124] that, to combat *S. aureus* infections, the therapeutic results can also be influenced by the sequence in which the therapeutic agents are administered: best results were obtained when phage therapy precedes antibiotic therapy. As interesting as this effect may be, methods for determining the best choice of phage(s) and antibiotic(s) are still lacking. Nevertheless, the reintroduction of phage therapy deserves to be approached with the idea that it could be not only an alternative, but also a complement, in circumstances where the diffusion of an antibiotic is weak, as is the case in bone tissue or in the presence of a biofilm for example [6].

The activity of phages against bacterial biofilms is yet another factor in support of phage therapy. The pathogenic role of biofilms appears fundamental in chronic infections, especially in the presence of foreign materials (i.e., prosthesis, catheter). The proteolytic enzymes of certain bacteriophages are capable of destroying polysaccharides in biofilms which allow bacteria to escape natural defenses and antibiotic treatments [125]. In addition to allowing the adhesion of bacteriophages on the bacterial surface, this action facilitates the diffusion of antibiotics. It should be noted that soluble degradation products of *S. aureus* biofilm components could have a deleterious role on osteoblasts [126] and thus limit the growth of bone callus, which would explain the rapid bone healing observed after bacteriophage treatment observed in the compassionate cases in Table 1.

The very interaction of phages with the surface of bacterial cells may itself have an additive effect for phage therapy. It has been shown that bacteriophages, by attaching themselves to the bacterial surface at particular sites, could block resistance mechanisms such as an efflux pump or impair the fitness or the virulence factor of a bacterium [127]. This would then make certain bacteria (i.e., *P. aeruginosa* or *K. pneumoniae*) more susceptible to traditional antibiotics and facilitate the healing of certain pathologies, such as endocarditis or vascular prosthesis infections.

Regarding the interaction with mammalian cells, facets that are directly linked to the bactericidal effects are further complemented by a larger understanding of phage interaction with human cells and physiology, particularly with the immune system. Studies indicate that, in addition to their well-known antibacterial action, bacteriophages have potent immunomodulatory properties. For some authors, the success of phage therapy, depending on the bacterial permissiveness of the phage, is related to the immunity of the subject. In particular, for Roach et al. [46], neutrophil–bacteriophage synergy demonstrated that it is essential for the cure of pneumonia. For Dabrowska [128], the impact on the immune system affects the final outcome of phage therapy. While antibody induction may play a role in eliminating bacteriophages, it has also been shown that they can induce cytokine production in mammalian immune cells.

In reference to phages and surrounding microbiota, bacteriophages are present in all micro-ecosystems found in nature, and their presence in human microbiota is becoming increasingly recognized. Human microbiomes are distinct for various anatomical niches of the body (digestive tract, vaginal cavity, mouth, airway, nares, skin, urine [129–135]) that house dense microbial communities containing not only bacteria, archaea and fungi, but also mainly viruses, of which bacteriophages are the majority and remain largely unexplored [135–137]. The notion of the microbiome must be borne in mind when a bacteriophage treatment is being considered [138]. Indeed, the introduction of a bacteriophage in a structured community is not without consequence, because it induces difficult-to-predict interactions that may facilitate or hinder the intended effect [139]. Interactions occur not only with the microbiome with which it comes in contact, but also eukaryotic tissue cells [140,141] and the immune system of the host organism, as mentioned above [142]. Consequently, a model consisting of this "ménage à trois" has been proposed and should be considered [143]. A good knowledge of these components could help improve treatment outcomes and should make phage therapy a more personalized therapy.

An interesting consideration of gastro-intestinal diseases is the interplay with surrounding gut microbiota. Indeed, many gastro-intestinal diseases are increasingly described as a dysbiosis in the microbial community rather than being caused by a discrete pathogen, and microbiome sequencing has been useful in revealing disease-associated microbial signatures [144,145]. Phages may be useful in restoring a proper balance, such as for Crohn's disease or ulcerative colitis, and a trial targeting Enteroaggregative *E. coli* (EAEC) has been initiated for Crohn's patients [146–148].

Currently in Western countries, *Clostridium difficile* is a major problem (regarding diarrhea and transmission within the community), against which conventional antibiotics are ineffective. Many authors in recent years have considered addressing this condition with bacteriophages [149–151], and a study has shown a strong adsorption of bacteriophages on human cells in vitro that would promote bacteriophage–bacterial interactions is important for treating such a condition [152]. Fecal microbiota transplantations (FMT) have been shown to be effective at treating *C. difficile*, and, more so, filtrates of FMT that are devoid of bacteria also retain therapeutic properties, which may be due to the presence or modulation of phages [153].

7. Conclusions

Noting a continuing increase in bacterial resistance to antibiotics and the scarcity of new antibiotic molecules, the World Health Organization declared in 2014 that a pre-antibiotic era was imminent [154] and that there was an urgent need worldwide to mobilize international cooperation. In view of the risk to public health, new strategies need to be considered without delay: phage therapy is one of the most successful options today, if not the most successful. The advancement of phage therapy will, however, require an entwinement of old and new, of science and medicine, of fundamental research and clinical application, that is unparalleled in other areas of medical research. A multidisciplinary approach is needed more than ever, bringing together microbiologists, ecologists, evolutionary biologists, infectious disease specialists, medical doctors, and public health professionals. The growing threat of antibiotic resistance is indeed a compelling motivation to include and evaluate as much historical, compassionate use, and pre-clinical information as possible to increase the likelihood of the effective implementation of phage therapy.

The limited knowledge of phages available when they were first used historically has been complemented by a wealth of scientific studies, and yet therapy remains largely as empiric today as it was then. Our own empirical compassionate experiences with phages have nevertheless resulted in good clinical outcomes and have led us to conclude that phage therapy has much to offer, particularly for osteoarticular infections. A clinical trial is now planned to treat osteoarticular infections as an extension of our empirical findings through compassionate treatment. Such observational evidence from individual treatments provides valuable information on how to refine treatment protocols and to guide effective clinical practice in the future.

The use of biological rather than chemical drugs, such as phages, is new and upsets conventional treatment paradigms. Moreover, this development is occurring in a more strictly regulated context than in the past, where therapeutic frameworks need to be navigated and financial support is lacking. If it is unlikely that phage therapy will ever replace antibiotic therapy, it would surely be best to combine available antimicrobial strategies to create an effective treatment, and more research is merited in this direction. In the interim of conclusive phage efficacy trials, compassionate use of phage therapy, in combination with appropriate antibiotics, should be continued to maximize positive treatment outcomes for patients suffering from antibiotic resistant or difficult-to-treat infections.

Funding: This research received no external funding.

Conflicts of Interest: The authors declare no conflict of interest.

References

1. D'Hérelle, F. Sur un microbe invisible antagoniste des bacilles dysentériques. *Acad. Sci. Paris* **1917**, *165*, 373–375.
2. D'Hérelle, F. *Le Bactériophage: Son Rôle dans L'immunité*; Masson et Cie: Paris, France, 1921.

3. Bruynoghe, R.; Maisin, J. Essais de thérapeutique au moyen du bactériophage. *CR Soc. Biol.* **1922**, *85*, 1120–1121.
4. WHO. Antimicrobial Resistance: Global Report on Surveillance. Available online: http://www.thehealthwell.info/node/763364 (accessed on 14 November 2018).
5. Wittebole, X.; Roock, S.; Opal, S.M. A historical overview of bacteriophage therapy as an alternative to antibiotics for the treatment of bacterial pathogens. *Virulence* **2013**, *5*, 226–235. [CrossRef] [PubMed]
6. Pires, D.; Melo, L.; Boas, V.D.; Sillankorva, S.; Azeredo, J. Phage therapy as an alternative or complementary strategy to prevent and control biofilm-related infections. *Curr. Opin. Microbiol.* **2017**, *39*, 48–56. [CrossRef] [PubMed]
7. Torres-Barceló, C.; Arias-Sánchez, F.I.; Vasse, M.; Ramsayer, J.; Kaltz, O.; Hochberg, M.E. A window of opportunity to control the bacterial pathogen *Pseudomonas aeruginosa* combining antibiotics and phages. *PLoS ONE* **2014**, *9*, e106628. [CrossRef] [PubMed]
8. Torres-Barceló, C.; Hochberg, M.E. Evolutionary rationale for phages as complements of antibiotics. *Trends Microbiol.* **2016**, *24*, 249–256. [CrossRef] [PubMed]
9. Jault, P.; Leclerc, T.; Jennes, S.; Pirnay, J.P.; Que, Y.-A.A.; Resch, G.; Rousseau, A.F.; Ravat, F.; Carsin, H.; Floch, R.; et al. Efficacy and tolerability of a cocktail of bacteriophages to treat burn wounds infected by *Pseudomonas aeruginosa* (PhagoBurn): A randomised, controlled, double-blind phase 1/2 trial. *Lancet Infect. Dis.* **2018**, *19*, 35–45. [CrossRef]
10. Leitner, L.; Sybesma, W.; Chanishvili, N.; Goderdzishvili, M.; Chkhotua, A.; Ujmajuridze, A.; Schneider, M.P.; Sartori, A.; Mehnert, U.; Bachmann, L.M.; et al. Bacteriophages for treating urinary tract infections in patients undergoing transurethral resection of the prostate: A randomized, placebo-controlled, double-blind clinical trial. *BMC Urol.* **2017**, *17*, 90. [CrossRef]
11. Sarker, S.; Sultana, S.; Reuteler, G.; Moine, D.; Descombes, P.; Charton, F.; Bourdin, G.; McCallin, S.; Ngom-Bru, C.; Neville, T.; et al. Oral phage therapy of acute bacterial diarrhea with two coliphage preparations: A randomized trial in children from bangladesh. *EBioMedicine* **2016**, *4*, 124–137. [CrossRef]
12. Wright, A.; Hawkins, C.; Anggård, E.; Harper, D. A controlled clinical trial of a therapeutic bacteriophage preparation in chronic otitis due to antibiotic-resistant *Pseudomonas aeruginosa*; a preliminary report of efficacy. *Clin. Otolaryngol. Allied Sci.* **2009**, *34*, 349–357. [CrossRef]
13. Ujmajuridze, A.; Chanishvili, N.; Goderdzishvili, M.; Leitner, L.; Mehnert, U.; Chkhotua, A.; Kessler, T.M.; Sybesma, W. Adapted bacteriophages for treating urinary tract infections. *Front. Microbiol.* **2018**, *9*, 1832. [CrossRef] [PubMed]
14. Pirnay, J.-P.; Verbeken, G.; Ceyssens, P.-J.; Huys, I.; de Vos, D.; Ameloot, C.; Fauconnier, A. The Magistral Phage. *Viruses* **2018**, *10*, 64. [CrossRef] [PubMed]
15. Sybesma, W.; Rohde, C.; Bardy, P.; Pirnay, J.-P.; Cooper, I.; Caplin, J.; Chanishvili, N.; Coffey, A.; de Vos, D.; Scholz, A.; et al. Silk route to the acceptance and re-implementation of bacteriophage therapy-Part II. *Antibiotics* **2018**, *7*, 2.
16. Górski, A.; Jończyk-Matysiak, E.; Łusiak-Szelachowska, M.; Międzybrodzki, R.; Weber-Dąbrowska, B.; Borysowski, J.; Letkiewicz, S.; Bagińska, N.; Sfanos, K.S. Phage therapy in prostatitis: Recent prospects. *Front. Microbiol.* **2018**, *9*, 1434. [CrossRef] [PubMed]
17. Weber-Dąbrowska, B.; Jończyk-Matysiak, E.; Żaczek, M.; Łobocka, M.; Łusiak-Szelachowska, M.; Górski, A. Bacteriophage procurement for therapeutic purposes. *Front. Microbiol.* **2016**, *7*, 1177. [CrossRef] [PubMed]
18. Weber-Dabrowska, B.; Mulczyk, M.; Górski, A. Bacteriophages as an efficient therapy for antibiotic-resistant septicemia in man. *Transplant. Proc.* **2003**, *35*, 1385–1386. [CrossRef]
19. Weber-Dabrowska, B.; Mulczyk, M.; Górski, A. Bacteriophage therapy for infections in cancer patients. *Clin. Appl. Immunol. Rev.* **2001**, *1*, 131–134. [CrossRef]
20. Weber-Dabrowska, B.; Mulczyk, M.; Górski, A. Bacteriophage therapy of bacterial infections: An update of our institute's experience. *Archivum Immunologiae Therapiae Experimentalis* **2000**, *48*, 547–551.
21. Schooley, R.T.; Biswas, B.; Gill, J.J.; Hernandez-Morales, A.; Lancaster, J.; Lessor, L.; Barr, J.J.; Reed, S.L.; Rohwer, F.; Benler, S.; et al. Development and use of personalized bacteriophage-based therapeutic cocktails to treat a patient with a disseminated resistant *Acinetobacter baumannii* infection. *Antimicrob. Agents Chemother.* **2017**, *61*, AAC-00954. [CrossRef]

22. LaVergne, S.; Hamilton, T.; Biswas, B.; Kumaraswamy, M.; Schooley, R.; Wooten, D. Phage therapy for a multidrug-resistant *Acinetobacter baumannii* craniectomy site infection. *Open Forum Infect. Dis.* **2018**, *5*, ofy064. [CrossRef]
23. Pirnay, J.-P.P.; de Vos, D.; Verbeken, G.; Merabishvili, M.; Chanishvili, N.; Vaneechoutte, M.; Zizi, M.; Laire, G.; Lavigne, R.; Huys, I.; et al. The phage therapy paradigm: Prêt-à-porter or sur-mesure? *Pharm. Res.* **2011**, *28*, 934–937. [CrossRef] [PubMed]
24. Phage Directory. Available online: https://phage.directory/ (accessed on 14 November 2018).
25. Sillankorva, S. Isolation of bacteriophages for clinically relevant bacteria. *Methods Mol. Biol.* **2018**, *1693*, 23–30. [PubMed]
26. EudraLex Good Manufacturing Practice (GMP) Guidelines. European Commission. 2010. Available online: https://ec.europa.eu/health/documents/eudralex/vol-4_en (accessed on 14 November 2018).
27. Verbeken, G.; Pirnay, J.-P.; de Vos, D.; Jennes, S.; Zizi, M.; Lavigne, R.; Casteels, M.; Huys, I. Optimizing the European Regulatory Framework for Sustainable Bacteriophage Therapy in Human Medicine. *Archivum Immunologiae Therapiae Experimentalis* **2012**, *60*, 161–172. [CrossRef] [PubMed]
28. Verbeken, G.; Pirnay, J.-P.; Lavigne, R.; Jennes, S.; de Vos, D.; Casteels, M.; Huys, I. Call for a dedicated european legal framework for bacteriophage therapy. *Archivum Immunologiae Therapiae Experimentalis* **2014**, *62*, 117–129. [CrossRef] [PubMed]
29. Pirnay, J.-P.P.; Blasdel, B.G.; Bretaudeau, L.; Buckling, A.; Chanishvili, N.; Clark, J.R.; Corte-Real, S.; Debarbieux, L.; Dublanchet, A.; de Vos, D.; et al. Quality and safety requirements for sustainable phage therapy products. *Pharm. Res.* **2015**, *32*, 2173–2179. [CrossRef] [PubMed]
30. Pirnay, J.-P.; Merabishvili, M.; Raemdonck, H.; de Vos, D.; Verbeken, G. Bacteriophage production in compliance with regulatory requirements. In Bacteriophage Therapy. *Methods Mol. Biol.* **2018**, *1693*, 233–252. [PubMed]
31. Fauconnier, A. Regulating phage therapy. *EMBO Rep.* **2017**, *18*, 198–200. [CrossRef]
32. Furfaro, L.L.; Payne, M.S.; Chang, B.J. Bacteriophage therapy: Clinical trials and regulatory hurdles. *Front. Cell. Infect. Microbiol.* **2018**, *8*, 376. [CrossRef]
33. Kutter, E.M.; Gvasalia, G.; Alavidze, Z.; Brewster, E. Phage Therapy. In *Biotherapy—History, Principles and Practice*; CRC Press: Boca Raton, FL, USA, 2013; pp. 191–231.
34. Fauconnier, A. Bacteriophage Therapy. *Methods Mol. Biol.* **2018**, *1693*, 253–268.
35. Skurnik, M.; Pajunen, M.; Kiljunen, S. Biotechnological challenges of phage therapy. *Biotechnol. Lett.* **2007**, *29*, 995–1003. [CrossRef]
36. Vandenheuvel, D.; Lavigne, R.; Brussow, H. Bacteriophage therapy: Advances in formulation strategies and human clinical trials. *Annu. Rev. Virol.* **2015**, *2*, 599–618. [CrossRef] [PubMed]
37. Rhoads, D.D.; Wolcott, R.D.; Kuskowski, M.A.; Wolcott, B.M.; Ward, L.S.; Sulakvelidze, A. Bacteriophage therapy of venous leg ulcers in humans: Results of a phase I safety trial. *J. Wound Care* **2009**, *18*, 237–243. [CrossRef] [PubMed]
38. Kutter, E.; Borysowski, J.; Miedzybrodzki, R.; Gorski, A.; Weber-Dąbrowska, B.; Kutateladze, M.; Alavidze, Z.; Goderdzishvili, M.; Adamia, R. Clinical Phage Therapy. In *Phage Therapy: Current Research and Applications*; Borysowski, J., Miedzybrodzki, R., Gorski, A., Eds.; Nova Science Publishers: Norfolk, UK, 2014; Chapter 11; pp. 257–288.
39. Speck, P.; Smithyman, A. Safety and efficacy of phage therapy via the intravenous route. *FEMS Microbiol. Lett.* **2016**, *363*. [CrossRef] [PubMed]
40. Międzybrodzki, R.; Borysowski, J.; Weber-Dabrowska, B.; Fortuna, W.; Letkiewicz, S.; Szufnarowski, K.; Pawełczyk, Z.; Rogoz, P.; Kłak, M.; Wojtasik, E.B.; et al. Clinical aspects of phage therapy. *Adv. Virus Res.* **2012**, *83*, 73–121. [PubMed]
41. Vandenheuvel, D.; Singh, A.; Vandersteegen, K.; Klumpp, J.; Lavigne, R.; Van den Mooter, G. Feasibility of spray drying bacteriophages into respirable powders to combat pulmonary bacterial infections. *Eur. J. Pharm. Biopharm.* **2013**, *84*, 578–582. [CrossRef] [PubMed]
42. Matinkhoo, S.; Lynch, K.; Dennis, J.J.; Finlay, W.H.; Vehring, R. Spray-dried respirable powders containing bacteriophages for the treatment of pulmonary infections. *J. Pharm. Sci.* **2011**, *100*, 5197–5205. [CrossRef] [PubMed]

43. Golshahi, L.; Lynch, K.H.; Dennis, J.J.; Finlay, W.H. In vitro lung delivery of bacteriophages KS4-M and PhiKZ using dry powder inhalers for treatment of *Burkholderia cepacia* complex and *Pseudomonas aeruginosa* infections in cystic fibrosis. *J. Appl. Microbiol.* **2011**, *110*, 106–117. [CrossRef]
44. Morello, E.; Saussereau, E.; Maura, D.; Huerre, M.; Touqui, L.; Debarbieux, L. Pulmonary bacteriophage therapy on *Pseudomonas aeruginosa* cystic fibrosis strains: First steps towards treatment and prevention. *PLoS ONE* **2011**, *6*, e16963. [CrossRef]
45. Abedon, S.T. Phage therapy of pulmonary infections. *Bacteriophage* **2015**, *5*, e1020260. [CrossRef]
46. Roach, D.R.; Leung, C.; Henry, M.; Morello, E.; Singh, D.; Santo, J.P.; Weitz, J.S.; Debarbieux, L. Synergy between the host immune system and bacteriophage is essential for successful phage therapy against an acute respiratory pathogen. *Cell Host Microbe* **2017**, *22*, 38–47. [CrossRef]
47. Semler, D.D.; Lynch, K.H.; Dennis, J.J. The promise of bacteriophage therapy for *Burkholderia cepacia* complex respiratory infections. *Front. Cell Infect. Microbiol.* **2012**, *1*, 27. [CrossRef] [PubMed]
48. Hoe, S.; Semler, D.D.; Goudie, A.D.; Lynch, K.H.; Matinkhoo, S.; Finlay, W.H.; Dennis, J.J.; Vehring, R. Respirable bacteriophages for the treatment of bacterial lung infections. *J. Aerosol Med. Pulm. Drug Deliv.* **2013**, *26*, 317–335. [CrossRef]
49. Abedon, S.T. Phage therapy: Eco-physiological pharmacology. *Scientifica* **2014**, *2014*, 581639. [CrossRef]
50. Parracho, H.M.; Burrowes, B.H.; Enright, M.C.; McConville, M.L.; Harper, D.R. The role of regulated clinical trials in the development of bacteriophage therapeutics. *J. Mol. Genet. Med.* **2012**, *6*, 279–286. [CrossRef]
51. McCallin, S.; Brüssow, H. Clinical Trials of Bacteriophage Therapeutics. In *Bacteriophages*; Harper, D., Abedon, S., Burrowes, B., McConville, M., Eds.; Springer: Cham, Switzerland, 2017; pp. 1–29.
52. Leszczynski, P.; Weber-Dabrowska, B.; Kohutnicka, M.; Luczak, M.; Gorecki, A.; Gorski, A. Successful eradication of methicillin-resistant *Staphylococcus aureus* (MRSA) intestinal carrier status in a healthcare worker-case report. *Folia Microbiol.* **2006**, *51*, 236–238. [CrossRef]
53. Marcuk, L.M.; Nikiforov, V.N.; Scerbak, J.F.; Levitov, T.A.; Kotljarova, R.I.; Naumgina, M.S.; Davydov, S.U.; Monsur, K.A.; Rahman, M.A.; Latif, M.A.; et al. Clinical studies of the use of bacteriophage in the treatment of cholera. *Bull. World Health Organ.* **1971**, *45*, 77–83. [PubMed]
54. Goodridge, L.D. Bacteriophages for managing *Shigella* in various clinical and non-clinical settings. *Bacteriophage* **2013**, *3*, e25098. [CrossRef]
55. Zhvania, P.; Hoyle, N.; Nadareishvili, L.; Nizharadze, D.; Kutateladze, M. Phage Therapy in a 16-Year-Old boy with Netherton Syndrome. *Front. Med.* **2017**, *4*, 94. [CrossRef]
56. Jonczyk-Matysiak, E.; Weber-Dabrowska, B.; Zaczek, M.; Miedzybrodzki, R.; Letkiewicz, S.; Lusiak-Szelchowska, M.; Gorski, A. Prospects of phage application in the treatment of acne caused by *Propionibacterium acnes*. *Front Microbiol.* **2017**, *8*, 164. [CrossRef]
57. Crutchfield, E.D.; Stout, B.F. Treatment of staphylococcic infections of the skin by the bacteriophage. *Arch. Derm. Syphilol.* **1930**, *22*, 1010–1021. [CrossRef]
58. Morozova, V.V.; Kozlova, Y.N.; Ganichev, D.A.; Tikunova, N.V. Bacteriophage treatment of infected diabetic foot ulcers. In *Bacteriophage Therapy: From Lab to Clinical Practice*; Azeredo, J., Sillankorva, S., Eds.; Humana Press: Clifton, NJ, USA; Springer: New York, NY, USA, 2018; Chapter 13; pp. 151–158.
59. Fish, R.; Kutter, E.; Wheat, G.; Blasdel, B.; Kutateladze, M.; Kuhl, S. Bacteriophage treatment of intransigent diabetic toe ulcers: A case series. *J. Wound Care* **2016**, *25* (Suppl. 7), S27–S33. [CrossRef]
60. Rose, T.; Verbeken, G.; de Vos, D.; Merabishvili, M.; Vaneechoutte, M.; Lavigne, R.; Jennes, S.; Zizi, M.; Pirnay, J.-P.P. Experimental phage therapy of burn wound infection: Difficult first steps. *Int. J. Burns Trauma* **2014**, *4*, 66–73.
61. Jennes, S.; Merabishvili, M.; Soentjens, P.; Pang, K.; Rose, T.; Keersebilck, E.; Soete, O.; François, P.-M.; Teodorescu, S.; Verween, G.; et al. Use of bacteriophages in the treatment of colistin-only-sensitive *Pseudomonas aeruginosa* septicaemia in a patient with acute kidney injury-a case report. *Crit. Care* **2017**, *21*, 129. [CrossRef]
62. Duplessis, C. Refractory *Pseudomonas* bacteremia in a 2-Year-Old sterilized by bacteriophage therapy. *J. Pediatr. Infect. Dis. Soc.* **2018**, *7*, 253–256. Available online: https://academic.oup.com/jpids/article/7/3/253/4004747 (accessed on 14 November 2018). [CrossRef]
63. Grimont, P.A.; Grimont, F.; Lacut, J.Y.; Issanchou, A.M.; Aubertin, J. Traitement d'une endocardite à *Serratia* par les bactériophages. *Nouvelle Presse Médicale* **1978**, *7*, 2251.

64. MacNeal, W.J.; Frisbee, F.C.; Blevins, A. Bacteriophage therapy of staphylococcic septic obstruction of cavernous sinus: II. Report of cases. *Arch. Ophthalmol.* **1943**, *29*, 341–368. [CrossRef]
65. MacNeal, W.J. The use of bacteriophages in wound infections and in bacteremias. *Am. J. Med. Sci.* **1932**, *184*, 805. [CrossRef]
66. Raiga, A. Septicémie à staphylocoque guérie par une inoculation intra-veineuse de bactériophage. *Bulletin et Mémoire de la Société des Chirurgiens de Paris* **1931**, *23*, 441–447.
67. Stroj, L.; Weber-Dabrowska, B.; Partyka, K.; Mulczyk, M.; Wojcik, M. Successful treatment with bacteriophage in purulent cerebrospinal meningitis in a newborn. *Neurologia i Neurochirurgia Polska* **1999**, *33*, 693–698.
68. Merril, C.R.; Biswas, B.; Carlton, R.; Jensen, N.C.; Creed, G.J.; Zullo, S.; Adhya, S. Long-circulating bacteriophage as antibacterial agents. *Proc. Natl. Acad. Sci. USA* **1996**, *93*, 3188–3192. [CrossRef]
69. Schless, R.A. *Staphylococcus aureus* meningitis: Treatment with specific bacteriophage. *Am. J. Dis. Child.* **1932**, *44*, 813–822. [CrossRef]
70. MacNeal, W.J.; Frisbee, F.C.; Blevins, A. Recoveries of staphylococcic meningitis following bacteriophage therapy. *Arch. Otolaryngol.* **1943**, *37*, 507–525. [CrossRef]
71. Martin, P. Méningite posttraumatique à pyocyaniques traitée par un bactériophage adapté intrarachidien. *Acta Chir. Belg.* **1959**, *58*, 85–90. [PubMed]
72. Sedallian, P.; Bertoye, A.; Gauthier, J.; Muller, M.; Courtieu, A.L. Méningite purulente à colibacilles traitée par un bactériophage adapté intrarachidien. *Lyon Med.* **1958**, *66*, 509–512.
73. Beckerich, A.; Hauduroy, P. Le traitement des infections urinaires à colibacille par le bactériophage de d'Hérelle. *Bull. Med.* **1923**, *37*, 273.
74. Schultz, E.W. Bacteriophage as a therapeutic agent in genito-urinary infections: Part I. *Calif. West. Med.* **1932**, *36*, 33–37.
75. Schultz, E.W. Bacteriophage as a therapeutic agent in genito-urinary infections: Part II. *Calif. West. Med.* **1932**, *36*, 91–96.
76. Wehrbein, H.; Nerb, L. Bacteriophage in the treatment of urinary infections: With an appendix on the technique of phage preparation. *Am. J. Surg.* **1935**, *29*, 48–53. [CrossRef]
77. Ujmajuridze, A.; Jvania, G.; Chanishvili, N.; Goderdzishvili, M.; Sybesma, W.; Managadze, L.; Chkhotua, A.; Kessler, T. Phage therapy for the treatment for urinary tract infection: Results of in-vitro screenings and in-vivo application using commercially available bacteriophage cocktails. *Eur. Urol. Suppl.* **2016**, *15*, e265. [CrossRef]
78. Khawaldeh, A.; Morales, S.; Dillon, B.; Alavidze, Z.; Ginn, A.; Thomas, L.; Chapman, S.; Dublanchet, A.; Smithyman, A.; Iredell, J. Bacteriophage therapy for refractory *Pseudomonas aeruginosa* urinary tract infection. *J. Med. Microbiol.* **2011**, *60*, 1697–1700. [CrossRef]
79. Melo, L.D.; Veiga, P.; Cerca, N.; Kropinski, A.M.; Almeida, C.; Azeredo, J.; Sillankorva, S. Development of a phage cocktail to control *Proteus mirabilis* catheter-associated urinary tract infections. *Front. Microbiol.* **2016**, *7*, 1024. [CrossRef] [PubMed]
80. Nzakizwanayo, J.; Hanin, A.; Alves, D.R.; McCutcheon, B.; Dedi, C.; Salvage, J.; Knox, K.; Stewart, B.; Metcalfe, A.; Clark, J.; et al. Bacteriophage can prevent encrustation and blockage of urinary catheters by *Proteus mirabilis*. *Antimicrob. Agents Chemother.* **2015**, *60*, 1530–1536. [CrossRef] [PubMed]
81. Letkiewicz, S.; Miedzybrodzki, R.; Fortuna, W.; Weber-Dabrowska, B.; Górski, A. Eradication of *Enterococcus faecalis* by phage therapy in chronic bacterial prostatitis-case report. *Folia Microbiol.* **2009**, *54*, 457–461. [CrossRef] [PubMed]
82. Letkiewicz, S.; Międzybrodzki, R.; Kłak, M.; Jończyk, E.; Weber-Dąbrowska, B.; Górski, A. The perspectives of the application of phage therapy in chronic bacterial prostatitis. *FEMS Immunol. Med. Microbiol.* **2010**, *60*, 99–112. [CrossRef] [PubMed]
83. Furr, C.L.; Lehman, S.M.; Morales, S.P.; Rosas, F.X.; Gaidamaka, A.; Bilinsky, I.P.; Grint, P.C.; Schooley, R.T.; Aslam, S. Bacteriophage treatment of multidrug-resistant *Pseudomonas aeruginosa* pneumonia in a cystic fibrosis patient. In 41st European Cystic Fibrosis Conference. *J. Cystic Fibrosis* **2018**, *17* (Suppl. 3), S1–S150. [CrossRef]
84. Kvachadze, L.; Balarjishvili, N.; Meskhi, T.; Tevdoradze, E.; Skhirtladze, N.; Pataridze, T.; Adamia, R.; Topuria, T.; Kutter, E.; Rohde, C.; et al. Evaluation of lytic activity of staphylococcal bacteriophage Sb-1 against freshly isolated clinical pathogens. *Microb. Biotechnol.* **2011**, *4*, 643–650. [CrossRef]

85. Pherecydes Pneumophage. Available online: https://www.pherecydes-pharma.com/pneumophage.html (accessed on 14 November 2018).
86. Zhang, G.; Zhao, Y.; Paramasivan, S.; Richter, K.; Morales, S.; Wormald, P.; Vreugde, S. Bacteriophage effectively kills multidrug resistant *Staphylococcus aureus* clinical isolates from chronic rhinosinusitis patients. *Int. Forum Allergy Rhinol.* **2018**, *8*, 406–414. [CrossRef]
87. Fong, S.A.; Drilling, A.; Morales, S.; Cornet, M.E.; Woodworth, B.A.; Fokkens, W.J.; Psaltis, A.J.; Vreugde, S.; Wormald, P.-J. Activity of bacteriophages in removing biofilms of *Pseudomonas aeruginosa* isolates from chronic rhinosinusitis patients. *Front. Cell. Infect. Microbiol.* **2017**, *7*, 418. [CrossRef]
88. Drilling, A.J.; Ooi, M.L.; Miljkovic, D.; James, C.; Speck, P.; Vreugde, S.; Clark, J.; Wormald, P.J. Long-term safety of topical bacteriophage application to the frontal sinus region. *Front. Cell. Infect. Microbiol.* **2017**, *7*, 49. [CrossRef]
89. Drilling, A.; Morales, S.; Jardeleza, C.; Vreugde, S.; Speck, P.; Wormald, P.J. Bacteriophage reduces biofilm of *Staphylococcus aureus* ex vivo isolates from chronic rhinosinusitis patients. *Am. J. Rhinol. Allergy* **2014**, *28*, 3–11. [CrossRef]
90. Town, A.E.; Frisbee, F.C. Bacteriophage in ophthalmology. *Arch. Ophthalmol.* **1932**, *8*, 683–689. [CrossRef]
91. Fadlallah, A.; Chelala, E.; Legeais, J.-M.M. Corneal infection therapy with topical bacteriophage administration. *Open Ophthalmol. J.* **2015**, *9*, 167–168. [CrossRef] [PubMed]
92. Górski, A.; Targonska, M.; Borysowski, J.; Weber-Dabrowska, B. The potential of phage therapy in bacterial infections of the eye. *Ophthalmologica* **2009**, *223*, 162–165. [CrossRef] [PubMed]
93. Shlezinger, M.; Houri-Haddad, Y.; Coppenhagen-Glazer, S.; Resch, G.; Que, Y.A.; Beyth, S.; Dorfman, E.; Hazan, R.; Beyth, N. Phage therapy: A new horizon in the antibacterial treatment of oral pathogens. *Curr. Top. Med. Chem.* **2017**, *17*, 1199–1211. [CrossRef] [PubMed]
94. Khalifa, L.; Shlezinger, M.; Beyth, S.; Houri-Haddad, Y.; Coppenhagen-Glazer, S.; Beyth, N.; Hazan, R. Phage therapy against *Enterococcus faecalis* in dental root canals. *J. Oral Microbiol.* **2016**, *8*, 32157. [CrossRef] [PubMed]
95. Ly, M.; Abeles, S.R.; Boehm, T.K.; Robles-Sikisaka, R.; Naidu, M.; Santiago-Rodriguez, T.; Pride, D.T. Altered oral viral ecology in association with periodontal disease. *mBio* **2014**, *5*, e01133-14. [CrossRef] [PubMed]
96. Rossitto, M.; Fiscarelli, E.V.; Rosati, P. Challenges and promises for planning future clinical research into bacteriophage therapy against Pseudomonas aeruginosa in cystic fibrosis. An argumentative review. *Front. Microbiol.* **2018**, *9*, 775. [CrossRef]
97. Matsuzaki, S.; Uchiyama, J.; Takemura-Uchiyama, I.; Ujihara, T.; Daibata, M. Isolation of bacteriophages for fastidious bacteria. In *Bacteriophage Therapy: From Lab to Clinical Practice, Humana Press ed.*; Azeredo, J., Sillankorva, S., Eds.; Springer: New York, NY, USA, 2018; Chapter 1; pp. 3–10.
98. Wan, X.Q.; Li, H.M.; Bai, Y. Advances in phage therapy of Helicobacter pylori infection. *World Chin. J. Digestol.* **2009**, *17*, 3623–3626. [CrossRef]
99. Shan, J.; Teulieres, L.; Clockie, M. Is there a place for bacteriophages in diagnosis and treatment of Lyme Disease? In Proceedings of the Lymes Disease Action Conference, International Lyme And Associated Diseases Society (ILADS), Philadelphia, PA, USA, 3–6 November 2016.
100. Zakowska, D.; Bartoszcze, M.; Niemcewicz, M.; Bielawska-Drozd, A.; Knap, J.; Cieslik, P.; Chomiczewski, K.; Kocik, J. Bacillus anthracis infections—New possibilities of treatment. *Ann. Agric. Environ. Med. AAEM* **2015**, *22*, 202–207. [CrossRef]
101. Filippov, A.A.; Kirill, V.S.; Mikeljon, P.N. Bacteriophages against biothreat bacteria: Diagnostic, environmental and therapeutic applications. *J. Bioterrorism Biodefense* **2013**, *S3*, 010. [CrossRef]
102. Kremers, H.M.; Nwojo, M.E.; Ransom, J.E.; Wood-Wentz, C.M.; Melton, L.J., III; Huddleston, P.M., III. Trends in the epidemiology of osteomyelitis: A population-based study, 1969 to 2009. *J. Bone Jt. Surg.* **2015**, *97*, 837–845. [CrossRef] [PubMed]
103. Grammatico-Guillon, L.; Baron, S.; Gettner, S.; Lecuyer, A.-I.; Gaborit, C.; Rosset, P.; Rusch, E.; Bernard, L. Surveillance hospitalière des infections ostéo-articulaires en France: Analyse des données médico-administratives, PMSI 2008. *Bull. Epidémiol. Hosp.* **2013**, *4–5*, 39–44.
104. Albee, F.H. The treatment of osteomyelitis by bacteriophage. *J. Bone Jt. Surg.* **1933**, *15*, 58–66.
105. Kutateladze, M. Experience of the Eliava Institute in bacteriophage therapy. *Virol. Sin.* **2015**, *30*, 80–81. [CrossRef] [PubMed]

106. Slopek, S.; Weber Dabrowska, B.; Dabrowski, M.; Kucharewicz-Krukowska, A. Results of bacteriophage treatment of suppurative bacterial infections in the years 1981–1986. *Archivum Immunologiae et Therapiae Experimentalis* **1987**, *37*, 369–383.
107. Chanishvili, N.; Sharp, R. Bacteriophage therapy: Experience from the Eliva Institute, Georgia. *Microbiol. Aust.* **2008**, *20*, 96–101.
108. Raiga, A. Considérations Générales sur L'ostéomyélite Aiguë et son Traitement par le Bactériophage de d'Hérelle. In Proceedings of the 52ème Congrès Français de Chirurgie, Paris, France, 4 October 1949.
109. Lang, G.; Kher, P.; Mathevon, H.; Clavert, J.M.; Sejourne, P.; Pointu, J. Bactériophages et chirurgie orthopédique—A propos de sept cas. *Rev. Chir. Orthop. Reparatrice Appar. Mot.* **1979**, *65*, 33–37.
110. Chanishvili, N. Phage Therapy—History from Twort and d'Herelle through soviet experience to current approaches. *Adv. Virus Res.* **2012**, *82*, 3–40.
111. Ferry, T.; Boucher, F.; Fevre, C.; Perpoint, T.; Chateau, J.; Petitjean, C.; Josse, J.; Chidiac, C.; L'hostis, G.; Leboucher, G.; et al. Innovations for the treatment of a complex bone and joint infection due to XDR *Pseudomonas aeruginosa* including local application of a selected cocktail of bacteriophages. *J. Antimicrob. Chemother.* **2018**, *73*, 2901–2903. [CrossRef]
112. Yilmaz, C.; Colak, M.; Yilmaz, B.C.; Ersoz, G.; Kutateladze, M.; Gozlugol, M. Bacteriophage therapy in implant-related infections: An experimental study. *J. Bone Jt. Surg.* **2013**, *95*, 117–125. [CrossRef]
113. Kishor, C.; Mishra, R.R.; Saraf, S.K.; Kumar, M.; vastav, A.K.; Nath, G. Phage therapy of staphylococcal chronic osteomyelitis in experimental animal model. *Indian J. Med. Res.* **2016**, *143*, 87–94. [PubMed]
114. Kaur, S.; Harjai, K.; Chhibber, S. In vivo assessment of phage and linezolid based implant coatings for treatment of methicillin resistant *S. aureus* (MRSA) mediated orthopaedic device related infections. *PLoS ONE* **2016**, *11*, e0157626. [CrossRef] [PubMed]
115. Zhang, L.; Sun, L.; Wei, R.; Gao, Q.; He, T.; Xu, C.; Liu, X.; Wang, R. Study of intracellular *Staphylococcus aureus* control by virulent bacteriophage within MAC-T bovine mammary epithelial cells. *Antimicrob. Agents Chemother.* **2017**, *61*, e01990-16. [CrossRef] [PubMed]
116. Pherecydes Phosa. Available online: https://www.pherecydes-pharma.com/phosa-collaborative-project.html (accessed on 14 November 2018).
117. Vieu, J.-F.; Guillermet, F.; Minck, R.; Nicolle, P. Données actuelles sur les applications thérapeutiques des bactériophages. *Bull. Acad. Nat. Méd.* **1979**, *1*, 61–66.
118. Houssaye, C. Evaluation In vitro D'une Suspension de Bactériophages Anti-Staphylococcique à Usage Thérapeutique. Ph.D. Thesis, Université Pierre et Marie Curie, Paris, France, 2004.
119. Comeau, A.; Tétart, F.; Trojet, S.; Prere, M.; One, K.H. Phage-antibiotic synergy (PAS): β-lactam and quinolone antibiotics stimulate virulent phage growth. *PLoS ONE* **2007**, *2*, e799. [CrossRef] [PubMed]
120. Kamal, F.; Dennis, J.J. *Burkholderia cepacia* complex phage-antibiotic synergy (PAS): Antibiotics stimulate lytic phage activity. *Appl. Environ. Microbiol.* **2015**, *81*, 1132–1138. [CrossRef] [PubMed]
121. Oechslin, F.; Piccardi, P.; Mancini, S.; Gabard, J.; Moreillon, P.; Entenza, J.M.; Resch, G.; Que, Y.A. Synergistic interaction between phage therapy and antibiotics clears *Pseudomonas aeruginosa* infection in endocarditis and reduces virulence. *J. Infect. Dis.* **2017**, *215*, 703–712. [CrossRef]
122. Valério, N.; Oliveira, C.; Jesus, V.; Branco, T.; Pereira, C.; Moreirinha, C.; Almeida, A. Effects of single and combined use of bacteriophages and antibiotics to inactivate *Escherichia coli*. *Virus Res.* **2017**, *240*, 8–17. [CrossRef]
123. Knezevic, P.; Curcin, S.; Aleksic, V.; Petrusic, M.; Vlaski, L. Phage-antibiotic synergism: A possible approach to combatting *Pseudomonas aeruginosa*. *Res. Microbiol.* **2013**, *164*, 55–60. [CrossRef]
124. Kumaran, D.; Taha, M.; Yi, Q.; Ramirez-Arcos, S.; Diallo, J.-S.; Carli, A.; Abdelbary, H. Does treatment order matter? Investigating the ability of bacteriophage to augment antibiotic activity against *Staphylococcus aureus* biofilms. *Front. Microbiol.* **2018**, *9*, 127. [CrossRef]
125. Chan, B.K.; Abedon, S.T. Bacteriophages and their enzymes in biofilm control. *Curr. Pharm. Des.* **2015**, *21*, 85–99. [CrossRef] [PubMed]
126. Sanchez, C.J.; Ward, C.L.; Romano, D.R.; Hurtgen, B.J.; Hardy, S.K.; Woodbury, R.L.; Trevino, A.V.; Rathbone, C.R.; Wenke, J.C. *Staphylococcus aureus* biofilms decrease osteoblast viability, inhibits osteogenic differentiation, and increases bone resorption in vitro. *BMC Musculoskeletal Disord.* **2013**, *14*, 187–198. [CrossRef] [PubMed]

127. Tevdoradze, E.; Kvachadze, L.; Kutateladze, M.; Stewart, C.R. Bactericidal genes of staphylococcal bacteriophage Sb-1. *Curr. Microbiol.* **2014**, *68*, 204–210. [CrossRef] [PubMed]
128. Dąbrowska, K. Interaction of bacteriophages with the immune system: Induction of bacteriophage-specific antibodies. *Methods Mol. Biol.* **2018**, *1693*, 139–150. [PubMed]
129. Aragón, I.M.; Herrera-Imbroda, B.; Queipo-Ortuño, M.I.; Castillo, E.; Moral, J.S.; Gómez-Millán, J.; Yucel, G.; Lara, M.F. The urinary tract microbiome in health and disease. *Eur. Urol. Focus* **2018**, *4*, 128–138. [CrossRef] [PubMed]
130. Oh, J.; Byrd, A.L.; Park, M.; Program, N.; Kong, H.H.; Segre, J.A. Temporal Stability of the Human Skin Microbiome. *Cell* **2016**, *165*, 854–866. [CrossRef]
131. Miller-Ensminger, T.; Garretto, A.; Brenner, J.; Thomas-White, K.; Zambom, A.; Wolfe, A.J.; Putonti, C. Bacteriophages of the urinary microbiome. *J. Bacterial.* **2018**, *200*, JB-00738. [CrossRef]
132. Gilbert, J.A.; Blaser, M.J.; Caporaso, G.J.; Jansson, J.K.; Lynch, S.V.; Knight, R. Current understanding of the human microbiome. *Nat. Med.* **2018**, *24*, 392–400. [CrossRef]
133. Xian, P.; Xuedong, Z.; Xin, X.; Yuqing, L.; Yan, L.; Jiyao, L.; Xiaoquan, S.; Shi, H.; Jian, X.; Ga, L. The Oral Microbiome Bank of China. *Int. J. Oral Sci.* **2018**, *10*, 16. [CrossRef]
134. Kong, H.H. Skin microbiome: Genomics-based insights into the diversity and role of skin microbes. *Trends Mol. Med.* **2011**, *17*, 320–328. [CrossRef]
135. Chu, D.M.; Ma, J.; Prince, A.L.; Antony, K.M.; Seferovic, M.D.; Aagaard, K.M. Maturation of the infant microbiome community structure and function across multiple body sites and in relation to mode of delivery. *Nat. Med.* **2017**, *23*, 314–326. [CrossRef] [PubMed]
136. Ravel, J.; Brotman, R.M. Translating the vaginal microbiome: Gaps and challenges. *Genome Med.* **2016**, *8*, 35. [CrossRef] [PubMed]
137. Manrique, P.; Bolduc, B.; Walk, S.T.; van der Oost, J.; de Vos, W.M.; Young, M.J. Healthy human gut phageome. *Proc. Natl. Acad. Sci. USA* **2016**, *113*, 10400–10405. [CrossRef] [PubMed]
138. Zarate, S.; Taboada, B.; Yocupicio-Monroy, M.; Arias, C.F. Human virome. *Arch. Med. Res.* **2017**, *48*, 701–716. [CrossRef] [PubMed]
139. Microbiome: Phage community in the gut. *Nat. Rev. Microbiol.* **2016**, *14*, 605. [CrossRef] [PubMed]
140. Kashyap, P.C.; Chia, N.; Nelson, H.; Segal, E.; Elinav, E. Microbiome at the frontier of personalized medicine. *Mayo Clin. Proc.* **2017**, *92*, 1855–1864. [CrossRef] [PubMed]
141. Sunderland, K.S.; Yang, M.; Mao, C. Phage-enabled nanomedicine: From probes to therapeutics in precision medicine. *Angew. Chem. Int. Engl.* **2017**, *56*, 1964–1992. [CrossRef]
142. Palmela, C.; Chevarin, C.; Xu, Z.; Torres, J.; Sevrin, G.; Hirten, R.; Barnich, N.; Ng, S.C.; Colombel, J.-F.F. Adherent-invasive *Escherichia coli* in inflammatory bowel disease. *Gut* **2018**, *67*, 574–587. [CrossRef]
143. Mirzaei, M.K.; Maurice, C.F. Ménage à trois in the human gut: Interactions between host, bacteria and phages. *Nat. Rev. Microbiol.* **2017**, *15*, 397–408. [CrossRef]
144. Waldschmitt, N.; Metwaly, A.; Fischer, S.; Haller, D. Microbial signatures as a predictive tool in IBD-Pearls and pitfalls. *Inflamm. Bowel Dis.* **2018**, *24*, 1123–1132. [CrossRef]
145. Knights, D.; Parfrey, L.W.; Zaneveld, J.; Lozupone, C.; Knight, R. Human-associated microbial signatures: Examining their predictive value. *Cell Host Microbe* **2011**, *10*, 292–296. [CrossRef] [PubMed]
146. Carding, S.; Davis, N.; Hoyles, L. Review article: The human intestinal virome in health and disease. *Aliment. Pharmacol. Ther.* **2017**, *46*, 800–815. [CrossRef] [PubMed]
147. Galtier, M.; Sordi, L.; Sivignon, A.; de Vallée, A.; Maura, D.; Neut, C.; Rahmouni, O.; Wannerberger, K.; Darfeuille-Michaud, A.; Desreumaux, P.; et al. Bacteriophages targeting adherent invasive *Escherichia coli* strains as a promising new treatment for Crohn's disease. *J. Crohn's Colitis* **2017**, *11*, 840–847. [CrossRef] [PubMed]
148. López, R.; Burgos, M.J.; Gálvez, A.; Pulido, R. The human gastrointestinal tract and oral microbiota in inflammatory bowel disease: A state of the science review. *APMIS Acta Pathol. Microbiol. Immunol. Scand.* **2017**, *125*, 3–10. [CrossRef] [PubMed]
149. Nale, J.Y.; Spencer, J.; Hargreaves, K.R.; Buckley, A.M.; Trzepiński, P.; Douce, G.R.; Clokie, M.R. Bacteriophage Combinations Significantly Reduce *Clostridium dificile* Growth In vitro and Proliferation In vivo. *Antimicrob. Agents Chemother.* **2016**, *60*, 968–981. [CrossRef] [PubMed]

150. Rea, M.C.; Alemayehu, D.; Ross, R.; Hill, C. Gut solutions to a gut problem: Bacteriocins, probiotics and bacteriophage for control of *Clostridium difficile* infection. *J. Med. Microbiol.* **2013**, *62*, 1369–1378. [CrossRef] [PubMed]
151. Thanki, A.M.; Taylor-Joyce, G.; Dowah, A.; Nale, J.Y.; Malik, D.; Clokie, M.R.J. Unravelling the links between phage adsorption and successful infection in *Clostridium difficile*. *Viruses* **2018**, *10*, 411. [CrossRef]
152. Shan, J.; Ramachandran, A.; Thanki, A.M.; Vukusic, F.B.; Barylski, J.; Clokie, M.R. Bacteriophages are more virulent to bacteria with human cells than they are in bacterial culture; insights from HT-29 cells. *Sci. Rep.* **2018**, *8*, 5091. [CrossRef]
153. Ott, S.J.; Waetzig, G.H.; Rehman, A.; Moltzau-Anderson, J.; Bharti, R.; Grasis, J.A.; Cassidy, L.; Tholey, A.; Fickenscher, H.; Seegert, D.; et al. Efficacy of sterile fecal filtrate transfer for treating patients with *Clostridium difficile* infection. *Gastroenterology* **2017**, *152*, 799–811. [CrossRef]
154. WHO. *Global Action Plan on Antimicrobial Resistance*; World Health Organization: Geneva, Switzerland, 2015; Available online: http://www.who.int/antimicrobial-resistance/global-action-plan/en/ (accessed on 14 November 2018).

Brief Report

Production of Bacteriophages by Listeria Cells Entrapped in Organic Polymers

Brigitte Roy [1,2,3], Cécile Philippe [1,3], Martin J. Loessner [4], Jacques Goulet [2] and Sylvain Moineau [1,3,*]

1 Département de Biochimie, de Microbiologie et de Bio-Informatique, Faculté des Sciences et de Génie, Université Laval, Québec, QC G1V OA6, Canada; brigitte.roy.12@ulaval.ca (B.R.); cecile.philippe.1@ulaval.ca (C.P.)
2 Département des Sciences des Aliments, Faculté des Sciences de L'agriculture et de L'alimentation, Université Laval, Québec, QC G1V OA6, Canada; Jacques.Goulet@fsaa.ulaval.ca
3 Félix d'Hérelle Reference Center for Bacterial Viruses and GREB, Faculté de Médecine Dentaire, Université Laval, Québec, QC G1V OA6, Canada
4 ETH Zurich, Institute of Food, Nutrition and Health, Schmelzbergstrasse, 7CH-8092 Zürich, Switzerland; martin.loessner@ethz.ch
* Correspondence: Sylvain.Moineau@bcm.ulaval.ca; Tel.: +1-418-656-3712

Received: 13 May 2018; Accepted: 8 June 2018; Published: 13 June 2018

Abstract: Applications for bacteriophages as antimicrobial agents are increasing. The industrial use of these bacterial viruses requires the production of large amounts of suitable strictly lytic phages, particularly for food and agricultural applications. This work describes a new approach for phage production. Phages H387 (*Siphoviridae*) and A511 (*Myoviridae*) were propagated separately using *Listeria ivanovii* host cells immobilised in alginate beads. The same batch of alginate beads could be used for four successive and efficient phage productions. This technique enables the production of large volumes of high-titer phage lysates in continuous or semi-continuous (fed-batch) cultures.

Keywords: *Listeria ivanovii*; bacteriophages; alginate; production; disinfection; phagodisinfection

1. Introduction

Listeria monocytogenes is responsible for fatal cases of listeriosis in humans via contaminated food products [1]. This bacterial species is ubiquitous in nature and can contaminate the food processing line at any critical point. The increasing resistance of these pathogens to disinfectants under certain conditions requires the use of higher concentrations of chemical products [2]. Furthermore, bacteria exposed to disinfectants may be more likely to develop antibiotic resistance [3,4]. Despite strict regulatory policies, the occurrence of *L. monocytogenes* still has detrimental consequences for the food industry.

The search for alternatives to overcome these challenges has rekindled interest for bacterial viruses (bacteriophages) in agriculture [5], aquaculture [6], food safety [7], and even in infectious diseases [8,9]. The use of strictly lytic (i.e., virulent) phages infecting *Listeria* as biosanitisers represents an ecological alternative that could reduce the use of chemical compounds and lower the concentrations of toxic residues in the environment [10]. Specific biodisinfectants consisting of suspensions of phages can provide a natural means to control pathogens in processed foods and on contact surfaces. For example, the virulent phage A511 has a very broad host range against several strains of *Listeria* spp. [11–13] and could be included in the formulation of this type of biodisinfectants.

Phage biocontrol of *L. monocytogenes* strains was first introduced in 2006 with the commercial product ListShield, which contained a cocktail of phages applicable to various foods. Another product is Phageguard Listex P100, which also aimed to reduce *L. monocytogenes* in a range of food products [14–18]. Moreover, it has been demonstrated that different virulent phages can reduce the *L. monocytogenes* population after adhesion to stainless steel or polypropylene surfaces, and a synergistic effect has been observed by combining phages with quaternary ammonium [19–21].

However, even if virulent phages have shown great potential for killing pathogenic or opportunistic food-borne bacteria [22], their production on a large scale often remains challenging. Phage production still involves traditional methods using tubes or Erlenmeyer flasks, or it is done in bioreactors as a batch process [23]. High phage titers can be obtained [24,25], but batch processes require significant manpower and non-operational periods of time that may be limiting [26].

Attempts have been made to overcome the disadvantages of the batch process with continuous phage production. Studies involving chemostats [25,27] have been conducted, as well as two-stage continuous processes or cellstat [25,28–30], which consists of culturing bacteria, separately, in the first stage to feed to a second stage when phages are produced. Although chemostat allows the cultivation of microorganisms at a physiological steady state [31], the bacterial culture may be less genetically stable, as mutations can occur [32]. Cellstat is recognised as a phage production system for strictly lytic phages [33] that avoids direct phage exposure and pressure but requires the use of two different bioreactors [34].

With the aim of reducing production time and costs, we investigated here a different phage production procedure that employs host bacteria entrapped in a porous gel matrix. Alginate gel was selected for the matrix because of its low cost and widespread use in a range of applications in medicine, pharmacy, biotechnology, and the food industry [35,36]. An alginate matrix with entrapped bacterial cells can be produced in a single-step process and has virtually no impact on the viability of the cells. Alginate can also form a gel in the presence of divalent cations, such as calcium, which are also often necessary as co-factors for phage multiplication [37,38].

Entrapped cells will still grow because nutrients diffuse through the gel matrix [39], and while microcolonies spread deeper in the beads, the bacterial density has been shown to be higher near to or at the surface of the beads [40–42]. Such a growth pattern leads to bacterial cell release in the medium by micro-fracture events in the matrix. Only bacterial cells released from the gel become infected and contribute to the propagation of virulent phages. Those cells remaining in the gel have been shown to be protected from the phages, as the bacterial viruses do not migrate into the beads because of their size [43–45]. Protein diffusion through the matrix is highly reduced when molecular weight is above 150 kDa [46].

Relevant advantages of using entrapped cells to produce strictly lytic phages are that the phage lysate can be easily recovered and the alginate beads can be reused for successive phage propagations. Multiple phage lytic cycles can also be favoured, because this protective system prevents the rapid decline of the phage-sensitive host population. This process also provides an opportunity to produce phages in continuous or semi-continuous (fed-batch) cultures. Taken together, the use of calcium alginate immobilised cells (in spheres or fibers) to produce phages is a bi-phasic technique that controls the bacterial population and preserves the integrity of the cells, as long as they remain entrapped in the matrix.

2. Materials and Methods

2.1. Bacteria, Phages, and Media

Listeria ivanovii WSLC 3009 and the broad-host-range virulent myovirus A511 [47] were obtained from the Institut für Mikrobiologie, ZIEL Institute for Food and Health, Technische Universität München (Germany). The siphovirus H387 [48,49] was obtained from the Félix d'Hérelle Reference Center for Bacterial Viruses (www.phage.ulaval.ca) of the Université Laval (Québec, Canada). Bacterial strains were grown in Trypticase soy broth (TSB) or plated on Trypticase soy agar (TSA) at 30 °C. Phage titration was done using the double-layer plating technique [50] on TSA. Phage stocks (>1 $\times$ 10^8 Plaque Forming Unit (PFU) mL^{-1}) were stored at 4 °C prior to use.

2.2. Alginate Gels and Cell Immobilisation by Entrapment

A 2–4% (*w*/*v*) aqueous solution of sodium alginate was prepared by suspending the polymer in distilled water. Solutions were sterilised by autoclaving (121 °C, 15 min). *L. ivanovii* cells were harvested by centrifugation (8000 rpm, 10 min) and resuspended in sterile TSB (3 mL). The cell suspensions were then mixed with sterile alginate [51]. Beads were formed by the dropwise addition of the alginate-cell mixtures into sterile $CaCl_2$ (200 mM) using a syringe and a 20 Gauge (G) needle. The cell-containing beads, 2 to 3 mm in diameter, were allowed to solidify for 1 to 2 h before $CaCl_2$ was replaced by fresh TSB containing 0.5 mM $CaCl_2$ to maintain the integrity of the alginate beads.

2.3. Morphology of Cells Immobilised in Beads

Alginate beads were observed by scanning electron microscopy (SEM) to visualise entrapped *Listeria* cells. The alginate beads were cut in half, and the specimens were fixed by immersion in glutaraldehyde (2.5% *v*/*v*) in 0.1 M sterile cacodylate buffer (pH 7.0) for 4 h. The samples were washed twice in 0.1 M sterile cacodylate for 20 min. Post-fixation was done in osmium tetroxide (2% *v*/*v*) in sterile cacodylate buffer for 30 min at 30 °C, and dehydration was completed using CO_2 in a critical point dryer (Model 3000 CPD, Bio-Rad, Mississauga, ON, Canada). The samples were mounted on stubs and covered with 15 nm of gold using a sputter coater (Emscope, Bio-Rad). A Nanolab LE 2100 (Vickers Instruments, Bausch and Lomb, Nepean, ON, Canada) scanning electron microscope operating at 15 hV was used to examine the bead surfaces.

2.4. Phage Adsorption

A set of alginate beads was made as described above but omitting the bacterial cells. Ten grams of pure alginate beads was transferred into TSB. Aliquots of phage suspensions (0.1 and 1 mL) were added and incubated at 30 °C for 12 h. The adsorption of phages on alginate beads was monitored by determining phage titers every 4 h. Two independent experiments were performed.

2.5. Biomass Concentration

To estimate the population of immobilised bacteria, 1 mL of alginate beads was dissolved in 9.0 mL of Na^+ citrate (50 mM), a sequestrant for Ca^{++}. The number of viable cells in the dissolved alginate gel was determined by direct plating on TSA for two independent experiments.

2.6. Phage Production

2.6.1. Free Cells

TSB (100 mL) was inoculated (5%) with an overnight culture of *L. ivanovii* 3009 from the Weihenstephan *Listeria* collection (WSLC) and grown to an optical density at 600 nm (OD_{600}) of 0.5–0.8. Phages were added at multiplicities of infection (MOIs) of 0.1 (1:10) and 1 (1:1), and the mixture was incubated for 16 h at 30 °C. Phage titers were measured every 4 h for two independent experiments.

2.6.2. Immobilised Cells Used for Single and Successive Phage Propagations

Beads containing entrapped microorganisms were transferred at least 2–4 times into prewarmed (30 °C) fresh TSB before phage production. Ten grams of beads containing *L. ivanovii* cells was added to 100 mL of TSB (OD_{600} of 0.5–0.8), and phage suspensions (at MOIs of 0.1 and 1) were added to the cultures. The flasks were incubated at 30 °C for 16 h, and phage titers were also determined as described above for two independent experiments. Between each successive production, the beads were stored overnight at 4 °C in sterile 2% (*w*/*v*) $CaCl_2$. The beads were then washed twice with sterile 2% $CaCl_2$ and reactivated as described above before each production.

2.7. Statistical Analysis

Mean and standard deviation values were calculated using Microsoft Excel (Microsoft, Redmond, WA, USA).

3. Results and Discussion

3.1. Morphological Observations

For the efficient use of alginate microbeads, morphological characteristics such as size and shape are important [52]. The produced alginate beads had proper sphericality and were typically 2 to 3 mm in size (Figure 1). Scanning electron micrographs of entrapped *Listeria* revealed no major changes in cell morphology (Figure 1). The mechanical constraints of the polymer did not seem to interfere with cell growth.

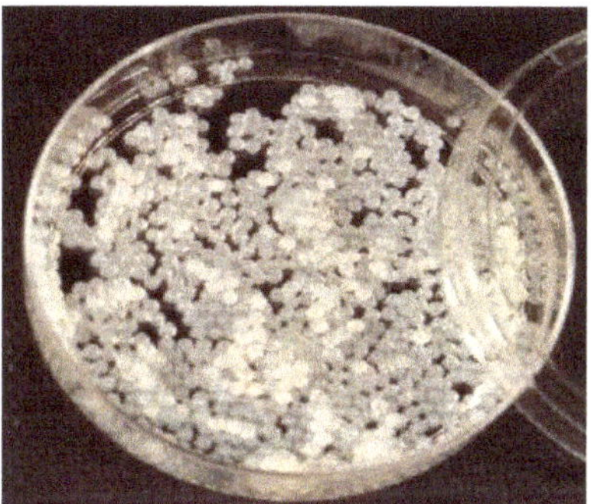

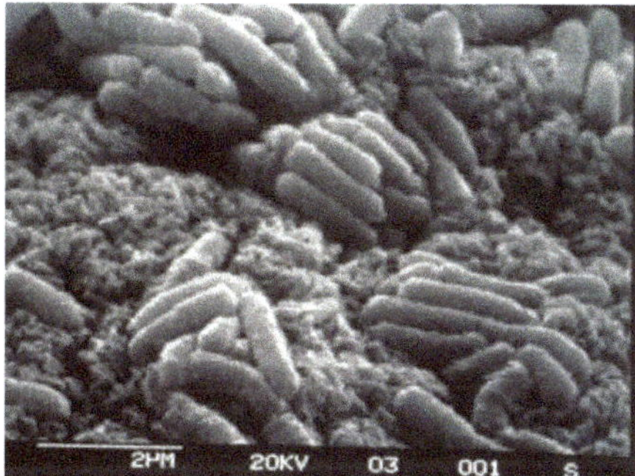

Figure 1. Observation of entrapped alginate bacteria. (**Left**) Visual appearance of alginate beads containing *Listeria ivanovii* WSLC 3009 (10^8 Colony Forming Unit (CFU) mL^{-1}) in a Petri dish. (**Right**) Scanning electron micrograph of *L. ivanovii* WSLC 3009 immobilised in alginate beads ($\times$10,000).

3.2. Phage Adsorption

Phage adsorption onto the gel matrix is a parameter that may impact the overall performance of the production system. The electrostatic adsorption of phages onto polymer beads could decrease the number and infectivity of phage particles in the medium. For this reason, the organic material selected for phage production should be tested for ionic attraction of viral particles. No decreases in the titers of phages A511 and H387 were observed in the medium after contact with the alginate beads. These results suggest that no major ionic interactions exist between the organic polymer and the phages.

3.3. Biomass Concentration

The concentration of entrapped cells in alginate beads has been studied for several types of bacteria [39,51,53,54]. Alginate is non-toxic to most living cells [55] and provides protection against external stresses such as temperature, pH, and toxic molecules. Figure 2 shows that three successive transfers (reactivations) of entrapped *L. ivanovii* cells in fresh TSB could raise the bacterial cell concentration inside the gel to almost 1×10^9 cells mL^{-1}, while five transfers increased the bacterial counts to almost 10^{10} cells mL^{-1}. Because the number of bacteria released into a medium is related to, among other parameters, the saturation level of the cells in the alginate structure, the yield of phage production will likely be influenced by the concentration of bacteria in the beads and at the bead surface.

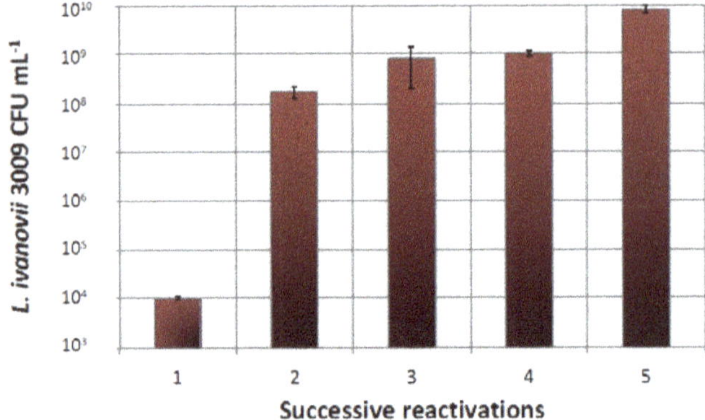

Figure 2. Transfers of *Listeria ivanovii* WSLC 3009 immobilised cells in fresh Trypticase soy broth (TSB) medium. Each reactivation was followed by an interval of 12 h. Bacterial concentrations were measured after 12 h of growth at 30 °C. Mean values were calculated from two independent experiments, and error bars correspond to standard deviations.

3.4. Phage Production

3.4.1. Free Cells

Phage productions in liquid medium were performed with both phages individually (Figure 3). All phage productions were characterized by a lag phase for the first 4 h, followed by a sharp increase in phage titers at 8 h. Maximal phage titers were close to 10^{10} PFU mL^{-1} of medium after 12 h. Very small variations in phage titers were observed at different MOIs. After 16 h, the titer of phage H387 decreased when using a 1:10 ratio. It is unclear at this time what caused this decrease in the phage titer, but it could have been due to phage adsorption to cell debris.

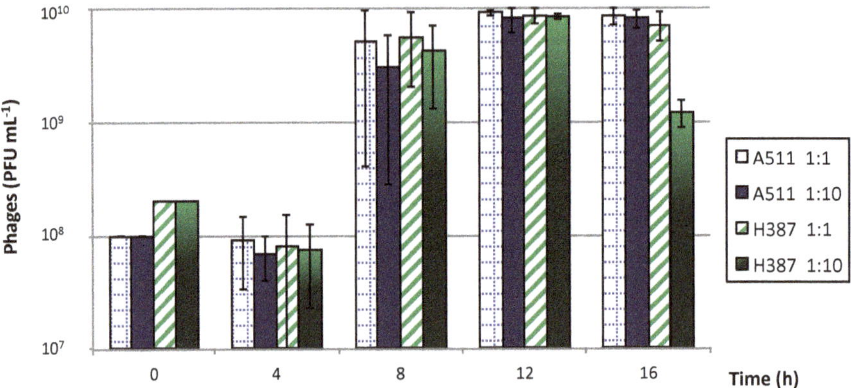

Figure 3. Production of phages A511 and H387 on *Listeria ivanovii* WSLC 3009 in liquid medium, using multiplicities of infection (MOIs) of 1 and 0.1. Phage counts were measured every 4 h. Mean values were calculated from two independent experiments, and error bars correspond to standard deviations.

3.4.2. Immobilised Cells Used for Single and Successive Phage Propagations

Microorganisms immobilised in polymers produce concentrated host bacteria that can be more easily and rapidly manipulated than free cells. Phage production using gel-entrapped host cells was compared to that of free cells in the same culture medium and under the same growing conditions. The highest production of virulent *Listeria* phages A511 and H387 was obtained after 12 h using a MOI of 1 (Figure 4). The maximum phage titers achieved using entrapped cells were slightly lower than for free cells. Some phage productions reached their maximum titer after 8 h of incubation, which was faster than for the free cells.

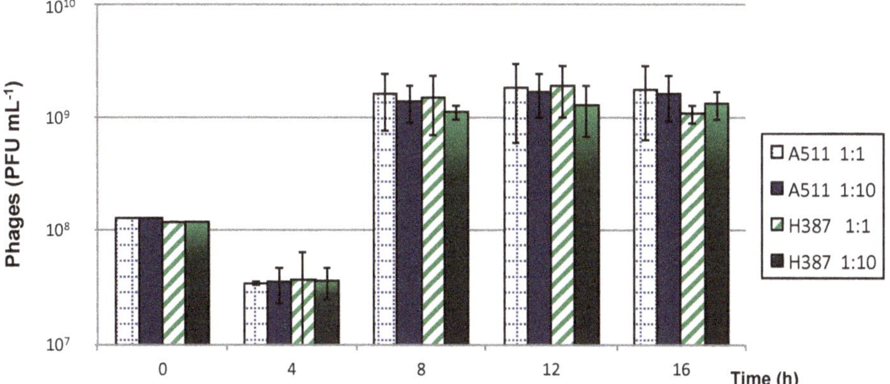

Figure 4. Production of phages A511 and H387 on *Listeria ivanovii* WSLC 3009 immobilised in alginate beads, using 1 and 0.1 multiplicities of infection (MOIs). Phage titers were measured every 4 h. Mean values were calculated from two independent experiments, and error bars correspond to standard deviations.

Two advantages of using gel-entrapped cells to produce virulent phages are that phage particles can be easily recovered by draining the culture medium (followed by centrifugation and filtration) and that phage propagation can be immediately resumed or pursued after a short or prolonged storage period. The same alginate beads with immobilised *L. ivanovii* cells were used for four successive phage productions. In all cases, phage titers were maintained at over 10^9 PFU mL^{-1} after the four productions (Figure 5). In general, the final phage titers of the virulent phage A511 were higher than for phage H387.

It has been shown previously that phages infecting some lactic acid bacteria cannot penetrate calcium alginate gels [43,44]. Because *Listeria* phages are the same size as dairy phages and even larger in the case of A511 [56,57], bacterial cells are well protected from phage infection as long as they remain entrapped in the gel. It is likely that this physical constraint, protecting the integrity of the bacterial population, allows the gel beads to be reused for successive phage production in new media. This advantage cannot be provided by free-cell amplification. Only small molecules can diffuse through the alginate matrix [43]. *L. ivanovii* cells entrapped in alginate beads are, therefore, protected against phages as well as against contamination by other bacteria. The production of phages after infection of the host bacteria likely only takes place on the beads' surface and in the medium after the cells have been released from the matrix. In fact, we noticed that the structure of the alginate gel was rather loose and easily broken up at the periphery of the beads, where cells usually most actively grow. These cells were likely released into the medium and infected by phages.

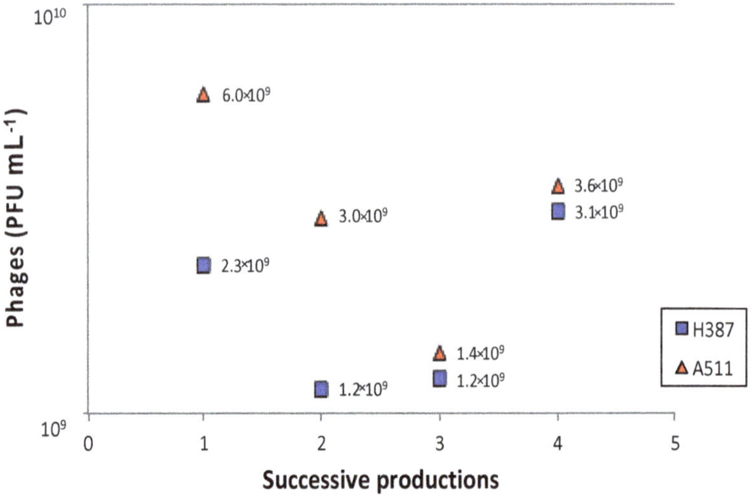

Figure 5. Successive productions of the two phages, A511 and H387, on *Listeria ivanovii* WSLC 3009 using a multiplicity of infection (MOI) of 1. Aliquots were withdrawn after 10 h of incubation. Mean values were calculated from two independent experiments.

While the process described here still requires optimisation, the gel entrapment of cells to produce specific phages offers the potential for the large-scale and rapid production of phages. Successive phage productions have shown that entrapped cells can be reused for at least four propagation cycles. Although the viral titer of lysate produced with entrapped cells was nearly 10-fold reduced compared to that of free-cell production, successive productions with the same beads should be globally seen as an interesting advantage. Continuous phage production using entrapped cells could be enhanced and applied to a large variety of phages.

Author Contributions: B.R. and J.G. conceived and designed the experiments; B.R. performed the experiments; B.R., M.J.L., J.G., and S.M. analysed the data; B.R., C.P., and S.M. wrote the paper.

Acknowledgments: We thank Alpha-Biotech Inc. and DAAD (Deutscher Akademischer Austauschdienst) for providing financial assistance for this study. We also thank Claude Champagne for discussion. S.M. holds the Canada Research Chair in Bacteriophages.

Conflicts of Interest: The authors declare no conflict of interest.

References

1. Radoshevich, L.; Cossart, P. *Listeria monocytogenes*: Towards a complete picture of its physiology and pathogenesis. *Nat. Rev. Microbiol.* **2018**, *16*, 32–46. [CrossRef] [PubMed]
2. Frank, F.J.; Koffe, A.R. Surface-adherent growth of *Listeria monocytogenes* is associated with increased resistance to surfactant sanitizers and heat. *J. Food Protect.* **1989**, *53*, 550–554. [CrossRef]
3. Directorate-General for Health and Consumers. *Assessment of the Antibiotic Resistance Effects of Biocides, 28th Plenary*; Commission Européenne, Scientific Committee on Emerging and Newly Identified Health Risks: Bruxelles, Belgium, 2009.
4. Pearce, H.; Messager, S.; Maillard, J.-Y. Effect of biocides commonly used in the hospital environment on the transfer of antibiotic-resistance genes in *Staphylococcus aureus*. *J. Hosp. Infect.* **1999**, *43*, 101–108. [CrossRef] [PubMed]
5. Buttimer, C.; McAuliffe, O.; Ross, R.P.; Hill, C.; O'Mahony, J.; Coffey, A. Bacteriophages and bacterial plant diseases. *Front. Microbiol.* **2017**, *8*, 34. [CrossRef] [PubMed]

6. Stalin, N.; Srinivasan, P. Efficacy of potential phage cocktails against *Vibrio harveyi* and closely related *Vibrio* species isolated from shrimp aquaculture environment in the south east coast of India. *Vet. Microbiol.* **2017**, *207*, 83–96. [CrossRef] [PubMed]
7. Bai, J.; Kim, Y.-T.; Ryu, S.; Lee, J.-H. Biocontrol and rapid detection of food-borne pathogens using bacteriophages and endolysins. *Front. Microbiol.* **2016**, *7*, 474. [CrossRef] [PubMed]
8. Abedon, S.T. Phage therapy of pulmonary infections. *Bacteriophage* **2015**, *5*, e1020260. [CrossRef] [PubMed]
9. Hagens, S.; Loessner, M.J. Phages of *Listeria* offer novel tools for diagnostics and biocontrol. *Front. Microbiol.* **2014**, *5*, 159. [CrossRef] [PubMed]
10. Lin, D.M.; Koskella, B.; Lin, H.C. Phage therapy: An alternative to antibiotics in the age of multi-drug resistance. *World J. Gastrointest. Pharmacol. Ther.* **2017**, *8*, 162–173. [CrossRef] [PubMed]
11. Loessner, M.J.; Busse, M. Bacteriophage typing of *Listeria* species. *Appl. Environ. Microbiol.* **1990**, *56*, 1912–1918. [PubMed]
12. Guenther, S.; Huwyler, D.; Richard, S.; Loessner, M.J. Virulent bacteriophages for biocontrol of *Listeria monocytogenes* in ready-to-eat foods. *Appl. Environ. Microbiol.* **2009**, *75*, 93–100. [CrossRef] [PubMed]
13. Klumpp, J.; Dorscht, J.; Lurz, R.; Bielmann, R.; Wieland, M.; Zimmer, M.; Calendar, R.; Loessner, M.J. The terminally redundant, non-permuted genome of *Listeria* bacteriophage A511: A model for the SPO1-like myoviruses of Gram-positive bacteria. *J. Bacteriol.* **2008**, *190*, 5753–5765. [CrossRef] [PubMed]
14. Silva, E.N.G.; Figueiredo, A.C.L.; Miranda, F.A.; de Castro Almeida, R.C. Control of *Listeria monocytogenes* growth in soft cheeses by bacteriophage P100. *Braz. J. Microbiol.* **2014**, *45*, 11–16. [CrossRef] [PubMed]
15. Chaitiemwong, N.; Hazeleger, W.C.; Beumer, R.R. Inactivation of *Listeria monocytogenes* by disinfectants and bacteriophages in suspension and stainless-steel carrier tests. *J. Food Protect.* **2014**, *77*, 2012–2020. [CrossRef] [PubMed]
16. Chibeu, A.; Agius, L.; Gao, A.; Sabour, P.M.; Kropinski, A.M.; Balamurugan, S. Efficacy of bacteriophage LISTEX™P100 combined with chemical antimicrobials in reducing *Listeria monocytogenes* in cooked turkey and roast beef. *Int. J. Food Microbiol.* **2013**, *167*, 208–214. [CrossRef] [PubMed]
17. Miguéis, S.; Saraiva, C.; Esteves, A. Efficacy of LISTEX P100 at different concentrations for reduction of *Listeria monocytogenes* inoculated in sashimi. *J. Food Protect.* **2017**, *12*, 2094–2098. [CrossRef] [PubMed]
18. Greer, G.G. Homologous bacteriophage control of *Pseudomonas* growth on beef spoilage. *J. Food Protect.* **1986**, *49*, 104–109. [CrossRef]
19. Ganegama Arachchi, G.J.; Cridge, A.G.; Dias-Wanigasekera, B.M.; Cruz, C.D.; McIntyre, L.; Liu, R.; Flint, S.H.; Mutukumira, A.N. Effectiveness of phages in the decontamination of *Listeria monocytogenes* adhered to clean stainless steel, stainless steel coated with fish protein, and as a biofilm. *J. Ind. Microbiol. Biotechnol.* **2013**, *10*, 1105–1116. [CrossRef] [PubMed]
20. Gray, J.A.; Chandry, P.S.; Kaur, M.; Kocharunchitt, C.; Bowman, J.P.; Fox, E.M. Novel biocontrol methods for *Listeria monocytogenes* biofilms in food production facilities. *Front. Microbiol.* **2018**, *9*, 605. [CrossRef] [PubMed]
21. Roy, B.; Ackermann, H.-W.; Pandian, S.; Picard, G.; Goulet, J. Biological inactivation of adhering *Listeria monocytogenes* by listeriaphages and a quaternary ammonium compound. *Appl. Environ. Microbiol.* **1993**, *59*, 2914–2917. [PubMed]
22. Lee, S.; Kim, M.G.; Lee, H.S.; Heo, S.; Kwon, M.; Kim, G. Isolation and characterization of *Listeria* phages for control of growth of *Listeria monocytogenes* in milk. *Korean J. Food Sci. Anim. Resour.* **2017**, *37*, 320–328. [CrossRef] [PubMed]
23. Agboluaje, M.; Sauvageau, D. Bacteriophage production in bioreactors. *Methods Mol. Biol.* **2018**, *1692*, 173–193.
24. Sargeant, K.; Yeo, K.G.; Lethbridge, J.H.; Shooter, K.V. Production of bacteriophage T7. *Appl. Microbiol.* **1968**, *16*, 1483–1488. [PubMed]
25. Chen, B.Y.; Lim, H.C. Bioreactor studies on temperature induction of the Qmutant of bacteriophage λ in *Escherichia coli*. *J. Biotechnol.* **1996**, *51*, 1–20. [CrossRef]
26. Sauvageau, D.; Cooper, D.G. Two-stage, self-cycling process for the production of bacteriophages. *Microb. Cell Fact.* **2010**, *9*, 81. [CrossRef] [PubMed]
27. Los, M.; Wegrzyn, G.; Neubauer, P. A role for bacteriophage T4 rI gene function in the control of phage development during pseudolysogeny and in slow growing host cells. *Res. Microbiol.* **2003**, *154*, 547–552. [CrossRef]

28. Park, S.H.; Park, T.H. Analysis of two-stage continuous operation of *Escherichia coli* containing bacteriophage λ vector. *Bioprocess Eng.* **2000**, *23*, 187–190. [CrossRef]
29. Chen, X.A.; Cen, P.L. A novel three-stage process for continuous production of penicillin G acylase by a temperature-sensitive expression system of *Bacillus subtilis* phage phi105. *Chem. Biochem. Q.* **2005**, *19*, 367–372.
30. Oh, J.S.; Cho, D.; Park, T.H. Two-stage continuous operation of recombinant *Escherichia coli* using the bacteriophage λ Q-vector. *Bioprocess Biosyst. Eng.* **2005**, *28*, 1–7. [CrossRef] [PubMed]
31. Ziv, N.; Brandt, N.J.; Gresham, D. The use of chemostats in microbial systems biology. *J. Vis. Exp.* **2013**, *80*, 50168. [CrossRef] [PubMed]
32. Chao, L.; Levin, B.R.; Stewart, F.M. A complex community in a simple habitat: An experimental study with bacteria and phage. *Ecology* **1977**, *58*, 369–378. [CrossRef]
33. Schwienhorst, A.; Lindemann, B.F.; Eigen, M. Growth kinetics of a bacteriophage in continuous culture. *Biotechnol. Bioeng.* **1996**, *50*, 217–221. [CrossRef]
34. Nabergoj, D.; Kuzmić, N.; Drakslar, B.; Podgornik, A. Effect of dilution rate on productivity of continuous bacteriophage production in cellstat. *Appl. Microbiol. Biotechnol.* **2018**, *102*, 3649–3661. [CrossRef] [PubMed]
35. Diviès, C.; Prévost, H.; Cavin, J.F. Les bactéries immobilisées dans l'industrie laitière. *Process. Mag.* **1989**, *1041*, 28–33.
36. Gacesa, P. Alginates. *Carbohydr. Polym.* **1988**, *53*, 161–182. [CrossRef]
37. Ackermann, H.-W.; Dubow, M.S. Phage multiplication. In *Viruses of Prokaryotes*; CRC Press, Inc.: Boca Raton, FL, USA, 1987; Volume 1, pp. 49–85.
38. Mahony, J.; Tremblay, D.M.; Labrie, S.J.; Moineau, S.; van Sinderen, D. Investigating the requirement for calcium during lactococcal phage infection. *Int. J. Food Microbiol.* **2015**, *201*, 47–51. [CrossRef] [PubMed]
39. Gosmann, B.; Rehm, H.J. Influence of growth behaviour and physiology of alginate-entrapped microorganisms on the oxygen consumption. *Appl. Microbiol. Biotechnol.* **1988**, *29*, 554–559. [CrossRef]
40. Cachon, R.; Catté, M.; Nommé, R.; Prévost, H.; Diviès, C. Kinetic behaviour of *Lactococcus lactis* ssp. *lactis* bv. *diacetylactis* immobilized in calcium alginate gel beads. *Process Biochem.* **1995**, *6*, 503–510.
41. Prévost, H.; Diviès, C.; Rousseau, E. Continuous yoghurt production with *Lactobacillus bulgaricus* and *Streptococcus thermophilus* entrapped in Ca-alginate. *Biotechnol. Lett.* **1985**, *4*, 247–252. [CrossRef]
42. Audet, P.; Paquin, C.; Lacroix, C. Immobilized growing lactic acid bacteria with κ-carrageenan—Locust bean gum gel. *Appl. Microbiol. Biotechnol.* **1988**, *1*, 11–18. [CrossRef]
43. Steenson, L.R.; Klaenhammer, T.R.; Swaisgood, H.E. Calcium alginate-immobilized cultures of lactic streptococci are protected from bacteriophages. *J. Dairy Sci.* **1987**, *70*, 1121–1127. [CrossRef]
44. Champagne, C.P.; Morin, N.; Couture, R.; Gagnon, C.; Jelen, P.; Lacroix, C. The potential of immobilized cell technology to produce freeze-dried, phage-protected cultures of *Lactococcus lactis*. *Food Res. Intern.* **1992**, *25*, 419–427. [CrossRef]
45. Champagne, C.P.; Moineau, S.; Lafleur, S.; Savard, T. The effect of bacteriophages on the acidification of a vegetable juice medium by microencapsulated *Lactobacillus plantarum*. *Food Microbiol.* **2017**, *63*, 28–34. [CrossRef] [PubMed]
46. Tanaka, H.; Matsumura, M.; Veliky, I.A. Diffusion characteristics of substrates in Ca alginate gel beads. *Biotechnol. Bioeng.* **1984**, *1*, 53–58. [CrossRef] [PubMed]
47. Zink, R.; Loessner, M.J. Classification of virulent and temperate bacteriophages of *Listeria* spp. on the basis of morphology and protein analysis. *Appl. Environ. Microbiol.* **1992**, *58*, 296–302. [PubMed]
48. Ackermann, H.-W.; Dubow, M.S. Description and identification of new phages. In *Viruses of Prokaryotes*; CRC Press, Inc.: Boca Raton, FL, USA, 1987; Volume 1, pp. 103–142.
49. Ortel, S.; Ackermann, H.-A. Morphologie von Neuen Listeriaphagen. *Zentralbl. Bakteriol. Mikrobiol. Hyg. Ser. A Med. Microbiol. Infect. Dis. Virol. Parasitol.* **1985**, *260*, 423–427.
50. Adams, M.H. Methods of study of bacterial viruses. In *Bacteriophages*; Interscience Publishers, Inc.: New York, NY, USA, 1959; pp. 443–457.
51. Boyaval, P.; Lebrun, A.; Goulet, J. Étude de l'immobilisation de *Lactobacillus helveticus* dans des billes d'alginate de calcium. *Le Lait* **1985**, *65*, 185–199. [CrossRef]
52. Börner, R.A.; Aliaga, M.T.A.; Mattiasson, B. Microcultivation of anaerobic bacteria single cells entrapped in alginate microbeads. *Biotechnol. Lett.* **2013**, *35*, 397–405. [CrossRef] [PubMed]

53. Kurosawa, H.; Matsumura, M.; Tanaka, H. Oxygen diffusivity in gel beads containing viable cells. *Biotechnol. Bioeng.* **1989**, *34*, 926–932. [CrossRef] [PubMed]
54. Chen, K.-C.; Huang, C.-T. Effects of the growth of *Trichosporon cutaneum* in calcium alginate gel beads upon bead structure and oxygen transfer characteristics. *Enzym. Microb. Technol.* **1988**, *10*, 284–292. [CrossRef]
55. Ogbonna, J.C.; Amano, Y.; Nakamura, K. Elucidation of optimum conditions for immobilization of viable cells by using calcium alginate. *J. Ferment. Bioeng.* **1989**, *67*, 92–96. [CrossRef]
56. Klumpp, J.; Loessner, M.J. *Listeria* phages: Genomes, evolution, and application. *Bacteriophage* **2013**, *3*, e26861. [CrossRef] [PubMed]
57. Ackermann, H.-W. Bacteriophage taxonomy. *Microbiol. Aust.* **2011**, *32*, 90–94.

Comment

Phage Therapy Faces Evolutionary Challenges

Clara Torres-Barceló

University of Reunion Island, UMR Plant populations and bio-agressors in tropical environment (PVBMT), Saint-Pierre 97410, Reunion, France; clara.torres@cirad.fr; Tel.: +262-262492727

Received: 18 May 2018; Accepted: 12 June 2018; Published: 12 June 2018

Abstract: Antibiotic resistance evolution in bacteria indicates that one of the challenges faced by phage therapy is that, sooner or later, bacteria will evolve resistance to phages. Evidently, this is the case of every known antimicrobial therapy, but here this is also part of a ubiquitous natural process of co-evolution between phages and bacteria. Fundamental evolutionary studies hold some clues that are crucial to limit the problematic process of bacterial resistance during phage applications. First, I discuss here the importance of defining evolutionary and ecological factors influencing bacterial resistance and phage counter-defense mechanisms. Then, I comment on the interest of determining the co-evolutionary dynamics between phages and bacteria that may allow for selecting the conditions that will increase the probability of therapeutic success. I go on to suggest the varied strategies that may ensure the long-term success of phage therapy, including analysis of internal phage parameters and personalized treatments. In practical terms, these types of approaches will define evolutionary criteria regarding how to develop, and when to apply, therapeutic phage cocktails. Integrating this perspective in antimicrobial treatments, such as phage therapy, is among the necessary steps to expand its use in the near future, and to ensure its durability and success.

Keywords: phage therapy; evolution; bacterial resistance; virulence

1. Introduction

Phage therapy is gradually becoming a reality in clinical, veterinary, and agricultural settings [1–3]. In order to avoid the past mistakes of chemical treatments, it is important to prevent bacterial resistance to phages where possible. A review and an expert comment in this same Special Issue advocate for a molecular and evolutionary combined basis in the selection of therapeutic phages [4,5]. Phage cocktails, in particular, are given special attention, and represent an excellent method to face bacterial genetic variability and prevent the evolution of resistance. In the past, a large host range was the main criterion to select different phages [6,7]. The newly published review suggests an expansion in phage choice to specifically include phages that target different bacterial receptors and phages with counter-defense abilities. Also considered are the phages' resistance to environmental factors (e.g., pH and temperature) and phages' viability, which could facilitate storage and production. In the expert opinion piece, the researchers suggested the use of pre-adapted, or "trained", phages to overcome the resistance capacity of bacteria, a method that has proven to be successful in the past. Both articles point out that continued exposure to phages may activate the immune system, increasing their elimination within the body. The experts bring their attention to the fact that overuse of phages could result in the evolutionary selection of resistance to phages in bacteria, which, in a similar light to antimicrobials, could be transferred to clinical situations via horizontal gene transfer. To avoid this, the authors consider the use of personalized medicine as a means to develop a successful and sustainable phage therapy strategy. Here, I present additional evolutionary concerns that can guide the design of therapeutic treatments with phages. Some concepts are open questions still in need of confirmation in the laboratory, clinic, or in vivo settings. Altogether, these are perspectives that may increase the benefits of any antimicrobial strategy.

2. Factors to Consider Regarding the Different Mechanisms of Resistance to Phages and Counter-Defense

The interaction between phages and bacteria stands as one of the fastest and more assorted evolutionary processes on Earth. Not only do bacteria and their parasites have short generation times, large population sizes, and high mutation rates, there are also bacteria and phages of all kinds and in all sorts of environments [8]. This co-evolution likely centers around resistance in bacteria and counter-resistance of phages. Among the resistance mechanisms active in bacteria facing phages are phenotypic shifts [9], point mutations in surface structures used by phages as receptors [10], or acquisition of CRISPR (Clustered Regularly Interspaced Short Palindromic Repeats) spacers [11]. Additionally, new and diverse bacterial defense mechanisms have been described lately [12]. Phages in turn can avoid bacterial defenses by, for instance, modifying their life cycle parameters (burst size, lysis time, etc.) [13], mutating receptor binding proteins [10], or recombining with other viruses [14].

Determining which resistance or counter-defense processes will take place at any given moment will depend on several factors, the understanding of which can help to predict the outcome of phage–bacteria interactions (Figure 1). First, there is the probability of occurrence of any resistance and counter-resistance mechanism, conditioned by mutation rate, and phage and bacterial population diversity. Besides internal microbial factors, parameters such as nutrient availability will determine the growth rate of bacteria (and their obligatory parasites), and thus the frequency of resistance generation. Second, there is the ecology of the infection environment, including the spatial structure. This is illustrated in the mammalian gut, where phages adhere to the mucus and attack invasive bacteria [15], whereas other body sites are less prone to phage–bacteria interaction [16]. Furthermore, the microbial communities present could both facilitate or impede phage therapeutic action. For example, their competitive interaction with the pathogenic strain will impose an additional fitness cost, which could enhance phage control, whereas a complex community can complicate the chance phages have to encounter their specific target. Third, a decisive parameter is the fitness cost imposed by each defense and counter-defense mechanism on phages and bacteria. For example, it has recently been shown that the CRISPR-Cas system as a resistance mechanism in *Pseudomonas aeruginosa* is more likely to be selected or maintained if the same host frequently faces the same phage [17]. In this example, the authors demonstrated that the absolute cost of receptor modification was lower than the inducible cost of adaptive immunity (provided by a CRISPR-Cas system), unless facing a static target. Thus, the presence of a phage defense mechanism (e.g., CRISPR-Cas) in a bacterium targeted for therapy does not always mean that it will be crucial for resistance to a therapeutic phage, especially if it is part of a diverse cocktail.

On the phage side, their small genome size may restrain their evolutionary capacities, favoring the selection of non-pleiotropic and less costly mutations. For example, modifications on genes coding for receptor binding proteins have been more frequently observed when compared to mutations in the phage polymerase, a core enzymatic activity [10,11]. Interestingly, a recent study proves that at least some phages are able to extend their genome size while exposed to bacteria in natural environments [11]. For therapeutic purposes, a bigger genome size and fewer overlapping genes in phage candidates may indicate potential for evolution and the capacity to overcome bacteria, although this idea remains to be tested. In conclusion, when exploring resistance/infectivity mechanisms in these microbes, one needs to consider the evolutionary forces favoring their selection (Figure 1). Also, the life-history of the pathogen and its past encounters with phages may determine the success of any therapeutic phages applied. Genomic tools are unveiling new and fascinating defense mechanisms, but without understanding their biological relevance, this knowledge is just a hint of its essence.

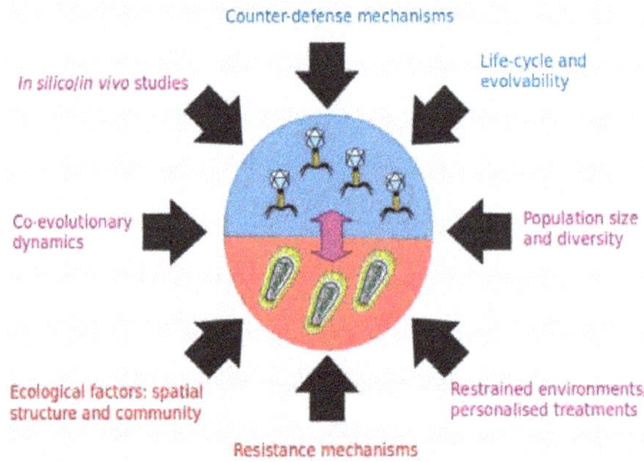

Figure 1. Different factors to consider in phage therapy applications in order to ensure the durability of the strategy against pathogenic bacteria. Color code associates the factors to therapeutic phages (blue), the targeted bacteria (red), or both microorganisms and the interaction between them (violet).

3. Co-Evolutionary Dynamics between Phages and Bacteria Influence the Therapeutic Outcome

Co-evolutionary dynamics between phages and bacteria can differ relative to time and genetic variability in the population of microbes, influencing the probability of resistance. Deeper understanding of these processes could point to the conditions that can ensure a successful treatment against pathogenic bacteria (Figure 1). It has been shown experimentally that an arms-race dynamics (ARD) type of evolutionary process can change to fluctuating selection dynamics (FSD), depending on the phase of the interaction [18]. This was observed in *P. fluorescens* and its phage phi2, a well-studied model of co-evolution, over a hundreds of generations (60 transfers) experiment. In the first stages of a bacteria-phage encounter (first 10 transfers), more types of resistance and infectivity alleles were available, which imposed lower fitness costs than in later transfers. Afterwards (between transfers 10 and 50), alleles of defense and attack were more limited, and their selection fluctuated depending on the most abundant genotypes of phages and bacteria in the environment, and likely imposed a higher fitness cost to both microbes. Co-evolutionary dynamics may be different for other phage-bacteria pairs but, in this particular case, it is plausible that an ARD type of interaction between the targeted bacteria and a "trained" therapeutic phage will be a very efficient strategy. In an FSD process, the probability of success is likely to be lower as the resistance/infectivity process is strongly frequency dependent. Conversely, the higher fitness costs confronted by bacteria during FSD may benefit therapeutic phages in the evolutionary race.

To recapitulate, in studying the evolution of phage resistance in bacteria, short- and long-term scales must be considered. Experimental evolution approaches encourage researchers to consider bacteria–phage interactions further than 24 h to account for the evolvability of both microorganisms [19]. Different effects have been observed experimentally when looking at short- or long-term co-evolving phages and bacteria. As stated before, evolutionary dynamics may differ over time, e.g., early arms-race *versus* late fluctuating dynamics [18]. It has also been shown that in combination therapies using phages and antibiotics, the control of pathogenic bacteria was more efficient in the first stages of the exposure (e.g., [20]). In practical terms, quantifying the capacity of the resistance of each strain, and performing co-evolutionary studies with candidate phages, could guide and improve phage therapy. Studies such as those just mentioned can direct efforts towards the characterization of the situations (e.g., timing, type of bacterial population, etc.) most likely to be controlled by phages.

Similarly, the genetic variability and population size (i.e., standing genetic variation) of the targeted bacteria will probably increase the diversity of the type of evolutionary dynamics and defense mechanisms selected [17]. In the course of a phage therapy treatment, input of new bacteria or conditions that favor a high microbial growth rate (e.g., immunocompromised patients, resource availability, etc.) may contribute to a higher probability of developing resistance to phages. In those conditions, a more varied phage cocktail and/or higher initial inoculum (MOI: multiplicity-of-infection) may be advised as a therapeutic strategy. Carrying on with the rationale of the expert comment published in this journal, in intensive veterinary or agricultural settings, the large population size of bacteria and the number of phages required for treatments will enhance the likelihood of resistance [5]. Notwithstanding, the extensive spread of phage resistance in bacteria (e.g., via horizontal gene transfer) in this type of natural framework remains to be decidedly proven. Smaller agricultural or veterinary frameworks and contained environments (e.g., plant nurseries, tool disinfection, etc.) are more adapted and likely to produce a successful outcome of the use of therapeutic phages. This is the reasoning behind, for example, the Listex and Biolyse preventive phage preparations against *Salmonella* and *Pectobacterium* pathogens currently used in packed products in the food industry [3,21]. Timely use of phages as a preventive treatment will target a smaller population of bacteria, since it is a more effective and less risky scenario. In conclusion, the epidemiology and ecology of bacteria and phages must certainly be integrated into disease management before any therapeutic incursion against rapidly evolving microbes (Figure 1).

4. Managing Disease: Towards Sustainability

Current knowledge suggests that the way antimicrobial treatments influence virulence parameters in bacteria are important to controlling infectious diseases in the longer term [22]. Increasingly, medical approaches include the evolutionary perspective, where the aim is not to eliminate pathogens completely, but to reduce or impair their population. This implies to apply a weaken selection pressure on the pathogens, or to target and dismantle virulence-specific mechanisms. In order to restrain disease, this approach often relies on the role of the host's immune system or the local microbiota.

In phage therapy, it is conceivable to alter or contain the virulence, and phage or antibiotic resistance emergence of bacterial pathogens. When selecting a therapeutic phage, the frequency of bacterial resistance induced by every particular phage is an advisable parameter to be examined. Several factors should be explored related to the facilitation of phage resistance in bacteria or the ability of phages to counteract it. Together, the phage genome size, mutation rate, and burst size could all be features determining the speed and potential adaptation of phages to bacteria. A large burst size increases the probability that phages contact target bacteria, the first step of infection. The amplification ability of phages could compensate the advantage of antibiotics at better diffusing in the body, and therefore reaching the site of infection. If phages can eliminate bacteria faster than they can replicate, a high burst size also results in a lower risk of selection for phage-resistant bacteria. At the same time, however, these phages imply a stronger selective pressure for bacteria, and could lead to faster or stronger resistance. An open question would then be trying to understand whether there is an evolutionary optimum of phage virulence that could constrain bacterial resistance evolution over longer periods of time.

The type of receptors desired for therapeutic phages is one of the most important parameters, and is frequently discussed (e.g., [4]). Are receptors such as lypopolysaccharides (LPS) more likely to be mutated (i.e., resistant) because of their intrinsic variability? Or are porins and pili structures easier to be modified, or their expression repressed in bacteria? It will likely depend on their environment and the fitness impact each mutation accrues. Interestingly, both LPS and pili are virulence factors that can be modified via selection, decreasing virulence in phage resistant bacteria. Several examples in humans, plants, and animals have demonstrated the role of phages in decreasing bacterial virulence [23–25]. Experimental evolution approaches may help elucidate which receptors favor phage treatment in the long term, i.e., enhance phages' adaptation. This is the case in a study from Betts and collaborators, where they find that *P. aeruginosa* phages targeting LPS receptors show ARD compared to phages targeting pili, which undertake FSD patterns [26]. Their interpretation is that pili are retractable structures, whereas the LPS is an essential component of the bacteria membrane whose complexity allows for ARD sequential changes. Consequently, phages attaching to LPS receptors may be interesting for their wide and efficient potential of evolution. In contrast, loss of pili is associated with a high fitness cost in bacteria depending on the environmental conditions and the bacteria are only temporarily modified. From a therapeutic perspective, conclusions are mixed but worth further exploration.

A much clearer case are phages targeting mechanisms associated with antibiotic resistance in bacteria. Although these phages can be engineered [27], they are also found in nature. The phage PRD1 of *Escherichia coli* uses proteins encoded by plasmids as receptors, selecting against bacteria that contain conjugative plasmids, which are mobile structures that carry and spread antibiotic resistance genes [28]. Another example of phages selecting against antibiotic resistant bacteria is that of phage OMKO1 of *P. aeruginosa*. This phage recognizes a cell surface protein that is part of the multi-drug efflux system as the bacterial receptor. Phage-resistant bacteria harbor mutated efflux pumps that are ineffective for antibiotic resistance [29]. At least one case of compassionate use of this last phage has been successful [30], prompting further treatments.

Different researchers and clinicians have advocated for applying phage therapy as a personalized medicine for different reasons, including a reduced selection for phage resistance in bacterial populations. First, a moderate (personalized versus generalized) use of phages could avoid strong evolutionary pressures on bacterial populations derived from high doses found with other antimicrobials [31,32]. Second, as stated in the recent articles of this Special Issue, sensitization of the immune system to phages and their elimination after repeated use should not be discarded. Third, a tailored phage strategy, such as the "magistral preparation" recently approved in Belgium [33], is low-priced and fast compared to the standard drug licensing pathway. The best way to ensure the durability of phage therapies' efficacy is the careful application of phage therapy, minimizing past mistakes with other antimicrobials; personalized phage therapy currently seems a favorable procedure to accomplish this.

5. In Silico and In Vivo Studies are Necessary to Understand Bacteria–Phage Evolution

In community ecology and evolutionary studies, the interaction of bacteria and their phages has become a cornerstone field [34]. However, the majority of these are in vitro studies; while in the wild, with diverse microbial communities, abiotic factors or host colonization by bacteria, co-evolutionary dynamics appear much more complex (e.g., [35]). Indeed, bacterial phage resistance is a condition that potentially implies metabolic and evolutionary costs, which are not always accounted for in vitro analyses [36–38]. Different studies have proved that a phage candidate's effect in vitro does not always relate to their capacity to control the disease in vivo (e.g., [39]). It has been suggested that different phage resistance mechanisms are selected in bacteria depending on the ecological conditions [17], although this remains to be largely explored in vivo. A recent study proved that phage evolution differed between dixenic mice and planktonic cultures [40]. Additionally, it has been demonstrated that a synergism between phages and the immune system is essential to wipe out pathogens from hosts [41]. Environmental complexity, the host immune system, differential bacterial gene expression, evolutionary trade-offs, or the interactions with diverse intra- and inter-microbial communities play a significant role in this regard. In order to understand evolutionary pressures, assays in both conditions are necessary.

The design and use of phage cocktails for therapy certainly allows for combating bacterial resistance evolution. The underlying idea is to deploy phages with complementary systems of attack for the same bacterium, while covering the entire (or as much as possible) target bacterial variability. Interactions between phages, either competition or facilitation, may exist in phage combinations (as in any ecological community), and make the efficiency of the cocktail deviate from the addition of the phage isolates' effects [42]. In addition to purely experimental approaches, computer simulations and algorithms of the interaction networks between phages and bacterial populations could help to guide the design of phage cocktails, defining characteristics of phage assemblages that are key to optimizing the cocktail stability and efficiency [43]. Understanding these complex dynamics may therefore help in choosing phages that increase the efficiency of a cocktail and aid in determining the frequency at which the cocktail should be applied in real systems.

6. Conclusions

It is essential in the design of effective antimicrobial strategies to consider the evolution of resistance in bacteria. Our understanding of phage–bacteria interactions can guide the selection of specific combinations that can help to minimize any possible impact of resistance development in an effective therapy. Including diverse approaches using different pathogenic bacteria will set up evolutionary principles to refine the selection of candidate phages. In other words, we must aim to detect potentially long-term effective phages to be used as durable control strategies. Among the evolutionary criteria and unsolved questions related to phages and their effect in bacteria are the following: timing of application, effects of bacterial population diversity and life-history related to phages, as well as phage features regarding their potential of adaptation to bacteria. All this information will help compile data on the capacity of each phage to select resistant bacteria and aid in choosing effective phages accordingly. Future research should provide much-needed results on the evolutionary and molecular consequences of phage therapy treatments in complex environments. Evolutionary approaches can provide insights into how to limit the evolution of bacterial resistance to phages and truly advance phage therapy, a potential solution to several worldwide problems.

Acknowledgments: I thank James Gurney and Pedro Gómez for fruitful discussions. This work was supported by the EPIBIO-OI project co-funded by the European Union: European regional development fund (ERDF, program INTERREG V), by the "Conseil Régional de la Réunion" and by the "Centre de Coopération internationale en Recherche agronomique pour le Développement" (CIRAD).

Conflicts of Interest: The author declares no conflict of interest. The founding sponsors had no role in the writing of the manuscript.

References

1. Kingwell, K. Bacteriophage therapies re-enter clinical trials. *Nat. Rev. Drug Discov.* **2015**, *14*, 515–516. [CrossRef] [PubMed]
2. Buttimer, C.; McAuliffe, O.; Ross, R.P.; Hill, C.; O'Mahony, J.; Coffey, A. Bacteriophages and Bacterial Plant Diseases. *Front. Microbiol.* **2017**, *8*. [CrossRef] [PubMed]
3. Moye, Z.; Woolston, J.; Sulakvelidze, A. Bacteriophage Applications for Food Production and Processing. *Viruses* **2018**, *10*, 205. [CrossRef] [PubMed]
4. Casey, E.; van Sinderen, D.; Mahony, J. In Vitro Characteristics of Phages to Guide "Real Life" Phage Therapy Suitability. *Viruses* **2018**, *10*, 163. [CrossRef] [PubMed]
5. Rohde, C.; Resch, G.; Pirnay, J.-P.; Blasdel, B.G.; Debarbieux, L.; Gelman, D.; Górski, A.; Hazan, R.; Huys, I.; Kakabadze, E.; et al. Expert Opinion on Three Phage Therapy Related Topics: Bacterial Phage Resistance, Phage Training and Prophages in Bacterial Production Strains. *Viruses* **2018**, *10*, 178. [CrossRef] [PubMed]
6. Mapes, A.C.; Trautner, B.W.; Liao, K.S.; Ramig, R.F. Development of expanded host range phage active on biofilms of multi-drug resistant *Pseudomonas aeruginosa*. *Bacteriophage* **2016**, *6*, e1096995. [CrossRef] [PubMed]
7. Ross, A.; Ward, S.; Hyman, P. More Is Better: Selecting for Broad Host Range Bacteriophages. *Front. Microbiol.* **2016**, *7*, 1352. [CrossRef] [PubMed]
8. Díaz-Muñoz, S.L. Viral coinfection is shaped by host ecology and virus-virus interactions across diverse microbial taxa and environments. *Virus Evol.* **2017**, *3*, vex011. [CrossRef] [PubMed]
9. Scanlan, P.D.; Buckling, A. Co-evolution with lytic phage selects for the mucoid phenotype of *Pseudomonas fluorescens* SBW25. *ISME J.* **2012**, *6*, 1148–1158. [CrossRef] [PubMed]
10. Scanlan, P.D.; Hall, A.R.; Lopez-Pascua, L.D.C.; Buckling, A. Genetic basis of infectivity evolution in a bacteriophage. *Mol. Ecol.* **2011**, *20*, 981–989. [CrossRef] [PubMed]

11. Laanto, E.; Hoikkala, V.; Ravantti, J.; Sundberg, L.R. Long-term genomic coevolution of host-parasite interaction in the natural environment. *Nat. Commun.* **2017**, *8*, 111. [CrossRef] [PubMed]
12. Kim, J.-S. Microbial warfare against viruses. *Science* **2018**, *359*, 993. [CrossRef] [PubMed]
13. Roychoudhury, P.; Shrestha, N.; Wiss, V.R.; Krone, S.M. Fitness benefits of low infectivity in a spatially structured population of bacteriophages. *Proc. R. Soc. B Biol. Sci.* **2013**, *281*, 20132563. [CrossRef] [PubMed]
14. De Paepe, M.; Hutinet, G.; Son, O.; Amarir-Bouhram, J.; Schbath, S.; Petit, M.A. Temperate Phages Acquire DNA from Defective Prophages by Relaxed Homologous Recombination: The Role of Rad52-Like Recombinases. *PLoS Genet.* **2014**, *10*, e1004181. [CrossRef] [PubMed]
15. Barr, J.J.; Auro, R.; Furlan, M.; Whiteson, K.L.; Erb, M.L.; Pogliano, J.; Stotland, A.; Wolkowicz, R.; Cutting, A.S.; Doran, K.S.; et al. Bacteriophage on mucus provide immunity. *Proc. Natl. Acad. Sci. USA* **2013**, *110*, 10771–10776. [CrossRef] [PubMed]
16. Costello, E.K.; Lauber, C.L.; Hamady, M.; Fierer, N.; Gordon, J.I.; Knight, R. Bacterial Community Variation in Human Body Habitats Across Space and Time. *Science* **2009**, *326*, 1694–1697. [CrossRef] [PubMed]
17. Chabas, H.; van Houte, S.; Høyland-Kroghsbo, N.M.; Buckling, A.; Westra, E.R. Immigration of susceptible hosts triggers the evolution of alternative parasite defence strategies. *Proc. R. Soc. B Biol. Sci.* **2016**, *283*, 20160721. [CrossRef] [PubMed]
18. Hall, A.R.; Scanlan, P.D.; Morgan, A.D.; Buckling, A. Host-parasite coevolutionary arms races give way to fluctuating selection. *Ecol. Lett.* **2011**, *14*, 635–642. [CrossRef] [PubMed]
19. Scanlan, P.D.; Buckling, A.; Hall, A.R. Experimental evolution and bacterial resistance: (Co)evolutionary costs and trade-offs as opportunities in phage therapy research. *Bacteriophage* **2015**, *5*, e1050153. [CrossRef] [PubMed]
20. Torres-Barceló, C.; Franzon, B.; Vasse, M.; Hochberg, M.E. Long-term effects of single and combined introductions of antibiotics and bacteriophages on populations of *Pseudomonas aeruginosa*. *Evol. Appl.* **2016**, *9*, 583–595. [CrossRef] [PubMed]
21. Chibeu, A.; Agius, L.; Gao, A.; Sabour, P.M.; Kropinski, A.M.; Balamurugan, S. Efficacy of bacteriophage LISTEX^{TM}P100 combined with chemical antimicrobials in reducing *Listeria monocytogenes* in cooked turkey and roast beef. *Int. J. Food Microbiol.* **2013**, *167*, 208–214. [CrossRef] [PubMed]
22. Allen, R.C.; Popat, R.; Diggle, S.P.; Brown, S.P. Targeting virulence: Can we make evolution-proof drugs? *Nat. Rev. Microbiol.* **2014**, *12*, 300–308. [CrossRef] [PubMed]
23. Seed, K.D.; Yen, M.; Jesse Shapiro, B.; Hilaire, I.J.; Charles, R.C.; Teng, J.E.; Ivers, L.C.; Boncy, J.; Harris, J.B.; Camilli, A. Evolutionary consequences of intra-patient phage predation on microbial populations. *eLife* **2014**, *3*, 1–10. [CrossRef] [PubMed]
24. Laanto, E.; Bamford, J.K.H.; Laakso, J.; Sundberg, L.-R. Phage-driven loss of virulence in a fish pathogenic bacterium. *PLoS ONE* **2012**, *7*, e53157. [CrossRef] [PubMed]
25. Evans, T.J.; Trauner, A.; Komitopoulou, E.; Salmond, G.P.C. Exploitation of a new flagellatropic phage of *Erwinia* for positive selection of bacterial mutants attenuated in plant virulence: Towards phage therapy. *J. Appl. Microbiol.* **2010**, *108*, 676–685. [CrossRef] [PubMed]
26. Betts, A.; Kaltz, O.; Hochberg, M.E. Contrasted coevolutionary dynamics between a bacterial pathogen and its bacteriophages. *Proc. Natl. Acad. Sci. USA* **2014**, *111*, 11109–11114. [CrossRef] [PubMed]
27. Pires, D.P.; Cleto, S.; Sillankorva, S.; Azeredo, J.; Lu, T.K. Genetically Engineered Phages: A Review of Advances over the Last Decade. *Microbiol. Mol. Biol. Rev.* **2016**, *80*, 523–543. [CrossRef] [PubMed]
28. Ojala, V.; Laitalainen, J.; Jalasvuori, M. Fight evolution with evolution: Plasmid-dependent phages with a wide host range prevent the spread of antibiotic resistance. *Evol. Appl.* **2013**, *6*, 925–932. [CrossRef] [PubMed]
29. Chan, B.K.; Sistrom, M.; Wertz, J.E.; Kortright, K.E.; Narayan, D.; Turner, P.E. Phage selection restores antibiotic sensitivity in MDR *Pseudomonas aeruginosa*. *Sci. Rep.* **2016**, *6*, 26717. [CrossRef] [PubMed]
30. Chan, B.K.; Turner, P.E.; Kim, S.; Mojibian, H.R.; Elefteriades, J.A.; Narayan, D. Phage treatment of an aortic graft infected with *Pseudomonas aeruginosa*. *Evol. Med. Public Heal.* **2018**, *2018*, 60–66. [CrossRef] [PubMed]
31. Day, T.; Read, A.F.; Shaw, M.; Hobbelen, P.; Oliver, R.; Evans, H. Does High-Dose Antimicrobial Chemotherapy Prevent the Evolution of Resistance? *PLOS Comput. Biol.* **2016**, *12*, e1004689. [CrossRef] [PubMed]
32. Trasta, A. Personalized medicine and proper dosage: Over- and undertreatment of chronic diseases endanger patients' health and strain public health systems. *EMBO Rep.* **2018**, *19*, e45957. [CrossRef] [PubMed]

33. Pirnay, J.P.; Verbeken, G.; Ceyssens, P.J.; Huys, I.; de Vos, D.; Ameloot, C.; Fauconnier, A. The magistral phage. *Viruses* **2018**, *10*, 64. [CrossRef] [PubMed]
34. Dennehy, J.J. What Can Phages Tell Us about Host-Pathogen Coevolution? *Int. J. Evol. Biol.* **2012**, *2012*, 396165. [CrossRef] [PubMed]
35. Gómez, P.; Bennie, J.; Gaston, K.J.; Buckling, A. The Impact of Resource Availability on Bacterial Resistance to Phages in Soil. *PLoS ONE* **2015**, *10*, e0123752. [CrossRef] [PubMed]
36. Meaden, S.; Paszkiewicz, K.; Koskella, B. The cost of phage resistance in a plant pathogenic bacterium is context-dependent. *Evolution* **2015**, *69*, 1321–1328. [CrossRef] [PubMed]
37. Koskella, B.; Thompson, J.N.; Preston, G.M.; Buckling, A. Local biotic environment shapes the spatial scale of bacteriophage adaptation to bacteria. *Am. Nat.* **2011**, *177*, 440–451. [CrossRef] [PubMed]
38. Gurney, J.; Aldakak, L.; Betts, A.; Gougat-Barbera, C.; Poisot, T.; Kaltz, O.; Hochberg, M.E. Network structure and local adaptation in co-evolving bacteria-phage interactions. *Mol. Ecol.* **2017**, *26*, 1764–1777. [CrossRef] [PubMed]
39. Henry, M.; Lavigne, R.; Debarbieux, L. Predicting in vivo efficacy of therapeutic bacteriophages used to treat pulmonary infections. *Antimicrob. Agents Chemother.* **2013**, *57*, 5961–5968. [CrossRef] [PubMed]
40. De Sordi, L.; Khanna, V.; Debarbieux, L. The Gut Microbiota Facilitates Drifts in the Genetic Diversity and Infectivity of Bacterial Viruses. *Cell Host Microbe* **2017**, *22*, 801–808.e3. [CrossRef] [PubMed]
41. Roach, D.R.; Leung, C.Y.; Henry, M.; Morello, E.; Singh, D.; Di Santo, J.P.; Weitz, J.S.; Debarbieux, L. Synergy between the Host Immune System and Bacteriophage Is Essential for Successful Phage Therapy against an Acute Respiratory Pathogen. *Cell Host Microbe* **2017**, *22*. [CrossRef] [PubMed]
42. Sanjuán, R. Collective Infectious Units in Viruses. *Trends Microbiol.* **2017**, *25*, 402–412. [CrossRef] [PubMed]
43. Weitz, J.S.; Poisot, T.; Meyer, J.R.; Flores, C.O.; Valverde, S.; Sullivan, M.B.; Hochberg, M.E. Phage-bacteria infection networks. *Trends Microbiol.* **2013**, *21*, 82–91. [CrossRef] [PubMed]

Comment

Phages Make for Jolly Good Stories

Thomas Häusler

Swiss Public Radio SRF, Redaktion Wissenschaft, Studio Basel, Novarastrasse 2, 4059 Basel, Switzerland; thomas.haeusler@srf.ch

Received: 12 March 2018; Accepted: 19 April 2018; Published: 20 April 2018

Abstract: Phage therapy has an intriguing history. It was widely used from the 1920s until the 1940s. After this period, it was nearly completely forgotten in the Western world, while it continued to be used in the Soviet part of the globe. The study of the history of phage therapy provides valuable input into the present development of the field. Science journalists uncovered much of this history and played an important role in the communication of phage therapy after the fall of the Soviet Union, when it came to the attention of Western researchers and doctors. This interest was fueled by the antibiotic resistance crisis. At this time, communication about phage therapy had a wide potential audience, that encompassed medical experts and researchers, as well as the public, because knowledge about this forgotten therapy was very limited. In such a situation, good communication had and still has the potential to catalyze important discussions among different groups; whereas, bad communication could have considerably hindered and still can hinder the possible renaissance of phage therapy.

Keywords: phage therapy; history of science; science communication

1. A Journalist's Paradise

It is a core competence of journalists who are worth their salary to spot a good story. The phage therapy story that presented itself in the mid-1990s surely was one. A few years before, the Soviet Union had collapsed, and, slowly, information about a therapy that could tackle bacteria resistant to antibiotics began to trickle westwards. In view of the looming antibiotic resistance crisis this, in itself, would have been worthy of many "exotic therapy" reports for Western eyes at the time. However, there was a lot more to it. The therapy in question—called phage therapy—had been pioneered more than 70 years prior. It had been used extensively in many parts of the world until about World War II, when it was forgotten in the Western part of the world. Phage therapy lived on behind the "Iron Curtain" until the Eastern bloc disintegrated. Additional lure lay in the way the therapy worked, by using viruses to fight pathogenic bacteria.

There were even more ingredients for a perfect story with a lot of human interest. One of the important centers of Soviet phage therapy practice and research had been based in Tbilisi, the capital of the now independent state of Georgia. After independence, Georgia went through a great deal of turmoil, a short civil war, and huge economic problems. As a consequence, the phage researchers and their institutions in Tbilisi were in very dire straits and the staff had to subsist on a very meager pay. Most buildings were run-down, and the equipment in labs and hospitals was old, sparse and often broken. The walls lacked paint, floors were cracked, and many lights did not work. Peter Radetzky was, to the author's best knowledge, the first Western journalist to travel to Tbilisi to report on all of this and must have felt in a journalist's paradise. His 1996 article in Discover magazine [1] opened the way for more reporting.

2. To Hype or Not to(o) Hype?

Hypothetical observers of this state of affairs could have asked themselves: What will be the optimal type of reporting in the following years if we want to maximize the benefit of this discovery to society? Journalism has many functions, but this one question was chosen for the sake of this exercise. A first quick answer could have been: report it in a way that helps phage therapy achieve a quick global comeback. This answer would have been wrong. Research uncovered that the evidence for phage therapy's effectiveness was not clear-cut at all.

Journalism textbooks provide ample messages of caution in such situations. Do not hype! Yet, there are quite a few examples from recent history where journalists, science communicators and scientists were guilty of exactly this. Think, for example, about the enthusiasm that greeted the first gene therapy attempts or the hypothesis that cancers could be cured by inhibiting angiogenesis. To be sure, both approaches eventually yielded important therapies and researchers continue to develop new ones. However, there were times when expectations were much higher than reality. These were fueled by over-optimistic assessments by scientists, and by journalists who uncritically featured them in their reports.

Several studies have shown that the main sources of exaggerations about scientific or medical findings are actually press releases issued by the institutions of the scientists that did the original studies [2,3]. Often, after inevitable roadblocks and setbacks have emerged, the reaction turns to the other extreme. The initial hype is followed by periods in which an approach is all but pronounced dead, as happened with both the examples mentioned above. It is a boom–bust cycle [4] that helps neither science nor journalism. How much hype and the boom–bust cycle actually damage the reputation of science as a whole or an individual discipline is debated in the literature. Many scholars argue that considerable damage can be done [5].

I would argue that phage therapy is especially vulnerable to such damage to its reputation due to its "exotic" history. However, there is a competition of ideas, research disciplines, and stories for the attention of the public, for funding and investments. To have a chance in this competition you need to make your case. Clearly, phage therapy has great potential, especially with regard to the antibiotics resistance crisis. Scientists and other actors in the field have to find the right balance between selling and overselling what they have to offer. The same holds true for reporters; they need to tell a good story and to give an accurate picture of the potential of their topic. How all actors have fared in this balancing act is discussed in different places in the text below. As a disclaimer, because I am an actor myself, I will mostly refrain from judgments and will instead try to highlight some of the aspects that I deem interesting.

3. Finding the Facts

Returning to phage therapy in the late 90s, it was not easy even for experienced reporters to find the fact base necessary to give an accurate picture. On the one hand, it seemed that if more than 70 years of research had not been able to prove the efficacy of the therapy beyond doubt, it was worth only very skeptical reporting. This seemed to be the order of the day. A skeptical journalist is, of course, a good journalist. However, perhaps depicting phage therapy as a fascinating yet utterly obsolete method was not the best way to go about it. Maybe the therapy did have the potential to help combat antibiotic resistance and reports that were too negative mounted additional barriers for a struggling, yet helpful, medical therapy?

For reporters, this difficult situation lasted well into the 2000s. At the time, little information was forthcoming. Most of the scientific literature was buried in pre-medline repositories, such as a long-forgotten, hand-collected literature list [6,7] that a German phage researcher collected and published in the 1950s and 1960s (11,405 references, painstakingly listed in two two-tome publications, and painstakingly read by the author of this article to be able to track down as many interesting articles as possible). Much of the phage literature concerning medical therapy was written in Russian or Georgian, which made it even more difficult to obtain and read. Ideally, a reporter would travel to

Georgia and see this research for themselves. However, not many could do this, given that they had to find an editor willing to produce the funds for such an investigation. Once in Georgia, the reporter could meet with phage researchers and doctors who warmly testify in favor of the method. Georgia's many troubles after independence cast most research institutes and hospitals in a sorry state. It was obvious that researchers and doctors did an admirable job to keep things running. Nevertheless, this situation did not instill a lot of conviction in a visitor who came looking for evidence for a therapy that was co-developed here.

The reporter's resolve to keep an open mind was further weakened by researchers from Western institutions uttering pungently negative views [8]. Often, these seemed influenced by the renowned phage researcher Gunter Stent. In his 1963 textbook "Molecular Biology of Bacterial Viruses" he offered the assessment that phage therapy was not working, because the human immune system removes phages, stomach acid destroys them, and bacteria become resistant to the phages applied in therapy. Stent was quite adamant in his negative opinion, as this quote from the book attests: "The strange bacteriophage therapy chapter of the history of medicine may now be fairly considered as closed" [9].

Another line of reasoning feeding into the negative view was the very skeptical opinion of science conducted in the Soviet Union [8,10]. This view might have been based on strong indications; however, it was not bona-fide proof that phage therapy did not work. Yet, all these factors cast their shadows into the 1990s and early 2000s, when Western scientists became aware of the vast practice of phage therapy in the Soviet Union. An interesting impression of how many Western scientists thought about the field at the time is given by a quote by Ry Young (Center for Phage Technology, Texas A&M). He stated, in 2017, in a phage workshop hosted by the US Food and Drug Administration (FDA): "In fact, since I was up until quite recently biased by my training in the phage group [a group of scientists that, starting in the 1940s, studied the molecular biology of phage], I was actually very anti-phage therapy" [11].

4. In the Age of Story Telling

Humans have probably told each other stories since they started to talk. However, it seems that we are living in an era in which story telling as a concept receives renewed and intense interest, in journalism [12], in science communication [12,13], and in science [14,15]. Journalism basically thrives on the millennia old wisdom that nothing captivates people more than a good story. However, even in science, journal editors and readers alike favor a good story; a set of experiments that fit nicely together into one narrative and that convey a coherent picture of a sub-set of reality. There is nothing amiss with this. The chemist and Nobel Laureate, Roald Hoffmann, even argues that using story-telling in a scientific paper can help to convey complex research findings or concepts to fellow scientists [15], and an analysis of scientific publications dealing with climate change has shown that publications with more narrative abstracts are cited more often [16]. However, if everybody is tuned to "good stories" and to coherent narratives, readers might have less patience for stories that are more complex or look to much into the underlying complex reality. They might even mistrust stories that do not fit a clear-cut narrative.

Does phage therapy work or does it not? Maybe it works, but only under certain circumstances. Scientists should be prepared for such a reality. There are plenty of established examples of drugs that work for patients with a certain genetic background, for example, but not for others. In the case of phage therapy, for instance, bacteria infecting a patient can develop resistance against a phage used for treatment [17]. In this case, an alternative phage or phage cocktail has to be used to continue treatment. Of course, scientists and doctors working in the field are well aware of this and other potential obstacles. However, every scientist is not an expert in all disciplines except their own. This may sound like a truism, but it is also a hint that all of us will not necessarily approach foreign disciplines with the usual scientific rigor and an open mind.

In the early days of renewed Western interest in phage therapy, at least one thing seemed clear: there was a huge amount of ignorance about the subject across most groups one could think of. This included the general public, most scientists and doctors, the pharmaceutical industry, the intellectual property community, investors, drug approval agencies, and so on. If phage therapy stands a chance at making a comeback, the education of all of these stakeholders is important. Scientists and doctors are necessary to create evidence that the therapy works. The pharmaceutical industry and investors are needed to fund the research and bring phage drugs to market. Since phages are natural entities that have been long-known to science, the patenting of drugs is not straightforward at all. Drug agencies need to explore the ways in which a preparation of viable viruses could be fed into the drug approval process. More will be discussed on this point later.

The researchers and practitioners of phage therapy in Tbilisi and other parts of the former Eastern bloc (namely Poland and Russia) knew, of course, a lot about the core topic but they needed knowledge about other matters, namely drug development, business development, intellectual property issues and many more.

How large the dangers of a bipartite information deficit was, became quickly and painfully clear. Radetzky's 1996 article catalyzed the first contact between Western investors and phage researchers in Tbilisi. A Canadian financier visited the city and quickly struck a deal with some of the researchers of the Eliava institute. However, things ended in bitter conflict. The US start-up set up to develop a phage drug for the Western market soon pulled out of the co-operation with Tbilisi. The reason for this, reportedly, was that the difficult circumstances in Georgia might taint the reputation of the future drug. The phage researchers in Tbilisi felt cheated. Additionally, the contract with the start-up created much tension among different groups of the Eliava Institute [18]. In 2003, a manager of the US start-up was quoted in the media with very negative statements about the work of the phage researchers in Tbilisi: "They made every mistake in the book and did some really stupid things. [...] They just took what was on the shelf and assumed it would work." [19]. At this still very early stage of rediscovery of phage therapy in Western countries, such a statement ran a very high risk of tainting the reputation of the whole field severely.

5. History's Worth

It turns out that much can be learned from extensive research into past literature. There are studies covering hundreds of thousands of subjects treated with phages, as well as many different types of infections and species of bacteria. Due to the fact that most of these studies were done before the 1960s, they do not hold up to the standards of the modern era, of double-blind, randomized, controlled clinical trials. There is a lot of information that can be found, nevertheless. For example, there is enough information to instill interest in potential investors and to help gauge the medical potential of the therapy.

Analysis of the past literature also provides reasons for why phage therapy was nearly completely forgotten after World War II in Western countries. One reason surely had to do with the fact that, even for the time, many practitioners and researchers were neglecting scientific rigor and standards significantly. This led to much-deserved criticism and skepticism, for instance in the form of several derogatory editorials in the Journal of the American Medical Association [10,20–22]. Another reason was that knowledge about the basic biology of phages was still rather limited.

Ironically, this kind of trouble had been foreshadowed with uncanny precision by the US writer and Nobel Laureate, Sinclair Lewis, in his 1925 bestseller, Arrowsmith [23]. In his novel, Lewis heavily criticized the way in which medical researchers and doctors were business-minded. At the center of his story, is a phage therapy trial in the Caribbean. His hero, Martin Arrowsmith, had travelled to an island where plague raged. He had all intents to do a properly controlled study by treating only half of the sick with a phage against *Yersinia pestis*. The other half was to receive a placebo. Then, Arrowsmith's wife died from the plague, and in his grief, he let slip all his resolve and discipline for protocol and treated all patients alike with the phage.

6. Communication in Times of Slow News

Lewis' pen was guided by the scientist-turned-science-writer, Paul de Kruif. Together, they crafted a story with a shadow that reached into the era of phage therapy's potential renaissance. After the "Iron Curtain" came down, the full comeback of phage therapy was, and still is, slow to materialize. What is needed are successful clinical trials. Some Phase I and II trials have been conducted but, to date, no Phase III trial has ever been initiated. Arrowsmith's great aim has still not yet been achieved. Medical progress is often slow in the making, and obstacles are part of the business. However, phage therapy seems to fight with more than its fair share of obstacles [24]. The reasons are manifold, but the most important is probably that the way phage therapy could be made to work does not fit within the current paradigms of the pharma market.

This poses special difficulties for the actors in the field of phage therapy, as far as communication is concerned. Slow progress means that little news is forthcoming that could be interesting to the public. In this situation, some actors could be tempted to hype their work to garner attention. Indeed, this has happened in some instances. In one case, the media department of a university issued a release about a study using phages against bacterial infections in Cystic Fibrosis (CF) lung infections [25]. The title read "Phage therapy shown to kill drug-resistant superbug". The release discussed the severe problem of lung infections in CF patients and stated: "Here for the first time, researchers have shown that phage therapy is highly effective in treating established and recalcitrant chronic respiratory tract infections caused by multi-drug resistant *Pseudomonas aeruginosa* strains. They show that phages are capable of killing the bacteria in long term infected lungs, such as those suffered by patients with the inherited disease Cystic Fibrosis, indicating a potential new therapeutic option for these hard to treat life threatening infections".

Careful readers will wonder if actual patients have indeed been treated, but the casual reader will probably presume so. The media release did not help readers perceive the research clearly, as it never mentioned that the experiments were done exclusively in mice and in an artificial model of a lung [26]. The plight of CF patients is well known. Many of them would jump at such a news and presume that help for their infection is just around the corner. Judging by the information contained in the original paper, this is not the case. However, the media department release was taken up virtually unchanged by a widely read internet site publishing popular science articles [27].

A quantitative survey of reports regarding phage therapy is outside of the scope of this article. I have surveyed, approximately, the first 50 articles that show up in a Google News search with the term "phage therapy". Based on this, it is fair to say that this kind of communication, by scientists or university media staff, is the minority. Many researchers seem to do an excellent job in providing a balanced picture. For example, in 2016, a diverse group of US researchers and doctors treated a patient that was suffering from a very severe *Acinetobacter baumannii* infection with phages. As is the nature of anecdotal cases, there was no definitive proof that the phages really cured the patient but the probability seemed quite high [17]. This publication triggered a host of reports in the press. Several aspects made the case intriguing. The patient was saved from what seemed to be sure death and he was treated with phages intravenously. Nevertheless, articles about the case were generally of a high quality.

Obviously, the scientists and media staff of their institutions did a good job. They provided extensive material for the media [28], and the scientists positioned the nature of a single case study very clearly when they were interviewed, stating, for example, "But for now, his case is just an anecdote—albeit a hopeful one—as all the physicians and scientists who worked on his case point out." [29]. This constructive communication was remarkable, since quite often the media overblow the significance of such anecdotal stories of healing, giving the impression that a single cured person constitutes proof of effectiveness.

This example shows the chance that lies in such a story with a human dimension if the science involved is communicated well. It triggered extensive news coverage, which is quite a success for a medical therapy that, in the eyes of a hard-nosed news editor, has not been making significant news for quite some years. The renewed media coverage may, in turn, have generated renewed interest from the grant and investment community. Intriguingly, the reports were quite long and detailed, not only in the aspects of human interest but also in the science. For instance, they mentioned the fact that the bacteria causing the infection developed resistance and that to continue treatment new phages had to be found. In this way, the reports conveyed a picture of the complexity of phage therapy, which helps to explain to the public why it is taking so long to develop this method into an established medical treatment.

Phage researchers could potentially learn still more from colleagues in other disciplines that have a lot of experience with communication. Climate science springs to mind because it is also a field that needs to communicate a lot with the public, and over a long period of time. Some climate scientists have devised a strategy of building a certain basic consensus among scholars to provide clearer communication to the public [30]. This has been taken up by the media quite extensively. Perhaps this concept could be adapted to the field of phage therapy.

7. Giving a Balanced View

Section 6 looked mainly at the role of scientists in the way facts about phage therapy are communicated. What about journalists? Ideally, journalists should be able to give a balanced view of the potential of phage therapy and the obstacles it faces. In the majority of cases that came out of the Google News search mentioned above, journalists did a good job. They usually mentioned that anecdotal cases are not general proof of efficacy, for example. However, some fail to provide this important aspect. Others offered general statements that were incorrect without proper specifications: "Even though bacteriophages [...] work better than antibiotics [...]" [31]. Many ran titles that seemed to talk of a miracle cure ("Bacteriophage therapy, the amazing cure for MRSA being ignored by mainstream medicine"), and this initial impression was never put into perspective or it was, but only very late into the article, when many readers probably had long escaped the text [28].

Whether the picture regarding communication quality overall is rather bright grey (as I think) or a bit darker, is up for personal judgment. Of course, there is always room for improvement. Platforms like healthnewsreview.org are trying to tap into that potential. Journalists provide critical appraisals of selected media articles and point out weaknesses. Up until now, they have done so with one article concerned with phage therapy [32].

8. Patience Is Still Required

What about the way forward? Again, a glance back in time might help. It turns out that the curative powers of phages were not completely forgotten in the West. Dedicated scientists, doctors and small companies kept it alive until the 1960s and 1970s, for example in France and Switzerland [33,34]. The way it was practiced there hinted at possible schemes that could be used today. Researchers were keeping a large and diverse bank of phages that could be used to search for effective variants when a patient presented with an infection. The concoction used was produced on demand. There was no or only a small profit involved.

This kind of scheme was again proposed by several actors in the early 2000s. In Georgia, it has been used for the last 80 years, and has a long tradition in Poland, as well. There is still some work required before it can be applied in other European countries or the US. However, after long attempts to bring phage therapy to market using more conventional approaches, there seems to be a willingness by several stakeholders to explore avenues that are unconventional relative to the regulatory framework in Western countries. In 2015, for instance, researchers, physicians, and industrial representatives of several countries met in Tbilisi to discuss how phage therapy could be used in Western health settings [35]. In the same year, the European drug regulatory agency EMA convened a workshop with representatives from industry, academia, regulatory authorities and legislators to discuss similar questions [36]. In 2017, the FDA held, as well, a workshop to discuss scientific and regulatory issues [37].

At this workshop, a representative of a French company that produces phages according to an officially regulated standard ("Good Manufacturing Practice"-production (GMP)) for clinical research, mentioned that these phages obtained clearance by the French authorities to be used outside the clinical trial in cases where patients are critically ill [38]. Belgian health authorities are preparing an approach that paves the way for the small-scale use of phage therapy [39]. The basis for this is a central phage bank, with controlled deposits that serve as a source for treatments.

There is still work to be done. Solid proof for phage therapy's effectiveness remains patchy [40]. Laudable initiatives like the EU-funded PhagoBurn study have been initiated, but have apparently run into difficulties [41]. If the fascinating history of phage therapy is any guide, this is not surprising. Yet, there seems to be, once again, renewed activity in the field. Amongst other developments, the US Navy is working on phage therapy [17], the German government is funding a phage project ("Phage4Cure"), and in the US, scientists have set up a virtual phage bank to help quickly source therapeutic phages for urgent needs [42]. If journalists, with their research and analysis [43,44], have had even a small catalytic function in this development, it would have been well worth the considerable investment.

Acknowledgments: Over the course of my journalistic and historical research into phage therapy, I have had the privilege to discuss this topic with many researchers and other actors in the field. I am deeply indebted to all of them. Additionally, I wish to thank the reviewers for their excellent suggestions and comments, which helped to improve this paper.

Conflicts of Interest: The author has written a book about phage therapy [40]. He has briefly consulted for the former biomedical company, Mondobiotech, on phage therapy. This engagement had no influence on the content of this review.

References

1. Radetzky, P. The good virus. *Discover Magazine*, 1 November 1996.
2. Sumner, P.; Vivian-Griffiths, S.; Boivin, J.; Williams, A.; Venetis, C.A.; Davies, A.; Ogden, J.; Whelan, L.; Hughes, B.; Dalton, B.; et al. The association between exaggeration in health related science news and academic press releases: retrospective observational study. *BMJ* **2014**, *349*, g7015. [CrossRef] [PubMed]
3. Sumner, P.; Vivian-Griffiths, S.; Boivin, J.; Williams, A.; Bott, L.; Adams, R.; Venetis, C.A.; Whelan, L.; Hughes, B.; Chambers, C.D. Exaggerations and caveats in press releases and health-related science news. *PLoS ONE* **2016**, *11*, e0168217. [CrossRef] [PubMed]
4. Rinaldi, A. To hype, or not to(o) hype. *EMBO Rep.* **2012**, *13*, 303. [CrossRef] [PubMed]
5. Master, Z.; Resnik, D. Hype and Public Trust in Science. *Sci. Eng. Ethics* **2013**, *19*, 321–335. [CrossRef] [PubMed]
6. Raettig, H.J. *Bakteriophagie 1917–1956*; Gustav Fischer: Stuttgart, Germany, 1958.
7. Raettig, H.J. *Bakteriophagie 1957–1965*; Gustav Fischer: Stuttgart, Germany, 1967.
8. Häusler, T.; Swiss Public Radio, Basel, Switzerland. Unpublished data from interviews with researchers. 2000–2006.
9. Stent, G. *Molecular Biology of Bacterial Viruses*; Freeman: San Francisco, CA, USA, 1963.
10. Summers, W.C. The strange history of phage therapy. *Bacteriophage* **2012**, *2*, 130–133. [CrossRef] [PubMed]
11. Young, R. Presentation at Workshop: Bacteriophage Therapy: Scientific and Regulatory Issues, Public Workshop. Page 9. Available online: https://www.fda.gov/downloads/BiologicsBloodVaccines/NewsEvents/WorkshopsMeetingsConferences/UCM579441.pdf (accessed on 7 April 2018).
12. Dahlstrom, M.F. Using narratives and storytelling to communicate science with nonexpert audiences. *Proc. Natl. Acad. Sci. USA* **2014**, *111*, 13614–13620. [CrossRef] [PubMed]
13. Martinez-Conde, S.; Macnik, S.L. Finding the plot in science storytelling in hopes of enhancing science communication. *Proc. Natl. Acad. Sci. USA* **2017**, *114*, 8127–8129. [CrossRef] [PubMed]
14. Science as Storytelling. Available online: http://serendip.brynmawr.edu/sci_cult/scienceis/bickmoregrandy.html (accessed on 7 April 2018).
15. Hoffmann, R. The Tensions of Scientific Storytelling. *Am. Sci.* **2014**, *102*, 250. [CrossRef]

16. Hillier, A.; Kelly, R.P.; Klinger, T. Narrative Style Influences Citation Frequency in Climate Change Science. *PLoS ONE* **2016**, *11*, e0167983. [CrossRef] [PubMed]
17. Schooley, R.T.; Biswas, B.; Gill, J.J.; Hernandez-Morales, A.; Lancaster, J.; Lessor, L.; Barr, J.J.; Reed, S.L.; Rohwer, F.; Benler, S.; et al. Development and use of personalized bacteriophage-based therapeutic cocktails to treat a patient with a disseminated resistant *Acinetobacter baumannii* infection. *Antimicrob. Agents Chemother.* **2017**, *61*, e00954-17. [CrossRef] [PubMed]
18. Häusler, T. *Viruses vs. Superbugs: A Solution to the Antibiotics Crisis?* Palgrave Macmillan: London, UK, 2006; pp. 179–184.
19. Marsa, L. Enlisting Viruses to Battle Bacteria. Los Angeles Times, 2003. Available online: http://articles.latimes.com/2003/mar/31/health/he-lab31 (accessed on 7 April 2018).
20. Anonymous. Editorial. *JAMA* **1931**, *96*, 693.
21. Anonymous. Editorial. *JAMA* **1932**, *98*, 1190.
22. Anonymous. Editorial. *JAMA* **1933**, *100*, 1431-2, 1603-4.
23. Lewis, S. *Arrowsmith*; P. F. Collier & Son: New York, NY, USA, 1925.
24. Brüssow, H. What is needed for phage therapy to become a reality in Western medicine? *Virology* **2012**, *434*, 138–142. [CrossRef] [PubMed]
25. University of Liverpool—News. Phage Therapy Shown to Kill Drug-Resistant Superbug. Available online: https://news.liverpool.ac.uk/2017/03/13/phage-therapy-shown-kill-drug-resistant-superbug/#comments (accessed on 15 April 2018).
26. Waters, E.M.; Neill, D.R.; Kaman, B.; Sahota, J.S.; Clokie, M.R.; Winstanley, C.; Kadioglu, A. Phage therapy is highly effective against chronic lung infections with *Pseudomonas aeruginosa*. *Thorax* **2017**, *72*, 666–667. [CrossRef] [PubMed]
27. Science Daily. Phage Therapy Shown to Kill Drug-Resistant Superbug. Available online: https://www.sciencedaily.com/releases/2017/03/170313134956.htm (accessed on 15 April 2018).
28. UC San Diego Health—Newsroom. Available online: https://health.ucsd.edu/news/topics/phage-therapy/Pages/default.aspx (accessed on 7 April 2018).
29. Weber, L. Sewage Saved This Man's Life. Someday It Could Save Yours. Huffpost, 2017. Available online: https://www.huffingtonpost.com/entry/antibiotic-resistant-superbugs-phage-therapy_us_5913414de4b05e1ca203f7d4 (accessed on 7 April 2018).
30. Cook, J.; van der Linden, S.; Maibach, E.; Lewandowsky, S. The Consensus Handbook. 2018. Available online: http://www.climatechangecommunication.org/all/consensus-handbook/ (accessed on 7 April 2018).
31. Benson, J. Bacteriophage Therapy, the Amazing Cure for MRSA Being Ignored by Mainstream Medicine. Natural News 2013. Available online: https://www.naturalnews.com/039226_bacteriophage_therapy_MRSA_cure.html (accessed on 7 April 2018).
32. Healthnewsreview.org. Available online: https://www.healthnewsreview.org/review/mail-order-viruses-are-the-new-antibiotics (accessed on 9 April 2018).
33. Häusler, T. *Viruses vs. Superbugs: A Solution to the Antibiotics Crisis?* Palgrave Macmillan: London, UK, 2006; pp. 167–169.
34. Häusler, T.; Swiss Public Radio SRF, Basel, Switzerland. Unpublished data from research in archives and from interviews with researchers. 2002–2004.
35. Sybesma, W.; Prinay, J.-P. Silk route to the acceptance and re-implementation of bacteriophage therapy. *Biotechnol. J.* **2016**, *11*, 595–600. [CrossRef]
36. Pelfrene, E.; Willebrand, E.; Cavaleiro Sanches, A.; Sebris, Z.; Cavaleri, M. Bacteriophage therapy: A regulatory perspective. *J. Antimicrob. Chemother.* **2016**, *71*, 2071–2074. [CrossRef] [PubMed]
37. U.S. Food and Drug Administration. Available online: https://www.fda.gov/BiologicsBloodVaccines/NewsEvents/WorkshopsMeetingsConferences/ucm544294.htm (accessed on 7 April 2018).
38. Gabard, J. Presentation at Workshop: Bacteriophage Therapy: Scientific and Regulatory Issues, Public Workshop. Page 93. Available online: https://www.fda.gov/downloads/BiologicsBloodVaccines/NewsEvents/WorkshopsMeetingsConferences/UCM579441.pdf (accessed on 7 April 2018).
39. Pirnay, J.-P.; Verbeken, G.; Ceyssens, P.-J.; Huys, I.; De Vos, D.; Ameloot, C.; Fauconnier, A. The Magistral Phage. *Viruses* **2018**, *10*, 64. [CrossRef] [PubMed]

40. Roach, D.R.; Debarbieux, L. Phage therapy: Awakening a sleeping giant. *Emerg. Top. Life Sci.* **2017**, *1*, 93–103. [CrossRef]
41. Servick, K. Beleaguered phage therapy trial presses on. *Science* **2016**, *352*, 1506. [CrossRef] [PubMed]
42. Phage Directory. Available online: phage.directory (accessed on 7 April 2018).
43. Häusler, T. *Viruses vs. Superbugs: A Solution to the Antibiotics Crisis?* Palgrave Macmillan: London, UK, 2006.
44. Kuchment, A. *The Forgotten Cure*; Springer: New York, NY, USA, 2012.

Conference Report

Expert Opinion on Three Phage Therapy Related Topics: Bacterial Phage Resistance, Phage Training and Prophages in Bacterial Production Strains

Christine Rohde [1,†,‡], Grégory Resch [2,†,‡], Jean-Paul Pirnay [3,†,‡], Bob G. Blasdel [4,†], Laurent Debarbieux [5], Daniel Gelman [6], Andrzej Górski [7,8], Ronen Hazan [6], Isabelle Huys [9], Elene Kakabadze [10], Małgorzata Łobocka [11,12], Alice Maestri [13], Gabriel Magno de Freitas Almeida [14], Khatuna Makalatia [10], Danish J. Malik [15], Ivana Mašlaňová [16], Maia Merabishvili [3,17], Roman Pantucek [16], Thomas Rose [3], Dana Štveráková [16,18], Hilde Van Raemdonck [3], Gilbert Verbeken [3] and Nina Chanishvili [10,*]

1. Department of Microorganisms, Leibniz Institute DSMZ—German Collection of Microorganisms and Cell Cultures, 38100 Braunschweig, Germany; Christine.Rohde@dsmz.de
2. Department of Fundamental Microbiology, University of Lausanne, 1015 Lausanne, Switzerland; Gregory.Resch@unil.ch
3. Laboratory for Molecular and Cellular Technology, Queen Astrid Military Hospital, 1120 Brussels, Belgium; jean-paul.pirnay@mil.be (J.-P.P.); maya.merabishvili@pha.ge (M.M.); thomas.rose@mil.be (T.R.); hilde.vanraemdonck@mil.be (H.V.R.); Gilbert.Verbeken@mil.be (G.V.)
4. Laboratory of Gene Technology, Department of Biosystems, 3000 Leuven, Belgium; blasdelb@gmail.com
5. Department of Microbiology, Institut Pasteur, 75015 Paris, France; laurent.debarbieux@pasteur.fr
6. Faculty of Dental Medicine, The Hebrew University of Jerusalem, Jerusalem 9112001, Israel; daniel.gelman@mail.huji.ac.il (D.G.); ronenh@ekmd.huji.ac.il (R.H.)
7. Bacteriophage Laboratory, Hirszfeld Institute of Immunology and Experimental Therapy, Polish Academy of Sciences, 53-114 Wroclaw, Poland; agorski@ikp.pl
8. Department of Clinical Immunology, Transplantation Institute, Medical University of Warsaw, 02-006 Warsaw, Poland
9. Department of Pharmaceutical and Pharmacological Sciences, KU Leuven, 3000 Leuven, Belgium; isabelle.huys@kuleuven.be
10. Eliava Institute of Bacteriophage, Microbiology and Virology, Gotua Street 3, 0160 Tbilisi, Georgia; elene.kakabadze@pha.ge (E.K.); khatuna.makalatia@pha.ge (K.M.)
11. Institute of Biochemistry and Biophysics, Polish Academy of Sciences, 00-901 Warsaw, Poland; lobocka@ibb.waw.pl
12. Autonomous Department of Microorganisms' Biology, Faculty of Agriculture and Biology, Warsaw University of Life Sciences—SGGW, 02-787 Warsaw, Poland
13. University of Turin, 10124 Turin, Italy; alice.maestri@edu.unito.it
14. Centre of Excellence in Biological Interactions, Department of Biological and Environmental Science, Nanoscience Center, University of Jyväskylä, Survontie 9C, FI-40014 Jyväskylä, Finland; gabriel.m.almeida@jyu.fi
15. Chemical Engineering Department, Loughborough University, Leicestershire LE11 3TU, UK; d.j.malik@lboro.ac.uk
16. Department of Experimental Biology, Faculty of Science, Masaryk University, 60000 Brno, Czech Republic; iva.maslanova@gmail.com (I.M.); pantucek@sci.muni.cz (R.P.); stverakova@mbph.cz (D.Š.)
17. Laboratory for Bacteriology Research, Faculty Medicine & Health Sciences, Ghent University, 9000 Ghent, Belgium
18. MB Pharma, 120 00 Prague 2 Vinohrady, Czech Republic

* Correspondence: nina.chanishvili@pha.ge

† These authors contributed equally to this work.

‡ Chair persons of the round tables.

Received: 31 March 2018; Accepted: 3 April 2018; Published: 5 April 2018

Abstract: Phage therapy is increasingly put forward as a "new" potential tool in the fight against antibiotic resistant infections. During the "Centennial Celebration of Bacteriophage Research" conference in Tbilisi, Georgia on 26–29 June 2017, an international group of phage researchers committed to elaborate an expert opinion on three contentious phage therapy related issues that are hampering clinical progress in the field of phage therapy. This paper explores and discusses bacterial phage resistance, phage training and the presence of prophages in bacterial production strains while reviewing relevant research findings and experiences. Our purpose is to inform phage therapy stakeholders such as policy makers, officials of the competent authorities for medicines, phage researchers and phage producers, and members of the pharmaceutical industry. This brief also points out potential avenues for future phage therapy research and development as it specifically addresses those overarching questions that currently call for attention whenever phages go into purification processes for application.

Keywords: Bacteriophage; phage therapy; resistance; adaptation; prophage; production; regulation

---

1. Foreword

This article is a reflection of three roundtable discussions in question–answer format held at the "Centennial Celebration of Bacteriophage Research" conference, which took place in Tbilisi, Georgia, on 26–29 June 2017. The goal was to elaborate a concerted expert opinion, based on clinical experience and scientific knowledge, on three commonly identified knowledge gaps with regard to phage therapy and the manufacturing of adequate phage therapy products: bacterial phage resistance, phage training and the presence of prophages in bacterial production strains. Whenever phages are foreseen for application, they need to undergo a careful pre-selection after intensive application-oriented biological investigation; only such phages should go into a purification process chain. However, the combination of single phages into cocktails creates additional more complex investigation. The goal of this brief is to inform phage therapy stakeholders from the academic, industrial, medical and regulatory areas on these three contentious issues in the context of an increasing demand for human and veterinary phage applications.

2. Bacterial Resistance to Phages

The antagonistic co-evolution between bacterial hosts and their infecting phages is considered to be an important driver of ecological and evolutionary processes in microbial communities [1]. In the light of a renewed interest in using phages to treat bacterial infections, in vitro studies indicate that bacteria–phage co-evolution could be an important factor in the success (or failure) of certain phage therapy applications. The evolution of bacterial resistance to individual phages is often (if not always) observed in vitro, but with considerable variance. However, phages have evolved multiple strategies to overcome the antiviral mechanisms they encounter when infecting bacterial cells [2], such as anti-CRISPR (Clustered Regularly Interspaced Short Palindromic Repeats) proteins [3]. On the other hand, BREX and DISARM are phage resistance systems widespread in bacterial genomes that have been recently discovered [4,5].

In experimental settings, phage-resistant bacteria are observed to emerge rapidly, but often at significant fitness costs, commonly including a reduced growth rate in the absence of phages [6]. Evolved (pre-adapted or "trained") phages were shown to be more effective in reducing the densities of chronic bacterial isolates [7]. The in vivo evolution of bacterial resistance to phages in human clinical practice seems inevitable, but this has been poorly documented in the scientific literature to date. When rats with aortic experimental endocarditis (EE) were treated with an anti-*Pseudomonas* phage cocktail, phage-resistant mutants with impaired infectivity were shown to emerge in vitro but not in vivo, presumably because resistance mutations in bacteria involved bacterial surface determinants necessary for infectivity (e.g., genes involved in pilus motility and lipopolysaccharide (LPS) formation) [8]. In a recent long-term study that followed co-evolution between phages and bacteria in a natural environment, *Flavobacterium columnare* isolates were found to be generally resistant to phages from the past and susceptible to phages isolated in years after bacterial isolation. Bacterial resistance had selected for increased phage infectivity and host range. Bacterial resistance was correlated to the appearance of new anti-phage spacers in CRISPR loci, and on several occasions the corresponding protospacer regions in the genome of phages isolated in the following samplings were found to be modified in response. This study shows that, in natural conditions (e.g., natural phage/bacteria ratios and diversities), phages and bacteria co-evolve in a continuous arms race [9].

One expert further notes that the in vivo growth rates as well as the metabolic status of hosts in a polymicrobial biofilm may typically be quite different compared to observations in in vitro studies where the host is in the log growth phase in a nutrient rich environment.

2.1. Strategies to Minimize Bacterial Phage Resistance

Most of the round table participants had no idea of the frequency of emergence of bacterial phage resistance in clinical practice. According to one group's experience, however, the number of patients in whom a pathogen acquired resistance to the phage used during therapy may vary from 17% (*Staphylococcus aureus* phages) to 85% (*Escherichia coli* phages) [10]. The majority of the participants feel that it is difficult to develop a phage cocktail to which bacteria would not be able to evolve resistance during therapy. In contrast, three participants, including two biopharmaceutical researchers, presume that it might, however, be possible to develop resistance-proof therapeutic phage cocktails, using phages with a broad host range and targeting highly conserved structures that are essential for bacterial survival and/or infectivity. Two of them note that the phage resistance problem is not caused by the de novo emergence of phage resistant clones, but by the selection of naturally present phage resistant isolates harboring antiviral mechanisms such as restriction modification systems and CRISPR/Cas (CRISPR associated proteins) systems. It is not hard to imagine that the spread of these mechanisms through horizontal gene transfer may indeed be the main driver of bacterial phage resistance occurrence in natural environments, with large population diversities and dynamics, but little is known if this is also the case in the patient's infection site. These phage-resistance-proof cocktails would need to be updated regularly to target newly selected phage resistant clones. One expert stresses that, in the experience at the Eliava Phage Therapy Center, even when a phage (cocktail) shows no in vitro lytic activity against an infecting bacterial strain (e.g., using the spot test), this phage (cocktail) might still be clinically effective in vivo. A reason for this might be that these phage resistant bacteria display an impaired virulence to support an ongoing infection and may be more easily managed by the immune system [8].

All participants do believe that it is possible to minimize the occurrence of bacterial phage resistance. Phages should be selected that belong to different families/groups and that individually show important infectious ability, such as a broad host range, high efficiency of plating (EOP), high adsorption rates, short latent periods, large burst sizes and a low inclination to select resistance (e.g., as determined by the Appelmans method [11]), and which act synergistically when mixed into one cocktail. Ideally, phage cocktails should be composed of phages that adsorb to different highly conserved bacterial cell wall structures or virulence factors and exert a selective pressure on different antiviral resistance mechanisms in the same target bacterium. Based on an extensive experience as phage researcher in the Eliava Institute, one expert stresses that there is a limit to the number of phages that can be successfully combined into a single phage cocktail, as some phages are bound to be incompatible or to compete for the same bacterial host. In addition, some experts feel that these cocktails might need to be tailored to a specific patient and when necessary adapted in vitro during therapy to reduce the risk of generating persistent bacterial phage resistance. Knowing which bacterial structures phages interact with, as well as a better understanding of the resistance mechanisms they elicit, are crucial for the success of this approach. One participant states that bacterial resistance against any phage cocktail will inevitably occur at some point during treatment, but that by then the amount of pathogenic bacteria might have been sufficiently reduced (the equilibrium is restored) for the patient's immune system (or other antibacterials) to resolve the infection. It was suggested earlier that synergy between phages and the patient's immune system might be required for the resolution of the bacterial disease in certain indications [12].

Should sequential strategies in which individual active phages are applied one after the other, so that treatment does not simultaneously select for broad resistance in the targeted bacteria, be considered [6]? Most participants believe that this approach could be (more) effective, but would be very difficult to implement in clinical practice. Especially in severe acute infections, this approach would require large collections of different lytic phage clones and rapid (automated) phage selection and adaptation techniques. Rapid sequencing technologies and algorithm based phage selection of phages from libraries may allow rational selection of phage cocktails targeting different conserved receptors. In addition, sequential strategies are only feasible for hospitalized patients or ambulant patients visiting the hospital on a regular basis (e.g., every day). It would also require the use of significantly more diagnostic tools than is currently the habit of medical doctors and veterinarians when administering broad-spectrum antibacterials. One expert notes that sequential approaches could also be achieved using burst release and time delayed release systems [13]. Another participant fears that the sequential approach will give bacteria the opportunity to develop resistance against one active phage at the time. Two participants assume that decreases of phage efficiency are partly due to the patient's immune response and that sequential approaches could therefore be more effective.

In vitro studies indicate that pre-adapting lytic phages to a pathogen leads to increased pathogen clearance and lowered resistance evolution [7], but will it lower the occurrence of bacterial phage resistance in clinical practice? Most participants presume that it may, as properly pre-adapted phages could harbor mutations (e.g., single-nucleotide polymorphisms (SNPs) or short deletions), which would allow them to escape antiviral mechanisms, but they would like to see clinical evidence of this. In addition, pre-adaptation could also result in phages with broader host ranges and increased infectious abilities (see Section 3). Two participants claim that pre-adapting phages will likely only result in a faster co-evolution process and will have no impact on bacterial phage resistance.

One participant points out that phage resistance is sometimes due to an interruption of the lytic cycle and therefore advocates the use of phage endolysins. Some experts propose to combine the use of phage cocktails and antibiotics, while choosing phages that interact with relevant antibiotic resistance determinants. As such, bacterial phage resistance could lead to (regained) increased susceptibility to antibiotics [14], leading to synergistic selective pressure. More studies are needed to investigate the significance of this synergistic effect in vivo.

2.2. *Phages in Agriculture, Fisheries and Food*

Phage products are on the market for the decontamination of food pathogens and phage probiotics as well as products for farms are in development [15]. However, what do we know about the impact of the large-scale and empirical use of phages in agriculture and food on bacterial resistance to phages or on the shape and diversity of bacterial populations in the environment or in field trials?

Most participants believe that, in a way analogous to antibiotics, the uncontrolled widespread use of phages in agriculture and food decontamination might become a contributor to phage resistant bacterial diseases if phage therapy is to be (re-)integrated in human medicine. Some participants fear that bacterial phage resistance determinants will spread (e.g., through horizontal gene transfer should this transfer pathway definitely play a major role) and persist in the environment. In addition, since complex interactions between phages and bacteria already play significant roles in the composition of environmental microbial communities (e.g., bacterial adaptation to stress via phage transduction), there could be important and unpredictable impacts on the ecosystem.

One expert in aquaculture-associated phage research suggests that phage resistant bacteria might not be able to persist in the environment in the absence of the applied phage, due to the fitness cost of typical mutations that confer phage resistance. It is, however, not clear how phage resistance would differ from antibiotic resistance in this perspective. Three participants argue that the use of phages in agriculture, fisheries and food will likely increase bacterial phage resistance, but they are confident that the host–parasite co-evolutionary arms race will always result in the emergence of successful phages [16]. One expert points out that many bacterial serotypes that cause cattle/fish/plant diseases are different from those causing human disease, but that this does not exclude the emergence of cross-resistance.

With the exception of two participants, who do not believe that bacterial phage resistance will persist in the environment, most experts suggest restricting the use of phages to a greater or lesser extent to prophylactically limit the potential spread of bacterial phage resistance in anticipation of relevant data. The majority proposes avoiding the empirical (without previous diagnosis) use of phages and to control and limit the scale of phage applications especially in agriculture, fisheries and food. A few participants would like to reserve the use of phages for serious (antibiotic resistant) human infections at first priority, with phages obtained only upon medical prescription. Finally, all participants feel that phage products should not be produced, marketed and used as if they were a new class of antibiotics and suggest that phage therapy should have its own regulatory platform, allowing flexible approaches including the timely production, composition, and adaptation of phage products. Two experts stress that it is important that the production of phage preparations should comply with Good Manufacturing Practices (GMP). To summarize, more fundamental research is required in order to differentiate between bacterial phage resistance mechanisms, between in vitro and in vivo resistance phenomena and to quantify these more precisely. This is necessary to better understand therapeutic phage efficacy. Finally, and in the context of the patient's immune response, such comparative data assessments will shed light on the truth of bacterial phage resistance.

3. Phage Training

3.1. *Phage Therapy and the Problem of Heterogeneity in Bacterial Populations*

It is well known that bacterial populations are heterogeneous and that this heterogeneity can originate either genetically or phenotypically [17–19]. Heterogeneity is already considered problematic from a therapeutic point of view when it concerns bacterial susceptibility to antibiotics [20–22]. Similarly, it could challenge phage therapy since it is known that phage-resistant variants pre-exist within bacterial populations and can be relatively easily selected in vitro through bet hedging [8,17,23]. A very recent case of a patient suffering from an *Acinetobacter baumannii* disseminated infection treated with phage therapy highlighted the clinical relevance of such phage-resistant variants [24]. Indeed, phage-resistant clones were selected in the patient during the treatment course, which necessitated adjustment of the phage cocktail composition twice. In this case, it stands to the credit of the involved teams that they were able to sequentially produce tailored cocktails of natural phages (i.e., non-trained) able to kill the resistant clones, within the very limited amount of time available to the patient.

3.2. *Phage Training?*

In parallel to the adaptation of the bacterial host to the attacking phage, phages in turn naturally adapt to their hosts during co-evolution in common habitats following an arms race or fluctuating selection processes [25–27], explaining why both bacteria (the prey) and phages (the predator) are still present on the surface of our planet after billions of years of co-habitation. As a result, phages that have been evolved to better fit the context of phage therapy can be selected for in vitro and in vivo. Various forms of phage training, also known as phage adaptation or phage pre-adaptation, have been developed to select for these evolved phages through experimental procedures performed in a laboratory. It is generally acknowledged that phage training protocols originated from the so-called Appelmans experiment reported in 1921 [11]. At first, Appelmans designed his classical eponymous experiment to titer a phage solution more precisely than d'Hérelle performed at that time. In order to do so, Appelmans was inspired by an approach of serial dilutions to quantify bacteria in water samples as described in chapter III of Miquel's "*Manuel Pratique d'analyse bactériologique des eaux*" published in 1891. The principle is relatively simple and is still used today with some modifications.

In this original study, Appelmans exposed a liquid culture of a susceptible bacterium to serial dilutions of the phage (up to 10^{-12}), an experiment very similar to what is currently done in the macro-dilution method for the determination of the Minimum Inhibitory Concentration (MIC) of antibiotics. After incubation (incubation time is only indicated in Appelmans' paper as "immediate" or "lately"), the tubes in which bacteria were able to grow were considered to be devoid of phages and the tubes in which no growth was observed were considered as containing phages. Taking into consideration the dilution factor, Appelmans was able to precisely determine the phage titer in the undiluted solution. This experiment also allowed him to further validate d'Hérelle's hypothesis about the nature of phage amplification on bacteria. Moreover, Appelmans decided to perform an additional series of experiments in which he serially diluted the phage into 50% alcohol or 5% phenol. Indeed, at that time it had already been published that the phage Appelmans used in his study was stable when exposed to both agents. However, stability was only tested in a highly concentrated phage solution. Surprisingly, Appelmans, with his dilution approach, highlighted the fact that not all phages in the solution were equally resistant to both agents. Indeed, while the non-exposed phage solution was still active at a 10^{-10} dilution, the corresponding phage solution exposed to 50% alcohol remained active only when diluted up to 10^{-6} independent of the incubation time (6 h, 24 h, 3 days, 10 days or 20 days). An additional dilution of this solution led to its inactivity, arguing for the presence of a fixed number of phages insensitive to the agent in the test tube. Appelmans made the same observation with 5% phenol except that the number of phages able to resist this treatment was much lower (dilution 10^{-3} still active). In other words, this experiment demonstrated selection through dilution of phage variants able to resist to some chemicals otherwise toxic for the majority of individuals in the phage population. This is indeed exactly the principle of phage training in which phage variants able to very efficiently lyse a bacterial population are selected through dilution. There are indications that Félix d'Herelle introduced the concept of serial passages that were not performed in the original Appelmans experiments [28].

Two primary protocols of phage training for expanded host range have been reported in the literature [7,29,30]. Firstly, a phage/bacteria mixture is simply diluted into fresh growth medium after a period of co-incubation [7,30]. In the second, a fixed concentration of bacteria is co-incubated with serial dilution of a phage stock in growth medium for 16–24 h. The next morning, the mixture in the tube in which lysis occurred at the lowest phage concentration is further chloroformed and filtered before being serially re-diluted and mixed with sample of a fresh culture of the ancestor bacteria [31]. In both protocols the procedure can be repeated for several "passages".

3.3. Outcome of Phage Training

Experts pointed out that while several mechanisms of phage adaptation ensuring phage propagation on co-evolving hosts were previously described [32]; it is only more recently that the benefits of experimental phage training started to be investigated in controlled assays. In a first study [33], the training *P. aeruginosa* phage PAK_P3, which initially showed only slight lysis on strain CHA, led to the selection of phage P3-CHA with significantly increased in vitro efficiency of plating. This in vitro result was confirmed in vivo in a mouse model of lung infection with 100% versus 20% survival rate achieved by P3-CHA and PAK_P3, respectively. Strikingly, this improved in vivo activity was reported to be due to only two single nucleotide changes in different putative open reading frames (ORFs). This result highlights how quickly a new therapeutic phage with tremendously increased infectivity can emerge in a natural co-evolution process, and could therefore be artificially selected for by phage training. In another study it has been shown that four evolved *P. aeruginosa* phages obtained within two passages over four days were more efficient than ancestral phages in reducing in vitro mean bacterial densities of ten *P. aeruginosa* strains [7]. Accordingly, *P. aeruginosa* phages LKD16 and 14/1 pre-adapted over six serial passages were shown to target clones within a bacterial population of the strain PAO1 that were originally resistant to the ancestor phages. Indeed, the proportion of susceptibility over 20 different clones increased from 80% to 85% for the ancestral phages to 100% for the evolved phages [26]. Therefore, increased infectivity of evolved over ancestral phages is usually attributed to a decreased capacity of the ancestral bacterial strains to evolve resistance towards the evolved phages.

3.4. Relevance and Implementation of Phage Training in the Clinic

While one expert described how phage training has been common practice for more than 80 years at the Eliava Institute, all tend to think that there is no doubt that relying on evolved phages able to sidestep bacterial heterogeneity would have been highly desirable for the case discussed above (see Section 3.1) and would therefore be relevant in the clinic in general. Although a pre-clinical study has reported the benefit of a trained phage relative to its original counterpart in vivo, comparable studies should be set up to shed additional light on this biological phenomenon [33].

The implementation of phage training in clinical protocols is appealing for at least two reasons. First, having access to phages covering 100% of the clones within the population of a given strain could dramatically increase the success of phage therapy in a given patient. For instance, *P. aeruginosa* populations in the lungs of patients with cystic fibrosis (CF) harbor a very high phenotypic diversity [34]. Therefore, training for phages that would render them able to cover this phenotypic diversity in CF patients could be clinically significant. Secondly, having access to single phages covering close to 100% of the circulating strains of a given pathogen would allow usage of a very limited number of phages (or, in rare cases, even a single phage) for many patients, thus simplifying the production process.

All experts further distinguished two situations, i.e., acute and chronic infections. If phage training is demonstrated in the future to be an efficient way to significantly improve therapeutic outcomes, several experts noticed that implementation in chronic situations where time is not such an issue would in principle be considerably easier. However, the addition of phage training steps to a treatment protocol would be time consuming and feasible only if sufficient qualified personnel were available. In such a situation, experts pointed out that phage training could be done by either following (i) a one-size-fits-all strategy by training phages on already available representatives of local strains, which will help with setting up and regularly updating specialized phage collections, or (ii) a tailored strategy by training phages on the patient's strain as soon as it became available in a form of highly personalized medicine. Of note, this latter strategy is applied at the Eliava Institute in the process of development of so-called "autophages".

Accordingly, many experts agreed that in intensive care units (ICUs), where patients need to be treated within minutes or hours, implementation of a tailored strategy would be difficult due to time limitations. Indeed, as discussed before, phage training protocols usually require a week to be performed or at minimum 24–48 h in case of a single passage [29]. Nevertheless, an expert pointed out that often the strain that will cause the life-threatening condition, often including sepsis, in ICU patients is known days before as the dominant colonizing strain. In such a situation, patients could be decolonized with available phages (see above) and autophages could then be developed to adjust the treatment and cover potential phage-resistant variants selected by the former phages, in a way that is similar to what occurred in the case discussed in Section 3.1). This strategy is very similar to the situation where patients are first treated with broad-spectrum antibiotics and treatment is adjusted later according to an antibiogram. However, in acute situations where the strain would not be available in advance, experts agree that the emergency use of broad-host range phage cocktails could be a viable strategy. As a conclusion, while all experts rather agree on the clinical importance of developing trained phages, some think that detailed pre-clinical and clinical studies still need to be performed to properly evaluate "cost vs. benefits" and decide if it is worthwhile to invest into the time- and resources-consuming development of trained phage collections. If such a strategy were shown to be a viable option, phage training could either be implemented as a one-size-fits-all or a tailored strategy depending on the patient's condition.

3.5. Regulatory Considerations Regarding Trained Phages

In answer to the question "should trained phages be considered as natural phages?" if one would use them in clinical trials, a large majority of experts answered "yes". Indeed, phage training is based on a naturally occurring event and does not rely on "human-guided" modification of the phage genome through for instance synthetic biology or any other tools. A trained phage is a phage in which random mutations have been introduced by co-evolution with bacteria as it happens in nature. In other words, a trained phage is a phage variant that pre-exists in nature and is selected and amplified in the laboratory through natural, but accelerated, co-evolution. Since adapted phages are variants selected from a population of natural phages, their status regarding regulatory agencies should be similar to the original phage.

4. Prophages in Bacterial Production Strains

4.1. Relevance of Prophages to the Production of Therapeutic Phages

Temperate phages are champions of evolution, but unwanted during pharmaceutical phage production. Why do we address the "prophage issue" here and discuss it in depth? It is a matter of course to consider all relevant questions and parameters before therapeutic phage preparations are produced. As discussed earlier by international experts, production of phage preparations for medical application must follow defined procedures and quality assessments [35–38]. It is important to understand the biology of the two different types of phages that are in fact two different forms of life: obligately lytic, virulent phages are attractive potent alternatives to antibacterial drugs as these phages kill their bacterial host cells upon infection. Temperate phages lysogenize their bacterial host cells, exist as prophages after integration into the bacterial chromosomes or as plasmid-like extra-chromosomal elements. When "induced", they then change their life cycle and behave like virulent phages while lysing bacterial hosts. The term "temperate phage" thus refers to the character of this form of phage life whereas "prophage" describes a status in the complex life cycle of such a phage. Temperate phages exist in most bacteria as prophages, often abundantly, being normal parts of bacterial genomes.

Indeed, prophages are extremely abundant elements in the biosphere. Consequently, they also belong to our own microbiome and are part of its virome, and can be described as a phageome [39]. It has even been found that prophages in our microbiome are communicating via signaling peptides though the "arbitrium code" system [40]. Lysogeny clearly plays an enormous evolutionary role, with temperate phages substantially involved in co-evolution processes of both bacteria and phages [41]. This is relevant also in the context of this article: we are carrying countless prophages. Both, the microbiome and the individual macro-organism represent the holobiont, this rather recent and Solomonic perception of life includes all forms of life that contribute to an individual's existence and health [42].

Prophage induction occurs when expression of the transcriptional repressor keeping host lethal genes involved in lytic infection shut off is impaired to such an extent that lytic infection begins. Classically, this impairment is known to be caused by stresses such as UV irradiation, mutagens, quorum-sensing signaling molecules, fluoroquinolones, oxidative stresses (such as hydrogen peroxide) or a temperature shift that generates sufficient transient damage to the bacterial cell. In addition, it has also been long known that infection by lytic phages, like happens as an inherent part of the lytic phage production process, also strongly induces the excision of prophages. Indeed, five of seven *P. aeruginosa* phages examined by Blasdel et al. (including lytic phages being currently used for therapy) induce the transcription of at least part of at least one prophage element in their PAO1 host [43]. If prophage induction successfully highjacks the phage infection during production, then the prophage will be released into the medium, and possibly contaminate the therapeutic preparation that should exclusively contain the lytic phage. Finally, the induction of active prophages can also occur spontaneously with various frequencies [44], typically one in 10^3–10^5 cells, but sometimes one in 10^2 cells [45], and certain toxin-encoding phages were shown to be induced spontaneously with higher frequency than their non-toxin-encoding relatives [46]. As a consequence, populations of lysogens are typically contaminated with free temperate phages [47].

According to pharmaceutical standards, such a contamination of therapeutic phages with temperate phages is out of the question. Selection of the strain and a carefully adapted experimental quality control procedure is crucial as both phage types, the therapeutic phage and a temperate contaminant phage, cannot be separated chemically or physically as they share biochemical and structural similarities. Prophages are common in probably all taxonomic groups of bacteria, and knowledge about them is growing as more strains are sequenced and bioinformatics tools are getting more sophisticated. Comparative genomics is a fundamentally important scientific development. Indeed, many prophages are associated with pathogenicity, such as in *E. coli*, *Streptococcus pyogenes*, *Salmonella enterica*, *Staphylococcus aureus* and can encode for exotoxins like in *E. coli* EHEC or *Vibrio cholera* as well as a rather broad spectrum of enzymes significant for bacterial virulence [48].

However, bacterial genomes often bear non-complete (cryptic) prophages that cannot be induced or released from the cell. Due to the intensive relatedness to the human microbiome, most clinically important bacterial isolates, especially those belonging to the ESKAPE group (Enterococci (VRE), *S. aureus* (MRSA), *Klebsiella pneumoniae* (Carbapenem resistant), *A. baumannii*, *P. aeruginosa*, *Enterobacteriaceae* (ESBL)) contain both intact and cryptic prophage sequences in their genomes. It is our own rich and dense microbiome habitat that inevitably causes constant genetic exchange and co-evolution of bacteria and their phages. Especially for clinical isolates, it is a selective advantage to carry many prophages, which often encode genetic cassettes that benefit their hosts such as virulence factors [49,50]. Extreme examples are the food pathogen *E. coli* EHEC O157:H7 strain Sakai carrying 18 prophages which amount to 16% of the total genome or *S. pyogenes* with up to six prophages comprising 12% of the bacterial genome [51,52]. Stable prophages that are typically observed in isolated bacterial cultures have a long-lasting bond with their hosts comparable to a symbiosis finally supporting duration of both. However, it is important to note that prophages can either cause loss of host fitness or, just the opposite by adding new functions to the cell via lysogenic conversion comparable to a "gain-of-function". Still another type of host–phage interaction is active lysogeny: prophages may integrate in critical bacterial genes such that they function as regulatory switches [50]. Stable prophages tend to protect their bacterial hosts until "no other choice but escape" is left. In this context, it is important to distinguish precisely between different genetic transfer mechanisms like general and specialized transduction and lysogenic conversion. Bacteria may benefit from lysogenic conversion by obtaining novel virulence factors (toxins, super-antigens, immune evasion, invasivity, adherence, resistance to phages) [53,54]. It might be postulated that cryptic prophages and active lysogeny are part of bacterial regulatory systems. The interaction between such prophages and lytic phages could be considered as interaction between bacterial genomes and phages or, between lysogenic and lytic phages. Such questions are required to get in focus of fundamental research. For the purpose of this article, it is important to understand the biology of obligately lytic and of temperate phages. There are many fascinating and as yet unexplored aspects of the microbial and phage world, however, conditions outside the laboratory or production facility are not a priority for this article.

4.2. Detection of Prophages in Bacterial Genomes

There is consensus among experts that host strain genome sequencing is an essential initial investigation before starting phage production processes as it allows the identification of prophage genes like integrases, repressors, excisases, recombinases, terminases. Having this sequence available thus makes predictions for prophage properties like virulence factors or for prophage incompleteness possible. Induction ability might be experimentally studied by using Mitomycin C or UV irradiation but these induction methods are not successful enough and require indicator strains that are not always available. Due to the increasing importance of phage applications, entering prophage genomes in databases is necessary when bacterial genomes are analyzed. Algorithms for finding prophages are available like PhiSpy [55], PHAST (http://phast.wishartlab.com) and PHASTER [56] and genome annotations will shed light on prophage properties and this is getting less complicated the more sequence data are entered into databases. We strongly encourage researchers to make such data publically available.

4.3. Prophages in Bacterial Strains Used to Produce Therapeutic Phages

Finding production strains that are completely free of prophages is generally difficult, especially in the case of some pathogens. Therefore, inducibility and the genetic outfit of a prophage should be considered, definitions are necessary to deal with this issue. Genome sequencing is indispensable as prophages might be carriers of pathogenicity factors, as well as antibiotic and phage resistance determinants. Isolation of new phages aiming at therapeutic preparations cannot initially exclude temperate phages but further characterization steps should definitely eliminate them. Regarding the production strain and production process, the individual situation has to be assessed: is a strain available (and effective!) that is lacking prophages? If not, do bioinformatic genome analyses describe the prophage's genetic outfit sufficiently to confirm a cryptic nature? Is there a potential of infectious particles being released into the medium during production and if so, how probable is this event? If all these evaluations conclude that the production strain does not present likelihood of a problem, the final point should be considered that in a special situation the preparation may be the only viable alternative left, this is especially the case for magistral purposes. It is also important to note that a phage production strain must efficiently produce phages, meaning the EOP has to be satisfactory. Expert consensus was that production strains should ideally not carry prophages or should no other suitable strain exist then the risk needs to be controlled to minimize prophage expression [35].

We suggest that the significance of this concern must be put into perspective and seen in comparison to the global multidrug resistance crisis and the human crisis in individual cases. Improvement of phage production is important but it is fortunate that (1) lytic phages for clinical isolates of the ESKAPE bacteria (see above) are not difficult to isolate and (2) a rather good number of production strains within the ESKAPE bacteria should be available. Careful monitoring of the production process and where possible, a pre-adaptation of the therapeutic phage candidates to suitable production strains is a way forward. Repeated phage genome sequencing during the production process seems mandatory. Gene modification of production strains is discussed, but may only be one way out in extremely challenging cases in the future whereas currently, experts agree that eliminating prophages from the bacterial genome should not become a prerequisite. Q-PCR or sequencing can confirm purity of the DNA of the lytic phage and exclude prophage DNA. According to one expert, one possible way to assess the capacity of prophage elements to successfully induce under production conditions involves the use of Q-RT-PCR to track the concentration of DNA for each prophage relative to other genomic DNA during infection with primers specific to each prophage element as was done by Ceyssens et al. [57]. If replication of prophage DNA relative to other genomic DNA does not occur, this would definitively exclude successful prophage induction.

As a conclusion, it is obvious that more phage research is required going hand-in-hand with applied therapy approaches since an easy solution to the "prophage problem" is not obvious. Indeed, this is even more problematic if a phage should be amplified on a freshly isolated patient strain (see magistral application [38]). If compared to a threshold endotoxin limit in phage preparations, it might be feasible to define such a limit for prophage presence. Mathematical modeling of phage–bacterium pharmacodynamics may allow simulation of whether such concentrations are significant in terms of propagation of the temperate phages; low concentration of temperate phages may bear a very low statistical probability of significant amplification as the phage binding process is concentration dependent. Such thoughts seem theoretical but might be key for the design of strategies in the near future and for negotiating with licensing authorities.

4.4. Prophages in Production Strains: What Does It Mean for the Licensing Pathway?

It must be stated that intact prophages are not necessarily induced and released at high frequency. If a production strain contains a frequently inducible intact prophage it should not bear antibiotic resistance genes, toxin genes or other virulence factors to reduce the risk of gene transfer. The presence of temperate phages in a therapeutic preparation should normally not affect its efficacy (see above). Lytic phages can in principal also contribute to horizontal gene transfer via generalized transduction, but a risk–benefit evaluation must be rational, and a pragmatic common-sense approach is needed; just look at the human microbiome and its virome where prophages are the most frequent inhabitants! For commercially available phage preparations in the future it should be compulsory that production strains are free of functional prophages whereas exceptions might be made for highly experimental treatments. Generally, functional-prophage-free production strains should be the primary choice, however, other parameters need consideration including determining the suitability of a production strain e.g., efficacy and prophage-independent pathogenicity factors etc. For regulatory pathways, some kind of "prophage acceptance threshold" should be defined, it would contribute to a constructive timely European approach to design the regulatory framework for safe phage therapy. Furthermore, the establishment of collections/banks of prophage-free potential production strains, especially those of the ESKAPE bacteria, is urgently needed. Such strains can sometimes be acquired by natural means, such as prophage induction and prophage-free cell selection, without the use of genetic modifications [58].

5. Concluding Remarks

5.1. Bacterial Phage Resistance

It is commonly accepted that combining phages with different infection strategies into cocktails will help reduce the selection of phage resistant bacterial clones, but this point has only been studied on a few occasions both in vitro and in vivo. Sequential strategies might be interesting when there is no sense of urgency (e.g., long-term phage therapy of chronic infections), but more research is needed here. In acute life-threatening infections, and in anticipation of rapid (automated) bacterial identification and phage selection and adaptation techniques, well-thought-out broad-spectrum phage cocktails are warranted. More in vivo studies are needed to document the impact of phage pre-adaptation on bacterial phage resistance reduction. More research is needed to determine if the theoretical bacterial phage resistance issue (if phage therapy would be applied intensively, which, even in Georgia, is not the case today) will be comparable to the persistent bacterial multidrug resistance problem we are facing today. Meanwhile, most experts suggest restricting the use of phages, to a greater or lesser extent, to limit the potential spread of bacterial phage resistance in anticipation of relevant data.

5.2. Phage Training

Phage training is an experimental co-evolution approach considerably accelerating the pace at which "increased" phages are selected, thanks to the short doubling time of many bacterial species. Ultimately, naturally pre-adapted phages could be very interesting alternatives to original phages if their improved abilities translate from the preclinical situation to the clinic. Therefore, we believe that trained phages should be included in products to be tested in clinical trials, provided that their safety and superior efficacy compared to the ancestral phage have been documented in in vitro pre-clinical studies. In addition to their potential increased efficacy, broad host and/or variant range of trained phages could offer a significant economic advantage regarding the costs of production in GMP, which is a pre-requisite to prospective clinical trials.

5.3. Prophages in Bacterial Production Strains

Temperate phages/prophages present a problem in phage production processes, but are so abundant in all habitats including our own microbiome that we should not blindly demand their elimination from phage production processes. Instead, we broadly advise that an intensive discussion be had on how to deal with situations when no prophage-free production strains are available. We have explained which properties prophages might carry and what their unwanted features are, how abundant they are in opportunistic common ESKAPE bacteria, how genome analyses precisely characterize them and what to calculate if a production strain carries a prophage. We advocate that researchers enter bacterial and (pro)phage genome data into public databases because rich databases will provide the platform that is needed to deal with these particular pressing questions and because phage therapy will be a potent alternative therapy in the global multidrug resistance threat. Resolute and concurrent activities are urgently needed and harmonized licensing pathways are desirable. They might include best practice guidance for those few gaps that cannot be completely solved when a biological therapy approach is used; this may include the prophage issue.

Acknowledgments: This work was supported through the PHAGEFORWARD project of the fourth European Joint Programming Initiative on Antimicrobial Resistance (JPIAMR) call: "AMR Networks/Working Groups".

Conflicts of Interest: The authors declare no conflict of interest.

References

1. Koskella, B.; Brockhurst, M.A. Bacteria-phage coevolution as a driver of ecological and evolutionary processes in microbial communities. *FEMS Microbiol. Rev.* **2014**, *38*, 916–931. [CrossRef] [PubMed]
2. Labrie, S.J.; Samson, J.E.; Moineau, S. Bacteriophage resistance mechanisms. *Nat. Rev. Microbiol.* **2010**, *8*, 317–327. [CrossRef] [PubMed]
3. Maxwell, K.L. Phages fight back: Inactivation of the CRISPR-Cas bacterial immune system by anti-CRISPR proteins. *PLoS Pathog.* **2016**, *12*, e1005282. [CrossRef] [PubMed]
4. Goldfarb, T.; Sberro, H.; Weinstock, E.; Cohen, O.; Doron, S.; Yoav Charpak-Amikam, Y.; Afik, S.; Ofir, G.; Sorek, R. BREX is a novel phage resistance system widespread in microbial genomes. *EMBO J.* **2015**, *34*, 169–183. [CrossRef] [PubMed]
5. Ofir, G.; Melamed, S.; Sberro, H.; Mukamel, Z.; Silverman, S.; Yaakov, G.; Doron, S.; Sorek, R. DISARM is a widespread bacterial defence system with broad anti-phage activities. *Nat. Microbiol.* **2018**, *3*, 90–98. [CrossRef] [PubMed]
6. Hall, A.R.; de Vos, D.; Friman, V.P.; Pirnay, J.P.; Buckling, A. Effects of sequential and simultaneous applications of bacteriophages on populations of *Pseudomonas aeruginosa* in vitro and in wax moth larvae. *Appl. Environ. Microb.* **2012**, *78*, 5646–5652. [CrossRef] [PubMed]
7. Friman, V.P.; Soanes-Brown, D.; Sierocinski, P.; Molin, S.; Johansen, H.K.; Merabishvili, M.; Pirnay, J.P.; de Vos, D.; Buckling, A. Pre-adapting parasitic phages to a pathogen leads to increased pathogen clearance and lowered resistance evolution with *Pseudomonas aeruginosa* cystic fibrosis bacterial isolates. *J. Evolut. Biol.* **2016**, *29*, 188–198. [CrossRef] [PubMed]
8. Oechslin, F.; Piccardi, P.; Mancini, S.; Gabard, J.; Moreillon, P.; Entenza, J.M.; Resch, G.; Que, Y.-A. Synergistic interaction between phage therapy and antibiotics clears *Pseudomonas aeruginosa* infection in endocarditis and Reduces Virulence. *J. Infect. Dis.* **2017**, *215*, 703–712. [CrossRef] [PubMed]
9. Laanto, E.; Hoikkala, V.; Ravantti, J.; Sundberg, L.R. Long-term genomic coevolution of host-parasite interaction in the natural environment. *Nat. Commun.* **2017**, *8*. [CrossRef] [PubMed]
10. Międzybrodzki, R.; Borysowski, J.; Weber-Dąbrowska, B.; Fortuna, W.; Letkiewicz, S.; Szufnarowski, K.; Pawełczyk, Z.; Rogóż, P.; Kłak, M.; Wojtasik, E.; et al. Clinical aspects of phage therapy. *Adv. Virus Res.* **2012**, *83*, 73–121. [PubMed]
11. Appelmans, R. Le dosage du bactériophage. *C. R. Soc. Biol. Fil.* **1921**, *89*, 1098.
12. Roach, D.R.; Leung, C.Y.; Henry, M.; Morello, E.; Singh, D.; Di Santo, J.P.; Weitz, J.S.; Debarbieux, L. Synergy between the host immune system and bacteriophage is essential for successful phage therapy against an acute respiratory pathogen. *Cell Host Microbe* **2017**, *22*, 38–47. [CrossRef] [PubMed]
13. Malik, D.J.; Sokolov, I.J.; Vinner, G.K.; Mancuso, F.; Cinquerrui, S.; Vladisavljevic, G.T.; Clokie, M.R.J.; Stapley, A.G.F.; Kirpichnikova, A. Formulation, stabilisation and encapsulation of bacteriophage for phage therapy. *Adv. Colloid Interface Sci.* **2017**, *249*, 100–133. [CrossRef] [PubMed]

14. Chan, B.K.; Sistrom, M.; Wertz, J.E.; Kortright, K.E.; Narayan, D.; Turner, P.E. Phage selection restores antibiotic sensitivity in MDR *Pseudomonas aeruginosa*. *Sci. Rep.* **2016**, *6*, 26717. [CrossRef] [PubMed]
15. Cooper, I.R. A review of current methods using bacteriophages in live animals, food and animal products intended for human consumption. *J. Microbiol. Meth.* **2016**, *130*, 38–47. [CrossRef] [PubMed]
16. Ormala, A.M.; Jalasvuori, M. Phage therapy: Should bacterial resistance to phages be a concern, even in the long run? *Bacteriophage* **2013**, *3*, e24219. [CrossRef] [PubMed]
17. Davis, K.M.; Isberg, R.R. Defining heterogeneity within bacterial populations via single cell approaches. *BioEssays* **2016**, *38*, 782–790. [CrossRef] [PubMed]
18. Magdanova, L.A.; Goliasnaia, N.V. Heterogeneity as an adaptive trait of the bacterial community. *Mikrobiologiia* **2013**, *82*, 3–13. [PubMed]
19. Veening, J.W.; Smits, W.K.; Kuipers, O.P. Bistability, epigenetics, and bet-hedging in bacteria. *Annu. Rev. Microbiol.* **2008**, *62*, 193–210. [CrossRef] [PubMed]
20. Woodford, N.; Ellington, M.J. The emergence of antibiotic resistance by mutation. *Clin. Microbiol. Infect.* **2007**, *13*, 5–18. [CrossRef] [PubMed]
21. Babouee Flury, B.; Ellington, M.J.; Hopkins, K.L.; Turton, J.F.; Doumith, M.; Loy, R.; Staves, P.; Hinic, V.; Frei, R.; Woodford, N. Association of novel nonsynonymous single nucleotide polymorphisms in *ampD* with cephalosporin resistance and phylogenetic variations in *ampC*, *ampR*, *ompF*, and *ompC* in *Enterobacter cloacae* isolates that are highly resistant to carbapenems. *Antimicrob. Agents Chempother.* **2016**, *60*, 2383–2390. [CrossRef] [PubMed]
22. Proctor, R.A.; Kahl, B.; von Eiff, C.; Vaudaux, P.E.; Lew, D.P.; Peters, G. Staphylococcal small colony variants have novel mechanisms for antibiotic resistance. *Clin. Infect. Dis.* **1998**, *27* (Suppl. 1), S68–S74. [CrossRef] [PubMed]
23. Beaumont, H.J.; Gallie, J.; Kost, C.; Ferguson, G.C.; Rainey, P.B. Experimental evolution of bet hedging. *Nature* **2009**, *462*, 90–93. [CrossRef] [PubMed]
24. Schooley, R.T.; Biswas, B.; Gill, J.J.; Hernandez-Morales, A.; Lancaster, J.; Lessor, L.; Barr, J.J.; Reed, S.L.; Rohwer, F.; Benler, S.; et al. Development and use of personalized bacteriophage-based therapeutic cocktails to treat a patient with a disseminated resistant *Acinetobacter baumannii* infection. *Antimicrob. Agents Chemother.* **2017**, *61*. [CrossRef] [PubMed]
25. Stern, A.; Sorek, R. The phage-host arms race: Shaping the evolution of microbes. *BioEssays* **2011**, *33*, 43–51. [CrossRef] [PubMed]
26. Betts, A.; Kaltz, O.; Hochberg, M.E. Contrasted coevolutionary dynamics between a bacterial pathogen and its bacteriophages. *Proc. Natl. Acad. Sci. USA* **2014**, *111*, 11109–11114. [CrossRef] [PubMed]
27. De Sordi, L.; Khanna, V.; Debarbieux, L. The gut microbiota facilitates drifts in the genetic diversity and infectivity of bacterial viruses. *Cell Host Microbe* **2017**, *22*, 801–808. [CrossRef] [PubMed]
28. D'Herelle, F. On an invisible microbe antagonistic toward dysenteric bacilli: Brief note by Mr. F. D'Herelle, presented by Mr. Roux. 1917. *Res. Microbiol.* **2007**, *158*, 553–554. [PubMed]
29. Merabishvili, M.; Pirnay, J.P.; de Vos, D. Guidelines to compose an ideal bacteriophage cocktail. *Methods Mol. Biol.* **2018**, *1693*, 99–110. [CrossRef] [PubMed]
30. Betts, A.; Vasse, M.; Kaltz, O.; Hochberg, M.E. Back to the future: Evolving bacteriophages to increase their effectiveness against the pathogen *Pseudomonas aeruginosa* PAO1. *Evolut. Appl.* **2013**, *6*, 1054–1063. [CrossRef]
31. Merabishvili, M.; Pirnay, J.P.; de Vos, D. Guidelines to compose an ideal bacteriophage cocktail. In *Bacteriophage Therapy: From Lab to Clinical Practice*, 1st ed.; Azaredo, J., Sillankorva, S., Eds.; Humana Press: New York, NY, USA, 2017; Volume 1, pp. 107–108, ISBN 9781493973941 1493973940.
32. Samson, J.E.; Magadan, A.H.; Sabri, M.; Moineau, S. Revenge of the phages: Defeating bacterial defences. *Nat. Rev. Microbiol.* **2013**, *11*, 675–687. [CrossRef] [PubMed]
33. Morello, E.; Saussereau, E.; Maura, D.; Huerre, M.; Touqui, L.; Debarbieux, L. Pulmonary bacteriophage therapy on *Pseudomonas aeruginosa* cystic fibrosis strains: First steps towards treatment and prevention. *PLoS ONE* **2011**, *6*, e16963. [CrossRef] [PubMed]
34. Clark, S.T.; Diaz Caballero, J.; Cheang, M.; Coburn, B.; Wang, P.W.; Donaldson, S.L.; Zhang, Y.; Liu, M.; Keshavjee, S.; Yau, Y.C.; et al. Phenotypic diversity within a *Pseudomonas aeruginosa* population infecting an adult with cystic fibrosis. *Sci. Rep.* **2015**, *5*, 10932. [CrossRef] [PubMed]

35. Pirnay, J.P.; Blasdel, B.G.; Bretaudeau, L.; Buckling, A.; Chanishvili, N.; Clark, J.R.; Corte-Real, S.; Debarbieux, L.; Dublanchet, A.; de Vos, D.; et al. Quality and safety requirements for sustainable phage therapy products. *Pharm. Res.* **2015**, *32*, 2173–2179. [CrossRef] [PubMed]
36. Expert Round Table on Acceptance and Re-implementation of Bacteriophage Therapy. Silk route to the acceptance and re-implementation of bacteriophage therapy. *Biotechnol. J.* **2016**, *11*, 595–600. [CrossRef]
37. Cooper, C.J.; Khan Mirzaei, M.; Nilsson, A.S. Adapting drug approval pathways for bacteriophage-based therapeutics. *Front. Microbiol.* **2016**, *7*, 1209. [CrossRef] [PubMed]
38. Pirnay, J.P.; Verbeken, G.; Ceyssens, P.-J.; Huys, I.; de Vos, D.; Ameloot, C.; Fauconnier, A. The magistral phage. *Viruses* **2018**, *10*, 64. [CrossRef] [PubMed]
39. Manrique, P.; Bolduc, B.; Walk, S.T.; van der Oost, J.; de Vos, W.M.; Young, M.J. Healthy human gut phageome. *Proc. Natl. Acad. Sci. USA* **2016**, *113*, 10400–10405. [CrossRef] [PubMed]
40. Erez, Z.; Steinberger-Levy, I.; Shamir, M.; Doron, S.; Stokar-Avihail, A.; Peleg, Y.; Melamed, S.; Leavitt, A.; Savidor, A.; Albeck, S.; et al. Communication between viruses guides lysis-lysogeny decisions. *Nature* **2017**, *541*, 488–493. [CrossRef] [PubMed]
41. Howard-Varona, C.; Hargreaves, K.R.; Abedon, S.T.; Sullivan, M.B. Lysogeny in nature: Mechanisms, impact and ecology of temperate phages. *ISME J.* **2017**, *11*, 1511–1520. [CrossRef] [PubMed]
42. Thomas, S.; Izard, J.; Walsh, E.; Batich, K.; Chongsathidkiet, P.; Clarke, G.; Sela, D.A.; Muller, A.J.; Mullin, J.M.; Albert, K.; et al. The host microbiome regulates and maintains human health: A primer and perspective for non-microbiologists. *Cancer Res.* **2017**, *77*, 1783–1812. [CrossRef] [PubMed]
43. Blasdel, B.G.; Ceyssens, P.J.; Chevallereau, A.; Debarbieux, L.; Lavigne, R. Comparative transcriptomics reveals a conserved Bacterial Adaptive Phage Response (BAPR) to viral predation. *bioRxiv* **2018**. [CrossRef]
44. Lwoff, A. Lysogeny. *Bacteriol. Rev.* **1953**, *17*, 269–337. [PubMed]
45. Nanda, A.M.; Heyer, A.; Krämer, C.; Grünberger, A.; Kohlheyer, D.; Frunzke, J. Analysis of SOS-induced spontaneous prophage induction in *Corynebacterium glutamicum* at the single-cell level. *J. Bacteriol.* **2014**, *196*, 180–188. [CrossRef] [PubMed]
46. Colon, M.P.; Chakraborty, D.; Pevzner, Y.; Koudelka, G.B. Mechanisms that determine the differential stability of Stx(+) and Stx(−) lysogens. *Toxins* **2016**, *8*, 96. [CrossRef] [PubMed]
47. Łobocka, M.; Hejnowicz, M.S.; Dąbrowski, K.; Izak, D.; Gozdek, A.; Głowacka, A.; Gawor, J.; Kosakowski, J.; Gromadka, R.; Weber-Dąbrowska, B.; et al. *Staphylococcus aureus* Strains for the Production of Monoclonal Bacteriophage Preparations Deprived of Contamination with Plasmid DNA. U.S. Patent WO 2016/030871 A1, 16 March 2016.
48. Fortier, L.C.; Sekulovic, O. Importance of prophages to evolution and virulence of bacterial pathogens. *Virulence* **2013**, *4*, 354–365. [CrossRef] [PubMed]
49. Colavecchio, A.; Cadieux, B.; Lo, A.; Goodridge, L.D. Bacteriophages contribute to the spread of antibiotic resistance genes among foodborne pathogens of the *Enterobacteriaceae* family—A review. *Front. Microbiol.* **2017**, *8*, 1108. [CrossRef] [PubMed]
50. Feiner, R.; Argov, T.; Rabinovich, L.; Sigal, N.; Borovok, L.; Herskovits, A.A. A new perspective on lysogeny: Prophages as active regulatory switches of bacteria. *Nat. Rev. Microbiol.* **2015**, *13*, 641–650. [CrossRef] [PubMed]
51. Canchaya, C.; Desiere, F.; Mcshan, W.M.; Ferretti, J.J.; Parkhill, J.; Brüssow, H. Genome analysis of an inducible prophage and prophage remnants integrated in the *Streptococcus pyogenes* strain SF370. *Virology* **2002**, *302*, 245–258. [CrossRef] [PubMed]
52. Touchon, M.; Bernheim, A.; Rocha, E.P. Genetic and life-history traits associated with the distribution of prophages in bacteria. *ISME J.* **2016**, *10*, 2744–2754. [CrossRef] [PubMed]
53. Maslanova, I.; Stribna, S.; Doskar, J.; Pantucek, R. Efficient plasmid transduction to *Staphylococcus aureus* strains insensitive to the lytic action of transducing phage. *FEMS Microbiol. Lett.* **2016**, *363*. [CrossRef] [PubMed]
54. Haaber, J.; Leisner, J.L.; Cohn, M.T.; Catalan-Moreno, A.; Nielsen, J.B.; Westh, H.; Penadés, J.R.; Ingmer, H. Bacterial viruses enable their host to acquire antibiotic resistance genes from neighbouring cells. *Nat. Commun.* **2016**, *7*, 13333. [CrossRef] [PubMed]
55. Akhter, S.; Aziz, R.K.; Edwards, R.A. PhiSpy: A novel algorithm for finding prophages in bacterial genomes that combines similarity- and composition-based strategies. *Nucleic Acids Res.* **2012**, *40*, e126. [CrossRef] [PubMed]

56. Arndt, D.; Grant, J.R.; Marcu, A.; Sajed, T.; Pon, A.; Liang, Y.; Wishart, D.S. PHASTER: A better, faster version of the PHAST phage search tool. *Nucleic Acids Res.* **2016**, *44*, W16–W21. [CrossRef] [PubMed]
57. Ceyssens, P.J.; Minakhin, L.; Van den Bossche, A.; Yakunina, M.; Klimuk, E.; Blasdel, B.G.; de Smet, J.; Noben, J.P.; Bläsi, U.; Severinov, K.; et al. Development of giant bacteriophage ϕKZ is independent of the host transcription apparatus. *J. Virol.* **2018**, *88*, 10501–10510. [CrossRef] [PubMed]
58. Łobocka, M.; Hejnowicz, M.S.; Gągała, U.; Weber-Dąbrowska, B.; Węgrzyn, G.; Dadlez, M. The first step to bacteriophage therapy—How to choose the correct phage. In *Phage Therapy: Current Research and Applications;* Borysowski, J., Międzybrodzki, R., Górski, A., Eds.; Caister Academic Press: Poole, UK, 2004; pp. 23–69.

Brief Report

Selection of Potential Therapeutic Bacteriophages that Lyse a CTX-M-15 Extended Spectrum β-Lactamase Producing *Salmonella enterica* Serovar Typhi Strain from the Democratic Republic of the Congo

Elene Kakabadze [1,†], Khatuna Makalatia [1,†], Nino Grdzelishvili [1], Nata Bakuradze [1], Marina Goderdzishvili [1], Ia Kusradze [1], Marie-France Phoba [2,3], Octavie Lunguya [2,3], Cédric Lood [4,5], Rob Lavigne [4], Jan Jacobs [6,7], Stijn Deborggraeve [8], Tessa De Block [8], Sandra Van Puyvelde [8,9], David Lee [10], Aidan Coffey [10], Anahit Sedrakyan [11], Patrick Soentjens [6,12], Daniel De Vos [13], Jean-Paul Pirnay [13] and Nina Chanishvili [1,*]

1. Eliava Institute of Bacteriophage, Microbiology & Virology and Tbilisi State University, Gotua Street 3, Tbilisi 0160, Georgia; elene.kakabadze@pha.ge (E.K.); khatuna.makalatia@pha.ge (K.M.); nino.grdzelishvili.3@iliauni.edu.ge (N.G.); nata.bakuradze@pha.ge (N.B.); mgoderdzishvili@pha.ge (M.G.); iakusradze@pha.ge (I.K.)
2. National Institute for Biomedical Research, University Teaching Hospital of Kinshasa, Kinshasa, Democratic Republic of the Congo; mfphoba@hotmail.com (M.-F.P.); octmetila@yahoo.fr (O.L.)
3. Department of Microbiology, University Teaching Hospital of Kinshasa, Kinshasa, Democratic Republic of the Congo
4. KU Leuven, Laboratory for Gene Technology, B-3000 Leuven, Belgium; cedric.lood@kuleuven.be (C.L.); rob.lavigne@kuleuven.be (R.L.)
5. Centre of Microbial and Plant Genetics, KU Leuven, B-3000 Leuven, Belgium
6. Department of Clinical Sciences, Institute of Tropical Medicine, B-2000 Antwerpen, Belgium; jjacobs@itg.be (J.J.); Patrick.Soentjens@mil.be (P.S.)
7. Department of Microbiology and Immunology, KU Leuven, B-3000 Leuven, Belgium
8. Department of Biomedical Sciences, Institute of Tropical Medicine, B-2000 Antwerpen, Belgium; sdeborggraeve@itg.be (S.D.); tdeblock@itg.be (T.D.B.); svanpuyvelde@itg.be (S.V.P.)
9. Wellcome Trust Sanger Institute, Hinxton, Cambridge CB10 1SA, UK
10. Department of Biological Sciences, Cork Institute of Technology, Cork T12 P928, Ireland; david.lee@mycit.ie (D.L.); aidan.coffey@cit.ie (A.C.)
11. Institute of Molecular Biology, 0014 Yerevan, Armenia; sedanahit@gmail.com
12. Center for Infectious Diseases ID4C, Queen Astrid Military Hospital, B-1120 Brussels, Belgium
13. Laboratory for Molecular and Cellular Technology, Queen Astrid Military Hospital, B-1120 Brussels, Belgium; DanielMarie.DeVos@mil.be (D.D.V.); jean-paul.pirnay@mil.be (J.-P.P.)

* Correspondence: nina.chanishvili@pha.ge

† These authors contributed equally to this work.

Received: 9 March 2018; Accepted: 2 April 2018; Published: 3 April 2018

Abstract: Recently, a *Salmonella* Typhi isolate producing CTX-M-15 extended spectrum β-lactamase (ESBL) and with decreased ciprofloxacin susceptibility was isolated in the Democratic Republic of the Congo. We have selected bacteriophages that show strong lytic activity against this isolate and have potential for phage-based treatment of *S.* Typhi, and *Salmonella* in general.

Keywords: typhoid fever; *Salmonella* Typhi; extended-spectrum beta lactamases (ESBL); Democratic Republic of the Congo; bacteriophages

---

Although effective oral antibiotics used to be readily available, today antimicrobial resistance in *Salmonella* Typhi is becoming an increasingly serious public health concern, especially in low- and middle-income countries (LMICs). Resistance to all first-line antimicrobials used in the treatment of S. Typhi infections emerged sequentially, leading to multidrug resistance (MDR) in the 1990s, and more recently, high levels of fluoroquinolone resistance in South Asia [1]. Recent data from the Typhoid Fever Surveillance in Africa Program (TSAP) indicates that the incidence rate for typhoid fever in Africa has been underestimated and is equal to, or even greater than, incidences reported in Asia [2]. Therefore, extended spectrum β-lactamase (ESBL) producing *Enterobacteriaceae* and fluoroquinolone resistant *Salmonellae* were included in the high-priority tier of the recent WHO priority list of antibiotic-resistant bacteria [3].

In general, it is acknowledged that global antimicrobial resistance (AMR) poses a fundamental long-term threat to human health, the production of food, and sustainable development. Based on scenarios of rising drug resistance for only six pathogens, experts estimated that by 2050, up to 10 million people could die every year from the effects of AMR and it could also impose an economic burden of US$100 trillion [4]. Recently, the UN committed to supporting the development of new antimicrobial agents and therapies [5].

Phage therapy is one of the promising "new" treatments that has been increasingly featured [6]. Bacteriophages (phages in short) are naturally occurring viruses of bacteria. Since the early phases of evolution, phages have controlled bacteria on our planet. In the early twentieth century, humans discovered them and immediately applied them to medicine. This was especially true in the former Soviet Union, where the use of phages continued after the advent of commercial antibiotics [7]. They can be selected to kill only certain bacteria of concern (e.g., bacteria causing infectious diseases) while leaving non-pathogenic bacteria and mammalian cells unharmed. As such, they can be effective against antibiotic-resistant bacteria and, in contrast to broad-spectrum antibiotics, spare the gut microbiota, which could particularly benefit malnourished and immunocompromised patients. In addition, phages can be easily isolated from environmental sources such as river or sewage water, using basic tools available in LMICs [8].

In 2017, a case of typhoid fever in a six-year-old boy in the Democratic Republic of the Congo (DRC) caused by an S. Typhi isolate producing CTX-M-15 extended spectrum β-lactamase (ESBL) and showing decreased ciprofloxacin susceptibility, was reported [9]. CTX-M-15 is part of the M1 group that includes six plasmid-mediated enzymes [10]. This isolate, named Typhi 10040_15, was sent to the Eliava Institute of Bacteriophage, Microbiology & Virology (EIBMV) in Tbilisi (Georgia) to determine phage susceptibility. The 14 *Salmonella* phage clones of the Eliava R&D collection and five batches of the commercial phage cocktail "INTESTI phage" were tested. Phage screening against this S. Typhi strain was performed at the EIBMV's BSL-2 Plus laboratory, meeting the required safety requirements. The fourteen phage clones were isolated from the river Mtkvari in Tbilisi, from the Black Sea (Batumi) and from the Tbilisi sewage water supply system in the period 2013–2017 (Table 1). The five tested INTESTI phage batches were #M 067 (produced in July 2017), #M2 901 (November 2017), #84 of (February 2017), #82 (January 2017), and #78 (December 2016).

As in many other pathogenic bacteria, temperate phages contribute to virulence in *Salmonella enterica* through the acquisition and exchange of virulence factors [11]. To assert the strictly lytic nature of these phages, high-resolution genome maps of 12 of the 14 individual phages were obtained using nanopore sequencing [12]. A pooled library consisting of barcoded genomic DNA of the phages was prepared using native barcodes and the 1D ligation kit from Oxford Nanopore Technology (ONT). The result was then sequenced on a MinION device, equipped with an R9.4 flowcell. For the data analysis, Albacore v2.1 (ONT, Oxford, UK) was used for base-calling the reads, followed by porechop v0.2.1 (https://github.com/rrwick/Porechop) in order to remove barcode sequences. Genome map assembly was performed with Canu v1.6 (https://github.com/marbl/canu) [13]. All the assembled genomes were subsequently processed with Racon v0.5 for better consensus sequences [14], and nanopolish v0.8.3 (https://github.com/jts/nanopolish) for higher accuracy of base-called nucleotides in the sequences.

Considering the intrinsic properties of nanopore sequencing, together with the run coverage (30× to 60×), we define these assemblies as high-resolution phage maps, rather than fully accurate genome sequences. Known homologous phage isolates were first located using the blastn tool on the NCBI nucleotide database [15]. For each of our new isolates, the closest match (highlighted in bold) in terms of query coverage and identity was identified and the corresponding genome downloaded. The genomic distance between all the pairs of phages was calculated using Mash [16] and the resulting distance matrix was used to build the clustering tree (Figure 1) with the hclust function found in the R stats package [17]. No known toxin genes were present and the genomes did not contain recognizable integrase genes, corroborating the lytic nature of these bacteriophages. Sequences were submitted to GenBank (Accession numbers: MG969404-15).

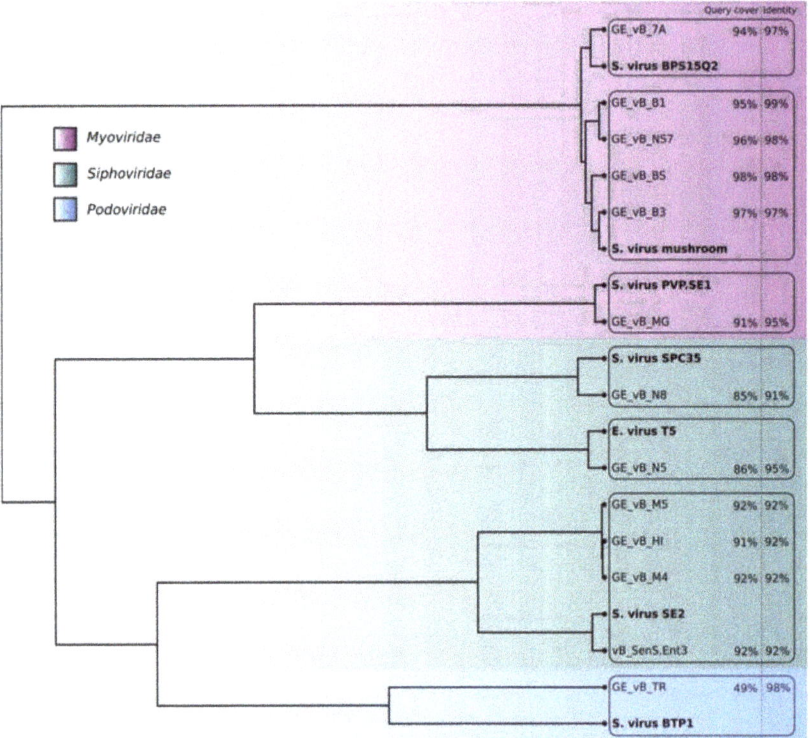

Figure 1. Clustering tree based on the genomic distance matrix generated for the *Salmonella* phages from the Eliava R&D collection and their closest matches (in bold) in the NCBI database. No genome maps were obtained for phages GE_vB_N3 (Siphovirus) and GE_vB_M1 (Podovirus).

Morphological analysis using Transmission Electron Microscopy (TEM) of the phage clones confirmed their general classification (Table 1). Purification and staining of the samples was performed according to Hans-W. Ackermann [18] and preparations were examined with JEOL-JEM-1400 TEM (Figure 2). Phages belonged to the families of the *Siphoviridae* (n = 6), *Myoviridae* (n = 6) and *Podoviridae* (n = 2). The genome map-based grouping allowed us to further assign these phage clones to individual phage species/genera, as indicated in Table 1.

The host range of the phages was assessed by screening their lytic activity and using the spot test against 118 clinical and 121 veterinary *Salmonella*. spp. isolates from Georgia (20), Armenia (71), Germany (7), and Ireland (141). These isolates belonged to the following serotypes:

S. Typhimurium (95), S. Enteritidis (45), S. Dublin (23), S. Anatum (11), S. Infantis (9), S. Newport (8), S. Derbey (8), S. Bredney (5), S. Branderburg (3), S. Germinara (2), S. Uganda (2), S. Senftenberg (2), S. Kentucky (2), S. Reading (2), S. Parat. B (2), S. Java (1), S. Bareilly (1), S. Virchow (1), S. Goldcost (1), S. Kottbus (1), S. Agona (1), and S. Poona (1). Thirteen isolates were not attributed to any known serotypes. Two hundred microlitres of *S. enterica* mid-log phase cultures were mixed with 5 mL of lukewarm 0.6% Lysogeny Broth (LB, Merck, Darmstadt, Germany) agar and overlaid on LB agar plates. The LB broth consisted of 10 g peptone from casein, 5 g yeast extract and 10 g NaCl in 1 L of deionized water. After the plates had cooled, 5 μL of the phage clones, with a titer of 10^7 plaque forming units (pfu)/mL was spotted on the lawn. Drops were air-dried and plates were incubated for 18 h at 37 °C. After the incubation, the plates were checked for zones of clearance resulting from phage activity [19]. The host range of the phages varied from 12 to 81% of the *S. enterica* strains (Table 1). It should be noted that the spot test is usually performed to determine bacterial susceptibility and host range using as many bacterial strains as possible because this method is simple, quick and inexpensive. A significant part of bacterial cell killing can be due to "lysis from without", i.e., the destruction of bacterial cells by the adherence of a sufficiently high number of phages to the bacterial cell, and the destruction of an essential cell wall structure by an extracellular lytic enzyme with subsequent lysis, but without phage replication.

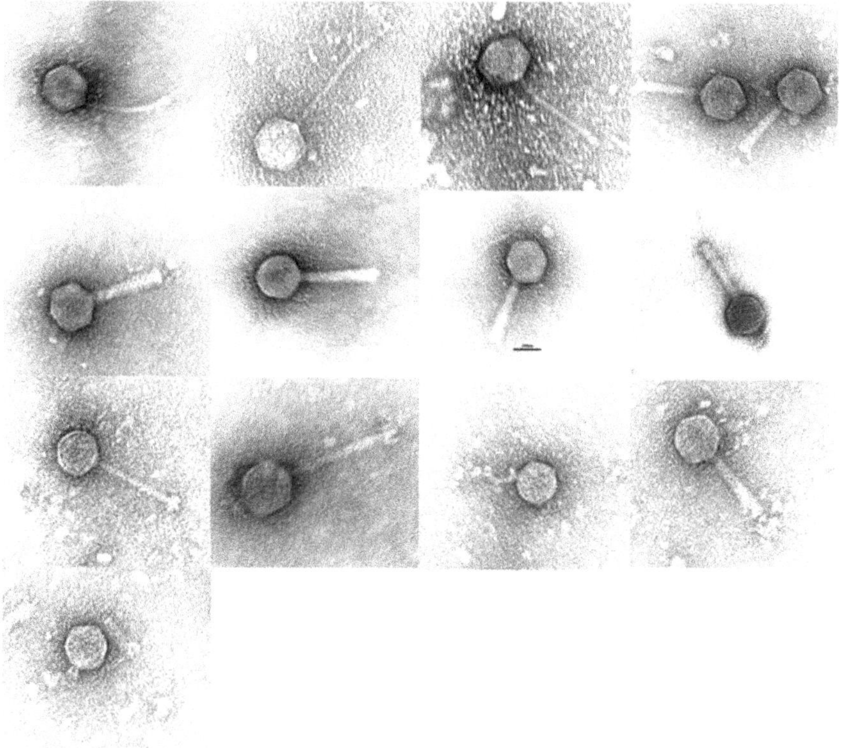

Figure 2. Tansmission electron micrographs of phages in left->right, top->bottom order: GE_vB_N3, GE_vB_N5, GE_vB_N8, GE_vB_MG, GE_vB_BS, GE_vB_B1, GE_vB_B3, GE_vB_NS7, GE_vB_M4, GE_vB_M5, GE_vB_TR, GE_vB_7A, GE_vB_M1. Scale bar, 100 nm.

The lytic activity of the 14 phage clones and of the five INTESTI phage batches against the Typhi 10040_15 isolate from the DRC was evaluated using the spot test and the streak method. For the streak method, mid-log phase S. Typhi was plated as a single line on an agar plate, air dried and 5 µL of single phage clones (titer 10^7 pfu/mL) was dropped on the bacterial line. Drops were air-dried and plates were incubated for 18 h at 37 °C. After incubation, plates were checked for zones of clearance resulting from phage activity [20]. Twelve out of 14 phage clones and three out of five batches of the commercial preparation INTESTI phage formed clear lysis zones on the CTX-M-15, producing an S. Typhi strain from the DRC. This occurred in the streak method, as well as in the spot test. The batches #M 067 and #M2 901 showed the strongest lytic activity with clear confluent zones, while batch #84 only developed weakly distinguishable discrete lytic zones. Only two phage clones, phage GE_vB_MG (*Myoviridae, Vequintavirinae, Se1virus, Salmonella virus SE1*) and phage GE_vB_TR, a potential lysogenic phage (*Podoviridae, P22virus*), showed no activity against the Typhi 10040_15 isolate (Table 1).

To assess the ability of phages to multiply inside the bacterial cells (creating phage plaques), the activity of the 14 phage clones (but not the INTESTI phage batches) was tested using the method of Gratia. A 200 µL mid-log phase culture of S. Typhi and 1 mL of a ten-fold diluted phage suspension, ranging from 10^6 to 10^{10} pfu/mL, were mixed with 5 mL of lukewarm 0.6% LB agar and overlaid on LB agar. Plates were incubated at 37 °C for 18 h and checked for plaque formation [21]. All phage clones, with the exception of three (GE_vB_N8, GE_vB_HIL and GE_vB_M1), were found to form plaques on the bacterial lawn (Table 1).

Finally, to confirm the ability of the phages to lyse the DRC S. Typhi strain in aqueous solutions, the lytic activity of the 14 phage clones was determined using Appelman's method. The 14 phage clones were diluted ten-fold to range from 10^6 to 10^{10} pfu/mL in 5 mL of LB broth and inoculated with 150 µL (10^9 cfu/mL) overnight S. Typhi culture. Mixtures were incubated at 37 °C without shaking and the turbidity of the samples was checked visually after 6, 18 and 24 h. As reference and control samples, phage-free bacterial culture and diluted phages without bacteria were tested under the same conditions [22]. Ten phage clones showed activity at different time points and concentrations, five of which showed the ability to lyse the CTX-M-15 producing S. Typhi isolate without forming phage-resistant mutants after 24 h of incubation, which is indicative of their inherent ability to limit the growth of phage-resistant mutants during phage therapy: GE_vB_N3, GE_vB_N5, GE_vB_N8, GE_vB_NS7 and GE_vB_HIL (Table 1). It should be noted that phages GE_vB_MG and GE_vB_TR are not active according to the spot and the streak tests, while still forming plaques according to the method of Gratia. The difference between these two methods is not that uncommon and could, for example, be caused by differences between phages' "lysis from without" (accentuated in the spot test) and phage infection, or "lysis from within" (accentuated in the Gratia test) capabilities. In other words, some phages could be less efficient in adhering to, and entering bacterial cells, while being very efficient once the normal lytic cycle is initiated. The spot test and the streak method are rapid ways to check whether a phage can infect a bacterium by trickling small droplets of the phage suspensions to be tested on a plate prepared with a bacterial isolate. Limitations of these tests compared with the Gratia and Appelmans methods are that clear zones on the bacterial lawn (a positive test), may be the result of abortive infection or lysis from without, both forming clear zones without new phages being produced.

The evolution of bacterial resistance to phages is often observed in vitro. Phages have evolved multiple strategies to overcome the antiviral mechanisms they encounter when infecting bacterial cells [23]. In experimental settings, phage-resistant bacteria emerged rapidly, but often at significant fitness costs, shown by the reduced growth rate in the absence of phages [24]. The in vivo evolution of bacterial resistance to phages in human clinical practice, however, was poorly documented until now. It has been suggested that phages could be combined with antibiotics to improve phage activity (synergy) [25].

Table 1. Characteristics of 14 *Salmonella* phages from the Eliava collection tested on the Typhi 10040_15_DRC_2015 isolate from the Democratic Republic of the Congo (DRC).

Nr	Name	GenBank Accession Numbers	Source	Isolation Year	Host Strain	Family	Homology to Other Phages	Host Range (%) *All Serovars/S. Enteritidis/S. Typhimurium/S. Dublin (Total Number of Tested Isolates)	Lytic Activity on the Typhi 10040_15_DRC_2015 Isolate							
									Streak Method	Spot Test	Gratia's Method	Appelmans' Method				
												Titer (pfu/mL)	Dilution	6 h	18 h	24 h
001	GE_vB_N3	ND	Mtkvari river water	2013	*S.* Enteritidis 3	Siphoviridae	ND	78/98/73/87 (239)	+	+	+	10^9	−1	−	−	+
												10^8	−2	−	−	+
												10^7	−3	−	+	+
002	GE_vB_N5	MG969412	Mtkvari river water	2013	*S.* Enteritidis 3	Siphoviridae	*E. coli T5 strain ATCC 11303-B5*	45/80/25/74 (239)	+	+	+	10^8	−1	−	+	+
												10^7	−2	−	+	+
												10^6	−3	−	+	+
003	GE_vB_N8	MG969413	Mtkvari river water	2013	*S.* Enteritidis 3	Siphoviridae	*phage SPC35*	65/84/67/78 (239)	+	+	−	10^9	−1	−	−	+
												10^8	−2	−	−	+
												10^7	−3	−	+	+
004	GE_vB_MG	MG969411	Tbilisi sewage water	2013	*S.* Enteritidis 3	Myoviridae	*S. phage PVP-SE1*	47/49/59/43 (239)	−	−	+	10^9	−1	−	−	−
												10^8	−2	−	−	−
												10^7	−3	−	−	−
005	GE_vB_BS	MG969407	Black Sea water	2013	*S.* Typhimurium 4	Myoviridae	*S SPT-1, partial genome*	81/96/93/83 (239)	+	+	+	10^{10}	−1	+	−	−
												10^9	−2	+	−	−
												10^8	−3	+	−	−
006	GE_vB_B1	MG969405	Mtkvari river water	2013	*S.* Typhimurium 6	Myoviridae	*S. phage Mushroom*	80/93/83/87 (239)	+	+	+	10^9	−1	+	−	−
												10^8	−2	+	−	−
												10^7	−3	+	−	−
007	GE_vB_B3	MG969406	Mtkvari river water	2013	*S.* Typhimurium 6	Myoviridae	*S. phage Mushroom*	81/98/73/87 (239)	+	+	+	10^9	−1	+	−	−
												10^8	−2	+	−	−
												10^7	−3	+	−	−
008	GE_vB_NS7	MG969414	Raw cow milk	2015	*S.* Typhimurium 6	Myoviridae	*S. phage Mushroom*	75/91/81/78 (239)	+	+	+	10^9	−1	+	−	−
												10^8	−2	−	+	−
												10^7	−3	−	+	+
009	GE_vB_M4	MG969409	Black Sea water	2016	*S.* Enteritidis 232	Siphoviridae	*S. phage vB_SenS-Ent3*	23/64/18/22 (218)	+	+	+	10^{10}	−1	+	−	−
												10^9	−2	+	−	−
												10^8	−3	−	−	−

Table 1. *Cont.*

Nr	Name	GenBank Accession Numbers	Source	Isolation Year	Host Strain	Family	Homology to Other Phages	Host Range (%) *All Serovars/S. Enteritidis/S. Typhimurium/S. Dublin (Total Number of Tested Isolates)	Lytic Activity on the Typhi 10040_15_DRC_2015 Isolate							
									Streak Method	Spot Test	Gratia's Method	Appelmans' Method				
												Titer (pfu/mL)	Dilution	6 h	18 h	24 h
010	GE_vB_M5	MG969410	Black Sea water	2016	*S.* Enteritidis 407	Siphoviridae	*S. phage vB_SenS-Ent3*	33/66/26/61 (218)	+	+	+	10^8	−1	+	–	–
												10^7	−2	+	–	–
												10^6	−3	+	–	–
011	GE_vB_TR	MG969415	Mtkvari river water	2017	*S.* Typhimurium 641	Podoviridae	*S. phage BTP1*	40/90/28/59 (141)	–	–	+	10^9	−1	–	–	–
												10^8	−2	–	–	–
												10^7	−3	–	–	–
012	GE_vB_HIL	MG969408	Mtkvari river water	2017	*S.* Enteritidis 765	Siphoviridae	*S. phage vB_SenS-Ent3*	58/81/75/77 (141)	+	+	–	10^{10}	−1	+	–	–
												10^9	−2	–	–	–
												10^8	−3	–	+	+
013	GE_vB_7A	MG969404	Mtkvari river water	2017	*S.* Typhimurium 1328	Myoviridae	*S. phage BPS15Q2*	37/62/28/23 (141)	+	+	+	10^8	−1	–	–	–
												10^7	−2	–	–	–
												10^6	−3	–	–	–
014	GE_vB_M1	ND	Black Sea water	2016	*S.* Enteritidis 104	Podoviridae	ND	12/20/19/0 (77)	+	+	–	10^9	−1	–	–	–
												10^8	−2	–	–	–
												10^7	−3	–	–	–

* The host range of the phages was determined for the total *Salmonella* strain collection (all serovars), with the total number of strains indicated between brackets, and for the three main serovars (*S.* Enteritidis/*S.* Typhimurium/*S.* Dublin) separately. ND, not done; "+", phage lytic activity; "–", no phage lytic activity; *S. phage*: *Salmonella phage*.

In conclusion, in a short time frame (two days), at least five phage clones from the Eliava collection were found to exhibit excellent in vitro lytic activity against the ESBL producing *S.* Typhi isolate from the DRC. Phages can be considered a potential additional tool for the treatment of MDR *Salmonella* infections and a (food) decontamination agent. Antimicrobials that address foodborne diseases are particularly important for LMICs as many of them lack reliable cold chain infrastructure and adequate hygiene practices [8]. In Western countries, several phage products are currently approved for the control of food pathogens, including *Salmonella.* In addition, phage preparations can be developed and produced faster and cheaper than conventional drugs. They can also be (freeze-)dried [26] so that they require no refrigeration [8].

Author Contributions: All authors contributed to the conception and writing of this paper.

Conflicts of Interest: The authors declare no conflict of interest.

Funding: This work was supported by the International Science and Technology Center (grant number ISTC-A 2140), the Science Foundation Ireland (12/R1/2335), the KU Leuven Orchestrated Research Action (Phage Biosystems to RL), the Fonds Wetenschappelijk Onderzoek Vlaanderen (1S64718N to CL), the Belgian Directorate of Development Cooperation (DGD) (Project 2.01 of the 3rd Framework Agreement between the Belgian DGD and the ITM Antwerp, Belgium), the Baillet-Latour Fund (The Bacterial Infections in the Tropics (BIT) research cluster at ITM Antwerp, Belgium) and the Department of Economy, Science and Innovation in Flanders (EWI funding ITM Antwerp).

References

1. Karkey, A.; Thwaites, G.E.; Baker, S. The evolution of antimicrobial resistance in *Salmonella Typhi. Curr. Opin. Gastroenterol.* **2018**, *34*, 25–30. [CrossRef] [PubMed]
2. Marks, F.; von Kalckreuth, V.; Aaby, P.; Adu-Sarkodie, Y.; El Tayeb, M.A.; Ali, M.; Aseffa, A.; Baker, S.; Biggs, H.M.; Bjerregaard-Andersen, M.; et al. Incidence of invasive *Salmonella* disease in sub-Saharan Africa: A multicentre population-based surveillance study. *Lancet Glob. Health* **2017**, *5*, e310–e323. [CrossRef]
3. Tacconelli, E.; Carrara, E.; Savoldi, A.; Harbarth, S.; Mendelson, M.; Monnet, D.L.; Pulcini, C.; Kahlmeter, G.; Kluytmans, J.; Carmeli, Y.; et al. Discovery, research, and development of new antibiotics: The WHO priority list of antibiotic-resistant bacteria and tuberculosis. *Lancet Infect. Dis.* **2017**, *18*, 234–236. [CrossRef]
4. Tackling Drug-Resistant Infections Globally: Final Report and Recommendations. The Review on Antimicrobial Resistance. 2016. Available online: https://amr-review.org/sites/default/files/160525_Finalpaper_withcover.pdf (accessed on 27 January 2018).
5. United Nations. Draft Political Declaration of the High-Level Meeting of the General Assembly on Antimicrobial Resistance (16-16108 (E)). Available online: http://www.un.org/pga/71/wp-content/uploads/sites/40/2016/09/DGACM_GAEAD_ESCAB-AMR-Draft-Political-Declaration-1616108E.pdf (accessed on 17 December 2017).
6. Watts, G. Phage therapy: Revival of the bygone antimicrobial. *Lancet* **2017**, *390*, 2539–2540. [CrossRef]
7. Chanishvili, N. Bacteriophages as Therapeutic and Prophylactic Means: Summary of the Soviet and Post Soviet Experiences. *Curr. Drug Deliv.* **2016**, *13*, 309–323. [CrossRef] [PubMed]
8. Nagel, T.E.; Chan, B.K.; de Vos, D.; El-Shibiny, A.; Kang'ethe, E.K.; Makumi, A.; Pirnay, J.P. The Developing World Urgently Needs Phages to Combat Pathogenic Bacteria. *Front. Microbiol.* **2016**, *7*, 882. [CrossRef] [PubMed]
9. Phoba, M.F.; Barbé, B.; Lunguya, O.; Masendu, L.; Lulengwa, D.; Dougan, G.; Wong, V.K.; Bertrand, S.; Ceyssens, P.J.; Jacobs, J.; et al. *Salmonella enterica* serovar *Typhi* Producing CTX-M-15 Extended Spectrum β-Lactamase in the Democratic Republic of the Congo. *Clin. Infect. Dis.* **2017**, *65*, 1229–1231. [CrossRef] [PubMed]
10. Bonnet, R. Growing group of extended-spectrum β-lactamases: The CTX-M enzymes. *Antimicrob. Agents Chemother.* **2004**, *48*, 1–14. [CrossRef] [PubMed]
11. Ho, T.D.; Figueroa-Bossi, N.; Wang, M.; Uzzau, S.; Bossi, L.; Slauch, J.M. Identification of GtgE, a novel virulence factor encoded on the Gifsy-2 bacteriophage of *Salmonella enterica* serovar *Typhimurium*. *J. Bacteriol.* **2002**, *184*, 5234–5239. [CrossRef] [PubMed]
12. Deamer, D.W.; Akeson, M. Nanopores and nucleic acids: Prospects for ultrarapid sequencing. *Trends Biotechnol.* **2000**, *18*, 147–150. [CrossRef]

13. Koren, S.; Walenz, B.P.; Berlin, K.; Miller, J.R.; Bergman, N.H.; Phillippy, A.M. Canu: Scalable and accurate long-read assembly via adaptive K-mer weighting and repeat separation. *Genome Res.* **2017**, *27*, 722–736. [CrossRef] [PubMed]
14. Vaser, R.; Sović, I.; Nagarajan, N.; Šikić, M. Fast and accurate de novo genome assembly from long uncorrected reads. *Genome Res.* **2017**, *27*, 737–746. [CrossRef] [PubMed]
15. Coordinators, N.R. Database resources of the national center for biotechnology information. *Nucleic Acids Res.* **2016**, *44*, D7.
16. Ondov, B.D.; Treangen, T.J.; Melsted, P.; Mallonee, A.B.; Bergman, N.H.; Koren, S.; Phillippy, A.M. Mash: Fast genome and metagenome distance estimation using MinHash. *Genome Biol.* **2016**, *17*, 132. [CrossRef] [PubMed]
17. R Core Team. *R: A Language and Environment for Statistical Computing*; R Foundation for Statistical Computing: Vienna, Austria, 2015; Available online: https://www.R-project.org/ (accessed on 24 March 2018).
18. Ackermann, H.W. Basic phage electron microscopy. In *Bacteriophages: Methods and Protocols*, 1st ed.; Clokie, M.R.J., Kropinski, A., Eds.; Humana Press: New York, NY, USA, 2009; Volume 1, pp. 113–126. ISBN 978-1-58829-682-5.
19. Kutter, E. Phage host range and efficiency of plating. In *Bacteriophages: Methods and Protocols*, 1st ed.; Clokie, M.R.J., Kropinski, A., Eds.; Humana Press: New York, NY, USA, 2009; Volume 1, pp. 141–149. ISBN 978-1-58829-682-5.
20. Matsuzaki, S.; Uchiyama, J.; Takemura-Uchiyama, I.; Ujihara, T.; Daibata, M. Isolation of Bacteriophages for Fastidious Bacteria. In *Bacteriophage Therapy*, 1st ed.; Azeredo, J., Sillankorva, S., Eds.; Humana Press: New York, NY, USA, 2018; pp. 3–10. ISBN 978-1-4939-7394-1.
21. Gratia, A. Des relations numeriques entre bacteries *lysogenes* et particules de bacteriophage. *Ann. Inst. Pasteur* **1936**, *57*, 652–676.
22. Appelmans, R. Le dosage du bactériophage. *C. R. Seances Soc. Biol.* **1921**, *85*, 1098.
23. Labrie, S.J.; Samson, J.E.; Moineau, S. Bacteriophage resistance mechanisms. *Nat. Rev. Microbiol.* **2010**, *8*, 317–327. [CrossRef] [PubMed]
24. Hall, A.R.; de Vos, D.; Friman, V.P.; Pirnay, J.P.; Buckling, A. Effects of sequential and simultaneous applications of bacteriophages on populations of *Pseudomonas aeruginosa* in vitro and in wax *moth larvae*. *Appl. Environ. Microbiol.* **2012**, *78*, 5646–5652. [CrossRef] [PubMed]
25. Kamal, F.; Dennis, J.J. *Burkholderia cepacia* complex Phage-Antibiotic Synergy (PAS): Antibiotics stimulate lytic phage activity. *Appl. Environ. Microbiol.* **2015**, *81*, 1132–1138. [CrossRef] [PubMed]
26. Merabishvili, M.; Vervaet, C.; Pirnay, J.P.; De Vos, D.; Verbeken, G.; Mast, J.; Chanishvili, N.; Vaneechoutte, M. Stability of *Staphylococcus aureus* phage ISP after freeze-drying (lyophilization). *PLoS ONE* **2013**, *8*, e68797. [CrossRef] [PubMed]

Conference Report

1st German Phage Symposium—Conference Report

Irene Huber [1], Katerina Potapova [1], Andreas Kuhn [1,2], Herbert Schmidt [1,3], Jörg Hinrichs [1,3], Christine Rohde [4] and Wolfgang Beyer [1,5,*]

1 Hohenheim Research Center for Health Sciences, University of Hohenheim, 70599 Stuttgart, Germany; irene.huber@uni-hohenheim.de (I.H.); Katerina.Potapova@uni-hohenheim.de (K.P.); andikuhn@uni-hohenheim.de (A.K.); Herbert.schmidt@uni-hohenheim.de (H.S.); j.hinrichs@uni-hohenheim.de (J.H.)
2 Institute of Microbiology, University of Hohenheim, 70599 Stuttgart, Germany
3 Institute of Food Science and Biotechnology, University of Hohenheim, 70599 Stuttgart, Germany
4 Leibniz-Institute DSMZ—German Collection of Microorganisms and Cell Cultures, 38124 Braunschweig, Germany; Chr@dsmz.de
5 Institute of Animal Sciences, University of Hohenheim, 70599 Stuttgart, Germany
* Correspondence: Wolfgang.Beyer@uni-hohenheim.de; Tel.: +49-711-459-22-429

Received: 13 March 2018; Accepted: 25 March 2018; Published: 29 March 2018

Abstract: In Germany, phage research and application can be traced back to the beginning of the 20th century. However, with the triumphal march of antibiotics around the world, the significance of bacteriophages faded in most countries, and respective research mainly focused on fundamental questions and niche applications. After a century, we pay tribute to the overuse of antibiotics that led to multidrug resistance and calls for new strategies to combat pathogenic microbes. Against this background, bacteriophages came into the spotlight of researchers and practitioners again resulting in a fast growing "phage community". In October 2017, part of this community met at the 1st German Phage Symposium to share their knowledge and experiences. The participants discussed open questions and challenges related to phage therapy and the application of phages in general. This report summarizes the presentations given, highlights the main points of the round table discussion and concludes with an outlook for the different aspects of phage application.

Keywords: bacteriophage; phage; horizontal gene transfer; co-evolution; phage therapy; industrial phage application; antimicrobial resistance (AMR); Germany

1. Introduction

Phage research and related business activities (e.g., BAYER's "Polyfagin" preparation) have some tradition in Germany [1,2]. However, with the advent of antibiotics, interest in phage applications declined and for a long time was focused on fundamental questions. Although noticeable efforts have been made to use phages as molecular tools [3,4], the application of phages in medicine [5], veterinary science [6], and for hygienic purposes along the food chain [7] only recently gained wider attention in Germany. The main reason for that is obvious: multiple drug resistant bacteria are a problem worldwide. In order to develop and implement AMR counter measures concerted actions have been initiated by researchers, stakeholders and regulatory authorities in Germany. Federal and State Ministries support these efforts with tailor-made funding programs and by joining international alliances (e.g., JPI-AMR). However, except for one major project, that is, Phage4Cure [8,9] funded by the Federal Ministry of Education and Research (BMBF), practical applications of phages in Germany are rare and still hampered by an unsatisfactory regulatory frame.

In addition, and similar to the situation in France some time ago [10], the German phage community is scattered and many did not know each other. It is in this context that all German phage researchers and stakeholders from industry, regulatory agencies, politics and society were invited to contribute to the 1st German Phage Symposium. Renowned guest speakers from all over the world enriched the conference that was to foster networking, exchange and new project ideas. As a visible result, the German Phage Forum ("Nationales Forum Phagen", www.nf-phagen.de) was founded; its first workshop will take place on 18 June 2018 at the University of Hohenheim in Stuttgart.

2. Summary of Scientific Sessions

During the 3 day conference, 170 participants from 20 countries contributed to five main topics, a plenary session and small workgroup discussions.

2.1. Structure-Function Relationship

The session was opened with Dennis Bamford's (University of Helsinki) perspective on the structural basis of the viral universe and functional consequences. By studying architectural principles, his group found that the variety of virus structures is restricted. Their experiments yielded two other major insights: (1) Host range cannot be used as primary criterion for the classification of viruses. Instead, all viruses on earth could be grouped into a small number of structure-based lineages. (2) Viruses may have existed before the three cellular domains of life have separated and therefore, a polyphyletic origin versus a monophyletic origin should be considered [11].

Johannes Wittmann (DSMZ—German Collection of Microorganisms and Cell Cultures) et al. approached the matter from a taxonomic point of view. First classifications of bacteriophages were based on morphology (capsid size and shape, tail existence and size) and genomic features (ss vs. ds DNA/RNA, genome size). Later, host-specificity provided another criterion for classification but could not bring convincing order into the growing number of newly discovered phages [12]. With new sequencing techniques and proteome-based tools, phage taxonomy has yet again undergone a number of changes. Wittmann and his colleagues from the Bacterial and Archaeal Viruses Subcommittee of the International Committee on Taxonomy of Viruses (ICTV) brought order into the taxonomy chaos [13] and in addition, create a tool box/method to classify higher numbers of phages simultaneously [14].

Tailed bacteriophages that infect gram-negative bacteria have developed a variety of strategies to overcome the rigid barrier of lipopolysaccharide-enforced outer membrane and to transfer their genome into the bacterial host. Pascale Boulanger (Centre National de la Recherché Scientifique) focused her talk on bacteriophage T5. After introducing its genomic composition, Boulanger pointed out how T5 architecture facilitates the sophisticated two-step infection mechanism [15,16]: After docking of T5 tail-tip proteins to the bacterial cell wall, "puncturing" occurs via interaction with *Escherichia coli* outer membrane protein FhuA and allows for limited phage DNA transfer (~8%) into the host cell. This step is accompanied by substantial destruction of bacterial DNA. The subsequent pause that occurs in vivo suggests a host factor mediated defence mechanism, before T5 completely takes over the infected cell and fully degrades the bacterial DNA by a phage-encoded, Mn^{2+}-dependent DNase. How this DNase activity is regulated (while maintaining T5 phage genome integrity) is still not known. Interestingly, in vitro infection experiments demonstrated that T5 phage is able to complete DNA transfer into proteoliposomes without interruption, a finding that further strengthens the hypothesis of host-specific defence mechanisms in vivo [16].

Stefanie Barbirz et al. (University of Potsdam) investigated a similar infection mechanism with a set of O-antigen specific dsDNA model phages that infect *Salmonella enterica*. Myovirus Det7, Podovirus P22, and Siphovirus 9NA differ in tail architecture but all anchor to receptors of the bacterial outer membrane and use their enzymatically active tail spike proteins (TSP) to hydrolyse the O-antigen polysaccharide. The group speculates that tail-protruding components are non-specifically pressed against the rigid outer cell membrane and this mechanical signal then transmits to the tail to proceed with the DNA transfer. Non-O-antigen specific phages may follow a different mechanism, but similar opening steps are conceivable [17].

Although tailed bacteriophage SPP1 performs a similarly fast (within 30 min after infection) takeover of host functions, Paulo Tavares et al. from Institut de biologie intégrative de la cellule (I2BC), France showed that mechanisms and proteins involved are quite different: After injection of viral DNA, the viral helicase gp40 seems to be key for high-jacking the host replisome, that is, recruitment of the host replication machinery to the phage DNA replication foci. Moreover, a major remodelling of the host cytoplasm occurs. Most likely, this reorganization optimizes not only the efficiency of phage replication but also the assembly of virions (warehouse model) which are in close proximity to the viral replisomes [18–20].

In contrast to lytic phages, filamentous phages are assembled in the inner membrane of their host and secreted across the bacterial envelope without killing its bacterial host [21]. Despite their minimalistic, plasmid-like genome, filamentous phage M13 has quite a complex life cycle. There are only 11 genes and respective proteins allow for infection of host bacteria (*E. coli*), reproduction, and assembly of new phages. Sebastian Leptihn et al. (University of Hohenheim) presented a model for the assembly of filamentous phages, detailing the molecular assembly motor (one gene, two open reading frames) and speculated on how DNA translocation and protein assembly are powered [21].

2.2. Host-Phage Interaction & Evolution of Microbial Communities

Bacteriophages regulate and drive the evolution of microbial communities in all the various ecosystems on earth. Part of this interaction is the steady evolvement of defence and anti-defence mechanisms of the host-phage system. Stan Brouns' (Delft University of Technology) keynote summarized the current knowledge on the diverse types and mechanisms of the Clustered Regularly Interspaced Short Palindromic Repeats (CRISPR) immune system and its application in genome engineering. For the latter, it is essential to know how upon selective pressure phages can circumvent bacterial defence systems. Several mechanisms have already been described, that is, the production of anti-CRISPR proteins [22–24], mutations in defence-specific target DNA sequences (https://www.ncbi.nlm.nih.gov/pubmed/10223975) [25,26], deletion of CRISPR target sites [27], genome recombination [28], and DNA glucosylation of target DNA binding sites [29,30]. The degree of CRISPR inhibition by sequence alterations or biochemical modification of DNA is strongly position-dependent. Brouns et al. propose that glucosylation-induced hydrogen bonds between the side groups of the glucosyl moieties and neighbouring bases may alter the topology of base pairs [31] which could impair R-loop formation and thereby prevent cleavage of the modified targets. However, bacteria can rapidly adapt their CRISPR defence systems by 'priming.' For this, Cas1–Cas2 protein complexes catalyse the addition of new spacers to the CRISPR memory bank, which in turn enables the bacterial host to recognize viral invaders even with changed sequences [32].

Interestingly, not all viral anti-defence strategies are similarly effective in fighting the different CRISPR-Cas types, implying that they have evolved independently [30,32]. The selective pressure posed upon bacterial defence by anti-defence mechanisms of bacteriophages leads to rapid co-evolution between host and phage [30,33]. Brouns also reflected on the consequences of these findings for genetic engineering, raised open questions and speculated on new applications.

Horizontal gene transfer (HGT) is an important driving force of evolution. It often involves temperate bacteriophages, which dependent on the surrounding conditions can choose between a lytic and lysogenic replication cycle. Once decided on the lysogenic state, most temperate phages [34,35] integrate their genomes as prophage into the bacterial chromosome and replicate vertically with their host. The acquisition of foreign DNA via prophages allows for selective advantages in the host cell, but also bears an additional risk of cell death if toxic phage genes are transferred to the host. Hence, the integration of prophage elements into the genome and into host regulatory circuits requires stringent regulation. Christiane Wolz et al. (University of Tübingen) presented work on such regulatory mechanisms and their consequences for *Staphylococcus aureus* and its host. *S. aureus* possesses a set of virulence factors necessary for infection of its human hosts. Some of these factors are encoded by temperate phages. Sa2 phages, for example, carry genes for Panton-Valentine leukocidin

(PVL), that is, one of the β-pore-forming toxins. PVL-enriched *S. aureus* perforate infected human cells and cause necrotic lesions that are very hard to treat. Similarly, Sa3 phages carrying several additional virulence factors invade *S. aureus* by integrating into the staphylococcal *hlb*-gene leading to the loss of β-haemolysin production. The excision of the phage restores the *hlb*-gene and increases the virulence of *S. aureus*. Sa phages also become highly mobile during chronic lung infections in cystic fibrosis patients but little is known about the triggers and molecular mechanisms involved [36]. In order to gain more insight into the role and interference of mobile genetic elements in *S. aureus*, the group of Wolz established molecular tools to analyse the molecular basis for strain specific phage transfer [37]. They could show that the genetic background of bacterial hosts has a substantial impact on lysogenization, induction, and phage gene expression.

In line with this notion are the studies presented by Julia Frunzke et al. (Helmholtz Research Center Jülich), who unravelled the role of xenogeneic silencing of prophages CGP1-3 in the Gram-positive soil bacterium *Corynebacterium glutamicum*, which is important for the industrial production of amino acids [38]. Her group could demonstrate that the small nucleoid-associated protein CgpS is key to the repression of AT-rich genes in horizontally acquired prophage sequences by forming protein-DNA complexes. Overexpressing the N-terminal oligomerization domain of CgpS disrupts the integrity of these protein-DNA complexes, as truncated CgpS proteins outnumber and compete with their native counterparts but cannot bind to their DNA targets anymore. Consequently, such counter silencing leads to the activation of CGP3 prophage sequences by allowing transcription factors to bind to the respective target sequences instead, resulting in bacterial growth defects and a highly increased frequency of CGP3-induced cell death [39,40]. Beyond this role in the control of gene expression, recent preliminary data of Frunzke's team suggest an even broader function for CgpS in chromosomal organization and replication. Bioinformatics revealed CgpS homologs in almost all actinobacterial species and, remarkably, also in the genomes of several actinobacteriophages and prophages. Despite low sequence conservation, the highly conserved secondary structure suggests an ancient function of these proteins in (pro-) phage-host interaction.

Another example of phage-mediated virulence, tripartite species interaction and phage/bacteria co-evolution was presented by Heiko Liesegang et al. (University of Göttingen). To study how bacterial resistance on phages impacts the bacterial virulence on a eukaryotic host, the group used a tripartite model system consisting of *Vibrio alginolyticus*, its phages and its host, that is, *Sygnatus typhle* (pipefish). When pipefish were challenged with three *Vibrio* strains (highly susceptible (hs), intermediate susceptible (is) and resistant (r) to phages), bacterial counts on the fish did not differ, but the gene expression of the fish clearly showed strain specific patterns. By challenging a culture of hs strains with phage solutions under co-evolutionary conditions, resistant mutants immediately emerged. Bacteria with a phage-susceptible phenotype show an increased virulence on their host fish while under appropriate conditions, phage-resistant and less virulent parasites evolve [41]. Hyperparasitism is thus an important factor for the virulence of bacterial pathogens. Comparative genome analysis shall reveal the genotypes under selection that are responsible for the acquired phage resistance as well as for the modified virulence on the eukaryotic host. In the second part of his talk, Liesegang took a closer look on *Inoviridae* genomics to identify prophages, determine their sequence composition and gene content. By this, they defined a scoring weight matrix consisting of 9 prominent features to identify new *Inoviridae*. The model for such scans and pilot data will be shared by this group on a website and further developed into a General Markov Model for (pro)-phage genomics [42,43].

The focus of Li Deng, Emmy Noether group leader at the Helmholtz Research Center München, is on the ecological role of viruses in the environment. Starting with impressive numbers of microorganisms decomposed by lytic phages, she provided a colourful picture how phages drive biogeochemical cycles, control microbial populations and maintain homeostasis. Employing viral tagging [44], novel purification procedures for environmental probes, microcosm experiments, metagenomics and bioinformatics, the group could assess, quantify and describe the ecological role of phages in the ocean, freshwater environments, groundwater aquifers, and other habitats [45].

When comparing contaminated versus unspoiled sites, an increased abundance of phages as compared to the number of their bacterial hosts was observed. One explanation how HGT contributes to this phenomenon might be the transfer of catabolic enzymes necessary for biodegradation within or across bacterial species [46]. In a strict sense, infected tissues of patients, for example, with Chronic Obstructive Pulmonary Disease (COPD), can also be defined as "contaminated environment." Based on analyses of a Human Virome Protein Cluster Database, preliminary data suggest that patients with COPD carry more virulence factors than healthy controls [46]. As pollution of our planet increases and leaves only a few unspoiled sites for control experiments, Deng heads towards experiments in the extra-terrestrial space soon. It will be interesting to see which experimental setups she chooses for the ISS, SpaceX and Mars missions, and it will certainly yield interesting findings and new insights.

Jacques Mahillon (Université Catholique de Louvain) and his colleagues focus on the lysogeny of Tectiviridae. This family of tail-less phages with a lipid membrane as the inner layer of their capsid has been found in less than 3% of bacterial isolates. Analysis of the Tectiviridae host range showed that no simple relationship could be established between the infection patterns of these phages and their diversity. However, data revealed that tectiviruses in the *Bacillus cereus* group clustered into two major groups: the ones infecting *Bacillus anthracis* and those isolated from other *B. cereus* group members [47]. Remarkably, tectiviral plasmid-related molecules with recombinant characteristics were also discovered by analyses of whole genome sequences. Additionally, Gillis and Mahillon demonstrated that tectiviral lysogeny had a significant influence on morphology, metabolic profile, growth kinetics, sporulation rate, biofilm formation, and swarming motility of their *Bacillus thuringiensis* host [48]. All these traits are involved in the survival and colonization of Bacillus strains in different environmental habitats. Overall, Mahillon's findings provide evidence that Tectiviridae are more diverse than previously thought and that they also have ecological roles in the already complex life cycle of *B. thuringiensis* and its kin [48,49]. Current research of the group is directed towards (1) the identification of phage-specific receptors in *B. thuringiensis*, (2) deciphering of phage-resistant mutant bacteria in order to identify and confirm the mutations involved in the resistance to tectiviruses. Preliminary results indicate changes in the sugar metabolism of the bacterial host and a slight difference in growth kinetics and swarming motility. Isolated mutants show twisted cell-chain morphology and highly increased biofilm production [50].

Josué L. Castro-Mejía et al. (University of Copenhagen) described new findings on phage-bacterium interaction in the gut and discussed the influence on intestinal and extra-intestinal disorders. Several environmental factors have been shown to contribute to imbalances in gut microbiome (GM), including diet, drugs, antibiotics and enteric pathogens [51,52]. Less is known about the impact of the virome on the GM composition, its functionality and interactions with age-related comorbidities in older adults. As part of the Danish Counterstrike Initiative [53], the group assessed probands with and without interventions (diet, exercise) in respect to their microbiome composition (prokaryotes, eukaryotes, phages), metagenome (subset), metabolome and physiological parameter. Co-abundance correlation analysis on metagenome data from faecal preparations demonstrated a large number of phage-bacterium interactions. Strikingly, the fluctuations in bacterial abundance (as a function of phage attack) resulted in dramatic variations of the global metabolic potential. It also triggered dysbiosis and influenced host renal function. Together, the data indicate that members of the gut virome are associated with age-related co-morbidities [54].

Health-relevant aspects of Shiga toxin producing *E. coli* and their phages were highlighted by Herbert Schmidt et al. (University of Hohenheim). Enterohaemorrhagic *Escherichia coli* (EHEC) are the causative agents of haemorrhagic colitis and the haemolytic-uremic syndrome (HUS). Shiga toxins (Stx) are responsible for pathogenicity and are encoded by lambdoid prophages at distinct positions in the EHEC chromosome. Other, non-Stx-encoding lambdoid prophages are integrated in the EHEC chromosomes in varying numbers. The foodborne EHEC O157:H7 strain EDL933 harbours 7–10 lambdoid prophages, two of which encode Stx1 and Stx2, respectively. The stx integration sites are close to the anti-terminator Q in the late transcribed region. Upon induction of prophages, stx is co-transcribed with the late transcribed phage genes [55].

Earlier studies demonstrated a large open reading frame 3′ adjacent to the stx genes. This ORF codes for a homologue of the chromosomal nanS gene, a well-described esterase cleaving an acetyl residue from 5-*N*-acetyl-9-*O*-acetyl-neuraminic acid (Neu5, 9Ac2) resulting in Neu5Ac as bacterial carbon source [56]. Bioinformatics revealed varying numbers of phage-encoded nanS homologs (designated nanS-p) in different EHEC strains, some of which already could be confirmed functional esterases [57,58]. Interestingly, growth ability of *E. coli* O157:H7 strain EDL933 and its isogenic mutants is dependent on the number of *nanS-p* genes; deletion of all nanS-p alleles inhibits growth on Neu5,9Ac2 completely. On the other hand, recombinant NanS-p proteins cleave acetyl residues in native mucin from *O*-acetylated neuraminic acids and *O*-acetylated glycolylic neuraminic acids [58]. Schmidt hypothesized that the prophage-encoded *nanS-p* genes represent a mobile gene pool, which ensures effective substrate utilization and therefore growth and preservation of pathogenic EHEC in the large intestine in spite of its thick mucus layer and high turnover rate. Therefore, Stx- and non Stx-prophages of EHEC should be considered a pathogenic principle of EHEC strains causing serious diseases.

2.3. Clinical Applications

The keynote by Mzia Kutateladze, Director of the famous G. Eliava Institute of Bacteriophages, Microbiology and Virology, started with a historical view on the use of bacteriophages for treatment of infectious diseases. This historic picture is partially based on the immense screening effort during ISTC Project G-1467: "Preparation of a detailed review article/monograph on the practical application of bacteriophages in medicine, veterinary, environmental research, based on old documents and publications." However, the data from old studies need to be treated with caution: Experimental setups were not as strict as today in terms of statistical significance, standardized methods, or the use of controls and placebos. This has changed and current data presented for phage treatments at the Eliava Phage Therapy Center are derived from randomized, placebo-controlled, double-blind trials. For example, urinary tract infections (e.g., *Streptococcus*, *E. coli*, *Enterococcus*, *Proteus*) [59] had a phage susceptibility of ~80% and no side effects have been noted in the patients treated [60].

Another focus of the Eliava Institute is the development and use of phages for prophylactic and therapeutic treatment of non-fermenting Gram-negative bacilli (NFGNB; for example, *Pseudomonas*, *Acinetobacter*, *Burkholderia* spp.), related infections as well as their elimination from hospital environment. In vitro screening of 467 multi-resistant *Streptococcus aureus* (MRSA) strains from the UK revealed 98.5% responsiveness to treatment with *Staphylococcus*-specific phages. Similar numbers were obtained on 54 MRSA/38 toxin-producing non-MRSA strains from Germany (99%), 56 MRSA strains from New York University (95%) and 100 MRSA strains from the Royal College of Surgeon in Ireland (97%) [61]. Analogous studies have been performed for β-lactamase producing *E. coli* and *Klebsiella* [62]. Attention is also given to especially dangerous pathogens, like *B. anthracis*, *Brucella* spp. [63], *Vibrio* spp. [64,65], or *Francisella tularensis*. Here, the goal is to find specific phages as well as to unravel their biology, ecology, stability, mechanisms of host interaction and other characteristics.

Research at the Eliava Institute is not solely focused on human phage therapy including wound care products [66], but also on animal health (e.g., phage therapy in aquaculture, against cattle mastitis), and for the purpose of environmental biocontrol (e.g., bacterial blight in cotton and rice, crown-gall disease in grape, diseases caused by *Ralstonia solanacearum*). Kutateladze concluded her talk with selected case reports for irritable bowel syndrome (IBS), cystic fibrosis, post-operational infections, chronic bacterial prostatitis, and chronic *Staphylococcus aureus* skin infections as a complication of Netherton syndrome [67].

"Can phage therapy provide an alternative to antibiotics?" was the initial question of Hans-Peter Horz's (RWTH Aachen University) presentation but at the end he would not give a general YES or NO because while phages can support overcoming the AMR crisis, they probably will not replace antibiotics. In fact, ongoing studies in Horz's lab showed remarkable synergism of phages and antibiotics (AB) against bacteria, which otherwise are resistant against the AB alone. When the group compared the efficiency of single phages versus phage cocktails to fight multi-drug resistant *Pseudomonas aeruginosa* strains, they observed that certain phages (e.g., SL2) were functioning equally effective as phage cocktails supplemented with additional phages (SL1–4), that is, neither additive nor multiplying effects could be detected [68,69]. The case suggests that a phage cocktail can only be as effective as its most effective single ingredient. Whether this is a general rule needs to be investigated. Horz also reflected on a number of antibacterial strategies involving phages or products encoded by them [70] and concluded with emerging perspectives on the human virome [71].

Key issues for the working and use of phages in experimental therapy were addressed by Thomas Rose et al. (Queen Astrid Military Hospital Brussels). He reported on the difficult first steps of experimental phage therapy in burn wound infections. As burn wounds are often colonized by multiple bacterial species, doubts about the efficacy of a 3-phage cocktail against *P. aeruginosa* and *S. aureus* were confirmed by a lack of clinical improvement in all 9 patients treated [72]. Moreover, general obstacles like (1) the limited number of evidence based studies according to modern standards, (2) the lack of a suitable regulatory frame [73], and (3) sufficient quantities of GMP-compliant phage products prevent rapid progress in human studies that meet ethical, regulatory and scientific standards. Despite these difficulties, a few European groups (members of the Bactériophages & Phagothérapie—Consortium (PHOSA) [74] and PneumoPhage [75]) strive for, or have already started, clinical studies with phages under special regulatory provisions (PhagoBurn [76] and Phage4Cure [9,10]). At the Queen Astrid Military Hospital, eight patients received phage therapy under the Declaration of Helsinki [77], and a clinical safety study in burn patients was conducted. Several cases were presented [78] and their treatment protocols outlined: urinary bladder infections were treated with phage solutions via a Foley catheter, decontamination of nose cavity and throat was achieved by phage spray application, or soft tissue treatment by washout and rinsing with phage fluid. As a partner of the European PHAGOBURN consortium and as a consequence of their practical experiences, the group also engages in the development of phage products [66] and supports regulatory approaches that will aid the clinical application of phages [79].

Andrzej Górski reported on the experiences, practices and results of the Phage Therapy Unit at the Institute of Immunology and Experimental Therapy (Polish Academy of Science) in Wrocław. In line with Truog's opinion, "While the most trustworthy advances come through the performance of well-designed trials, sometimes experimental treatments based on theoretical considerations alone may lead to major breakthroughs" [80], Gorski and colleagues performed more than 280 treatments according to current administrative, legislative and ethical requirements. The outcome has been evaluated meticulously and classified into 7 categories. The 40% success rate for patients that were beyond any other treatment refers to complete pathogen eradication or sustainable clinical improvement in the patients. The remaining 60% showed either questionable outcomes, transient responses, no therapeutic effects, or even deterioration as a result of phage therapy [81]. Moreover, efficacy, side effects, resistance and immune response have been assessed dependent on the route of phage administration. Immune responses were shown to be low in patients receiving phages orally and not necessarily adversely affect therapy outcome. Remarkably, phages (and their proteins) downregulated proinflammatory cytokines in mice and reactive oxygen species [82,83]. Similar findings have been noted in patients on phage therapy [83] suggesting that phage therapy in addition to its well-known antibacterial action may also have anti-inflammatory and immunomodulatory effects which may be of use in clinical medicine [84–86].

The efficiency of bacteriophage therapy in children with diarrheal diseases was assessed in a clinical trial by Karaman Pagava and co-workers from Tbilisi State Medical University and JSC Biochimpharm. The double-blind randomized, placebo-controlled study included 71 hospitalized children from the age of 6 months to 6 years with moderate to severe diarrhoea. The improvement of symptoms and period of hospitalization after treatment with a polyvalent phage cocktail (SEPTAPHAGE, JSC Biochimpharm) or a placebo was compared. Results were summarized as follows: Phage therapy (1) significantly shortened the hospital stay (on average for 1.9 ± 0.6 days); (2) prevented clinical deterioration (especially in case of positive test on calprotectin) and switching to antibiotic therapy; (3) alleviated the severity of symptoms; (4) did not elicit any measurable side effects. Pagava also reflected on possible immunological reactions which might decrease the efficacy of phage therapy. Taking the results of this study and literature data together, Pagava reasoned that phage therapy for the treatment of diarrheal diseases can be beneficial for both in-patients and out-patients of children of all ages.

Experiences from Phase I and II clinical trials at the International Center for Diarrheal Diseases Research (ICDDRB) in Dhaka were presented by Harald Brüssow (Nestlé Research Center Lausanne). About 60% of the children hospitalized suffered from *E. coli* diarrhoea, as microbiologically determined at local laboratories. Although enterotoxigenic *E. coli* (ETEC) were supposed to be the etiological agent, other pathogens later were found to contribute to the clinical signs. Phage therapy included and compared two phage products: a cocktail of T4-like phages produced on *E. coli* strain K803 (a prophage-free K-12 derivative) and the commercial Microgen product Coli-Proteus, that is, a phage cocktail of 18 distinct phage types [87]. After a series of safety tests, no elevated safety risk was detected for the two phage products in comparison to a placebo [88]. The double-blinded, placebo-controlled, randomized phage therapy trial at ICDDRB included 120 patients, a third of which were either treated with the aforementioned cocktail of T4-like phages, the commercial Coli-Proteus product or a placebo. No adverse effects attributable to oral phage application were observed. Although the faecal coliphage titre was increased in treated over control children, phage therapy did not outperform standard rehydration/zinc treatment. Various factors might have contributed to the therapeutic failure of the phage trial: mixed infections with other pathogens causing the diarrhoea phenotype being most likely. Indeed, subsequent stool microbiota analyses revealed a correlation between diarrhoea and increased levels of Streptococcus [89].

Bacteriophage therapy for the treatment of lung infections was introduced by Martin Witzenrath (Charité—University Hospital Berlin). Multidrug resistant bacteria are a severe threat for patients with nosocomial pneumonia, as there are no therapeutic options. Experiments with phage-derived endolysins [90] demonstrated that inhalative [5] or intraperitoneal application of Cpl-1 [91] was efficient in treating severe pneumococcal pneumonia in mice, but increased inflammation markers IL-1b and IL-6 [5]. In conclusion, Witzenrath recommended the unequivocal identification of the pathogen prior to any endolysin treatment to ensure their efficacy. He also stressed the need for more studies that (1) confirm the therapeutic safety of lysins in humans (e.g., ContraFect Phase IIa Study endolysins against *S. aureus* bacteraemia), (2) address possible risks for the development of resistance, and (3) improve pharmacokinetic and pharmacodynamic properties of therapeutic lysin products.

Whereas *Streptococcus pneumoniae* plays a major role in community-acquired pneumonia, *Acinetobacter baumannii* is causal in many hospital-acquired pneumonia, and often multidrug resistant. Mice infected with *A. baumannii* and treated with a purified phage Acibel004 preparation intratracheally showed significantly reduced bacterial load in broncho alveolar lavage fluid and lung as well as a significantly improved clinical outcome and lung permeability. Neither cellular nor humoral adverse effects were observed [92]. Chronic lung diseases are increasing and complicated by airway infections. Patients suffering from pre-impaired lungs are often chronically infected with *P. aeruginosa*, especially patients with cystic fibrosis or 'non-CF' bronchiectasis. The recently started BMBF-funded project "Phage4Cure" [9,10] will realize the development of phage preparations and a clinical trial to test safety, tolerability and efficacy in healthy volunteers and patients with chronic pulmonary *P. aeruginosa* infections.

The European Project PhagoBurn [76] started in June 2013 and aims at evaluating the efficacy of phage therapy for the treatment of burn wounds infected with bacteria *Escherichia coli* and *Pseudomonas aeruginosa*. The coordinator of the consortium, Patrick Jault (Percy Military Hospital of Clamart), introduced the design of the first randomized, single-blinded, multi-centric and controlled clinical trial in human phage therapy in Europe. Initially planned for 3 years, the implementation of the project plan needed much more time because the setup and quality control of a GMP-compliant phage bioproduction chain (CLEAN CELLS, France) and its approval by the French Medicine Agency (ANSM) needed 24 instead of the scheduled 7 months. The clinical trial started when the first patient was treated in July 2015 and ended in December 2016 after phage therapy in a total of 27 patients. Although analyses and discussion of the results are still in progress, some valuable results can already be summarized: (I) The first European GMP-compliant production chain is operating and approved by 3 national regulators (FR, BE, CH); (II) Design and prerequisites of randomized multi-centric clinical trials have been developed and validated; (III) More lessons will be learned and conclusions drawn after the trial results are publicly available and hopefully will shed more light on questions regarding the reliable shelf life of phage cocktails, potential development of phage resistance in patients, immune reactions and possible precautions and interventions.

Christine Rohde (Leibniz Institute DSMZ) summarized the pros and cons of human phage therapy and set it in a broader context [93]. As "stable survivalists" phages co-evolve with or drive the evolution of their bacterial hosts and modulate natural and artificial microbial ecosystems. Their potential as antimicrobials and/or modulator of microbial communities is widely recognized and our knowledge steadily increases.

"Pro phage" criteria included: (1) host specificity avoids dysbiosis in treated environments, (2) no toxic side effects if purified, (3) self-replication/self-limitation, (4) occurrence of resistance to one phage does not cause generalized phage resistance, (5) phage-resistant bacteria are often less fit/virulent, (6) inexhaustible reserves, (7) effective regardless of MDR/ESKAPE bacteria, (8) isolated phage lysins as alternative, (9) comparatively inexpensive, thus public health costs can be reduced, (10) flexible application practices, (11) phages can replace last resort antibiotics once production, purification standards and regulatory pathway are defined, (12) availability of tailor-made phage preps could be done in realistic time. In line with the concept of personalized medicine, Rohde envisages phage therapy as a flexible concept reacting to specific settings, that is, single phage preparations or mixtures may be adapted to fit changing host spectra.

On the "contra phage" side, that is, a list of possible obstacles, Rohde addressed: (I) causative bacterial pathogens must be identified unequivocally, also in mixed infections, (II) bacterial phage resistance/immune system of probands can cause problems, (III) phages hardly reach intracellular pathogens, (IV) shelf life/stability might vary from phage to phage, and requires regular titre controls, (V) phage therapy needs physicians' know-how, and (VI) some human body targets might be challenging for phage application, for example, bones/joints, deep wounds.

Rohde also addressed current challenges that prevent wider use and safe application of phages: (a) phage preparations need to be highly specific for the bacterial target, fully characterized, and produced according to GMP standards; (b) protection of intellectual property rights (IPR) and infrastructures are yet to be established, because phage banks, pharmaceutical production facilities, diagnostic laboratories and hospitals conversant with phage use are rare. In a multi-stakeholder approach, model licensing pathways for phage preparations should be developed which are accepted and streamlined by regulatory authorities across national borders [94]. Such prerequisites provided, Rohde suggests that phage therapy should not only be considered in life-threatening situations but could routinely avoid antibiotic treatments.

2.4. Application of Phages for Veterinary Practices, in the Food and Environmental Sector

The use of phages as antimicrobials has spread to other sectors, for example, food production, environmental protection and agrobioindustry. "Despite a century of bacteriophage application in medicine and as biocontrol agent, our understanding of the molecular details of the phage infection process is still limited. Few model phages have been studied in detail, but the majority of potentially useful phage-encoded resources remain untapped in this respect." After this introductory line, Jochen Klumpp's (ETH Zürich) keynote highlighted how investigation of molecular and structural aspects of the phage infection process can be used and transferred to efficient biotechnological applications. Using long tail fibre proteins of *Salmonella* phage S16 for immobilization, Klumpp and his colleagues developed a rapid detection assay for *Salmonella* [95], which can significantly improve food safety. Likewise, *Listeria* phage A511 [96] was employed as a model for the large family of SPO1-related phages, which attack bacteria involved in foodborne and non-foodborne diseases. Knowledge of the structure of the distal tail apparatus and deduced information about the infection process helped to establish a specific *Listeria* assay that can be easily adapted for other SPO1-related phage applications. For instance, *Erwinia amylovora* phage Y2 has its main use as an efficient biocontrol agent against fire blight in apple and pear trees, but it is also the basis for a low-cost detection assay, which will serve as an environmental monitoring tool for *Erwinia* [97].

Effects of phages in dairy fermentation processes were addressed by Horst Neve (Max Rubner-Institute Kiel). Starter cultures can be inhibited by bacteriophages and subsequently even stop fermentation completely. *Lactococcus lactis* and *Streptococcus thermophilus*, prominent bacteria in mesophilic and thermophilic starter cultures, can be attacked by a broad range of bacteriophages. For *L. lactis* at least 10 different phage groups are known [98], and many strains do also contain prophages. In dairies, there are also hybrid *S. thermophilus* phages with genome regions derived from both, lactococcal and *Streptococcus thermophilus* phages [99], as well as phages attacking only flavour-producing *Leuconostoc* strains. Interestingly, a pilot study shows that phage types differ when comparing raw milk with whey or other dairy products [100], the reasons for which remain to be elucidated. In order to prevent economic losses and maintain process safety, Neve et al. recommend monitoring of all starter cultures and a special focus on thermo-resistant and new phage types. To this end, he and other colleagues focus on fast and reliable detection assays as well as on non-thermal removal processes (e.g., membrane filtration immobilization/UV-C irradiation [101,102]) that can significantly reduce phage load in fermentation processes and products thereof.

Campylobacter is an important food-borne pathogen. The family comprises 25 species of which thermophilic *C. jejuni* and *C. coli* are the most common causes of acute bacterial enteritis. In his talk, Stefan T. Hertwig (German Federal Institute for Risk Assessment) focused on the characterization of *Campylobacter* phages and their application along the food chain. Although *Campylobacter* phages have been used to fight this bacterium in chicken since 2003, not much is known about the genetics of these phages, to some extent because of unusual DNA modifications of the *Campylobacter* phage genome. According to genome size, these phages are classified: Group II (180–190 kb) and group III (130–140 kb) are the most common, while group I (320 kb) is rare. Data base searches and genome organization revealed a close relationship of phages belonging to each group. Based on the genome organization of group II phages (4 modules separated by long repeat regions), Hertwig et al. developed a multiplex PCR system [103].

While group II phages lyse strains of *C. jejuni* and *C. coli*, group III phages exclusively infect *C. jejuni*. However, group III phages generally lysed more *C. jejuni* strains than group II phages. In addition, the in vitro kinetics of cell lysis diverged in the two groups, probably caused by the different burst size of phages [104].

Sophie Kittler (University of Veterinary Medicine Hannover) and colleagues presented a practical approach of using phages to reduce *Campylobacter* load in broiler chickens. Risk assessment considers the reduction of *Campylobacter* in primary production to be most beneficial for human health [105,106]. Based on various pilot studies [6,107,108], a phage cocktail of four *Campylobacter* phages was tested during three *in vivo* trials under experimental conditions and in commercial broiler houses. Significant reduction of *Campylobacter* counts was confirmed for all trials under experimental conditions as well as in the two field studies. Reduction of up to $\log_{10}$ 3.2 CFU in *Campylobacter* load at slaughter was demonstrated in one field trial; one day after phage application *Campylobacter* numbers in another experimental group were reduced under the detection limit (<50 CFU/g) in faecal samples. Resistance analyses with re-isolates yielded three major results: (1) Phage-susceptible *Campylobacter* overgrow resistant isolates, (2) resistance of *Campylobacter* against phages stabilizes at a low level after an initial increase, and (3) different mechanisms of resistance seem to affect different phages. The latter seems quite plausible considering recent findings published by Doron et al. [109].

In light of modern food technologies, an interesting question was raised by Meike Samtlebe (University of Hohenheim) et al.: Do hygienic measures and subsequent reduced microbial load of our food (and thus also reduced number of phages) influence our intestinal performance and microbial balance? The human gut contains about 10^{15} individual phage particles, but little is known about their impact on gut microbiota [110], health and diseases. It is obvious that phages influence their bacterial hosts in various ways and hence, could specifically be employed to modulate the microbial composition of the gastrointestinal tract to maintain a healthy balance. However, targeted application of phages by integration into food matrixes faces numerous challenges, for example, bacterial resistance, manufacturing issues, suitable delivery systems and the adaptation to gastrointestinal conditions. In vitro experiments with encapsulated *Lactococcus lactis* phage P008 were carried out to test phage viability under various conditions (encapsulation techniques, enzymes/pH of surrounding fluids). In comparison to free phages (surviving pH > 2.5), encapsulated lactococcal phages are protected during their transit through the stomach and are released effectively under intestine conditions. This result could be confirmed in a dynamic in vitro gastrointestinal model (TIM-1). The study also demonstrated a protective effect of dairy matrices resulting in significant higher phage survival rates after undergoing acid gastric conditions. In conclusion, phages may be suitable modulators of human gut microbiota when applied through dairy food matrices [111].

The potential use of bacteriophages in honeybees was presented by Hannes Beims (Lower Saxony State Office for Consumer Protection and Food Safety) and targets *Paenibacillus larvae* as the causative agent of American foulbrood. To combat this most serious bacterial disease in honey bees, the team isolated and characterized *P. larvae*-specific bacteriophages from infected beehives. Whole-genome analysis of the phages allowed for a detailed safety profile and uncovered their lysogenic nature [112]. The bacteriolytic activity of phages HB10c2 and HBχ (*Siphoviridae*) was tested in plaque assays and growth inhibition was found for all genotypes of *P. larvae* tested (ERIC I–IV), as well as for 40 field isolates of the genotypes ERIC I and II. In vivo bioexposure assays showed that the feeding of bee larvae with bacteriophages has no negative effect on the development of the brood. In fact, mortality of bee larvae was reduced by phage application. Therapeutic effect could be improved by daily application.

2.5. Phage Lysins and Commercial Perspectives

In some cases, the use of whole phages to eliminate bacterial pathogens is inhibited either by active defence systems of the host [109] or physico-chemical conditions of the surrounding environment [113]. The use of phage-specific enzymes can circumvent such restrictions and moreover, may bypass the limited host range of most phages. Aidan Coffey (Cork Institute of Technology) presented a successful example of this approach by using phage-derived peptidoglycan hydrolases to target MRSA and antibiotic resistant *Clostridium difficile* [114]. After genomic characterization of three anti-staphylococcal phages (DW2, K, and CS1), their genes for peptidoglycan-degrading hydrolases were cloned. One of the phage endolysins displays a modular organisation with three domains: a Cys/His-dependent amido hydrolase peptidase (CHAPk), an amidase, and a cell-wall binding domain [115,116]. The latter facilitates attachment of the enzyme to the bacterial cell wall, while the other two domains catalyse the degradation of the peptidoglycan to mediate rapid bacterial cell death. Deletion analysis showed that full lytic activity against antibiotic-resistant staphylococci was retained, even when truncated to its CHAPk (peptidase) domain. X-ray crystallography and site-directed mutagenesis allowed further insight and modelling of its enzymatic mechanism. CHAPk was successfully tested in vitro on MRSA cultures and in vivo eliminated MRSA colonization in mouse without adverse effects. Ex vivo application showed no inflammatory response in primary human umbilical vascular endothelial cells (HUVECs) but immunogenicity in peripheral blood mononuclear cells (PBMCs) was detected in some subjects [115,117]. Analogous experiments were presented for the amidase endolysin from *Clostridium difficile* bacteriophage CD6356. In order to deliver the designer endolysins to their targets, a host-specific secretion and expression system was developed for dairy application (*L. lactis*) and successfully tested. Additionally, nanoparticle gels and adhesive dressings with anchored CHAPk-nanoparticles were developed for skin application. A thermal trigger concept (activation of CHAPk at 37 °C) was successfully implemented.

Wolfgang Mutter et al. from HYpharm reported on a similar approach using designer lysins to target MRSA. He stated that lysins are as efficient as antibiotics, have a comparable minimum inhibitory concentration (MIC) and work much faster. Their application on surfaces (skin, nasal mucosa) is possible, but requires optimization of the proteins with respect to stability and expression rate. Moreover, one has to keep in mind that in vivo applications can trigger immune responses in some cases, that is, production of anti-lysin antibodies in the host [118]. Based on their unique phage recombinant protein technology, HyPharm holds eight patent families.

One designer lysin HY-133 which is directed against *Staphylococcus aureus* has successfully passed laboratory and animal (cotton rat model) tests. In comparison to PRF-119, another recombinant chimeric bacteriophage endolysin [119], HY-133 displays the same activity and specificity but a significant higher stability. The molecule is currently in GMP production; clinical phase I trials are planned for 2019. The studies are conducted by an interdisciplinary public-private consortium (Fraunhofer ITEM, Coreolis Pharma, Center for Clinical Trials Tübingen, University Hospital Münster, German Center for Infection Research) and get support from German regulatory authorities (Federal Institute for Drugs and Medical Devices—BfArM and Paul-Ehrlich-Institute—PEI) and funders.

The representative of Micreos BV, Steven Hagens, pointed out that not all phages are suitable for bio-controlling. Which selection criteria need to be met was discussed on the basis of two examples, a single phage product against *Listeria* (Listex P100 = PhageGuard Listex) and a two-phage cocktail targeting *Salmonella* (S16 + FO1 = PhageGuard S). Favourable for PG Listex is (a) the extremely broad host range within the *Listeria* genus (b) the high efficacy (94–100% reduction) for various *Listeria*-contaminated food sources after storage for 6 days, and (c) its speed of action (1 min seems sufficient) [119].

Because of its different receptors, phages in PhageGuard S can attach to cell receptors present on all *Salmonella* serovars. Additional DNA modifications protect the phage product against varying bacterial defence mechanisms. An industrial trial consisting of 7 s dip treatment/24 h hold for various meat products resulted in a 1–3 log reduction of *Salmonella* counts [120]. Regulatory authorities have acknowledged safety and efficacy of both products by approving their use in the US, Canada, the Netherlands, Australia and New Zealand. Nevertheless, food manufacturers need to comply with hygienic rules; phages can neither mask poor hygiene nor replace it.

3. Plenary Session with Panel Discussion "Quo Vadis, German Bacteriophage Research?"

One of the conference highlights was a panel discussion with participants from several sectors. It mainly covered the phage application in human medicine but also dealt with other related fields. Representatives from two main regulatory bodies the Federal Institute for Drug and Medical Devices (BfArM) and Paul-Ehrlich-Institute (PEI, Federal Institute for Vaccines and Biomedicines), academia and industry engaged in an in-depth discussion around the current state of affairs and the most pressing problems in the field of phage application. The discussion started with the acknowledgement that there is an extensive amount of basic research taking place in Germany. A national phage bank has been set up at Leibniz Institute DSMZ-German Collection of Microorganisms and Cell Cultures in Braunschweig about 25 years ago and is further expanding [121]. Its main aim is to stock phages against the most common pathogens, for example, ESKAPE. The bank is now involved as a pool for the first ever government-funded clinical trial with Charité University Hospital in Berlin (Phage4Cure) [9,10]. Still, a range of problems has been identified during the discussion, such as:

- Lack of a clear regulatory framework;
- Lack of a clearly pre-defined phage product;
- Lack of clinical trials;
- Lack of financial incentives, particularly for start-ups and small companies;

- Lack of involvement of the pharma industry due to the low return on investment [122] as well as liability and reimbursement issues.

Thus, the situation in Germany is similar to most Western countries and lags behind the developments in some EU member states like Poland, Belgium, France or The Netherlands. The current situation in Germany has been described as one where everyone appears to be waiting for the other side to take action and break the vicious circle of a missing approved phage product. Academia pointed to the lack of a clear European regulatory framework for phages and the rather confusing responsibilities regarding the application of phages throughout the various settings like human and animal health, food, plant and environmental protection as well as hygiene.

The regulatory bodies stated that there is no clearly defined product out there prompting an immediate licensing activity. While achievements have been made in the basic research and all the needed technology is available there are still no clinical trials running except for the aforementioned "Phage4Cure" project. It was suggested that efforts should be made to develop pilot like clinical trials targeting clearly defined disease syndromes with limited numbers of patients instead of huge time- and money-consuming multi-centre studies. Medical practitioners, under the umbrella of their respective medical organizations, will not use phages as there is no approved medicinal product available and no health insurance company would cover the costs. Some patients still travel to Georgia or Poland for help and their number is expected to increase in the future. Two more points have been identified as hurdles: (a) there are no patient organisations in Germany who would advocate for phage therapy, and (b) the public awareness of the topic is still low, although the media coverage has been on the rise for the last few years.

Both regulatory bodies declared their general openness for dialogue and willingness to support companies who plan to develop a product and apply for an approval. They urged the interested companies to be more proactive and approach the authorities in advance in order to discuss the issues in detail and avoid costly procedures and prevent errors which could jeopardize the approval process. They reported on their experiences with the European Medicines Agency (EMA), which employs a SME Office, that is, a unit advising small and medium-sized enterprises [123]. Unfortunately, this service is rarely being used. A promising instrument is the so-called PRIME scheme, which allows for an accelerated approval of medicinal products where an urgent unmet medical need exists [124]. Concerted efforts to set up a European regulatory frame are desirable and could benefit from lessons learned during similar initiatives, for example, the regulation of genetically modified plants. The speakers concluded with a statement that any regulatory attempt can be a reactive one *per se*, and thus encouraged companies to be more proactive and engage in an early and intensive exchange with regulatory bodies.

Both small company representatives complained about the lack of interest from the pharma industry. Hansjörg Lehnherr (PTC Phage Technology Center GmbH) pointed out that it is not financially feasible to develop a GMP-compatible phage product for a small business and pleaded for more support from the government. Wolfgang Mutter from Hyglos GmbH reported on their successful cooperation with the authorities and with three university hospitals, all three partners of the German Center for Infection Research (DZIF). The cooperation project aims at testing of HY-133 in three German clinics, an active component against *S. aureus*, which has been designed by the company together with the University Hospital Münster [125,126]. Mutter highlighted the essential role of the DZIF in development and research of phage products, as a government-funded body comprising 35 leading research organisations in the field of infection research, with the needed finances, expertise and flexibility. Both company representatives referred to successful approval cases in the US as a potential model for Germany or the EU. However, there has also been some controversial debate on how to get approval for endolysin products.

All participants agreed that there is an urgent need to establish a dialogue between all stakeholders including regulatory bodies, legal experts, industry (pharma, biotech, food), insurance companies, researchers and academia to define possible solutions. Such a dialogue could lead to a more favourable climate for research and development of phage products throughout all sectors.

Furthermore, general funding possibilities were discussed. A representative of VDI/VDE Innovation + Technology GmbH (a project management agency for BMBF) informed that the German Federal Government was to allocate up to 500 million Euros (i.e., a two-third more than before) in the coming decade to fight antimicrobial resistance [127]. That framework which was agreed upon during the G20 Health Ministers meeting in Berlin [128] could offer the needed support for phage research and application consortia.

In conclusion, it was established that bacteriophages have become a source of hope in the face of ever-increasing AMR problems and that more coordinated efforts are needed to engage all stakeholders in a dialogue and to raise public awareness. Any phage application and regulation efforts should be in line with the WHO's One Health approach which addresses all settings as one single system [129,130]. Pharma and health insurance companies, medical practitioners as well as patient organisations are expected to be the promotors of innovation in phage therapy. An important key role is placed with the DZIF as the only organisation in Germany today capable of introducing the desired product. Some of the issues are regulated nationally (e.g., blood products, tissue preparations, vaccines under the responsibility of the PEI) [131], however, in the long-run effective regulatory measures should best be aimed at the supra-national level [94].

4. Conclusions and Perspectives

The diversity of phages, their properties and functional interactions in various settings are enormous. With the increasing number of studies, our knowledge about the structure, function and interaction of bacteriophages becomes richer in detail and further substantiated. The current "Omics" repertoire and other methods also help to revisit "old" findings and put them into new context. Extra potential comes with the yet "hypothetical proteins" that could complete our picture of structural elements and unravel their functional significance and relations. Interaction of phages within their specific environments opens up exciting new fields of research and application. For instance, the role of phages in modulating the human microbiome has been addressed in several talks and lively discussions during the 1st German Phage Symposium. Since then, a number of papers have shed more light on this topic in humans [132–135], animals [136–138], and the environment [139] making it safe to predict that phage therapy will conquer this emerging field of application rather sooner than later.

One century of ground breaking and experimental phage research has opened new perspectives and set the stage for multifaceted applications of bacteriophages. Many methodical obstacles were removed, "teething problems" of GMP-compliant phage production have been addressed, and some regulatory hurdles have already been taken. And yet, reality does not keep pace with scientific progress. IP protection, licensing and other regulatory issues must be adapted to the new world of personalized medicine and other fields of phage application, not vice versa. Open access to data and suitable infrastructure (phage repositories, GMP-compliant production facilities, diagnostic units and clinics experienced with phage use) is needed to tap the full potential of phage applications. What's more, close international cooperation can compensate for the still limited number of phage applications worldwide. Conferences and other exchange platforms offer a suitable forum for that. It is in this context, that we would like to thank all participants of the 1st German Phage Symposium for their contributions and we hope to continue this exchange at the 2nd German Phage Symposium in 2019 and at other occasions.

Supplementary Materials: All abstracts of oral presentations and posters of the 1st German Phage Symposium are available online at http://www.mdpi.com/1999-4915/10/4/158/s1 as supplement (S1).

Acknowledgments: The organization of the 1st German Phage Symposium was kindly supported by the German Research Foundation (DFG), Boehringer Ingelheim Foundation, cc pharma GmbH, BASF, HYpharm GmbH, and 'Friends and Supporters of the University of Hohenheim' eV.

Conflicts of Interest: The authors declare no conflict of interest.

Disclaimer: This review has been prepared by the authors to the best of knowledge and was cross-checked with abstracts and presentations with the greatest care. However, errors or misinterpretations cannot be completely ruled out. The authors assume no responsibility for third party content. All liability is excluded.

References

1. Ruska, H. Die Sichtbarmachung der Bakteriophagen Lyse im Übermikroskop. *Naturwissen* **1940**, *28*, 45–46. [CrossRef]
2. Rohde, C.; Sikorski, J. Bakteriophagen. Vielfalt, Anwendung und ihre Bedeutung für die Wissenschaft vom Leben. *Nat. Rundsch.* **2011**, *64*, 5–14.
3. Ray, D.S.; Bscheider, H.P.; Hofschneider, P.H. Replication of the single-stranded DNA of the male-specific bacteriophage M13. Isolation of intracellular forms of phage-specific DNA. *J. Mol. Biol.* **1966**, *21*, 473–483. [CrossRef]
4. Georgieva, Y.; Konthur, Z. Design and screening of M13 phage display cDNA libraries. *Molecules* **2011**, *16*, 1667–1681. [CrossRef] [PubMed]
5. Doehn, J.M.; Fischer, K.; Reppe, K.; Gutbier, B.; Tschernig, T.; Hocke, A.C.; Fischetti, V.A.; Löffler, J.; Suttorp, N.; Hippenstiel, S.; et al. Delivery of the endolysin Cpl-1 by inhalation rescues mice with fatal pneumococcal pneumonia. *J. Antimicrob. Chemother.* **2013**, *68*, 2111–2117. [CrossRef] [PubMed]
6. Kittler, S.; Fischer, S.; Abdulmawjood, A.; Glünder, G.; Klein, G. Effect of bacteriophage application on Campylobacter JEJUNI loads in commercial broiler flocks. *Appl. Environ. Microbiol.* **2013**, *79*, 7525–7533. [CrossRef] [PubMed]
7. Hertwig, S.; Hammerl, J.A.; Appel, B.; Alter, T. Post-harvest application of lytic bacteriophages for biocontrol of foodborne pathogens and spoilage bacteria. *Berliner und Münchener Tierärztliche Wochenschrift* **2013**, *126*, 357–369. [CrossRef] [PubMed]
8. Charité Research Organization. Bacteriophages Join the Fight against Infection. Available online: https://www.charite-research.org/en/bacteriophages-join-fight-against-infection (accessed on 2 March 2018).
9. Phage4Cure—Developing Bacteriophages as Approved Therapy against Bacterial Infection, Project Homepage. Available online: http://phage4cure.de/en/ (accessed on 16 February 2018).
10. Torres-Barceló, C.; Kaltz, O.; Froissart, R.; Gandon, S.; Ginet, N.; Ansaldi, M. "French Phage Network"—Second Meeting Report. *Viruses* **2017**, *9*, 87. [CrossRef] [PubMed]
11. Bamford, D.H. Do viruses form lineages across different domains of life? *Res. Microbiol.* **2003**, *154*, 231–236. [CrossRef]
12. Nelson, D. Phage taxonomy: We agree to disagree. *J. Bacteriol.* **2004**, *186*, 7029–7031. [CrossRef] [PubMed]
13. Adriaenssens, E.M.; Wittmann, J.; Kuhn, J.H.; Turner, D.; Sullivan, M.B.; Dutilh, B.E.; Jang, H.B.; van Zyl, L.J.; Klumpp, J.; Lobocka, M.; et al. Taxonomy of prokaryotic viruses: 2017 update from the ICTV Bacterial and Archaeal Viruses Subcommittee. *Arch. Virol.* **2017**, *162*, 1153–1157. [CrossRef] [PubMed]
14. Wittmann, J. Phage genomics and taxonomy—Bringing order into chaos. In Proceedings of the 1st German Phage Symposium, Stuttgart, Germany, 9–11 October 2017.
15. Arnaud, C.-A.; Effantin, G.; Vivès, C.; Engilberge, S.; Bacia, M.; Boulanger, P.; Girard, E.; Schoehn, G.; Breyton, C. Bacteriophage T5 tail tube structure suggests a trigger mechanism for Siphoviridae DNA ejection. *Nat. Commun.* **2017**, *8*, 1953. [CrossRef] [PubMed]
16. Boulanger, P. Singularities of bacteriophage T5 structure and infection mechanism. In Proceedings of the 1st German Phage Symposium, Stuttgart, Germany, 9–11 October 2017.
17. Broeker, N.K.; Barbirz, S. Not a barrier but a key: How bacteriophages exploit host's O-antigen as an essential receptor to initiate infection. *Mol. Microbiol.* **2017**, *105*, 353–357. [CrossRef] [PubMed]
18. Tavares, P. Genome Replication and Assembly of the Bacteriophage SPP1 Particle in vivo. In Proceedings of the 1st German Phage Symposium, Stuttgart, Germany, 9–11 October2017.
19. Fernandes, S.; Labarde, A.; Baptista, C.; Jakutytė, L.; Tavares, P.; São-José, C. A non-invasive method for studying viral DNA delivery to bacteria reveals key requirements for phage SPP1 DNA entry in Bacillus subtilis cells. *Virology* **2016**, *495*, 79–91. [CrossRef] [PubMed]
20. Djacem, K.; Tavares, P.; Oliveira, L. Bacteriophage SPP1 PAC Cleavage: A Precise Cut without Sequence Specificity Requirement. *J. Mol. Biol.* **2017**, *429*, 1381–1395. [CrossRef] [PubMed]

21. Loh, B.; Haase, M.; Mueller, L.; Kuhn, A.; Leptihn, S. The Transmembrane Morphogenesis Protein GP1 of Filamentous Phages Contains Walker A and Walker B Motifs Essential for Phage Assembly. *Viruses* **2017**, *9*, 73. [CrossRef] [PubMed]
22. Pawluk, A.; Staals, R.H.J.; Taylor, C.; Watson, B.N.J.; Saha, S.; Fineran, P.C.; Maxwell, K.L.; Davidson, A.R. Inactivation of CRISPR-Cas systems by anti-CRISPR proteins in diverse bacterial species. *Nat. Microbiol.* **2016**, *1*, 16085. [CrossRef] [PubMed]
23. Pawluk, A.; Bondy-Denomy, J.; Cheung, V.H.W.; Maxwell, K.L.; Davidson, A.R. A new group of phage anti-CRISPR genes inhibits the type I-E CRISPR-Cas system of *Pseudomonas aeruginosa*. *mBio* **2014**, *5*, e00896. [CrossRef] [PubMed]
24. Bondy-Denomy, J.; Pawluk, A.; Maxwell, K.L.; Davidson, A.R. Bacteriophage genes that inactivate the CRISPR/Cas bacterial immune system. *Nature* **2013**, *493*, 429–432. [CrossRef] [PubMed]
25. Semenova, E.; Jore, M.M.; Datsenko, K.A.; Semenova, A.; Westra, E.R.; Wanner, B.; van der Oost, J.; Brouns, S.J.J.; Severinov, K. Interference by clustered regularly interspaced short palindromic repeat (CRISPR) RNA is governed by a seed sequence. *Proc. Natl. Acad. Sci. USA* **2011**, *108*, 10098–10103. [CrossRef] [PubMed]
26. Deveau, H.; Barrangou, R.; Garneau, J.E.; Labonté, J.; Fremaux, C.; Boyaval, P.; Romero, D.A.; Horvath, P.; Moineau, S. Phage response to CRISPR-encoded resistance in *Streptococcus thermophilus*. *J. Bacteriol.* **2008**, *190*, 1390–1400. [CrossRef] [PubMed]
27. Pyenson, N.C.; Gayvert, K.; Varble, A.; Elemento, O.; Marraffini, L.A. Broad Targeting Specificity during Bacterial Type III CRISPR-Cas Immunity Constrains Viral Escape. *Cell Host Microbe* **2017**, *22*, 343–353. [CrossRef] [PubMed]
28. Paez-Espino, D.; Sharon, I.; Morovic, W.; Stahl, B.; Thomas, B.C.; Barrangou, R.; Banfield, J.F. CRISPR immunity drives rapid phage genome evolution in *Streptococcus thermophilus*. *mBio* **2015**, *6*. [CrossRef] [PubMed]
29. Bryson, A.L.; Hwang, Y.; Sherrill-Mix, S.; Wu, G.D.; Lewis, J.D.; Black, L.; Clark, T.A.; Bushman, F.D. Covalent Modification of Bacteriophage T4 DNA Inhibits CRISPR-Cas9. *mBio* **2015**, *6*, e00648. [CrossRef] [PubMed]
30. Vlot, M.; Houkes, J.; Lochs, S.J.A.; Swarts, D.C.; Zheng, P.; Kunne, T.; Mohanraju, P.; Anders, C.; Jinek, M.; van der Oost, J.; et al. Bacteriophage DNA glucosylation impairs target DNA binding by type I and II but not by type V CRISPR-Cas effector complexes. *Nucleic Acids Res.* **2018**, *46*, 873–885. [CrossRef] [PubMed]
31. El Hassan, M.A.; Calladine, C.R. Propeller-twisting of base-pairs and the conformational mobility of dinucleotide steps in DNA. *J. Mol. Biol.* **1996**, *259*, 95–103. [CrossRef] [PubMed]
32. Jackson, S.A.; McKenzie, R.E.; Fagerlund, R.D.; Kieper, S.N.; Fineran, P.C.; Brouns, S.J.J. CRISPR-Cas: Adapting to change. *Science* **2017**, *356*. [CrossRef] [PubMed]
33. Tao, P.; Wu, X.; Rao, V. Unexpected evolutionary benefit to phages imparted by bacterial CRISPR-Cas9. *Sci. Adv.* **2018**, *4*. [CrossRef] [PubMed]
34. Heinrich, J.; Velleman, M.; Schuster, H. The tripartite immunity system of Phages P1 and P7. *FEMS Microbiol. Rev.* **1995**, *17*, 121–126. [CrossRef] [PubMed]
35. Ravin, V.; Ravin, N.; Casjens, S.; Ford, M.E.; Hatfull, G.F.; Hendrix, R.W. Genomic sequence and analysis of the atypical temperate bacteriophage N15. *J. Mol. Biol.* **2000**, *299*, 53–73. [CrossRef] [PubMed]
36. Kahl, B.C. Impact of *Staphylococcus aureus* on the pathogenesis of chronic cystic fibrosis lung disease. *Int. J. Med. Microbiol.* **2010**, *300*, 514–519. [CrossRef] [PubMed]
37. Tang, Y.; Nielsen, L.N.; Hvitved, A.; Haaber, J.K.; Wirtz, C.; Andersen, P.S.; Larsen, J.; Wolz, C.; Ingmer, H. Commercial Biocides Induce Transfer of Prophage Φ13 from Human Strains of *Staphylococcus aureus* to Livestock CC398. *Front. Microbiol.* **2017**, *8*, 2418. [CrossRef] [PubMed]
38. Kalinowski, J.; Bathe, B.; Bartels, D.; Bischoff, N.; Bott, M.; Burkovski, A.; Dusch, N.; Eggeling, L.; Eikmanns, B.J.; Gaigalat, L.; et al. The complete *Corynebacterium glutamicum* ATCC 13032 genome sequence and its impact on the production of l-aspartate-derived amino acids and vitamins. *J. Biotechnol.* **2003**, *104*, 5–25. [CrossRef]
39. Helfrich, S.; Pfeifer, E.; Krämer, C.; Sachs, C.C.; Wiechert, W.; Kohlheyer, D.; Nöh, K.; Frunzke, J. Live cell imaging of SOS and prophage dynamics in isogenic bacterial populations. *Mol. Microbiol.* **2015**, *98*, 636–650. [CrossRef] [PubMed]
40. Pfeifer, E.; Gätgens, C.; Polen, T.; Frunzke, J. Adaptive laboratory evolution of *Corynebacterium glutamicum* towards higher growth rates on glucose minimal medium. *Sci. Rep.* **2017**, *7*, 16780. [CrossRef] [PubMed]

41. Wendling, C.C.; Piecyk, A.; Refardt, D.; Chibani, C.; Hertel, R.; Liesegang, H.; Bunk, B.; Overmann, J.; Roth, O. Tripartite species interaction: Eukaryotic hosts suffer more from phage susceptible than from phage resistant bacteria. *BMC Evol. Biol.* **2017**, *17*, 98. [CrossRef] [PubMed]
42. Research Group Heiko Liesegang. Group Website. Available online: http://appmibio.uni-goettingen.de/index.php?sec=agl (accessed on 2 March 2018).
43. Wendling, C.C.; Chibani, C.; Hertel, R.; Bunk, B.; Dietrich, S.; Overmann, J.; Liesegang, H.; Roth, O. Tripartite Species Interaction: The Impact of Phage/Bacteria Co-Evolution within a Eukaryotic Host on the Genome Structure of Inoviridae. In Proceedings of the 1st German Phage Symposium, Stuttgart, Germany, 9–11 October 2017.
44. Deng, L.; Ignacio-Espinoza, J.C.; Gregory, A.C.; Poulos, B.T.; Weitz, J.S.; Hugenholtz, P.; Sullivan, M.B. Viral tagging reveals discrete populations in *Synechococcus* viral genome sequence space. *Nature* **2014**, *513*, 242–245. [CrossRef] [PubMed]
45. Feichtmayer, J.; Deng, L.; Griebler, C. Antagonistic Microbial Interactions: Contributions and Potential Applications for Controlling Pathogens in the Aquatic Systems. *Front. Microbiol.* **2017**, *8*, 2192. [CrossRef] [PubMed]
46. Deng, L. Assess the role of viruses in contaminant biodegradation through metagenomics. In Proceedings of the 1st German Phage Symposium, Stuttgart, Germany, 9–11 October 2017.
47. Gillis, A.; Mahillon, J. Phages preying on *Bacillus anthracis*, *Bacillus cereus*, and *Bacillus thuringiensis*: Past, present and future. *Viruses* **2014**, *6*, 2623–2672. [CrossRef] [PubMed]
48. Gillis, A.; Mahillon, J. Influence of lysogeny of Tectiviruses GIL01 and GIL16 on *Bacillus thuringiensis* growth, biofilm formation, and swarming motility. *Appl. Environ. Microbiol.* **2014**, *80*, 7620–7630. [CrossRef] [PubMed]
49. Bolotin, A.; Gillis, A.; Sanchis, V.; Nielsen-LeRoux, C.; Mahillon, J.; Lereclus, D.; Sorokin, A. Comparative genomics of extrachromosomal elements in *Bacillus thuringiensis* subsp. israelensis. *Microbiol. Res.* **2017**, *168*, 331–344. [CrossRef] [PubMed]
50. Gillis, A.; Mahillon, J. Tectiviruses infecting members of the *Bacillus cereus* group. In Proceedings of the 1st German Phage Symposium, Stuttgart, Germany, 9–11 October 2017.
51. Tomasello, G.; Mazzola, M.; Leone, A.; Sinagra, E.; Zummo, G.; Farina, F.; Damiani, P.; Cappello, F.; Gerges Geagea, A.; Jurjus, A.; et al. Nutrition, oxidative stress and intestinal dysbiosis: Influence of diet on gut microbiota in inflammatory bowel diseases. *Acta Univ. Palacki. Olomuc. Fac. Med.* **2016**, *160*, 461–466. [CrossRef] [PubMed]
52. Weiss, G.A.; Hennet, T. Mechanisms and consequences of intestinal dysbiosis. *CMLS* **2017**, *74*, 2959–2977. [CrossRef] [PubMed]
53. COUNTERSTRIKE: COUNTERacting Sarcopenia with ProTeins and ExeRcise—Screening the CALM Cohort for Lipoprotein Biomarkers, Project Website. Available online: http://counterstrike.ku.dk/ (accessed on 3 February 2017).
54. Castro-Mejía, J.L.; Jacobsen, R.; Krych, L.; Kot, W.; Hansen, L.H.; Reitelseder, S.; Holm, L.; Engelsen, S.B.; Vogensen, F.K.; Nielsen, D.S. Elucidating phage-bacterium interactions that trigger changes in bacterial composition and functional profile in the gut of older adults. Presented at the 1st German Phage Symposium, Stuttgart, Germany, 9–10 October 2017. Unpublished work.
55. Wagner, P.L.; Livny, J.; Neely, M.N.; Acheson, D.W.K.; Friedman, D.I.; Waldor, M.K. Bacteriophage control of Shiga toxin 1 production and release by *Escherichia coli*. *Mol. Microbiol.* **2002**, *44*, 957–970. [CrossRef] [PubMed]
56. Nubling, S.; Eisele, T.; Stober, H.; Funk, J.; Polzin, S.; Fischer, L.; Schmidt, H. Bacteriophage 933 W encodes a functional esterase downstream of the Shiga toxin 2a operon. *Int. J. Med. Microbiol.* **2014**, *304*, 269–274. [CrossRef] [PubMed]
57. Saile, N.; Voigt, A.; Kessler, S.; Stressler, T.; Klumpp, J.; Fischer, L.; Schmidt, H. *Escherichia coli* O157:H7 Strain EDL933 Harbors Multiple Functional Prophage-Associated Genes Necessary for the Utilization of 5-*N*-Acetyl-9-*O*-Acetyl Neuraminic Acid as a Growth Substrate. *Appl. Environ. Microbiol.* **2016**, *82*, 5940–5950. [CrossRef] [PubMed]

58. Schmidt, H. Bacteriophages of Shiga toxin producing *E. coli*—small molecules with high impact. Presented at the 1st German Phage Symposium, Stuttgart, Germany, 9–11 October 2017; University of Hohenheim, 2017. Available online: https://1st-german-phage-symposium.uni-hohenheim.de/fileadmin/einrichtungen/1st-german-phage-symposium/Presentations/Schmidt_1st-German-Phage-Symposium_web.pdf (accessed on 16 February 2018).
59. Leitner, L.; Sybesma, W.; Chanishvili, N.; Goderdzishvili, M.; Chkhotua, A.; Ujmajuridze, A.; Schneider, M.P.; Sartori, A.; Mehnert, U.; Bachmann, L.M.; et al. Bacteriophages for treating urinary tract infections in patients undergoing transurethral resection of the prostate: A randomized, placebo-controlled, double-blind clinical trial. *BMC Urol.* **2017**, *17*, 90. [CrossRef] [PubMed]
60. Kutadeladze, M. Bacteriophages for Treatment of Infectious Diseases. Presented at the 1st German Phage Symposium, Stuttgart, Germany, 9–11 October 2017. Unpublished work.
61. Fitzgerald-Hughes, D.; Bolkvadze, D.; Balarjishvili, N.; Leshkasheli, L.; Ryan, M.; Burke, L.; Stevens, N.; Humphreys, H.; Kutateladze, M. Susceptibility of extended-spectrum- β-lactamase-producing *Escherichia coli* to commercially available and laboratory-isolated bacteriophages. *J. Antimicrob. Chemother.* **2014**, *69*, 1148–1150. [CrossRef] [PubMed]
62. Sybesma, W.; Zbinden, R.; Chanishvili, N.; Kutateladze, M.; Chkhotua, A.; Ujmajuridze, A.; Mehnert, U.; Kessler, T.M. Bacteriophages as Potential Treatment for Urinary Tract Infections. *Front. Microbiol.* **2016**, *7*, 465. [CrossRef] [PubMed]
63. Tevdoradze, E.; Farlow, J.; Kotorashvili, A.; Skhirtladze, N.; Antadze, I.; Gunia, S.; Balarjishvili, N.; Kvachadze, L.; Kutateladze, M. Whole genome sequence comparison of ten diagnostic brucellaphages propagated on two *Brucella abortus* hosts. *Virol. J.* **2015**, *12*, 66. [CrossRef] [PubMed]
64. Kokashvili, T.; Whitehouse, C.A.; Tskhvediani, A.; Grim, C.J.; Elbakidze, T.; Mitaishvili, N.; Janelidze, N.; Jaiani, E.; Haley, B.J.; Lashkhi, N.; et al. Occurrence and Diversity of Clinically Important Vibrio Species in the Aquatic Environment of Georgia. *Front. Public Health* **2015**, *3*, 232. [CrossRef] [PubMed]
65. Whitehouse, C.A.; Baldwin, C.; Sampath, R.; Blyn, L.B.; Melton, R.; Li, F.; Hall, T.A.; Harpin, V.; Matthews, H.; Tediashvili, M.; et al. Identification of pathogenic Vibrio species by multilocus PCR-electrospray ionization mass spectrometry and its application to aquatic environments of the former soviet republic of Georgia. *Appl. Environ. Microbiol.* **2010**, *76*, 1996–2001. [CrossRef] [PubMed]
66. Merabishvili, M.; Monserez, R.; van Belleghem, J.; Rose, T.; Jennes, S.; de Vos, D.; Verbeken, G.; Vaneechoutte, M.; Pirnay, J.-P. Stability of bacteriophages in burn wound care products. *PLoS ONE* **2017**, *12*, e0182121. [CrossRef] [PubMed]
67. Zhvania, P.; Hoyle, N.S.; Nadareishvili, L.; Nizharadze, D.; Kutateladze, M. Phage Therapy in a 16-Year-Old Boy with Netherton Syndrome. *Front. Med.* **2017**, *4*, 94. [CrossRef] [PubMed]
68. Latz, S.; Krüttgen, A.; Häfner, H.; Buhl, E.M.; Ritter, K.; Horz, H.-P. Differential Effect of Newly Isolated Phages Belonging to PB1-Like, phiKZ-Like and LUZ24-Like Viruses against Multi-Drug Resistant *Pseudomonas aeruginosa* under Varying Growth Conditions. *Viruses* **2017**, *9*, 315. [CrossRef] [PubMed]
69. Horz, H.-P. Phage therapy: An alternative to antibiotics? In Proceedings of the 1st German Phage Symposium, Stuttgart, Germany, 9–11 October 2017.
70. Viertel, T.M.; Ritter, K.; Horz, H.-P. Viruses versus bacteria-novel approaches to phage therapy as a tool against multidrug-resistant pathogens. *J. Antimicrob. Chemother.* **2014**, *69*, 2326–2336. [CrossRef] [PubMed]
71. Wahida, A.; Ritter, K.; Horz, H.-P. The Janus-Face of Bacteriophages across Human Body Habitats. *PLoS Pathog.* **2016**, *12*, e1005634. [CrossRef] [PubMed]
72. Rose, T.; Verbeken, G.; de Vos, D.; Merabishvili, M.; Vaneechoutte, M.; Lavigne, R.; Jennes, S.; Zizi, M.; Pirnay, J.P. Experimental phage therapy of burn wound infection: Difficult first steps. *Int. J. Burns Trauma* **2014**, *4*, 66–73. [PubMed]
73. Verbeken, G.; Huys, I.; de Vos, D.; de Coninck, A.; Roseeuw, D.; Kets, E.; Vanderkelen, A.; Draye, J.P.; Rose, T.; Jennes, S.; et al. Access to bacteriophage therapy: Discouraging experiences from the human cell and tissue legal framework. *FEMS Microbiol. Lett.* **2016**, *363*. [CrossRef] [PubMed]
74. PHOSA. Project Website. Available online: http://www.phosa.eu/ (accessed on 16 February 2018).
75. Pherecydes Pharma. PneumoPhage: A Collaborative Research Project for the Development of an Effective Phage Therapy Treatment against Respiratory Tract Infections. Available online: http://www.pherecydes-pharma.com/pneumophage.html (accessed on 16 February 2018).
76. PhagoBurn. Project Website. Available online: http://www.PhagoBurn.eu/ (accessed on 16 February 2018).

77. Rose, T. Key issues in phage therapy. In Proceedings of the 1st German Phage Symposium, Stuttgart, Germany, 9–11 October 2017.
78. Jennes, S.; Merabishvili, M.; Soentjens, P.; Pang, K.W.; Rose, T.; Keersebilck, E.; Soete, O.; François, P.-M.; Teodorescu, S.; Verween, G.; et al. Use of bacteriophages in the treatment of colistin-only-sensitive *Pseudomonas aeruginosa* septicaemia in a patient with acute kidney injury—A case report. *Crit. Care* **2017**, *21*, 129. [CrossRef] [PubMed]
79. Pirnay, J.-P.; Verbeken, G.; Ceyssens, P.-J.; Huys, I.; de Vos, D.; Ameloot, C.; Fauconnier, A. The Magistral Phage. *Viruses* **2018**, *10*, 64. [CrossRef] [PubMed]
80. Truog, R.D. The United Kingdom Sets Limits on Experimental Treatments: The Case of Charlie Gard. *JAMA* **2017**, *318*, 1001–1002. [CrossRef] [PubMed]
81. Międzybrodzki, R.; Borysowski, J.; Weber-Dąbrowska, B.; Fortuna, W.; Letkiewicz, S.; Szufnarowski, K.; Pawełczyk, Z.; Rogóż, P.; Kłak, M.; Wojtasik, E.; et al. Clinical aspects of phage therapy. *Adv. Virus Res.* **2012**, *83*, 73–121. [CrossRef] [PubMed]
82. Zimecki, M.; Artym, J.; Kocieba, M.; Weber-Dabrowska, B.; Borysowski, J.; Górski, A. Effects of prophylactic administration of bacteriophages to immunosuppressed mice infected with *Staphylococcus aureus*. *BMC Microbiol.* **2009**, *9*, 169. [CrossRef] [PubMed]
83. Przerwa, A.; Zimecki, M.; Switała-Jeleń, K.; Dabrowska, K.; Krawczyk, E.; Łuczak, M.; Weber-Dabrowska, B.; Syper, D.; Miedzybrodzki, R.; Górski, A. Effects of bacteriophages on free radical production and phagocytic functions. *Med. Microbiol. Immunol.* **2006**, *195*, 143–150. [CrossRef] [PubMed]
84. Górski, A.; Jończyk-Matysiak, E.; Łusiak-Szelachowska, M.; Międzybrodzki, R.; Weber-Dąbrowska, B.; Borysowski, J. Phage therapy in allergic disorders? *Exp. Biol. Med.* **2018**. [CrossRef] [PubMed]
85. Górski, A.; Dąbrowska, K.; Międzybrodzki, R.; Weber-Dąbrowska, B.; Łusiak-Szelachowska, M.; Jończyk-Matysiak, E.; Borysowski, J. Phages and immunomodulation. *Future Microbiol.* **2017**, *12*, 905–914. [CrossRef] [PubMed]
86. Górski, A.; Jończyk-Matysiak, E.; Łusiak-Szelachowska, M.; Międzybrodzki, R.; Weber-Dąbrowska, B.; Borysowski, J. The Potential of Phage Therapy in Sepsis. *Front. Immunol.* **2017**, *8*, 1783. [CrossRef] [PubMed]
87. McCallin, S.; Alam Sarker, S.; Barretto, C.; Sultana, S.; Berger, B.; Huq, S.; Krause, L.; Bibiloni, R.; Schmitt, B.; Reuteler, G.; et al. Safety analysis of a Russian phage cocktail: From metagenomic analysis to oral application in healthy human subjects. *Virology* **2013**, *443*, 187–196. [CrossRef] [PubMed]
88. Sarker, S.A.; Sultana, S.; Reuteler, G.; Moine, D.; Descombes, P.; Charton, F.; Bourdin, G.; McCallin, S.; Ngom-Bru, C.; Neville, T.; et al. Oral Phage Therapy of Acute Bacterial Diarrhea with Two Coliphage Preparations: A Randomized Trial in Children from Bangladesh. *EBioMedicine* **2016**, *4*, 124–137. [CrossRef] [PubMed]
89. Sarker, S.A.; Brüssow, H. From bench to bed and back again: Phage therapy of childhood *Escherichia coli* diarrhea. *Ann. N. Y. Acad. Sci.* **2016**, *1372*, 42–52. [CrossRef] [PubMed]
90. Loeffler, J.M.; Nelson, D.; Fischetti, V.A. Rapid killing of *Streptococcus pneumoniae* with a bacteriophage cell wall hydrolase. *Science* **2001**, *294*, 2170–2172. [CrossRef] [PubMed]
91. Witzenrath, M.; Schmeck, B.; Doehn, J.M.; Tschernig, T.; Zahlten, J.; Loeffler, J.M.; Zemlin, M.; Müller, H.; Gutbier, B.; Schütte, H.; et al. Systemic use of the endolysin Cpl-1 rescues mice with fatal *pneumococcal* pneumonia. *Crit. Care Med.* **2009**, *37*, 642–649. [CrossRef] [PubMed]
92. Witzenrath, M. Bacteriophage therapy in lung infections. In Proceedings of the 1st German Phage Symposium, Stuttgart, Germany, 9–11 October 2017.
93. Rohde, C. Pro's and con's of the phage applications. In Proceedings of the 1st German Phage Symposium, Stuttgart, Germany, 9–11 October 2017.
94. Debarbieux, L.; Pirnay, J.; Verbeken, G.; de Vos, D.; Merabishvili, M.; Huys, I.; Patey, O.; Schoonjans, D.; Vaneechoutte, M.; Zizi, M.; et al. A bacteriophage journey at the European Medicines Agency. *FEMS Microbiol. Lett.* **2016**, *363*. [CrossRef] [PubMed]
95. Denyes, J.M.; Dunne, M.; Steiner, S.; Mittelviefhaus, M.; Weiss, A.; Schmidt, H.; Klumpp, J.; Loessner, M.J. Modified Bacteriophage S16 Long Tail Fiber Proteins for Rapid and Specific Immobilization and Detection of Salmonella Cells. *Appl. Environ. Microbiol.* **2017**, *83*. [CrossRef] [PubMed]
96. Klumpp, J.; Loessner, M.J. Listeria phages: Genomes, evolution, and application. *Bacteriophage* **2013**, *3*, e26861. [CrossRef] [PubMed]

97. Rombouts, S.; Volckaert, A.; Venneman, S.; Declercq, B.; Vandenheuvel, D.; Allonsius, C.N.; van Malderghem, C.; Jang, H.B.; Briers, Y.; Noben, J.P.; et al. Characterization of Novel Bacteriophages for Biocontrol of Bacterial Blight in Leek Caused by *Pseudomonas syringae* pv. porri. *Front. Microbiol.* **2016**, *7*, 279. [CrossRef] [PubMed]
98. Luján Quiberoni, A.D.; Reinheimer, J.A. (Eds.) *Bacteriophages in Dairy Processing*; Nova Science Publishers Inc.: Hauppauge, NY, USA, 2012; ISBN 978-1-61324-517-0.
99. Szymczak, P.; Janzen, T.; Neves, A.R.; Kot, W.; Hansen, L.H.; Lametsch, R.; Neve, H.; Franz Charles, M.A.P.; Vogensen, F.K. Novel Variants of *Streptococcus thermophilus* Bacteriophages Are Indicative of Genetic Recombination among Phages from Different Bacterial Species. *Appl. Environ. Microbiol.* **2017**, *83*. [CrossRef] [PubMed]
100. Wagner, N.; Brinks, E.; Samtlebe, M.; Hinrichs, J.; Atamer, Z.; Kot, W.; Franz, C.M.A.P.; Neve, H.; Heller, K.J. Whey powders are a rich source and excellent storage matrix for dairy bacteriophages. *Int. J. Food Microbiol.* **2017**, *241*, 308–317. [CrossRef] [PubMed]
101. Ghugare, G.S.; Nair, A.; Nimkande, V.; Sarode, P.; Rangari, P.; Khairnar, K. Membrane filtration immobilization technique—A simple and novel method for primary isolation and enrichment of bacteriophages. *J. Appl. Microbiol.* **2017**, *122*, 531–539. [CrossRef]
102. Atamer, Z.; Samtlebe, M.; Neve, H.; Heller, K.J.; Hinrichs, J. Review: Elimination of bacteriophages in whey and whey products. *Front. Microbiol.* **2013**, *4*. [CrossRef] [PubMed]
103. Jäckel, C.; Hammerl, J.A.; Rau, J.; Hertwig, S. A multiplex real-time PCR for the detection and differentiation of *Campylobacter* phages. *PLoS ONE* **2017**, *12*, e0190240. [CrossRef] [PubMed]
104. Hertwig, S. Characterization of Campylobacter phages and their application. In Proceedings of the 1st German Phage Symposium, Stuttgart, Germany, 9–11 October 2017.
105. Moore, J.E.; Corcoran, D.; Dooley, J.S.G.; Fanning, S.; Lucey, B.; Matsuda, M.; McDowell, D.A.; Mégraud, F.; Millar, B.C.; O'Mahony, R.; et al. Campylobacter. *Vet. Res.* **2005**, *36*, 351–382. [CrossRef] [PubMed]
106. Silva, J.; Leite, D.; Fernandes, M.; Mena, C.; Gibbs, P.A.; Teixeira, P. *Campylobacter* spp. as a Foodborne Pathogen: A Review. *Front. Microbiol.* **2011**, *2*, 200. [CrossRef] [PubMed]
107. Ludwig, M.Z.; Kittler, R.; White, K.P.; Kreitman, M. Consequences of eukaryotic enhancer architecture for gene expression dynamics, development, and fitness. *PLoS Genet.* **2011**, *7*, e1002364. [CrossRef] [PubMed]
108. Kittler, S.; Fischer, S.; Abdulmawjood, A.; Glünder, G.; Klein, G. Colonisation of a phage susceptible *Campylobacter jejuni* population in two phage positive broiler flocks. *PLoS ONE* **2014**, *9*, e94782. [CrossRef] [PubMed]
109. Doron, S.; Melamed, S.; Ofir, G.; Leavitt, A.; Lopatina, A.; Keren, M.; Amitai, G.; Sorek, R. Systematic discovery of antiphage defense systems in the microbial pangenome. *Science* **2018**, *359*. [CrossRef] [PubMed]
110. Gilbert, R.A.; Kelly, W.J.; Altermann, E.; Leahy, S.C.; Minchin, C.; Ouwerkerk, D.; Klieve, A.V. Toward Understanding Phage: Host Interactions in the Rumen; Complete Genome Sequences of Lytic Phages Infecting Rumen Bacteria. *Front. Microbiol.* **2017**, *8*, 2340. [CrossRef] [PubMed]
111. Samtlebe, M.; Denis, S.; Chalancon, S.; Atamer, Z.; Wagner, N.; Neve, H.; Franz, C.; Schmidt, H.; Blanquet-Diot, S.; Hinrichs, J. Bacteriophages as modulator for the human gut microbiota. *LWT* **2018**, *91*, 235–241. [CrossRef]
112. Beims, H.; Wittmann, J.; Bunk, B.; Spröer, C.; Rohde, C.; Günther, G.; Rohde, M.; von der Ohe, W.; Steinert, M. Paenibacillus larvae-Directed Bacteriophage HB10c2 and Its Application in American Foulbrood-Affected Honey Bee Larvae. *Appl. Environ. Microbiol.* **2015**, *81*, 5411–5419. [CrossRef] [PubMed]
113. O'Flaherty, S.; Coffey, A.; Meaney, W.J.; Fitzgerald, G.F.; Ross, R.P. Inhibition of bacteriophage K proliferation on *Staphylococcus aureus* in raw bovine milk. *Lett. Appl. Microbiol.* **2005**, *41*, 274–279. [CrossRef] [PubMed]
114. O'Flaherty, S.; Ross, R.P.; Coffey, A. Bacteriophage and their lysins for elimination of infectious bacteria. *FEMS Microbiol. Rev.* **2009**, *33*, 801–819. [CrossRef] [PubMed]
115. Keary, R.; Sanz-Gaitero, M.; van Raaij, M.; O'Mahony, J.; Fenton, M.; McAuliffe, O.; Hill, C.; Paul Ross, R.; Coffey, A. Characterization of a Bacteriophage-Derived Murein Peptidase for Elimination of Antibiotic-Resistant *Staphylococcus aureus*. *CPPS* **2016**, *17*, 183–190. [CrossRef]
116. Fenton, M.; Casey, P.G.; Hill, C.; Gahan, C.G.; Ross, R.P.; McAuliffe, O.; O'Mahony, J.; Maher, F.; Coffey, A. The truncated phage lysin CHAP(K) eliminates *Staphylococcus aureus* in the nares of mice. *Bioeng. Bugs* **2010**, *1*, 404–407. [CrossRef] [PubMed]

117. Coffey, A. Characterization and applications of bacteriophage-derived peptidoglycan hydrolase enzymes targeting MRSA and antibiotic resistant *Clostridium difficile*. In Proceedings of the 1st German Phage Symposium, Stuttgart, Germany, 9–11 October 2017.
118. Love, M.J.; Bhandari, D.; Dobson, R.C.J.; Billington, C. Potential for Bacteriophage Endolysins to Supplement or Replace Antibiotics in Food Production and Clinical Care. *Antibiotics* **2018**, *7*. [CrossRef] [PubMed]
119. Idelevich, E.A.; von Eiff, C.; Friedrich, A.W.; Iannelli, D.; Xia, G.; Peters, G.; Peschel, A.; Wanninger, I.; Becker, K. In vitro activity against *Staphylococcus aureus* of a novel antimicrobial agent, PRF-119, a recombinant chimeric bacteriophage Endolysin. *Antimicrob. Agents Chemother.* **2011**, *55*, 4416–4419. [CrossRef] [PubMed]
120. Hagens, S. Phages to combat Listeria and Salmonella. In Proceedings of the 1st German Phage Symposium, Stuttgart, Germany, 9–11 October 2017.
121. Leibniz Institute DSMZ–German Collection of Microorganisms and Cell Cultures. FAQ Phage Therapy. Available online: https://www.dsmz.de/home/news-and-events/faq-phage-therapy.html (accessed on 16 February 2018).
122. Cooper, C.J.; Khan Mirzaei, M.; Nilsson, A.S. Adapting Drug Approval Pathways for Bacteriophage-Based Therapeutics. *Front. Microbiol.* **2016**, *7*, 1209. [CrossRef] [PubMed]
123. European Medicines Agency (EMA). SME Office, Set up by the Article 11 of the Commission Regulation (EC) No 2049/2005. Available online: http://eur-lex.europa.eu/legal-content/EN/TXT/?uri=CELEX%3A32005R2049 (accessed on 16 February 2018).
124. European Medicines Agency (EMA). PRIME—Priority Medicines. Available online: http://www.ema.europa.eu/ema/index.jsp?curl=pages/regulation/general/general_content_000660.jsp&mid=WC0b01ac05809f8439 (accessed on 16 February 2018).
125. DZIF—German Center for Infection Research. Project Start: New Active Substance Targeting Dreaded Hospital Pathogens. Available online: http://www.dzif.de/en/about_us/dzif_people/view/detail/artikel/project_start_new_active_substance_targeting_dreaded_hospital_pathogens/ (accessed on 16 February 2018).
126. Idelevich, E.A.; Schaumburg, F.; Knaack, D.; Scherzinger, A.S.; Mutter, W.; Peters, G. The Recombinant Bacteriophage Endolysin HY-133 Exhibits in Vitro Activity against Different African Clonal Lineages of the *Staphylococcus aureus* Complex, Including *Staphylococcus schweitzeri*. *Antimicrob. Agents Chemother.* **2016**, *60*, 2551–2553. [CrossRef] [PubMed]
127. Bundesministerium für Bildung und Forschung (BMBF). Globale Gesundheitskrisen Verhindern, Pressemitteilung. Available online: https://www.bmbf.de/de/globale-gesundheitskrisen-verhindern-4506.html (accessed on 15 February 2018).
128. Berlin Declaration of the G20 Health Ministers. Together Today for a Healthy Tomorrow. Available online: https://www.bundesgesundheitsministerium.de/ministry/g20-health-ministers-meeting.html (accessed on 16 February 2018).
129. World Health Organization 2015: Global Action Plan on Antimicrobial Resistance. Available online: http://www.who.int/antimicrobial-resistance/publications/global-action-plan/en/ (accessed on 16 February 2018).
130. Kittler, S.; Wittmann, J.; Mengden, R.; Klein, G.; Rohde, C.; Lehnherr, H. The use of bacteriophages as One-Health approach to reduce multidrug-resistant bacteria. *Sustain. Chem. Pharm.* **2017**, *5*, 80–83. [CrossRef]
131. Paul-Ehrlich-Institut. Medicinal Products. Available online: https://www.pei.de/EN/medicinal-products/medicinal-products-node.html (accessed on 16 February 2018).
132. Lerner, A.; Matthias, T.; Aminov, R. Potential Effects of Horizontal Gene Exchange in the Human Gut. *Front. Immunol.* **2017**, *8*, 1630. [CrossRef] [PubMed]
133. Miller-Ensminger, T.; Garretto, A.; Brenner, J.; Thomas-White, K.; Zambom, A.; Wolfe, A.J.; Putonti, C. Bacteriophages of the Urinary Microbiome. *J. Bacteriol.* **2018**. [CrossRef] [PubMed]
134. Yutin, N.; Makarova, K.S.; Gussow, A.B.; Krupovic, M.; Segall, A.; Edwards, R.A.; Koonin, E.V. Discovery of an expansive bacteriophage family that includes the most abundant viruses from the human gut. *Nat. Microbiol.* **2018**, *3*, 38–46. [CrossRef] [PubMed]
135. Kieser, S.; Sarker, S.A.; Berger, B.; Sultana, S.; Chisti, M.J.; Islam, S.B.; Foata, F.; Porta, N.; Betrisey, B.; Fournier, C.; et al. Antibiotic Treatment Leads to Fecal Escherichia coli and Coliphage Expansion in Severely Malnourished Diarrhea Patients. *Cell Mol. Gastroenterol. Hepatol.* **2018**, *5*. [CrossRef]
136. Clavijo, V.; Flórez, M.J.V. The gastrointestinal microbiome and its association with the control of pathogens in broiler chicken production: A review. *Poult. Sci.* **2018**, *97*, 1006–1021. [CrossRef] [PubMed]

137. Kim, M.S.; Bae, J.W. Lysogeny is prevalent and widely distributed in the murine gut microbiota. *ISME J.* **2018**. [CrossRef] [PubMed]
138. Leigh, B.A.; Djurhuus, A.; Breitbart, M.; Dishaw, L.J. The gut virome of the protochordate model organism, Ciona intestinalis subtype A. *Virus Res.* **2018**, *244*, 137–146. [CrossRef] [PubMed]
139. Morella, N.M.; Gomez, A.L.; Wang, G.; Leung, M.S.; Koskella, B. The impact of bacteriophages on phyllosphere bacterial abundance and composition. *Mol. Ecol.* **2018**. [CrossRef] [PubMed]

MDPI
St. Alban-Anlage 66
4052 Basel
Switzerland
Tel. +41 61 683 77 34
Fax +41 61 302 89 18
www.mdpi.com

Viruses Editorial Office
E-mail: viruses@mdpi.com
www.mdpi.com/journal/viruses

www.ingramcontent.com/pod-product-compliance
Lightning Source LLC
Chambersburg PA
CBHW051702210326
41597CB00032B/5339